ENZYME ENGINEERING 7

ANNALS OF THE NEW YORK ACADEMY OF SCIENCES
Volume 434

ENZYME ENGINEERING 7

Edited by A. I. Laskin, G. T. Tsao, and L. B. Wingard, Jr.

The New York Academy of Sciences
New York, New York
1984

Cover: The cover shows a cross-section (SEM) of immobilized *Saccharomyces cerevisiae* cells in Ca-alginate gel bead. (Photographed by Valio Laboratory.)

SP
Printed in the United States of America
ISBN 0-89766-262-8 (cloth)
ISBN 0-89766-263-6 (paper)
ISSN 0094-8500

ANNALS OF THE NEW YORK ACADEMY OF SCIENCES

Volume 434

December 31, 1984

ENZYME ENGINEERING 7[a]

Editors

A. I. LASKIN, G. T. TSAO, and L. B. WINGARD, JR.

CONTENTS

Foreword. *By* A. I. LASKIN, G. T. TSAO, and L. B. WINGARD, JR. xiii

First Enzyme Engineering Award xvii

Part I. Stability and Stabilization of Enzymes

Strategies for Increasing the Stability of Enzymes. *By* EARLE STELLWAGEN 1

Thermal and Chemical Deactivation of Soluble and Stabilized Enzymes. *By* GUIDO GRECO, JR., GIUSEPPE MARRUCCI, NINO GRIZZUTI, and LILIANA GIANFREDA 7

Mechanisms of Irreversible Thermoinactivation of Enzymes. *By* STEPHEN E. ZALE and ALEXANDER M. KLIBANOV 20

Stabilization of Subunit Enzymes by Intramolecular Crosslinking with Bifunctional Reagents. *By* V. P. TORCHILIN and V. S. TRUBETSKOY 27

Kinetic and EPR Spectroscopy Studies of Immobilized Chymotrypsin Deactivation. *By* DOUGLAS S. CLARK and JAMES E. BAILEY 31

Enzyme Stability and Glucose Inhibition in Cellulose Saccharification. *By* M. CANTARELLA, A. GALLIFUOCO, V. SCARDI, and F. ALFANI 39

Gramicidin S Synthetase Stabilization *in Vivo*. *By* S. N. AGATHOS and A. L. DEMAIN 44

Effect of Water Activity on Enzyme Action and Stability. *By* PIERRE MONSAN and DIDIER COMBES 48

Effect of Polyhydric Alcohols on Invertase Stabilization. *By* DIDIER COMBES and PIERRE MONSAN 61

Series-type Enzyme Deactivation Kinetics: Influence of Immobilization, Chemical Modifiers, and Enzyme Aging. *By* JAMES P. HENLEY and AJIT SADANA 64

Part II. Enzyme Processes and Processing Steps of Industrial Interest

The Production of α-Cyclodextrin by Enzymatic Degradation of Starch. *By* E. FLASCHEL, J.-P. LANDERT, D. SPIESSER, and A. RENKEN 70

[a]The papers in this volume were presented at the Seventh International Enzyme Engineering Conference, which was sponsored by the Engineering Foundation and held in White Haven, Pennsylvania on September 25–30, 1983.

Scale-up of an Enzyme Membrane Reactor Process for the Manufacture of L-Enantiomeric Compounds. *By* WOLFGANG LEUCHTENBERGER, MICHAEL KARRENBAUER, and ULF PLÖCKER 78

Enzymatic Production of Optically Active 2-Hydroxy Acids. *By* R. WICHMANN, C. WANDREY, W. HUMMEL, H. SCHÜTTE, A. F. BÜCKMANN, and M.-R. KULA 87

L-Amino Acids from a Racemic Mixture of α-Hydroxy Acids. *By* C. WANDREY, E. FIOLITAKIS, U. WICHMANN, and M.-R. KULA 91

Enzymatic Production of Aspartame. *By* KIYOTAKA OYAMA, SHIGEAKI IRINO, TSUNEO HARADA, and NORIO HAGI 95

Kinetically Controlled Semisynthesis of β-Lactam Antibiotics and Peptides. *By* V. KASCHE, U. HAUFLER, and L. RIECHMANN 99

Improved Steroid-1-Dehydrogenation Using Heat-dried Bacterial Cells. *By* HOLLY J. WOLF and LEO A. KOMINEK 106

Production of Microbial Enzymes from Agroindustrial By-products. *By* C. HUITRÓN, S. SAVAL, and M. E. ACUÑA 110

Aqueous Two-Phase Systems for Producing α-Amylase Using *Bacillus subtilis*. *By* ELIS ANDERSSON, ANN-CHRISTIN JOHANSSON, and BÄRBEL HAHN-HÄGERDAL 115

Enzymatic Hydrolysis of Lactose in a Hollow-Fiber Reactor. *By* C. K. S. JONES, E. T. WHITE, and R. Y. K. YANG 119

Characterization of an Enzymatic Capillary Membrane Reactor. *By* G. IORIO, G. CATAPANO, E. DRIOLI, M. ROSSI, and R. RELLA 123

Immobilization of Purified Penicillin Acylase in a Polarized Ultrafiltration Membrane Reactor. *By* FRANCESCO M. VERONESE, ENRICO BOCCÙ, ODDONE SCHIAVON, GUIDO GRECO, JR., and LILIANA GIANFREDA 127

The Influence of Pretreatments on Rice Husk Utilization by *Sporotrichum pulverulentum*. *By* JOSÉ M. C. DUARTE, L. RODA-SANTOS, AND A. CLEMENTE 131

Enzymatic Whey Hydrolysis in a Pilot-Plant "Lactohyd." *By* J. E. PRENOSIL, E. STUKER, T. HEDIGER, and J. R. BOURNE 136

Kinetics of Enzymatic Starch Liquefaction. *By* JAMES E. ROLLINGS and ROBERT W. THOMPSON 140

Continuous Conversion of Starch to Ethanol Using a Combination of an Aqueous Two-Phase System and an Ultrafiltration Unit. *By* MATS LARSSON and BO MATTIASSON 144

Continuous Alcohol Fermentation of Starchy Materials with a Novel Immobilized Cell/Enzyme Bioreactor. *By* SUSUMU FUKUSHIMA and KAZUHIRO YAMADE 148

Shifting Product Formation from Xylitol to Ethanol in Pentose Fermentation with *Candida tropicalis* by Altering Environmental Parameters. *By* ELKE LOHMEIER-VOGEL, BIRGITTA JÖNSSON, and BÄRBEL HAHN-HÄGERDAL 152

Economics of Recycling Cellulases. *By* HENRY R. BUNGAY 155

A Model for Hydrolysis of Microcrystalline Cellulose by Cellulase. *By* R. MATSUNO, M. TANIGUCHI, M. TANAKA, and T. KAMIKUBO 158

Solid Superacids for Hydrolyzing Oligo- and Polysaccharides. *By* BÄRBEL HAHN-HÄGERDAL, KERSTEIN SKOOG, and BO MATTIASSON 161

Kinetics of Acid-catalyzed Cellulose Hydrolysis under Low Water Conditions. *By* S. K. SONG and Y. Y. LEE 164

Effects of Compaction on the Effectiveness Factor of a Gel Particle Containing Enzyme. *By* K. UEYAMA and S. FURUSAKI 168

Part III. Enzymes in Preparative Chemistry

Biohydrogenation and Electromicrobial and Electroenzymatic Reduction Methods for the Preparation of Chiral Compounds. *By* H. SIMON, J. BADER, H. GÜNTHER, S. NEUMANN, and J. THANOS 171

Enantioselective Reductions of β-keto-Esters by Bakers' Yeast. *By* CHARLES J. SIH, BING-NAN ZHOU, ARAVAMUDAN S. GOPALAN, WOAN-RU SHIEH, CHING-SHIH CHEN, GARY GIRDAUKAS, and FRANK VANMIDDLESWORTH ... 186

New Enzymes for the Synthesis of Chiral Compounds. *By* W. HUMMEL, H. SCHÜTTE, and M.-R. KULA 194

Enzymatic Synthesis of Optically Pure 1-Carbacephem Compounds. *By* Y. HASHIMOTO, S. TAKASAWA, T. OGASA, H. SAITO, T. HIRATA, and K. KIMURA 206

Synthesis of Phenoxymethylpenicillin Using α-Acylamino-β-lactam Acylhydrolase from *Erwinia aroideae*. *By* D. H. NAM and DEWEY D. Y. RYU 210

Hydroxylation by Hemoglobin-containing Systems: Activities and Regioselectivities. *By* DIDIER GUILLOCHON, LAURENT ESCLADE, BERNARD CAMBOU, and DANIEL THOMAS 214

Unusual Catalytic Properties of Usual Enzymes. *By* BERNARD CAMBOU and ALEXANDER M. KLIBANOV 219

Part IV. New Methods, Enzymes, and Approaches

Rotating Ring-Disc Electrode: Its Use in Studying Mass Transport Resistances through Immobilized Enzyme Support Matrices. *By* LEMUEL B. WINGARD, JR., JAMES F. CASTNER, OSATO MIYAWAKI, and CARL MARRESE 224

Studies of Protein Function by Various Mutagenic Strategies: β-Lactamase. *By* G. DALBADIE-MCFARLAND, J. NEITZEL, A. D. RIGGS, and J. H. RICHARDS 232

New Immobilization Techniques and Examples of Their Application. *By* KLAUS MOSBACH 239

Chromophoric Sulfonyl Chloride Agarose for Immobilizing Bioligands. WILLIAM H. SCOUTEN and WILL VAN DER TWEEL 249

New Synthetic Carriers for Enzymes. *By* O. MAUZ, S. NOETZEL, and K. SAUBER 251

New Methods for Activation of Polysaccharides for Protein Immobilization and Affinity Chromatography. *By* MEIR WILCHEK, TALIA MIRON, and JOACHIM KOHN 254

Continuous Regeneration of ATP for Enzymatic Syntheses. *By* W. BERKE, M. MORR, C. WANDREY, and M.-R. KULA 257
Study and Use of Spontaneous Cofactor Regeneration in an Immobilized Enzyme-Coenzyme System. *By* PHILIPPE MINARD, MARIE-DOMINIQUE LEGOY, and DANIEL THOMAS 259
Immobilization of NAD Kinase. *By* XINSONG JI, HUIXIAN LI, ZHONGYI YUAN, and SHUHUANG LIU 264
Production and Purification of Dextransucrase from *Leuconostoc mesenteroides,* NRRL B 512 (F). *By* F. PAUL, D. AURIOL, E. ORIOL, and P. MONSAN 267
Characterization of a New Class of Thermophilic Pullulanases from *Bacillus acidopullulyticus. By* MARTIN SCHÜLEIN and BIRGITTE HØJER-PEDERSEN 271
Selection of a β-Amylase-producing Strain and Its Fermentation Conditions. *By* HE BINGWANG, GUO JUNJUN, JIANG ZHAOYUAN, and ZHANG SHUZHENG 275
Improvement of β-Amylase Production and Characterization of Polypepton Fraction Effective for β-Amylase Production. *By* R. SHINKE, H. YAMANAKA, T. NANMORI, K. AOKI, H. NISHIRA, and S. YUKI 278
Microbial Glycosidases: Research and Application. *By* ZHANG SHUZHENG, GE SUGUO, and YANG SHOUJUN 282
Enzyme Recovery by Adsorption from Unclarified Microbial Cell Homogenates. *By* PER HEDMAN and JAN-GUNNAR GUSTAFSSON 285
Immobilized Thrombolytic Enzymes Possessing Increased Affinity toward Substrate. *By* V. P. TORCHILIN, A. V. MAKSIMENKO, E. G. TISCHENKO, G. V. IGNASHENKOVA, and G. A. ERMOLIN 289
A New Viologen Mediator for Hydrogenase-catalyzed Reactions. *By* E. ZIOMEK, W. G. MARTIN, and R. E. WILLIAMS 292
Expression of Hydrogenase Activity in Cereals. *By* V. TORRES, A. BALLESTEROS, V. M. FERNÁNDEZ, and M. NÚÑEZ 296
A Novel Method of Affinity Isolation of Acetylcholine Receptor with Nylon Affinity Tubes. *By* P. V. SUNDARAM 299

Part V. Modified Enzymes and Enzyme-like Compounds

Synthetic Polymers with Enzyme-like Activities. *By* IRVING M. KLOTZ 302
Semisynthetic Enzymes. *By* E. T. KAISER 321
Molecular Imprinting. *By* GÜNTER WULFF 327
Synzymes: Synthetic Hydrogenation Catalysts. *By* G. P. ROYER, K. S. HATTON, and W. S. CHOW 334
Amino Acid and NMR Analysis of Oxidized *Mucor miehei* Rennett. *By* SVEN BRANNER, PETER EIGTVED, MOGENS CHRISTENSEN, and HENNING THØGERSEN 340
New Properties of Chymotrypsin Modified by Fixation with a Hydrophobic Molecule: Hexanal. *By* M. H. REMY, C. BOURDILLON, and D. THOMAS 343

Part VI. Immobilized Microbial and Plant Cells and Organelles

Reproductive Immobilized Cells. *By* A. CONSTANTINIDES and G. K. CHOTANI .. 347

Microbiological Measurements by Immobilization of Cells within Small-Volume Elements. *By* JAMES C. WEAVER, PETER E. SEISSLER, STEVEN A. THREEFOOT, JEFFREY W. LORENZ, TIMOTHY HUIE, RONALD RODRIGUEZ, and ALEXANDER M. KLIBANOV 363

Growth of Procaryotic Cells in Hollow-Fiber Reactors. *By* HARVEY W. BLANCH, T. BRUCE VICKROY, and CHARLES R. WILKE 373

Immobilized Viable Plant Cells. *By* PETER BRODELIUS 382

Semicommercial Production of Ethanol Using Immobilized Microbial Cells. *By* H. SAMEJIMA, M. NAGASHIMA, M. AZUMA, S. NOGUCHI, and K. INUZUKA 394

Applications of Immobilized Lactic Acid Bacteria. *By* P. LINKO, S.-L. STENROOS, Y.-Y. LINKO, T. KOISTINEN, M. HARJU, and M. HEIKONEN 406

Gel Entrapment of Whole Cells and Enzymes in Crosslinked, Prepolymerized Polyacrylamide Hydrazide. *By* AMIHAY FREEMAN 418

Large-Scale Bacterial Fuel Cell Using Immobilized Photosynthetic Bacteria. *By* ISAO KARUBE, HIDEAKI MATSUOKA, HIDEKI MURATA, KAZUHITO KAJIWARA, SHUICHI SUZUKI, and MITSUO MAEDA 427

The Implication of Reaction Kinetics and Mass Transfer on the Design of Biocatalytic Processes with Immobilized Cells. *By* J. KLEIN, K.-D. VORLOP, and F. WAGNER 437

Continuous Production of L-Alanine: Successive Enzyme Reactions with Two Immobilized Cells. *By* TETSUYA TOSA, SATORU TAKAMATSU, MASAKATSU FURUI, and ICHIRO CHIBATA 450

2,3-Butanediol Production by Immobilized Cells of *Enterobacter* sp. H. KAUTOLA, Y.-Y. LINKO, and P. LINKO 454

Transformation of Steroids by Immobilized Microbial Cells. *By* YANG LIAN-WAN and ZHONG LI-CHAN 459

Malolactic Fermentation and Secondary Product Formation in Wine by *Leuconostoc oenos* Cells Immobilized in a Continuous-flow Reactor. *By* PAOLO SPETTOLI, MARCO P. NUTI, ANGELO DAL, BELIN PERUFFO, and ARTURO ZAMORANI 461

Immobilization of Microorganisms. *By* JOHN L. GAINER and DONALD J. KIRWAN 465

Production, Purification, and Immobilization of *Bacillus subtilis* Levan Sucrase. *By* PATRICE PERLOT and PIERRE MONSAN 468

Oxygen Supply to Immobilized Cells Using Hydrogen Peroxide. *By* OLLE HOLST, HANS LUNDBÄCK, and BO MATTIASSON 472

Metabolic Behavior of Immobilized Cells—Effects of Some Microenvironmental Factors. *By* BO MATTIASSON, MATS LARSSON, and BÄRBEL HAHN-HÄGERDAL 475

Various Applications of Living Cells Immobilized by Prepolymer Methods. *By* ATSUO TANAKA, KENJI SONOMOTO, and SABURO FUKUI 479

Immobilization of Yeast Cells on Transition Metal-activated Pumice Stone. *By* J. M. S. Cabral, M. M. Cadete, J. M. Novais, and J. P. Cardoso 483

Immobilization of Plant Protoplasts. *By* L. Linse and P. Brodelius 487

Growth and Cardenolide Production by *Digitalis lanata* in Different Fermentor Types. *By* P. Markkanen, T. Idman, and V. Kauppinen 491

Noninvasive ^{31}P NMR Studies of the Metabolism of Suspended and Immobilized Plant Cells. *By* Peter Brodelius and Hans J. Vogel 496

Comparison of Immobilization Methods for Plant Cells and Protoplasts. *By* J. M. S. Cabral, P. Fevereiro, J. M. Novais, and M. S. S. Pais 501

Part VII. Enzyme Sensors and Assays

A Simple Inulin Assay for Renal Clearance Determination Using an Immobilized β-Fructofuranosidase. *By* D. F. Day and W. E. Workman.... 504

Hybrid Biosensor for Clinical and Fermentation Process Control. *By* Isao Karube, Izumi Kubo, and Shuichi Suzuki 508

Oxalate-sensing Enzyme Electrode. *By* Leland C. Clark, Jr., Linda K. Noyes, Thomas A. Grooms, and Pamela E. Moore 512

Direct Rapid Electroenzymatic Sensor for Measuring Alcohol in Whole Blood and Fermentation Products. *By* Leland C. Clark, Jr., Linda K. Noyes, Thomas A. Grooms and Pamela E. Moore 515

FAD and Glucose Oxidase Immobilized on Carbon. *By* Osato Miyawaki and Lemuel B. Wingard, Jr 520

Enzyme Electrode Based on a Fluoride Ion Selective Sensor Coupled with Immobilized Enzyme Membrane. *By* Mat H. Ho and Tai-Guang Wu 523

Use of Immobilized Enzymes in Flow Injection Analysis. *By* Mat H. Ho and Moore U. Asouzu 526

Multifunctional Biosensor for the Determination of Fish Meat Freshness. *By* Estuo Watanabe, Kenzo Toyama, Isao Karube, Hideaki Matsouka, and Shuichi Suzuki 529

Oxidase Enzyme: Enzyme and Immunoenzyme Sensor. *By* J. L. Romette and J. L. Boitieux 533

Current Status of Activity Assays for Tissue Plasminogen Activator. *By* John C.-T. Tang, Shirley Li, Paula McGray, and Andrew Vecchio 536

Part VIII. Enzymes and Hydrocarbon Substrates and Environments

Oxidation of Gaseous Hydrocarbons by Methanotrophs: Heterogeneous Bioreactor. *By* Ching T. Hou 541

Hydrocarbon Micellar Solutions in Enzymatic Reactions of Apolar Compounds. *By* P. L. Luisi, P. Lüthi, I. Tomka, J. Prenosil, and A. Pande 549

Continuous Synthesis of Glycerides by Lipase in a Microporous Membrane Bioreactor. *By* Tsuneo Yamané, Mohammad Mozammel Hoq, Shoichi Shimizu, Shiro Ishida, and Tadashi Funada 558

The Enzymatic Synthesis of Esters in Nonaqueous Systems. *By* I. L. Gatfield 569

Cholesterol Degradation by Polymer-entrapped *Nocardia* in Organic Solvents. *By* JOSÉ M. C. DUARTE and M. D. LILLY 573

Catalysis by Enzymes Entrapped in Reversed Micelles of Surfactants in Organic Solvents. *By* I. V. BEREZIN and KAREL MARTINEK 577

Protein Immobilization on Liposomes. *By* A. L. KLIBANOV, N. N. IVANOV, A. A. BOGDANOV, and V. P. TORCHILIN 580

Index of Contributors 583

Subject Index 587

Foreword

Enzyme engineering, which deals with the efficient utilization of enzymes, microorganisms, and cultured plant and animal cells as organic catalysts, continues to be one of the major thrusts in the burgeoning area called biotechnology. This volume contains most of the papers presented at the Seventh International Enzyme Engineering Conference, which was held September 25–30, 1983 at White Haven, Pennsylvania in the USA. Papers from previous conferences in this biannual series were published by Wiley (first conference) and by Plenum Press (second–sixth conferences). Highlighted at this seventh conference were papers on the preparation (by genetic or chemical means), stabilization, and application of enzymes either free in solution or immobilized within cells or on solid supports.

The seventh conference, like the earlier ones, was sponsored by the Engineering Foundation. Registration was limited to 204, with participants this time from 20 countries. The conference was organized by an Executive Committee that consisted of the following members:

Allen I. Laskin .... Exeuctive Chairman
George T. Tsao .... Program Chairman
Shuichi Suzuki .... Japan Member
Gunter Schmidt-Kastner .... Europe Member
Daniel Thomas .... Europe Member
Howard H. Weetall .... USA Member
Michael K. Weibel .... USA Member
Lemuel B. Wingard, Jr. .... Permanent Member
Sanford S. Cole .... Director of Conferences

Additional input was provided by the Advisory Board, composed as follows:

Francesco Bartoli	Ephraim Katchalski-Katzir
Sven Branner	Joachim Klein
George Broun	Jan Konecny
Robert P. Chambers	Maria R. Kula
Thomas M. S. Chang	Pekka Linko
Ichiro Chibata	George Manecke
Enrico Drioli	Bo Mattiasson
Peter Dunnill	Allan S. Michaels
Horbour Filippusson	Saul L. Neidleman
David J. Fink	A. Hiro Nishikawa
Saburo Fukui	Hirosuki Okada
Daniele C. Gautheron	Bob Poldermans
Hans U. Geyer	Garfield P. Royer
Manfred Gloger	Fritz Wagner
Leon Goldstein	Meir Wilchek
James J. Hamsher	Hideaki Yamada
Hakuai Inoue	Shu-zheng Zhang
J. Bryan Jones	

The success of the conference was aided markedly by the help provided by many of our colleagues and friends. We particularly want to acknowledge the contributions of (1) Ching T. Hou, who organized the poster paper sessions and helped in editing manuscripts, (2) Dolores M. Pirnik, Michael Pirnik, Nancy Barnabe, Kathy Greaney, and Copper Haith, all of whom helped in conference organization and in manuscript editing, and (3) Sanford S. Cole, Harold A. Comerer, and the staff at the Engineering Foundation.

We are deeply grateful to the following organizations that contributed financial support for the conference:

A C Biotechnics, Sweden
Ajinomoto Company Inc., Japan
Alfa-Laval, Sweden
Association of Microbial Food & Enzyme Producers, Belgium
BASF Wyandotte Corporation, USA
Bayer AG, FRG
Beecham Pharmaceuticals, UK
Biochemie GmbH, Austria
Boehringer Mannheim, FRG
B. Braun Melsungen AG, FRG
Cetus Corporation, USA
Chemap AG, Switzerland
Ciba-Geigy, Switzerland
Corning Glass Works, USA
D.S.M., the Netherlands
Edwards Kniese & Co., FRG
Enka AG, FRG
Exxon Research and Engineering Company, USA
Giovanola Freres SA, Switzerland
Gist Brocades NV, the Netherlands
Haarmann & Reimer GmbH, FRG
Henkel KG AA, FRG
Hoechst-Roussel Pharmaceuticals Inc., USA
Japanese Society of Enzyme Engineering, Japan
Kyowa Hakko Kogyo Company, Japan
Membrana, FRG
E. Merck, FRG
Merck Sharp & Dohme, USA
Miles Kali-Chemie GmbH, FRG
Miles Laboratories Inc., USA
Monsanto Company, USA
National Science Foundation, USA
New York Academy of Sciences, USA
Novo Laboratories Inc., USA
Röhm GmbH, FRG
Schering AG, FRG
Süddeutsche Zucker AG, FRG
Westfalia Separator AG, FRG
Worthington Diagnostic Systems Inc., USA

The Eighth International Enzyme Engineering Conference, sponsored by the Engineering Foundation, is scheduled to be held September 22–27, 1985 at Helsingör, Denmark. Information can be obtained from the Engineering Foundation, 345 East 47th Street, New York, NY 10017, USA. The Executive Committee, charged with organizing the eighth conference, is composed as follows:

Klaus Mosbach	Executive Chairman
Daniel Thomas	Program Chairman
Ilia Berezin	Member
Villy J. Jensen	Member
Joachim Klein	Member
Allen I. Laskin	Member
Hirosuke Okada	Member
Günter Schmidt-Kastner	Member
Howard H. Weetall	Member
Meir Wilchek	Member
Lemuel B. Wingard, Jr.	Member
Harold A. Comerer	Director, Engineering Foundation

A. I. Laskin
G. T. Tsao
L. B. Wingard, Jr.

First Enzyme Engineering Award

In 1982 the Board of Directors of the Engineering Foundation approved the creation of a new major award to honor individuals who have made outstanding contributions of major importance in the field of enzyme engineering. The Corning Glass Works agreed to act as the initial financial sponsor. The award is to be presented at the biannual enzyme engineering conferences that are sponsored by the Engineering Foundation.

Dr. Ichiro Chibata of Tanabe Seiyaku Co., Ltd. of Osaka, Japan was the recipient of the first international Enzyme Engineering Award. He was presented with an engraved piece of Steuben glass by the Corning Glass Works and a hand engrossed certificate by the Engineering Foundation. The presentation was made at the Seventh Enzyme Engineering Conference in September 1983. Dr. Chibata was selected for his engineering, scientific, and managerial leadership in bringing about the first industrial scale applications of immobilized enzymes and immobilized microbial cells in 1969 and 1973, respectively. He received his Ph.D. in biochemistry and agricultural chemistry from Kyoto University in 1959 and then joined the Tanabe Seiyaku Co., where he is Managing Director and Research and Development Executive. He is the author of many scientific papers and several books.

The awardee was selected by the conference Executive Committee upon the recommendation of an Awards Committee that consisted of the following members: G. Broun, S. Fukui, G. Manecke, E. K. Pye, H. H. Weetall, and L. B. Wingard, Jr. (chairman).

Strategies for Increasing the Stability of Enzymes

EARLE STELLWAGEN

Department of Biochemistry
University of Iowa
Iowa City, Iowa 52242

The economics of a continuous-flow, enzyme-catalyzed chemical reaction are dependent in large measure on the lifetime of the catalytically efficient form of the enzyme under the reaction conditions employed. Unfortunately, the catalytic lifetime desired for many flow reactions exceeds that of available enzyme catalysts. Accordingly, enzyme catalysts with the desired stability must either be obtained from exotic natural sources adapted to more stressful conditions or be manufactured either by genetic engineering or by chemical modification of available enzymes.

In this discussion, a two-state reversible equilibrium between a globular catalytically functional native conformation and a randomly coiled, catalytically inactive denatured conformation will be considered. Enhanced stability will mean a shift in the conformational equilibrium toward the native conformation at 25° at neutral pH in aqueous solvents. Such a shift will be denoted as an increase in the numerical value of the free energy of the native conformation measured as outlined in FIGURE 1. Such a shift will increase the lifetime of the native conformation by decreasing the rate of denaturation, increasing the rate of renaturation, or both. In the case of the cytochromes *c*, the rate of renaturation is increased.[1] Enhanced thermostability implies a shift in the conformational equilibrium toward the native form at temperatures in excess of 50°. However, most denaturation reactions at these higher temperatures are coupled with irreversible reactions leading to rapid precipitation or covalent modification of the denatured protein. In these cases, an increase in the lifetime of the native conformation will necessarily result from a decrease in the rate of denaturation.

Among life forms adapted to stress, the extreme thermophilic bacteria have been most thoroughly studied.[2] Such bacteria are commonly obtained in the runoff streams from hot springs or in commercial hot water storage tanks and are capable of growth at temperatures in excess of 70°C. The native conformation of enzymes purified from extreme thermophilic bacteria are usually stable at temperatures in excess of 90°C, as demonstrated in the work of Stellwagen *et al.*[3] and Saiki *et al.*[4] These enzymes not only exhibit enhanced thermostability but also enhanced stability to protein denaturants, such as ionic detergents, pH extremes, chaotropic agents, guanidine hydrochloride, urea, and organic solvents when compared with the same enzyme purified from a normal or mesophilic organism, as shown by Stellwagen *et al.*[3] and Kagawa *et al.*[5] TABLE 1). Accordingly, purification of a desired catalytic activity from an extreme thermophilic organism will likely yield an enzyme of enhanced stability to a range of rigorous *in vitro* conditions. Samples of thermophilic organisms can be obtained from bacterial culture banks or from individual investigators and grown in relatively simple media maintained at elevated temperatures. The plasma membranes of the cells are easily ruptured and the cytoplasmic enzymes purified in reasonable yield using standard purification procedures, including affinity chromatography at room temperature.

For some applications, the desired enzyme may not occur in thermophilic bacteria

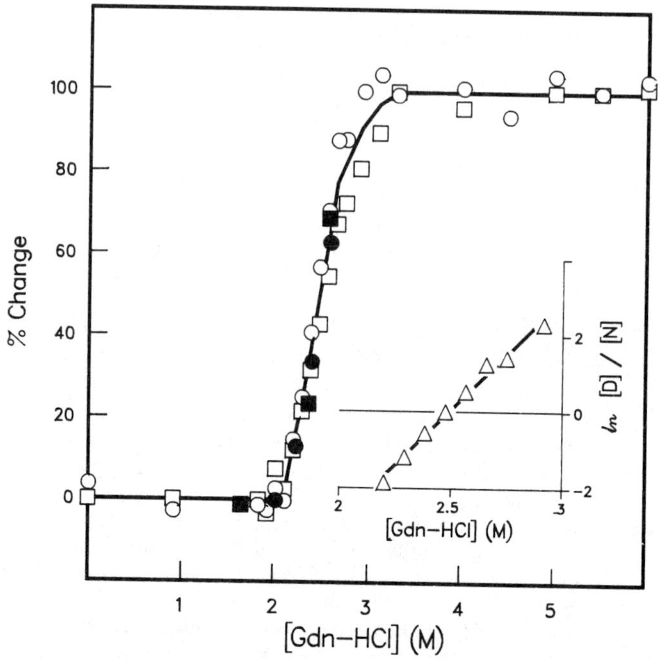

FIGURE 1. Analysis of a protein conformational transition. The main body of the figure depicts the reversible cooperative conformational transition of the protein thioredoxin generated by the protein denaturant guanidine hydrochloride (Gdn-HCl). The circles indicate changes in the globular tertiary structure as detected by the increase in the fluorescence emission amplitude of the two tryptophan residues observed at 350 nm using 295-nm excitation. The squares indicate changes in the secondary structural content, which includes several α-helixes, a β-sheet and several reverse turns, as detected by circular dichroism at 219 nm. Open symbols represent values obtained beginning with the native protein and filled symbols represent values obtained beginning with the dentured protein. The experimental values describe three zones: a native baseline zone encompassing 0–2 *M* denaturant, a conformational transition zone 2–3 *M* in denaturant, and a denatured baseline zone above 3 *M* denaturant. Assuming a two-state transition, experimental values in the transition zone can be resolved into the fractional amounts of native and denatured forms present. The insert illustrates a logarithmic plot of the transition zone so resolved. Assuming no binding of denaturant, the slope of the line multiplied by the [Gdn–HCl] at the transition midpoint (i.e. ln D/N = 0) multiplied by −RT is the $\Delta G°$ of the N $\rightleftharpoons$ D equilibrium.[20]

TABLE 1. Comparative Stability of ATPase F_1 to Preincubation[5]

Denaturant	[Denaturant] Causing 50% Inactivation: Bovine	Thermophilic Bacterium PS3
KCNS	0.07 *M*	0.53 *M*
Urea	0.85 *M*	5.52 *M*
Sodium dodecyl sulfate	0.005%	0.03%
Acetone	13%	63%

or if present may not have adequately enhanced stability for the conditions employed. Accordingly, it would be instructive to determine the structural basis for the enhanced stability of enzymes from thermophilic bacteria and to apply this information to the modification of available enzymes to enhance their stability. Surprisingly, an enzyme obtained from a thermophile invariably possesses the same structural features such as polypeptide chain length, secondary structural elements, globular domains, and degree of polymerization as does the same enzyme obtained from a mesophile. While such paired enzymes can be considered homologous proteins, their polypeptide chains when aligned contain numerous amino acid sequence differences. These involve all categories of amino acid side chains, including apolar, polar, ionic, linear, branched, and aromatic. Pairwise comparisons of a number of polypeptide sequences, having the same catalytic function and enhanced and ordinary thermostability, have failed to identify a particular kind of amino acid side chain as responsible for enhanced stability. At present one can only conclude that enhanced stability results from the incremental contributions, both positive and negative, of a variety of interactions which *in toto* result in an increased stability.[6] In principle it should be possible to identify by energy calculations those amino acid replacements mainly responsible for enhanced thermostability in enzymes of known three-dimensional structure, but in practice such calculations are not adequately precise.

Part of the difficulty in identification of amino acid replacements responsible for enhanced thermostability using naturally occurring enzymes lies in the concurrence of replacements resulting from genetic drift, which may have little effect on stability, and from replacements responsible for enhanced stability. Accordingly, it would be ideal to evaluate the stability changes of single controlled amino acid replacements, a capability recently made feasible by genetic engineering and chemical oligonucleotide synthesis procedures. A single amino acid replacement judiciously positioned in the native structure can make a significant contribution to enzyme stability as shown by comparison of the two naturally occurring forms of the enzyme tyrosinase from the organism *Neurospora crassa*.[7] The sole replacement of an asparagine at position 201 with an aspartate residue increased the stabilization free energy of the catalytic form of the enzyme from 2.4 kcal/mole to 4.9 kcal/mole, a substantive enhancement that can be readily evaluated using the procedures outlined in FIGURE 1.

The tools of contemporary molecular biology generate a bewildering prospect in that an astronomical number of single amino acid replacements can be generated in even a small enzyme containing only about 100 amino acid residues. Since generation of each replacement is no small task given possession of the structural gene, some perception of replacements likely to enhance stability must be operative. I suggest that replacement of amino acid residues having an internal or buried location in the bifunctional enzyme structure be avoided. Analyses of crystallographic structures of biofunctional proteins has indicated that their interiors are optimally packed and that all buried polar atoms are hydrogen bonded.[8] Accordingly, altering the size, shape, or polarity of a single internal side chain would very likely destabilize an enzyme structure. By contrast, the mobility of water molecules affords no barrier to the packing or hydrogen bonding alterations of amino acid side chains residing on the surface of enzymes. Analyses[9,10] of the surfaces of crystallographic enzyme structures indicate that about 50% of the atoms in contact with the aqueous solvent are apolar, a surprising departure from the oil drop model for proteins. Among the exposed apolar atoms are to be found some side chains of leucine, isoleucine, valine, phenylalanine, tyrosine, tryptophan, methionine, and proline residues largely if not totally in contact with water. These apolar residues are often exposed to solvent to serve as aggregation sites leading to formation of intracellular assemblies, an activity of no consequence to the function of an isolated enzyme *in vitro*. Transfer free energy values[11,12] indicate

TABLE 2. Residue Parameters

Residue	Transfer Free Energy[11,12] (kcal/mole)	Secondary Structural Probabilities[15]		
		α-Helix	β-Strand	Reverse Turn
Alanine	0.6	1.4	0.8	0.6
Arginine	0.6	1.0	0.9	1.1
Asparagine	0	0.7	0.9	1.6
Aspartate	0	1.0	0.5	1.6
Cyteine	0	0.7	1.2	0.9
Glutamate	0	1.5	0.4	0.8
Glutamine	0	1.1	1.1	0.8
Glycine	0	0.6	0.8	1.6
Histidine	0	1.0	0.9	0.8
Isoleucine	2.5	1.1	1.6	0.3
Leucine	2.5	1.2	1.3	0.4
Lysine	1.5	1.2	0.7	1.1
Methionine	1.5	1.5	1.0	0.5
Phenylalanine	2.5	1.1	1.4	0.6
Proline	2.5	0.6	0.5	2.0
Serine	0	0.8	0.8	1.5
Threonine	0	0.8	1.2	1.0
Tryptophan	2.5	1.1	1.4	0.5
Tyrosine	2.5	0.7	1.5	1.1
Valine	1.5	1.1	1.7	0.4

that between 0.6 and 2.5 kcal/mole of stabilization energy could be obtained per apolar residue removed from contact with water, as shown in TABLE 2. Accordingly, their solvation by water leads to a *de*stabilization of between 0.6 and 2.5 kcal/mole residue. Replacement by genetic engineering of any surface apolar residue by a small polar or ionic residue having a transfer free energy of zero should then enhance the stability of the enzyme structure by between 0.6 and 2.5 kcal/mole residue replaced.

While no deliberate effort has been made to date to test this strategy, a naturally occurring example suggests that it is promising. Bovine heart and horse heart cytochromes *c* differ at the three sequential positions (indicated in TABLE 3) out of the total of 104 residues. Analysis of the reversible denaturation of each of these proteins indicates that the native structure of the bovine protein is more stable by about 1.1 kcal/mole.[13] This result was initially considered surprising since all three residues involved in the replacement are located on the surface. The transer free energy values for threonine, serine, and glycine are all zero; this generates no energetic changes in replacement at positions 47 and 89. However, exposure of the apolar side chain portion of the lysine residue at position 60 costs about 1.5 kcal/mole. Accordingly, replacement of the exposed lysine in the horse protein by a glycine in the bovine protein

TABLE 3. Comparative Stability of Cytochromes *c*[13]

Sequence Position	Horse	Bovine
47	Threonine	Serine
60	Lysine	Glycine
89	Threonine	Glycine
ΔG° (kcal/mole)	−7.3	−8.4

recovers this deficit consistent with the observed enhanced stability of the bovine protein.

Analysis of the surface of the crystallographic models of most proteins should target at least one surface residue whose replacement would substantially increase the stability of the protein. For example, the side chains of one isoleucine, one leucine, and two valines in the small protein thioredoxin[14] are each at least two thirds as exposed to the solvent. If the residue position targeted for replacement is part of an α-helix, a β-strand, or a reverse turn, a residue having both a transfer free energy of zero *and* a high probability for residence in the concerned secondary structural element[15] should be selected. For example, the values listed in TABLE 2 suggest that an exposed methionine participating in an α-helix might usefully be replaced by a glutamate, an exposed leucine in a β-strand by a threonine, and an exposed proline in a reverse turn by a glycine.

Unfortunately, most enzymes do not presently have a crystallographic model available for identification of one or more residues having exposed apolar side chains. However, the exposure of aromatic apolar side chains of tyrosine and tryptophan residues can be established using the relatively simple technique of solvent perturbation difference absorbance spectroscopy.[16] Selective chemical modification procedures using nondenaturing conditions have been described[17] for detection of exposed arginine, lysine, methionine, tryptophan, and tyrosine residues. The position(s) of such exposed residues in the amino acid sequence can be established using a radiolabeled reagent, proteolytic cleavage, peptide fractionation, and amino acid analysis. In the absence of any exposure information, lysine residues that have an exposure probability of 0.96,[8] can be replaced by histidine residues. This maintains the side chain net charge over at least a portion of the pH range and has no preference for particular secondary structural elements, while decreasing the transfer free energy. Indeed, several lysine residues could likely be replaced by neutral side chains with little perturbation of either catalytic activity or native conformation. In executing these chemical modifications, it will be important to occupy the catalytic site with a competitive inhibitor so as not to target a residue participating directly in catalysis for replacement.

For investigators having neither a crystallographic model of the enzyme of interests or a facility for obtaining and engineering the structural gene, use of a cross-linking reagent is recommended to increase stability. This approach is based on diminishing the polypeptide entropy that is the principal thermodynamic quantity stabilizing the denatured form. The chain entropy is diminished by the equation.[18]

$$\Delta S = 0.75(Rc)\ln(r + 3)$$

where c is the number of intracrosslinks and r is the average number of residues encompassed by the crosslinks. A recent review[19] lists an abundance of protein cross-linking reagents that are targeted for various side chain groups and have variable spacer lengths. Since carboxyl and amino groups constitute the most abundant reactive groups on protein surfaces, I recommend use of cross-linking agents designed to attack either of these two groups. Use of a flexible spacer between the reactive ends of the crosslinker would increase the probability of linking residues remote from each other in the polypeptide sequence, a consideration advantageous to the entropic reduction as shown in the above equation. Cross-linking should be done in the presence of a substrate or substrate analogue to preserve the catalytic function. Some caution must be exercised in this undertaking in that cross-linking can be continued to an unproductive extreme in which the flexibility of conformational changes required of catalysis can be compromised.

REFERENCES

1. BREMS, D. N., R. CASS & E. STELLWAGEN. 1982. Conformational transitions of frog heart ferricytochrome *c*. Biochemistry **21**(7): 1488–1493.
2. FRIEDMAN, S. M. 1978. Biochemistry of Thermophily. Academic Press. New York.
3. STELLWAGEN, E., M. M. CRONLUND & L. D. BARNES. 1973. A thermostable enolase from the extreme thermophile, *Thermus aquaticus* YT-1. Biochemistry **12**(8): 1552–1558.
4. SAIKI, T., S. IIJIMA, K. TOHDA, T. BEPPU & K. ARIMA. 1978. Purification and properties of malate dehydrogenase and isocitrate dehydrogenase from the extreme thermophile, *Thermus flavus* AT-62. *In* Biochemistry of Thermophily. S. M. Friedman, Ed. Academic Press. New York. pp. 287–303.
5. KAGAWA, Y., N. SONE, M. YOSHIDA, H. HIRATA & H. OKAMOTO. 1976. Proton translocating ATPase of a thermophilic bacterium. J. Biochem. (Tokyo) **80**(1): 141–151.
6. ARGOS, P., M. G. ROSSMAN, U. M. GRAU, H. ZUBER, G. FRANK & J. D. TRATSCHIN. 1979. Thermal Stability and Protein Structure. Biochemistry **18**(25): 5698–5703.
7. RUEGG, C., D. AMMER & K. LERCH. 1982. Comparison of amino acid sequence and thermostability of tyrosinase from three wild type strains of *Neurospora crassa*. J. Biol. Chem. **257**(11): 6420–6426.
8. CHOTHIA, C. 1975. Structural invariants in protein folding. Nature **254**(5498): 304–308.
9. LEE, B. & F. M. RICHARDS. 1971. The interpretation of protein structures: Estimation of static accessibility. J. Mol. Biol. **55**(3): 379–400.
10. SHRAKE, A. & J. A. RUPLEY. 1973. Environment and exposure to solvent of protein atoms. Lysozyme and insulin. J. Mol. Biol. **79**(2): 351–371.
11. BRANDTS, J. F. 1964. The thermodynamics of protein denaturation. II. A model of reversible denaturation and interpretations regarding the stability of chymotyrpsinogen. J. Am. Chem. Soc. **86**(20): 4302–4314.
12. BRANDTS, J. F. & L. HUNT. 1967. The thermodynamics of protein denaturation. III. The denaturation of ribonuclease in water and in aqueous urea and aqueous methanol mixtures. J. Am. Chem. Soc. **89**(19): 4826–4838.
13. KNAPP, J. A. & C. N. PACE. 1974. Guanidine hydrochloride and acid denaturation of horse, cow and *Candida krusei* cytochromes *c*. Biochemistry **13**(6): 1289–1294.
14. HOLMGREN, A., B. O. SODERBERG, H. EKLUND & C. I. BRANDEN. 1975. Three-dimensional structure of *Escherichia coli* thioredoxin-S_2 to 2.8-Å resolution. Proc. Natl. Acad. Sci. USA **72**(6): 2305–2309.
15. CHOU, P. Y. & G. D. FASMAN. 1978. Prediction of the secondary structure of proteins from their amino acid sequence. Adv. Enzymol. **47:** 45–148.
16. DONOVAN, J. W. 1969. Changes in ultraviolet absorption produced by alteration of protein conformation. J. Biol. Chem. **244**(8): 1961–1967.
17. GLAZER, A. N. 1976. The chemical modification of proteins by group-scientific and site-specific reagents. *In* The Proteins, 3rd edition. H. Neurath & R. L. Hill, Eds. Vol. **II:** 2–103.
18. FLORY, P. J. 1956. Theory of elastic mechanisms in fibrous proteins. J. Am. Chem. Soc. **78**(20): 5222–5235.
19. JI, T. H. 1983. Bifunctional reagents. Methods Enzymol. **91:** 580–609.
20. SCHELLMAN, J. A. Solvent denaturation. Biopolymers **17**(5): 1305–1322.

Thermal and Chemical Deactivation of Soluble and Stabilized Enzymes

GUIDO GRECO, JR., GIUSEPPE MARRUCCI, AND NINO GRIZZUTI

Instituto di Principi di Ingegneria Chimica
Facoltà di Ingegneria
Università di Napoli
piazzale V. Tecchio
I-80125 Naples, Italy

LILIANA GIANFREDA

Facoltà di Farmacia
Università di Napoli
I-80125 Naples, Italy

INTRODUCTION

The activity of most enzymes decays rapidly with time (especially when they operate in noncellular environments), thus ruling out possible applications in industrial conversions. Traditional immobilization techniques essentially consist of binding the protein to an insoluble support. The introduction of additional bonds does not necessarily enhance the stability of the protein structure (increases in enzyme half-life upon immobilization have been observed episodically.[13] In any case, the overall costs grow, both directly and indirectly (because of losses in catalytic activity produced by the chemical manipulations undergone by the enzyme).

Recently, a new stabilization technique has been proposed, in which use is made of a polarized ultrafiltration (UF) cell.[2,3] The procedure consists of injecting the enzyme into the system together with a soluble, linear-chain, high-molecular-weight polymer. Both macromolecules are rejected, and accumulate in a very narrow region immediately upstream from the membrane surface, because of concentration polarization phenomena. Depending on operating conditions (enzyme and polymer amounts injected, permeating flow rates), extremely high overall concentrations are attained. As a consequence, entanglements are formed among the polymeric chains and a network is produced that surrounds the protein. The synthetic macromolecule thus exerts a mechanical constraint onto the enzyme that results in considerable reductions in the rate of unfolding of the protein structure.

These effects have been shown to hold for a wide variety of different enzymes and of stabilizing polymers as well. However, the results refer to single-step thermal deactivation kinetics:

$$N \rightarrow D$$

where N is the native enzyme and D is a totally inactive structure (possibly random coil).

Deactivation kinetics other than irreversible first-order are followed quite commonly. Indeed, nonlinear log(activity) versus time curves have been often described in the literature.[4,5] Since enzyme deactivation should be an intrinsically monomolecular process, overall kinetics other than first-order clearly suggest a multistep reaction scheme. Recently a two-step, in-series model has been proposed[6] that assumes that the native enzyme undergoes an irreversible transition to an equilibrium distribution of at least two intermediate structures of entirely different stability and kinetic behavior. These, in turn, degrade to a totally inactive, random-coil situation. Schematically, the pattern proposed can be depicted as:

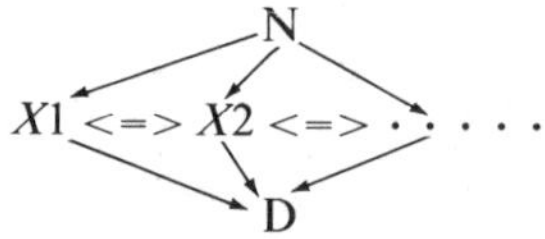

Where N is the native enzyme, $X1$, $X2$, . . . are the intermediate structures, D is random-coil. On the assumption that both steps follow first-order, irreversible kinetics,

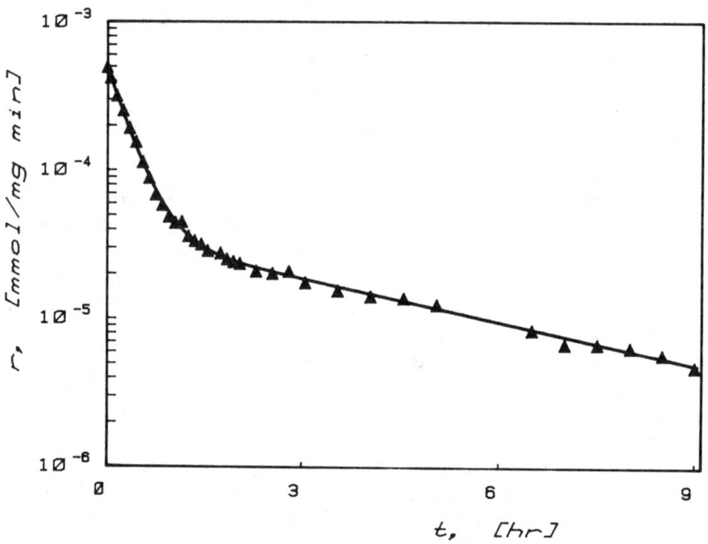

FIGURE 1. *p*-Nitro-phenyl phosphate hydrolysis by acid phosphatase. Specific enzyme activity versus deactivation time. Experimental conditions: soluble enzyme, T = 45°C, pH = 8.50 in 200 mM Tris-maleate, substrate concentration S° = 3 mM.

the overall specific activity (referred to the total amount N° of enzyme initially charged in a batch system) can be predicted to decay with time according to:

$$r = [r1 - r2\,k1/(k1 - k2)]\exp(-k1\,t) + [r2\,k1/(k1 - k2)]\exp(-k2\,t) \quad \textbf{(1)}$$

where r1 and r2 are the specific activities (both referred to N°) of N and of the particular equilibrium distribution of intermediate structures pertaining to the operating conditions adopted, respectively, and k1 and k2 are the two corresponding deactivation constants. If $0 < [r2\,k1/(k1 - k2)] < r1$, a two-slope log(activity) versus time curve is predicted by Equation 1 like the one depicted in FIGURE 1, which refers to acid-phosphatase thermal deactivation at pH = 8.50, T = 45°C, soluble enzyme.

Experimental runs have been carried out to check whether the stabilization techniques proposed[2,3] are still effective when the deactivation pattern agrees with the model described above. The results are discussed in the following. A few features of the deactivation process undergone by the soluble enzyme working at high dilution will be illustrated.

A further point is discussed in the present communication, that is, chemical denaturation by urea. When the deactivation is produced by a chemical, the points to be investigated are:

(1) the modifications in the deactivation kinetics, as compared to thermal denaturation, in terms of mechanism, kinetic constants, characteristics of the distribution of intermediate structures; and
(2) the extent of the enzyme stabilization by linear-chain polymers, if any.

Preliminary results are presented for both the homogeneous-phase enzyme and for the polymer-stabilized one.

MATERIALS AND METHODS

The enzyme, the deactivation kinetics of which are analyzed in detail, is acid phosphatase (E.C. 3.1.3.2, from potato, Boheringer Biochemia, Mannheim, FRG). The reaction used to monitor its activity as a function of time is hydrolysis of *p*-nitro-phenyl phosphate.

The linear-chain, water-soluble macromolecule employed as a stabilizing polymer is SEPARAN AP273 (Anionic poly-acrylamide, molecular weight >1,000,000, Dow Chemical Co., Lexington, MA).

All other chemicals used, reagent grade, are produced by B.D.H. (Poole, UK) or by Carlo Erba (Milano, Italy).

Experimental runs have been performed with soluble native enzyme. At the beginning of each test, the lyophilized, commercial enzyme preparation is dissolved in 200 m*M* buffer solution (Tris-maleate, pH 8.50, pH 5.60, or Na-citrate, pH 3.77), preheated at the deactivation temperature. This is taken as zero deactivation time. At predetermined time intervals, suitable samples of the enzymatic solution are drawn and injected into 1.5 ml of 3 m*M* *p*-nitro-phenyl phosphate solution in the same buffer. Substrate is saturating if reference is made to the native enzyme kinetics.[7] After an incubation time interval, which is always negligible as compared to the characteristic time scales of the deactivation process, the reaction is stopped with 1 *M* NaOH and the *p*-nitro-phenol concentration is subsequently read at 405 nm (extinction coefficient of 18.5 1/m*M*/cm). The enzyme amounts injected are such as to insure 10% conversion, at most. Therefore, for all practical purposes, all the runs have been carried out at the same constant substrate concentration of 3 m*M*. Each data point is referred to a deactivation time that is equal to the sample collection time plus the duration of the kinetic assay. The same urea concentration is present in both the buffer where the enzyme is dissolved and in the assay mixture, in order to apply this procedure to chemical deactivation tests also.

The stabilized enzyme runs have been performed in an ultrafiltration membrane reactor described elsewhere in detail.[8] Its basic features are:

(1) small internal volume (less than 3 ml, I.D. = 4 cm);
(2) plane membrane (GR 41, molecular weight cut-off of 20,000, The Danish Sugar Co., Nakskov, Denmark);

(3) protection of the region where the polarization layer develops, by a porous polyethylene disc;
(4) radial distribution of the feed within the reactor, in order to minimize preferential flow patterns.

The reactor is connected with a pressurized feed vessel containing the substrate solution. Permeate is collected automatically at predetermined time intervals. The injection of the macromolecular solution containing acid phosphatase and AP273 is performed via a multiport gas-chromatographic-type valve, inserted upstream from the UF cell, thus preventing perturbations in the flow rate and pressure regimes. Permeate flow rate and enzyme amounts fed are such that the reactor operates in a differential situation in all the experimental runs performed.

By inspection of Equation 1, it can be seen that the deactivation process is completely characterized once the four parameters r1, r2, k1, and k2 have been determined. The procedure adopted is described in Gianfreda *et al.*[6] Briefly, it consists of obtaining k2 and

$$[r2\,k1/(k1 - k2)]$$

by linear regression of the log(r) versus t experimental data at "high" deactivation time, that is, when N has disappeared. The differences between the experimental overall specific reaction rate r data and the corresponding values of

$$[r2\,k1/(k1 - k2)]\exp(-k2\,t)$$

calculated at the same deactivation time, yield

$$[r1 - r2\,k1/(k1 - k2)]\exp(-k1\,t)$$

values, which eventually produce k1 and $[r1 - r2\,k1/(k1 - k2)]$, by further linear regression. A variance-minimizing computer routine has been set up that optimizes the choice of the "final" portion of the deactivation curve.

RESULTS AND DISCUSSION

An initial set of experimental runs was devoted to the characterization of the soluble enzyme deactivation kinetics, at pH 3.77. Two-slope log(activity) versus time curves are still produced by soluble acid phosphatase, like the one shown in FIGURE 1, which refers to pH 8.50. In FIGURE 2, the values of the specific activity r2 of the intermediate structure distribution are reported as a function of deactivation temperature. A typical S-shaped pattern is followed, which recalls previous experimental results obtained for other proteins, plotting some structure-related parameters such as intrinsic viscosity, optical rotation, UV and absorbance (see Tanford[9] for a review). Needless to say, r1 obeys the Arrhenius law (activation energy 8 kcal/mole,[6]) which is quite obvious since N is a well-defined, unique protein structure. Within the framework of the model proposed, the temperature dependence of r2 should result from two opposing effects: (1) the rearrangement in the equilibrium distribution of intermediate structures, which presumably yields more deteriorated, less active situations, with increasing temperature, and (2) the temperature-increasing rate of hydrolysis, since the chemical reaction is an intrinsically activated process. By inspection of FIGURE 2, it can be seen that the first phenomenon is largely dominating, since, at least in an intermediate temperature range, r2 decreases with increasing temperature. Above the upper and below the lower boundary of the temperature interval explored, r2 becomes an increasing function of T. This agrees with the hypothesis that the intermediate structures are related to each other by an equilibrium condition. Indeed, at sufficiently high and at sufficiently low temperature, the equilibrium is totally biased toward one of the two limiting structures. Therefore, only one single protein species is present outside an intermediate temperature range and, hence, in both these regions, r2 increases exponentially with deactivation temperature, as usual.

In FIGURE 3, another diagram is shown that still refers to the temperature dependence of r2 for soluble acid phosphatase, now operating at pH 8.50. This is an Arrhenius plot only formally, since the specific rates refer to a set of different protein structures that change in composition by varying the temperature. The curve is not

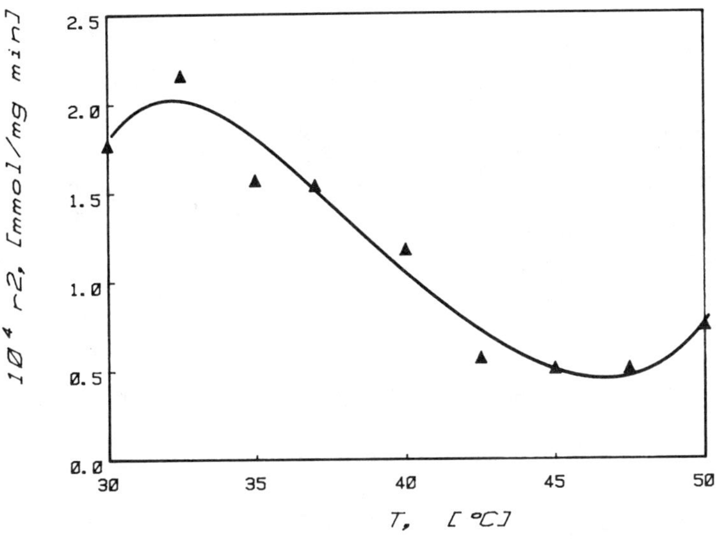

FIGURE 2. *p*-Nitro-phenyl phosphate hydrolysis by acid phosphatase. Specific activity of the intermediate structure distribution versus T. Experimental conditions: soluble enzyme, pH = 3.77 in 200 m*M* Na-citrate, S° = 3 m*M*.

S-shaped; nonetheless, the two contradicting effects described previously are still present, as shown by the strong reduction in apparent activation energy (E less than 2 kcal/mole) as compared to the values obtained for r1. This indicates that, at this pH, the decrease in r2 associated with the modifications in the equilibrium distribution of intermediate structures, is virtually compensated for by a corresponding increase in the rate of hydrolysis.

Since the soluble enzyme deactivation model has been further validated, experimental results can now be discussed concerning the polymer-stabilized situation. FIGURE 4 shows an activity versus time curve for AP273-stabilized acid phosphatase, operating at pH = 8.50, $T = 45$°C. These are the same experimental conditions as the ones adopted in the soluble enzyme run of FIGURE 1. It can be noticed that stabilization

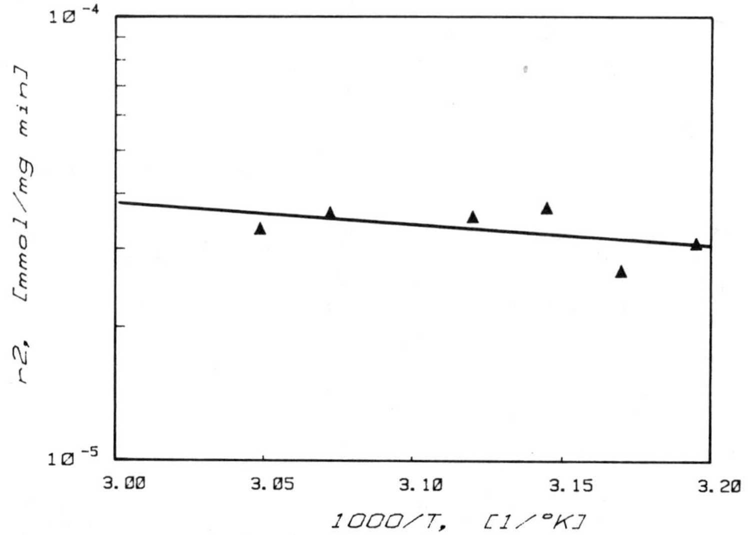

FIGURE 3. *p*-Nitro-phenyl phosphate hydrolysis by acid phosphatase. Specific activity of the intermediate structure distribution versus 1/T. Experimental conditions: soluble enzyme, pH = 8.50 in 200 m*M* Tris-maleate, S° = 3 m*M*.

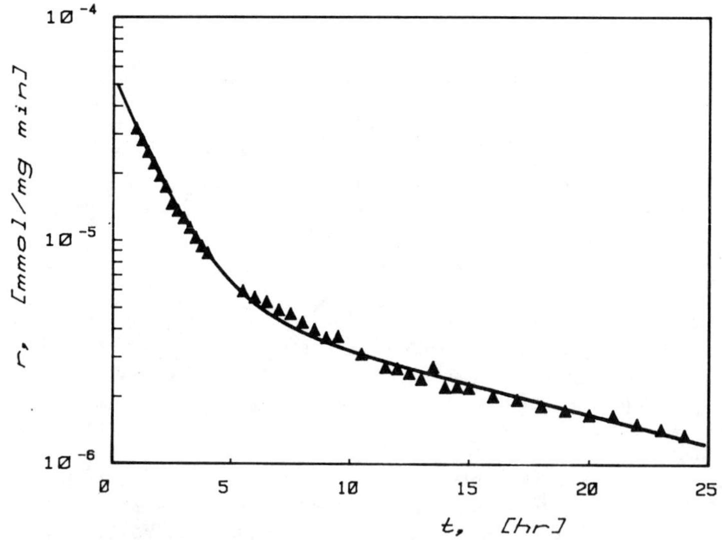

FIGURE 4. *p*-Nitro-phenyl phosphate hydrolysis by acid phosphatase. Specific enzyme activity versus deactivation time. Experimental conditions: AP273-stabilized enzyme, T = 45°C, pH = 8.50 in 200 m*M* Tris-maleate, S° = 3 m*M*.

does not affect the deactivation mechanism, since a two-slope behavior is still present. The kinetic parameters r1, r2, k1, and k2 have been calculated from both curves and are reported in TABLE 1, for comparison purposes.

It can be seen that:

(1) stabilization decreases the rate of the first step in the deactivation process (N → intermediate structure distribution) by a factor of more than 5;
(2) the rate of the second step (intermediate structure distribution → D) is decreased by a factor of approximately 4;
(3) the stabilized enzyme specific activity r1 (native enzyme N) is ninefold lower, as compared to the soluble enzyme ones;
(4) the stabilized enzyme initial specific activity r2 (intermediate structure distribution) is more than fivefold lower, as compared to that of the soluble enzyme.

As regards points (1) and (2), enzyme stabilization effects are obtained, when a complex, two-step deactivation mechanism is followed, which are comparable to the one observed when dealing with single-step (N → random coil) denaturation patterns.[2,3] Hence, the hypothesis that the reduction in the rate of unfolding of the protein structure takes place because of a relatively mild mechanical constraint exerted onto the enzyme by the polymeric network, seems to apply also to two-step deactivation mechanisms. It follows that the structural variations undergone by the enzymatic protein in the step N → intermediate structure distribution are comparable in extent to the ones that yield random-coil, starting from a native structure, although a residual enzymatic activity is still retained.

TABLE 1. *p*-Nitro-phenyl Phosphate Hydrolysis by Acid Phosphatase: Comparison between Soluble and Stabilized Enzyme[a]

Status[b]	10^5 r1	10^6 r2	10 k1	10^2 k2
Soluble	49.4	33.6	31.5	22.2
Stabilized	5.46	5.17	6.14	6.20

[a]pH = 8.50, T = 45°C.
[b]r1, r2 = mmol/mg min; k1, k2 = 1/h.

Points (3) and (4) indicate that the reductions in specific activity undergone by N and by the intermediate structure distribution are not equal. This could mean that the stabilization procedures affect also the thermodynamics of the deactivation process. On the contrary, if these observed reductions were a consequence of substrate mass-transfer resistances being significant, the stabilization effects measured would be spurious, at least partially. Indeed, effectiveness factors of less than one, associated with time-decaying enzyme activities, do result in apparent reductions in the rate of the deactivation process. However, this does not seem to be the case for the UF membrane reactor, since, within the polarization layer, where the enzymatic reaction takes place, substrate is transferred by an essentially convective mechanism. Furthermore, r2 is one order of magnitude lower than r1 and hence, the assumption that the overall reaction rate data are affected by substrate mass-transfer limitations even when N has disappeared does not seem realistic. Nonetheless, further experimental runs have been performed to rule out definitely this possibility.

In order to simplify the problem, an optimal acid phosphatase pH of 5.60 has been chosen. Indeed, at this pH, the deactivation process is single step[10] and the rate of

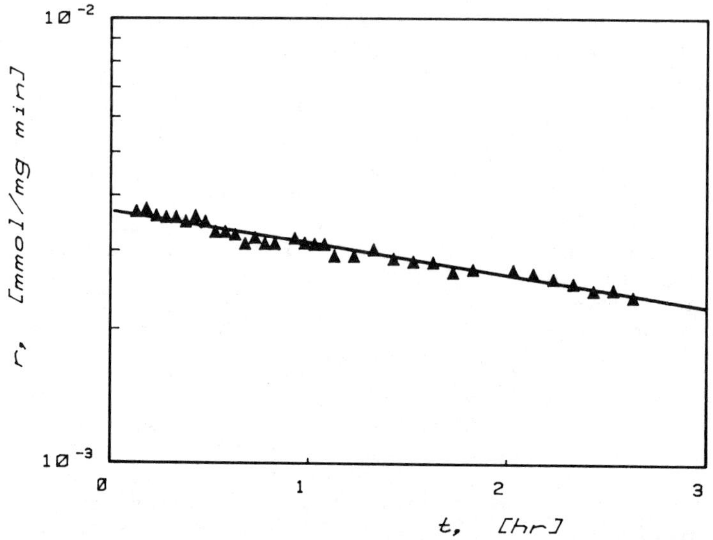

FIGURE 5. *p*-Nitro-phenyl phosphate hydrolysis by acid phosphatase. Specific enzyme activity versus deactivation time. Experimental conditions: soluble enzyme, T = 45°C, pH = 5.60 in 200 m*M* Tris-maleate, S° = 3 m*M*.

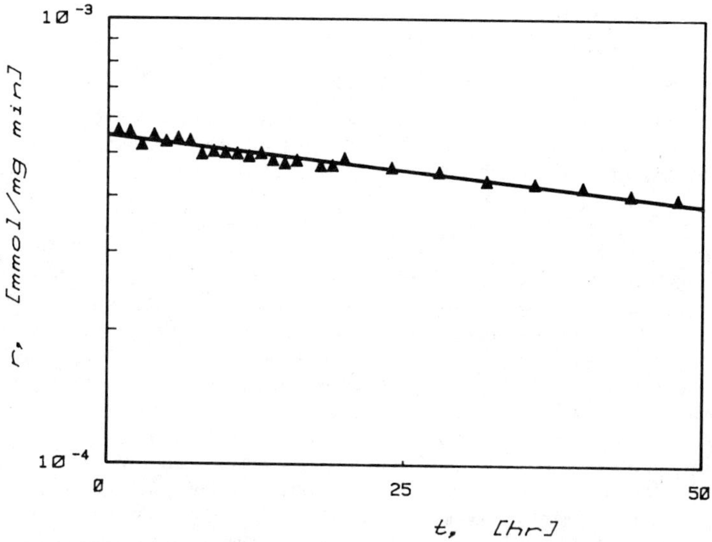

FIGURE 6. *p*-Nitro-phenyl phosphate hydrolysis by acid phosphatase. Specific enzyme activity versus deactivation time. Experimental conditions: AP273-stabilized enzyme, T = 45°C, pH = 5.60 in 200 m*M* Tris-maleate, S° = 3 m*M*.

TABLE 2. *p*-Nitro-phenyl Phosphate Hydrolysis by Acid Phosphatase: Comparison between Soluble and Stabilized Enzyme[a]

Status[b]	10^4 r°	10^3 k
Soluble	36.9	165
Stabilized	5.53	7.24

[a]pH = 5.60, T = 45°C.
[b]r° = mmol/mg min; k = 1/h.

reaction is much higher than that obtained at a pH of 8.50. The first step is to verify if at a pH of 5.60, reductions in activity also take place for the stabilized enzyme situation, as compared to that with the soluble enzyme. In FIGURES 5 and 6, log (activity) versus time data are reported, obtained with soluble and stabilized enzyme, respectively, at T = 45°C, pH = 5.60. The linearity of the data confirms that a single-step deactivation pattern is followed (N → D) in both cases. Therefore, only two parameters completely characterize the deactivation process, that is, the initial enzyme activity r° and the kinetic constant k. The values obtained by linear regression of the data are reported in TABLE 2. An approximately sevenfold reduction in enzyme activity can be observed, which agrees with the results obtained at pH = 8.50. Incidentally, it can be seen that an even better stabilization is achieved, since the rate of deactivation is reduced by 22.8.

Since the general behavior of the data at a pH of 5.60 is similar to that observed at a pH of 8.50, a further set of experimental runs has been devoted to the determination of the activation energy of the hydrolysis reaction at the former pH, both with the soluble and the stabilized enzyme. Only the final results, in terms of initial enzyme activity r° versus reciprocal absolute temperature are shown in FIGURE 7. The

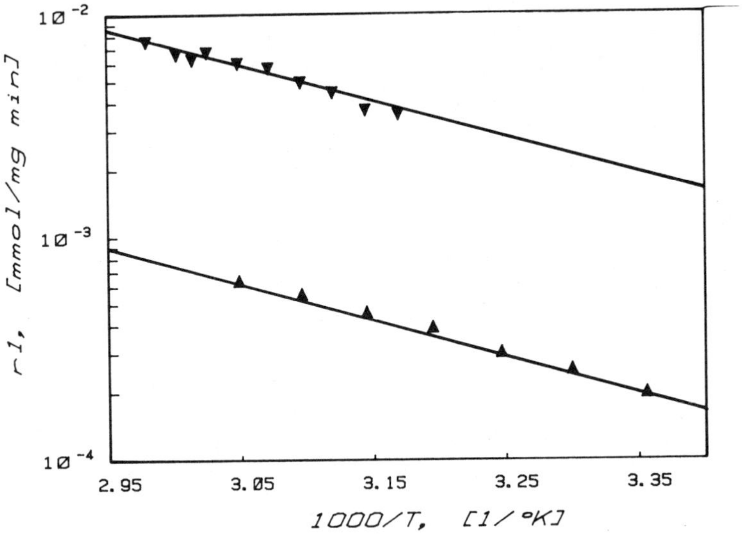

FIGURE 7. *p*-Nitro-phenyl phosphate hydrolysis by acid phosphatase. Specific activity of the native enzyme versus 1/T. Experimental conditions: pH = 5.60 in 200 m*M* Tris-maleate, S° = 3 m*M*, ▼ curve: soluble enzyme, ▲ curve: AP273-stabilized enzyme.

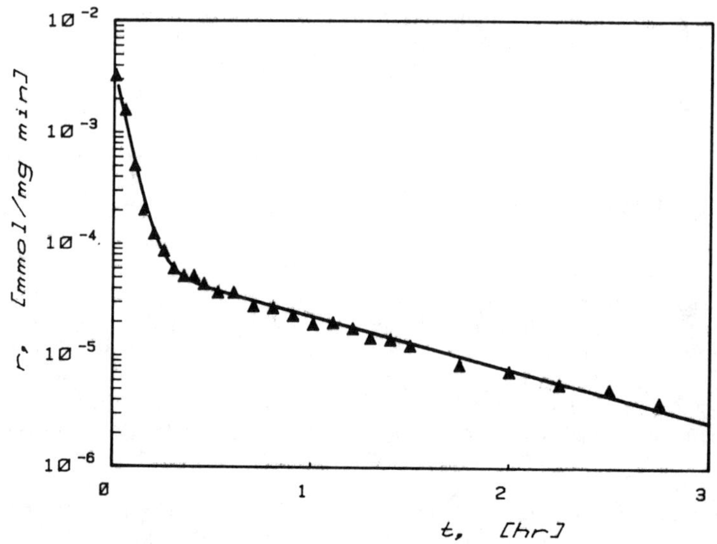

FIGURE 8. *p*-Nitro-phenyl phosphate hydrolysis by acid phosphatase. Specific enzyme activity versus deactivation time. Experimental conditions: soluble enzyme, T = 45°C, pH = 5.60 in 200 m*M* Tris-maleate, S° = 3 m*M*. Urea concentration = 6 M.

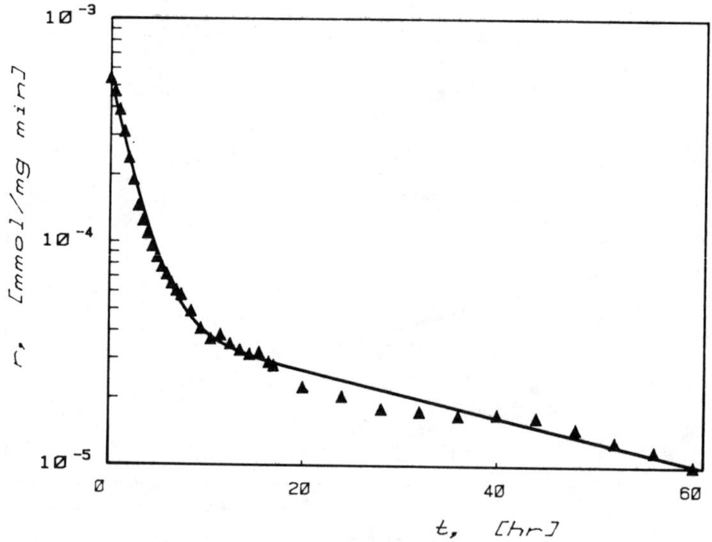

FIGURE 9. *p*-Nitro-phenyl phosphate hydrolysis by acid phosphatase. Specific enzyme activity versus deactivation time. Experimental conditions: AP273-stabilized enzyme, T = 45°C pH = 5.60 in 200 m*M* Tris-maleate, S° = 3 m*M*. Urea concentration = 6 *M*.

calculated values of the activation energy are: 8 kcal/mole (soluble enzyme) and 7.6 kcal/mole (stabilized enzyme). In other words, the activation energies coincide, well within the limits of experimental error. Therefore:

(1) no mass-transfer limitations affect the stabilized enzyme data at a pH of 5.60, and even more so at pH of 8.50, because of the reduced enzyme activity;
(2) an actual enzyme stabilization has been achieved at both pH values; and
(3) the reduction in overall rate of reaction upon stabilization stems from an actual decrease in enzyme activity and presumably depends on a specific interaction between the enzyme and its peculiar environment (tight network produced by the electrically charged polymer within the polarization layer).

In previous work, situations have been reported where no appreciable variation occurred in activity as a consequence of polymer stabilization,[11] but there have been reports where even stronger reductions took place than the ones discussed in this paper.[12]

A final set of runs has been carried out on chemical deactivation. The experimental conditions adopted are: pH = 5.60, T = 45°C, urea concentration = 6 *M*. The usual two experiments have been performed, namely, with soluble and with AP273-stabilized

TABLE 3. *p*-Nitro-phenyl Phosphate Hydrolysis by Acid Phosphatase, Chemical Denaturation by 6 *M* Urea: Comparison between Soluble and Stabilized Enzyme[a]

Status[b]	10^4 r1	10^5 r2	10 k1	10^2 k2
Soluble	37.2	6.28	185	109
Stabilized	5.22	4.04	4.44	2.39

[a]pH = 5.60, T = 45°C.
[b]r1, r2 = mmol/mg min; k1, k2 = 1/h.

enzyme. The results appear in FIGURES 8 and 9. The kinetic parameters are reported in TABLE 3. By inspection of TABLE 3, it can be seen that:

(1) a two-step mechanism holds at a pH of 5.60, when a chemical deactivation process takes place (whereas, at the same pH, single-step, irreversible first-order kinetics seem to be followed for thermal inactivation);
(2) both steps in the deactivation pattern are stabilized by the presence of the linear-chain polymer approximately to the same extent; the first, N → intermediate structure distribution, by a factor of 42, the second, leading to random coil, by a factor of 46;
(3) both specific activities r1 agree with the corresponding ones reported in TABLE 2 (this could be predicted, since each refers to the same situation: soluble native enzyme and stabilized native enzyme, pH = 5.60, T = 45°C, respectively); hence, the same, approximately sevenfold reduction in the specific activity of N takes place; and
(4) on the contrary, r2 is reduced by a factor of only 1.5, because of the presence of the stabilizing polymer network; again, this discrepancy can be interpreted as an effect of stabilization on the thermodynamics of the equilibrium among the intermediate protein structures.

CONCLUSIONS

The basic conclusions drawn can be summarized briefly as follows:

(1) The two-step thermal deactivation mechanism proposed by Gianfreda,[6] which postulates the existence of an equilibrium distribution of intermediate protein structures, reached by the native enzyme before its final denaturation to random-coil, is consistent with S-shaped specific activity versus temperature experimental curves obtained with soluble acid phosphatase.

(2) When operating in a polymer-stabilizing enzyme situation, reductions in the rate occur for both thermal deactivation steps; if the stabilization occurs by means of a purely mechanical constraint exerted onto the protein by the polymeric network, this implies that the structural modifications undergone by the enzyme in the transition to the intermediate structures are not minor.

(3) When the enzyme deactivation is produced by urea, a two-step mechanism is followed even at a pH value at which thermal denaturation is a single-step phenomenon.

(4) Polymer stabilization is quite effective in countering both chemical deactivation steps.

(5) Stabilization is accompanied by reductions in the specific activities of all the protein structures present; for this phenomenon, substrate mass transfer limitations have been shown to play no role.

(6) The specific activity reductions undergone by the native enzyme N are not equal to the ones experienced by the intermediate structures. This seems to imply that modifications are produced by the stabilizing polymer on the thermodynamics of the distribution of the latter.

SUMMARY

The thermal and chemical stability of enzymes is usually analyzed by measuring variations in initial rates of the enzymatic reaction, or in some protein structure-related physicochemical property, as a function of operating conditions (temperature, pH, denaturant, concentration, etc.)

Little, if any, attention is devoted to the long-term behavior, as regards deactivation phenomena. Hence, reliable predictions in terms of specific activity versus time cannot be made. This is quite unsatisfactory, mainly when an industrial application is envisaged.

In the present communication, experimental results are discussed that describe some complex mechanisms followed in thermal and chemical enzyme denaturation processes. The dependence of the kinetic parameters on operating conditions is analyzed and comparisons are made between the results obtained when the enzyme operates under soluble and "stabilized" conditions. Stabilization is achieved within a completely polarized ultrafiltration reactor in the presence of high molecular weight, linear-chain macromolecules.

REFERENCES

1. Goldman, R., L. Goldstein & E. Katchalski. 1971. *In* Biochemical Aspects of Reactions on Solid Supports. G. R. Stark, Ed. Academic Press. New York. pp. 1–78.

2. GRECO, G. JR. & L. GIANFREDA. 1981. Enzyme stabilization by linear-chain polymers in U.F. membrane reactors. Biotechnol. Bioeng. **23:** 2199–2210.
3. GIANFREDA, L. & G. GRECO, JR. 1981. The stabilizing effect of soluble macromolecules on enzyme performance. Biotechnol. Lett. **3:** 33–38.
4. BOUIN, J. C. & H. O. HULTIN. 1982. Stabilization of glucose oxidase by immobilization/modification as a function of pH. Biotechnol. Bioeng. **24:** 1225–1231.
5. SADANA A. 1982. Deactivation model involving a grace period for immobilized and soluble enzymes. Enzyme Microb. Technol. **4:** 44–46.
6. GIANFREDA, L., G. MARRUCCI, N. GRIZZUTI & G. GRECO, JR. Acid phosphatase deactivation by a series mechanism. Biotechnol. Bioeng. In press.
7. CANTARELLA, M., M. H. REMY, V. SCARDI, F. ALFANI, G. IORIO & G. GRECO, JR. 1979. Kinetic behavior of acid phosphatase/albumin copolymers in homogeneous phase and under gel-immobilized conditions. Biochem. J. **179:** 15–20.
8. MODAFFERI, M. & P. FORMISANO. 1983. Chemical Engineering Thesis. University of Naples, Naples, Italy.
9. TANFORD C. 1968. Protein denaturation. Adv. Protein Chem. **23,** 121–282.
10. GRECO, G. JR., A. M. LIVOLSI, F. MANSI, M. R. SCARFI & L. GIANFREDA. 1981. UF membrane enzymatic reactors. Polarized versus stirred reactor performance. Eur. J. Appl. Microbiol. Biotechnol. **13:** 251–253.
11. GIANFREDA, L., A. M. LIVOLSI, M. R. SCARFI & G. GRECO, JR. 1982. β-D-Glucosidase stabilization in a polarized UF reactor. Enzyme Microb. Technol. **4:** 322–326.
12. GRECO, G. JR., F. VERONESE, R. LARAGAJOLLI & L. GIANFREDA. 1983. Purified penicillin acylase performance in a stabilized membrane reactor. Eur. J. Appl. Microbiol. Biotechnol. In press.

Mechanisms of Irreversible Thermoinactivation of Enzymes

STEPHEN E. ZALE AND ALEXANDER M. KLIBANOV

Laboratory of Applied Biochemistry
Department of Nutrition and Food Science
Massachusetts Institute of Technology
Cambridge, Massachusetts 02139

The usefulness of enzymes as industrial catalysts is severely limited by the fact that, at elevated temperatures, they irreversibly inactivate. In order to overcome this limitation, rational approaches to thermostabilization of enzymes must be developed.[1,2] The latter, in turn, require a mechanistic understanding of why and how enzymes irreversibly inactivate upon heating.

For this study, bovine pancreatic ribonuclease A has been selected as a model enzyme, with the expectation that the findings obtained will be applicable to other enzymes as well. Ribonuclease is an ideal model because it is a small, simple protein whose structure and conformational dynamics have been thoroughly characterized (see Richards and Wyckoff[3] and Blackburn and Moore[4]).

In a previous publication,[5] we have demonstrated that ribonuclease undergoes irreversible thermoinactivation in accordance with the following two-step kinetic scheme, first proposed by Lumry and Eyring[6]:

$$\begin{array}{c}\text{Native} \\ \text{Enzyme}\end{array} \xrightarrow{\quad K \quad} \begin{array}{c}\text{Reversibly} \\ \text{Thermounfolded} \\ \text{Enzyme}\end{array} \xrightarrow{\quad \mathrm{k} \quad} \begin{array}{c}\text{Irreversibly} \\ \text{Thermoinactivated} \\ \text{Enzyme}\end{array}$$

where irreversible thermoinactivation (governed by rate constant k) is preceded by reversible thermal unfolding (governed by equilibrium constant *K*). The aim of our current research is to determine the exact nature of the irreversible step[7] as a function of pH within the range from pH 3 to 9, where most enzymes function and are expected to find practical utility. The experiments presented here are focused on irreversible thermoinactivation of ribonuclease at two representative values of the pH, 4 and 8, which are within this range, but are sufficiently far apart so that different mechanisms may contribute to irreversible thermoinactivation.

COVALENT VERSUS CONFORMATIONAL IRREVERSIBLE THERMOINACTIVATION PROCESSES

For a single polypeptide chain protein with no prosthetic groups, such as ribonuclease, possible mechanisms of irreversible thermoinactivation can be divided into two categories:[1] (1) those in which enzyme inactivation is the result of a covalent change in the protein molecule,[8] and (2) those in which no covalent change has occurred, that is, the cause of irreversible inactivation is purely conformational.[7] An important goal of this study was to assess the relative contributions of these two types of inactivation processes toward the irreversible thermoinactivation of ribonuclease.

In contrast to that of covalent mechanisms, the irreversibility of conformational processes (e.g., formation of incorrect structures[7]) must be due only to noncovalent forces (i.e., hydrophobic and electrostatic interactions, H-bonding, etc.). Noncovalent interactions of the same nature are responsible for maintenance of the correct tertiary structure of native proteins,[9] hence it is likely that reagents that disrupt the tertiary structure of native proteins, that is, denaturants such as GuHCl and acetamide, will also stabilize enzymes against formation of incorrect structures. Such stabilization has been reported against irreversible thermoinactivation of both trypsin and α-chymotrypsin.[10] Given that denaturants will prevent conformational irreversible inactivation processes, determination of the effects of such reagents on the kinetics of irreversible thermal inactivation of an enzyme should provide a means to distinguish between conformational and covalent thermoinactivation processes.

A second criterion for differentiating conformational and covalent inactivation mechanisms is whether or not the irreversibly inactivated enzyme can be reactivated by complete unfolding to a random coil, followed by refolding under physiological conditions, as shown below:

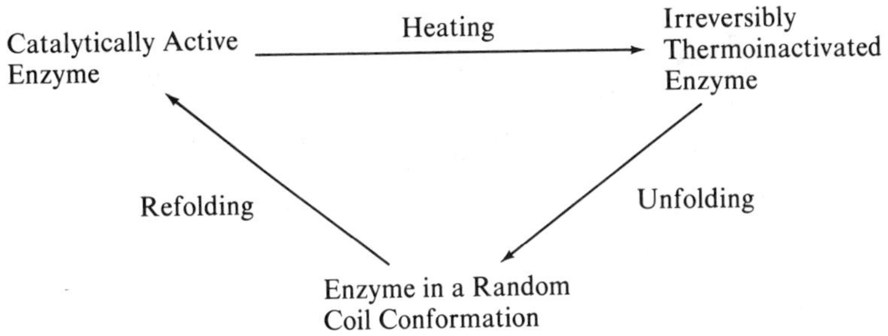

It is expected that reduction of disulfide bonds in the presence of a denaturant (conditions under which native proteins are completely unfolded) should be sufficient to overcome the kinetic barriers responsible for the apparent irreversibility of inactivation by incorrect structure formation. This has been shown to be the case for thermoinactivated trypsin, which was completely reactivated by this method.[7] On the other hand, a covalently altered thermoinactivated enzyme is not expected to be amenable to reactivation by this procedure.

IRREVERSIBLE THERMOINACTIVATION OF RIBONUCLEASE AT pH 8 AND 90°C

The two criteria described in the previous section (i.e., the effect of denaturants on thermoinactivation kinetics and the reactivation of the thermoinactivated enzyme) were applied to ribonuclease thermoinactivation at pH 8, in order to assess the relative contributions of covalent and conformational inactivation processes.

The effect of 6 *M* GuHCl on the time course of ribonuclease thermoinactivation at pH 8 and 90°C is illustrated in FIGURE 1, curves a and b. The stabilizing effect of GuHCl is evident, resulting in a sixfold reduction of the rate constant of inactivation. Addition of another denaturant, 8 *M* acetamide, resulted in a similar stabilization. The observation that denaturants stabilize the enzyme indicates that at pH 8, ribonuclease inactivates via a conformational mechanism. The fact that stabilization is only partial suggests that covalent inactivation processes also occur, the kinetics of which are reflected in the time course of inactivation in the presence of denaturant (where conformational processes are prevented).

The results of attempts to reactivate ribonuclease irreversibly thermoinactivated at pH 8 were consistent with these conclusions. Reduction of the inactivated enzyme in 8 *M* urea, followed by reoxidation in the absence of denaturant yielded a recovery of over one-third of the specific activity of the native enzyme. Ribonuclease inactivated at pH 8 in 6 *M* GuHCl (where incorrect structure formation does not occur) could not be reactivated when subjected to this procedure.

Thus, the results on ribonuclease thermoinactivation at pH 8 and 90°C presented so far can be summarized as follows:

(1) Conformational irreversible inactivation processes (i.e., incorrect structure formation) contribute to RNase thermoinactivation, as evidenced by the stabilizing effect of denaturants and by the partial reactivation of the irreversibly inactivated enzyme.

(2) Competing covalent mechanism(s) also occur, as indicated by the observations that both stabilization and reactivation are only partial, and that ribonuclease inactivated in the presence of denaturant is not amenable to reactivation at all.

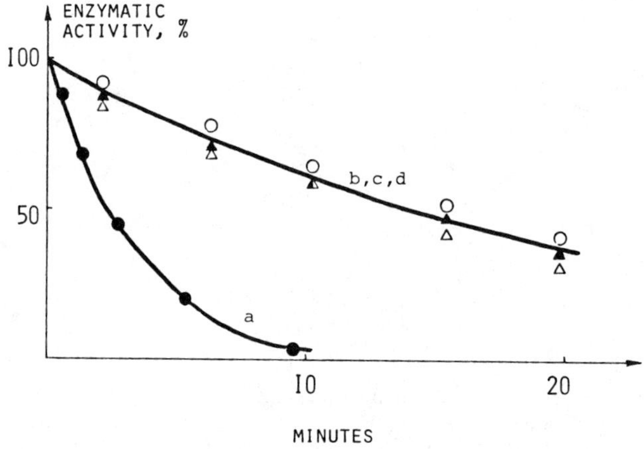

FIGURE 1. The time course of irreversible thermoinactivation of ribonuclease at 90°C and pH 8. (a) (●) Ribonuclease; (b) (○) ribonuclease in 6 *M* GuHCl; (c) (▲) succinyl-ribonuclease; (d) (△) succinyl-ribonuclease in 6 *M* GuHCl. All experiments were carried out in 0.01 *M* phosphate buffer at 50 μM enzyme concentration.

In considering why ribonuclease readily inactivates via an incorrect folding mechanism at pH 8, it occurred to us that perhaps this was related to the fact that pH 8 is close to the isoelectric point of the enzyme (pH 9.6[3]). Thus, at pH 8, the reversibly thermounfolded protein is relatively uncharged, and may therefore be able to assume an incorrect structure without hindrance by electrostatic repulsions. To test this hypothesis, we altered the isoelectric point of ribonuclease by chemical modification with succinic anhydride. This procedure converts positively charged amino groups to negatively charged carboxylates, as shown below:

$$Enz-NH_3^+ + \begin{array}{c} H_2C-C(=O) \\ | \quad\quad\quad O \\ H_2C-C(=O) \end{array} \longrightarrow Enz-NH-\overset{O}{\overset{\|}{C}}-CH_2-CH_2-\overset{O}{\overset{\|}{C}}-O^-$$

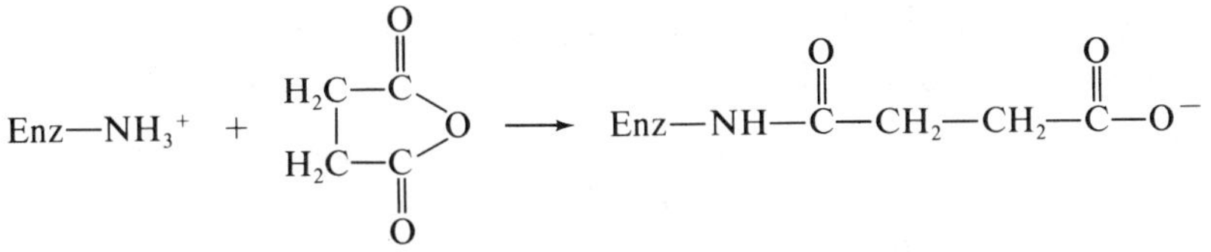

Succinylation of ribonuclease in accordance with a method adapted from Giese and Vallee[11] resulted in modification of 5–7 of the 11 amino groups present with no loss of catalytic activity.

The kinetics of thermoinactivation of succinylated ribonuclease at 90°C and pH 8 were then measured both in the absence and presence of 6 M GuHCl. The results of these experiments are shown in FIGURE 1, curves c and d, respectively. The two major findings obtained are that (1) succinylation of ribonuclease stabilizes the enzyme to the same degree as addition of 6 M GuHCl, indicating that succinylation also prevents inactivation by incorrect structure formation and that (2) addition of 6 M GuHCl has no effect on the kinetics of inactivation of the succinylated enzyme, suggesting that the succinylated enzyme inactivates via a covalent process. Furthermore, the fact that the unmodified enzyme in the presence of GuHCl, and the succinylated enzyme both the absence and presence of GuHCl (FIG. 1, curves b, c, and d, respectively) all have identical inactivation kinetics suggests that the same covalent process is responsible for thermoinactivation in all three cases.

The final problem we addressed concerning ribonuclease thermoinactivation at pH 8 was the determination of the exact nature of the covalent process that inactivates ribonuclease under conditions where the conformational process is prevented. It has been shown[12] that at alkaline pH, cystine residues of ribonuclease, and other proteins as well, are labile to a base-catalyzed β-elimination reaction. This results in the formation of thiocysteine and dehydroalanine, both of which can undergo a variety of subsequent chemical transformations.[13] Conceivably, β-elimination of cystine might also occur at high temperature, under less extreme alkaline conditions, and hence contribute to enzyme thermoinactivation.

To evaluate this possibility, ribonuclease was heated in 6 M GuHCl (to exclude conformational processes) at pH 8 and 90°C, and both the loss of catalytic activity and the destruction of cystine residues were simultaneously measured. Unreacted disulfides were conveniently determined by reducing the enzyme, and then titrating free sulfhydryls using Ellman's reagent.[14] The results of this experiment are shown in FIGURE 2A. The kinetics of the two processes are consistent with the destruction of a single cystine residue in ribonuclease resulting in the loss of catalytic activity.

In order to ascertain that the β-elimination of cystine is not a consequence of the presence of GuHCl, the kinetics of cystine destruction were measured both in the absence and presence of the denaturant. The two curves are shown in FIGURE 2B. Clearly, the covalent process is not affected by GuHCl—an assumption implicit in the use of stabilization by denaturants as a criterion for distinguishing between covalent and conformational mechanisms. Importantly, the figure also indicates that heating the enzyme beyond the time required for complete thermoinactivation results in the eventual destruction of all four cystine residues. That all four residues, which occur in totally different bonding environments within the ribonuclease molecule, are more or less equally labile to thermal disruption strongly suggests that this process is not a specific feature of ribonuclease, but is a general one for all proteins containing cystine residues. In fact, it was found that the rates of cystine destruction in other proteins (soybean trypsin inhibitor, α-chymotrypsinogen) under these conditions (pH 8, 90°) were similar to those observed for ribonuclease.

IRREVERSIBLE THERMOINACTIVATION OF RIBONUCLEASE AT pH 4 AND 90°C

To determine the contribution of conformational mechanisms toward thermoinactivation of ribonuclease at pH 4 and 90°C, the criteria outlined in a previous section

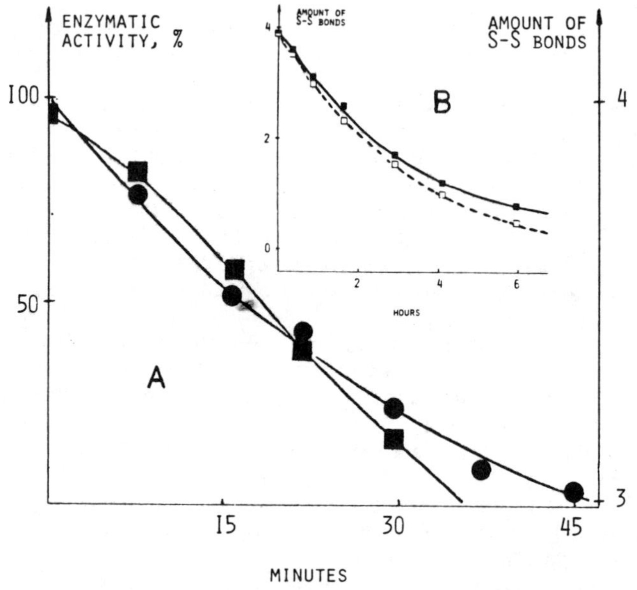

FIGURE 2. A. The time course of irreversible thermoinactivation of ribonuclease and of destruction of cystine residues at 90°C and pH 8 in 6 *M* GuHCl. (●) Percent of initial enzyme activity; (■) number of cystine residues. B. Destruction of cystine residues of ribonuclease at 90°C and pH 8 over a longer period of time. (□) Ribonuclease; (■) ribonuclease in 6 *M* GuHCl. Experimental conditions the same as in FIGURE 1.

were applied to this process. It was found that addition of either GuHCl or acetamide had no appreciable effect on the rate constant of inactivation, and that the thermoinactivated enzyme could not be reactivated using the approach described earlier. These results indicate that incorrect structure formation is *not* involved in ribonuclease inactivation at pH 4, and hence, inactivation must be due to one or more covalent mechanisms.

In order to identify the covalent change(s) responsible for ribonuclease thermoinactivation at pH 4, the thermoinactivated enzyme was first subjected to complete acid hydrolysis and its amino acid composition was compared to that of the native enzyme. No change in amino acid composition was observed. This is not surprising, since any amino acid stable under the conditions of acid hydrolysis (6 *N* HCl, 110°, 20 hours) would also be expected to be stable under the much milder conditions required to inactivate ribonuclease at pH 4 (90°, 20 hours).

There are, however, several possible covalent processes that may contribute to enzyme inactivation, but are not detectable by amino acid analysis of an acid hydrolysate. These include (1) deamidation of glutamine and asparagine, (2) racemization of amino acid residues, (3) cleavage of peptide bonds, and (4) formation of new peptide bonds.

The kinetics of deamidation of glutamine and asparagine residues of ribonuclease at pH 4 and 90°C were determined by measuring the time course of the appearance of ammonia, which was conveniently assayed using glutamate dehydrogenase.[15] We found that deamidation does occur under these conditions, but at too slow a rate to contribute significantly toward the thermoinactivation of ribonuclease.

In order to assess the likelihood of racemization of amino acid residues as a cause of ribonuclease inactivation, a number of N-acetyl amino acids and N-acetyl amino acid amides were employed as models of amino acid residues in a protein molecule. Solutions of these compounds were heated, and the kinetics of racemization determined by polarimetry. The rates observed were extremely slow relative to the rate of inactivation of ribonuclease. This result indicates that racemization of amino acids probably does not contribute appreciably toward thermoinactivation.

The possibility that irreversible thermoinactivation of ribonuclease at pH 4 is the result of peptide bond hydrolysis is currently under investigation. Preliminary data strongly suggest that cleavage of the polypeptide backbone is a major cause of thermoinactivation. Amino acid analysis of low-molecular-weight products of ribonuclease thermoinactivation at pH 4 and 90°C, isolated by ultrafiltration, revealed the presence of alanyl-seryl-valine, the COOH-terminal tripeptide of ribonuclease, presumably the product of the hydrolysis of the peptide bond linking Asp-121 to Ala-122. The rate of this process was measured, and was found to be approximately 20% of the overall thermoinactivation rate.

It is known from the literature[16] that peptide bonds involving aspartic acid are preferentially hydrolyzed under conditions of high temperature and dilute acid. It is conceivable that this process may occur at other aspartic acid residues of the ribonuclease molecule (of which there are a total of five) in addition to Asp-121. Results of analysis of thermoinactivated ribonuclease using SDS-polyacrylamide gel electrophoresis are consistent with this hypothesis. Further experiments, including measurement of the appearance of new amino and carboxyl termini during thermoinactivation are currently in progress.

CONCLUSIONS

The nature of the irreversible step in the thermoinactivation of ribonuclease at 90°C has been fully characterized at pH 8 and tentatively identified at pH 4.

At pH 8, the major thermoinactivation process is the formation of incorrect structures. A second process, the destruction of cystine residues, also occurs at pH 8, and is the cause of inactivation under conditions where the conformational mechanism is excluded.

Thermoinactivation of ribonuclease at pH 4 does not involve conformational irreversible processes, and hence must be due only to covalent mechanisms. One such mechanism has been identified as the scission of the peptide bond linking Asp-121 and

Ala-122, resulting in the release of the COOH-terminal tripeptide. The possibility that other peptide bonds involving aspartic acid residues are also labile to hydrolysis is under investigation.

The practical reason for identification of the mechanisms of irreversible inactivation of enzymes is to provide a rational basis for their thermostabilization. Prevention of the incorrect folding of ribonuclease at pH 8 by succinylation of the enzyme is an example of this approach put into practice. Manipulation of the pI of an enzyme may prove to be a general method for stabilization against conformational thermoinactivation processes.

The possibility that at mildly acidic pH, peptide bonds involving aspartic acid are potential "weak links" in the peptide backbone of enzymes suggests a general approach toward enzyme thermostabilization utilizing recent developments in genetic engineering. Replacement of aspartic acid residues with glutamic acid via directed mutation is likely to increase the thermal stability of an enzyme without altering its desirable catalytic properties. Again, the approach toward stabilization is dependent upon knowledge of the nature of the thermoinactivation mechanism.

REFERENCES

1. Klibanov, A. M. 1983. Stabilization of enzymes against thermal inactivation. Adv. Appl. Microbiol. **29:** 1–28.
2. Schmid, R. D. 1979. Stabilized soluble enzymes. Adv. Biochem. Eng. **12:** 41–118.
3. Richards, F. M. & H. W. Wyckoff. 1970. Bovine pancreatic ribonuclease. *In* The Enzymes. 3rd Edition. P. D. Boyer, Ed. **4:** 647–806.
4. Blackburn, P. & S. Moore. 1982. Pancreatic ribonuclease. *In* The Enzymes, 3rd edition. P. D. Boyer, Ed. **15:** 317–433.
5. Zale, S. E. & A. M. Klibanov. 1983. On the role of reversible denaturation (unfolding) in the irreversible thermal inactivation of enzymes. Biotechnol. Bioeng. **25:**2221–2230.
6. Lumry, R. & H. Eyring. 1954. Conformational changes of proteins. J. Phys. Chem. **58:** 110–120.
7. Klibanov, A. M. & V. V. Mozhaev. 1978. On the mechanism of irreversible thermoinactivation of enzymes and possibilities for reactivation of "irreversibly" inactivated enzymes. Biochem. Biophys. Res. Commun. **83:** 1012–1014.
8. Feeney, R. E. 1980. Overview on the chemical deteriorative changes of proteins and their consequences. *In* Chemical Deterioration of Proteins. J. R. Whitaker & M. Fujimaki, Eds. ACS. Washington, D.C. pp. 1–47.
9. Kauzmann, W. 1959. Some factors in the interpretation of protein denaturation. Adv. Protein Chem. **14:**1–59.
10. Martinek, K., V. S. Goldmacher, A. M. Klibanov & I. V. Berezin. 1975. Denaturing agents (urea, acetamide) protect enzymes against irreversible thermoinactivation: A study with native and immobilized α-chymotrypsin and trypsin. FEBS Lett. **51:** 152–155.
11. Giese, R. W. & B. L. Vallee. 1972. A novel class of reagents for protein modification. I. Maleic anhydride—iron tetracarbonyl. J. Am. Chem. Soc. **94:** 6199–6200.
12. Bohak, Z. 1964. N^{ϵ}-(DL-2-amino-2-carboxyethyl)-L-lysine, a new amino acid formed on alkaline treatment of proteins. J. Biol. Chem. **239:** 2878–2887.
13. Nashef, A. S., D. T. Osuga, H. S. Lee, A. I. Ahmed, J. R. Whitaker & R. E. Feeney. 1977. Effects of alkali on proteins. Disulfides and their products. J. Agric. Food Chem. **25:** 245–251.
14. Taniuchi, H. 1970. Formation of randomly paired disulfide bonds in des-(121–124)-ribonuclease after reduction and reoxidation. J. Biol. Chem. **245:** 5459–5468.
15. Kun, E. & E. B. Kearney. 1974. Ammonia. *In* Methods of Enzymatic Analysis, 2nd edition, Vol. 4. H. U. Bergmeyer, Ed. Academic Press. New York and London. pp. 1802–1806.
16. Partridge, S. M. & H. F. Davis. 1950. Preferential release of aspartic acid during the hydrolysis of proteins. Nature **165:** 62–63.

Stabilization of Subunit Enzymes by Intramolecular Crosslinking with Bifunctional Reagents

V. P. TORCHILIN AND V. S. TRUBETSKOY

USSR Cardiology Research Center
Institute of Experimental Cardiology
Moscow, USSR

There exist many cases when the use of a stabilizing polymeric carrier can dramatically affect the biological activity of enzymes: for example, their interaction with high-molecular-weight substrates or with cell surface receptors. Methods of enzyme stabilization also are available that do not require carriers, but instead depend on chemical modification or intramolecular crosslinking with bifunctional reagents (for a review see Torchilin & Martinek[1]). In the case without the carriers, the crosslinkages hamper the unfolding of the native protein globule that would otherwise occur in the presence of various denaturating agents. This was shown earlier by us for one-chain enzymes.[2] The approach is even more promising for subunit enzymes, wherein conformational inactivation is a more complicated two-step process.[3] In the first step, the enzyme dissociates reversibly into nonactive subunits, while in the second step, irreversible unfolding of the individual subunits takes place (FIG. 1). Intersubunit linkages may in this case hinder or impede the dissociation of the protein into subunits, which therefore results in the stability increase.

We have studied the possibility of stabilizing subunit enzymes against different dissociative conditions. The enzymes we studied were glyceraldehyde-3-phosphate dehydrogenase (GAPD) and asparaginase. Each enzyme consisted of four identical subunits; and the intramolecular intersubunit crosslinking was tested with different bifunctional reagents, namely dicarboxylic acids of $HOOC\text{-}(CH_2)_n\text{-}COOH$ type, which can interact with the protein after preliminary activation with water-soluble carbodiimide, and diimidoesters. We have used a homologous series of bifunctional reagents, which permitted us to choose for each enzyme the optimal length of the crosslinking reagent. This length corresponded to the distance between the points to be linked on different subunits. Diacids with "n" varying from 0 to 12 and a set of diimidoesters were used. In each group of experiments, control tests were made with monofunctional carboxylic acid and imidoester to be sure that the effects observed were not conditioned by one-point modification. The experimental details of the protein modification with dicarboxylic acids are described in Torchilin *et al.*[4] and with diimidoesters in Torchilin *et al.*[5] Published methods for measurement of enzyme activity of GAPD[6] and asparaginase[7] were used.

Thermoinactivation experiments were performed in the following way: $2 \times 10^{-6} M$ solutions of native or modified enzyme in 0.05 *M* phosphate buffer, pH 7.5, were incubated at 60 °C. Samples were taken at appropriate time intervals; and the residual catalytic activity was measured. Analytical electrophoresis in polyacrylamide gel in the presence of SDS was performed as described by Laemmli.[8] The data are summarized in TABLE 1 and FIGURE 2. From the results several conclusions can be drawn. First, the interaction of the subunit enzyme with bifunctional reagents leads to the formation of a crosslinked fraction of the enzyme preparation; this fraction consists

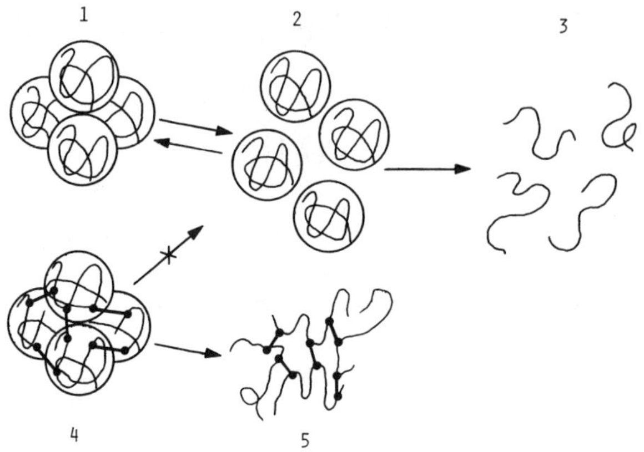

FIGURE 1. Conformational inactivation of tetrameric enzyme and its stabilization with inter-subunit cross-linking: (1) native enzyme, (2) reversibly dissociated subunits, (3) irreversibly unfolded subunits, (4) crosslinked enzyme, (5) irreversibly denaturated crosslinked enzyme.

of covalently crosslinked dimers, trimers, tetramers and a small quantity of higher oligomers, according to the SDS-electrophoresis data. Second, the amount of cross-linked fraction depends on the type of bifunctional reagent, with only one or two optimal reagents giving the highest amount of crosslinked fraction for each series. Third, the intersubunit crosslinking always leads to an increase in enzyme thermostability. The degree of thermostabilization correlates strictly with the amount of crosslinked fraction. Greater amounts of the nondissociated SDS-electrophoresis fraction agreed with the higher thermostability of the preparation. Fourth, monofunctional reagents do not cause any stabilization effects. Thus, the observed stabilization cannot be explained by simple chemical modification of the protein. Fifth, for GAPD, succinic acid and dimethyl pimelimidate are optimal crosslinking reagents. For asparaginase, dimethyl suberimidate is optimal. Even among a small set of homologous

TABLE 1. Properties of Native and Modified Oligomeric Enzymes

Enzyme	Modifier	Crosslinked Fraction (%)	Incubation time for Inactivation (min)
Glyceraldehyde-3-phosphate dehydrogenase	none	0	0.5
	4-hydroxybenzimidate	0	0.5
	dimethyladipimidate	ca. 40	2.5
	dimethylpimelimidate	ca. 50	7
	carbodiimide-activated succinic acid	ca. 30	6.5
	carbodiimide-activated adipic acid	ca. 5	1.5
L-Asparaginase	none	0	6
	dimethyladipimidate	ca. 25	8
	dimethylsuberimidate	ca. 50	40

bifunctional reagents, an optimal one can be found that gives the highest amount of crosslinked fraction and the highest increase in thermostability.

In thermoinactivation kinetic experiments, it can be easily shown that the mechanism of the observed stabilization is the elimination of the first step of the inactivation, that is, elimination of enzymes dissociation into individual subunits, and the increase of the activation energy of the whole inactivation process. It has also been demonstrated that if the first inactivation step for native GAPD is reversible, then the modified GAPD undergoes one-step unfolding without dissociation and cannot be reactivated (FIG. 1).

The crosslinked subunit enzymes acquire not only increased thermostability but also increased stability against other denaturating factors, such as urea, guanidine chloride, detergents, and different solvents.[4] It is interesting to note that similar stabilization results have been observed in nature, in the form of thermophilic

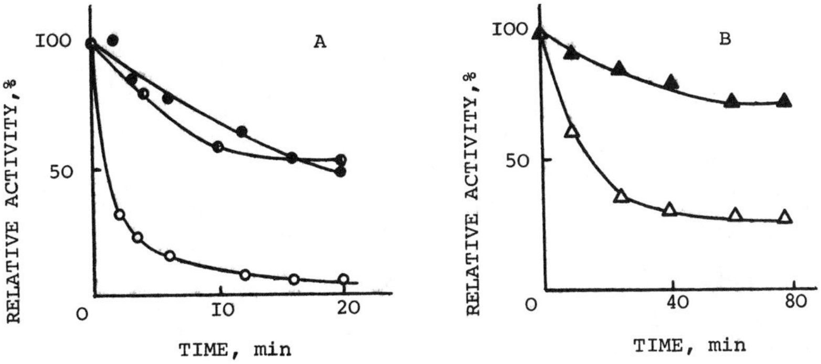

FIGURE 2. Thermoinactivation of native and modified oligomeric enzymes at 60°C. Part A used GAPD: (O) native enzyme, (O) GAPD modified with activated succinic acid, and (●) GAPD modified with dimethyl pimelimidate. Part B used asparaginase: (△) native enzyme and (▲) asparaginase modified with dimethyl suberimidate. With the crosslinked enzymes, the data are presented only for the maximally stabilized preparations.

organisms. The enzymes in these organisms subunit enzymes included, are extremely stable against denaturation by the surrounding medium, particularly by heat. Detailed analysis of the properties of GAPD from thermophilic microorganisms[9] has shown minor changes in primary structure which result in the formation of additional intramolecular intersubunit electrostatic or hydrophobic bonds. Even a small number of these bonds can drastically increase the stability of subunit enzymes against dissociative inactivation.

REFERENCES

1. TORCHILIN, V. P. & K. MARTINEK. 1979. Enzyme stabilization without carriers. Enzyme Microb. Technol. **1**: 74–82.
2. TORCHILIN, V. P., A. V. MAKSIMENKO, V. N. SMIRNOV, I. V. BEREZIN, A. M. KLIBANOV & K. MARTINEK. 1978. The principles of enzyme stabilization. III. The effect of the length of intramolecular cross-linkages on thermostability of enzymes. Biochim. Biophys. Acta **522**: 277–283.

3. POLTORAK, O. M. & E. S. CHYKHRAY. 1979. Kinetics and mechanism of catalytic inactivation of enzymes with subunit structure. Moscow State Univ. Proc. Ser. 2. Chemistry. **20**: 195–211.
4. TORCHILIN, V. P., V. S. TRUBETSKOY, V. G. OMEL'YANENKO & K. MARTINEK. 1983. Stabilization of subunit enzyme by intersubunit cross-linking with bifunctional reagents: Studies with glyceraldehyde-3-phosphate dehydrogenase. J. Mol. Catal. **19**: 291–301.
5. TORCHILIN, V. P. & V. S. TRUBETSKOY. 1984. Artificial and natural thermostabilization of subunit enzymes. Do they have similar mechanisms? Submitted for publication.
6. FERDINAND, W. 1964. The isolation and specific activity of rabbit muscle glyceraldehyde-3-phosphate dehydrogenase. Biochem. J. **92**: 578–585.
7. HOWARD, J. B. & F. H. CARPENTER. 1972. L-Asparaginase from *Ewinia carotevora.* Substrate specificity and enzymatic properties. J. Biol. Chem. **247**: 1020–1030.
8. LAEMMLI, V. K. 1976. Cleavage of structural proteins during the assembly of the head of bacteriophage T4. Nature **227**: 680–685.
9. WALKER, J. E., A. J. WONACOTT & J. I. HARRIS. 1980. Heat stability of a tetrameric enzyme, D-glyceraldehyde-3-phosphate dehydrogenase. Eur. J. Biochem. **108**: 581–586.

Kinetic and EPR Spectroscopy Studies of Immobilized Chymotrypsin Deactivation[a]

DOUGLAS S. CLARK AND JAMES E. BAILEY

Department of Chemical Engineering
California Institute of Technology
Pasadena, California 91125

In spite of many experimental investigations of enzyme deactivation, little information is available at the molecular level on the changes that occur in a population of enzyme molecules as a result of exposure to denaturing conditions. This is especially true for immobilized enzyme systems in which opaque supports prevent use of many of the available analytic methods for characterizing protein structural features. In this research, the deactivation of α-chymotrypsin on CNBr-Sepharose 4B in the presence of a solution containing 50% *n*-propanol has been investigated. Although considerable detail on this particular system will be provided, the major objective of this work is to illustrate the application of a combination of characterization methods to better understand the fundamental properties of immobilized enzyme catalysis and deactivation.

The catalytic activity of α-chymotrypsin-CNBr-Sepharose 4B conjugates was determined by withdrawing samples of the immobilized enzyme catalyst from a slurry of catalyst in 50% *n*-propanol after different exposure times. The samples were washed and then contacted with standard mixtures of ATEE (N-acetyl-L-tyrosine ethyl ester) to determine the overall activity in terms of μmoles ATEE converted per second. It is known from previous investigations of this system that diffusion limitations have a significant effect on the catalysts under the conditions employed for these assays.[1] Accordingly, a procedure previously described for extracting intrinsic activity information from the overall activity measurement was applied to evaluate the intrinsic activities. In these calculations, the diffusivity of ATEE in these types of conjugates determined from previous investigations, $D_e = 3.8 \times 10^{-6}\text{cm}^2\text{sec}^{-1}$, was employed.

An extremely useful technique for immobilized enzyme characterization is active-site titration.[1,2] This method uses the nonfluorescent suicide inhibitor MUTMAC, designated by S in Equation 1, which upon reaction yields a fluorescent product in solution, P_1, according to

$$E + S \underset{k_{-1}}{\overset{k_1}{\rightleftharpoons}} ES \xrightarrow{k_2} EP_2 + P_1 \xrightarrow{k_3} E + P_2 \tag{1}$$

where EP_2 is the acylenzyme intermediate. Under typical reaction conditions $k_2 \gg k_3$, and a single molecule of fluorescent product is produced for each active molecule of α-chymotrypsin. The extreme sensitivity of fluorescent measurements enables this method to be used for direct assay of the quantity of active enzyme in a relatively small sample. For example, in these experiments, the number of active protein molecules employed in any single assay was of the order of 10^{15}. Another advantage of active-site titration measurements is their complete insensitivity to diffusion effects. The active-site titration measurement determines the amount of active enzyme, not the catalytic activity of that enzyme.

Immobilized enzyme samples withdrawn from *n*-propanol solution were also

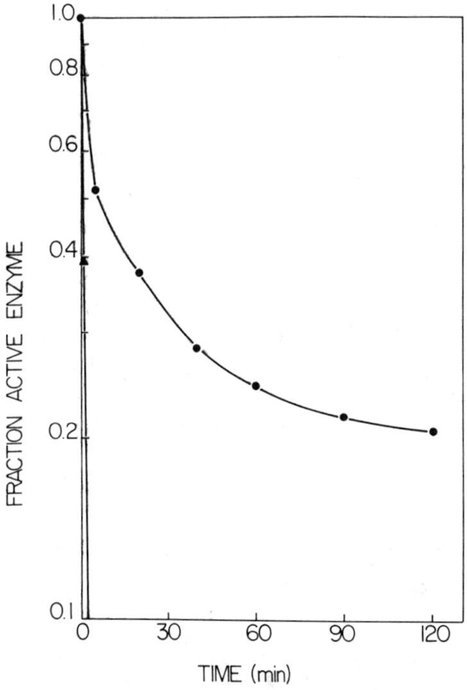

FIGURE 1. Deactivation of α-chymotrypsin free in solution (▲) and coupled to CNBr-Sepharose 4B(●) in 50% *n*-propanol solution. Active enzyme content was determined by active-site titration with MUTMAC.

assayed for the total amount of active enzyme using active-site titration. The results for a 120-minute experiment are shown in FIGURE 1. Also shown for comparison is the time trajectory of the quantity of active enzyme when a parallel experiment is done with enzyme in solution. The soluble enzyme loses all of its activity extremely rapidly under these conditions, while the immobilized enzyme is substantially stabilized. Interestingly, the kinetics of change in the quantity of active enzyme deviate significantly from first order (note the ordinate log scale). Also, the active enzyme trajectory apparently approaches an asymptote, indicating that some portion of the immobilized enzyme completely resists deactivation under these conditions.

Combining intrinsic activity and the quantity of active enzyme measurements at different times of exposure to *n*-propanol, the specific activity of active enzyme is experimentally determined. These results are shown in TABLE 1. Surprisingly, the enzyme remaining active after exposure to *n*-propanol possesses a higher specific

TABLE 1. Change in the Specific Activity of Active Immobilized Chymotrypsin during Exposure to 50% *n*-Propanol

Exposure Time (min)	Specific Activity of *Active* Enzyme (μmol ATEE/sec · μmol active CT)
0	35.5
5	39.7
65	38.4
120	39.8

activity than does the initial active enzyme. This counterintuitive result can be explained by considering the results of a third class of measurements.

EPR SPECTROSCOPY OF SPIN-LABELED α-CHYMOTRYPSIN

In order to provide direct experimental information on the active-site structure of immobilized chymotrypsin, spin label m-IV was attached to the serine-195 residue in the immobilized chymotrypsin active site.[1,3] This spin label is a substrate analogue with a stable nitroxide on its free terminus. A number of investigations of this label in solution as well as experimental evidence in this laboratory shows that this spin label interacts with and labels essentially only those enzyme molecules that are catalytically active. A schematic diagram illustrating the label location in the active site is provided in FIGURE 2.

The label motion is conditioned by the nearby active-site structure. As illustrated clearly in FIGURE 3, the active-site structure of chymotrypsin is modified substantially

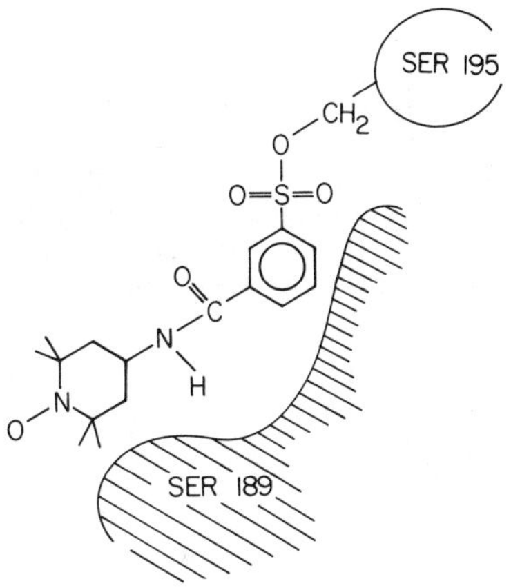

FIGURE 2. Schematic diagram of spin label m-IV attached to the serine-195 residue in the α-chymotrypsin active site.

by immobilization.[1] Both of the spectra in FIGURE 3 were measured in the presence of 10.5 m*M* indole solution. Addition of this competitive inhibitor of the enzyme during the EPR analysis provides greatly enhanced resolution as a clear distinction between the immobilized and soluble forms of the enzyme.

Further examination of the EPR spectra for five immobilized chymotrypsin preparations with different initial active enzyme contents showed that all initial indole spectra could be represented extremely accurately as linear combinations of the same two spectra:[4]

$$f(G) = \alpha_A f_A(G) + \alpha_B f_B(G) \quad (2)$$

$f_B(G)$ is the indole spectrum of chymotrypsin in solution. $f_A(G)$ is a spectrum indicating greatly restricted label motion. This spectrum was determined both by spectral analysis based on Equation 2, using the measured overall spectra $f(G)$ and the known B spectrum, and by chemical subtraction of the EPR spectrum for the more mobile label by addition of potassium ferricyanide.[4] The success of Equation 2 in fitting all of

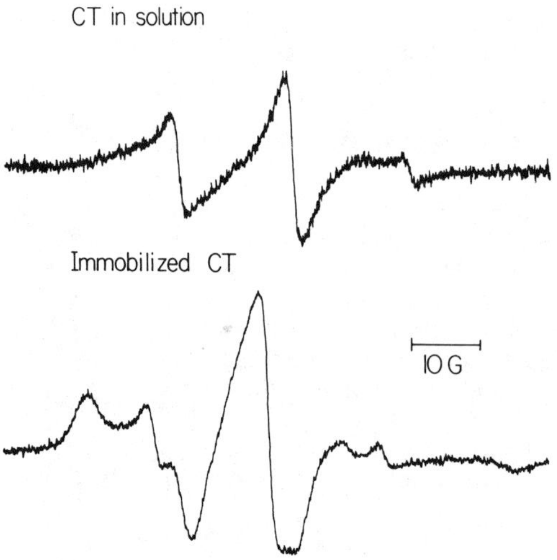

FIGURE 3. EPR spectra in 10.5 m*M* indole solution, pH 4.0, of spin-labeled α-chymotrypsin (CT) in solution (*top*) and coupled to CNBr-Sepharose 4B (*bottom*). The immobilized chymotrypsin preparation analyzed here contains 1.27 μmol active chymotrypsin per g catalyst.

the EPR spectra for five different preparations before exposure to *n*-propanol supports the hypothesis that two distinct active forms of immobilized chymotrypsin exist in these preparations. The fractions of active enzyme consisting of form A and form B are equal to α_A and α_B, respectively. Consequently, by analysis of overall EPR spectra for any preparation, the fractions of A and B forms may be determined. FIGURE 4 illustrates the subpopulation spectra and the percentages of each that must be added in

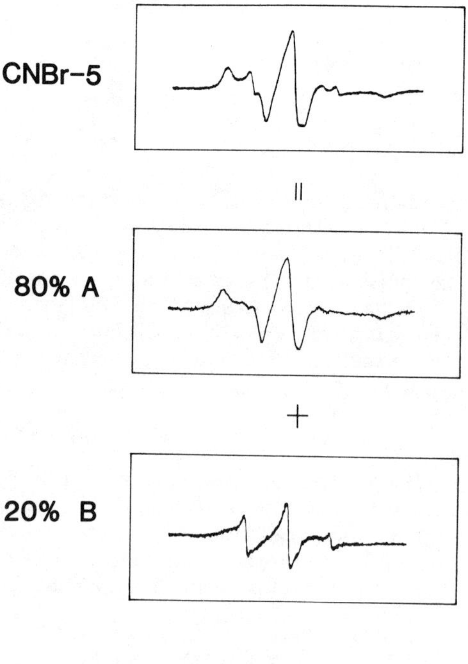

FIGURE 4. Resolution of the overall indole EPR spectrum for immobilized chymotrypsin preparation CNBr-5 into contributions from immobilized enzymes of forms A and B.[4] This catalyst contains 80% chymotrypsin A and 20% chymotrypsin B (spectra added to give that for CNBr-5).

order to fit the overall spectrum for a representative chymotrypsin-Sepharose conjugate designated CNBr-5.[4]

This information on the distribution of A and B forms combined with overall activity measurements can be used to estimate the intrinsic catalytic activities of immobilized chymotrypsin forms A and B. Using overall specific activity measurements v_{1_0} and v_{2_0} for two different immobilized chymotrypsin preparations, and determining the fractions of A and B forms for each of these same two conjugates, the following two equations may be written

$$v_{1_0} = \alpha_{A_1} v_{A_0} + \alpha_{B_1} v_{B_0} \tag{3}$$

$$v_{2_0} = \alpha_{A_2} v_{A_0} + \alpha_{B_2} v_{B_0} \tag{4}$$

where v_{A_0} and v_{B_0} are the intrinsic specific activities of the A and B forms of chymotrypsin, respectively.

Basing these calculations on the first two support entries listed in TABLE 2, the subpopulation specific activities given in the table caption were determined.[4] These

TABLE 2. Properties of Five Different Preparations of Immobilized α-Chymotrypsin[a]

Support	Active CT Loading (μmol active CT / g catalyst)	α_B	Activity (μmol ATEE / sec · μmol active CT) Measured	Calculated
I	0.192	0.342	49.2	49.3
I	0.541	0.275	41.2	41.3
I	1.27	0.200	32.2	32.3
II	0.365	0.366	52.7	50.9
II	0.717	0.283	42.9	42.3

[a]The fraction of enzyme form B was determined by decomposition of indole EPR spectra. Calculated activity values based on Equation 3 using $v_{A_0} = 8.3$, $v_{B_0} = 128$ μmol ATEE (sec · μmol active CT)$^{-1}$.[4]

values were then used with the α_A and α_B values determined by EPR spectroscopy of the remaining three samples to calculate the specific activity of those samples, using equations such as Equations 3 and 4. The success of this procedure is clear from comparison of the measured and calculated values in TABLE 2. The ability of a representation of the form of Equation 3 to fit consistently activity data for a number of different preparations with different loadings and different overall specific activities provides confirming evidence for the two active subpopulations model for this immobilized enzyme system.

SUBPOPULATION DEACTIVATION

Since the catalyst contains two different active enzyme forms before exposure to alcohol, a reasonable conceptual view of the deactivation process and of the time dependence of overall specific activity of active enzyme is indicated schematically in FIGURE 5. Both enzyme forms may become completely inactive, and both enzyme

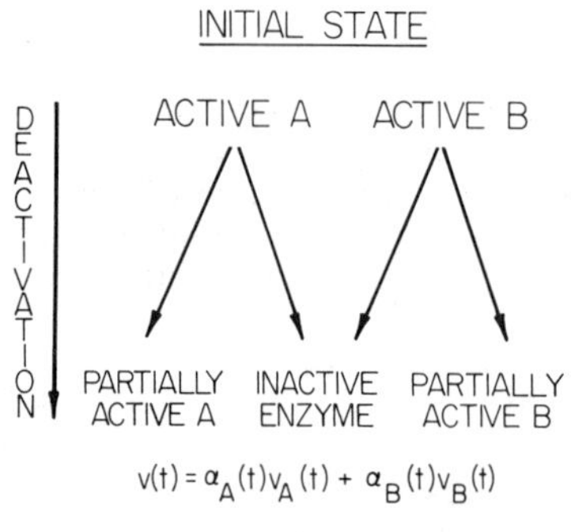

FIGURE 5. During deactivation, each of the two immobilized enzyme forms may become completely inactivated or may lose some portion of its initial activity.

forms may remain active but with reduced specific activity relative to the initial value. Several partially inactivated forms may exist; these are lumped together here because they could not be distinguished experimentally. Because the A and B forms may inactivate completely at different rates, the subpopulation fractions α_A and α_B must be viewed in general as time dependent. Because the average activity of the A form and of the B form may also change with the time of exposure to alcohol solution, both of these subpopulation specific activities are also time dependent. The EPR spectral decomposition method, active-site titration data, and intrinsic activity data have been used in concert to determine the fate of the two enzyme forms during exposure to *n*-propanol.

The basic strategy for determining $\alpha_A(t)$ and $\alpha_B(t)$ will be described next.[5] After exposure to alcohol solution for time t, a sample is withdrawn and spin labeled immediately. It is important to note at this step that the label attaches only to active enzyme. Then, the labeled, immobilized enzyme preparation is washed for sixteen hours, allowing time for reconstitution of the immobilized enzyme to its original state before exposure to alcohol. This procedure allows the A and B enzyme forms to return to the states at which their EPR spectra are known. Finally, the EPR spectral decomposition method previously developed for analysis of the initial enzyme preparations[4] is applied to determine $\alpha_A(t)$ and $\alpha_B(t)$.

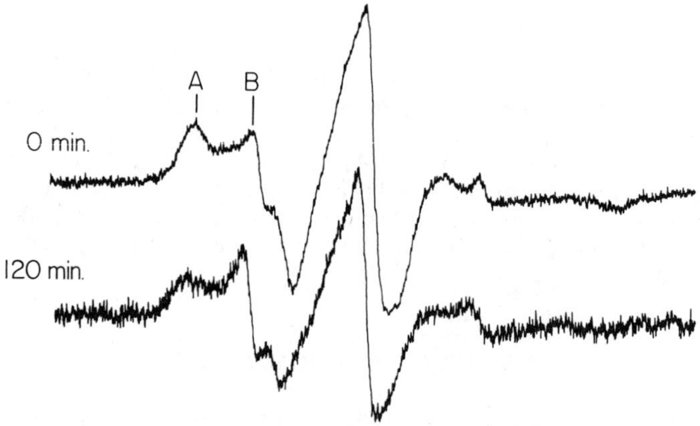

FIGURE 6. Indole EPR spectra of immobilized chymotrypsin initially and after a 120-min exposure to 50% *n*-propanol. The alcohol-deactivated sample was extensively washed to allow refolding of labeled active enzyme. Modes characteristic of enzyme forms A and B are marked.

Before considering the calculated results, it is useful to examine the EPR spectra themselves. Shown in FIGURE 6 are the indole EPR spectra of the initial immobilized enzyme population and that of the active enzyme after exposure to 50% *n*-propanol for 120 min after refolding. The relative heights of the modes labeled A and B, which may be taken as approximately indicative of the relative quantities of these two forms, changes in a clear fashion. The amount of active A decreases relative to the amount of active B. Another representation of this redistribution in active enzyme forms caused by deactivation is shown by the difference spectrum in FIGURE 7. This was obtained by subtracting the spectrum of the active enzyme after deactivation from the initial spectrum for the same sample. This is the EPR indole spectrum of the deactivated enzyme population. The predominance of chymotrypsin A in this completely deactivated enzyme is clear.

Spectral decomposition calculations indicate that the B fraction increased from 0.20 at time 0 to 0.33 after a 120-min exposure to alcohol. This result helps to explain the trend noted above in TABLE 1. The intrinsic specific activity of active enzyme

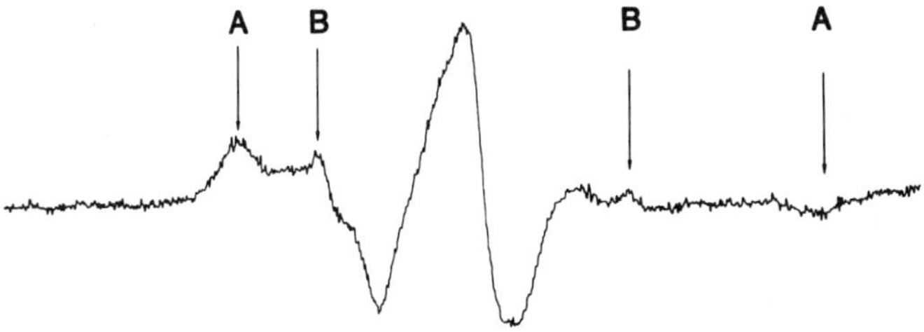

FIGURE 7. Indole EPR spectrum of immobilized chymotrypsin completely inactivated by 120 min in 50% *n*-propanol. The spectrum was obtained by subtraction of the bottom spectrum from the top spectrum in FIGURE 6. A: Spectral components of less active subpopulation; B: Spectral components of more active subpopulation. The deactivated enzyme is primarily form A.

increased as a result of exposure to *n*-propanol because the percentage of the relatively more active B form increased during the deactivation process.

CONCLUSIONS

This research has shown that immobilization greatly enhances the stability of α-chymotrypsin in 50% *n*-propanol. The deactivation kinetics are complex and cannot be described by a simple first-order model. Understanding the change in active enzyme specific activity requires detailed analysis of the catalyst before and during deactivation. This analysis, based on EPR spectroscopy of spin-labeled immobilized enzyme, shows that deactivation causes redistribution among the three immobilized enzyme states (A, B, and catalytically inactive forms).

The methodology applied here has significant potential for enhancing molecular-level understanding of immobilized enzyme catalyst formulation and utilization properties. Further efforts to probe and to measure directly molecular level properties of immobilized enzymes are crucial to eventual understanding of the influence of immobilization on proteins and the influence of denaturing environments on immobilized enzymes.

REFERENCES

1. Clark, D. S. & J. E. Bailey. 1983. Structure-function relationships in immobilized chymotrypsin catalysis. Biotechnol. Bioeng. **25:** 1027–1047.
2. Jameson, G. W., D. V. Roberts, R. W. Adams, W. S. A. Kyle & D. T. Elmore. 1973. Spectrofluorimetric titration of serine proteases. Biochem. J. **131:** 107–117.
3. Berliner, L. J. & S. S. Wong. 1974. Spin-labeled sulfonyl fluorides as active site probes of protease structure. J. Biol. Chem. **249:** 1668–1677.
4. Clark, D. S. & J. E. Bailey. 1984. Characterization of heterogeneous immobilized enzyme subpopulations using EPR spectroscopy. Biotechnol. Bioeng. **26:** 231–238.
5. Clark, D. S. & J. E. Bailey. 1984. Deactivation kinetics of immobilized α-chymotrypsin subpopulations. Biotechnol. Bioeng. In press.

Enzyme Stability and Glucose Inhibition in Cellulose Saccharification[a]

M. CANTARELLA,[b] A. GALLIFUOCO,[c] V. SCARDI,[b]
AND F. ALFANI[c]

[b]*Istituto di Fisiologia Generale*
Università di Napoli
via Mezzocannone 8
I-80134 Naples, Italy

[c]*Istituto di Principi di Ingegneria Chimica*
Università di Napoli
piazzale V. Tecchio
I-80125 Naples, Italy

Pure cellulose as well as cellulosic wastes can be hydrolyzed by an enzyme complex generally named cellulase, of which two main sources are known, the microfungi *Trichoderma viride* and *Aspergillus niger*. The complex exhibits three major activities: an endo- and an exo-glucanase (C_1 and C_x) synergistically acting in the breakdown of both crystalline and amorphous regions of the cellulose chain, and a third component, β-glucosidase, which catalyzes the hydrolysis of cellobiose to glucose. At present, however, various constraints still limit the enzymatic saccharification of cellulose. In the present study, our attention has been focused on thermal stability and on glucose inhibition.

The reactor used was an ultrafiltration (UF) stirred cell (Amicon Mod. 52) equipped with a YM 10 membrane; the nominal molecular weight cutoff (10,000) ensures both a 100% rejection of enzymes and cellulose and complete permeability towards low molecular weight products.

The flow rate and enzyme concentration were chosen so as to avoid the formation of a protein gel layer on the UF membrane by polarization concentration.

The characteristic features of the reactor and the advantages of using membrane reactors in enzymatic processes involving insoluble substrates, such as cellulose, have been discussed elsewhere.[1]

Samples of the outflowing reactor stream were collected automatically at regular time intervals and glucose concentration was determined using the Boehringer Glucose Kit (GOD-Perid), modified by the addition of 10 mM δ-gluconolactone, an inhibitor of the β-glucosidase that usually contaminates the assay kit. Reaction rates were determined according to the equation of Alfani *et al.*[1]

Different cellulase preparations, as listed in TABLE 1, were tested in 50 mM Na-acetate buffer, pH 4.8. All other experimental conditions were varied in the different runs.

Experiments were set up to measure the rate of thermal deactivation of the three enzymic components, C_1, C_x, and β-glucosidase. Each enzyme was allowed to react in the presence of its specific substrate, crystalline cellulose (Avicel), a soluble cellulose derivative, carboxymethylcellulose (CMC) and cellobiose, respectively. Typical plots of reaction rate versus process time are shown in FIGURE 1. During the first 10 hours,

[a]This research was funded under a CEE grant (Energy from Biomass, ESE-R-41-I).

TABLE 1. Thermal Stability and Specific Activity of the Enzymatic Components in Different Commercial Cellulases

Cellulase	Substrate	C°_E (μg/ml)	$10^2 \cdot K_d$ (h^{-1})	r_{RG} (μmoles/mg$_E$h)	r_G (μmoles/mg$_E$h)
T. viride					
(BDH)	Avicel	32.1	0.57	4.78	0.96
	CMC	32.1	0.52	4.26	1.35
	Cellobiose	32.1	1.52		32.50
T. virdide					
(Miles)	Avicel	34.9	2.43	7.88	1.16
	CMC	34.9	0.79	3.99	1.17
	Cellobiose	34.9	1.13		23.08
Celluclast 200					
(NOVO)	Avicel	28.3	0.24	3.54	0.73
	CMC	28.3	1.69	6.17	1.43
	Cellobiose	28.3			
Cellulase AP					
(Assoreni)	Avicel	22.0	6.50	4.14	2.76
	CMC	22.0	0.12	4.03	3.01
	Cellobiose	22.0	0.35		122.00
Cellulase AP3					
(Assoreni)	Avicel	9.5	3.36	6.63	1.30
	CMC	9.5	1.16	17.51	2.48
	Cellobiose	9.5	0.67		103.00
A. niger					
(Sigma)	Avicel	7.2	3.94	2.49	0.83
	CMC	7.2	0.76	8.44	2.10
	Cellobiose	7.2	0.59		199.20

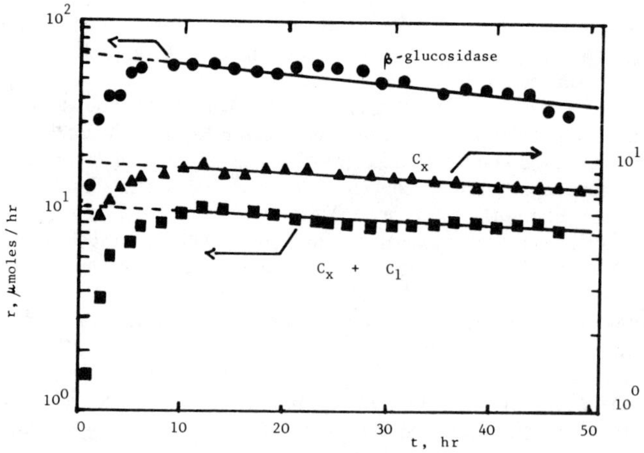

FIGURE 1. Thermal stability of cellulase components: specific reaction rate versus process time.

an increase in enzyme activity was observed. This increase is apparent, since the initial transient behavior is related to the response of the UF-membrane stirred reactor to a step input in glucose production,[2] while the remaining portion of the curve, if a decay is observed, shows the existence of a deactivation phenomenon. Since in the cellulose hydrolysis the activity of the three enzyme components fits a linear decay in a semilog plot, the result indicates that:

$$C_E = C_E^o \exp(-K_d t)$$

where C_E^o and C_E^o are the enzyme concentrations at time *t* and zero, respectively, and k_d is the rate constant of the deactivation process. In FIGURE 1, data refer to runs performed with the *T. viride* cellulase (BDH). The rate of glucose production is a measure of β-glucosidase activity, while the instantaneous concentration of C_1 and C_x is assumed to be proportional to the formation of reducing sugars. In fact, the experiments were always performed at saturating substrate concentration: Avicel, 33 mg/ml; CMC, 0.1% (wt/vol), and cellobiose, 10 m*M*. The reaction temperature was 45°C, the enzyme concentrations were those reported in TABLE 1.

The body of experimental results indicates that the thermal stability of the three enzymic components varies largely in the different cellulase preparations. High specific activity of a cellulase component is usually combined with a good thermal stability. The best β-glucosidase component is that present in the cellulase from *A. niger,* while satisfactory balanced activity and thermal stability were observed for the endo- and exo-glucanases from *T. viride*. The data of TABLE 1 enables us to draw another important conclusion. Glucose is present to a lesser extent, but it is still produced even in the presence of δ-gluconolactone used for total inhibition of the β-glucosidase component. Therefore, this evidence supports the proposed mechanism,[3] according to which glucose is mainly produced in cellobiose hydrolysis, but monomer units of glucose are also directly liberated by the attack of C_1 and C_x at the nonreducing end of the cellulose chain.

Another part of this study was devoted to the identification of the mechanism of glucose inhibition. In previous works,[4,5] a progressive activity decay of the β-glucosidase in the presence of increasing glucose concentrations was shown. However, the mechanism and the constant values were not determined since the data obtained in a UF membrane reactor, with high conversion levels, were not suitable to be plotted in the forms that allow these evaluations. On the other hand, a mathematical model to describe cellobiose hydrolysis and glucose inhibition in a batch reactor was studied[6] and a diagram clearly indicated that the difference between competitive and noncompetitive mechanisms are generally small and comparable with the average scattering of experimental data. Therefore, a true identification of the mechanism in a batch reactor is extremely difficult and the uncertainties present in the literature may depend mainly on the use of a differential batch reactor.

In order to overcome these difficulties, a mathematical model of cellobiose hydrolysis in a UF membrane stirred reactor was set up and the results were plotted as in FIGURE 2, where I is the glucose (inhibitor) concentration, X_S the degree of conversion, k_3 the kinetic constant (h^{-1}) and *r* the reaction rate (mg/ml · h). The curves of a noncompetitive mechanism are for different values of the second inhibition constant, k_{i2}, well distinct from the straight line of a competitive mechanism. This behavior was always observed, independently of the values of the Michaelis constant, K_m, the substrate concentration, S_o, and the first inhibition constant, k_{i1}. This result was considered very promising and experiments were set up to confirm model forecasts. The reaction was performed at 25°C, which is a good temperature to prevent thermal deactivation of the biocatalyst, and with enzyme concentrations ranging from 0.197 to

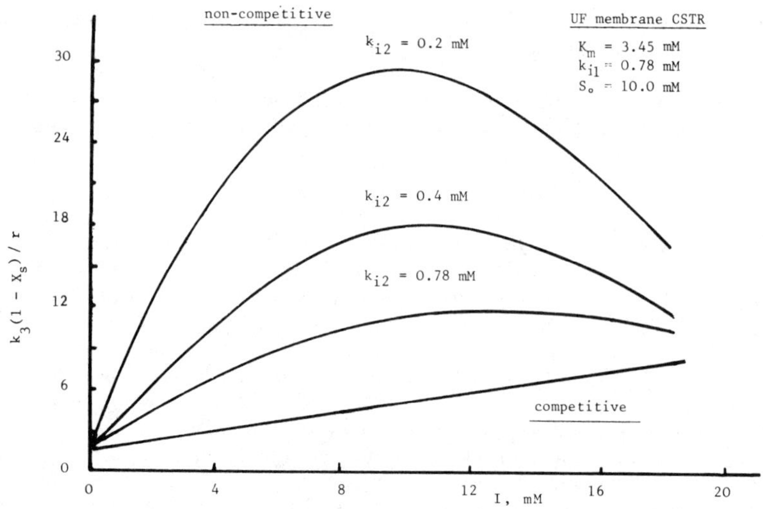

FIGURE 2. Mathematical modeling of product inhibition in a stirred membrane reactor.

19.7 μg/ml in order to explore a wide range of glucose concentration. Enzyme activity is strongly dependent on glucose concentration for values lower than 2 m*M* and decreases very smoothly for higher product concentrations. The data have been worked out according to the mathematical model and plotted in FIGURE 3. The diagram clearly shows that glucose inhibition follows a noncompetitive mechanism. The following values of the two inhibition constants and the Michaelis constant have been evaluated by regression analysis of the experimental data: k_{i1} = 0.33 m*M*, k_{i2} = 1.19 m*M* and K_m = 0.59 m*M*.

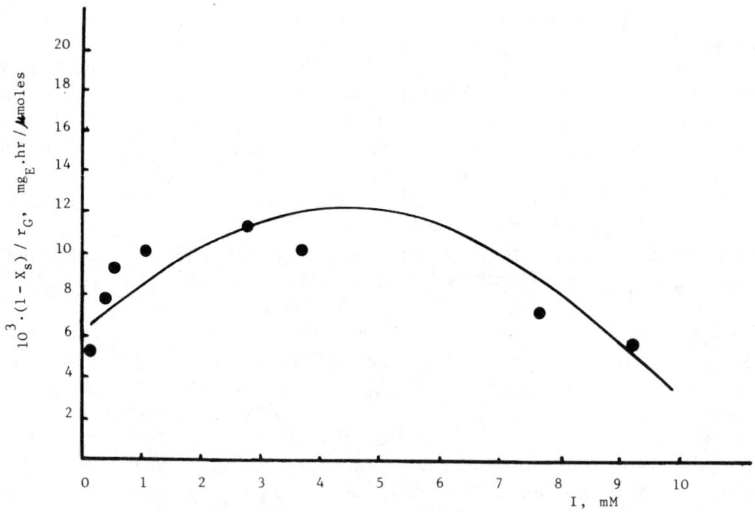

FIGURE 3. Experimental identification of the inhibition mechanism.

In conclusion, it seems possible to affirm that UF membrane reactors are very helpful for kinetic investigations of enzymatic reactions on a laboratory scale as compared to other reactor configurations.

REFERENCES

1. ALFANI, F., M. CANTARELLA & V. SCARDI. 1983. Use of membrane reactor for studying enzymatic hydrolysis of cellulose. J. Membr. Sci. **16:** 407–416.
2. GRECO, G., F. ALFANI, G. IORIO, M. CANTARELLA, A. FORMISANO, L. GIANFREDA, R. PALESCANDOLO & V. SCARDI. 1979. Theoretical and experimental analysis of a soluble enzyme membrane reactor. Biotechnol. Bioeng. **21:** 1421–1438.
3. LADISCH, M. R., K. W. LIN, M. VOLOCH & G. T. TSAO. 1983. Process considerations in the enzymatic hydrolysis of biomass. Enzyme Microbiol. Technol. **5:** 82–102.
4. ALFANI, F., M. CANTARELLA & V. SCARDI. 1982. Cellulose saccharification: Reaction kinetics and cellulase characterization. *In* Energy from Biomass. G. Grassi & W. Palz, Eds. Vol. **3:** 253–259. Reidel Publishing Company. Dordrecht, the Netherlands.
5. ALFANI, F., M. CANTARELLA, L. ERTO & V. SCARDI. 1982. Enzymatic saccharification of cellulose and cellulosic materials. *In* Energy from Biomass. A. Strub, P. Chartier & G. Schleser, Eds. Applied Science Publishers. London, England. pp. 1000–1005.
6. ALFANI, F., M. CANTARELLA & A. GALLIFUOCO. 1983. Utilizzazione di catalizzatori biologici: Immobilizzazione e loro impiego nella conversione enzimatica di residui lignocellulosici. *In* Sviluppo Della Ricerca Di Ingegneria Chimica. G.R.I.C.U., Eds. FAST. Milano, Italy. pp. 175–181.

Gramicidin S Synthetase Stabilization *in Vivo*

S. N. AGATHOS

Department of Chemical and Biochemical Engineering
Faculty of Engineering Science
The University of Western Ontario
London, Ontario, Canada N6A 5B9

A. L. DEMAIN

Department of Nutrition and Food Science
Massachusetts Institute of Technology
Cambridge, Massachusetts 02139

We recently proposed a novel way of approaching the problem of optimal microbial production of secondary metabolites, such as antibiotics, namely increasing the *in vivo* stability of enzymes (synthetases) that catalyze the biosynthesis of these important chemicals.[1] This suggestion reflects our current awareness that *in vivo* inactivation of microbial enzymes appears to be a general regulatory phenomenon irreversibly affecting the catalytic activity of many enzymes in the cell.[2] The rapid *in vivo* inactivation of antibiotic synthetases is widespread and occurs during the idiophase of many batch fermentations.[3] As a model system, we have attempted to understand and control the *in vivo* inactivation of gramicidin S synthetase, the enzyme complex responsible for the formation of gramicidin S (GS), a cyclic decapeptide antibiotic, in *Bacillus brevis* ATCC 9999. The pattern of appearance and disappearance of the synthetase[4] is typical of a large number of antibiotic fermentations: after the appearance of the enzyme complex in late exponential growth phase and the attainment of a peak value in specific activity, it rapidly disappears (half-life $\simeq$ 1–2 hours) as the cell population moves into stationary phase.[5] Previous work[6] had shown that inactivation of GS synthetase *in vivo* is oxygen-dependent and irreversible.

MATERIALS AND METHODS

The culture used in this work was *Bacillus brevis* ATCC 9999. Spore and inoculum preparation, determination of growth and assays of GS were as previously described.[7] Bulk sulfhydryl (SH) content of cells was measured according to Ellman.[8] In order to study the *in vivo* disappearance of GS synthetase, we have used a system that exhibits inactivation kinetics in short-term experiments. The system utilizes frozen and thawed cells of *B. brevis* harvested after growth in yeast extract–peptone (YP) medium in a 180-liter fermentor and frozen to -20°C immediately after harvesting. Small batches of cells were thawed just before performing *in vivo* inactivation studies according to the methodology of Friebel and Demain[6] as modified by Agathos.[3] Briefly, these cells were agitated under air in buffer for various periods of time (baffled flasks, 250 rpm, 37°C) in the presence of appropriate additives versus nitrogen-covered controls, crude cell-free extracts were subsequently prepared using lysozyme and the extracts were assayed by the overall GS synthetase assay (incorporation of L-[^{3}H]ornithine into GS[3]).

RESULTS AND DISCUSSION

Our initial biochemical characterization of the oxygen-dependent *in vivo* inactivation of GS synthetase was performed through the use of inhibitors of aerobic metabolism and of protein synthesis. Addition of such inhibitors to aerated cell suspensions failed to prevent enzyme inactivation, thus suggesting that the loss of synthetase activity *in vivo* did not require aerobic energy-yielding metabolism and also that the mediation of an induced enzymatic step in the inactivation process was unlikely.

We further hypothesized that a possible target of the oxygen-dependent inactivation *in vivo* could be labile sulfhydryl (SH) active groups on the synthetase surface, given an estimated minimum of seven SH groups required for catalytic activity per mole of synthetase complex[9] and possibly additional SH groups that might be involved in the structural integrity of the native enzyme. The following evidence tends to confirm this hypothesis. Firstly, long-term anaerobic incubations of *B. brevis* cells in

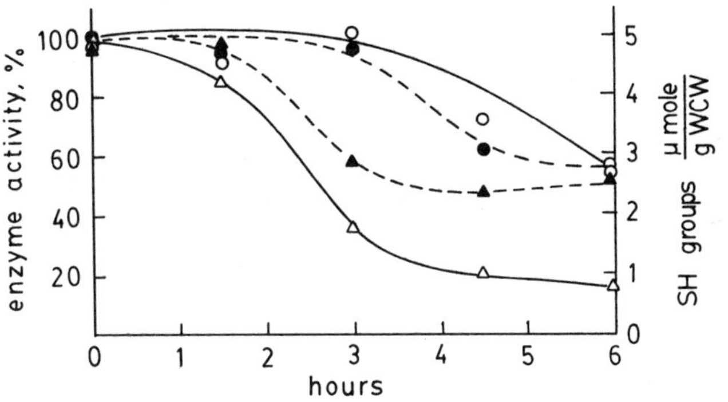

FIGURE 1. Time course of intracellular sulfhydryl group concentration and of synthetase activity under air in the presence of glycerol.

▲ SH with buffer only; ● SH with glycerol (20 g/l); △ enzyme activity with buffer only; ○ enzyme activity with glycerol (20 g/l).

the presence of organic thiols resulted in significant retardation of the inactivation process. For example, L-cysteine at 50 m*M* preserved more than 80% of the synthetase activity during 12 hours of incubation, whereas by that time less than 30% of enzyme activity had remained in the control vessel. Secondly, further short-term anaerobic incubations of preaerated cells (having suffered various degrees of O_2-dependent inactivation) with organic thiols resulted in partial reactivation of the synthetase. L-Cysteine at 30–50 m*M* was the most effective thiol in this respect.

Another series of experiments addressed the possible influence of carbon sources upon GS synthetase inactivation *in vivo,* since a shift in C metabolism is one of the physiological conditions most commonly associated with *in vivo* disappearance of microbial enzyme activity.[2] We found that inactivation could be retarded upon addition of utilizable carbon sources in aerated suspensions of *B. brevis* cells, and that the degree of stabilization achieved, in decreasing order of effectiveness, is glycerol, fructose, inositol, which correlates with the order of effectiveness for growth. It was further determined that increasing concentrations of C-source (between 1 and 4%)

provided increasing retardation of the inactivation (between 1½- and 3-fold increase in synthetase half-life), and it was confirmed that the degree of enzyme protection was directly related to the rate of C-source uptake. The apparent metabolic basis of this C-source-mediated effect was further elucidated by following the concentration of total SH groups in cell extracts from cell suspensions aerated under standard inactivation conditions in the presence and absence of 2% glycerol. As can be seen from FIGURE 1, the drop in enzyme activity in both vessels was accompanied by a similar drop in the intracellular SH concentrations and, furthermore, the [SH] remained higher in the glycerol-supplemented cells during the period of retardation of inactivation. These observations support the idea that, while SH groups on the synthetase are susceptible to oxidation during aeration, the active catabolism of a C-source could lead to a low intracellular redox potential that could maintain intracellular cysteine residues in the -SH (i.e. reduced) state. For example, it is possible that reduced

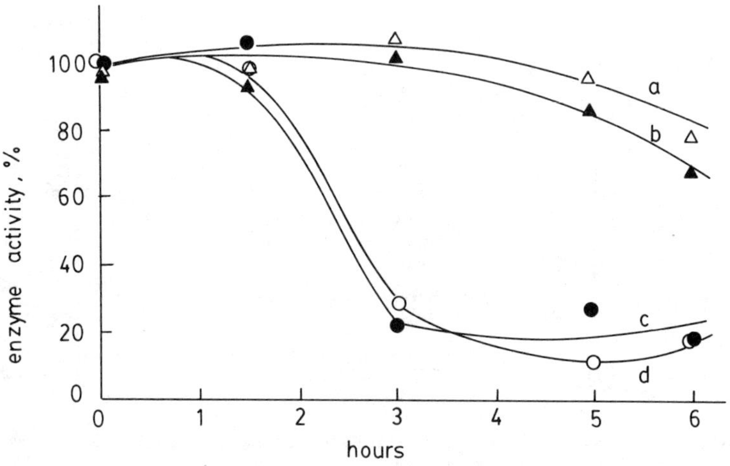

FIGURE 2. Effect of five amino acids on synthetase stability under aeration in the presence and absence of chloramphenicol (CAP).

△ 5 amino acids (6 m*M*); ▲ 5 amino acids (6 m*M*) + CAP (500 μg/ml); ● CAP only (500 μl/ml); ○ no addition.

pyridine nucleotides produced under these conditions help regenerate SH groups on the enzyme via thiol interchange with intracellular thiols such as glutathione, dihydrolipoamide, thioredoxin, or glutaredoxin.[1]

A final series of experiments examined the influence of the five natural amino acid substrates of the synthetase upon its stability *in vivo,* since it has been well established that substrates, cofactors and other natural ligands often increase enzyme stability.[10] A mixture of L-phenylalanine, L-proline, L-leucine, L-ornithine, and L-valine at 6 m*M* each, added to standard aerated suspensions of *B. brevis* cells provided the best *in vivo* stabilization of GS synthetase to date; at the end of 6 hours of aeration, more than 80% of synthetase activity still remained. The effect was not due to cell growth or protein synthesis, since an excess of chloramphenicol (CAP) used with the five amino acids did not abolish the protective effect of the amino acids on the synthetase (FIG. 2). In addition, when growing cells of *B. brevis* were supplemented with the same mixture of five amino acids, they exhibited increases by 50% in their biosynthetic capacity, a behavior compatible with a substrate-mediated stabilization of the synthetase *in vivo*.

ACKNOWLEDGMENTS

We thank the Greek Ministry of National Economy for a NATO Science Fellowship for S. N. Agathos. The technical assistance of AiQi Fang is gratefully appreciated.

REFERENCES

1. AGATHOS, S. N. & A. L. DEMAIN. 1983. Optimization of fermentation processes through the control of *in vivo* inactivation of microbial biosynthetic enzymes. *In* Foundations of Biochemical Engineering: Kinetics and Thermodynamics in Biological Systems. H. W. Blanch, E. T. Papoutsakis & G. Stephanopoulos, Eds. ACS Symp. Ser. No. 207. ACS. Washington, DC. pp. 53–67.
2. SWITZER, R. L. 1977. The inactivation of microbial enzymes *in vivo*. Ann. Rev. Microbiol. **31:** 135–157.
3. AGATHOS, S. N. 1983. Studies on the *in vivo* inactivation of gramicidin S synthetase. Ph.D. Thesis. M.I.T. Cambridge, MA.
4. DEMAIN, A. L., A. POIRIER, S. AGATHOS & O. NIMI. 1981. Appearance and disappearance of gramicidin S synthetases. Dev. Ind. Microbiol. **22:** 233–244.
5. TZENG, C. H., K. D. THRASHER, J. P. MONTGOMERY, B. K. HAMILTON & D. I. C. WANG. 1975. High productivity tank fermentation for gramicidin S synthetases. Biotechnol. Bioeng. **17:** 143–151.
6. FRIEBEL, E. T. & A. L. DEMAIN. 1977. Oxygen-dependent inactivation of gramicidin S synthetase in *Bacillus brevis*. J. Bacteriol. **130:** 1010–1016.
7. MATTEO, C. C., M. GLADE, A. TANAKA, J. PIRET & A. L. DEMAIN. 1975. Microbiological studies on the formation of gramicidin S synthetases. Biotechnol. Bioeng. **17:** 129–142.
8. ELLMAN, G. L. 1959. Tissue sulfhydryl groups. Arch. Biochem. Biophys. **82:** 70–77.
9. LALAND, S. G. & T.-L. ZIMMER. 1973. The protein thiotemplate mechanism of synthesis for the peptide antibiotics produced by *Bacillus brevis*. Essays Biochem. **9:** 31–57.
10. GRISOLIA, S. 1964. The catalytic environment and its biological implications. Physiol. Rev. **44:** 657–712.

Effect of Water Activity on Enzyme Action and Stability

PIERRE MONSAN AND DIDIER COMBES

Departement de Genie Biochimique et Alimentaire
ERA-CNRS 879, INSA
Avenue de Rangueil
F-31077 Toulouse Cedex, France

Until now, there has been a great difference between fundamental enzymology research and its technological applications. The first case is more concerned with the characterization of kinetic behavior and establishing structure-function relationships of enzymes in dilute solutions. The second case is concerned more with the use of these catalysts, usually in concentrated media (starch/protein hydrolysis, etc.) or even in solids (bread/cheese production, etc.).

In order to allow technological applications of biocatalysts to be exploited fully, it has become necessary to understand their operating conditions in concentrated or solid media and to improve the effects of these reaction conditions on enzyme activity and stability. As well as being of purely technological interest, such research should also lead to a greater understanding of enzyme action *in vivo,* at the level of cell structure (membranes, organelles, etc.) and enzyme stability in low-water media (seeds, spores, etc.).

The development of an industrial process, for hydrolysis of concentrated sucrose solutions, using immobilized invertase,[1] has led to the study of the action and stability of this enzyme in concentrated media. In turn this led to modeling of the action mechanism of invertase and research into the causes of its increased stability in the presence of high levels of substrate.

EFFECT ON ENZYME ACTION

The hydrolysis of sucrose, also called inversion, by invertase is a classical enzymatic reaction that has been studied by hundreds of investigators since the end of the nineteenth century. The apparent activity of invertase declines in sucrose solutions of concentration above 0.4 *M*. Although this phenomenon is generally attributed to inhibition by excess substrate, the curve relating reaction rate and substrate concentration thus obtained cannot be mathematically described by the corresponding classical equation:

$$V = \frac{V_{max}\,(S)}{K_m + (S) + \dfrac{(S)^2}{K_s}}$$

where: V is hydrolysis reaction rate; V_{max} is maximal reaction rate; (S) is sucrose concentration; K_m is Michaelis constant; and K_s is constant of inhibition by excess substrate. In fact, invertase activity decreases faster than would result from a simple excess substrate inhibition.[2-4] In a first attempt to explain this difference, we determined the effect of the addition of viscosity-producing macromolecular derivatives. Medium viscosity is greatly increased with increasing sucrose concentration: a 2.67 *M* sucrose solution has a viscosity equal to 43 mPa · sec at 53°C. It may be seen from FIGURE 1 that the addition of polyacrylamide to a 0.4 *M* sucrose solution does not significantly modify invertase activity, while medium viscosity is increased from 1 mPa · sec to 106 mPa · sec when polyacrylamide concentration is increased from 0% to 1.5% (wt/vol). Similarly, invertase activity was not decreased by the addition of carboxymethylcellulose.[5] This clearly demonstrates that the increase in sucrose solution viscosity, obtained when sucrose concentration is increased, is not responsible for the observed decrease in invertase efficiency. This is consistent with the absence of any significant mass transfer limitation in such conditions of homogeneous catalysis.

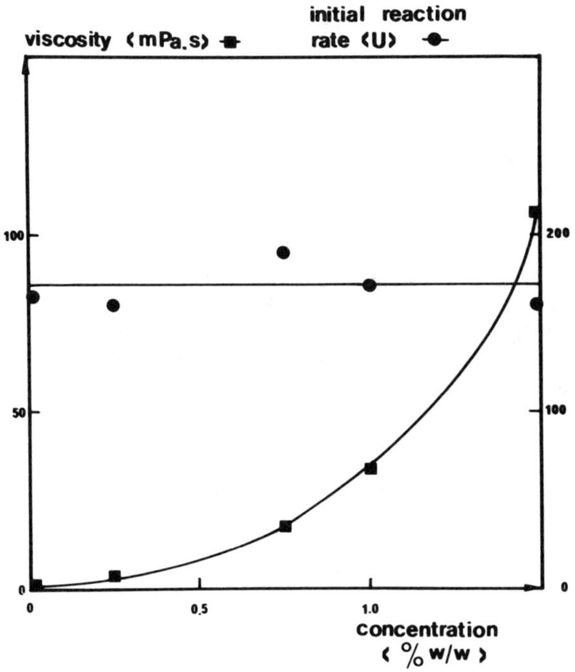

FIGURE 1. Effect of polyacrylamide concentration on invertase activity. Polyacrylamide was added at various concentrations to a 0.4 *M* sucrose solution in 0.1 *M* acetate buffer, pH 4.5 (20-ml final volume). The reaction was started by the addition of a 2 g/l invertase solution (0.2 ml) in the same buffer, at 53°C.

Study of the kinetics of invertase as a function of sucrose concentration brought out irregularities in the decreasing part of the activity/sucrose concentration curve (FIG. 2): when sucrose concentration was increased above 0.4 *M*, invertase activity did not decrease regularly. Two discontinuities were observed at 1 *M* and 1.8 *M* sucrose concentration, at 25°C. Such discontinuities, which until now have not been taken into account, appear in previously reported data.[2,6] Furthermore, the concentration ranges corresponding to these discontinuities coincide with discontinuities observed in the Laser-Raman[7] and x-ray diffraction[8] studies of solute-solvent interactions in aqueous

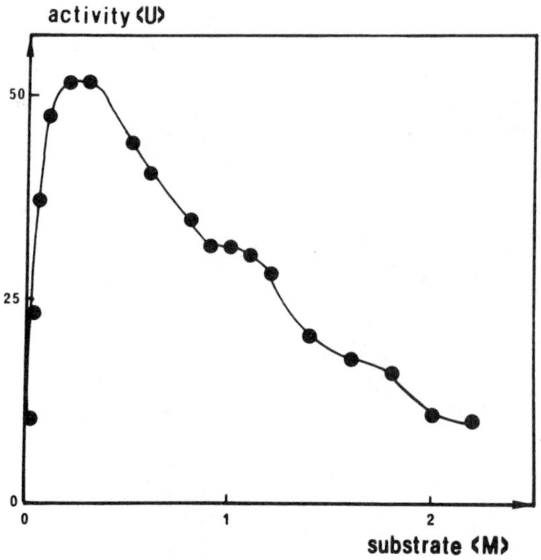

FIGURE 2. Effect of sucrose concentration on invertase activity at 25°C. Initial reaction rate was determined after addition of a 2 g/l invertase solution (0.2 ml) to 20 ml of a sucrose solution at different concentration in 0.1 *M* acetate buffer, pH 4.5.

solutions of sucrose, which were attributed to a modification of the conformation of the sucrose molecule.[8] In fact, when sucrose concentration is increased, the attendant decrease in water activity results in a modification in the distribution of hydrogen bonds: the water-sucrose intermolecular hydrogen bonds are increasingly replaced by (1) sucrose intramolecular hydrogen bonds (one or two) and (2) sucrose-sucrose intermolecular hydrogen bonds. This gives a folded and aggregated structure for sucrose molecules, which may finally yield the sucrose crystal structure. We thus postulated that the decreased invertase activity observed with increasing sucrose concentrations may result not only from inhibition by excess substrate, but also from the appearance of folded and aggregated sucrose molecules that are not recognized by invertase. To test this hypothesis, we studied the influence of compounds such as magnesium ions or poly(1,2-ethanediol) that modify water activity but do not affect invertase activity. In fact, the addition of either 0.1% (wt/vol) magnesium chloride or 1% (wt/vol) poly(1,2-ethanediol) results in a significant shift of the position of the discontinuities towards lower sucrose concentrations.[5] This may be attributed to the fact that in the presence of reagents that interact strongly with water molecules, the distribution of intra- and intermolecular hydrogen bonds within the sucrose molecule (i.e., sucrose folding and aggregation) can occur at lower sucrose concentrations. On the other hand, when temperature is increased, sucrose solubility also increases, and sucrose folding and aggregation occur at higher sucrose concentration. It can be seen from FIGURE 3 that an increase in temperature shifts the position of the discontinuities towards higher sucrose concentrations. These results indicate that effects that can alter the water activity of reaction mixtures can modify the shape of the curve relating invertase activity to sucrose concentration, without implying any modification in invertase activity.

In a second attempt to test our hypothesis, we applied our rationale to the mathematical description of invertase kinetics. In order to describe the effect of sucrose concentration on the initial reaction rate of sucrose hydrolysis, we made the following assumptions:

(1) The discontinuities observed in FIGURE 1 can be attributed to folding of the

sucrose molecule, involving one or two intramolecular hydrogen bonds.[8] If the folded sucrose molecules are no longer susceptible to invertase action, an increase in total sucrose concentration does not then result in a similar increase in the amount of sucrose actually available for invertase hydrolysis. In consequence, it is necessary to substitute for the total sucrose concentration (S) a corrected sucrose concentration (S^*), taking into account this decrease in the enzyme-available sucrose concentration.[5]

(2) In addition to sucrose folding, sucrose molecules undergo an aggregation phenomenon. The resulting sucrose clusters are not susceptible to invertase action, and combine with the enzyme to give unproductive complexes.[2] If only the clusters resulting from the association of two sucrose molecules are taken into account, invertase action may be more simply described as follows:

$$E + S \rightleftharpoons ES \rightarrow E + P \qquad (1)$$

$$ES + S \rightleftharpoons ESS \qquad (2)$$

$$ES + S - S \rightleftharpoons ESSS \qquad (3)$$

Excess substrate inhibition is characterized by the constant

$$K_s = (ES)\,(S)/(ESS) \qquad (4)$$

while sucrose aggregation is characterized by the constant

$$K_s' = (ES)\,(S)/(ESSS) \qquad (5)$$

The substitution of the corrected sucrose concentration (S^*) for the total sucrose

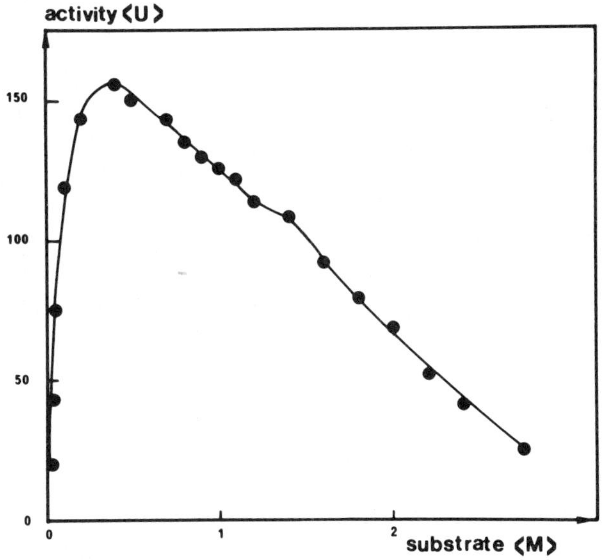

FIGURE 3. Effect of sucrose concentration on invertase activity at 53°C. Experimental conditions were as indicated in FIGURE 2.

concentration gives the following equation:

$$V = \frac{V_{\max}(S^*)}{K_m + (S^*) + \frac{(S^*)^2}{K_s} + \frac{(S^*)^3}{K_s'}} \tag{6}$$

Equation 6 allows a very good description of the effect of corrected sucrose concentration on the initial rate of sucrose hydrolysis by invertase.[5]

EFFECT ON ENZYME STABILITY

It may be seen from FIGURE 4 that sucrose concentration has a very marked effect on the stability of invertase covalently grafted onto corn stover.[9] At 60°C, for example, the half-life of immobilized invertase is 3 days when the sucrose concentration is lower than 1.5 *M*. For sucrose concentrations higher than 1.5 *M*, invertase half-life increases in a linear way. A tenfold increase in half-life is found for invertase in a 2.75 *M* sucrose concentration. Understanding such stabilization is important, as it may result in a general approach to enzyme stabilization. The question then is: is the increase in stability observed for immobilized invertase with increasing sucrose concentration a specific phenomenon due to substrate protection or a more general effect? We determined the effect of several compounds containing hydroxylic groups [polyhydric alcohols, poly(1,2-ethanediol), dextran] on invertase stability. For this purpose, we took as a criterion for enzyme stability the half-life, that is, the time necessary to lose

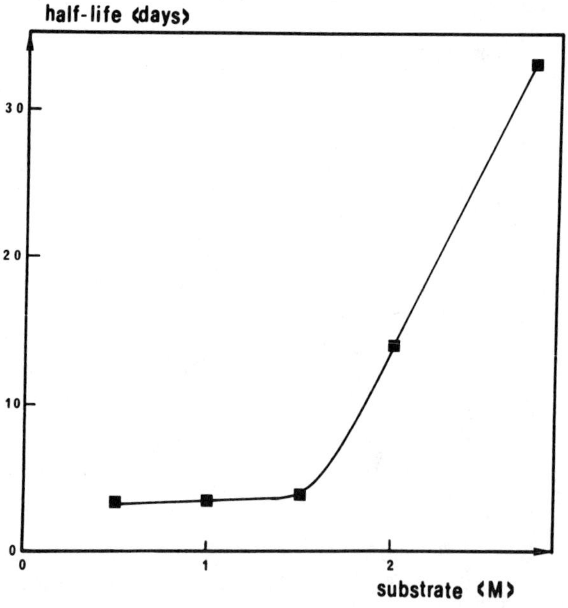

FIGURE 4. Effect of sucrose concentration on half-life of immobilized invertase. Temperature was 60°C and pH was 4.5.

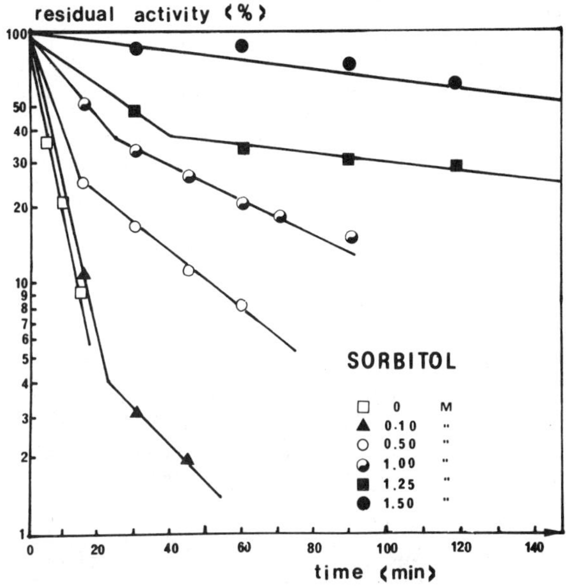

FIGURE 5. Effect of sorbitol concentration on invertase residual activity. Denaturation temperature was 60°C; residual activity of invertase was measured at 40°C using 0.4 *M* sucrose solution in 0.1 *M* acetate buffer, pH 4.5.

fifty per cent of the initial activity at 60°C. The stabilizing factor was defined as the ratio of the free invertase half-life in the presence of the hydroxyl group molecule, to the half-life of free invertase without any additive. A 2 g/l invertase solution in a 0.1 *M* acetate buffer, pH 4.5, was kept at 60°C with and without additive. Samples were withdrawn at intervals, and residual invertase activity was determined as previously described.[5] The invertase half-life was then determined assuming an exponential decay of enzyme activity as a function of time (FIG. 5) and used to calculate the stabilizing factor previously defined.

Effect of Polyhydric Alcohols

The stabilizing effect of different polyhydric alcohols, containing two to six carbon atoms, was determined. Ethylene glycol (C2), glycerol (C3), erythritol (C4), xylitol (C5), and sorbitol (C6) were used for this study. The kinetics of invertase denaturation in the presence of sorbitol is given in FIGURE 5. Free invertase has a half-life of 4.5 min at 60°C. The addition of sorbitol, at concentrations ranging from 0.1 *M* to 1.5 *M*, results in improved enzyme stability. A protective effect corresponding to a stabilizing factor of 13.9 is obtained using a 1.5 *M* sorbitol solution (TABLE 1).

It may be noticed in FIGURE 5 that the exponential decay of invertase activity is not regular and presents a change of slope at a denaturation time corresponding to varying sorbitol concentration. This irregularity may be attributed either to heterogeneous enzyme or to a denaturation mechanism involving different consecutive steps.

Ethylene glycol does not increase invertase stability and even has a slightly negative effect on the enzyme half-life.[10] This may be due to the short length of the ethylene glycol molecule (two carbon atoms) that allows a possible interaction with the internal structure of the invertase molecule.

The protective effect of polyols increases with their concentration and the length of the carbon chain.[10]

Effect of Poly(1,2-Ethanediol)

The poly(1,2-ethanediol) (PEG) molecules contain two hydroxyl groups per molecule, according to the following general structure:

$$OH—CH_2—CH_2—(O—CH_2—CH_2)_n—O—CH_2—CH_2—OH$$

The effect of PEG addition on invertase stability has been determined under the same conditions as described for polyhydric alcohols, using PEG preparations with an average molecular weight ranging from 200 to 20,000.

PEG 200 (MW 200) does not modify invertase stability in the range 0–250 g/l (FIG. 6). The other PEG preparations increase the half-life of free invertase at 60°C.

TABLE 1. Effect of Sorbitol Addition on Free Invertase Half-life and Protective Effect

Sorbitol Concentration (*M*)	Invertase Half-life (min)	Protective Effect
0	4.5	1.0
0.1	5	1.2
0.5	8	1.9
1.0	18	4.2
1.25	30	9.3
1.5	183	13.9

The protective effect increases with PEG concentration and PEG molecular weight. It can be seen in FIGURE 6 that the only exception is PEG 1500. But PEG 1500 was supplied by Merck, while all the other PEG preparations were supplied by Sigma. This underlines the problem of the origin and the heterogeneity in molecular weight distribution of the PEG fractions used for such a study.

For a constant PEG concentration, FIGURE 7 shows that the PEG of molecular weight below 600 has an important effect on invertase stabilization. This effect is less important for higher molecular weight PEG. But if the protective effect is corrected by the molecular concentration of the PEG preparations, it would appear that there is a direct relation between the PEG protective effect per mole and the PEG molecular weight (FIG. 8). According to the structure of PEG molecules, which, regardless of molecular weight, contain two hydroxyl groups, one at each end of the molecule, and bearing in mind that these hydroxylic groups are the main points of interaction between PEG molecules and water, the hypothesis may be made that the stabilizing effect of PEG cannot be simply attributed to a modification of the water activity of the medium. It should be noted also that the rigid PEG molecules with a molecular weight above 1500 do not modify the stabilization of hemoglobin molecules by glucose and sucrose during lyophilization.[11] This is not the case for shorter PEG molecules (MW 300–600), which decrease the protective effect of glucose and sucrose. This varying

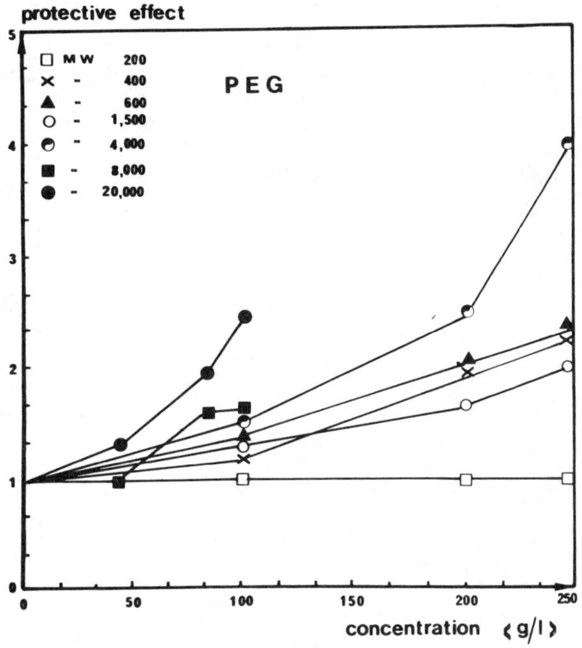

FIGURE 6. Effect of poly(1,2-ethanediol) concentration on invertase stability. Protective effect is defined as the ratio of invertase half-life with stabilizer to invertase half-life without stabilizer. Experimental conditions were as indicated in FIGURE 5.

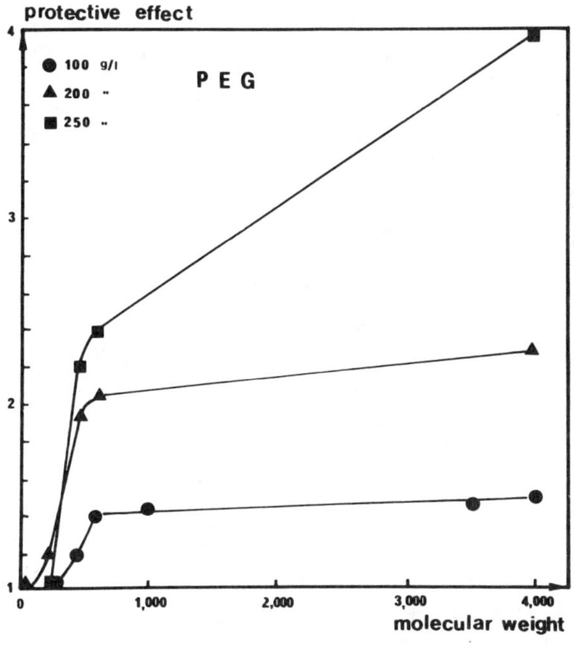

FIGURE 7. Effect of poly(1,2-ethanediol) molecular weight on invertase stability. Experimental conditions were as indicated in FIGURE 5.

behavior of PEG molecules cannot be ascertained in the case of the stabilization of invertase activity at 60°C (FIG. 8). This difference may be attributed to the fact that during lyophilization, the loss of the active conformation results from a loss of hydration water molecules which can be substituted by low molecular weight PEG but not by the high molecular weight derivatives. On the other hand, during thermal inactivation, the loss of the active conformation results from the collision between enzyme molecules and water molecules.

Effect of Dextrans

Dextran is a D-glucose polymer containing 95% α (1 → 6) linear bonds and 5% α (1 → 3) branching bonds. This polymer is obtained by enzymatic synthesis from sucrose, using dextransucrase, an exocellular enzyme from *Leuconostoc mesenteroides.*[12] The protective effect of dextran on free invertase activity at 60°C has been

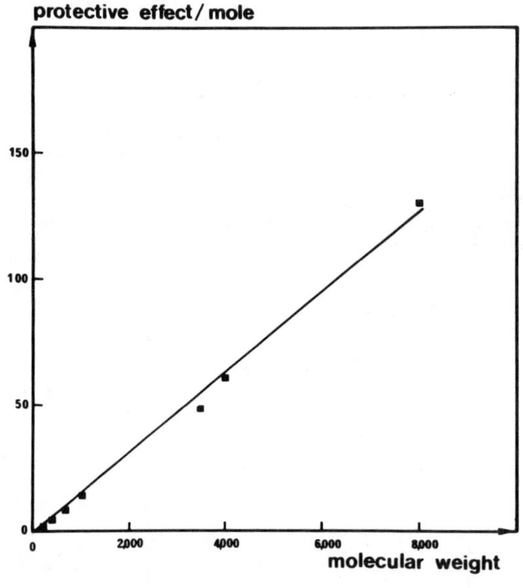

FIGURE 8. Effect of poly(1,2-ethanediol) molecular weight on invertase protective effect/mole. Poly(1,2-ethanediol) concentration was 100 g/l. Experimental conditions were as indicated in FIGURE 5.

determined using dextran preparations (Sigma) with an average molecular weight ranging from 10,500 to 2,000,000. It may be seen in FIGURE 9 that dextran's protective effect increases with molecular weight to maximum stabilization for dextran 18,100. Above this value, the protective effect decreases as dextran molecular weight increases. Furthermore, the degree of stabilization obtained in the case of dextran is relatively less than when using either PEG or polyhydric alcohols.

Effect of Stabilizing Mixtures

To determine the protective effect of mixtures of stabilizers, two of the compounds, sorbitol and PEG 1500, which combine an important increase in invertase stability with good solubility, were used.

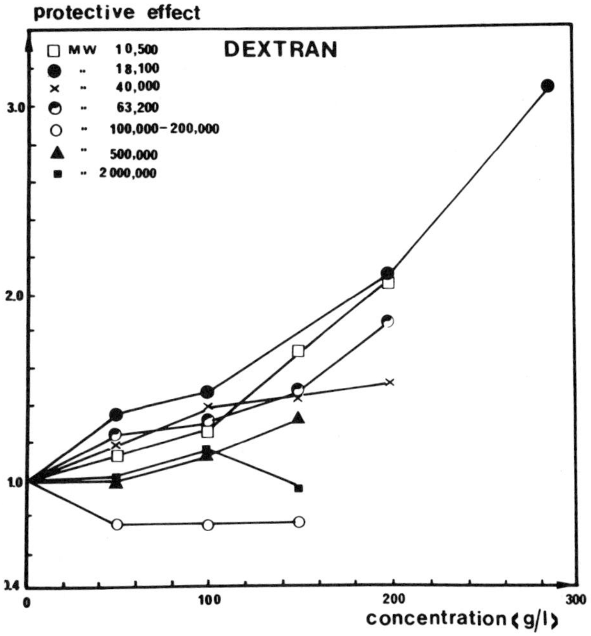

FIGURE 9. Effect of dextran concentration on invertase stability. Experimental conditions were as indicated in FIGURE 5.

Two mixtures containing sorbitol and PEG 1500 were tested for their protective effect (TABLE 2). The protective effect corresponding to the amount of PEG used was determined from FIGURE 6 and expressed as the amount of sorbitol necessary to obtain the same protecting effect (FIG. 5). The sum of the actual sorbitol concentration present in mixture 1 and 2 and of the sorbitol-equivalent concentration corresponding to PEG 1500 was used to determine from FIGURE 10 the predicted protective effect of the mixture. It may be seen from TABLE 2 that the experimental protective effect obtained for each mixture shows a slight synergistic effect, corresponding to an increase of about 25%. This may be attributed to a possible molecular interaction between sorbitol and PEG molecules in such concentrated solutions. The use of highly concentrated sorbitol solution (2.4 *M*) results in a protective effect of up to 57-fold, that is, the invertase half-life in a 2.4 *M* sorbitol solution at 60°C reaches 4.3 hours, being about sixty times higher than the half-life obtained in a pure water solution.

TABLE 2. Invertase Stabilization with Mixtures of Sorbitol and PEG

Mixture	Sorbitol Concentration (*M*)	PEG Concentration ($g \cdot l^{-1}$)	Total Stabilizers Concentration ($g \cdot l^{-1}$)	Experimental Protective Effect	Predicted Protective Effect
1	1.5	150	423	38	31
2	0.9	250	414	16,5	13

CONCLUSION

The use of invertase in the catalysis of concentrated sucrose solution hydrolysis led us to the study of the influence of water activity on the action and stability of this enzyme.

As far as invertase action is concerned, it may be observed that enzyme activity greatly decreases when substrate concentration is increased above 0.4 *M*. This phenomenon may be explained neither by simple excess substrate inhibition, nor by an increase in medium viscosity (FIG. 2). Close examination of the results obtained for the influence of sucrose concentration on hydrolysis initial reaction rate indicates the presence of irregularities in the decreasing part of the curve shown in FIGURE. 1. The

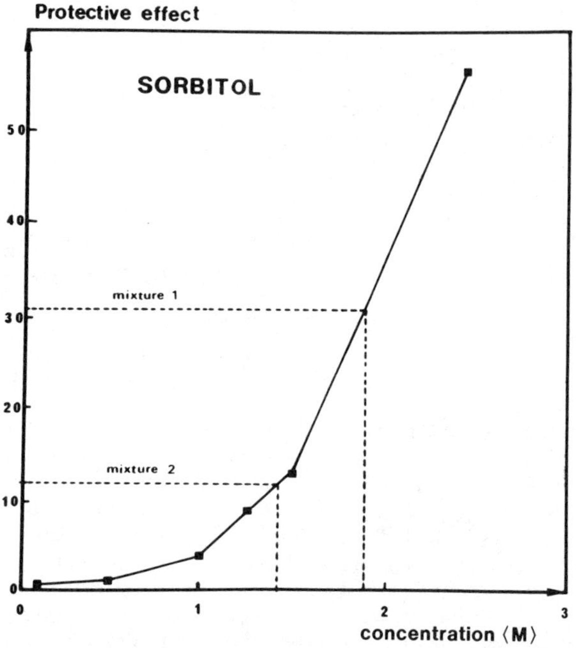

FIGURE 10. Effect of sorbitol concentration on invertase stability. Mixtures 1 and 2 are defined in TABLE 2. Experimental conditions were as indicated in FIGURE 5.

position of these plateaus coincides with the zone of sucrose concentration where the molecule undergoes a conformation change, due to the establishment of one and then two intramolecular hydrogen bonds,[7,8] and an aggregation phenomenon, due to a decrease in free water activity. It was observed that free water activity-lowering agents brought about such irregularities at lower sucrose concentrations, whereas an increase in temperature caused a shift towards the higher sucrose concentration values (FIG. 3). These two phenomena led to a mathematical model that gave an excellent description of the influence of sucrose concentration on invertase activity (Eq. 6).

While in general, the enzyme molecule itself has been considered, our results with invertase show that a decrease in free water activity does modify enzymatic behavior, directly, by a change in ionization state and indirectly, by a substrate conformational change. It may thus be envisaged that enzyme action on a given substrate may be regulated not only by a change in the catalytic properties of the enzyme molecule, but also by a conformational change of the substrate, the intensity of which would depend on free water activity in the medium. It is likely that such a phenomenon, as we have shown in the case of carbohydrases, may likewise play a very important role in the case of proteases. For example, the regulation of proteolysis reactions in cell structures or in solid media may be connected with water activity.

The change in water activity affects not only enzyme activity but stability, in particular, resistance to thermal denaturation. Enzyme stabilization against thermal denaturation has recently been discussed in greater detail.[14] Generally speaking, enzyme stability increases in concentrated media (FIG. 4), that is, when water activity decreases. It has been found that enzymes remain very stable when stored into the form of dry powder.[13] We have studied the influence of various water-activity-decreasing compounds (polyols, PEG, dextran) on the residual activity of free invertase at 60°C, taking enzyme half-life as the denaturation criterion.

In the case of polyols, the protective effect increases with polyol concentration (FIG. 5) and chain length. However, comparison of the results obtained with glycerol and sorbitol shows that this effect is not related simply to changes in water activity. In fact, while glycerol has a greater depressive effect on water activity than sorbitol, it is sorbitol that has the greater stabilizing effect.[10] In the case of PEG, the protective effect increases with molecular weight and concentration (FIGS. 6 and 7). Furthermore, a direct relationship may be observed between protective effect per mole and PEG molecular weight (FIG. 8). This would seem to suggest that the protective effect is due not only to the interaction of hydroxyl groups with water molecules, as PEG still only contains two hydroxyl groups per molecule.

Finally, dextran shows less of a protective effect, and an optimum is obtained with dextran of average molecular weight 18,100. High-molecular-weight dextrans have practically no stabilizing effect.

In view of the results obtained from the study of the stabilizing effect of polyols, PEG, and dextrans on invertase enzymatic activity, a schematic description of the action of such compounds can be proposed. For this purpose, it must first of all be considered that the catalytic activity of an enzyme is directly related to a precise three-dimensional conformation, the only one that allows the transformation of substrate into product. Any serious modification in this structure results in a loss of catalytic activity.

The three-dimensional conformation of enzyme molecules is stabilized by low-energy bonds that are broken mainly as a result of an increase in vibration energy and of collision with water molecules occurring following a rise in temperature. As the energy caused by water-protein hydrogen bonds is of the same magnitude as enzyme intramolecular bonds, compounds having a strong interaction with the enzyme (i.e., at an energy level greater than that of water) can stabilize unfolded structures and favor denaturation. A similar result has been obtained during the study of the denaturation of various synthetic polymers and polysaccharides.[15] However, compounds having a low interaction with the enzyme but interacting strongly with water molecules will result in the formation of water clusters having a structure comparable to that of ice.[15] They will thus have a protective role by reducing the amount of free water. At the intermediate level, compounds that interact with the enzyme molecule at an energy level similar to that of water, will have no noticeable effect on stability.

It may thus be assumed, in the case of polyols, that ethyleneglycol, with a slightly negative effect on invertase stability,[10] interacts with internal enzyme structure due to its small molecule size. This is impossible for polyols of greater chain length. The protective effect increases with molecular weight not only for polyols but also for PEG molecules, which have only one hydroxyl group at each end of the chain. This may be explained by the fact that the capacity of the molecules to mobilize and structure water in the form of clusters depends directly on their size. This may be observed for polyhydric molecules as well as for hydrophobic molecules.[15] Such a reinforcement of water molecule organization, directly related to PEG chain length, may be the cause of the direct relationship observed between the protective effect per mole and PEG molecular weight (FIG. 8). However, the stabilizing effect of dextrans does not increase with size of molecule for molecular weight greater than 18,100 (FIG. 9). Such behavior may be attributed to the fact that dextran chain structure does not allow, in the case of high-molecular-weight molecules, a high degree of water molecule organization.

Furthermore, such dextran molecules have a tendency toward strong interaction with one another.

It would thus appear that water activity exerts a strong influence on enzyme activity and stability. This phenomenon must be fully appreciated in order to analyze the effect of a compound on enzyme efficiency, and so as not to misinterpret as direct an indirect effect caused by a modification of water activity. Furthermore, from a technological viewpoint, a better understanding of this phenomenon may allow rationalization of the operation of enzyme catalysts and their improved efficiency.

REFERENCES

1. MONSAN, P. & D. COMBES. 1984. Application of immobilized invertase to continuous hydrolysis of concentrated sucrose solutions. Biotechnol. Bioeng. **26:** 347–351.
2. BOWSKY, L., R. SAINI, D. Y. RYU & W. R. VIETH. 1971. Kinetic modeling of the hydrolysis of sucrose by invertase. Biotechnol. Bioeng. **13:** 641–656.
3. BESSERDICH, H., E. KAHRIG, R. KRENZ & D. KIRSTEIN. 1977. Kinetics studies on substrate inhibition of free and bound yeast invertase. J. Mol. Catal. **2:** 361–367.
4. COMBES, D., P. MONSAN & M. MATHLOUTHI. 1981. Enzymic hydrolysis of sucrose. Carbohydr. Res. **93:** 312–316.
5. COMBES, D. & P. MONSAN. 1983. Sucrose hydrolysis by invertase. Characterization of products and substrate inhibition. Carbohydr. Res. **117:** 215–228.
6. MAC LAREN, A. D. 1963. Enzymes reactions in structurally restricted systems. III Yeast-fructofuranosidase (Invertase). Activity in concentrated sucrose solution. Enzymologia **26:** 1–11.
7. MATHLOUTHI, M., C. LUU, A. M. MEFFROY-BIGET & D. V. LUU. 1980. Laser-Raman study of solute-solvent interactions in aqueous solutions of D-fructose, D-glucose, and sucrose. Carbohydr. Res. **81:** 213–223.
8. MATHLOUTHI, M. 1981. X-ray diffraction study of the molecular association in aqueous solutions of D-fructose, D-glucose, and sucrose. Carbohydr. Res. **91:** 113–123.
9. MONSAN, P., D. COMBES & I. ALEMZADEH. 1984. Invertase covalent grafting onto corn stover. Biotechnol. Bioeng. **26:** In press.
10. COMBES, D. & P. MONSAN. 1983. Effect of polyhydric alcohols on invertase stabilization. Ann. N. Y. Acad. Sci. This volume.
11. PRISTOUPIL, T. I., S. ULRYCH & M. KRAMLOVA. 1981. Haemoglobin stabilization during lyophilization with saccharides. Perturbation effect of polyethylene glycols. Collect. Czek. Chem. Commun. **46:** 1856–1859.
12. JEANES, A. 1968. Dextran. *In* Encyclopoedia of Polymer Science and Technology. Interscience Pubishers. Vol. **4:** 805–829. John Wiley and Sons. New York.
13. SCHMID, R. D. 1979. Stabilized soluble enzymes. *In* Advances in Biochemical Engineering. T. K. Ghose, A. Fiechter & N. Blakebrough, Eds. Vol. **12:** 41–118. Springer-Verlag. New York.
14. KLIBANOV, A. M. 1983. Stabilization of enzymes against thermal inactivation. *In* Advances in Applied Microbiology. A. I. Laskin, Ed. Vol. **29:** 1–24. Academic Press, New York.
15. DOBBINS, R. J. 1973. Solute-solvent interactions in polysaccharides systems. *In* Industrial Gums. 2nd edition. R. L. Whistler, Ed. Academic Press. New York. pp. 19–25.

Effect of Polyhydric Alcohols on Invertase Stabilization

DIDIER COMBES AND PIERRE MONSAN

Department de Genie Biochimique et Alimentaire
ERA-CNRS 879, INSA
Avenue de Rangueil
F-31077 Toulouse Cedex, France

Further development of enzyme engineering requires improved enzyme stability. In order to be suitable for various industrial applications, enzymes have to be stable under operational conditions over prolonged periods of time. The most important factors leading to protein denaturation and the principles for the stabilization of water-soluble enzymes have recently been reviewed.[1] From a practical point of view, thermal inactivation is the most important mode of enzyme denaturation. The mechanisms of thermal inactivation have been described by Klibanov,[2] who also presents general considerations on the stabilization of enzymes by sugars and other polyols.

Previous studies of invertase action in concentrated sucrose solutions (up to 1 kg/1) have shown an important increase in enzyme stability (half-life) when sucrose concentration is varied from 1.5 *M* to 2.75 *M*.[3] For example, at 60°C, the half-life of invertase in the presence of 2.75 *M* sucrose solution was ten times greater than in the presence of 1.5 *M* sucrose solution. This increase in enzyme stability may be attributed either to a specific—or to an unspecific—protective effect due to the presence of the substrate, the latter resulting either from the preferential interaction of invertase with sucrose[4] or from the influence of sucrose concentration on water activity.

The effect of water-activity-lowering agents on invertase stabilization, therefore, has been studied. For example, polyhydric alcohols such as ethylene glycol, glycerol, or sorbitol have already been used to obtain low water-activity systems for the study of their enzyme behavior.[5]

FIGURE 1 shows the influence of the concentration of different polyhydric alcohols (ethylene glycol, glycerol, erythritol, xylitol, and sorbitol) on the stabilizing factor of invertase. The stabilizing factor is defined as follows:

Stabilizing factor = invertase half-life with polyol / invertase half-life without polyol.

Heat denaturation was performed at 60°C and residual activity was measured at 40°C with a 0.4 *M* sucrose solution in an acetate buffer (0.1 *M*, pH 4.5). The variation in concentration of the reducing sugars resulting from invertase activity was measured by the 2,4-dinitro-salicylic acid method.[6]

It may be seen from FIGURE 1 that ethylene glycol has no positive effect on invertase stability. On the other hand, all the other polyhydric alcohols used protect invertase against heat denaturation. The protective effect increases with the rising concentration of the polyhydric alcohol.

In FIGURE 2 the effect of the length of the polyhydric alcohol molecule on the stabilizing factor is given and related to the number of hydroxyl groups per molecule. The stabilizing effect increases with the number of hydroxyl groups contained in the molecule.

Similar results have been obtained using the transition temperature (i.e., the conformation change in the enzyme molecule) determined either by UV spectropho-

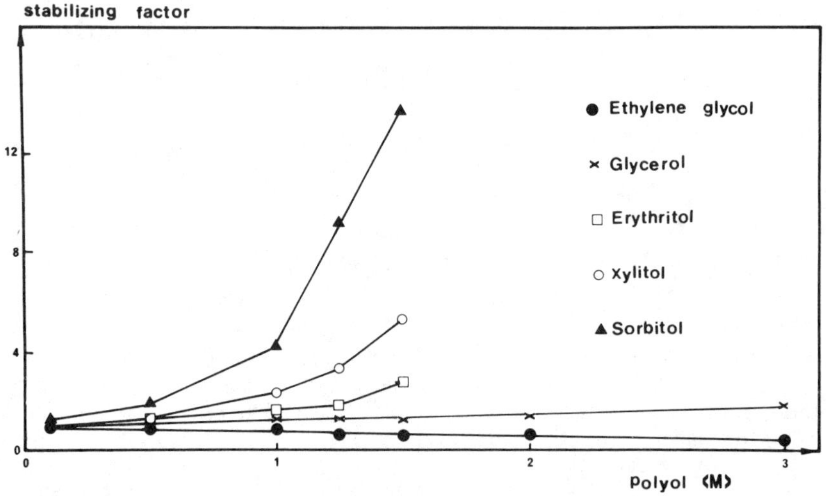

FIGURE 1. Influence of polyhydric alcohol concentration on stabilizing factor of invertase. Denaturation temperature: 60°C. Residual activity was measured using 0.4 *M* sucrose solution pH 4.5 at 40°C.

tometry[7] or by differential scanning calorimetry,[8,9] instead of the activity decrease, as a denaturation criterion.

It may be seen from our results that in the range of polyhydric alcohols used, there is no direct relation between invertase stabilization, taking as a criterion the modification of the half-life of the enzyme, and the number of hydroxyl groups per polyol

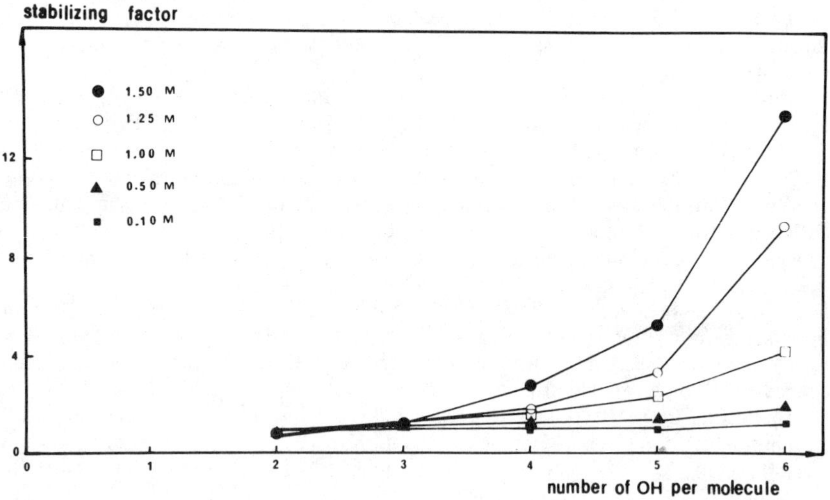

FIGURE 2. Influence of length of polyhydric alcohol molecule on stabilizing factor of invertase. Denaturation temperature: 60°C. Residual activity was measured using 0.4 *M* sucrose solution pH 4.5 at 40°C.

molecule. Furthermore, glycerol, which has a slightly greater water-activity-depressing effect than sorbitol,[10] has no marked stabilizing effect on invertase activity (FIG. 2). The addition of sorbitol, however, results in an important increase in enzyme stability. It may be concluded that the protective effect of polyhydric alcohols is not simply related to a modification in the water activity of the medium.

In conclusion, the increased stability of invertase observed in concentrated sucrose solutions is not due to a specific substrate protective effect, as similar stabilization may be obtained using different polyhydric alcohols. This protective effect increases with polyol concentration and molecule length and cannot be simply attributed to the polyol modification of water activity.

REFERENCES

1. SCHMID, R. D. 1979. Stabilized soluble enzymes. *In* Advances in Biochemical Engineering. T. K. GHOSE, A. FIECHTER & N. BLAKEBROUGH, Eds. Vol. **12:** 41–118. Springer-Verlag. New York.
2. KLIBANOV, A. M. 1983. Stabilization of enzymes against thermal inactivation. *In* Advances in Applied Microbiology. A. I. LASKIN, Ed. Vol. **29:** 1–24. Academic Press. New York.
3. MONSAN, P., D. COMBES & I. ALEMZADEH. 1984. Invertase covalent grafting onto corn stover. Biotechnol. Bioeng. **26:** In press.
4. ARAKAWA, T. & S. N. TIMASHEFF. 1982. Stabilization of protein structure by sugars. Biochemistry **21:** 6536–6544.
5. TOME, D., J. NICOLAS & R. DRAPRON. 1978. Influence of water activity on the reaction catalyzed by polyphenoloxidase (E.C.1.14.18.1.) from mushrooms in organic liquid media. Lebensm. Wiss. Technol. **11:** 38–41.
6. SUMNER, J. B. & S. F. HOWELL. 1935. A method for determination of invertase activity. J. Biol. Chem. **108:** 51–54.
7. GERLSMA, S. Y. 1968. Reversible denaturation of ribonuclease in aqueous solutions as influenced by polyhydric alcohols and some other additives. J. Biol. Chem. **243:** 957–961.
8. GEKKO, K. 1982. Calorimetric study on thermal denaturation of lysozyme in polyol-water mixtures. J. Biochem. **91:** 1197–1204.
9. FUJITA, Y., Y. IWASA & Y. NODA. 1982. The effect of polyhydric alcohols on the thermal denaturation of lysozyme as measured by differential scanning calorimetry. Bull. Chem. Soc. Jpn. **55:** 1896–1900.
10. GUILBERT, S., O. CLEMENT & J. C. CHEFTEL. 1981. Efficacité comparée d'agents depresseurs de l'aw en solution et dans des aliments a humidité intermediaire. Lebensm. Wiss. Technol. **14:** 245–251.

Series-type Enzyme Deactivation Kinetics: Influence of Immobilization, Chemical Modifiers, and Enzyme Aging

JAMES P. HENLEY AND AJIT SADANA

Chemical Engineering Department
University of Mississippi
University, Mississippi 38677

One method of explaining enzyme deactivations that do not follow simple first-order kinetics is to use the following series-deactivation scheme

$$E \xrightarrow{k_1} E_1 \xrightarrow{k_2} E_d \quad (1)$$

where k_1 and k_2 are first-order deactivation rate coefficients, E is the specific activity of the initial enzyme state, E_1 is the specific activity of the single active intermediate that is as active as the initial enzyme state, and the specific activity E_d of the deactivated enzyme is zero. The activity, $\underset{\approx}{a}$, is defined by Equation 2 if we assume that E and E_1 equal zero at time, $t = 0$.

$$\underset{\approx}{a} = (E + E_1)/E_o = [k_2 \exp(-k_1 t) - k_1 \exp(-k_2 t)]/(k_2 - k_1) \quad (2)$$

Since the intermediate will have a different spatial configuration from the initial enzyme state, it is very reasonable to assume that the enzyme intermediate may differ in specific activity from the initial enzyme. Then, the activity, $\underset{\approx}{a}$, is equal to $(E + \alpha_1 E_1)/E_o$ and yields Equation 3.

$$\underset{\approx}{a} = [1 + (\alpha_1 k_1/(k_2 - k_1))] \exp(-k_1 t) - [\alpha_1 k_1/(k_2 - k_1)] \exp(-k_2 t) \quad (3)$$

The activity, $\underset{\approx}{a}$ actually is a weighted function of the specific activities of the two enzyme states; and $\alpha_1 = E_1/\mathrm{E}$ is the ratio of the specific activities of the intermediate and the initial enzyme states. Equation 3 permits an initial rise in activity with time, or an "activation" if we assume that α_1 is greater than one.[1]

Classically, it has been assumed that the deactivated enzyme state, E_d suffers a complete loss of activity. There is no reason to believe that this should hold for all enzymes. If we assume partial deactivation for, at least, some enzymes, and if we let the specific activity of the final state be E_2 rather than E_d, then Equation 1 may be rewritten as Equation 4.

$$E \xrightarrow{k_1} \overset{\alpha_1}{E_1} \xrightarrow{k_2} \overset{\alpha_2}{E_2} \quad (4)$$

where $\alpha_2 = E_2/E$ is the ratio of the specific activities of the final and initial enzyme states. Then, the activity is expressed by Equation 5.

$$\underset{\approx}{a} = (E + \alpha_1 \mathrm{E}_1 + \alpha_2 E_2)/E_o = \alpha_2 + [1 + ((\alpha_1 k_1 - \alpha_2 k_2)/(k_2 - k_1))] \exp(-k_1 t) - [k_1(\alpha_1 - \alpha_2)/(k_2 - k_1)] \exp(-k_2 t) \quad (5)$$

where at time, $t = 0$ the specific activities of the intermediate and final enzyme states E_1 and E_2 are zero. If $\alpha_2 = 0$, then Equation 5 reduces to Equation 3 as it should. The α_2 parameter estimates the residual or stabilized activity of the enzyme.

Enzyme deactivation data were taken from the literature and modeled using Equation 5. The SAS NLIN (nonlinear regression) procedure[2] was used along with the Marquadt method of interative convergence. The fit of the curves obtained by the SAS procedure[2] performed on experimental data on immobilized and soluble enzymes using Equation 5 is shown in FIGURES 1 and 2. The estimated values of k_1, k_2, α_1, and α_2 are

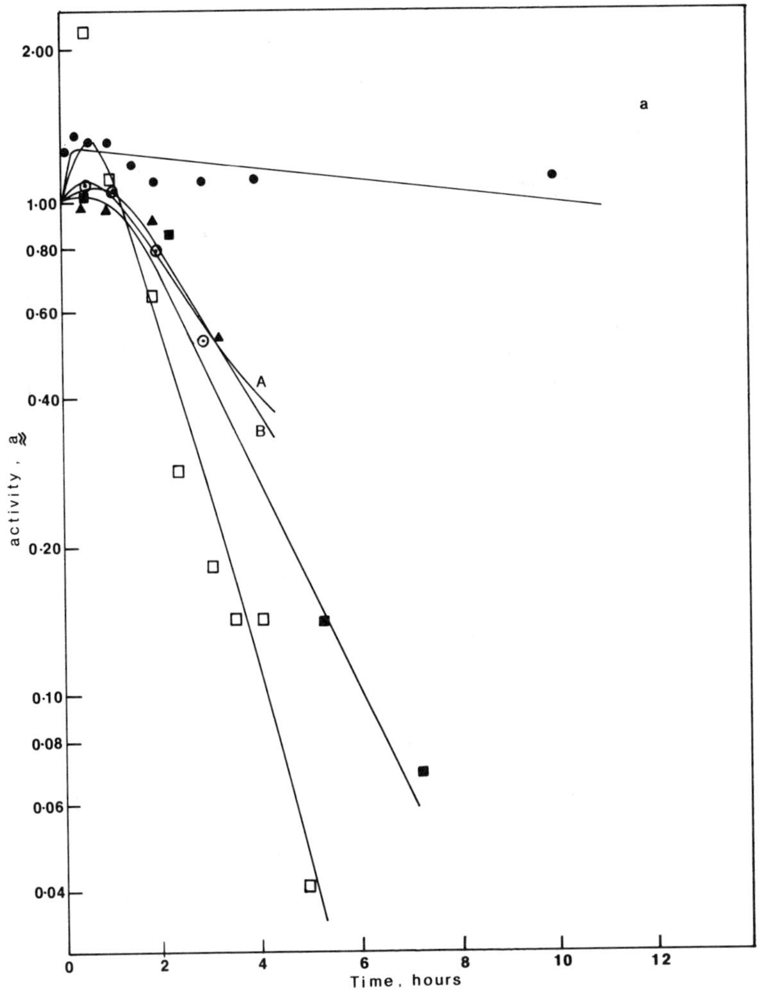

FIGURE 1. Nonlinear regression fit of the proposed model for enzymes exhibiting the "compensation-like" or "trade-off" effect. Different forms of mushroom tyrosinase: ■ native; ⊙ immobilized on collagen (A); ● collagen + glutaraldehyde; □ collagen + ethyl acetimidate; ▲ collagen + dimethyl adipimidate (B).

given in TABLES 1 and 2 for different enzymes. The reliability of the adjustable constants has been demonstrated earlier.[1] The values of the rate constants were purposely changed to determine a sensitivity analysis. Note that since four parameters are used to fit a single curve, the model provides a reasonable fit for the experimental data taken from the literature for the various deactivating immobilized and soluble enzymes. Out of these four parameters, two are required to model the initial activation and the residual activity of the enzyme.

Data are presented in TABLE 1 on the effect of the reticulating agents (the diimidates) on the deactivation kinetics of mushroom tyrosinase (FIG. 1), the alkaline phosphatases, and a mixture of glyceraldehyde-3-phosphate dehydrogenase/phosphoglycerate kinase. It is of interest to note that if k_2 for the immobilized enzyme (with or without chemical modifiers) is higher or lower than k_2 for the soluble enzyme, then α_1 is also higher or lower, respectively. This behavior seems to be a "compensation-like" effect or a "trade-off," wherein you lose a little enzyme specific activity to get a little

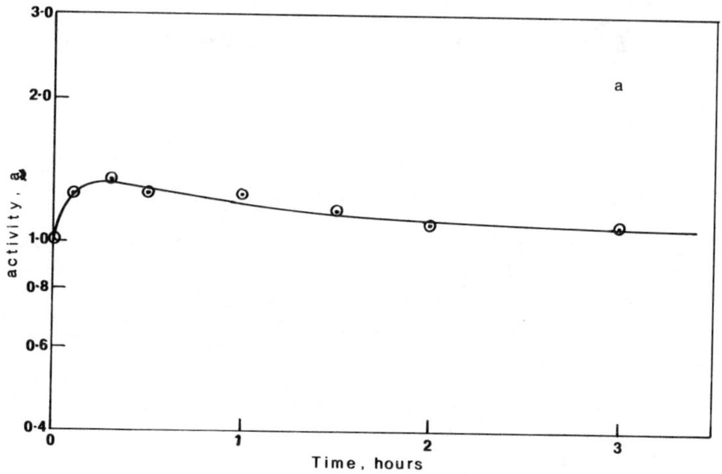

FIGURE 2. Stabilization of activity, $\underset{\approx}{a}$, of mushroom tyrosinase immobilized on collagen and chemically modified by glutaraldehyde at a level higher than the initial level.

increased enzyme stability. In these cases, a chemical modification of the enzyme structure apparently decreases the conformational flexibility of the native configuration that yields an enzyme with lower catalytic activity but enhanced stability.

Data are presented in TABLE 2 for enzymes that exhibit a stabilized or residual activity above (FIG. 2) or below the initial level. The chemical modification of immobilized mushroom tyrosinase by glutaraldehyde yields an estimated value of α_2 equal to 1.05. In this case, the crosslinks provided by glutaraldehyde do effectively stabilize the native conformation at an eventual activity level even slightly higher than the initial level. The unmodified enzyme exhibits a residual activity level below the initial level.

In conclusion, it is appropriate to indicate that since the proposed model fits the data adequately the presumed mechanism is consistent with the data. The fit does not prove that the mechanism is correct. Nevertheless, wherever the suggested model

TABLE 1. Soluble and Immobilized Enzymes Following the Deactivation Model: Illustration of the "Compensation-like" Effect

Enzyme	T	pH	Buffer	Support	Chemical Modifier	k_1 h^{-1}	k_2 h^{-1}	α_1	Reference
Mushroom tyrosinase	Room temperature	7	0.1 *M* Phosphate	Native	None	0.50	1.4	1.9	Letts & Chase[3]
Mushroom tyrosinase	Room temperature	7	0.1 *M* Phosphate	Collagen	None	0.60	0.80	1.6	Letts & Chase[3]
Mushroom tyrosinase	Room temperature	7	0.1 *M* Phosphate	Collagen	Glutaraldehyde	260	0.026	1.2	Letts & Chase[3]
Mushroom tyrosinase	Room temperature	7	0.1 *M* Phosphate	Collagen	Dimethyl-adipimidate	0.51	0.47	1.2	Letts & Chase[3]
Mushroom tyrosinase	Room temperature	7	0.1 *M* Phosphate	Collagen	Ethyl acetimidate	0.90	1.8	2.8	Letts & Chase[3]
Glyceraldehyde-3-phosphate Dehydrogenase (GAPDH)/phosphoglycerate kinase (PGK) (crude mix)	N/A[a]	7	0.05 *M* Imidazole	Sepharose	None	0.036	0.038	1.1	Marshall[4]
GAPDH/PGK (pure yeast enzyme)	N/A[a]	7	0.05 *M* Imidazole	Sepharose	None	0.20	0.18	1.3	Marshall[4]
Alkaline phosphatase	37°C	10.5	N/A[a]	Native	None	0.021	0.022	1.06	Galski *et al.*[5]
Alkaline phosphatase	37°C	10.5	N/A[a]	Native	None	0.022	0.024	1.09	Galski *et al.*[5]

[a]N/A—Not available

TABLE 2. Enzymes Following the Deactivation Model and Exhibiting a Residual Activity

Enzyme	T, °C	pH	Buffer	Support	Chemical Modifier	k_1 day^{-1}	k_2 day^{-1}	α_1	α_2	Reference
Mushroom tyrosinase	Room temperature	7	0.1 *M* phosphate	Collagen	Glutaral-dehyde	264	23	1.4	1.05	Letts & Chase[3]
LDH	0–5	7.2	0.1 *M* Tris	PTFE (polytetra-fluoroethylene)	None	4.9	5.2	1.3	0.95	Danielson & Siergiej[6]
Calf alkaline phosphatase	5	8	0.1 *M* NH_4 Tris-HCO_3/CO_3; 1 *M* NaCl	Tritylagarose	None	0.038	0.096	1.4	0.69	Cashion *et al.*[7]
UDP glucose: ceramide glucosyl transferase	37	6.35	100 μmoles cacodylate HCl	None	25 nmoles lactosyl ceramide and 100 nmoles CMP-sialic acid	0.48	0.48	1.6	0.86	Dreyfus *et al.*[8]
Cytidine-5′-mono-phospho-N-acetyl-neuraminic acid galactosyl glucosylceramide sialyltransferase	37	6.35	100 μmoles cacodylate HCl	None	25 nmoles lactosyl ceramide and 100 nmoles CMP-sialic acid	144	144	3.8	0.33	Dreyfus *et al.*[8]

correlates experimental data successfully, it does provide a plausible explanation along with valuable insights into the enzyme deactivation mechanisms and structure.

REFERENCES

1. HENLEY, J. P. & A. SADANA. 1984. Series-type enzyme deactivations. Influence of intermediate activity on deactivation kinetics. Enzyme Microb. Tech. **6**(1): 35–41.
2. SAS Institute, Inc. 1982. SAS User's Guide: Statistics Edition. Cary, NC.
3. LETTS, D. & T. CHASE, JR. 1974. Chemical modification of mushroom tyrosinase for stabilization to reaction inactivation. *In* Advances in Experimental Medicine and Biology. R. Bruce Dunlap, Ed. **42:** 317–28. Plenum. New York.
4. MARSHALL, D. L. 1974. Use of immobilized enzymes for synthetic purposes. *In* Advances in Experimental Medicine and Biology. R. Bruce Dunlap, Ed. **42:** 345–68. Plenum. New York.
5. GALSKI, H., S. E. FRIDOVIC, D. WEINSTEIN, N. DEGROOT, S. SEGAL, R. FOLMAN & A. A. HOCHBERG. 1981. Synthesis and secretion of alkaline phosphatase *in vitro* from first-trimester and term human placentas. Biochem. J. **194:** 857–66.
6. DANIELSON, N. D. & R. W. SIERGIEJ. 1981. Immobilization of enzymes on polytetrafluoroethylene particles packed in HPLC columns. Biotechnol. Bioeng. **23:** 1913–17.
7. CASHION, P., A. JAVED, D. HARRISON, J. SEELEY, V. LENTINI, & G. SATHE. 1982. Enzyme immobilization on tritylagarose. Biotechnol. Bioeng. **24:** 403–23.
8. DREYFUS, H., S. HARTH, A. PRETI, P. F. URBAN & P. MANDEL. 1974. Studies of Retinal Ganglioside Metabolism. *In* Advances in Experimental Medicine and Biology. R. Bruce Dunlap, Ed. **42:** 655–65. Plenum. New York.

The Production of α-Cyclodextrin by Enzymatic Degradation of Starch[a,b]

E. FLASCHEL, J.-P. LANDERT, D. SPIESSER, AND A. RENKEN

Institute of Chemical Engineering
Swiss Federal Institute of Technology
CH-1015 Lausanne, Switzerland

Enzymes known as cyclodextrin-glycosyltransferases (EC 2.4.1.19) degrade starch, liberating cyclic amyloses mainly composed of 6, 7, or 8 glucose units called α-, β- and γ-cyclodextrin, respectively. These compounds are able to form inclusion complexes with hydrophobic substances. This behavior has been studied extensively and has led to numerous industrial applications for cyclodextrins.[1]

Most of the application studies have been undertaken with β-cyclodextrin. Owing to its low solubility and the selectivity pattern of most of the enzymes known, β-cyclodextrin can be produced much easier in a pure state than can its homologs. β-Cyclodextrin is produced by Nihon Shokuhin Kako Co. on a 500 t/a scale.

α- And γ-cyclodextrins are produced by chromatographic separation,[2,3] a means far too expensive to allow general application of these compounds.

At least seven different bacteria are known as sources for cyclodextrin-glycosyltransferases.[4-6] It seems that five of these produce predominantly β-cyclodextrin.[4] The enzyme from *Bacillus macerans* has a poor selectivity for α-cyclodextrin[7] whereas the enzyme from *Klebsiella pneumoniae* M5 al produces α-cyclodextrin with a very high selectivity when the reaction is stopped at the maximum α-cyclodextrin level.[7-10]

A severe drawback for the preparation of α-cyclodextrin is that, under normal reaction conditions, its maximum concentration from potato starch is limited to about $13.5\ g \cdot l^{-1}$,[10] which is less than a tenth of its solubility at 25°C. For facilitating its purification, α-cyclodextrin can be selectively precipitated after[11] or during the reaction[12] by complexing reagents, preferentially by higher alcohols.

Decanol seems to fulfill the requirements for such a complexing agent, since it is essentially insoluble in water, it can be stripped off by steam distillation and it is approved to a certain extent as a food additive.[7] Preliminary studies have shown that decanol does not affect the enzyme of *K. pneumoniae* M5 al, that an α-cyclodextrin yield of 50% can be obtained, and that decanol and α-cyclodextrin seem to undergo complexation in an equimolar ratio.[10]

The displacement of the reaction equilibrium by addition of decanol in the presence of the highly active and selective enzyme from *K. pneumoniae* M5 al seemed to be a promising method for the production of α-cyclodextrin. Therefore, the influence of the main reaction variables on process performance has been studied.

[a] The cyclodextrin project is supported by the Swiss National Science Foundation.

[b] An explanation of the nomenclature and abbreviations used in this paper appears after the reference section.

MATERIALS AND METHODS

The cyclodextrin-glycosyltransferase from *K. pneumoniae* M5 al was obtained by continuous culture with a 1.5% starch medium.[13] The biomass was collected by semicontinuous centrifugation (Sharples T1P). The enzyme in the liquid phase was concentrated by ultrafiltration (Amicon H10 P10) and precipitated with 40% ammonium sulfate. The precipitate was collected by centrifugation and resuspended in a 30% ammonium sulfate solution.

Potato starch and white dextrin A-332 were purchased from Blattmann, Wädenswil, CH. Potato starch Noredux N 150 and N 150 mod. were obtained from Henkel, Düsseldorf, D. White dextrin A-332, Noredux N 150 and N 150 mod. showed a dextrose equivalent of 4.7, 0.7, and 2.7%, respectively.

All other reagents were of analytical grade. The activity of the cyclodextrin-glycosyltransferase was determined according to the cyclization test described by Landert *et al.*[14] The dextrose equivalent (DE) was measured by a method based on the ferricyanide/iodine reaction combined with a thiosulfate titration.[15]

Cyclodextrins were analyzed by HPLC on a reversed-phase column, Lichrosorb-NH_2, 10 μm (Merck, Mannheim, D). An aqueous acetonitrile solution of 65 vol % served as eluant with a flow rate of 2 ml · min^{-1} at 40°C. Detection was performed with a RI detector (model 61, Knauer, Berlin, D).

The enzymatic reactions were carried out in a jacketed glass vessel, equipped with a thermostat and a mechanical stirrer. A starch suspension of 1 liter was supplemented with 5 mmol $CaCl_2$. Sodium hydroxide was added to establish a pH of 6.8 if not stated otherwise. The starch suspension was heated to 90°C for 5 min to gel the starch. The reaction temperature was 40°C if not stated otherwise. During the reaction without decanol, samples were taken with a pipette, diluted with water to obtain an appropriate concentration for the HPLC analysis and heated for 15 min in a water bath of 90°C to inactivate the enzyme.

During reactions in the presence of decanol, samples were taken through a bottom valve of the vessel to assure that the sample would contain a representative composition, since the reaction medium is a slurry of the crystalline decanol/α-cyclodextrin complex. Before the HPLC analysis, decanol was stripped off by steam distillation. By this treatment, α-cyclodextrin is redissolved and the enzyme is denatured.

If not stated otherwise, decanol was added to the gelled starch (0.1 l · kg^{-1} starch) at the beginning of the reaction simultaneously with the enzyme. Potato starch Noredux N 150 mod. was employed at a concentration of 100 g · l^{-1} when no specific data are given.

RESULTS AND DISCUSSION

The pH- and the temperature-activity profile of the cyclodextrin-glycosyltransferase from *K. pneumoniae* M5 al measured by following the coupling of α-cyclodextrin to glucose[10] were reexamined by measuring the cyclization reaction from starch to α-cyclodextrin.

The pH-activity profile can be estimated from FIGURE 1. The evolution of the α-cyclodextrin (CD_6) concentration is given as a function of time for different pH values. The enzyme exhibits a broad pH optimum, at least from pH 6 to pH 8. Even at pH 9.5 it is still active.

The thermal denaturation of the enzyme was examined by carrying out batch

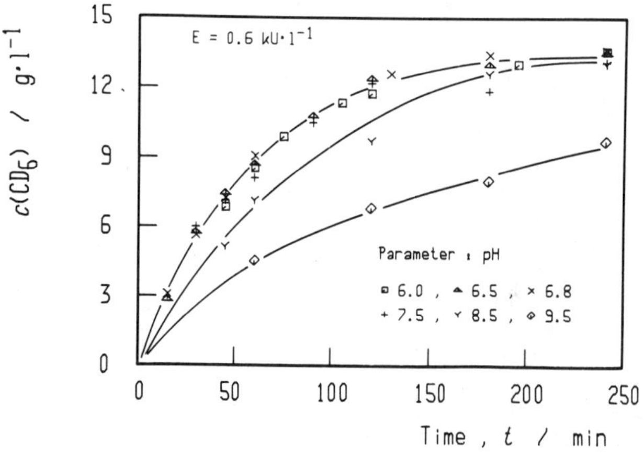

FIGURE 1. The influence of pH on enzymatic activity (potato starch, without decanol).

reactions in the presence of decanol as shown in FIGURE 2. The curves in this graph combine the ambiguous influence of the reaction temperature: an increase in temperature increases the enzymatic activity but enhances simultaneously the rate of deactivation. The enzyme is sufficiently stable up to 50°C for reaction periods up to 6 h.

Since there is no significant difference in activity from 40 to 50°C, the temperature can be chosen within this range. These two figures reveal a broader pH optimum and a better thermal stability compared to the results previously reported.[10] The enhanced thermal stability might be due to the presence of starch.

Since the final α-cyclodextrin yield in the presence of decanol ($0.1\ l \cdot kg^{-1}$ starch) is found to be independent of the temperature in the range of 30 to 50°C, the solubility of the decanol/α-cyclodextrin complex ($3\ g \cdot l^{-1}$ at 25°C, and $6.5\ g \cdot l^{-1}$ at 40°C) seems not to limit the α-cyclodextrin yield. This finding could indicate that the soluble

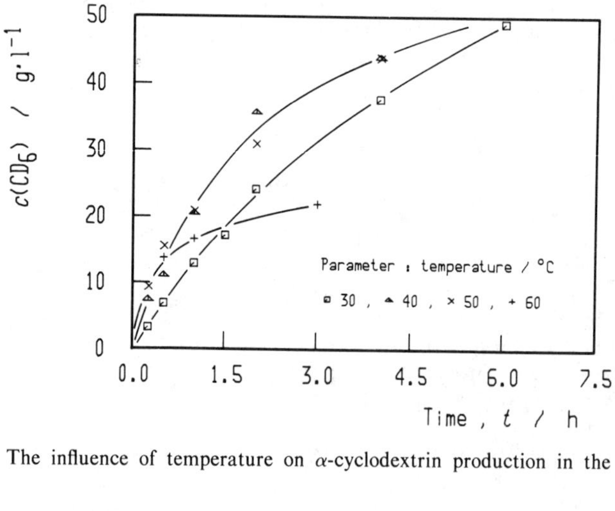

FIGURE 2. The influence of temperature on α-cyclodextrin production in the presence of decanol.

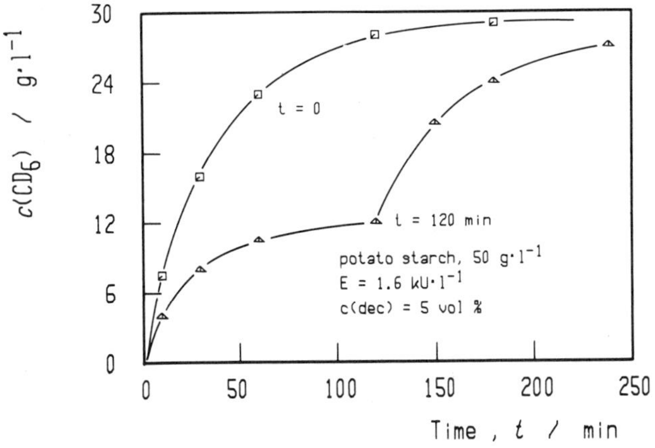

FIGURE 3. The influence of decanol addition at different times.

part of α-cyclodextrin exists almost completely in the complexed form and that, in this form, it is no longer available for the reverse reaction. This is confirmed by studying the strategy of decanol addition as shown in FIGURE 3 for two extreme cases. Decanol has been added at the beginning of the reaction (t = 0) or after having attained the maximum α-cyclodextrin concentration without decanol (t = 120 min). The rate of reaction is considerably enhanced by the addition of decanol even at the beginning. Therefore, it will be most advantageous to add decanol when the reaction is started.

The main variables determining the economy of the decanol process are the enzyme concentration, the substrate concentration and the quality of the substrate. The last two variables are not independent. With starch of low mean molar mass, quite high substrate concentrations can be established without running into technical problems of high viscosity. In the case of ordinary potato starch, the concentration would be limited

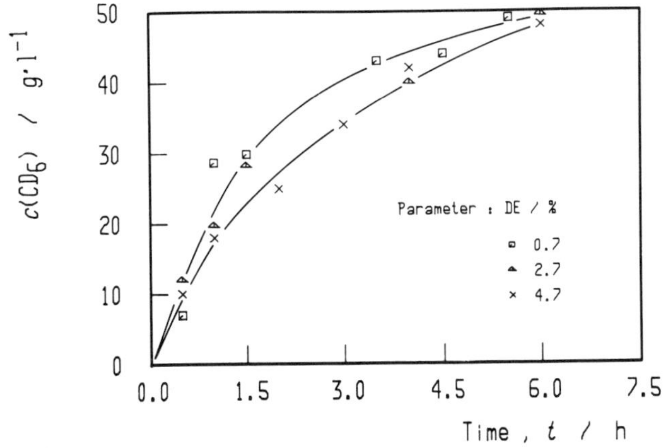

FIGURE 4. The influence of the dextrose equivalent of starch on α-cyclodextrin production. (DE = 0.7: Noredux N 150, DE = 2.7: Noredux N 150 mod., DE = 4.7: white dextrin A-332).

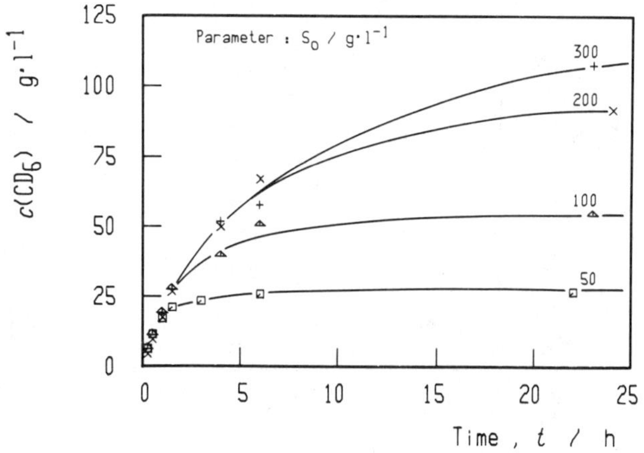

FIGURE 5. The dependence of the α-cyclodextrin production on starch concentration.

to about 100 g · l^{-1}. For starch of the type Noredux N 150 mod., this limit would be at about 200 g · l^{-1} and for dextrins even beyond 300 g · l^{-1}.

The reaction without decanol depends very much on the mean molar mass of the substrate, especially with respect to the maximum α-cyclodextrin concentration. The performance of the decanol process, however, does not depend significantly on this variable as shown in FIGURE 4. When the dextrose equivalent is increased considerably, the rate of reaction slows somewhat but only at an intermediate phase. Therefore, the choice of an appropriate substrate depends mainly on, besides its price, the criterion of whether or not it can be employed at a high concentration.

The aim of the decanol process should be to obtain maximum α-cyclodextrin in the form of the precipitating crystalline decanol complex. When starch can be applied in high concentrations, the soluble fraction of the complex is of minor importance. The

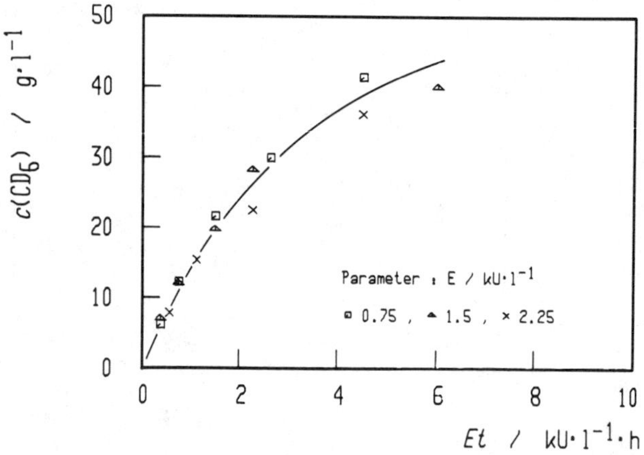

FIGURE 6. α-Cyclodextrin production as a function of normalized reaction time.

time course of the evolution of α-cyclodextrin in the presence of different initial concentrations of starch is shown in FIGURE 5. When the time of reaction is properly adjusted, a final α-cyclodextrin yield of about 50% is obtained and product concentrations of higher than 100 g · l^{-1} can be achieved. Since the solubility of α-cyclodextrin in the presence of decanol is only 6.5 g · l^{-1} at 40°C, the product recovery is almost complete without special precautions.

It has been shown that the reaction without decanol seems to agree with the hypothesis of quasi–steady-state kinetics.[10] Adding decanol introduces a mass transfer term that could result in a retarding effect at high enzyme concentrations. In a range of enzyme concentrations applicable in a production process, no such effect can be detected, as shown in FIGURE 6. The product concentration is given as a function of the normalized time of reaction. Since the measurements with three different enzyme concentrations can be described by a simple function, the enzyme concentration and the time of reaction are interchangeable quantities. This facilitates the reactor design considerably.

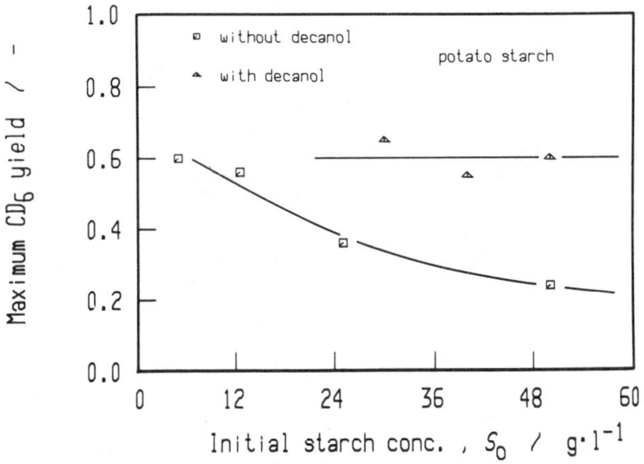

FIGURE 7. Comparison of the maximum α-cyclodextrin yields in the absence and in the presence of decanol.

The maximum α-cyclodextrin yield obtained in the absence of decanol amounts to about 60% in the case of potato starch. But, as shown in FIGURE 7, this yield can only be achieved at very low substrate concentrations. The same yield is obtained by the use of decanol as complexing agent also at high substrate concentrations. This might indicate that approximately 35% of the starch cannot be used by the enzyme.

CONCLUSION

The use of decanol as a complexing agent for α-cyclodextrin has shown that the equilibrium of the reaction system can be shifted towards α-cyclodextrin. The product is obtained in crystalline form, facilitating its recovery. The over-all yield amounts to about 50% and seems not to vary significantly with the starch quality used. The degradation of starch by the highly active cyclodextrin-glycosyltransferase from *K.*

pneumoniae M5 al in the presence of decanol seems to be a very promising process for the production of α-cyclodextrin.

SUMMARY

The maximum concentration of α-cyclodextrin for the enzymatic degradation of starch is limited to about 13.5 g · l^{-1}. By addition of decanol, the equilibrium of the reaction system can be shifted towards an α-cyclodextrin yield of 50% even at high substrate concentrations. The main variables of the decanol process—pH, temperature, substrate quality, substrate, and enzyme concentration—have been studied. The cyclodextrin-glycosyltransferase from *Klebsiella pneumoniae* M5 al can preferentially be employed at pH 6 to 8, temperatures of 40 to 50°C and a decanol concentration of 0.1 kg^{-1} starch. The dextrose equivalent of starch is important with respect to the maximum achievable starch concentration, but not with respect to the reaction. Under process conditions, the rate of α-cyclodextrin evolution is limited by the enzymatic reaction and not by mass transfer of decanol into the aqueous phase.

ACKNOWLEDGMENT

The aid of Dr. H. Bender, University of Freiburg, West Germany, is greatly appreciated. The cyclodextrin-glycosyltransferase was prepared in his laboratories owing to his kind advice and help.

REFERENCES

1. SZEJTLI, J. 1982. Cyclodextrins and their inclusion complexes. Akadémiai Kiado. Budapest.
2. HORIKOSHI, K., N. NAKAMURA, N. MATSUZAWA & M. YAMAMOTO. 1982. Industrial production of cyclodextrins. *In* Proceedings of the First International Symposium on Cyclodextrins. J. Szejtli, Ed. Akadémiai Kiado. Budapest. pp. 25–39.
3. HORIKOSHI, K., M. YAMAMOTO, N. NAKAMURA, M. OKADA, M. MATSUZAWA, O. UESHIMA & T. NAKAKUKI. 1982. Process for producing cyclodextrins. EP 0,045,464 A1.
4. BENDER, H. 1982. Enzymology of the cyclodextrins. *In* Proceedings of the First International Symposium on Cyclodextrins. J. Szejtli, Ed. Akadémiai Kiado. Budapest. pp. 77–87.
5. HORIKOSHI, K. 1979. Production and industrial applications of β-cyclodextrin. Proc. Biochem. May 1979. pp. 26–30.
6. YOSHIAKI, Y., K. KOUNO & T. INUI. 1980. A process for producing cyclodextrins. EP 0,017,242 A1.
7. LANDERT, J.-P. 1982. Herstellung von α-Cyclodextrin mittels der Cyclodextrin-glycosyltransferase aus *Klebsiella pneumoniae* M5 al. Ph.D. Thesis No 460. Swiss Federal Institute of Technology Lausanne. Lausanne.
8. BENDER, H. 1980. Kinetische Untersuchungen der durch die Cyclodextrin-glycosyltransferase katalysierten (1-4)–α-D-Glucopyranosylketten (durchschnittlicher Polymerisationsgrad von 16) als Substrate. Carbohydr. Res. **78:** 133–145.
9. BENDER, H. 1980. Kinetische Untersuchungen der durch die Cyclodextrin-glycosyltransferase katalysierten (1-4)–α-D-Glucopyranosyltransferreaktionen, insbesondere der Zyklisierungsreaktion, mit Amylosen, Amylopektin und Gesamtstärke als Substrate. Carbohydr. Res. **78:** 147–162.

10. FLASCHEL, E., J.-P. LANDERT & A. RENKEN. 1982. Process development for the production of α-cyclodextrin. *In* Proceedings of the First International Symposium on Cyclodextrins. J. Szejtli, Ed. Akadémiai Kiado. Budapest. pp. 41–49.
11. ARMBRUSTER, F. C. 1970. Method of preparing pure alpha-cyclodextrin. U.S. Patent 3,541,077.
12. ARMBRUSTER, F. C. 1972. Procedure for production of alpha-cyclodextrin. U.S. Patent 3,640,847.
13. BENDER, H. 1977. Cyclodextrin-glucanotransferase von *Klebsiella pneumoniae*. 1. Synthese, Reinigung und Eigenschaften des Enzyms von *K. pneumoniae* M5 al. Arch. Microbiol. **111:** 271–281.
14. LANDERT, J.-P., E. FLASCHEL & A. RENKEN. 1982. A photometric test for the cyclization activity of cyclodextrin-glycosyltransferases. *In* Proceedings of the First International Symposium on Cyclodextrins. J. Szejtli, Ed. Akadémiai Kiado. Budapest. pp. 89–94.
15. SMITH, J. 1967. Characterization and analysis of starches. *In* Starch: Chemistry and Technology. R. L. Whistler & E. F. Paschall, Eds. Vol. II. Chap. 25. Academic Press. New York.

NOMENCLATURE

SYMBOLS AND UNITS

c	$g \cdot l^{-1}$	concentration
DE	%	dextrose equivalent
E	$kU \cdot l^{-1}$	enzyme concentration
Et	$kU \cdot l^{-1} \cdot h$	normalized time of reaction
S	$g \cdot l^{-1}$	starch concentration
t	h (min)	time

INDICES

o	initial
max	maximum

ABBREVIATIONS

CD_6	α-cyclodextrin (cyclohexaamylose)
dec	decanol

Scale-up of an Enzyme Membrane Reactor Process for the Manufacture of L-Enantiomeric Compounds

WOLFGANG LEUCHTENBERGER,
MICHAEL KARRENBAUER, AND ULF PLÖCKER

Degussa AG
FC-O, VTC–I
D-6450 Hanau 1, Federal Republic of Germany

INTRODUCTION

In principle, there are four processes that are suitable for the manufacture of enantiomeric compounds used as fine chemicals or pharmaceuticals:

(1) extraction, which means isolating the product from natural sources such as plants or proteins;
(2) chemical asymmetric synthesis using sophisticated chiral catalysts;
(3) fermentation; and
(4) enzymatic catalysis utilizing whole cells or active cell components as biocatalysts.

For the industrial production of L-amino acids, each of these processes has its importance.

TABLE 1 shows an overview of the processes that are practiced for the production of individual amino acids, which are used as flavoring agents, feed additives, and in infusion solutions. Enzymatic catalysis is advantageous for cases where L-amino acids can be produced with high specificity from inexpensive precursors in a continuously operating process.

The acylase-catalyzed stereospecific hydrolysis of N-acetyl-DL-amino acids is employed industrially for the production of L-alanine, L-valine, L-methionine, L-phenylalanine, and L-tryptophan. Enzymatic resolution can only be competitive when coupled with an efficient enzyme recycling or immobilization system and an easy racemization of the undesired acetyl-D-amino acids (FIG. 1).

CONTINUOUS RECYCLING OF ENZYMES IN A MEMBRANE REACTOR

The current state of the art for the production of L-amino acids is utilization of immobilized acylase in a fixed-bed reactor.[1–3]

In cooperation with Prof. Wandrey's group from the Kernforschungsanlage Jülich and Dr. Kula's group from the Gesellschaft für Biotechnologische Forschung Braunschweig-Stöckheim, Degussa has developed an alternative process using an enzyme membrane reactor (EMR) that combines enzyme recycling with the advantages of homogeneous enzyme catalysis.[4]

FIGURE 2 shows a schematic representation of this enzyme membrane reactor. Substrate is pumped into a loop reactor that contains the dissolved enzyme. The

TABLE 1. Production of L-Amino Acids

Amino Acid	Chemical Synthesis	Extraction	Fermentation	Enzymatic Catalysis
L-Alanine		+		+
L-Arginine		+	+	
L-Aspartic acid		+		+
L-Cystine		+		
L-Cysteine[a]	+			
L-Glutamic acid (Na)		(+)	+	
L-Histidine (·HCl)		+	+	
L-Isoleucine		+	+	
L-Leucine		+		
L-Lysine (·HCl)			+	+
L-Methionine				+
L-Phenylalanine	(+)	(+)	(+)	+
L-Proline		+	(+)	
L-Serine		+	+	
L-Threonine		+	+	
L-Tryptophan			+	+
L-Tyrosine		+		
L-Valine		+	(+)	+

[a]Reduction of L-cystine.

product solution that accumulates behind the membrane is free from enzyme and ready for work-up.

Most of the synthetic ultrafiltration membranes presently on the market largely fulfill the criteria for use in a membrane reactor, such as molecular cutoff, chemical and biological stability, temperature and pressure tolerance, pore size distribution, and, finally, price per square meter.

Asymmetric membranes have proved themselves for the scale-up of the membrane reactor. These allow higher flow rates at lower pressure, are less susceptible to clogging and easier to clean. For dilute aqueous substrate solutions, capillary membranes are recommended. They exhibit advantageous fluid dynamic properties with acceptable surface-to-volume and surface-to-price ratios.

The economic operation of an EMR plant requires that the sterile reactor be operated for months in order to utilize the enzyme activity to the greatest possible extent. Precipitates on the membrane caused by concentration polarization in the boundary layer (FIG. 3) act as a second membrane; the filtration performance falls. This may be minimized in capillary membranes by increasing the flow rate of circulated reaction solution, but backflushing is eventually unavoidable. Cyclic backflushing is a method that has proved quite successful in prolonging membrane life.

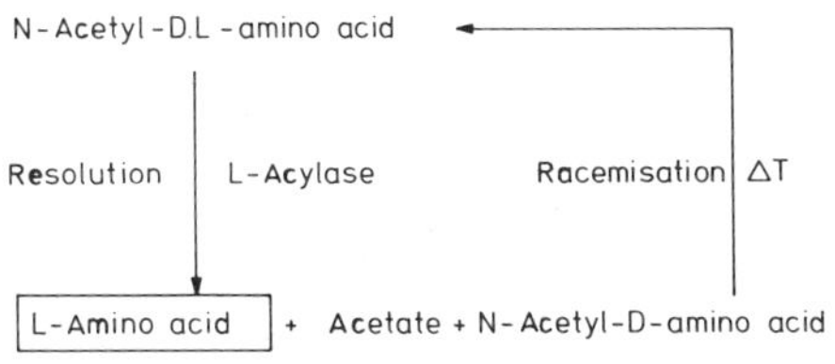

FIGURE 1. L-Amino acids from N-acetyl-DL-amino acids by enzymatic synthesis.

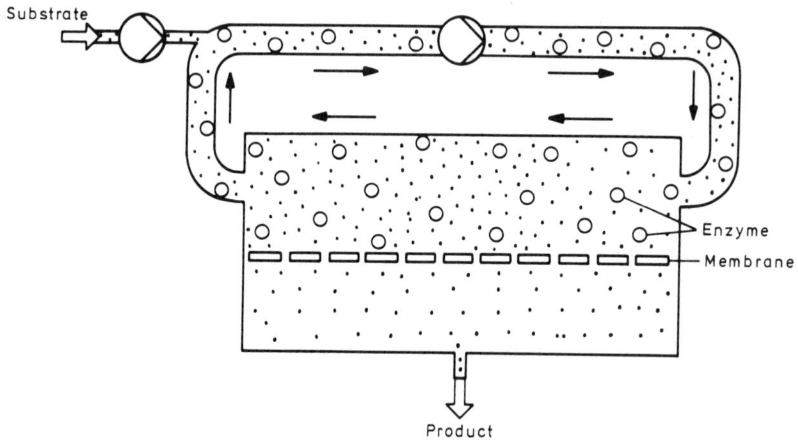

FIGURE 2. Enzyme membrane reactor.

The versatility of the enzyme membrane reactor is demonstrated by the different substrate/enzyme systems that are compiled in TABLE 2.

L-AMINO ACIDS PREPARED FROM N-ACETYL-DL-DERIVATIVES BY MEANS OF ENZYMATIC RESOLUTION

The production of L-methionine and L-phenylalanine from the corresponding N-acetyl-DL-derivatives using kidney or microbial acylase was chosen as a model to test the enzyme membrane reactor. An EMR bench-scale plant was constructed so that the important operation parameters such as residence time and enzyme concentra-

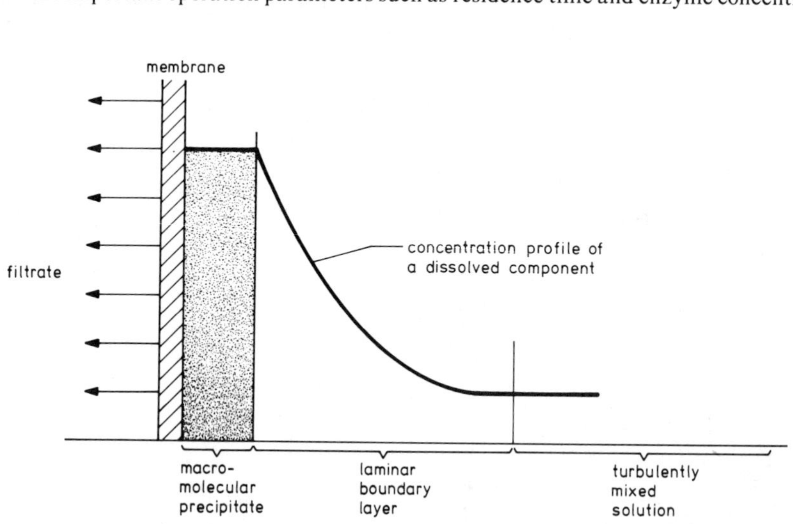

FIGURE 3. Concentration polarization during membrane filtration.

TABLE 2. Substrate-enzyme Systems Tested in the Enzyme Membrane Reactor

Substrate	Enzyme	Product
N-Acetyl-DL-alanine	Acylase[a]	L-Alanine
N-Acetyl-DL-methionine	Acylase[a,b]	L-Methionine
N-Acetyl-DL-valine	Acylase[a,b]	L-Valine
N-Acetyl-DL-phenylalanine	Acylase[b]	L-Phenylalanine
N-Acetyl-DL-tryptophan	Acylase[b]	L-Tryptophan
Fumaric acid	Fumarase	L-Malic acid
Fumaric acid	Aspartase	L-Aspartic acid
Pyruvic acid	L-Alanine dehydrogenase	L-Alanine
α-Keto isocaproic acid	L-Leucine dehydrogenase	L-Leucine
Phenyl-pyruvic acid	L-Lactate dehydrogenase	L-Phenyl lactic acid

[a]Kidney acylase (Serva, Medimpex).
[b]Acylase from aspergillus oryzae (Amano).

tion could be varied. The enzyme requirement was determined using a special enzyme-metering regulator.

FIGURE 4 shows a simplified flow chart of the EMR-plant. Substrate solution is prepared and pumped by way of a buffer tank into the substrate container, from which it is pumped through bubble trap, sterile filter, and heat exchanger into the loop reactor. The reactor essentially consists of a circulation pump, a membrane unit and a residence time tank. The product solution leaves the reactor through the membrane, passes through a flow-through polarimeter to monitor the conversion, and is collected for work-up. The continuous polarimetric measurement of the product concentration detects any alteration, such as lowered conversion caused by enzyme deactivation, and corrects by adding enzyme.

The conditions shown in TABLE 3 were maintained during a 1000-hour experiment. A 0.8 molar solution of N-acetyl-D,L-methionine adjusted to pH 7 was used as substrate. Kidney acylase with an activity of 15 IU/mg and a concentration of 0.02-0.1% served as enzyme. A conversion of almost 70% was achieved with a reaction temperature of 37°C, a residence time of 4–7 hours and a pressure of 1-1.5 bar. A polyamide membrane with a cutoff of 10,000 daltons was used.

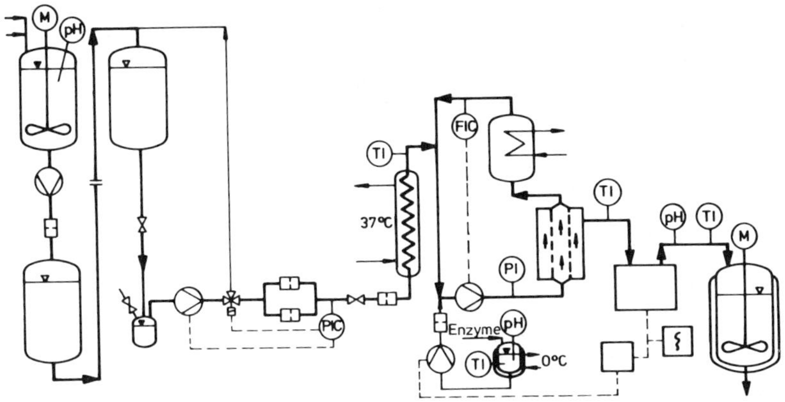

FIGURE 4. Flow chart of EMR bench-scale plant.

TABLE 3. Model Conditions for the N-Acetyl-DL-Methionine/Kidney Acylase System

Substrate	0.8 mol N-Acetyl-DL-methionine/liter
Enzyme	Kidney acylase (15 IU/mg)
Membrane	Polyamide cutoff 10,000 dalton
Temperature	37 ± 0.5 °C
pH	7 ± 0.2
Conversion	68%
Enzyme concentration	0.02–0.1%
Residence time	4–7 hours
Pressure	1–1.5 bar

The results in terms of enzyme consumption and productivity over 1000 hours can be seen in FIGURE 5. The specific enzyme consumption is less than 2100 IU/kg L-methionine.

Two 0.5-square-meter hollow-fiber modules and a 40-liter loop reactor formed the heart of a pilot plant with essentially the same flow scheme as that of the bench-scale plant. In this pilot plant, a maximum flow rate of 10 liters per hour was attained, corresponding to approximately 10 kg L-methionine per day. Compared with the laboratory results, the 1400-hour pilot plant experiment delivered much better results with regard to module life and enzyme consumption (TABLE 4). The most important results of the long-term experiments using L-phenylalanine/mold acylase are the discovery that enzyme consumption can be minimized by increasing the enzyme concentration and shortening the residence time and the confirmation that the membrane surface is not a limiting factor for scale-up. In a 2300-hour experiment, 779,000 IU acylase were used to prepare consecutively 366 kg of L-methionine and 179 kg of L-phenylalanine, corresponding to an enzyme consumption of 1437 IU/kg amino acid. Conversions of 82-86% for both L-phenylalanine and L-methionine were attained. In a performance test, resolution rates of about 85% were also achieved using a

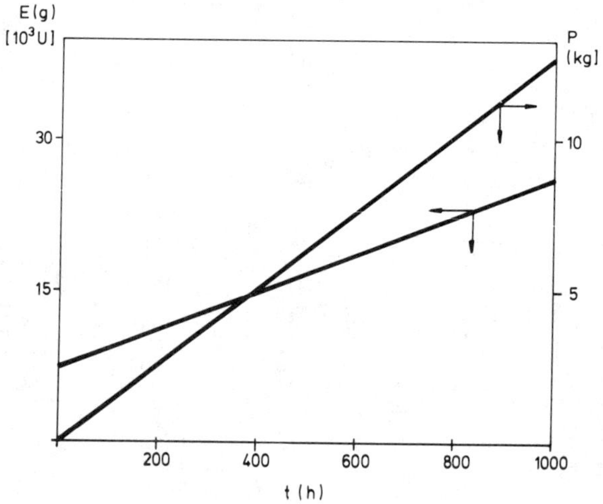

FIGURE 5. Continuous bench-scale production of L-methionine.

TABLE 4. Enzymatic Cleavage of N-Acetyl-DL-Methionine

		Laboratory	Pilot
Substrate concentration	[Mole]	0.8	0.8
Enzyme concentration in reactor[a]	[%]	0.1	0.02
Conversion	[%]	69	69
Module life[b]			
(without regeneration)	[h]	1700	3200
Initial deactivation of the enzyme	[%/d]	8	9
Enzyme consumption[c]	[IU/kg L-Met]	2100	1350
Maximum flow rate	[l/m^2]	5	10

[a]Kidney acylase with an activity of 15,000 IU/g.
[b]Extrapolated.
[c]Based on 1000 hours of operation.

substrate consisting of equal proportions of fresh and racemized N-acetyl-DL-phenylalanine.

L-MALIC ACID FROM FUMARIC ACID

The fumarase-catalyzed stereoselective addition of water to fumaric acid resulting in L-malic acid was also successfully tested in the membrane reactor. Different fumarases exhibited considerably different stabilities in the reaction solution. A fumarase extracted from pig heart lost activity quite rapidly in solution. A microbial fumarase, however, exhibited a relatively high stability, which encouraged us to utilize this enzyme in the membrane reactor.

Dr. Kula's group from the Gesellschaft für Biotechnologische Forschung succeeded in optimizing the isolation of fumarase from *Brevibacterium ammoniagenes* to the point that an economic production of L-malic acid in the membrane reactor could be carried out using a very stable fumarase.

Because the product is strongly inhibiting, a two-step process was tested using two

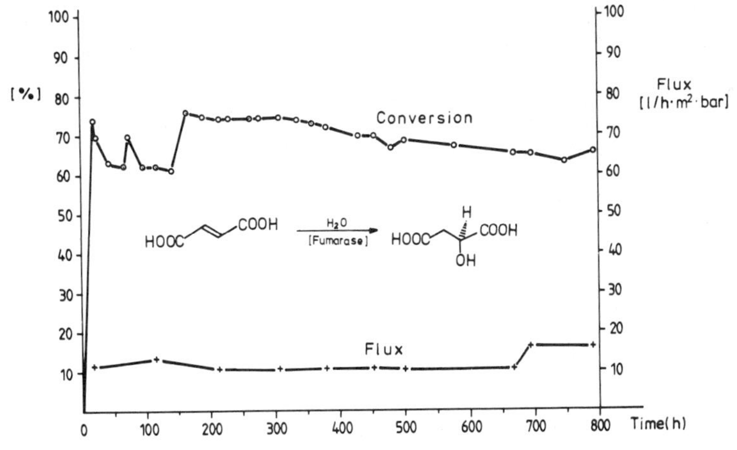

FIGURE 6. EMR pilot experiment. L-Malic acid from fumaric acid.

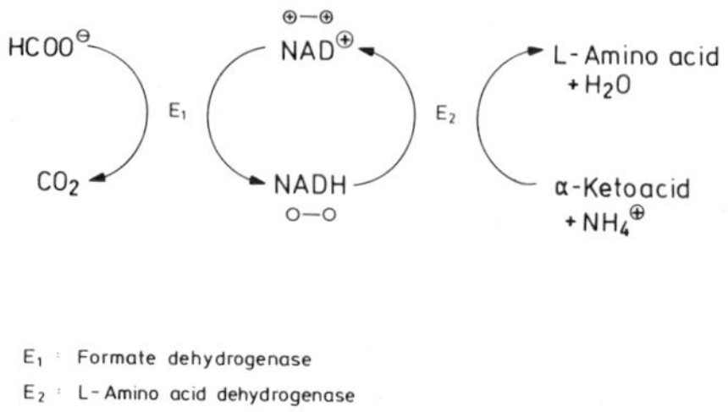

FIGURE 7. Cofactor regeneration.

enzyme membrane reactors in series. Measures such as increasing the reactor size, adding substances to stabilize the enzyme and periodic exchange of a portion of the enzyme within the reactor, however, allowed the one-step concept to be successfully pursued in pilot experiments.

FIGURE 6 shows an 800-hour experiment in a 10-liter EMR pilot plant. The conversion was maintained at approximately 70% by continuous addition of fresh fumarase. As can be seen from the lower curve, the flux and thus the membrane performance remained fairly constant. This experiment required 339,000 IU fumarase for the production of 300 kg of L-malic acid (640 IU/ml, 17 mg/ml protein), corresponding to an enzyme consumption of 1130 IU/kg product.

L-AMINO ACIDS FROM α-KETO ACIDS USING MULTIENZYME SYSTEMS WITH COFACTOR REGENERATION

In addition to simple enzyme systems, the use of more complicated multienzyme systems has been investigated. These allow redox reactions to be carried out in the enzyme membrane reactor. The concept of the membrane reactor should prove very promising for these conversions.[5,6]

The enzymatic conversion of α-keto acids to L-amino acids or to the corresponding hydroxy analogs is an ambitious but elegant process. Carrying out this stereoselective reductive amination or reduction is an ambitious undertaking because the required enzyme, the product-specific dehydrogenase, only functions in the presence of the coenzyme NADH. A continuous reaction process thus requires cofactor regeneration. The process is elegant, however, because the desired optically active compounds are obtained directly from prochiral precursors, eliminating the resolution, derivatization, and racemization steps.

FIGURE 7 describes the cofactor regeneration during the production of L-amino

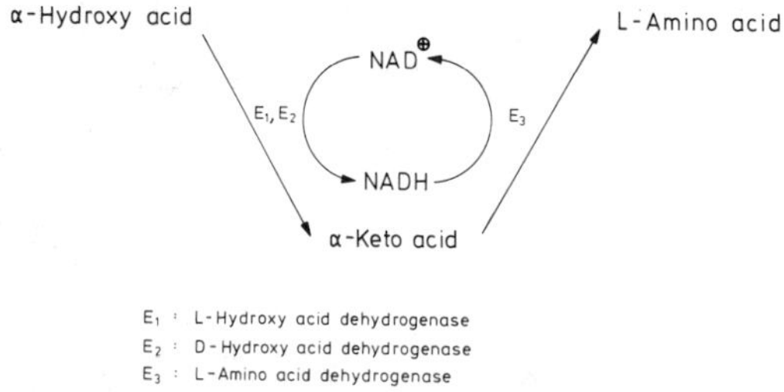

FIGURE 8. L-Amino acids from α-hydroxy acids.

acids from α-keto acids. The NAD^+, which is formed during the reductive amination of the keto acid by a specific dehydrogenase E_2, is reduced back to NADH by a second enzyme E_1, the formate dehydrogenase[7] for example, which oxidizes formate to carbonate.

If formate is added as a cosubstrate to the keto acid substrate and steps are taken to insure that the coenzyme is reusable, redox reactions may be carried out continuously in the enzyme membrane reactor.

The coenzyme is maintained in the system by binding it to a water-soluble polymer such as polyethylene glycol (PEG-10,000 to PEG-20,000). This increased-molecular-weight coenzyme developed by the GBF has proven itself in the membrane reactor. It is synthesized from PEG, succinic anhydride, ethylene imine, and NADH.[8]

In a 250-hour, bench-scale experiment, α-ketoisocaproate was converted by means of leucine dehydrogenase[9] to L-leucine with a productivity of 0.5 kg/day. Since this experiment, Prof. Wandrey's group has investigated a number of further systems and has attained up to 70,000 cycles.

The production of L-amino acids from α-keto acids that are enzymatically produced from α-hydroxy acids (FIG. 8) is a very promising variation that does not require the formate/formate dehydrogenase regeneration system. The NADH that is

TABLE 5. Advantages and Disadvantages of the EMR Process

Advantages	Disadvantages
• No enzyme fixation costs	• Large membrane surface
• No immobilization losses	• Efficient circulation pump required to prevent concentration polarization
• Enzyme dosage insures constant conversion rate	• Sterile filtration (0.2 μ) required
• Interchangeability of substrate/enzyme systems	• Possible deactivation by shear force
• Use of multienzyme systems	
• Sterilizability of the plant	
• Product solution is pyrogen free	
• No diffusion limitation	

consumed during the reductive amination is regenerated during the oxidation of the hydroxy acid to α-keto acid.

Thus DL-lactate may be converted to L-alanine via pyruvate or DL-α-hydroxy-isocaproate to L-leucine.

CONCLUSION

The concept of the enzyme membrane reactor offers the opportunity not only to utilize simple enzyme systems such as acylase, fumarase, and aspartase in homogeneous solution in a continuous process but also to carry out coenzyme-dependent redox reactions. The advantages and disadvantages of the process are compared in TABLE 5.

Based on the results of laboratory and pilot experiments, a production plant for the acylase-catalyzed resolution of acetyl-DL-amino acids was conceived and constructed with an initial capacity of 5 tons per month. The production plant has since been expanded to 15–20 tons per month of L-amino acids.

Complicated enzymatic reactions should play an increasingly significant role in the future. The progress that has been made in immobilizing whole cells has already given impetus to further work in the area.

The current results from investigations of cofactor regeneration using isolated enzymes offer hope that enzyme membrane technology using multienzyme systems will allow the large-scale realization of enzymatic transformations such as the enzymatic production of L-amino acids from α-keto or α-hydroxy acids.

REFERENCES

1. CHIBATA, J. & T. TOSA. 1976. Industrial application of immobilized enzymes and immobilized microbial cells. *In* Immobilized Enzyme Principles. Applied Biochemistry and Bioengineering. L. B. Wingard, E. Katchalski-Katzir & L. Goldstein, Eds. Vol. **1:** 329–357. Academic Press. New York.
2. CHIBATA, J. 1978. Immobilized Enzymes. Kodansha Ltd. Halsted Press. Tokyo. New York. pp. 168–178.
3. CHIBATA, J. *et al.* 1980. Immobilized aminoacylase. U.S. Patent 4, 224, 411.
4. WANDREY, C. & E. FLASCHEL. 1979. Process development and economic aspects in enzyme engineering. Acylase L-methionine system. *In* Advances in Biochemical Engineering. T. K. Ghose, A. Fiechter & N. Blakebrough, Eds. Vol. **3:** 147–218. Springer-Verlag. Berlin, Heidelberg, New York.
5. WANDREY, C., R. WICHMANN, W. LEUCHTENBERGER, M. -R. KULA & A. BÜCKMANN. 1982. Process for the continuous enzymatic change of water-soluble α-ketocarboxylic acids into the corresponding amino acids. U.S. Patent 4, 304, 858.
6. WANDREY, C., R. WICHMANN, W. LEUCHTENBERGER, M. -R. KULA & A. BÜCKMANN. 1982. Process for the continuous enzymatic change of water-soluble α-ketocarboxylic acids into the corresponding α-hydroxycarboxylic acids. U.S. Patent 4, 326, 031.
7. KRONER, K. H., H. SCHÜTTE, W. STACH & M. -R. KULA. 1982. Scale-up of formate dehydrogenase by partition. J. Chem. Tech. Biotechnol. **32:** 130–137.
8. BÜCKMANN, A., M.-R. KULA, R. WICHMANN & C. WANDREY. 1981. An efficient synthesis of high-molecular-weight NAD(H) derivatives suitable for continuous operation with coenzyme-dependent enzyme systems. J. Appl. Biochem. **3:** 301–315.
9. HUMMEL, W., H. SCHÜTTE M. -R. KULA. 1981. Leucine dehydrogenase from *Bacillus sphaericus*. Eur. J. Appl. Microbiol. Biotechnol. **12:** 22–27.

Enzymatic Production of Optically Active 2-Hydroxy Acids

R. WICHMANN AND C. WANDREY

Institut für Biotechnologie
Kernforschungsanlage Jülich GmbH
D-5170 Jülich, Federal Republic of Germany

W. HUMMEL, H. SCHÜTTE, A. F. BÜCKMANN, AND M.-R. KULA

Gesellschaft für Biotechnologische Forschung mbH
D-3300 Braunschweig-Stöckheim, Federal Republic of Germany

Optically active 2-hydroxycarbonic acids can be synthesized from the corresponding 2-oxocarbonic acids by catalysis with a stereospecific 2-hydroxycarbonic acid dehydrogenase like lactate dehydrogenase, or with new enzymes like L-2-hydroxyisocaproate dehydrogenase (L-HicDH) from *Lactobacillus confusus* or D-2-hydroxyisocaproate dehydrogenase (D-HicDH) from *Lactobacillus casei*.[1] For continuous production of these optically active 2-hydroxycarbonic acids, an ultrafiltration membrane is used in a membrane reactor to retain the enzyme in the reactor.

As NADH is used as the reductive agent for this reaction on a mole per mole product basis, a method for NADH regeneration is necessary to reduce the amount of coenzyme needed to a catalytic level. The enzymatic oxidation of formate to carbon dioxide catalyzed by formate dehydrogenase (FDH) from *Candida boidinii* was chosen for NADH regeneration. This reaction has the advantage that the thermodynamic equilibrium very strongly favors the reduction of NAD+ to NADH. The by-product can easily be removed from the product solution. If the coenzyme is to be retained by an ultrafiltration membrane, the pore size or the molecular cut-off range of the membrane must be very low due to the molecular weight of NAD(H). Such membranes usually have a low permeability even for water as well as for the low-molecular-weight substrates and products. To overcome this disadvantage, NAD(H) covalently bound to the water-soluble polymer, polyethylene glycol, was used instead of native NAD(H).[2] Polyethylene glycol (PEG-20000) with a molecular weight of 20,000 g/mole was used. This NAD(H) derivative is coenzymatically active with the enzymes used. The application of the principle of this reaction system has been shown for the production of D-phenyl lactic acid[3,4] and for the production of L-alanine and L-leucine using a D-lactate dehydrogenase or L-amino acid dehydrogenases,[5,6] respectively.

A sterile filter is used to maintain sterility in the reactor after sterilization with diluted peracetic acid. The concentration of the product is measured with a polarimeter as the produced 2-hydroxycarbonic acid is the only compound in the product solution that is optically active. The state of reduction of the coenzyme is checked by measurement of the extinction at 340 nm of the reaction solution in a bypass stream of the reactor. The enzymes and the coenzyme are fed to the reactor by injection into the substrate solution stream between the pump and the sterile filter. The reactor contents are stirred with a magnetic stirring bar to reduce the concentration polarization of the

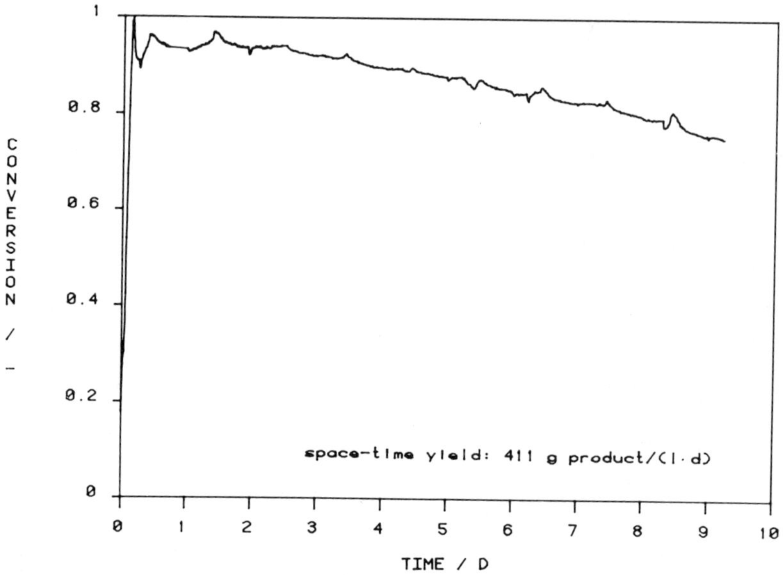

FIGURE 1. Continuous production of L-2-hydroxyisocaproic acid.

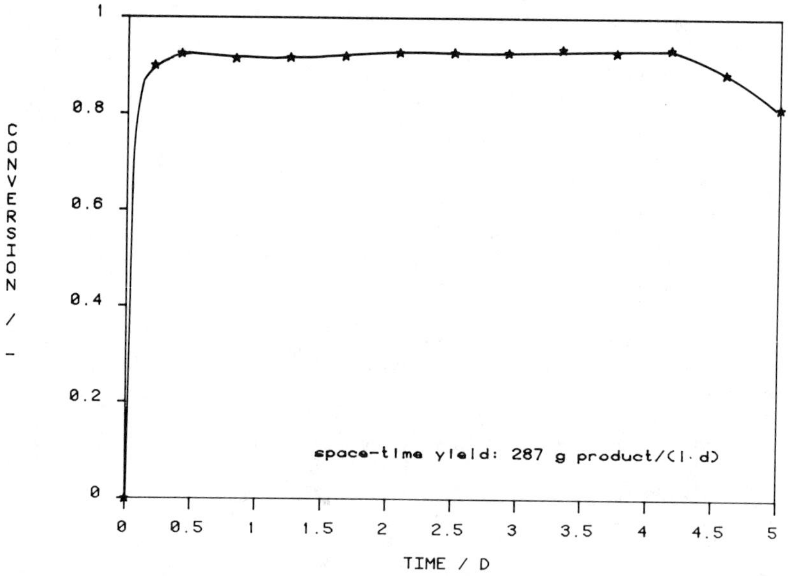

FIGURE 2. Continuous production of D-2-hydroxyisocaproic acid.

catalysts in front of the ultrafiltration membrane. An ultrafiltration membrane with a molecular cut-off range of 5000 (YM5, Amicon) was used.

In FIGURE 1, the conversion as a function of time is shown for the production of L-2-hydroxyisocaproic acid catalyzed by L-HicDH. The corresponding results of an experiment for continuous production of D-2-hydroxyisocaproic acid catalyzed by D-HicDH are shown in FIGURE 2.

The parameters and results of both experiments are shown in TABLE 1. In the experiment shown in FIGURE 1, a coenzyme cycle number of 69,400 moles of product per mole of coenzyme lost due to deactivation or insufficient membrane retention was reached on the average. The coenzyme cycle number for the experiment shown in FIGURE 2 was 22,800.

TABLE 1. Experimental Data and Results

		L-HicDH	D-HicDH
pH		7.0	7.0
Temperature	°C	25	25
V_R	ml	11.3	11.3
τ	min	40	50
2-ketoisocaproate	mmol/l	100	100
Na formate	mmol/l	400	400
Phosphate buffer	mmol/l	50	50
PEG-20000-NAD(H)	mmol/l	0.5	1.0
FDH	U/ml	3.8	5.8
HicDH	U/ml	5.5	5.3
Deactivation:			
FDH (k_{dea})	%/d	5.1	2.9
HicDH (k_{dea})	%/d	5.6	negligible within operating time
Coenzyme (k_{dea})	%/d	9.4	8.5
Catalyst consumption:			
FDH (U/g product)		0.46	0.44
HicDH (U/g product)		0.73	negligible within operating time
Coenzyme (μmol/g product)		0.10	0.33

ACKNOWLEDGMENTS

The authors express their appreciation to Mrs. U. Mackfeld and Mrs. H. Haase for their expert technical assistance.

REFERENCES

1. HUMMEL, W., H. SCHÜTTE & M.-R. KULA. 1983. New enzymes for the synthesis of chiral compounds. Ann. N.Y. Acad. Sci. This volume.
2. BÜCKMANN, A. F. 1979. German Patent No. 28.41.414.
3. WANDREY, C., R. WICHMANN, W. LEUCHTENBERGER, M.-R. KULA & A. F. BÜCKMANN. 1982. U.S. Patent No. 4.326.031.

4. WANDREY, C., R. WICHMANN, A. F. BÜCKMANN & M.-R. KULA. 1980. Immobilization of biocatalysts using ultrafiltration techniques. *In* Enzyme Engineering. H. H. Weetall & G. P. Royer, Eds. Vol. **5**: 453–456. Plenum Press. New York.
5. WICHMANN, R., C. WANDREY, A. F. BÜCKMANN & M.-R. KULA. 1981. Continuous enzymatic transformation in an enzyme membrane reactor with simultaneous NADH (H) regeneration. Biotechnol. Bioeng. Vol. **23**(issue): 2789–2802.
6. WANDREY, C., R. WICHMANN, W. LEUCHTENBERGER, M.-R. KULA & A. F. BÜCKMANN. 1981. U.S. Patent No. 4.304.858.

L-Amino Acids from a Racemic Mixture of α-Hydroxy Acids

C. WANDREY, E. FIOLITAKIS, AND U. WICHMANN

Institut für Biotechnologie
Kernforschungsanlage Jülich
D-5170 Jülich, Federal Republic of Germany

M.-R. KULA

Gesellschaft für Biotechnologische Forschung mbH
D-3300 Braunschweig-Stöckheim, Federal Republic of Germany

A racemic mixture of α-hydroxy acids can be transformed to the desired optically active L-amino acid by means of a reaction route that goes through the intermediate formation of the corresponding α-keto acid. The continuous realization of this process is possible in a multienzyme membrane reactor, which has been described earlier as to its technical characteristics.[1]

The feasibility of the process is demonstrated by the transformation of (LD)-lactate via pyruvate to L-alanine, as first suggested by Mosbach (See FIG. 1).[2]

In the process, a form of NAD covalently bound to water-soluble polyethylene glycol[3] (PEG-20000-NAD) together with the three enzymes L-LDH, D-LDH and ALADH is retained by an ultrafiltration membrane. A necessary condition for the steady-state continuous realization of the process is that the concentration of the intermediate (pyruvate) in the reactor has to be held at some level above zero by continuous feeding. If this is not done, the soluble intermediate will be washed out from the reactor and the reaction path will be shifted entirely to the formation of the reduced coenzyme form, NADH, that is, the reaction will be stopped. On the other hand, it is conceivable that large intermediate concentrations will favor the reverse reaction to lactate and the oxidized coenzyme form, NAD^+; thus, an optimal feed concentration of the intermediate will be expected.

The dependence of the rates of the four individual reaction steps on substrate and product concentrations has been investigated by measuring initial rates and subsequently verified by measuring in the entire range of conversion. The data-fitting has been accomplished by a nonlinear optimization procedure.[4] The theoretical analysis below is based on the initial rate kinetic data.

On the basis of the assumption that the sorption capacities of the enzymatic complexes are practically too low to have any measurable dynamic effect, the following model differential equations can be established for the dynamics of the reaction system under consideration:

$$\frac{d(\mathrm{lac})}{dt} = \frac{\mathrm{lac}_0 - \mathrm{lac}}{\tau} - R_1 + R_2 \quad (1)$$

$$\frac{d(\mathrm{pyr})}{dt} = \frac{\mathrm{pyr}_0 - \mathrm{pyr}}{\tau} + R_1 - R_2 - R_3 + R_4 \quad (2)$$

$$\frac{d(\mathrm{ala})}{dt} = \frac{\mathrm{ala}_0 - \mathrm{ala}}{\tau} + R_3 - R_4 \quad (3)$$

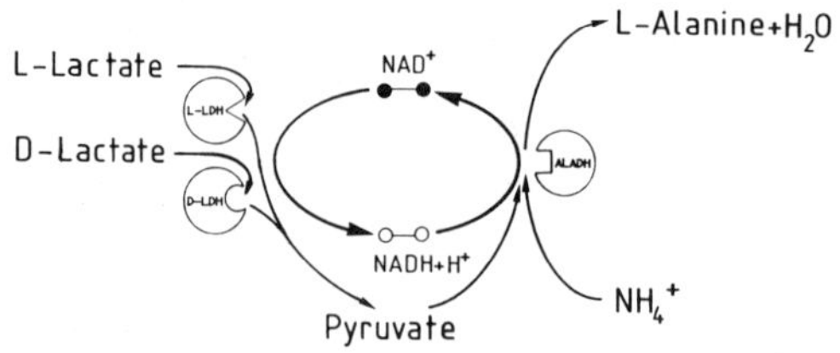

FIGURE 1. The transformation of LD-lactate via pyruvate to L-alanine.

$$\frac{d(NADH)}{dt} = + R_1 - R_2 - R_3 + R_4 \tag{4}$$

$$\frac{d(NAD^+)}{dt} = - R_1 + R_2 + R_3 - R_4 \tag{5}$$

(Index o means feed: τ is the mean residence time; the rate terms R_i (see Eqs. 1–5) have the following significance:

R_1 corresponds to the reaction step lactacte to pyruvate;
R_2 corresponds to the reaction step pyruvate to lactate;
R_3 corresponds to the reaction step pyruvate to alanine; and
R_4 corresponds to the reaction step alanine to pyruvate.)

In order to perform an engineering analysis of the continous process, for example, establishing optimal operating conditions or optimal reactor size, the steady-state equations (i.e., the above equations with the left side equal to zero) have been solved by using the optimization routine.[4] In FIGURE 2, a result of such calculations is depicted. It

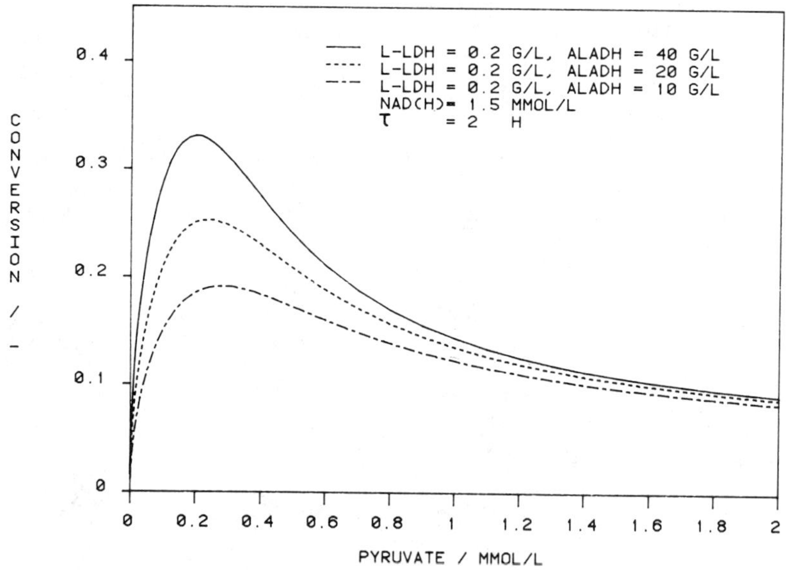

FIGURE 2. The dependence of steady-state conversion on the pyruvate concentration in the feed; calculated on initial rate data.

is clearly seen that conversion depends on the level of pyruvate in the feed, and that an optimal pyruvate feed concentration exists. A surplus of alanine dehydrogenase must be used owing to severe product inhibition. The character of NADH as cosubstrate is manifested by the nonlinear dependence of the maximal conversion on the coenzyme concentration level; this can be understood as the coenzyme saturation effect of enzymes.

The experimental verification of the feasibility of the continuous process as well as of the above-mentioned characteristics of the theoretical analysis are depicted in FIGURE 3. A continuous operation of more than one month has been achieved; the space time yield amounted to 134 g of alanine/(1.d). The cycle number was estimated at 19,200 mol L-alanine produced/mol NADH deactivated or lost across the membrane.

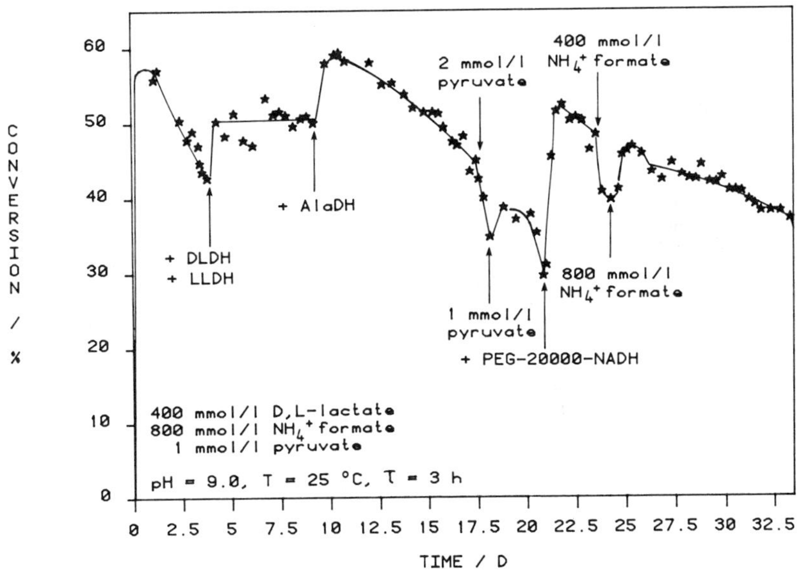

FIGURE 3. Experimental run: conversion of both L-LDH, D-LDH to L-alanine.

FIGURE 3 demonstrates that:

(1) the observed partial deactivation of enzymes in the initial phase has been compensated for by addition of lactate dehydrogenase;
(2) additional amounts of alanine dehydrogenase increase the conversion;
(3) increasing the pyruvate feed from 1 to 2 mmol/l lowers the conversion;
(4) lowering the pyruvate feed from 2 to 1 mmol/l restores the earlier conversion (deactivation has been considered);
(5) increasing the PEG-NADH level increases the conversion;
(6) lowering the NH_4^+ level lowers the conversion; and
(7) increasing the NH_4^+ level increases the conversion.

REFERENCES

1. WANDREY, C. 1979. Continuous homogeneous catalysis in enzyme engineering. Ann N.Y. Acad. Sci. **326:** 87–96.
2. MOSBACH, K. & P. -O. LARSSON. 1978. Immobilized cofactors and cofactor fragments in general ligand affinity chromatography and as active cofactors. Enzyme Engineering, Vol. **3:** 291–298. E. K. Pye & H. H. Weetall, Eds. Plenum Press. New York.
3. BÜCKMANN, A. F., M.-R. KULA, R. WICHMANN & C. WANDREY. 1981. An efficient synthesis of high molecular weight NAD(H) derivatives suitable for continuous operation with coenzyme depending enzyme systems. J. Appl. Biochem. **3:** 301–315.
4. DENNIS, J. E., D. M. GAY & R. E. WELSCH. 1981. Algorithm 573, NL2SOL, an adaptive nonlinear least squares algorithm. ACM Trans. Math. Software **7:** 369.

Enzymatic Production of Aspartame

KIYOTAKA OYAMA, SHIGEAKI IRINO,
TSUNEO HARADA, AND NORIO HAGI

*Research Laboratories, Toyo Soda Manufacturing Co., Ltd.
Tonda, Shin-Nanyo
Yamaguchi 746, Japan*

Aspartame (L-aspartyl-L-phenylalanine methyl ester, APM) is the methyl ester of the terminal dipeptide of gastrin, the hormone that is released from the antral mucosa during digestion. It is about 200 times as sweet as sucrose, and has a pleasant sweetness without a bitter aftertaste. Recently, it was approved in many countries as a food additive, and is receiving much attention as a low-calorie sweetener.

Proteinases catalyze the hydrolysis of peptide bonds. This is an equilibrium reaction, and thus a peptide bond may be formed in a significant yield if one can choose the proper enzyme and conditions. We thought that aspartame might be synthesized enzymatically. After screening many enzymes, we found that some metalloproteinases that hydrolyze an amine of hydrophobic amino acids can catalyze the following reactions:[1,2]

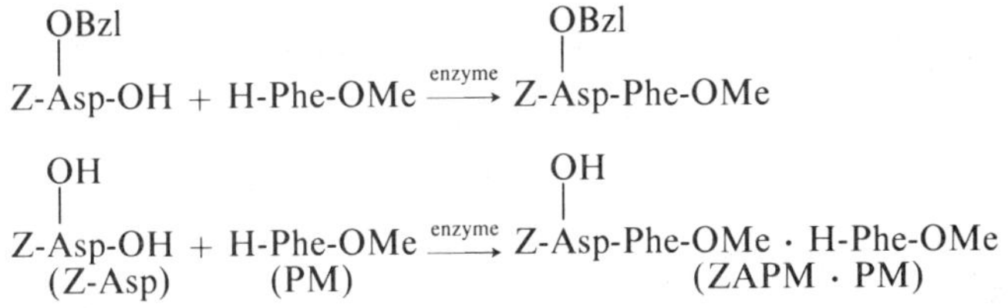

Since Z and Bzl groups can be removed easily by catalytic hydrogenation, these reactions are very useful for the production of aspartame, especially because in the latter reaction, only α-linkages are formed, even though the side-chain carboxylic acid is not protected. Furthermore, when racemic amino acids are used, only L-isomers react to give the L-L-dipeptide, while Z-D-Asp remains unreacted. D-PM forms a very insoluble addition compound with ZAPM, which precipitates out from the reaction mixture, thus forcing the equilibrium toward synthesis in almost quantitative yield. Since this enzymatic method appears to be very economical as compared with the known chemical methods, we performed basic studies as well as process development.

KINETICS

As we have already reported,[3] the reaction is first order with respect to the enzyme and L-PM concentrations. On the other hand, with respect to Z-L-Asp, the saturation of the rate, typical Michaelis-Menten behavior, can be seen (FIG. 1). From these facts, the

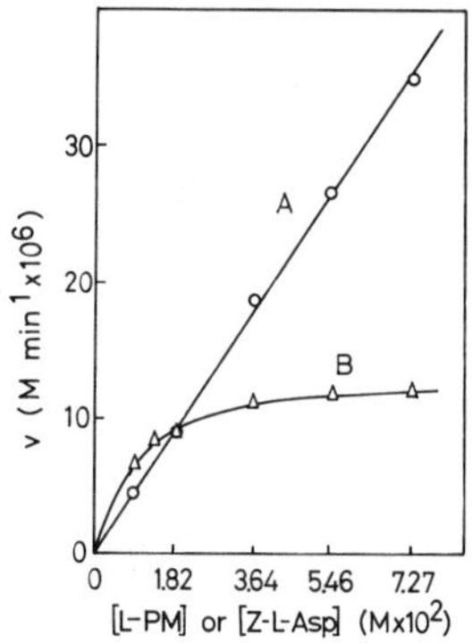

FIGURE 1. The effects of the substrate concentrations on the rate of the formation of ZAPM in the thermolysin-catalyzed condensation reaction of Z-L-Asp with L-PM at 40°C ($[E_0] = 4.85 \times 10^{-6} M$, $\mu = 3.64$). The straight line (A) represents the plot of v versus [L-PM] ([Z-L-Asp] = $1.82 \times 10^{-2} M$), and the curve (B) represents the plot of v versus [Z-L-Asp] ([L-PM] = $1.82 \times 10^{-2} M$).

reaction mechanism and the rate law can be expressed as follows:

$$\text{Z-L-Asp} + \text{E} \underset{k_{-1}}{\overset{k_{+1}}{\rightleftharpoons}} \text{Z-L-Asp, E}$$

$$\text{Z-L-Asp, E} + \text{L-PM} \xrightarrow{k_2} \text{ZAPM} + \text{E}$$

$$v = \frac{k_2[\text{E}_0]\,[\text{Z-L-Asp}]\,[\text{L-PM}]}{K + [\text{Z-L-Asp}]};\ \left(K = \frac{k_{-1}}{k_{+1}};\ k_{-1} \gg k_2\,[\text{L-PM}]\right)$$

It was also found that Z-D-Asp acts as a competitive inhibitor, while D-PM does not affect the reaction at all.[3]

EQUILIBRIUM

For the equilibrium, Z-L-Asp + L-PM $\overset{K}{\rightleftharpoons}$ ZAPM, the following equation can be given.

$$K = \frac{\text{S}}{[\text{A}(1-\alpha) - \text{S}]\,[\text{B}(1-\alpha) - \text{S}]} \tag{1}$$

where K is the equilibrium constant; S is the solubility of ZAPM; A, B are the initial concentrations of Z-L-Asp and L-PM, respectively; and α is the conversion to ZAPM.

FIGURE 2. The dependence of the yield of the equilibrium, Z-L-Asp + DL-PM $\rightleftharpoons$ ZAPM · D-PM, on the solvent volume ($K = 1.5\ M^{-1}$, S = 0.005 M, Z-L-Asp = 0.2 mole, DL-PM = 0.5 mole). The line is calculated by Equation 1, while the dots are obtained experimentally.

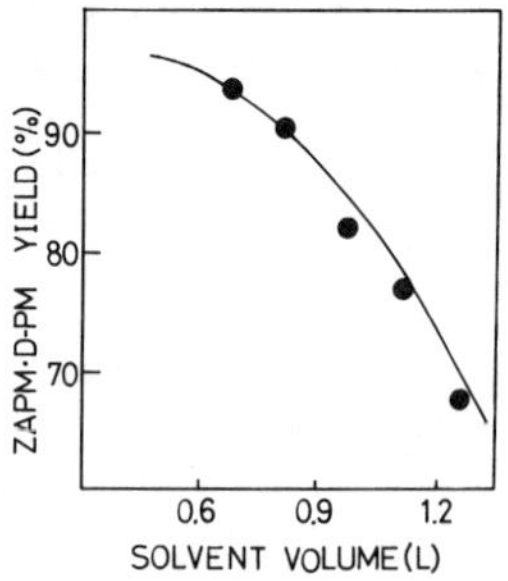

The equation indicates that the yield is higher with higher initial substrate concentrations and a lower solubility of the product. When DL-PM is used, the equilibrium constant (K) and the solubility of ZAPM · D-PM (S) are found to be 1.5 mol^{-1} liter and 5×10^{-3} mol $liter^{-1}$, respectively. Using these values, theoretical yields of ZAPM · D-PM are calculated according to Equation 1 and compared with the yields actually obtained; these values agree very well (FIG. 2).

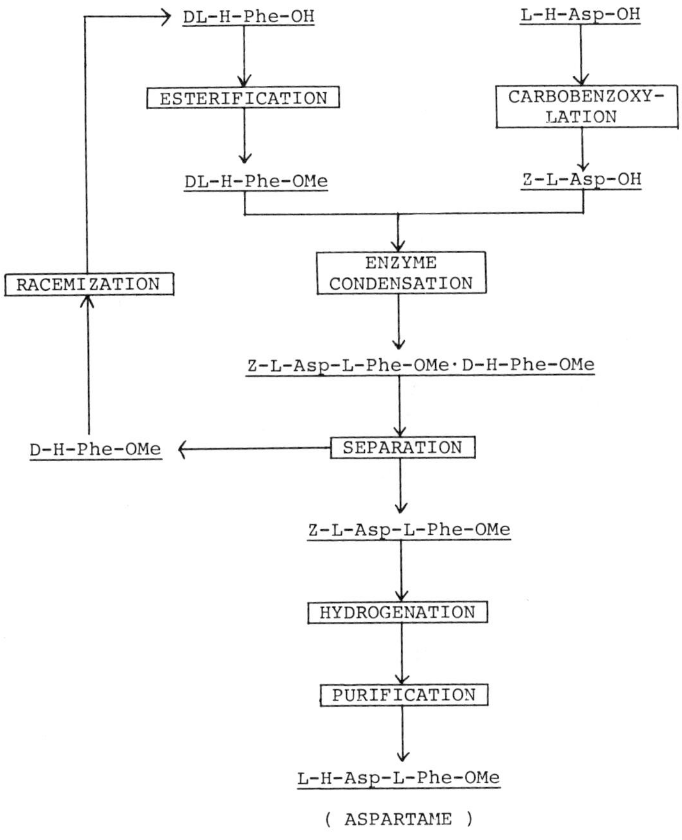

FIGURE 3. The enzymatic process for producing aspartame.

PROCESS DEVELOPMENT

Based on the above results and others, we established an industrial process to produce aspartame. In our process, we chose L-Asp and DL-Phe as the raw materials. The choice was made for the following reasons: (1) Z-D-Asp acts as a competitive inhibitor, while D-PM does not; (2) L-Asp can be available very cheaply, whereas L-Phe is much more expensive than DL-Phe; and (3) racemization of Z-D-Asp without destroying the expensive protecting group is very difficult, whereas D-Phe can be racemized easily. The outline of the process is shown in FIGURE 3.

Major characteristic features of our enzymatic method may be summarized as follows: (1) Inexpensive racemic raw materials can be used. (2) Only α-peptide linkages are formed due to the enzyme's strict specificity. On the contrary, as shown below, all of the chemical methods with an industrial potential produce 20–40% of the β-isomer, which has a bitter taste. (3) The enzymatic reaction can be done under very mild conditions in an aqueous solution. (4) The enzymes used in this method are thermolysin and its family of enzymes. They are highly stable against heat, organic solvents and pH. Therefore, unlike many other enzymes, they can be used rather easily as industrial catalysts.

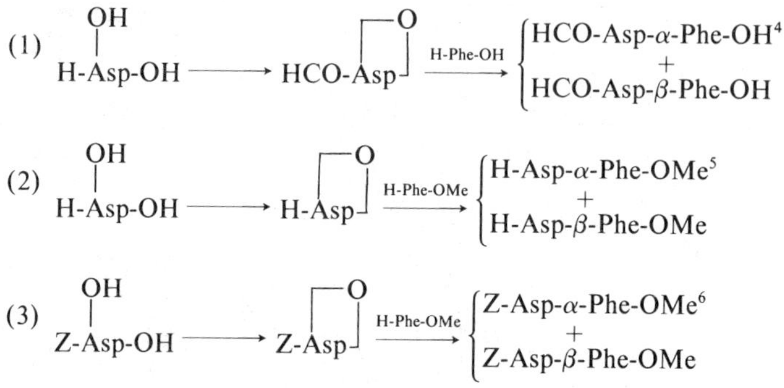

CONCLUSION

In this process peptide synthesis and optical resolution of amino acid are carried out simultaneously. Furthermore, only α-carboxylic acids are involved in the peptide bond formation. Therefore it can be said that this enzymatic method is utilizing the enzyme's outstanding properties fully, and thus we could greatly reduce the production cost of aspartame.

REFERENCES

1. ISOWA, Y., K. OYAMA & T. ICHIKAWA. 1976. Abstract of the 35th Annual Meeting of the Japanese Chemical Society. 4B23.
2. ISOWA, Y., M. OHMORI, T. ICHIKAWA, K. MORI, Y. NONAKA, K. KIHARA, K. OYAMA, H. SATOH & S. NISHIMURA. 1979. The thermolysin-catalyzed condensation reactions of N-substituted aspartic and glutamic acids with phenylalanine methyl ester. Tetrahedron Lett. **1979:** 2611–2622.
3. OYAMA, K., Y. NONAKA & K. KIHARA. 1981. On the mechanism of the action of thermolysin: Kinetic study of the thermolysin-catalyzed condensation reaction of *N*-benzyloxycarbonyl-L-aspartic acid with L-phenylalanine methyl ester. J. Chem. Soc. Perkin Trans. 2. **1981:** 356–360.
4. U. S. Patent. 1976. 3933781.
5. U. S. Patent. 1974. 3833553.
6. U. S. Patent. 1974. 3786039.

Kinetically Controlled Semisynthesis of β-Lactam Antibiotics and Peptides[a]

V. KASCHE, U. HAUFLER, AND L. RIECHMANN

Biology Department
Bremen University
D-28 Bremen 33, Federal Republic of Germany

Condensation reactions, like the synthesis of semisynthetic penicillins and cephalosporins, peptides, and oligosaccharides, can be catalyzed by the hydrolases that also catalyze the hydrolysis of these compounds. These condensation reactions can be carried out as equilibrium or kinetically controlled processes.[1] For the semisynthesis of ampicillin, the equilibrium-controlled process is:

$$\text{D-phenyl-glycine} + \text{6-APA} \xrightarrow{\text{penicillin amidase}} \text{ampicillin} \qquad (1)$$

where the yield of ampicillin is independent of the properties of the enzyme. The kinetically controlled process is:

$$\text{D-phenyl-glycine-ester} + \text{6-APA} \xrightarrow{\text{penicillin amidase}} \text{ampicillin} + \text{alcohol} \qquad (2)$$

where an activated substrate must be used, and the enzyme carries out a group transfer reaction. The yield of the desired product in this case is often much larger than the yield for the equilibrium-controlled process.[2-5] This requires that the catalyst be removed when the maximum is reached, and this is easily performed with immobilized enzymes.

The detailed reaction mechanism for the kinetically controlled process[2] must be known for a rational analysis of the factors controlling yield. Two different mechanisms have been proposed here. They are given in TABLE 1. The mechanism without binding of the nucleophile N (6-APA in Eq. 2, or an amino-acid derivative in peptide semisynthesis), before the deacylation of the acylenzyme E-A, is the most frequently used mechanism in the literature in this field. The other mechanism where the nucleophile N is bound to the acylenzyme before the deacylation may be inferred from the sequence specificity of the hydrolases and the principle of microscopic reversibility. The two mechanisms can be distinguished by measuring the initial rates of formation of AN (v_{AN}) and A (v_A) as functions of nucleophile concentration. The expressions for this and the yield at the kinetically controlled maximum are given in TABLE 1.[7]

Once the mechanism has been established, the yield-controlling factors can be rationally analyzed to evaluate the biotechnological potential of such processes. Some data pertinent to the kinetically controlled, enzyme-catalyzed synthesis of semisynthetic penicillins and peptides are presented here.

Free and immobilized penicillin amidase (EC 3.5.1.11) from *E. coli*, the substrates and 6-APA used to prepare semisynthetic penicillins were gifts from Dr. Sauber (Hoechst AG) and Dr. Krämer (Röhm), respectively. Phenyl-acetyl-glycine was

[a]This work has been supported by Bundesministerium für Forschung und Technologie (PTB-8478), Deutsche Forschungsgemeinschaft (Ka 501/3), and E. Röhm Gedächtnisstiftung.

TABLE 1. Quantities Used to Distinguish between Two Mechanisms for the Kinetically Controlled Synthesis[a]

Mechanism	Measured Quantity	
	$(k_3'/k_3)_{app} = \frac{v_{AN}}{v_{AN}} \cdot \frac{[H_2O]}{[N]_o}$ [b,c]	Maximum of an AN the Kinetically Controlled Maximum (Reverse Reaction with B Neglected) $= \alpha \cdot [N]_o/(1 + \alpha)$
No nucleophile binding to E-A	$\frac{k_3'}{k_3}$	$\alpha = \frac{k_3'}{k_3} \cdot \frac{\left(\frac{k_{cat}}{K_m}\right)_{AB}}{\left(\frac{k_{cat}}{K_m}\right)_{AN}} \cdot \frac{[AB]}{[H_2O]}$
Nucleophile binding to E-A	$\frac{k_{3,N}'}{k_3 \cdot K_N + k_{3,N} \cdot [N]_o}$	$\alpha = \frac{k_{3,N}'}{k_{3,N} \cdot K_N} \cdot \frac{\left(\frac{k_{cat}}{K_m}\right)_{AB,N}}{\left(\frac{k_{cat}}{K_m}\right)_{AN,N}} \cdot \frac{[AB]}{[H_2O]}$

^{a}Without:

$$AB + E \rightleftharpoons E-A \xrightarrow{N, k'_3} E + AN$$
$$H_2O \downarrow k_3$$
$$E + A$$

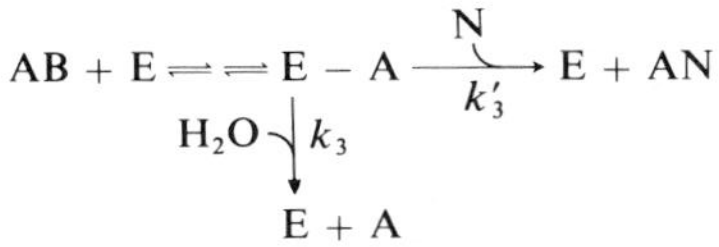

and with:

$$AB + E \rightleftharpoons^{B} E-A \underset{K_N}{\overset{N}{\rightleftharpoons}} N\cdot\cdot E-A \underset{}{\overset{k'_{3,N}}{\rightleftharpoons}} E\cdot\cdot\cdot A \rightleftharpoons AN + E$$
$$H_2O \downarrow k_3 \quad\quad H_2O \downarrow k_{3,N} \quad\quad\quad\quad N$$
$$E + A \quad\quad E\cdot\cdot\cdot N + A$$

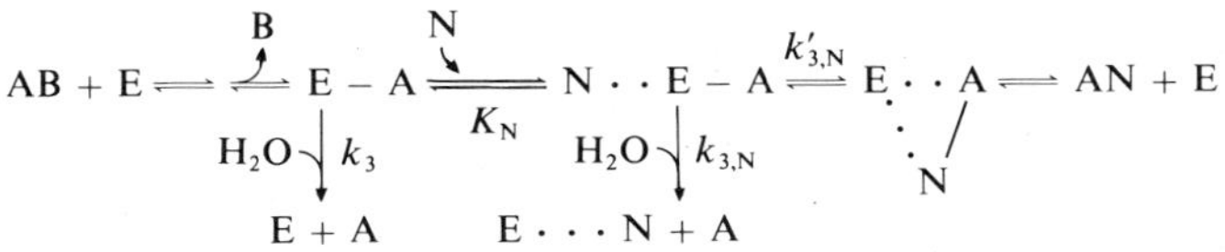

Binding of the nucleophile N to the acyl-enzyme E-A before the deacylation step.
^{b}Subscripts: o denotes initial condentrations; N denotes properties of E · · N.
^{c}N = nucleophile.

prepared as described in Kasche & Galunsky.[2] Bovine α-chymotrypsin (EC 3.4.21.1.) from Worthington was practically homogeneous and used as purchased. Bovine β-trypsin (EC 3.4.21.4) was prepared by affinity chromatography using soybean trypsin inhibitor sepharose[10] from trypsin (Merck 24579). The substrate N-acetyl-tyrosine-ethyl-ester and the nucleophiles (Val-NH_2, Ala-HN_2, Gly-NH_2) were obtained from Serva and Sigma, respectively.

The concentration of the reactants and products after mixing the catalyst and the reactant was monitored by HPLC using a Spectra Physics Chromatograph and detector.[2] The pH and temperature was kept constant during the reaction.

The apparent ratio of the deacylation rate constants was found to depend on nucleophile content for the penicillin amidase-catalyzed synthesis of penicillins and the protease (α-chymotrypsin, β-trypsin) catalyzed synthesis of peptides (FIG. 1). From TABLE 1, we conclude that the mechanism with nucleophile binding applies here. This conclusion can be drawn only when the enzyme preparations used are homogeneous. This applies only for the proteases used here. The maximum in the rate of ampicillin formation in FIGURE 1A, however, shows that this conclusion applies also for penicillin amidase.

From the data in TABLE 1, it follows that the yield at the kinetically controlled maximum should be independent of the enzyme content. This is verified in FIGURE 2. The time to reach the maximum yield is inversely proportional to enzyme content. This figure also demonstrates that the yield at the kinetically controlled maximum is much larger than the maximal yield in the equilibrium-controlled process and is also obtained much more rapidly than for the latter process.

For the kinetically controlled semisynthesis of ampicillin, after mixing 100 m*M*

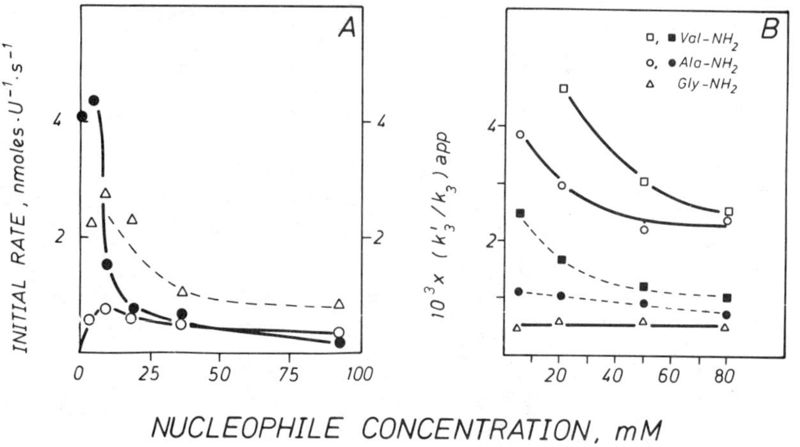

FIGURE 1. Initial rates of product formation and the apparent ratio of deacylation rates in the kinetically controlled semisynthesis of ampicillin (A) and peptides (B) with different nucleophile concentrations.

A: Penicillin amidase-catalyzed ampicillin synthesis from 20 m*M* D-phenyl-glycine-methyl-ester and 6-APA at pH 7.0 and 25°C; ionic strength I = 0.2; ○ rate of ampicillin formation; ● D-phenyl-glycine formation; △ (k_3'/k_3) app.

B: Dipeptide synthesis from 6 m*M* N-acetyl-l-tyrosine-ethyl-ester and different nucleophiles (given in the figure) at pH 10 and 25°C; ionic strength I = 0.2. catalyst: — α-chymotrypsin; --- β-trypsin.

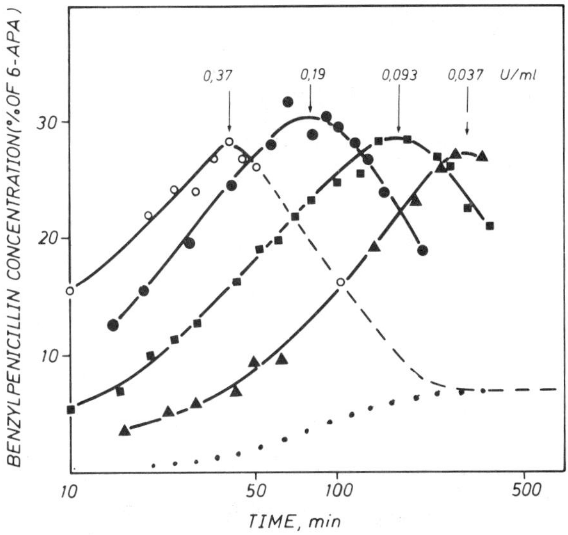

FIGURE 2. The influence of the enzyme content on the kinetically controlled semisynthesis of benzylpenicillin. Initial conditions: 10 m*M* phenyl-acetyl–glycine and 10 m*M* 6-APA at pH 6.0 and 25°C (ionic strength I = 0.2). The enzyme concentrations in U/ml are given above the maxima of each curve (. . . equilibrium-controlled synthesis).

6-APA and activated side chain at pH 7.0, the apparent ratio for the deacylation rates was found to decrease more rapidly with increasing ionic strength than the maximal yield. These results show that in the former case only the ionic-strength dependence of the binding of the negative ion 6-APA to the positively charged active site of the acyl-enzyme ion accounts for the observed results. For the maximum yield, the ionic strength dependence of (k_{cat}/K_m), a measure for the substrate specificity, for the different substrates must also be considered. The ionic strength dependence thus reduces the yield at technical concentrations (~100 m*M*). For peptide synthesis using amino acids with unprotected carboxyl groups and endoproteases as catalysts, an opposite ionic strength effect on the yield was observed. In this case, both the nucleophile and the active site have the same (−) charge.[7]

The effect of enzyme immobilization is shown in TABLE 2. The particle size dependence shows that mass transfer effects, on the nucleophile (6-APA-amino acid), must be considered.[6] At low concentrations, the 6-APA content in the support is lower than in the bulk solution. The effects can be neglected at higher (>100 m*M*) concentrations that are of technical interest, or reduced by using positively charged supports where the 6-APA content near the support is increased.

When condensation reactions are carried out as kinetically controlled processes with enzymes as catalysts, the yields are much higher than for the equilibrium-controlled process. For ampicillin, the former yield is >50%, the latter <1% of the 6-APA used for the synthetic reaction. Whether the enzyme-catalyzed process can compete with the chemical processes now used for the synthesis of semisynthetic penicillins, cephalosporins, and peptides, can be analyzed only when all yield-controlling factors are analyzed. For this aim, the mechanism with nucleophile binding and the relation for the maximum yield in TABLE 1 must be used.

TABLE 2. Effects of Enzyme Immobilization on Yields at the Kinetically Controlled Equilibrium (25°C)

Biocatalyst	Support	Support Diameter (μm)	pH	Concentration of Activated *Side* Chain/Nucleophile (m*M*)	Product	Yield (% of 6-AP)
Penicillin amidase	—	—	6.0	10/10	Benzylpenicillin	30
"	Lichrospher	10	6.0	10/10	"	29
"	Eupergit C	~100	6.0	10/10	"	16
"	—	—	7.0	20/20	Ampicillin	15
"	Eupergit C	~100	7.0	20/20	"	13
"	Eupergit C (with + charges)	~100	7.0	20/20	"	14
α-chymotrypsin	—	—	9.0	10/80	N-acetyl-tyrosine-	60[a]
"	Sepharose 4B	~100	9.0	10/80	Valin-amide	90[a]

[a] % of substrate AB.

The kinetically controlled synthesis process analyzed here may also influence hydrolysis reactions catalyzed by hydrolases. It gives rise to an apparent product inhibition that is due to the hydrolysis of transient products formed by the mechanism (Eq. 2), like formation and hydrolysis of tri- and tetrasaccharides in lactose hydrolysis.[8,9] In these cases, the synthetic processes must be minimized.

REFERENCES

1. ZHANG, Y. S. 1983. Trends Biochem. **8:** 16–18.
2. KASCHE, V. & B. GALUNSKY. 1982. Biochem. Biophys. Res. Commun. **104:** 1215–1222.
3. FLASCHEL, E., J.-P. LANDERT, D. SPIESSER & A. RENKEN. 1984. Ann. N.Y. Acad. Sci. **434:** This volume.
4. KONECNY, J., A. SCHNEIDER & M. SIEBER. 1983. Biotechnol. Bioeng. **25:** 461–472.
5. COLE, M. 1969. Biochem. J. **115:** 757–767.
6. KASCHE, V. 1983. Enzyme Microb. Technol. **5:** 2–13.
7. KASCHE, V. In preparation.
8. NAKANISHI, K., R. MATSUNO, K. TONI, K. YAMAMOTO & T. KANIKUBO. 1983. Enzyme Microb. Technol. **5:** 115–120.
9. PRENOSIL, J., E. STUKER, T. HEDIGER & J. R. BOURNE. 1984. Ann. N.Y. Acad. Sci. **434:** This volume.
10. KASCHE, V. 1976. Arch. Biochem. Biophys. **173:** 269-272.

Improved Steroid-1-Dehydrogenation Using Heat-dried Bacterial Cells

HOLLY J. WOLF AND LEO A. KOMINEK

Fermentation Research & Development
The Upjohn Company
Kalamazoo, Michigan 49001

Many steroids used for the treatment of corticoid-responsive diseases are unsaturated between the first and second carbon atoms in the A ring of the molecule. This chemical configuration confers on the molecule enhanced potency and results in less drug-induced salt retention than the corresponding 1,2-dihydro compounds. Microbial bioconversions are commonly used for industrial production of 1-dehydrogenated steroid intermediates.[1]

Arthrobacter simplex (Corynebacterium simplex) is a bacterium that is well known in the art for its steroid-1-dehydrogenase (Δ^1-dehydrogenase) activity. This enzyme catalyzes one step in the catabolism of steroids by the microorganism. Two types of commercially important steroid-1-dehydrogenations utilize *A. simplex*. Fermentation bioconversions are run by the direct addition of a steroid substrate to a grown culture in a fermentor.[2] However, the activity of other steroid-modifying enzymes in the microorganism can result in reduced yields or complete loss of the desired product. Acetone drying of *A. simplex* cells can eliminate some of these undesirable activities.[3,4] The Δ^1-dehydrogenation can then be performed by exposure of these acetone-dried cells to the steroid substrate and an added electron carrier in a buffered aqueous medium. We have discovered an improved method for steroid-1-dehydrogenation bioconversions that involves the use of heat-dried *A. simplex* cells as the microbial catalyst.

RESULTS AND DISCUSSION

Arthrobacter simplex was grown in a fermentor under conditions favorable for cell production and induction of the steroid-1-dehydrogenase. When assays indicated high cell yields and good enzyme activity, the cells were harvested by centrifugation. The cell cake was dried in an oven at elevated temperatures (55°C) under reduced pressure until a moisture content of 1–5% was reached. The dried cells were stored at 5°C until used as a catalyst. The bioconversion was performed by the exposure of rehydrated cells to a steroid substrate in the presence of an artificial electron acceptor. The mixture was aerated by high agitation and incubated until the 1-dehydrogenation was complete. The steroid product was recovered from the bioconversion mixture using standard techniques such as solvent extraction and precipitation or crystallization from the solvent.

Heat drying most successfully eliminated the undesirable steroid-modifying enzyme activities in the *A. simplex* cells. Alternative methods of treatment gave poor results. Bioconversions run in fermentation broths exhibited good steroid-1-dehydrogenation but complete degradation of the steroid nucleus followed. Washed, isolated cells showed slower degradation. Modification of the steroid rings and side chain cleavage or modification (20-keto reduction) were observable. Freeze-drying the cells further reduced undesired reactions to side chain modification or degradation. Cells dried at 45°C showed none of these detrimental activities on 6-methyl hydrocortisone.

Although acetone-dried cells did not exhibit the side reactions mentioned above, the amount of Δ^1-dehydrogenase activity was considerably lower than in untreated *A. simplex* cells. Good activity was preserved in the heat-dried cells. Although no added electron acceptors were required for enzyme activity, the addition of several such compounds stimulated the reaction. The addition of 0.05 g of menadione or 1,4-naphthoquinone to bioconversion mixtures containing 0.5 g of substrate and 0.05 g dried cells in one liter resulted in the formation of 0.48 g product/liter compared to 0.04 g product/liter in the control. Bioconversions containing 2,6-dichlorophenol indophenol, or phenazine methosulfate accumulated 0.14 g product/liter. The addition of phenazine methosulfate to heat-dried cells resulted in an increase in the steroid-1-dehydrogenase specific activity, measured spectrophotometrically as μmoles of electron acceptor reduced per minute per mg protein, to an average level of 1.8-fold greater than that observed in those same cells before they were dried.

The higher specific activity permitted an increase in the substrate concentration that could be successfully bioconverted (see TABLES 1 and 2). The heat-dried cells demonstrated broad substrate specificity for 3-keto-Δ^4-steroids, including several androstene and pregnene compounds that are modified at carbons 6, 9, 11, 16, 17, or 21 by hydroxylation, methylation, fluorination, acetylation, epoxidation, or unsaturation. The Δ^1-dehydrogenase activity in the heat-dried cells has proven to be completely stable to storage at 5°C for more than 3 years.

The steroid-1-dehydrogenase was stable to drying at high temperatures. Cells prepared by overnight drying at temperatures from 22°C to 90°C catalyzed good bioconversions of 10 g of androsta-4,9(11)-diene-3,17-dione/liter in shake flasks. Results varied from greater than 97% conversion (22°–59°C cell preps) to 49% (90°C cell prep). The highest rates of 1-dehydrogenation were observed for cells dried at temperatures of 40–60°C. The specific activity of the steroid-1-dehydrogenase in the cells (μmoles electron acceptor reduced/min/mg protein) showed similar effects of drying temperature with a much narrower optimum drying temperature range (approximately 53–55°C).

TABLES 1 and 2 demonstrate the improvements observed in bioconversions catalyzed by heat-dried cells. TABLE 1 describes bioconversions of two different substrates using fermentation broth and heat-dried cells. Elimination of steroid degradation, faster rates of 1-dehydrogenation and better isolation yields were obtained in the bioconversions with heat-dried cells. TABLE 2 compares bioconversions using acetone-dried and heat-dried cells. The heat-dried cells exhibited higher steroid-1-dehydrogenase activity and converted a higher substrate level of steroid.

SUMMARY

The use of heat-dried *A. simplex* cells for steroid-1-dehydrogenation bioconversions provides several advantages over previously described methods. Heat-dried cells possess high levels of Δ^1-dehydrogenase activity and convert higher substrate levels. Heat-drying changes the nature of the cells to a more simple microbial catalyst, by the elimination of undesirable steroid-modifying activities. The cells can be stored until it is convenient to run the bioconversion. The conditions for the bioconversion can be optimized for each steroid substrate without the necessity of maintaining conditions for

TABLE 1. Comparison of Fermentation Bioconversions by *A. simplex* and Bioconversions Using Heat-dried Cells

Type of Bioconversion		Incubation (Hours)	Remaining Substrate (g/l)	Product Accumulated (g/l)	Chemical Yield of Isolated Useful Steroid[a] (%)	Δ^1 Compound (%)
1. 8 g/liter 21-acetoxypregna-4,9(11),16-triene-3,20-dione						
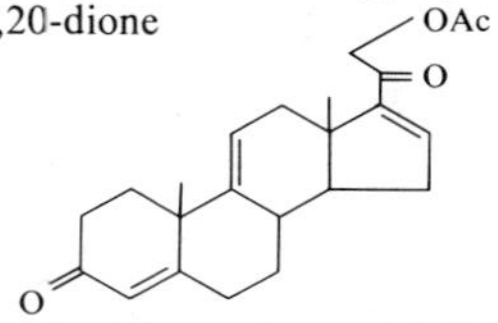	Fermentation	27	3.1	5.0		
		52	1.8	2.1		
		75	0.8	1.5	18.1[b]	59.1
	Dried Cell	4	1.7	6.3		
		23	0.3	7.7		
		27.5	0.2	7.8	86.5[c]	97.8
2. 10 g/liter androsta-4,9(11)-diene-3,17-dione						
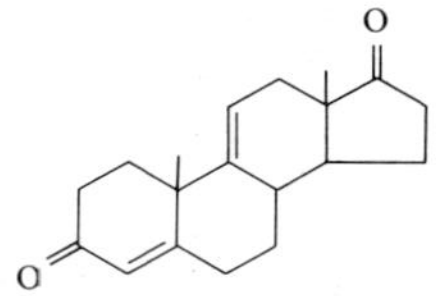	Fermentation	20	9.5	0.5		
		45	8.5	0.4		
		65	6.0	0.1		
		88	5.0	trace	80.0[d]	2.3
	Dried Cell	4	2.3	7.7		
		24	0.4	9.6		
		72	0.3	9.7	92.5	92.0

[a]Useful steroid is defined as either the Δ^1-compound that can be used in further synthesis or the 1,2-dihydro substrate that can be recycled into another bioconversion to make Δ^1 product.

[b]Isolated material contained approximately 80% impurities, including the 9,11-epoxides of the substrate and product and other hydroxylated forms of the compounds.

[c]Isolated material contained 100% substrate and product.

[d]Isolated material contained unconverted substrate, plus some 9,11-epoxides of the Δ^1 product.

TABLE 2. Comparison of Bioconversion Capacity[a] of *A. simplex* Cells Dried by Different Methods[b]

Time of Sampling (Hours)	% Unconverted Substrate	
	Acetone-dried Cells	Heat-dried Cells
1	98.7	85.1
4	95.5	23.2
24	84.1[c]	7.2[c]

[a]Bioconversions were run in 0.05 *M* potassium phosphate buffer, pH 7.5 and contained 10 g dried cell cake/liter, 10 g $\Delta^{9,11}$-androstenedione/liter, and 86 mg menadione/liter. 100 ml of the bioconversion mixture was incubated in 500-ml Erlenmeyer shaker flasks on a rotary shaker at 28°C.

[b]A 5-liter *A. simplex* culture was harvested by centrifugation. The recovered cell paste was divided in half. One portion was dried under reduced pressure at 55°C. The second was treated with acetone and dried at 5°C under reduced pressure as described in U.S. Patent No. 3,360,439.

[c]Residual levels did not decrease with further incubation.

cell viability or other problems associated with fermentations. Isolation of the product is improved because of simpler reaction mixtures versus more complex fermentation media or more complex mixtures of steroid molecules.

ACKNOWLEDGMENTS

We would like to thank Mary Lou Krumme, Dave Swan, Bob Perschon, Don VanOverloop, Del Gibbons, Tom Fleck, Tom Patt, and Tom Miller for their support, enthusiasm, and assistance.

REFERENCES

1. CHARNEY, W. & H. HERZOG. 1967. Introduction, first dehydrogenations. *In* Microbial Transformation of Steroids. Academic Press, Inc. New York. pp. 6–9.
2. NOBILE, A. 1958. Process for production of dienes by *Corynebacterium*. U.S. Patent No. 2,837,464.
3. FELDMAN, L. I., C. E. HOLMLUND & N. L. BARBACCI. 1963. Process for 1,2-dehydrogenation of steroids by the use of dried thalli. U.S. Patent No. 3,087,864.
4. ERICKSON, R.C., W. E. BROWN & R. W. THOMA. 1967. Process for preparing 1-dehydro steroids. U.S. Patent No. 3,360,439.

Production of Microbial Enzymes from Agroindustrial By-products

C. HUITRÓN, S. SAVAL, AND M. E. ACUÑA

Department of Biotechnology
Institute of Biomedical Research
UNAM, National University of Mexico,
A.P. 70228, Mexico, D.F., 04510, Mexico

Most studies on the use of agroindustrial wastes in fermentations have been concerned with the production of single cell protein or nutritional supplements.[1–4] There have been very few reports on the use of these by-products as starting material for the production of enzymes of industrial interest.[5,6] This aspect is important since agro-industrial by-products contain large amounts of compounds such as cellulose, hemicellulose, and pectin, which could serve as inducers for the production of extracellular enzymes such as cellulases, xylanases, and pectinases. Therefore, it would be interesting to develop processes for the use of these by-products.

In our laboratory, a strain identified as *Aspergillus* sp. has been isolated that is capable of producing extracellular pectinases in the fermentation of untreated henequen pulp.[7,8] Also, a yeast-like fungus has been isolated that has been identified as *Aureobasidium* sp.,[9,10,12] which produces extracellular cellulases and xylanases when it is cultured on microcrystalline cellulose or untreated sugar cane bagasse pith as the sole carbon source. When grown on henequen pulp, both in shake flasks[8] and in 14-l fermenters,[11] *Aspergillus* sp. produces sevenfold more pectinolytic activity (expressed as mg reducing groups per ml) (fifteenfold more by viscosimetric assay) than does *Aspergillus niger* ATCC 20107. These pectinases clarify apple juice in a way similar to commercial preparations.[7]

In this report, we present some of the results of experiments to find simple and economical culture media and the optimal inocula of *Aureobasidium* sp. and *Aspergillus* sp. for the production of pectinases, cellulases, and xylanases by submerged fermentation of either untreated henequen pulp or untreated sugar cane bagasse pith.

The strain of *Aspergillus* sp. was isolated from soil in the area where the pulp from a henequen defibering plant had been discarded.[7] It grows well at 37°C and needs pectin or henequen pulp to produce pectinase activity. *Aureobasidium* sp. is a cellulolytic strain that was isolated at 29°C.[9] Both strains were maintained on potato dextrose-agar plates. As the inoculum, either spores or mycelial culture (72 h for *Aspergillus* sp. and 24 h for *Aureobasidium* sp.) were used. The same medium used to obtain mycelia was used for fermentation. The inoculum for fermenters was 5% (vol/vol) for *Aspergillus* sp. and 10% (vol/vol) for *Aureobasidium* sp. Fermentations of *Aspergillus* sp. were carried out in 14-l fermenters (New Brunswick Scientific) containing 10 l of culture medium, maintained at 37°C, agitated at 200 rpm, and aerated at 1.0 vvm. Those of *Aureobasidium* sp. were carried out both in 500-ml shake flasks (180 rpm) and in 14-l fermenters and were maintained at 29°C, agitated at 200 rpm and aerated at 0.4 vvm. The composition of each culture medium is indicated in the figure legends. The pH value was initially adjusted to 4.5 for both strains. Samples were removed aseptically and were centrifuged at 5,000 rpm to obtain cell-free filtrates.

Pectinase activity was determined (1) by quantifying the reducing groups (as galacturonic acid) that were produced in the hydrolysis of pectin by using the 3,5-dinitrosalicyclic acid method,[13] and (2) by measuring the relative change in viscosity of the aqueous pectin solutions in an Ostwald viscosimeter as described previously.[7] Cellulase activity was determined by using a filter paper assay.[14] Xylanase activity was determined by quantifying the reducing sugars (as xylose) using the 3,5-dinitrosalicylic acid method.[12]

We found previously that *Aspergillus* sp. and *Aureobasidium* sp., isolated from soil in which henequen pulp and sugar cane bagasse pith, respectively, were discarded, are capable not only of rapidly adapting to a fermentation medium containing these agricultural by-products but also of producing high levels of extracellular cellulases, pectinases, and xylanases. In the studies on enzyme production carried out to date,[7,8,10,12] analytical-grade salts and distilled water have been used. However, for industrial production, it would be advantageous to cut costs not only by simplifying the medium but also by using industrial-grade salts and tap water.

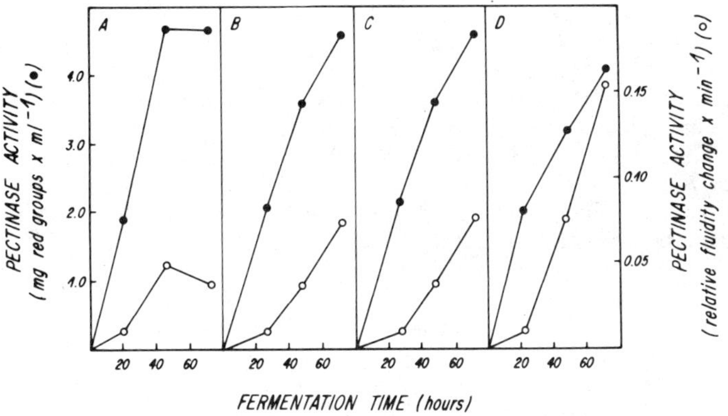

FIGURE 1. Production of extracellular pectinolytic activity by fermentation of 2% (wt/vol) henequen pulp by *Aspergillus* sp. 14-1 fermenters inoculated with spores were used. (A) medium containing 0.2% $(NH_4)_2HPO_4$, 0.2% KH_2PO_4, 0.05% $MgSO_4 . 7\ H_2O$, trace elements, and distilled water; (B) as in (A) except tap water was used; (C) medium containing 0.2% $(NH_4)_2HPO_4$, 0.2% KH_2PO_4, and tap water; (D) medium containing 0.2% $(NH_4)_2SO_4$, 0.2% K_2HPO_4, 0.2% KH_2PO_4 (all industrial grade), and well water.

We have found that the use of tap water has no negative effect on the production of pectinolytic activity (FIG. 1B) with respect to that produced with distilled water (FIG. 1A). The elimination of magnesium sulfate or of trace elements did not reduce the activity when tap water was used (FIG. 1C). The use of industrial-grade salts and well water reduced the pectinolytic activity, as determined by reducing groups, by 11%; however, as shown by the reduction in viscosity, this same activity was increased by 100% (FIG. 1D). Although we do not have direct data that can explain this effect, we think that the pectinolytic activity that we measured was due to the action of both endoenzymes and exoenzymes in such a way that, under the culture conditions used, the individual enzymatic components may be differentially affected. At any rate, from the practical point of view, the increased pectinolytic activity of the endo type is a desirable characteristic for industrial application.[15]

The cellulase production of *Aureobasidium* sp. in medium containing microcrystalline cellulose, analytical-grade salts, Tween 80, and distilled water was 2.0 mg reducing sugars per ml (FIG. 2A). When four of the components of the medium had been eliminated, the activity produced was practically the same, 1.9 mg/ml (FIG. 2B); under these conditions, not only were the number of components reduced but also tap water and sugar cane bagasse pith were used in place of distilled water and microcrystalline cellulose.

To determine the optimal inoculum type, spore suspensions and mycelia from *Aspergillus* sp. and *Aureobasidium* sp. were tested. When mycelia of *Aspergillus* sp. were used instead of spores, the pectinolytic activity, as measured by liberation of reducing groups, was less (TABLE 1). When measured by reduction in viscosity, this difference was greater, probably due to a greater effect on the endoenzymes. Since in this experiment industrial-grade salts were used (FIG. 1D), the spore inoculum seemed to be optimal for production of pectinolytic activity, particularly of the endo type.

In the case of *Aureobasidium* sp. the xylanolytic activity produced with the mycelial inoculum was 62% greater than that obtained with the spore inoculum, while the cellulolytic activity increased slightly (TABLE 1). This indicates that the mycelial inoculum is more convenient to use since this inoculum requires less of the costly solid culture medium than spores do and uses the same medium both for culturing the inoculum and for fermentation.

The synthesis of xylanases by *Aureobasidium* sp. is due to the presence of xylans in the structure of sugar cane bagasse pith.[6] However, in another study,[12] we found that this yeast-like fungus also produces xylanase activity when it is grown on microcrystalline cellulose.

The fact that *Aspergillus* sp. and *Aureobasidium* sp. are well adapted for growth and produce enzymes such as pectinases, cellulases, and xylanases in simple media and

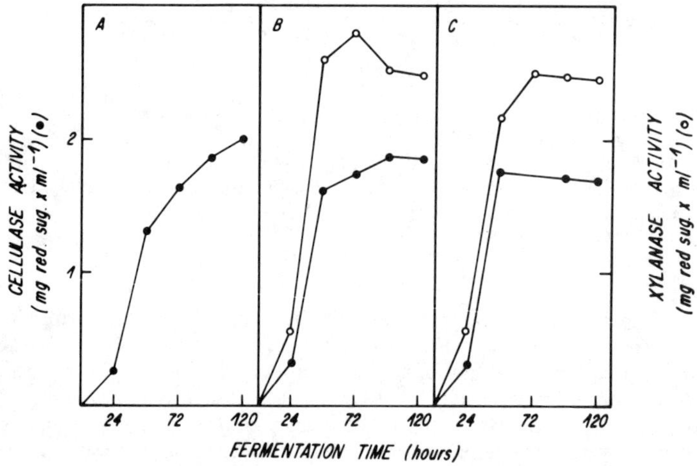

FIGURE 2. Production of extracellular cellulases and xylanases by *Aureobasidium* sp. Shake flasks inoculated with spores were used. (A) medium containing 0.75% microcrystalline cellulose, 0.2% KH_2PO_4, 0.03% urea, 0.03% $CaCl_2$, 0.14% $(NH_4)_2SO_4$, 0.03% $MgSO_4 \cdot 7\,H_2O$, 0.2% Tween 80, trace elements, and distilled water; (B) medium containing 2.0% sugar cane bagasse pith, 0.2% KH_2PO_4, 0.03% urea, 0.14% $(NH_4)_2SO_4$, and tap water; (C) same as in B, but using industrial-grade salts.

TABLE 1. Effect of Type of Inoculum on the Production of Extracellular Enzymes from Agroindustrial By-Products

Microorganism	Substrate	Enzymatic Activity	Inoculum	Enzyme (A)	Activities[a] (B)
Aspergillus sp.	henequen pulp	pectinase	spores	3.63	0.167
Aspergillus sp.	henequen pulp	pectinase	micelial culture	2.89	0.072
Aureobasidium sp.	sugar cane bagasse pith	cellulase	spores	1.22	
Aureobasidium sp.	sugar cane bagasse pith	cellulase	micelial culture	1.38	
Aureobasidium sp.	sugar cane bagasse pith	xylanase	spores	2.57	
Aureobasidium sp.	sugar cane bagasse pith	xylanase	micelial culture	4.17	

[a]Enzyme activities were determined at 72 hours of fermentation in a 14-liter fermenter. A: mg reducing groups × ml^{-1} cell-free filtrate. B: relative fluidity change × min^{-1}.

in the presence of untreated henequen pulp or untreated sugar cane bagasse pith is congruent with the fact that they were isolated from areas where, for several decades, these agricultural by-products had been discarded.

We find these results promising and are continuing to study the possibility of large-scale production of enzymes by these two microorganisms.

REFERENCES

1. BLANCAS, A., L. ALPIZAR, G. LARIOS, S. SAVAL & C. HUITRÓN. 1982. Conversion of henequen pulp to microbial biomass by submerged fermentation. Biotechnol. Bioeng. Symp. No. **12:** 171–175.
2. MOO-YOUNG, M., D. S. CHAHAL & D. VLACH. 1978. Single cell protein from various chemically pretreated wood substrates using *Chaetomium cellulolyticum*. Biotechnol. Bioeng. **20:** 107–118.
3. ZADRAZIL, F. 1977. The conversion of straw into feed by *Basidiomycetes*. Eur. J. Appl. Microbiol. **4:** 273–281.
4. CHANG, W. T. H., W. H. HSU, M. P. LAI & P. P. CHANG. 1979. Production of single cell protein from rice hulls for animal feed. Dev. Ind. Microbiol. **21:** 313–325.
5. ZETELAKI, K. 1976. Optimal carbon source concentration for the pectolytic enzyme formation of *Aspergilli*. Proc. Biochem. July/August: 11–18.
6. GARCÍA, M. D. V., T. OGAWA, SHINMYO & T. ENATSU. 1974. Hydrolytic degradation of bagasse by enzymes produced by *Penicillium variabile*. J. Ferment. Technol. **52:** 378–386.
7. SAVAL, S., R. SOLÓRZANO, L. ALPIZAR, A. CEA & C. HUITRÓN. 1983. Producción de pectinasas a partir de pulpa de henequén. *In* Biotecnología de Enzimas. C. Huitrón, Ed. México, D.F., Mexico pp. 203–215.
8. SAVAL, S., R. SOLÓRZANO, L. ALPIZAR, A. CEA & C. HUITRÓN. 1982. Production of pectinases by submerged fermentation of henequen pulp. Proc. Int. Symp. on the Use of Enzymes in Food Technology. Versailles, France. pp. 531–535.
9. GILBÓN, A., G. LARIOS & C. HUITRÓN. 1981. Aislamiento y selección de hongos que degradan celulosa cristalina. Tecnol. Aliment. **16**(5): 24–30.
10. LARIOS, G. & C. HUITRÓN. 1981. Celulasas extracelulares que produce un hongo levaduriforme. Rev. Tecnol. Aliment. (Mexico) **16**(5): 24–29.
11. SAVAL, S. & C. HUITRÓN. 1983. Microbial pectinases from henequen pulp. Dev. Ind. Microbiol. **24:** 547–551.

12. LARIOS, G., A. GILBÓN, Y. LARA & C. HUITRÓN. 1982. Extracellular cellulases produced by a yeast-like fungus. Enzyme Eng. **6:** 353–354.
13. MILLER, G. L. 1959. Use of dinitrosalicylic acid reagent for determination of reducing sugar. Anal. Chem. **31:** 426–428.
14. MANDELS, M., R. ANDREOTTI & C. ROCHE. 1976. Measurement of saccharifying cellulase. Biotechnol. Bioeng. Symp. No. **6:** 21–34.
15. FOGARTY, W. M. & O. P. WARD. 1974. Pectinases and pectic polysaccharides. Prog. Ind. Microbiol. **13:** 59–119.

Aqueous Two-Phase Systems for Producing α-Amylase Using *Bacillus subtilis*[a]

ELIS ANDERSSON, ANN-CHRISTIN JOHANSSON, AND
BÄRBEL HAHN-HÄGERDAL

Applied Microbiology
Chemical Center, Lund University
P.O. Box 740
S-220 07 Lund, Sweden

Since water is biocompatible, aqueous two-phase systems are suitable for use with biological materials. These systems have been used successfully for large-scale separations and purifications of intracellular enzymes.[1,2] It is also possible to culture living microbial cells,[3-5] as well as to perform enzymatic conversions[6] in aqueous two-phase systems where the product is later isolated from one of the two phases. The present communication describes the utilization of an aqueous two-phase system for the production of α-amylase using *Bacillus subtilis*.

FIGURE 1 shows a photograph of *B. subtilis* cells partitioned in a droplet of the dextran-rich bottom phase in a poly(ethylene glycol) (PEG)/dextran phase system. All cells were found in the bottom phase, which facilitates the upgrading of a product (α-amylase) from the cell-free top phase.

For the production of α-amylase, an aqueous two-phase system was selected in which the partition coefficient, K (enzymatic activity in the top phase over enzymatic activity in the bottom phase) for α-amylase was 0.5. The composition of the phase system was:

PEG 600	8% (wt/wt)
PEG 3350	5% (wt/wt)
Dextran T 500	2% (wt/wt)

The top phase contains mainly the two poly(ethylene glycol) components, while the bottom phase contains mainly dextran.

The volume ratio between the top and bottom phase was 9. The combined effect of the partition coefficient and volume ratio gave a yield of α-amylase in the top phase of approximately 80%.

In the next step, *B. subtilis* was cultured in this two-phase system (FIG. 2) and compared to a batch fermentation without the phase system added (FIG. 3). In the phase system, the α-amylase production was of a repeated batch type. After every production period, the stirring and aeration was stopped to allow the phase system to separate for 30 minutes. The top phase was then removed and thc production started again with addition of fresh medium.

α-Amylase activity measured in the initial and subsequent top phases was then compared to the activity found in the reference batch fermentation. Results showed

[a]This work was supported by the Swedish Board for Technical Development (STU).

that 30% more α-amylase was produced in the first cycle of the aqueous two-phase system. The reason for this is not yet known, but one possibility could be that the polymers interact with the surface of the cells, thus facilitating the formation and release of the enzyme.

Cell growth in the repeated batch fermentations proceeded well; however, there was a tendency for decreasing α-amylase production in the subsequent steps. The

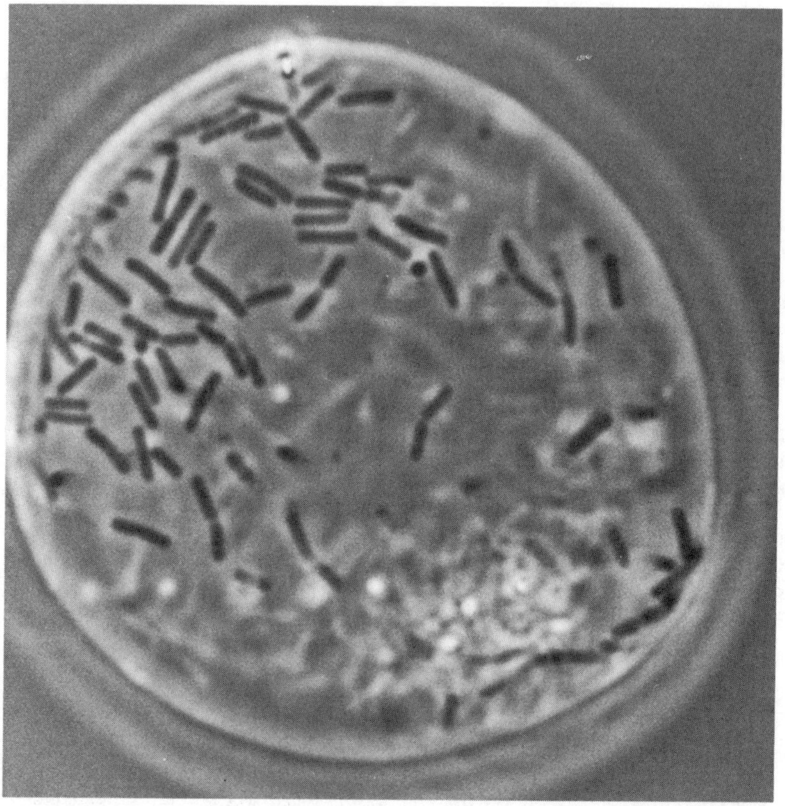

FIGURE 1. Photograph of *Bacillus subtilis* cells in a PEG-dextran aqueous two-phase system. The cells are confined in a droplet of the dextran-rich bottom phase. Magnification ×1250. (As shown, reduced in size to 65% of original.)

reason for this decrease is still unknown and subject to further investigation. Factors that might be of importance are the supply of oxygen as well as the ratio between the carbon and nitrogen sources during the enzyme production phase.

The results show that aqueous two-phase systems can be used for the production of extracellular products with living bacterial cells where the product can be recovered in a cell-free top phase. An improved α-amylase production was observed when the phase system was used, indicating that the environment of the bacterial cells also can influence the product formation.

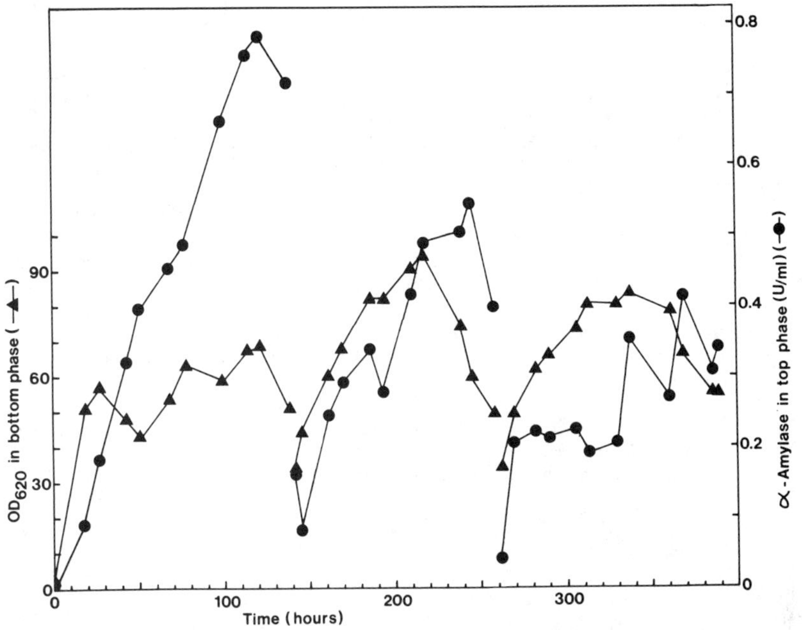

FIGURE 2. α-Amylase production with *B. subtilis* in a repeated batch fermentation in an aqueous two-phase system. Phase system: PEG 600, 8% (wt/wt), PEG 3350, 5% (wt/wt), Dextran T 500, 2% (wt/wt). Top phase composition: PEG 600, 8.4% (wt/wt), PEG 3350, 5.2% (wt/wt), Dextran T 500, 0.8% (wt/wt). Medium: (g/l) soluble starch, 50; peptone, 10; yeast extract, 2; NaCl, 1.5; $MgSO_4 \cdot 7\ H_2O$, 0.5; KH_2PO_4, 0.5; and $CaCl_2$, 0.1. Temperature: 37°C. Stirring speed: 300 rpm. Aeration: 1 vvm. Volume: 2.5 l; pH 7.0. α-Amylase analysis: Phadebas (Pharmacia Diagnostics, Uppsala, Sweden).

FIGURE 3. α-Amylase production with *B. subtilis* in a batch fermentation without phase system. Medium, growth conditions, and analysis as in FIGURE 2.

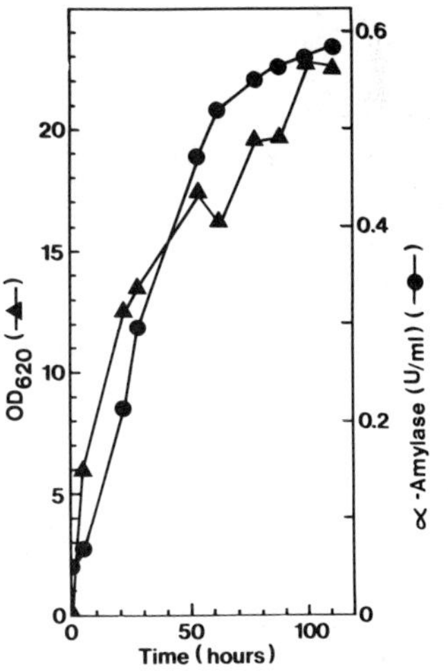

REFERENCES

1. KRONER, K. H., H. HUSTEDT & M.-R. KULA. 1982. Evaluation of crude dextran as phase-forming polymer for the extraction of enzymes in aqueous two-phase systems in large scale. Biotechnol. Bioeng. **24:** 1015–1045.
2. KULA, M.-R., K. H. KRONER & H. HUSTEDT. 1982. Purification of enzymes by liquid-liquid extraction. *In* Advances in Biochemical Engineering. A. Fiechter, Ed. Vol. **24:** 73–118. Springer-Verlag. Berlin.
3. PUZISS, M. & C.-G. HEDÉN. 1965. Toxin production by *Clostridium tetani* in biphasic liquid cultures. Biotechnol. Bioeng. **7:** 355–366.
4. MATTIASSON, B., M. SOUMINEN, E. ANDERSSON, L. HÄGGSTRÖM, P.-Å. ALBERTSSON & B. HAHN-HÄGERDAL. 1982. Solvent production by *Clostridium acetobutylicum* in aqueous two-phase system. *In* Enzyme Engineering. I. Chibata, S. Fukui, L. B. Wingard, Jr., Eds. Vol **6:** 153–155. Plenum Publishing Corporation. New York.
5. MATTIASSON, B. & B. HAHN-HÄGERDAL. 1983. Utilization of aqueous two-phase system for generating soluble immobilized preparations of biocatalysts. *In* Immobilized Cells and Organelles. B. Mattiasson, Ed. Vol **1:** 121–134. CRC Press. New York.
6. ANDERSSON, E., B. MATTIASSON & B. HAHN-HÄGERDAL. Enzymatic conversion in aqueous two-phase systems: Deacylation of benzylpenicillin to 6-aminopenicillanic acid with penicillin acylase. Enzyme Microb. Technol. In press.

Enzymatic Hydrolysis of Lactose in a Hollow-Fiber Reactor

C. K. S. JONES AND E. T. WHITE

Department of Chemical Engineering
University of Queensland
Brisbane, Australia

R. Y. K. YANG

Department of Chemical Engineering
West Virginia University
Morgantown, West Virginia 26506-6101

The use of hollow fibers for enzyme immobilization offers considerable advantages over other types of immobilization techniques. It allows enzymes to be immobilized under mild conditions without causing significant alteration of their kinetic properties and permits easy removal and replacement of the immobilized enzymes. For some enzyme-substrate combinations, with proper choice of hollow fibers and operation strategies, the method also allows simultaneous execution of reaction and separation within the same reactors.

These considerations have prompted considerable research efforts over the past decade to investigate the use of commercially available hollow-fiber cartridges as reactors for carrying out enzymatic reactions (For example: Breslau and Kilcullen,[2] Chambers *et al.*,[3] Davis,[4] Georgakis *et al.*,[5] Henley *et al.*,[6] Jones and Yang,[8] Kohlwey and Cheryan,[9] Korus and Olson,[10] Robertson *et al.*,[14] Rony,[15] Waterland *et al.*[16,17]) However, with the exception of the works by Breslau and Kilcullen involving liquid substrate and by Henley *et al.* and Jones and Yang involving solid substrates, all the investigated reactor schemes have used only diffusion as the major means of contacting substrates and enzymes. The backflush scheme proposed by Breslau and Kilcullen, in which bulk flow is expected to play a role in the enzymatic conversion of liquid substrates, has been investigated further in this work. Major findings are summarized here; details are described elsewhere.

Lactase (β-galactosidase) was chosen as the enzyme to be immobilized in the sponge layer of the hollow fibers. Four commercially available lactases derived from different microbial sources (*Escherichia coli, Kluyveromyces lactis, K. fragilis* and *Aspergillus niger*) were evaluated with respect to their suitability for this type of immobilization. The activity and the stability of the free and the immobilized enzymes, as well as their compatibility with Amicon hollow fibers of different molecular weight cut-offs, were the major factors of consideration. "Lactase N", derived from *A. niger* (G. B. Fermentation), was chosen as the best candidate for the study.

This lactase showed a broad range of pH tolerance with an optimum pH of 4.4 at 37°C. The temperature at which maximum activity was displayed at a pH of 4.4 was 58°C. Further kinetic studies using both the initial rate method (Allison and Purich[1]) and the progress curve method (Orsi and Tipton[13]), together with nonlinear regression analysis, confirmed that the rate expression associated with this enzyme was that of competitive product inhibition (by galactose). The values for the kinetic parameters associated with the rate expression were determined and, together with the rate expression, are shown in TABLE 1. The relatively high value (approximately 30 to 80) of the ratio of Michaelis constant to inhibition constant indicated strong product inhibition. Although the molecular weight of lactase has been reported to be over 1,000,000 (Leuba *et al.*[12]), it was found necessary to use Amicon H1P5 hollow fibers, which had a molecular weight cut-off as small as 5,000, to prevent enzyme leakage and to ensure total retention of Lactase N.

In the backflush mode of operating a hollow-fiber reactor, the feed was fed to the shell side of the hollow-fiber cartridge at one end and collected from the lumen exit port at the other end. The enzyme was immobilized by the pressure difference between the shell and the lumen. Since the rate of enzymatic conversion was directly related to the concentration of the active enzyme being retained in the sponge layer of the fibers, a reliable procedure for loading the lactase to the reactor was essential. Such a procedure was indeed developed, and a detailed description can be found elsewhere (Jones[7]). This immobilization procedure was found to cause no detectable gelling of enzyme, breakage of fiber, or presence of enzyme in the reactor shell surrounding the fiber bundle. Above all, it was very reproducible.

Although several different types of Amicon hollow-fiber cartridges were used for permeability measurements and residence time distribution studies, only the H1P5 cartridge was used as the reactor to evaluate the performance of this type of reactor scheme. During all the performance runs, the cartridge was operated with the shell exit port closed; no axial pressure drop was detected. Care was also taken to make sure that no enzyme deactivation occurred during the period of each run. Enzyme loadings of between 20 mg and 2200 mg, and lactose feed (in sodium citrate buffer, pH 4.4) varying between 0.5 and 50 g/l were used in this study. Flow rates between 0.4 and 15 ml/min were used. Experiments were performed at 35°C, 45°C, and 55°C.

The conversion data collected indicated that the reactor behaved more like a CSTR than a plug-flow reactor, particularly at low flow rates and high lactose concentrations. This appeared to confirm the conclusion derived from the residence time distribution study that a large degree of dispersion along the direction normal to the length of the fibers existed in the sponge layer and the value of the dispersion coefficient was of the same order of magnitude as that of the molecular diffusivity. This was consistent with our understanding that molecular diffusion strongly affects the rate of dispersion in laminar flow and implied that the substrate solution flowed slowly and evenly from the shell side through the sponge layer and the thin membrane to the lumen side. A mathematical model that related the performance of a backflush hollow-fiber reactor to the intrinsic kinetic and physical parameters of the system was developed and solved

TABLE 1. Kinetic Parameters of Lactase N at Optimum pH (4.4)

Method Used	Temperature	K_m(mM)	K_i(mM)	$K^a \times 10^3$
Progress curve (Foster-Niemann plot)	35°C	91 ± 30	1.1 ± 0.4	6.5 ± 0.7
Initial Rate (nonlinear regression)	35°C	95 ± 25	1.2 ± 0.3	6.8 ± 0.4
Initial rate (nonlinear regression)	45°C	62 ± 15	1.6 ± 0.3	9.3 ± 0.5
Initial rate (nonlinear regression)	55°C	91 ± 1.5	2.9 ± 0.4	14.9 ± 0.7

[a]K in unit of moles/mg enzyme · min as defined by the competitive product inhibition rate expression:

$$kE_0S/(S + K_m + \frac{K_m}{K_i}P)$$

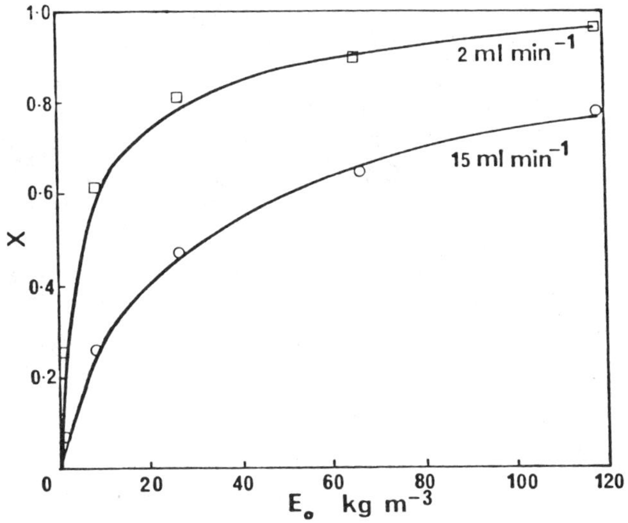

FIGURE 1. Lactose conversion X versus lactase loading E_0 at 35°C: S_0 = 5.85 m*M*; solid lines represent model predictions.

numerically.[7] Excellent agreement between model predictions and experimental data was found, suggesting the possibility of choosing *a priori* an enzyme loading and feed flow rate combination to attain a desired conversion in such a reactor. FIGURE 1, where lactose conversions versus lactase loadings are plotted for different flow rates, is typical of the results obtained in this study.

This work has also shown that the reactor is capable of achieving conversions in excess of 75% using a lactose feed of 50 g/l and at a flow rate corresponding to about 40 ml per minute per square meter of active membrane surface. This compares favorably with most of the conversions reported in the literature (*ca.* 20%), using much lower flow rates and feed concentrations, and should be a significant factor in any considerations regarding potential industrial applications of this type of enzyme reactor.

REFERENCES

1. ALLISON, R. D. & D. L. PURICH. 1979. Practical considerations in the design of initial velocity enzyme rate assays. Methods Enzymol. **63:** 3.
2. BRESLAU, B. R. & B. H. KILCULLEN. 1975. Hollow-fiber enzymatic reactors—an engineering approach. 3rd Int. Conf. Enzyme Eng. Portland, OR.
3. CHAMBERS, R. P., W. COHEN & W. H. BARICOS. 1976. Physical immobilization of enzymes by hollow-fiber membranes. Methods Enzymol. **44:** 291.
4. DAVIS, J. C. 1974. Kinetic studies in a continuous steady-state hollow-fiber membrane enzyme reactor. Biotechnol. Bioeng. **16:** 1113.
5. GEORGAKIS, C., P. C. CHAN & R. ARIS. 1975. The design of stirred reactors with hollow-fiber catalysts for Michaelis-Menten kinetics. Biotechnol. Bioeng. **17:** 99.
6. HENLEY, R. G., R. Y. K. YANG & P. F. GREENFIELD. 1980. Enzymatic saccharification of cellulose in membrane reactors. J. Enzyme Microb. Technol. **2:** 206–208.

7. JONES, C. K. S. 1983. Performance of a radial flow/backflush hollow-fiber enzymic reactor. Ph.D. Thesis. University of Queensland. Brisbane, Australia.
8. JONES, C. K. S. & R. Y. K. YANG. 1980. Simultaneous enzymatic hydrolysis of cellulose and sucrose. Chem. Eng. Commun. **6:** 283–291.
9. KOHLWEY, D. E. & M. CHERYAN. 1981. Performance of a β-D-galactosidase hollow-fiber reactor. Enzyme Microb. Technol. **3:** 64.
10. KORUS, R. A. & A. C. OLSON. 1977. The use of α-galactosidase and invertase in hollow-fiber reactors. Biotechnol. Bioeng. **19:** 1.
11. KORUS, R. A. & A. C. OLSON. 1978. Use of α-galactosidase, β-galactosidase, glucose isomerase, and invertase in hollow-fiber reactors. *In* Enzyme Engineering. **3:** 543. Plenum Press. New York.
12. LEUBA, J. L., F. WIDMER & D. MAGNOLATO. 1978. Purification and immobilization of a fungal β-galactosidase (lactase). *In* Enzyme Engineering **4:** 57. Plenum Press. New York.
13. ORSI, B. A. & K. F. TIPTON. 1979. Kinetic analysis of progress curves. Methods Enzymol. **63:** 159.
14. ROBERTSON, C. R., A. S. MICHAELS & L. R. WATERLAND. 1976. Molecular separation barriers and their application to catalytic reactor design. Sep. Purif. Method **5**(2): 301.
15. RONY, P. R. 1971. Multiphase catalysis II: Hollow-fiber catalysts. Biotechnol. Bioeng. **13:** 431.
16. WATERLAND, L. R., A. S. MICHAELS & C. R. ROBERTSON. 1974. A theoretical model for enzymatic catalysis using asymmetric hollow-fiber membranes. AIChE J. **20**(1): 50.
17. WATERLAND, L. R., L. R. ROBERTSON & A. S. MICHAELS. 1975. Enzymatic catalysis using asymmetric hollow-fiber membranes. Chem. Eng. Commun. **2:** 37.

Characterization of an Enzymatic Capillary Membrane Reactor

G. IORIO AND G. CATAPANO

Department of Chemical Engineering
University of Naples
Piazzale Tecchio
80125 Naples, Italy

E. DRIOLI

Department of Chemical Engineering
University of Calabria
Arcavacata di Rende (CS), Italy

M. ROSSI

Department of Enzymology
School of Science
University of Naples
Via Mezzocannone 16
80100 Naples, Italy

R. RELLA

International Institute of Genetics and Biophysics
CNR
Viale Marconi 10
80125 Naples, Italy

Many wines, for example red wines, attain their ultimate characteristics only if the initial malic acid content is reduced.[1] The decomposition of the L-malic acid is, therefore, a fundamental step towards the achievement of good organoleptic properties. The use of spontaneously growing fungi makes the process itself extremely uncertain and difficult to control.

Our group has been concerned with the decomposition of the malic acid catalyzed by the malic enzyme (E.C. 1.1.1.40.), from chicken liver cytosol.

The decomposition of the L-malic acid is accomplished by the enzyme according to the following scheme:

$$\text{L-malate} + 2\,NADP^+ \xrightarrow[Mn^{2+} \text{ or } Mg^{2+}]{\text{M.E.}} CO_2 + \text{pyruvate} + 2\,NADPH$$

The action of the enzyme is $NADP^+$-dependent, and a divalent cation, for example Mn^{2+} or Mg^{2+}, is needed to realize the oxidative decarboxylation of L-malate. The enzyme binds the coenzyme $NADP^+$ by means of Mn^{2+} or Mg^{2+} ions, producing an active complex, E-Mn(Mg)-$NADP^+$, which, in turn, reacts with the substrate. The products are released in order, according to the above scheme, and the enzymatic protein is eventually regenerated.[2]

We have developed an immobilization procedure that consists of a dynamic gelation of the malic enzyme onto the active side of capillary ultrafiltration (UF) membranes. Polyamide capillary membranes, with a molecular weight cut-off equal to 50,000, inner diameter 1 mm, thickness 0.1 mm, were used to support the enzymatic layer; 12 fibers were fitted in a cylindrical P.M.M.A. tube and shell reactor, sealed at both ends by means of epoxy resins. Approximately 1.2 enzymatic units (EU) per fiber, with a 3 EU per mg protein ratio, were charged under a constant $\Delta P = 1$ atm, at $T = 10°C$. Owing to concentration polarization phenomena, a thin layer, highly concentrated in malic enzyme, is formed on the active side of the capillaries. Spectrophotometric assays of both the permeate and the axial stream indicate that the rejection of the membrane towards the enzyme is close to 100% and that from 83% to 100% of the charged enzyme is retained in the gel. The enzymatic system has been operated continuously, feeding the substrate and the cofactor under pressure with no recycle, and collecting the product downstream of the membrane. A 0.02 *M* Tris-HCl buffer, pH 7.4, 0.5 m*M* $NADP^+$, 0.1 m*M* $MnCl_2$, 0.5 m*M* β-mercaptoethanol was used in all experiments.

The activity of the enzyme was tested initially in homogeneous solution, to establish a reference for its kinetic behavior; the enzyme exhibits its maximum activity at pH 7.4, $K_m = 0.06$ m*M*, at its optimum pH.[3]

The physical variables that affect the system behavior at constant temperature are essentially the axial flow rate and the applied pressure. As far as the former is concerned, its value has to be suitably balanced between conditions of intolerable shear stresses on the gel layer, due to high flow rates, and of product inhibition, which could be enhanced by low flow rates.

Experiments carried out at axial flow rates ranging up to 1.7 ml/min exhibit an axial conversion that diminishes and a poorer mechanical stability of the gel layer at

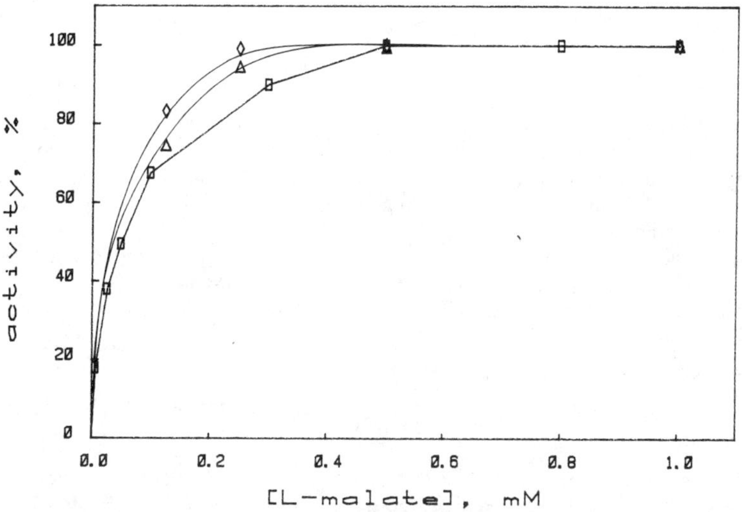

FIGURE 1. Percent activity versus L-malate concentration plot, at 25°C. □—native malic enzyme in homogeneous solution. △—malic enzyme gelled on capillary membranes, operated at $\Delta P = 1.2$ atm. ◇—malic enzyme gelled on capillary membranes, operated at $\Delta P = 1.4$ atm.

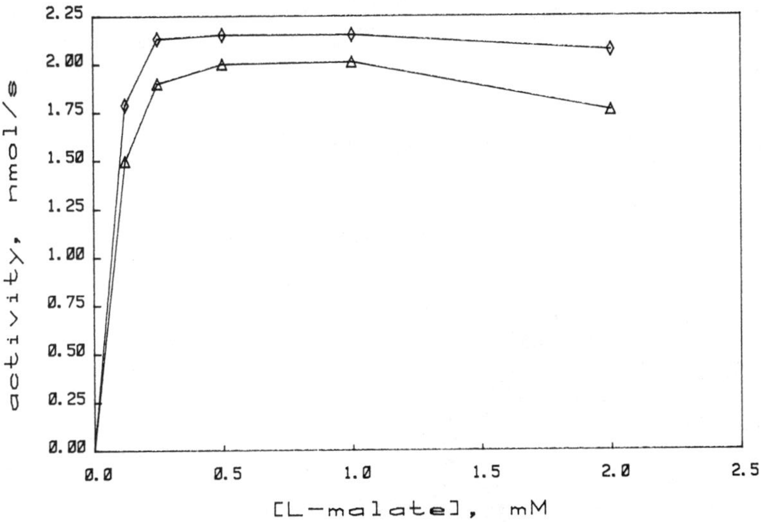

FIGURE 2. Substrate inhibition at saturating substrate concentrations, at T = 25°C. △—ΔP = 1.2 atm; ◇—ΔP = 1.4 atm.

higher axial flow rates. The effects of the applied pressure have been investigated in several runs at different ΔPs, at a constant temperature of 25°C. As shown in FIGURE 1, the enzyme under immobilized conditions still exhibits Michaelis-Menten behavior. In particular, it can be noticed that the apparent Michaelis constant, K_m, decreases from a value of 0.1 m*M* to 0.02 m*M* as the pressure increases from 1 atm to 1.4 atm.

This result is in good agreement with the physical meaning usually attributed, in homogeneous solution, to the Michaelis constant, that is, of an inverse measure of the affinity of the enzyme towards the substrate. Under immobilized conditions, with a kinetic behavior still analyzed in terms of the Michaelis-Menten rate equation, the apparent K_m becomes more generally an inverse measure of the enzyme availability also from a mass transfer point of view;[3] its decrease can be reasonably expected as a result of the pressure increase, which leads to higher permeate fluxes and, therefore, to lower diffusional resistances. When the ΔP is increased, a slight increase in the reaction rate at saturating substrate concentrations is also observed.

At higher substrate concentrations, apparently, substrate inhibition phenomena occur (FIG. 2), even though only in the permeate stream. They might be attributed to a diminution, in the enzyme layer, of the activating effects of the Mn^{2+} ions in the presence of high concentrations of substrate, which, in turn, acts as an inhibitor of the enzyme kinetic activity. Work is under way to explain these results.

The kinetic stability of the enzymatic system is higher than that of the enzyme in homogeneous solution: the activity falls exponentially, and asymptotically approaches the 60% of the initial value, at least up to 36 hours. No fouling occurs and a marked increase in the permeate flow rate is observed as the applied pressure increases. Yields up to 20% have been obtained: these values are almost insensitive to ΔP changes.

The reactor is inexpensive and easy to assemble and control.[4] A procedure to regenerate the reactor *in situ,* when it is exhausted, is under study. Once the continuous loss of the cofactor in the permeate stream can be avoided, and the substrate inhibition controlled, this kind of system appears to be very promising from an industrial point of view, as a continuous enzymatic reactor, for the oxidative decarboxylation of L-malic acid.

SUMMARY

This paper deals with the results obtained in a flow membrane enzymatic reactor. The enzyme adopted, the malic enzyme, catalyzes the oxidative decarboxylation of L-malic acid to pyruvic acid.

The reaction has interesting implications from an industrial point of view, being an intermediate step in the malolactic fermentation of wine.

The enzyme has been immobilized by gelation onto the active side of polyamide capillary membranes assembled in a tube and shell reactor configuration.

The results we have obtained seem quite promising and promote further reactor developments.

REFERENCES

1. OTTAVI, O. 1923. Classificazione dei vini. *In* Enologia Teorico-pratica. O. Ottavi, Ed. Casalmonforte, Bari. pp. 799–888.
2. HSU, R. V. 1982. Pigeon liver malic enzyme. Mol. Cell. Biochem. **43:** 3–26.
3. IORIO, G., G. CATAPANO, E. DRIOLI, M. ROSSI & R. RELLA. 1983. Malic enzyme immobilization in continuous capillary membrane reactors. Synthetic Membranes in Science and Industry. Tubingen, 7.9.1983. Report in press.
4. STRATHMANN, H. 1981. Membrane separation processes. J. Membr. Sci. **9:** 121–189.

Immobilization of Purified Penicillin Acylase in a Polarized Ultrafiltration Membrane Reactor

FRANCESCO M. VERONESE, ENRICO BOCCÙ, AND ODDONE SCHIAVON

Istituto di Chimica Farmaceutica
Università di Padova
Centro di Studio sulla Chimica del Farmaco del C.N.R.
Padua, Italy

GUIDO GRECO, JR.

Istituto di Principi di Ingegneria Chimica
Università di Napoli
Naples, Italy

LILIANA GIANFREDA

Facoltà di Farmacia
Università di Napoli
Naples, Italy

Penicillin acylase from *E. coli* is widely used in the pharmaceutical industry for the mild hydrolysis of penicillin G, obtained by fermentation. The product, 6-aminopenicillanic acid (6-APA), is the starting material in the production of new, semisynthetic penicillins.[1] Usually, crude penicillin acylase preparations are employed (either bound to insoluble matrices, or as soluble enzyme). This is the reason why few purification to homogeneity procedures have been described so far, and why little is known about acylase physico-chemical characteristics.

In this paper, three items are briefly discussed: (1) a purification technique with a final affinity chromatography step; (2) an analysis of the enzyme thermal deactivation kinetics; and (3) the results of preliminary acylase stabilization tests, performed in a polarized ultrafiltration membrane reactor, where the enzyme is injected together with linear-chain polyelectrolytes.

The procedure for penicillin acylase purification, usually employed in our laboratories,[2] involves the following steps: (1) enzyme extraction by washed-cell paste freezing and subsequent thawing in distilled water; (2) batch adsorption of acidic components, from the cell-free extract by DEAE-cellulose; (3) batch acylase adsorption by CM-cellulose at pH 5, followed by elution with 0.1 *M* acetate buffer; (4) ammonium sulfate cut between 50% and 70% saturation; and (5) adsorption on a hydroxylapatite column and elution with a linear phosphate buffer gradient. With some stocks of *E. coli* cells, however, this method did not yield a homogeneous enzyme. An affinity resin was therefore prepared, which employs 6-APA as a ligand and Sepharose as a support. The activation agent is tosyl chloride.[3]

The procedures for resin preparation are the following: (1) dry Sepharose (corresponding to 130 g of wet resin) is added to a solution of tosyl chloride (18.5 g in 37 ml acetone); (2) the mixture is agitated while 18 ml of pyridine are added dropwise; (3) after 1 hour at room temperature, the resin is washed with 0.1 *M* borate buffer, pH

9.3; and (4) the ligand (6-APA, 20 g) is added and, after 20 hours mixing, the unbound material is removed by filtration. The amount of bound 6-APA (0.05 mg/ml of settled resin) is estimated by titration. In acylase affinity chromatography, the enzyme solution (in 0.01 *M* phosphate buffer, pH 5.7) is applied to the column and eluted with a 0.02 *M* 6-APA solution, in the same buffer.

A different resin was also prepared with phenylacetic acid as a ligand (*p*-aminophenylacetic acid was bound to the tosylated resin through its amino group), but acylase was retained less effectively. Resins were also prepared with the same two ligands, using CNBr-activated or epoxy-activated Sepharose. In these cases, the results were not as satisfactory.

Tests were performed to characterize soluble, native acylase deactivation kinetics at pH 7 (in 200 m*M* phosphate buffer), T = 55°C.[5] Two different experimental procedures were adopted: (1) enzyme incubation in substrate-free buffer (both with

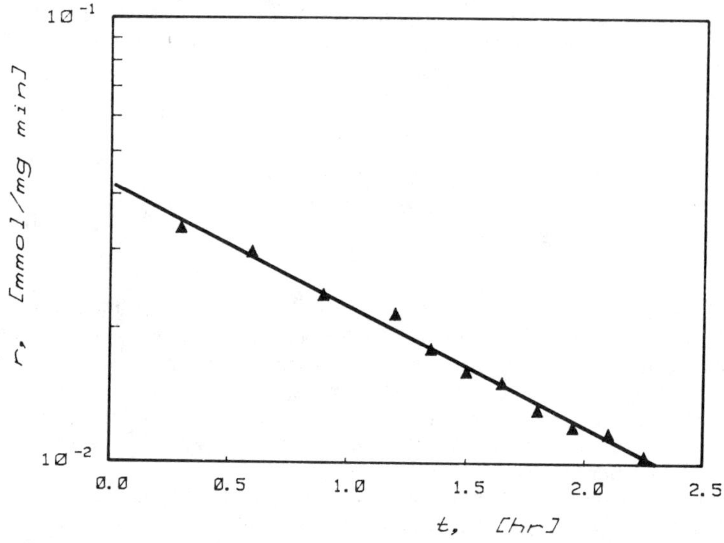

FIGURE 1. Typical soluble enzyme results.

and without reaction products), with initial rate measurements performed at predetermined time intervals, to monitor enzyme activity; (2) batch reactor runs (in which deactivation and enzymatic reaction take place simultaneously), where 6-APA concentration was determined as a function of deactivation time. 6-APA concentration was measured according to the experimental technique proposed in Balasingham *et al.*[4]

Basically, the following conclusions were reached: (1) the deactivation of acylase was a straightforward transition from active to inactive enzyme; (2) it obeyed irreversible first-order kinetics (therefore, the deactivation was completely characterized by two parameters: the kinetic constant kd and the initial acylase specific activity r°); (3) in the concentration range explored, substrate had no stabilizing effect on enzyme activity; (4) the results were not affected by product inhibition, owing to the low conversion levels attained; (5) the kinetics of acylase deactivation were not modified by the presence of either reaction product. Typical soluble enzyme results are

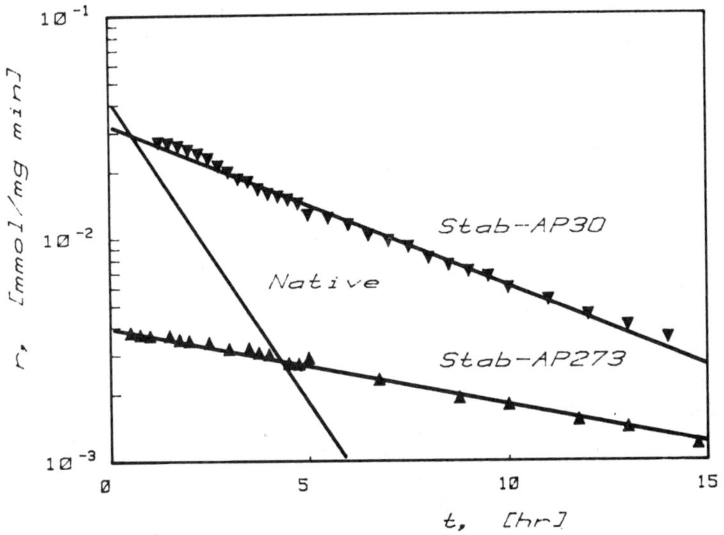

FIGURE 2. Typical results obtained with the polyacrylamides AP273 and AP30.

reported in FIGURE 1, in terms of enzyme activity versus time. The data refer to acylase deactivation and were obtained following the enzyme incubation procedure described above.

Details of the experimental procedure adopted to stabilize acylase within an unstirred ultrafiltration reactor are reported elsewhere in this book.[6] Briefly, the method consists of injecting the enzyme into a polarized, plane-membrane ultrafiltration (UF) cell, together with a linear-chain, high-molecular-weight polyelectrolyte (polyacrylamide, SEPARAN, The Dow Chemical Co., Lexington, MA). Both macromolecules accumulate immediately upstream from the UF membrane and reach extremely high concentration levels. As a consequence, the linear-chain polymer yields a fairly tight network surrounding the protein, and preventing the unfolding of the latter, to some extent, thus reducing kd. Different polyacrylamides have been tested. Typical results obtained with two of them, AP273 and AP30, are reported in FIGURE 2. Both polyelectrolytes are slightly anionic, the molecular weight of AP273 is three times that of AP30, both exceeding 1,000,000. A comparison among the three situations is given in TABLE 1, where the kinetic parameters of acylase deactivation are reported. It can be seen that AP273 amd AP30 increase enzyme half-life by roughly a factor of 8 and 4, respectively, as compared to native acylase. On the other hand, AP273 yields an

TABLE 1. Kinetic Parameters of Acylase Deactivation

Status	10^3 r[a]	kd[b]
Native	42.3	0.626
AP273-stabilized	3.78	0.079
AP30-stabilized	32.3	0.166

[a] r° = mmoles/mg min.
[b] kd = 1/hr.

elevenfold decrease in acylase initial activity, while with AP30 there is a 30% reduction in r°. It can be concluded that, in choosing the appropriate stabilizing polymer, a compromise must be found between effective enzyme stabilization and high residual catalytic activity.

REFERENCES

1. VANDAMME, E. J. 1981. Penicillin acylase and β-lactamase. *In* Economic Microbiology. A. N. Rose, Ed. **2:** 467–481. Academic Press. New York.
2. VERONESE, F. M., D. FRANCHI, E. BOCCÙ, A. GUERRATO & P. ORSOLINI. 1981. *E. coli* penicillin acylase: Extraction, purification and kinetic characterization by improved methods. Il Farmaco **7:** 663–670.
3. NILSSON, K. & K. MOSBACH. 1980. *p*-Toluensulphonyl chloride as an activating agent of agarose for the preparation of immobilized affinity ligands and proteins. Eur. J. Biochem. **112:** 397–402.
4. BALASINGHAM, K., D. WARBURTON, P. DUNNILL & M. D. LILLY. 1972. The isolation and purification of penicillin amidase from *E. coli.* Biochim. Biophys. Acta **284:** 250–256.
5. GRECO, G. JR., F. M. VERONESE, R. LARGAJOLLI & L. GIANFREDA. 1983. Purified penicillin acylase performance in a stabilized ultrafiltration membrane reactor. Eur. J. Appl. Microbiol. Biotechnol. **18:** 333–338.
6. GRECO, G. JR., G. MARRUCCI, N. GRIZZUTI & L. GIANFREDA. 1984. Thermal and chemical deactivation of soluble and stabilized enzymes. Ann. N.Y. Acad. Sci. **434:**. This volume.

The Influence of Pretreatments on Rice Husk Utilization by *Sporotrichum pulverulentum*[a]

JOSÉ M. C. DUARTE,[b] L. RODA-SANTOS,
AND A. CLEMENTE

Department of Chemical Industries
Laboratório Nacional de Engenharia e Tecnologia Industrial
DT1Q/LNETI
2745 Queluz, Portugal

Rice husks are among the most difficult agro-industrial wastes to dispose of. The husk is the protective envelope of the rice grain and constitutes about 20% by weight of the paddy rice.

Annual world paddy rice production is calculated to be larger than 300 million tons, which gives an idea of the amount of the husk left over. In spite of a recent report[1] advocating the conversion of rice husk into energy, mainly by combustion, there are difficulties that make it unattractive for these uses.

Biotechnological utilization of rice husk has been practically nonexistent. Fermentation of its pentose sugars was referred to in the USDA-NRRL process[2] for production of liquid fuels or solvents. Among others, acid and enzymatic hydrolysis to produce fermentable sugars is also a possibility. The husks are mainly of lignocellulosic nature, with variable composition: cellulose, the main component (40–50%), hemicellulose (15–25%) and lignin (10–20%). Growth of mushroom-like fungi have been recommended as an alternative to the disposal of several agro-industrial wastes. On the other hand, Zadrasil[3] found that solid-state fermentation of rice husk by a number of *Basidiomycetes* decreased the digestibility of the husk. This was attributed to the fact that rice husks contain large amounts of silica; silica may account for as much as 20% of the husk.

Therefore, we attempted to use rice husk in liquid heterogeneous fermentations as a substrate for the white-rot fungus *Sporotrichum pulverulentum.*[4]

However, to increase the husk susceptibility to the fungal attack, several pretreatments were tried.

MATERIALS AND METHODS

Growth Conditions

Sporotrichum pulverulentum, var. Novobranova was grown on 50 ml of a modified Norkrans medium (pH 5.5) in 250-ml shake flasks, and incubated for 10 days on a orbital shaker at 150 rpm with temperature control at 30°C. Rice husk concentration

[a]The Calouste Gulbenkian Foundation (Lisboa) and the Engineering Foundation (New York) provided support to J. M. C. Duarte.

[b]To whom correspondence should be addressed.

was chosen to be 2%.[4] The medium and substrate were sterilized by autoclaving at 120°C for 20 minutes. The flasks were inoculated with 5 ml of an eight-day-old culture grown in 2% cellulose (Av*i* cel).

Analytical Methods

After the 10 days of growth, the flask's contents were filtered and the residue was dried for 6 hours at 105°C. Protein content of the dried samples was determined by an automated Kjeldahl method. CMCase activity was determined in the culture filtrates by measuring the rate of formation of reducing sugars at 25°C by a method based on Witt.[5] Total reducing sugars were determined by the dinitrosalicyclic acid method.

Husk Pretreatments

Size reduction: The husk was milled on a Max Fritsch knife mill with a 1.5-mm screen. Several granulometries were separated and assayed.

Steaming: Pressure cooking by exposing the husk to different times of autoclaving was also used.

γ-Radiation: Rice husks were irradiated with a ^{60}Co source of 10 kCi (1 Curie is equivalent to 3.7×10^{10} Becquerel). Two different doses were tried (50 and 100 Mrad) on wet and dry samples of the husk.

Sodium hydroxide: Samples were treated with 5 N NaOH for 6 and 24 hours, respectively. Treated samples were either neutralized *in situ* or washed, and the insoluble residue and the filtrate were tested separately after neutralization with sulfuric acid.

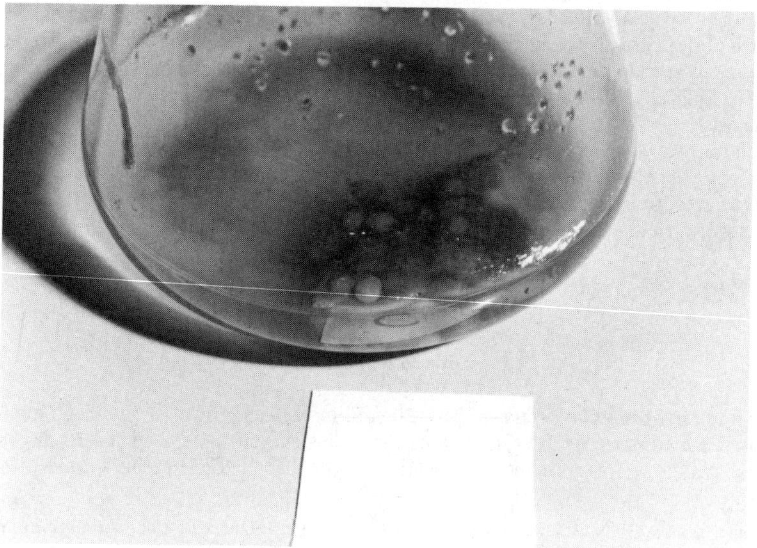

FIGURE 1. Aspect of a shake-flask culture of *S. pulverulentum* grown on rice husk pretreated for 6 hours with sodium hydroxide (solid residue).

FIGURE 2. Shake-flask cultures of *S. pulverulentum* grown on the filtrate liquor from the sodium hydroxide pretreated husk. Left: 6-hour treatment. Right: 24-hour treatment.

RESULTS AND DISCUSSION

The effect of pretreatments on husk attack by *S. pulverulentum* as measured by the reduction in the dry weight and protein content of the residue of the fermentation is shown in TABLE 1.

Growth of the fungus in commercial cellulose (Avicel) is also shown. It shows that milling of the husk alone seems not to have a noticeable effect on increasing the growth of the *Sporotrichum* since most of the protein value is accounted for by the inoculum. The 1.0-mm-sized sample showed a larger reduction in dry weight, and it was chosen for use with the other treatments. Autoclaving the samples (pressure cooking) for different periods did not increase significantly the protein content of the final sample. However when γ-irradiation was used, a significant increase was observed, especially with the 100 Mrad dose. The dry sample gave a better result than the wet sample. Han *et al.*[6] also used γ-radiation to pretreat lignocelullosic materials; they were able to completely solubilize a lignocellulosic (cane bagasse) sample. However, the degradation products contained little glucose owing to the combined effect of alkali and high dosage of γ-radiation.

Sodium hydroxide has been widely used for chemical pretreatments of lignocellulosics. By using sodium hydroxide, Chahal *et al.*[7] increased the protein content of the sample by 17.3% with a cellulose utilization of 78.5%. However, they used a high-temperature treatment for 30 minutes. In our work, we used a sodium hydroxide treatment at room temperature. As shown in TABLE 1, a large reduction in the dry weight was obtained (almost 70% for the 24-hr-treated sample). The protein content of the dried samples after fermentation was 15.5%, an increase of about eightfold compared with the original husk. The fungus could still be grown on the effluent liquor, which gave a dry residue of about 66% of the previous one but with a somewhat lower

TABLE 1. The Influence of Several Pretreatments on the Degradability of the Rice Husk by *S. pulverulentum*

Substrate/Treatment	Sample Size (mm)	Dry Weight Reduction (%)	Protein (% wt/wt)
Cellulose			
/none	—	53	28.3
Husk/Milling	0.5	−3.3	4.7
	1.0	7.4	3.2
Steaming			
0.5 h	1.0	5.9	3.1
1 h	1.0	10.3	3.4
2 h	1.0	0	3.7
γ-Irradiation			
mrad			
50 dry	1.0	3.6	7.1
wet	1.0	6.0	5.4
100 dry	1.0	17.0	12.1
wet	1.0	7.7	7.0
NaOH (5N)			
Residue 6 h	1.0	56.4	14.0
24 h	1.0	68.0	15.5

protein content (8.2%). Therefore, with this treatment, a material was obtained that had 55% of the protein when grown on pure commercial cellulose. Of all the other treatments, only γ-irradiation caused a relatively good increase of the protein content, about 43% of the Avicel preparation.

These differences must be mainly due to different ways by which the treatments affect the structure of the lignocellulosic material. Whereas autoclaving seems to be relatively inefficient, γ-radiation seems to expose the cellulose fiber to the fungal attack, probably by destroying the lignin barrier. It may also render the cellulose fibers more susceptible to the enzymatic action. However, γ-radiation still seems to maintain intact the apparent structure of the material, which was reflected by the low dry weight reduction obtained.

On the contrary, sodium hydroxide acts by solubilizing lignin and hemicelluloses[7] leaving mostly a cellulosic material, more easily degraded by the *Sporotrichum*.

TABLE 2 shows the activity of cellulases in the fermentation medium, using an assay with carboxymethyl cellulose (CMC). It is noteworthy that the γ-irradiated samples display the highest activity, followed by the sodium hydroxide-treated samples. Reducing sugars in solution also follow this pattern. Notice that γ-irradiation

TABLE 2. Effect of Husk Pretreatments on CMCase Activity of *S. pulverulentum* Cultures

Treatment	CMCase IU/ml	Increase[a] of CMCase Activity	Reducing Sugars (mg/l)
Milling (1.0 mm)	22	3.5-fold	116
Steaming (1 h)	26	3.2-fold	104
Radiation (100 mrad/dry)	54	2.5-fold	666
Sodium hydroxide (24 h)	36	3.2-fold	143

[a]Compared with the value at the beginning of fermentation.

causes a much larger increase of reducing sugars in solution owing probably to physical destruction of some of the cellulose fibers.

These results seem to indicate that the cellulase complex is relatively inefficient. Other results (not shown) also indicated that it may be due to a low exoglucanase activity. This could be due either to poor induction and conditions for enzyme action (the pH increases from 5.5 to 6.5) or most probably to inhibition of the exo-1,4-β-glucanase by products of the lignin degradation by the *S. pulverulentum.*[8] Therefore we believe that further studies on the enzymatic activities of the fungus and the effects of the pretreatments may yield even better results, which may in turn lead to a set of conditions for the successful exploitation of rice husks as a source of carbohydrate for the growth of *S. pulverulentum.* The solid product obtained could be used as an inexpensive material for animal feeding.

REFERENCES

1. BEAGLE, E. C. 1978. Rice husk conversion to energy. FAO Agric. Serv. Bull. No. 31. Rome.
2. TSUCHIYA, H. M., J. M. VAN LANEN & A. F. LANGLYKKE. 1949. U.S. Patent 2,481,263. Sept. 6.
3. ZADRAZIL, F. 1980. Conversion of different plant waste into feed by *Basidiomycetes*. Eur. J. Appl. Microbiol. Biotechnol. **11:** 183–188.
4. DUARTE, J. M. C., A. CLEMENTE, A. T. DIAS & M. E. ANDRADE. 1983. Biotechnological upgrade of rice husk. I. The influence of substrate pretreatments on the growth of *S. pulverulentum. In* Biomass Utilization. Wilfred A. Côté, Ed. Plenum Press. New York. Vol. **67**(Nato ASI Series): 393–410.
5. WITT, W. D. 1980. Utilization of cellulose and hemicellulose of pig faeces by *T. viride*. Meded. Landbouwhogesch. Wageningen **80**(2): 1–76.
6. HAN, Y. W., J. TIMPA, A. CIEGLER, J. COURTNEY, W. CURRY & E. LAMBREMONT. 1981. γ-Ray-induced degradation of lignocellulosic materials. Biotechnol. Bioeng. **23:** 2525–2535.
7. CHAHAL, D. S., M. MOO-YOUNG & D. VLACH. 1981. Effect of physical and physicochemical pretreatments of wood for SCP production with *Chaetomium cellulolyticum*. Biotechnol. Bioeng. **23:** 2417–2420.
8. ERIKSSON, K. E. & B. PETTERSSON. 1975. Extracellular enzyme system utilized by the fungus *S. pulverulentum* for breakdown of cellulose. Eur. J. Biochem. **51:** 213–218.

Enzymatic Whey Hydrolysis in a Pilot-Plant "Lactohyd"

J. E. PRENOSIL, E. STUKER, T. HEDIGER, AND J. R. BOURNE

Technisch-Chemisches Laboratorium
ETH Zurich
CH-8092 Zurich, Switzerland

The problem of whey disposal or its meaningful utilization has plagued the dairy industry for many decades. In some countries, like Switzerland and Norway, where the art of cheese making in small plants still prevails, whey is disposed of locally by pig feeding or by manufacture of whey cheese.

Worldwide, however, with modern dairy plants sometimes producing over 1 million liters of whey daily, traditional means of whey disposal are no longer acceptable. The same can be said about UF permeate from the manufacture of cheese or whey protein by ultrafiltration, which is becoming increasingly popular. Because whey and UF permeate, having BOD values of about 50,000 mg O_2/l, represent a significant environmental problem, they cannot be discharged without expensive treatment. Therefore, methods of whey or UF-permeate upgrading are being intensively studied. The enzymatic hydrolysis of lactose to glucose/galactose by β-galactosidase with simultaneous protein recovery is one of the most promising methods[1] currently being investigated.

Basic research in our laboratory on β-galactosidase and its immobilization[2,3] reached a stage where a scale-up of the enzyme reactor was required. A bigger reactor needed large quantities of substrate and the use of pure lactose was no longer economical. Hence, sweet whey, which was provided by the Swiss dairy industry, was chosen as a substrate. This called for other unit operations for whey processing before enzymatic hydrolysis as well as for final product tailoring. Therefore a complete pilot plant had to be constructed. It also gave us an opportunity to test the stability of our catalyst in the environment of a whey processing plant. An economic analysis of enzymatic whey hydrolysis, based on the pilot plant results, was attempted.

The plant was built around 3-l, fixed-bed reactors with the enzyme catalyst developed earlier.[2,3] A capacity of 800–1000 l of whey per day was chosen as a feasible plant size. A production cycle of 20 or 40 hr and a 4-hr sanitation cycle was used. Total automation was required as in an industrial plant. "Quickfit" glass was chosen for a construction material allowing easy plant construction and visual inspection under operating conditions. The products, a glucose/galactose syrup and a whey protein concentrate, had to be of a quality enabling them to be offered to the food industry for application and market tests. A glucose analyzer, was used for the analysis of glucose; HPLC for lactose, glucose, galactose, and oligosaccharides. A simplified flow diagram of the "Lactohyd" is given in FIGURE 1.

The enzyme catalyst, developed and optimized earlier,[2,3] was at first based on β-galactosidase from *Aspergillus niger*. This was expensive and its use required a new selective lactose crystallization step.[4] Four fixed-bed reactors were installed so that every combination of their connections was possible. Three of them were used for hydrolysis with the fourth one kept on stand-by, waiting to be exchanged with any

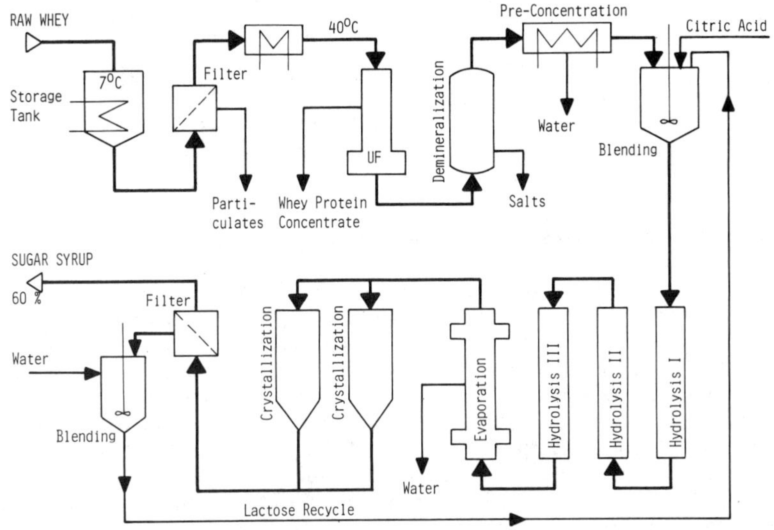

FIGURE 1. Flow diagram.

TABLE 1. Catalyst

Enzyme	β-Galactosidase *A. niger* (1) and *A. oryzae* (2)
Carrier	Ion exchange resin, Duolite S-761
Immobilization	Adsorption, crosslinking by glutaraldehyde
Particle size	0.4–0.8 mm
Activity	(1) 150–200 AU/g, (2) 300–450 AU/g
Half-life time (2)	31 days, T = 40°C, deionized whey
Volumetric weight	*ca.* 0.4 g/ml
Approximate costs related to product = 0.5¢ per kg s/rup	

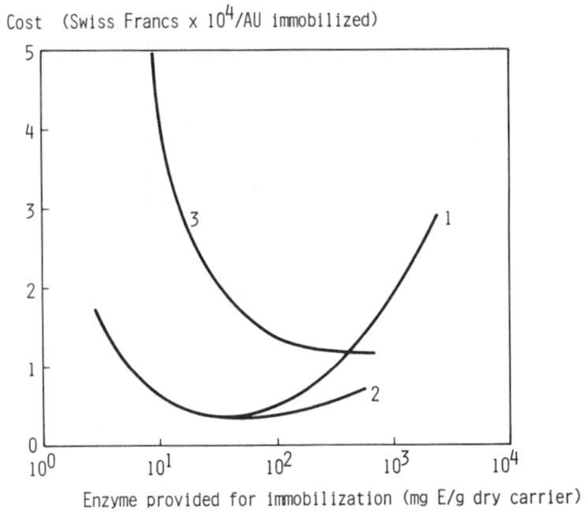

FIGURE 2. Price of immobilized AU in various catalysts as a function of enzyme provided for immobilization. (1) Ion-exchange resin Duolite S-761, 7.80 SFr./lt; (2) Ion-exchange resin Duolite A-7, 8.20 SFr./lt; (3) Controlled-pore glass (Corning Glass Co.) 100.— SFr./lt (assumed price).

TABLE 2. Products

Whey protein concentrate =	75–120 l/day
	8–10 wt% protein in first stage
Glucose/galactose syrup =	50–70 l/day
	75–95% hydrolysis
	60–70 wt% dry mass

reactor in need of *in situ* regeneration. This reactor in turn was put on stand-by after its regeneration.

Later, a catalyst based on β-galactosidase from *A. oryzae* was used. It was more active, but slightly less stable. This enzyme is comparatively inexpensive and the crystallization step was no longer needed. Furthermore, only one reactor, instead of the three used with the enzyme from *A. niger,* was required to achieve 90–95% lactose conversion. Catalyst data and its costs related to product are summarized in TABLE 1. The catalyst activity with respect to enzyme loading was investigated for various carriers. In FIGURE 2, the price per unit activity, including the carrier price and enzyme deactivation is plotted against the amount of enzyme consumed during its immobilization. The optimal price corresponds to about 20 mg enzyme/g Duolite, whereas for the Corning Glass controlled-pore glass carrier, the optimum is located at much higher amounts of enzyme and its price per unit activity is still substantially higher. Typical product characteristics are presented in TABLE 2.

Owing to its transgalactosylic action, β-galactosidases produce certain, variable amounts of oligosaccharides (OS) in the course of lactose hydrolysis. Various β-galactosidases of yeast and fungal origin were investigated and the enzyme from *A. oryzae* was found to produce most OS. The amount of OS depends strongly on the initial lactose concentration and degree of hydrolysis (FIG. 3). If a conversion of 80–85% is considered to be most economical, then the UF permeate must not be preconcentrated. Otherwise, the level of OS in the syrup would be too high.

The fully automated pilot plant "Lactohyd" was successfully operated for more than half a year with catalyst regeneration and reimmobilization *in situ;* no problems were encountered with microbial contamination.

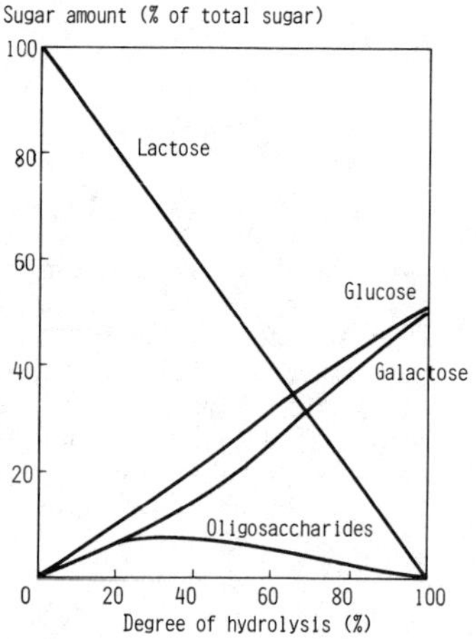

FIGURE 3. Relative amounts of sugars as a function of disaccharide concentration. 2 g of resin catalyst (A. oryzae) in 50 ml of 5 wt% lactose in buffer. T = 50°C; pH = 4.5.

REFERENCES

1. JELEN, P. 1983. Reprocessing of whey and other dairy wastes for use as food ingredients. Food Technol. (February): 81–84.
2. PETER, J. 1980. Ueber die enzymatische Hydrolyse von Lactose mittels traegergebundener β-Galactosidase: Immobilisierung des Enzymes und Anwendung in der Molkeverwertung. Diss. ETH Nr. 6628. Zurich.
3. BETSCHART, H. 1983. Traegergebundene β-Galactosidase bei der Lactosehydrolyse in Molke unter Beruecksichtigung der Oligosaccharide. Diss. ETH Nr. 7360. Zurich.
4. BOURNE, J. R., M. HEGGLIN & J. E. PRENOSIL. 1983. Solubility and selective crystallization of lactose from solutions of its hydrolysis products glucose and galactose. Biotechnol. Bioeng. **25:** 1625–1639.
5. PRENOSIL, J. E., J. PETER & J. R. BOURNE. 1980. Hydrolytische Spaltung des Milchzuckers der Molke durch immobilisierte Enzyme im Festbettreaktor. Verfahrenstechnik **14**(6): 392–396.

Kinetics of Enzymatic Starch Liquefaction

JAMES E. ROLLINGS AND ROBERT W. THOMPSON

Chemical Engineering Department
Worcester Polytechnic Institute
Worcester, Massachusetts 01609

Over the past two decades, considerable interest has persisted in both the academic and industrial communities for utilization of renewable polysaccharide materials in providing alternative sources of foods and petrochemicals. Currently, enzymatic starch hydrolysis processing technologies significantly contribute to sweetener production in the United States and in Western Europe. However, our understanding of the overall system dynamics is far from complete, and consequently appropriate kinetic modeling of such depolymerizations has been hindered. This paper presents a simplified model of starch liquefaction that may be useful in understanding the complexities of this process.

The first stage of starch processing (referred to as liquefaction) converts native starch to small oligosaccharides, lowering paste viscosity, and initiating sweetener production. Published studies concerning liquefaction have dealt mainly with end product production and for the large part have ignored the polymeric intermediates.[1-6] Early attempts at examining depolymerization products were generally limited to observations of approximately two decades of the molecular weight distribution for these materials;[7-9] this is the limitation imposed by use of a single size exclusion (SEC) chromatographic resin.[10] As the estimates of the initial molecular weight for starch are of the order of 10^9 daltons,[11,12] these data clearly represent only a small fraction of the overall picture. Yau and coworkers[13] described a method for extending the SEC separation range, which was exploited by us[14] for the explicit purpose of examining starch liquefaction products. By using this method,[15] certain trends in the molecular weight distribution pattern emerged that could not have been predicted a priori. Rollings *et al.*[15] concluded that starch depolymerization does not proceed in a random manner. In general, it was observed that four distinct groups of molecular-sized products are produced. Potentially, this situation could exist and therefore be attributed to phenomena associated with the enzyme and its action pattern,[2,16] the substrate's bonding pattern,[17-27] and/or physical chemical properties of the substrate such as the formation of polymeric complexes.[14,28-30] The first two areas have in part been examined by other investigators; here we present data to support the latter case.

A purified wet-milled 1% corn starch substrate was hydrolyzed with commercial α-amylase (Novo Laboratories) in a batch reactor at pH 6.0 at temperatures both above and below gelatinization (melting) range of the starch. Aqueous SEC was used to monitor the molecular weight distribution of the hydrolysate products. Half-normal aqueous alkaline solvents were used for the analytical experiments. In order to quantitate the molecular weight scale, the intrinsic viscocity data of Bendetskii and Yarovenko[9] was employed. This procedure is described elsewhere.[31]

Molecular weight distributions of corn starch hydrolyzed under all conditions tested were always multimodal. At temperatures below gelatinization, these distributions were primarily in two well-defined classes; one greater than the five-million-dalton upper limit of the linear SEC separation range and a second less than the lower 2000-dalton limit. Four distinct chromatographic peaks exist in the molecular weight distribution when starch is hydrolyzed above the gelatinization temperature; the peaks mentioned above and two others with average molecular weights at 5×10^5 daltons and 2×10^4 daltons. The latter peak is within the range associated with Bruchard's dissolving gap.[30] In a separate report,[31] we have demonstrated that this material within the dissolving gap region is at least partially insoluble in the reaction medium, forms aggregates that may be crystalline, and is not directly created as a consequence of native granule structure.

The essential features of these observations are incorporated into a mathematical description of starch depolymerization. Differential equations for the hydrolysate groups, labeled A, B, C, and D, having the peak average molecular weights of 10^7, 5×10^5, 2×10^4, and 1.5×10^3, respectively, are written with allowance for all possible transformations as:

$$\frac{d\eta_A}{dt} = -k_A \cdot E(t) \cdot A \cdot \eta_A \tag{1}$$

$$\frac{d\eta_B}{dt} = k_A \cdot E(t) \cdot B \cdot \eta_A - k_B E(t) \cdot C\eta_B \tag{2}$$

$$\frac{d\eta_C}{dt} = k_A \cdot E(t) \cdot D \cdot \eta_A + k_B \cdot E(t) \cdot F \cdot \eta_B - k_C \cdot E(t) \cdot G \cdot \eta_C \tag{3}$$

$$\frac{d\eta_D}{dt} = k_A \cdot E(t) \cdot \alpha_{AB} \cdot \eta_A + k_B \cdot E(t) \cdot \alpha_{BD} \cdot \eta_B + k_C \cdot E(t) \cdot G \cdot \eta_B \tag{4}$$

In these expressions, η_i (i = A, B, C, D) denotes the mass fraction of carbohydrate within each of the groups, α_{jk} denotes the relative molar mass of species j to k, and the constant terms A, B,.....,G represent specific collections of the $\alpha_{jk's}$ that account for all possible transformations between these groups. For example, it is possible to cleave off a hydrolysate product from group A, producing two polymers of group B, or one in group D and one in group A (note: each group is actually a range of products), but it is not possible to produce two polymeric products in group C from a group A substrate molecule. In addition to these equations, an expression for thermal enzymatic denaturation is required. Here we assumed a first-order dependency to be adequate as shown in Eq. 5:

$$E(t) = E_o e^{-k_E t} \tag{5}$$

where E_0 is the initial active enzyme concentration and k_E is its deactivation constant.

The differential equations were solved subject to the initial condition that η_A is unity and the other η's are zero for optimal values of the three parameters of interest (k_A, k_B, k_C, and k_E). Reasonable agreement was found between simulation and experiment when the relative rate of reactions k_A:k_B:k_C are 100:10:2.2 and a value for k_E of 3.5×10^{-3} min^{-1} was employed. For further information on this, the reader is referred to our more complete publication.[31]

REFERENCES

1. Roybt, J. F. & D. French. 1963. Action pattern and specificity of an amylase from Bacillus subtilis. Arch. Biochem. Biophys. **100:** 451–467.
2. Roybt, J. F. & D. French. 1970. The action pattern of porcine pancreatic α-amylase in

relationship to the substrate binding site of the enzyme. Arch. Biol. Chem. **245:** 3917–3927.
3. Schoch, T. J. 1969. Mechano-chemistry of starch. Wallerstein Lab. Commun. **32:** 149.
4. Roybt, J. F. & D. French. 1967. Multiple attack hypothesis of α-amylase action. Arch. Biochem. Biophys. **122:** 8–16.
5. Weetall, H. H. & N. B. Havenwala. 1962. Continuous production of dextrose from cornstarch. Biotechnol. Bioeng. Symp. **3:** 241–266.
6. Roels, J. A. & R. Vantilbury. 1979. Kinetics of reactions with amyloglucosidose and their relevant industrial application. Starke **31:** 338–344.
7. Henriksnas, H. & H. Bruun. 1978. Molecular weight distribution in starch solution hydrolyzed with α-amylase and when oxidized with NaH hypochlorite. Starke **30:** 233–237.
8. Henriksnas, H. & T. Lovgren. 1978. Chain-length distribution of starch hydrolyzed with α- or β-amylase action. Biotechnol. Bioeng. **20:** 1303–1307.
9. Benetskii, K. M. & V. L. Yarovenko. 1978. Change in the molecular weight parameters of starch and amylase on their hydrolysis. Bioorgan. Khim. **1:** 614–620.
10. Yau, W. W., J. J. Kirkland & D. D. Bly. 1979. Modern Size Exclusion Liquid Chromatography. John Wiley & Sons. New York.
11. Banks W., R. Geddes, C. T. Greenwood & E. G. Jones. 1972. Physicochemical studies on starches, Part 63. Starke **24:** 245–251.
12. Greenwood, C. T. 1956. Aspects of the physical chemistry of starch. Adv. Carbohydr. Chem. **11:** 335–343.
13. Yau, W. W., G. R. Ginnard & J. J. Kirkland. 1978. Broad-ranged linear calibration in high-performance SEC using column packings with bimodal pores. J. Chromatogr. **149:** 465–487.
14. Rollings, J. E., A. Bose, M. R. Okos & G. T. Tsao. 1982. System development for size exclusion chromatography of starch hydrolysis. J. Appl. Polym. Sci. **27:** 2281–2296.
15. Rollings, J. E., M. R. Okos & G. T. Tsao. 1983. Molecular size distribution of starch during enzymatic hydrolysis. ACS Symp. Ser. **207:** 443.
16. Abdulla, M., D. French & J. F. Roybt. 1966. Action pattern of amylases. Arch. Biochem. Biophys. **114:** 595.
17. French, D. 1972. Structure af amylopectin. J. Jpn. Soc. Starch Sci. **21:** 91–98.
18. Banks, W. & C. T. Greenwood. 1975. Starch and its components. Edinburgh University Press. Edinburgh, Scotland.
19. Robin, J. P., Mercier, F. Duprat, R. Charonniose & A. Guildot. 1975. Amidon lintnerises. Starch **27:** 36–45.
20. Banks W. & D. D. Muir. 1980. Chemistry of starch granule. *In* The Biochemistry of Plants: A Comprehensive Treatise. Vol. 3. J. Preiss, Ed. Academic Press. New York.
21. Hood, L. F. & C. Mercier. 1978. Molecular structure of unmodified and chemically modified mantioc starch. Carbohydr. Res. **61:** 53–66.
22. Ikawa, Y., D. V. Glover, Y. Siegimoto & H. Puna. 1981. Some structure characteristics of starches of maize having a special grain background. Starch **33:** 9–13.
23. French, D. 1975. Chemistry and biochemistry of starch. M.T.P. Int. Rev. Sci. Biochem. Ser. One **5:** 267–335.
24. Abdulla, M. & D. French. 1970. Substrate specificity of pullulanase. Arch. Biochem. Biophys. **137:** 483–493.
25. Hall, R. S. & D. J. Manners. 1980. α-(1 → 4)-D-glucans. Part XXIII. The structural analysis of some amylodextrins. Carbohydr. Res. **83:** 93–101.
26. Umeki, D. & K. Kainuma. 1981. Fine structure of Nageli amylodextrin obtained by acid treatment of defatted waxy-maize starch—structural evidence to support the double-helix hypothesis. Carbohydr. Res. **96:** 143–159.
27. Borovsky, D., E. E. Smith & W. J. Whelan. 1975. Purification and properties of potato 1,4-α-D-glucan: 1,4-α-D-glucan, (-α-(1,4-α-glucano)-transferase. Evidence against a dual catalytic function in amylose-branching enzyme. Eur. J. Biochem. **59:** 615–625.
28. Rollings, J. E. 1981. Development of aqueous size exclusion chromatography for the study of substrate structure and susceptability in enzymatic starch hydrolysis. Ph.D. thesis. Purdue University.

29. KODOMA, M., H. NODA & T. KAMATA. 1978. Conformation of amylose in water. Biopolymers **17:** 985–1002.
30. BURCHARD, W. 1963. Viscosity behavior of amylose in various solvents. Makromol. Chem. **64:** 110–125.
31. ROLLINGS, J. E. & R. W. THOMPSON. Kinetics of enzymatic starch liquefaction: Simulation of the high-molecular-weight product distribution. Biotechnol. Bioeng. In press.

Continuous Conversion of Starch to Ethanol Using a Combination of an Aqueous Two-Phase System and an Ultrafiltration Unit[a]

MATS LARSSON AND BO MATTIASSON

Pure and Applied Biochemistry
Chemical Center, University of Lund
P. O. Box 740
S-220 07 Lund, Sweden

Separation processes based on the use of ultrafiltration membranes or aqueous two-phase systems have both been predicted to grow in importance in biotechnology.[1,2] Both methods are generally used in downstream processing for purification of a product from a bioconvertive reaction, a fermentation broth, or a biological fluid. The methods both offer the possibility of integration of bioconversion and a first upgrading step. Integrating a separation step in the production stage may be advantageous since it is thus possible to retain substrates and catalytically active cells and enzymes while products are removed. If a first separation step was used before the membrane step, this would prevent or at least greatly reduce the risks of polarization of the membranes in a subsequent ultrafiltration step.

The bioconversion of starch-containing material to fermentable sugars traditionally includes a cooking step at 130–150°C. In this highly energy-intensive step, the starch particles are gelatinized and in some cases also partly hydrolyzed by alpha-amylase. The solution is cooled and a second enzyme, glucoamylase, is added to break down the starch to glucose. In this conventional process the enzymes are not reused.

The process described in this paper is a continuous extractive bioconversion of native starch granules to glucose or ethanol. An aqueous two-phase system was used in a mixer-settler reactor to make it possible to obtain a clean product stream to the ultrafiltration unit. The conversion of starch was carried out at 35°C and the yeast cells and/or the hydrolytic enzymes were retained and reused in the reactor.

When aqueous solutions of two different polymers are mixed, the mixture may become turbid and when left for a while, phase separation will occur. Each of the two phases will consist mainly of a solution of one of the polymers.[3] Aqueous two-phrase systems are characterized by containing 85–95% water and they therefore provide mild conditions for cells and proteins, that is, they are biocompatible. The interfacial tension is very low, less than 0.1 dyne/cm^2 and this requires only gentle mixing to create and maintain an emulsion. The basis for separation in an aqueous two-phase system is the nonuniform distribution of substances between the phases. Biological macromolecules, cells or particles distribute according to their surface properties. The aqueous two-phase system used in these experiments consisted of 5% (wt/wt) poly(ethylene glycol) (PEG 20M, Union Carbide, USA) and 3% (wt/wt) dextran (crude dextran, Sorigona, Sweden). The volume ratio between the top and the bottom phase was 4 : 1.

[a]This project was supported by The National Swedish Board for Technical Development and the Swedish Biotechnology Research Foundation.

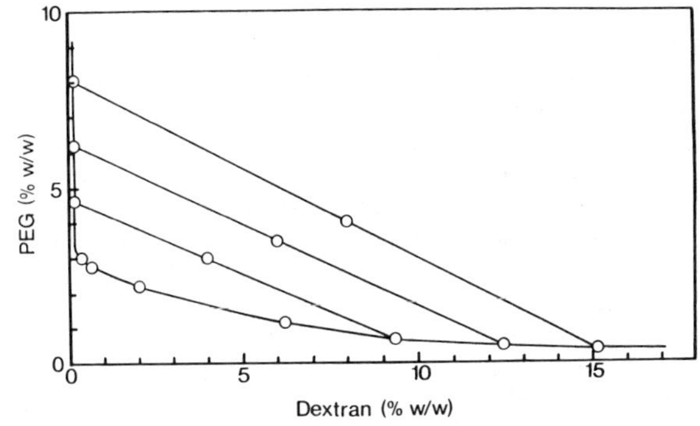

FIGURE 1. Phase diagram of the system crude dextran—PEG 20M—water at 20°C Experimental binodial and tie lines.[3]

A phase diagram for the two-phrase system is shown in FIGURE 1.[4] Mixtures of polymers in concentrations above the curve result in phase systems, whereas polymer mixtures below the curve give homogeneous solutions.

EXPERIMENTAL SET-UP AND EXPERIMENTAL CONDITIONS

The mixer-settler reactor consisted of a box of Plexiglass (500 x 200 x 150 mm). It was divided into two parts by a plate of Plexiglass with 3 vertical 5-mm slits, (see FIG. 2). The reactor was filled with the two-phase system and mixing was achieved by using a propeller in one of the chambers, in which an immersion heater acted as a thermostat. In the other chamber, the two-phase system was allowed to settle. The clear top phase from this chamber was pumped (pump A) to a beaker and then pumped (pump B, flow rate 75 ml/min) to the mixing chamber in the reactor. The level in the beaker was kept constant with an inductive sensor controlling pump A. From the beaker, the top phase was also pumped with a lobe rotor pump (pump C, 2000 ml/min) through the ultrafiltration unit (Pellicon cassette system) and back to the beaker. This arrangement was made to keep tangential flow in the membrane unit and still avoid stirring of

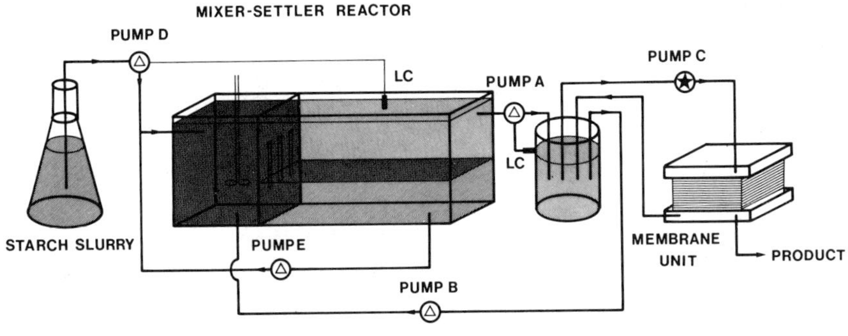

FIGURE 2. Experimental set-up. LC = level control. Experimental details are given in the text.

the phases in the settling chamber with the high flow from the lobe rotor pump. In the ultrafiltration unit, the poly(ethylene glycol) and the glucoamylase present were retained on the upstream side of the membrane (PTGC 10000MW, Millipore) while products, salt, and water were passed through the membrane.

The reactor was fed with substrate (native starch) with pump D controlled by the level in the reactor. The bottom phase, containing unreacted substrate, was recycled to the mixing chamber with pump E (flow rate 75ml/min).

Initially, native starch was continuously converted to glucose in 50 m*M* acetate buffer, pH 4.8, at 35°C in a reactor of a total volume of 21 liters. The reactor was fed with a 20% (wt/wt) starch slurry (waxy maize, Stadex, Sweden) in the same buffer, continuously for 38 hours. The starch was partitioned to the bottom phase. The enzymes, alpha-amylase, Termamyl (total 2400 KNU), and the glucoamylase, spritamylase (total 6000 AGL), were added to the system. Termamyl was adsorbed to the starch particles and thus partitioned to the bottom phase and spritamylase was present in both phases. The enzymes were generous gifts from NOVO A/S, Denmark. The total amount of starch added was 5.0 kg.

In a second series of experiments, yeast was added to produce ethanol from the glucose. The phase system of a total volume of 15 liters contained initially 10% (wt/wt) starch and an additional amount of 1.2 kg was added during the experiment. After 7 hours, 65 grams of yeast cells (dry weight, commercial bakers yeast) were added. The yeast cells were partitioned to the bottom phase.

Ethanol was analyzed by GLC and glucose by HPLC according to previous published procedures.[4]

RESULTS

The outcome of an experiment with enzymatic conversion of starch to glucose using the reactor configuration shown in FIGURE 2 is shown in FIGURE 3. During the first 38 hours, starch is added and the glucose content in the effluent increases to more than 120 g/l. The total yield of glucose calculated from the starch added was 94%.

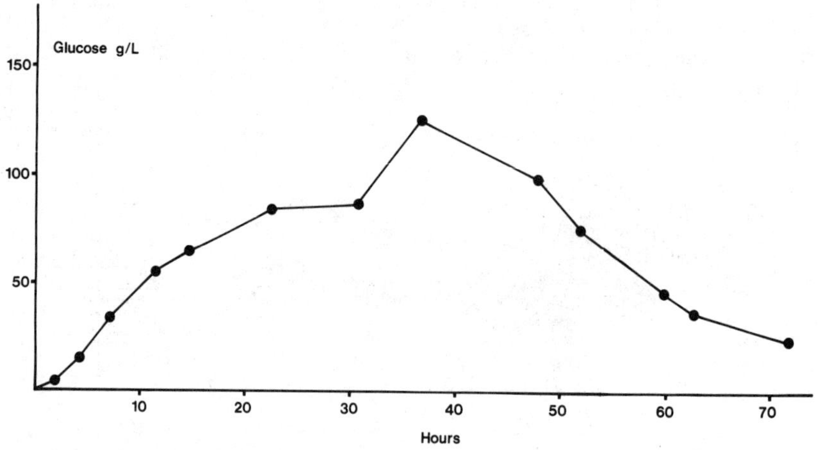

FIGURE 3. Glucose produced from native starch as a function of time. Details in the text.

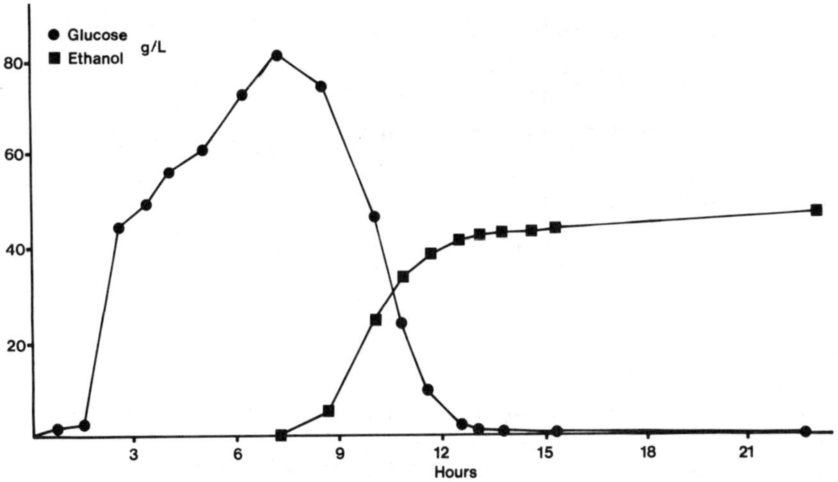

FIGURE 4. Glucose and ethanol produced from native starch as a function of time. Details in the text.

In another set of experiments in which yeast was also added, a profile according to FIGURE 4 was obtained. During the initial 7 hours, enzymatic hydrolysis of starch forms a pool of glucose, which upon addition of yeast is converted to ethanol with a concomitant decrease in glucose content.

The level of ethanol in the effluent stabilized close to the theoretical value calculated from added starch and 100% conversion.

Leakage of PEG from the membrane was tested by flushing the membrane with an 8% (wt/wt) PEG M20 solution (120 l/h). Permeate flux was 2.5 l/h. During a 100-h period, a leakage of less than 0.25 g PEG/h was observed. No tendency for fouling could be seen during the same period.

The process described here for starch hydrolysis eliminates the energy-demanding step of gelatinization. The yields observed are quite comparable with those obtained in conventional processes. One characteristic of the method is that the macromolecular substrate is kept in the reactor. Furthermore, the biocatalysts are recycled in the process, and this provides the possibility for improved economy of the process.

The integration of extraction in aqueous two-phase systems and separation in ultrafiltration units gives a reactor design with great applicability in biotechnology.

REFERENCES

1. MICHAELS, A. S. 1980. Membrane technology and biotechnology. Desalination **35:** 329–351
2. MATTIASSON, 1983. Application of aqueous two-phase systems in biotechnology. Trends Biotechnol. **1: 15–20.**
3. ALBERTSSON, P-Å. 1971. Partition of Cell Particles and Macromolecules. Almqvist & Wiksel. Uppsala, Sweden.
4. WENNERSTEN, R., F. TJARNELD, M. LARSSON & B. MATTIASSON. 1983. Extractive bioconversions. Hydrolysis of starch in an aqueous two-phase system. International Solvent Extraction Conference. Denver, Colorado.

Continuous Alcohol Fermentation of Starchy Materials with a Novel Immobilized Cell/Enzyme Bioreactor

SUSUMU FUKUSHIMA AND KAZUHIRO YAMADE

Department of Chemical Engineering
Faculty of Engineering
Kansai University
Suita, Osaka 564, Japan

The purpose of this work was to obtain rapid production of high concentrations of ethanol from liquefied starchy materials through parallel operation of saccharification and fermentation in an immobilized-cell enzyme bioreactor. The novel bioreactor that was designed was packed with both Al alginate biocatalyst A (entrapping adsorbed-glucoamylase on γ-alumina), and Al alginate biocatalyst B (entrapping a mixture of different yeast strains).

The bioreactor should meet the following criteria to produce high concentrations of ethanol on a large scale for a long period nonstop:

(1) Biocatalyst gel particles should allow only a low leakage of cells in the bioreactor, in addition to having mechanical properties such that they are neither cracked nor crushed.

(2) The bioreactor should not accumulate dextrin, sugar, CO_2 gas bubbles, ethanol, or by-products locally and should not cause shearing stress on the biocatalyst gel particles.

EXPERIMENTS

Biocatalyst A

One-eighth inch γ-alumina pellets were pulverized to a 100–145 mesh. The resulting powder was washed with toluene and acetone or ethanol and ethyl ether. The powder was dried at 100°C for one day. The refined powder was adjusted to pH 4.8 in $^1/_{15}$ kmol/m^3 phosphate buffer solution. One part of the powder, adsorbing glucoamylase, was mixed with three parts of 1.5% (wt/vol) Na-alginate solution. This slurry was put into 166 kg/m^3 $CaCl_2$ solution through the nozzle to make particles, entrapping adsorbed glucoamylase with Ca alginate. These particles were immersed in 10–20 kg/m^3 $Al_2(SO_4)_3 14–18H_2O$ solution for exchanging Ca to Al ion. The glucoamylases used were Amano Seiyaku GNL-2000 and Novo AMG.

Biocatalyst B[1]

The cells used were *Saccharomyces cerevisiae Hakken No. 1,* which is often used for ethanol production from starchy materials, and *Saccharomyces cerevisiae OC-2,* which is popular for wine production. The cells of the two strains were separately incubated in YM medium in rotating flasks for 24 to 36 hours. An equal amount of wet cells of each yeast strain, obtained after centrifugation, was mixed with 1.5 parts of 1.5% (wt/vol) Na alginate. Biocatalyst B, entrapping a mixture of different strains with Al alginate, was obtained using the method described above. The content of cells was 220 kg (dry base)/m^3.

Liquefied Starchy Materials

Starchy materials used were potatoes such as Goto (Japan) dried sweet potatoes, and grains such as Japanese dried barley and American dried corn. These were pulverized to less than 100 mesh in a ball mill. The powders were liquefied with commercial α-amylase by known methods. The feed solution was the supernatant after centrifugation. One or two kg/m^3 $Al_2(SO_4)_3 14–18H_2O$ and 1 kg/m^3 $(NH_4)_2SO_4$ were added to the feed solution. The pH of the feed solution was adjusted by adding H_2SO_4.

Bioreactor and Its Operation

A novel multi-stage, 3-phase (gas, liquid, solid) bioreactor was designed by the authors.[2] The units are arranged vertically, and sintered plate is installed at the bottom of the lowest unit for dispersing an inert gas such as N_2 or CO_2. The ratios of biocatalyst A and B to working volume were 0.15–0.30 and 0.05–0.10, respectively. The inert gas was introduced into the bioreactor through the sintered plate at 0.01 vol/vol/m and the liquefied starchy materials were continuously fed into the lowest unit of the bioreactor at pH 2.8–3.2, 30°C. The exhaust gas and effluent solution were discharged from an outlet located at the top of the bioreactor. The operating conditions are illustrated in TABLE 1, where F_L, V, V_S, τ, and τ_S indicate volumetric feed rate, working volume of bioreactor, biocatalyst volume, liquid residence time based on working volume, and liquid residence time based on biocatalyst volume, respectively.

RESULTS

At a residence time of 30 hours and pH 2.8, 10–11% (vol/vol) ethanol solution was produced from liquefied starchy materials of sweet potatoes over 30 days. From sweet potatoes, 12% (vol/vol) ethanol solution was produced at a residence time of 20 hours over 7 days. In this case, with high packing of biocatalyst, most of the dextrin was converted to sugar and ethanol. Seven percent (vol/vol) ethanol solution was obtained from barley for 50 days, and 7.6% (vol/vol) ethanol solution was produced from corn for 7 days.

These results demonstrated that this novel biosystem is capable of producing high concentrations of ethanol from liquefied starchy materials in a nonstop operation without contamination hazard.

SUMMARY

In a novel bioreactor that was packed with both Al alginate biocatalyst A entrapping glucoamylase and biocatalyst B entrapping a mixture of different strains of yeast cells, 7–12% (vol/vol) ethanol solution was continuously produced through

TABLE 1. Operating Conditions for the Bioreactor

Feed Starchy Materials		Total Volume of Biocatalyst	Biocatalyst A		Biocatalyst B		Residence Time	
Series	Dextrin Content in Feed Solution (kg/m^3)	$(V_S/V)_{A+B}$ $(-)^a$	$(V_S/V)_A$ (−)	$(\tau_S)_A$ (h)	$(V_S/V)_B$ (−)	$(\tau_S)_B$ (h)	τ (h)	pH (−)
Goto (Japan) dried sweet potatoes	185–200	0.20	0.15	4.5	0.05	1.5	30	2.8
		0.40	0.30	6.0	0.10	2.0	20	2.8
Japanese dried barley	130	0.24	0.17	3.4	0.07	1.4	20	3.2
American dried corn	130–140	0.21	0.16	3.2	0.05	1.0	20	3.2

[a] (−) indicates dimensionless.

parallel operation of saccharifying and fermenting liquefied starchy materials for a long period without contamination hazard. The nonstop operation was anaerobically performed at pH 2.8–3.2, 30°C, when it was maintained at liquid residence time of 20 to 30 hours over 30 days.

REFERENCES

1. FUKUSHIMA, S. & H. HATAKEYAMA. 1983. *In* Energy from Biomass, 2nd E.C. Conference. A. Strub, P. Chartier & G. Schleser, Eds. Applied Science, pp. 989–993. London and New York.
2. FUKUSHIMA, S., H. MUNENOBU & K. YAMADE. 1982. U.S. Patent 4337315.

Shifting Product Formation from Xylitol to Ethanol in Pentose Fermentation with *Candida tropicalis* by Altering Environmental Parameters[a]

ELKE LOHMEIER-VOGEL, BIRGITTA JÖNSSON, AND BÄRBEL HAHN-HÄGERDAL

Applied Microbiology
Chemical Center, Lund University
P. O. Box 740
S-220 07 Lund, Sweden

One problem in the utilization of lignocellulosic biomass for the production of liquid fuels such as ethanol is the fermentation of the hydrolyzed pentosan fraction, which can constitute as much as 37% of the raw material[1] and is composed mainly of the pentose, xylose. Several solutions for this problem have been suggested: the enzymatic conversion of xylose to xylulose followed by fermentation with *Saccharomyces cerevisiae*,[2] the use of thermophilic anaerobic bacteria such as *Thermoanaerobacter ethanolicus*[3] which can directly ferment xylose to ethanol, or the use of several yeast strains, for example, *Candida tropicalis*, which under "semiaerobic" conditions are also capable of carrying out this process.[4] Each of these routes, however, has a flaw of one kind or another. For example, xylose isomerase is of bacterial origin and has a pH and temperature optimum different from yeast, making a one-step process difficult. *T. ethanolicus* is substrate-inhibited, resulting in product concentrations too low for economical use. Finally, the xylose-fermenting yeasts tend to produce xylitol as a major by-product, thus not utilizing xylose efficiently.

This paper deals with the latter problem. Using *Candida tropicalis* ATCC 32113 (generously supplied by AC Biotechnics, Arlöv, Sweden), an attempt was made to shift product formation from xylitol to ethanol by altering aeration levels, with the use of the respiratory inhibitor, azide, and by decreasing the water activity in the medium.

"Semiaerobic" (or oxygen-limited) conditions have been found to enhance ethanol production from xylose in some yeast strains.[5] However, xylitol production is also enhanced in an oxygen-limited culture.[6] Azide has previously been shown to enhance ethanol production in *S. cerevisiae*[7] and the same was observed with a decreased water activity.[8]

TABLE 1 summarizes the results with *C. tropicalis*. In agreement with previous reports, the yield of ethanol is highest under "semiaerobic" conditions, 50% higher than under aerobic conditions. However, the xylitol production increases as well under these conditions. The semiaerobic conditions were chosen for further experiments with azide and with decreased water activities because *C. tropicalis* functions poorly anaerobically (little or no growth and poor utilization of xylose).

Azide was tested at two concentrations. 0.2 m*M* and 4.6 m*M* (TABLE 1). At the higher concentration, cell growth was inhibited and only 30% of the available xylose

[a]This study was supported by the Swedish Natural Research Council and the National Swedish Board for Energy Source Development.

was utilized, approximating the anaerobic situation. However, the ethanol yield was about the same as for the semiaerobic control, whereas xylitol production was repressed by 83%.

At an azide concentration of 0.2 m*M*, the available xylose was completely utilized. Ethanol production was enhanced by a factor of two and xylitol formation was repressed by 95%. Since cell growth was partially inhibited by this level of azide, the ethanol production per gram of cell dry weight increased by a factor of 2.5.

The present knowledge of xylose metabolism in yeast suggests that it is converted to xylitol via reduction by an NADPH-dependent D-xylose reductase, followed by oxidation to xylulose with an NAD-dependent xylitol dehydrogenase.[9] The oxygen requirement for ethanol production has been proposed to be necessary for cofactor regeneration.[10] The present results could therefore indicate that when respiration is slightly inhibited, the regeneration of NAD, which occurs through ethanol fermentation, is used to metabolize xylitol. However, further investigations are needed to fully understand xylose metabolism in *C. tropicalis*.

TABLE 1. Xylose Fermentation with *Candida tropicalis,* ATCC 32113, Cultured with Altered Concentrations of the Respiratory Inhibitor Azide, and with Decreased Water Activity (a_w)

Conditions	g EtOH / g Sugar	g EtOH / g Dry Weight	g Xylitol / g Xylose
Aerobic[a]	0.05	0.092	0.48
Semi-aerobic[b]	0.075	0.191	0.55
Anaerobic[c]	0.01	0.057	0.112
0.2 m*M* azide	0.141	0.480	0.032
4.6 m*M* azide	0.073	0.090	0.095
a_w			
6% PEG 0.994	0.109	0.312	0.424
12% PEG 0.992	0.126	0.335	0.425
18% PEG 0.989	0.131	0.353	0.378
21% PEG 0.987	0.139	0.407	0.337

[a]Aerobic conditions: 50 ml medium containing 1.3 g dry weight cells (initial) in a 1-l baffled flask; 200 rpm at 30°C.

[b]Semiaerobic conditions: 50 ml medium containing 5 g dry weight cells (initial) in a 250-ml baffled flask; 160 rpm at 30°C.

[c]Anaerobic conditions: 100 ml medium containing 1.3 g dry weight cells (initial) in a stoppered capped flask flushed with N_2 gas; grown at 30°C with slow stirring.

In the final experiments, the water activity in the fermentation medium was decreased by adding increasing concentrations of poly(ethylene glycol) 1540 (TABLE 1). The ethanol yield increased by about 80%, in agreement with what was found for *S. cerevisiae* in glucose fermentations.[8] The decreased water activity also seemed to switch the metabolism from cell growth to ethanol production as was seen by the almost twofold increase in ethanol output per gram of cell dry weight. This is also in agreement with what was seen for 0.2 m*M* azide (above) and with what has been reported for *Klebsiella pneumoniae*.[11] Xylitol formation is increasingly repressed as the water activity is decreased, but not as dramatically as with azide (38% is the best value obtained).

In conclusion, it is possible to shift the metabolism from xylitol formation to ethanol production, thereby increasing the yield of ethanol, by altering environmental parameters such as by adding a respiratory inhibitor or by lowering the water activity. For large-scale production of ethanol, however, these fermentation conditions do not seem realistic. The result may serve rather as a guide for which features to look for in a strain selection program.

REFERENCES

1. LADISCH, M. R., K. W. LIN, M. VOLOCH & G. T. TSAO. 1983. Process considerations in the enzymatic hydrolysis of biomass. Enzyme Microb. Technol. **5:** 82–102.
2. CHIANG, L-C., C-S. GONG, L-F. CHEN & G. T. TSAO. 1981. D-Xylulose fermentation to ethanol by *Saccharomyces cerevisiae.* Appl. Environ. Microbiol. **42:** 284–289.
3. CARREIRA, L. H. & L. LJUNGDAHL. Production of Ethanol from Biomass Using Anaerobic Thermophilic Bacteria. CRC Press. In press.
4. DU PREEZ, J. C. 1983. Fermentation of D-xylose to ethanol by a strain of *Candida shehatae.* Biotechnol. Lett. **5**: 357–362.
5. WANG, P. Y., C. SHOPSIS & H. SCHNEIDER. 1980. Fermentation of a pentose by yeast. Biochem. Biophys. Res. Commun. **94:** 248–254.
6. BAILLARGEON, M. W., N. B. JANSEN, C-S. GONG & G. T. TSAO. 1983. Effect of oxygen uptake rate on ethanol production by a xylose-fermenting yeast mutant, *Candida sp.* XF217. Biotechnol. Lett. **5:** 339–344.
7. HAHN-HÄGERDAL, B. & B. MATTIASSON. 1982. Azide sterilization of fermentation media. Eur. J. Appl. Microbiol. Biotechnol. **14:** 140–143.
8. HAHN-HÄGERDAL, B., M. LARSSON & B. MATTIASSON. 1982. Shift in metabolism towards ethanol production in *Saccharomyces cerevisiae* using alterations of the physical-chemical microenvironment. Biotechnol. Bioeng. Symp. **12:** 99–102.
9. BARNETT, J. A. 1968. *In* The Fungi. G. C. Ainsworth & A. S. Sussman, Eds. Vol. 3. Academic Press. New York.
10. HORECKER, B. L. 1962. *In* Pentose Metabolism in Bacteria. John Wiley & Sons. New York.
11. ESENER, A. A., G. BOL, N. W. F. KOSSEN & J. A. ROELS. 1981. *In* Advances in Biotechnology. M. Moo-Young, C. W. Robinson & C. Vezina, Eds. Vol. **1:** 339–344. Pergamon Press. Toronto.

Economics of Recycling Cellulases

HENRY R. BUNGAY

Rensselaer Polytechnic Institute
Troy, New York 12181

Two main considerations in obtaining sugars from lignocellulosic biomass are feedstock cost and the cost of enzymes. An excellent analysis by Ladisch *et al.*[1] expresses each cost per unit weight of glucose produced, and it may be inferred that enzymatic hydrolysis is uneconomic at the present cost of the enzymes. Yield is a function of time and enzyme dosage. If the concentration of enzyme is too low, hydrolysis is slow, and this increases the hazard of microbial contamination as well as raising equipment costs because larger tanks are needed for a given throughput.

Our group has had some success with recovering enzymes after hydrolysis, and it is of interest to estimate the economic impact. The assumptions of Ladisch *et al.*[1] were used, although the actual cost of cellulases will become much lower as better strains and better fermentation conditions are developed.

Cellulases can be recovered from the solid residue after hydrolysis by washing with dilute buffer.[2] Additional enzymes in the filtrate from hydrolysis can be adsorbed on fresh wood. Recycle is discussed by Sinitsyn *et al.*[3] One of the most important points is the ability to attain enzyme titers in the hydrolysis reactor that are higher than the titer of the fresh feed. Typical data for enzyme recovery after hydrolysis of steam-exploded wood from the Iotech process are in TABLE 1. Note that only two activities, endoglucanase and β-glucosidase are listed. We expect that exoglucanase would behave in a similar fashion to endoglucanase. In any event, preliminary tests of recycled enzymes show an unexpectedly high rate of hydrolysis of steam-exploded wood, so a detailed analysis of the kinetics of the various components and of blends of their activities with this substrate would have great value.

TABLE 1. Fate of Enzymes after Cellulose Hydrolysis[a]

Fraction	Percentage of Starting Activity
Hydrolysis filtrate	
Endoglucanase	30
β-Glucosidase	17
Extracted from residue	
Endoglucanase	47
β-Glucosidase	31
Missing from filtrate after contact with fresh exploded wood	
Endoglucanase	14
β-Glucosidase	6

[a]Conditions: 130-h hydrolysis, 10 percent of washed steam-exploded wood, about 2 IU/ml filter paper activity.[2]

FIGURE 1 shows that the hydrolysis step uses fresh feedstock, fresh enzyme solution, and recovered enzyme in two forms: a solution extracted from spent residue, and enzymes on fresh feedstock that has been contacted with spent filtrate. Enzyme

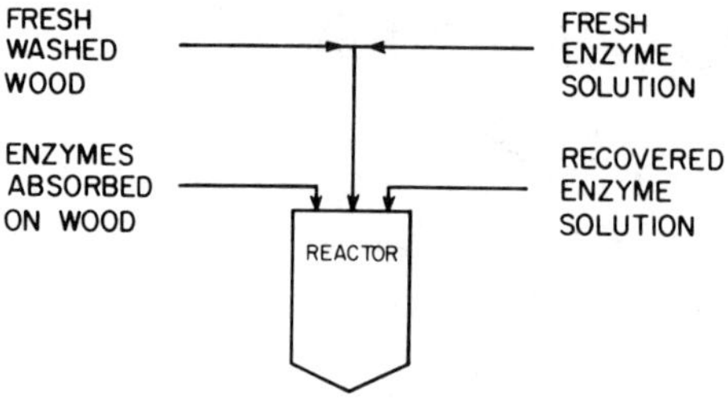

FIGURE 1. Streams entering hydrolysis reactor.

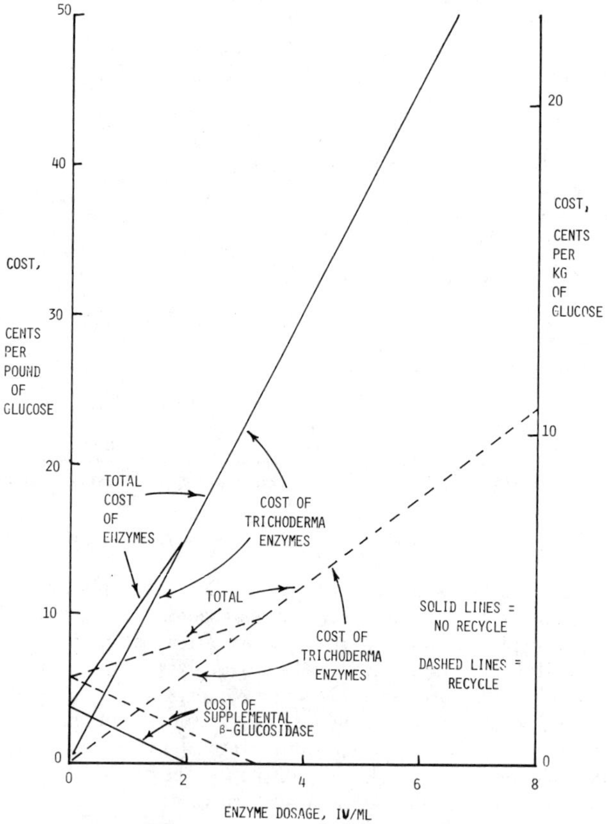

FIGURE 2. Enzyme costs versus dosage.

recycle causes a shift in the relative proportions of enzymatic activities in the reactor because the recoveries are different. For example, β-glucosidase is not recovered very well and will be in lowered proportions in the blend of fresh and recycled enzymes.

A simple mass balance for a single type of enzymatic activity is:

$$\text{feed} + \text{recycle} = \text{enzyme in reactor}$$

$$F + YC = C$$

where F = enzyme in fresh feed; Y = recovery yield for enzyme; and C = enzyme in reactor. Solving for C:

$$C = F/(1 - Y)$$

Using the recovery data in TABLE 1:

$$\text{endoglucanase in reactor} = 2.56 \text{ times fresh feed}$$

$$\beta\text{-glucosidase in reactor} = 1.59 \text{ times fresh feed}$$

Although there is scant information on the level of β-glucosidase that is needed, we will assume that this activity must be supplemented with additional enzyme from another source, such as a separate fermentation with the mold *Aspergillus phonecius*.

Rough calculations following the approach of Ladisch *et al.*[1] gave the results shown in FIGURE 2. The range of enzyme dosages is about the same as has been used by various investigators, but the dosage used in the analysis by Ladisch *et al.*[1] would be close to the origin on this plot.

The bases for the calculations were: cellulase costs $20 per million units; β-glucosidase costs half as much; at least 0.5 IU/ml of β-glucosidase is essential; and glucose yield is 120 g/l. There is β-glucosidase in with the cellulase complex from a *Trichoderma reesei* fermentation, and a ratio of 1 unit per 2 IU of filter paper activity was assumed.

Although recycle significantly reduces the cost of enzymes, supplementation with β-glucosidase from a separate fermentation is relatively expensive at low cellulase dosages. It is obvious that cellulase at a cost of $20 per million units is too expensive for any chance of profitability unless the dosage is unrealistically low. A company that produces its own cellulases will have a cost much lower than this, and better pretreatment of the cellulosic feedstock may permit lower dosages of enzymes. In any event, recycle will lower costs and make it easy to achieve the required dosage even if the enzyme titer in the fresh feed is inadequate.

REFERENCES

1. LADISCH, M. R., K. W. LIN, M. VOLOCH & G. T. TSAO. 1983. Process considerations in the enzymatic hydrolysis of biomass. Enzyme Microb. Technol. **5:** 82–102.
2. SINITSYN, A. P., M. L. BUNGAY, L. S. CLESCERI & H. R. BUNGAY. 1983. Recovery of enzymes from the insoluble residue of hydrolyzed wood. Appl. Biochem. Biotechnol. **8:** 25–29.
3. SINITSYN, A. P., H. R. BUNGAY & L. S. CLESCERI. 1983. Enzyme management in the Iotech process. Biotechnol. Bioeng. **25:** 1393–1399.

A Model for Hydrolysis of Microcrystalline Cellulose by Cellulase

R. MATSUNO, M. TANIGUCHI, M. TANAKA, AND T. KAMIKUBO

Department of Food Science and Technology
Faculty of Agriculture
Kyoto University
Sakyo-Ku, Kyoto 606, Japan

The kinetics of the heterogeneous cellulase reaction with microcrystalline cellulose (MCC) is closely related to the adsorption phenomenon[1-4] of the enzyme onto MCC. The rapid decrease in the reaction rate with reaction time might also be ascribable to the adsorption phenomenon. To elucidate the mechanism of the slowing down of the reaction rate, the rate of solubilization and the behavior of adsorption of enzymes to the substrate were observed during the course of hydrolysis of cellulose under various experimental conditions by purified and crude enzymes from *Pellicularia filamentosa,*[5] *Eupenicillium javanicum,*[6] and *Trichoderma viride.*

When purified enzyme having low catalytic activity (because of the absence of synergistic action such as B-4 component) from *E. javanicum*[6] was used, the enzyme adsorbed almost instantaneously on MCC, was desorbed gradually until 10 h and adsorbed again after 10 h at an extremely slow rate as shown in FIGURE 1. The ordinate represents the amount of total sugar solubilized and the degree of adsorbed enzyme calculated from the amount of protein remaining in the supernatant solution of the reaction mixture. Essentially the same results were obtained for other purified enzymes from three microorganisms used in this study. The slow adsorption found in the later period suggests that the enzyme is adsorbed not only on the surface of the cellulose fibrils, but that it diffuses into the inside of the fibrils and is adsorbed there also.

Contrary to the result with pure enzymes, when a crude enzyme preparation, Meicelase from *T. viride* (Meiji Seika, Japan) having high catalytic activity because of synergistic actions, was used, the enzyme adsorbed instantaneously and desorbed gradually during the entire course of the reaction with decreasing reaction rate as shown in FIGURE 2. Experiments were devised to make clear that a part of the enzymes preparation diffuses into the inside of the fibrils even in the hydrolysis by the crude enzyme. Based on the fact that the dependency of the rate of diffusion on the temperature is much lower than that of the rate of enzyme reaction, two experiments were performed. In one experiment with 2.0% MCC and 0.4 mg/ml Meicelase, the hydrolysis reaction proceeded at 30°C while in the other, at 3°C for 96 h when the temperature was shifted up to 30°C. The amount of total sugar solubilized reached 10 mg/ml after 96 h with 20% enzyme adsorption for the former experiment and after 144 h with 50% for the latter. The high degree of enzyme adsorption for the latter experiment may indicate that the crude enzyme diffused into the cellulose fibrils and adsorbed while the reaction temperature was kept low. The soluble products of hydrolysis might affect the rate of hydrolysis and the degree of adsorption. To elucidate the effect of soluble products, the hydrolysis was carried out in a vessel equipped with an ultrafiltration membrane of cutting molecular weight of 10K (Amicon, PM 10), by supplying buffer solution and removing the solution containing

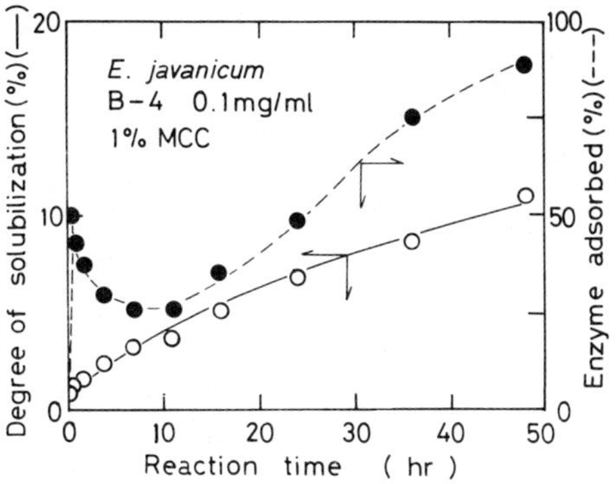

FIGURE 1. Time courses of solubilization of MCC and adsorption of enzyme on MCC during hydrolysis by purified cellulase. Experimental conditions: 30°C and pH 5.0 (acetate buffer).

only soluble products. The result (Exp. 2) is compared in TABLE 1 with that of a batch experiment taken as a control (Exp. 1). Comparing the amount of total sugar solubilized, it is obvious that product inhibition is significant. The rapid decrease in the activity per amount of enzyme adsorbed with reaction time in both experiments indicates that the enzyme adsorbed inside the fibril shows low activity. The amount of enzyme adsorbed per remaining cellulose increases with reaction time for experiment 2 while it decreases slightly in the control experiment. This suggests that the soluble products promote the desorption of enzyme.

On the basis of these experimental results, we proposed the following model for the kinetics: (1) Cellulose fibril is a cylindrical bundle of highly crystallized microfibrils. (2) Cellulase is adsorbed specifically on the surface of the bundle to form enzyme-substrate complex and hydrolyzes cellulose at a high rate. (3) Cellulase diffuses very slowly into the narrow spaces between the microfibrils and is adsorbed specifically on the surface of microfibrils inside the bundle. However, the rate of hydrolysis is low. (4) Cellulase once adsorbed inside the fibrils is desorbed as the reaction proceeds, since the adsorption sites are lost by hydrolysis. (5) Soluble products not only inhibit the hydrolysis but also promote the desorption by specific binding to cellulase. From this

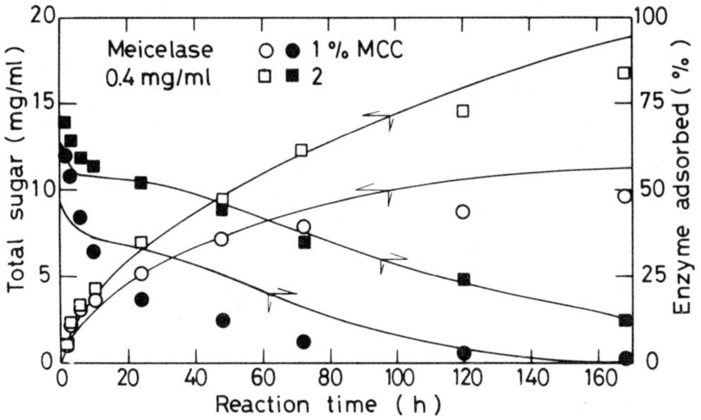

FIGURE 2. Time courses of solubilization of MCC and adsorption of enzyme on MCC during hydrolysis by crude cellulase. Solid curves show the calculated results by the model proposed. Experimental conditions: 30°C and pH 5.0 (acetate buffer).

TABLE 1. Effect of Soluble Products on Rate of Hydrolysis and Adsorption of Enzyme[a]

Time (hours):		1	3	6	10	24	48	72
Total sugar (mg/ml)	Exp. 1	1.12	1.97	2.74	3.40	5.30	6.76	7.98
	Exp. 2	2.00	2.10	4.30	5.1	9.7	15.0	17.7
Enzyme adsorbed/remaining cellulose (μg/mg)	Exp. 1	12.6	12.7	12.8	12.1	10.8	10.7	10.1
	Exp. 2	13.0	12.8	14.1	13.6	14.5	20.9	30.1
Enzyme activity/protein adsorbed (mg/h/mg protein adsorbed)	Exp. 1	2.15	—	0.9	0.7	0.5	0.35	0.25
	Exp. 2	3.65	—	1.75	1.30	1.30	0.9	0.6

[a]Experiment 1: Control experiment. Experiment 2: Experiment with supplying buffer solution and removing solution of soluble products. Experimental conditions: 30°C and pH 5.0 (acetate buffer).

model, contradictory results with respect to adsorption of enzyme found in pure and crude enzyme systems can be explained as follows. In the pure enzyme system, mechanism (3) predominates, while in the crude enzyme system, the effects of (4) and (5) are greater than that of (3). The rapid decrease in reaction rate is attributable to the effects of (3) and (5). The results of preliminary calculations based on this model are shown by solid curves in FIGURE 2. The model suggests that the rapid decrease in the reaction rate might be reduced by avoiding diffusion of enzyme into the cellulose fibril. The crude enzymes from *P. filamentosa* were crosslinked by glutaraldehyde treatment to increase the molecular size and to reduce the diffusion into the fibril. Although the initial reaction rate was decreased by crosslinking, the decrease in reaction rate was reduced and finally the amount of total sugar solubilized by the crosslinked enzyme was even higher than that by free enzyme.

REFERENCES

1. CASTANON, M. & C. R. WILKE. 1980. Adsorption and recovery of cellulases during hydrolysis of newspaper. Biotechnol. Bioeng. **22:** 1037–1053.
2. BELTRAME, P. L., P. CARNITI, B. FOCHER, A. MARZETTI & M. CATTANEO. 1982. Cotton cellulose: Enzyme adsorption and enzymatic hydrolysis. J. Appl. Polym. Sci. **27:** 3493–3502.
3. LEE, S. B., H. S. SHIN, D. D. Y. RYU & M. MANDELS. 1982. Adsorption of cellulase on cellulose: Effect of physicochemical properties of cellulose on adsorption and rate of hydrolysis. Biotechnol. Bioeng. **24:** 2137–2153.
4. LEE, Y.-H. & L. T. FAN. 1982. Kinetic studies of enzymatic hydrolysis of insoluble cellulose: Analysis of the initial rates. Biotechnol. Bioeng. **24:** 2383–2406.
5. TANAKA, M., S. TAKENAWA, R. MATSUNO & T. KAMIKUBO. 1977. Purification and properties of cellulases from *Pellicularia filamentosa*. J. Ferment. Technol. **55:** 137–142.
6. TANAKA, M., M. TANIGUCHI, R. MATSUNO & T. KAMIKUBO. 1981. Purification and properties of cellulases from *Eupenicillium javanicum*. J. Ferment. Technol. **59:** 177–183.

Solid Superacids for Hydrolyzing Oligo- and Polysaccharides[a]

BÄRBEL HAHN-HÄGERDAL AND KERSTIN SKOOG

Applied Microbiology
Chemical Center, Lund University
P. O. Box 740
S-220 07 Lund, Sweden

BO MATTIASSON

Pure and Applied Biochemistry
Chemical Center, Lund University
P. O. Box 740
S-220 07 Lund, Sweden

The current interest in utilizing carbohydrates as raw materials for producing chemicals and fuels via various fermentation routes has focused interest on cheap and efficient hydrolysis processes for the production of fermentable sugars. The hydrolysis of cellulose, especially, has received much attention, since cellulose is the most abundant source of renewable carbohydrates.[1]

The present technology essentially offers two ways for cellulose hydrolysis: acid and enzymatic. The former is fast but has the drawback of low yields. Furthermore, the corrosion problem has not yet been fully solved. The latter can give theoretical yields, but is slow and subject to severe product inhibition, particularly when the intermediate cellobiose accumulates.

This paper evaluates a third possibility: the utilization of solid superacids, perfluorinated resin–sulfonic acid catalysts, NAFION[R] 501, generously supplied by E. I. du Pont de Nemours & Company, Wilmington, DE.

Solid superacids have been successfully utilized in a number of reactions such as alkylations[2] and esterifications.[3] Their utilization in hydrolysis resembles an acid hydrolysis, but the fact that they are solid makes it possible to remove them as well as to operate the hydrolysis process continuously in a plug-flow-type reactor.

The hydrolysis of cellulose, starch, and cellobiose were first compared in a batch reaction (FIG. 1). The production of glucose from cellobiose and starch followed zero-order kinetics. No tendency for product inhibition was observed. However, cellulose was barely hydrolyzed. The difference in the results between starch and cellulose is probably due to the fact that starch is gelatinized at the experimental temperature used, 95°C, thus opening up the molecules to the catalyst more than for cellulose. In this way, steric hindrances imposed by both substrate and catalyst appear to prevent the hydrolysis of cellulose with solid superacids.

The manufacturer recommends that the solid superacids be used in the presence of sodium chloride in order to enhance the reaction rate. In repeated batch hydrolysis experiments, it was found that the presence of sodium chloride released hydrogen ions into the medium, the effect being that the superacids lost their catalytic activity after

[a]This study was supported by the Swedish Natural Research Council and the National Swedish Board for Energy Source Development.

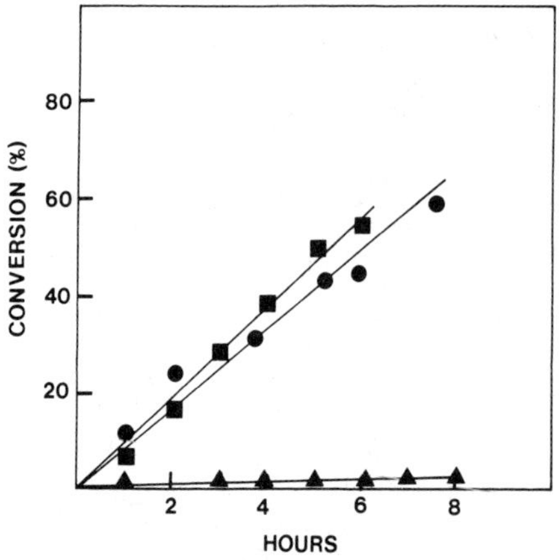

FIGURE 1. Hydrolysis of (–●–) D-(+)-cellobiose (Sigma Chemical Company, St. Louis, MO, (–■–) starch (KEBO, Stockholm, Sweden), (–▲–) cellulose, Solka floc BW 200, (a generous gift from James River Corp., Berlin, NH). 100 g/l substrate, 95°C, 10 g/l sodium chloride. Conversion expressed as glucose concentration over initial substrate concentration. Glucose determined enzymatically.[4]

one or two batch experiments depending on the concentration of sodium chloride. Therefore, in the continuous hydrolysis experiment, no sodium chloride was added.

A small plug-flow-type reactor containing solid superacids was continuously operated for one month (FIG. 2). It was fed cellobiose at two concentrations and sucrose at one concentration. Sucrose was included in the study because cellobiose has a limited solubility in water. Irrespective of concentrations, a 50% yield was obtained with all three substrates. The fact that even at the highest substrate concentration used during the last period of operation the same yield was obtained indicates that the superacids do not lose their activity under the operational conditions chosen in this investigation. The consistency in yield indicates that neither the size of the superacid bed nor the flow rate seem to be the reason for the incomplete conversion of the substrate. The solid superacids appear to have an inherent kinetic feature that results in only half the substrate being hydrolyzed. The mechanism for this remains to be understood.

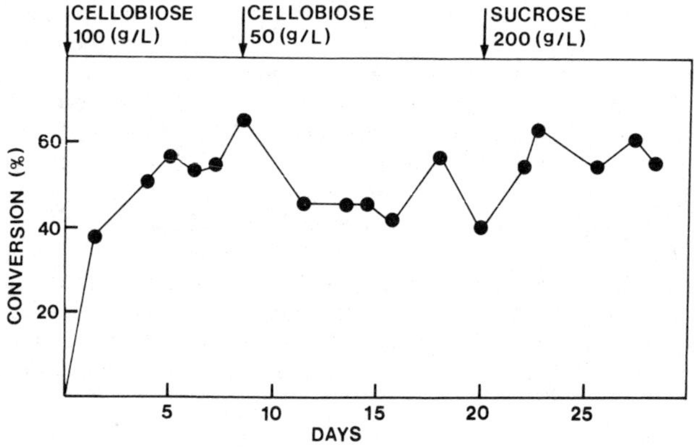

FIGURE 2. Continuous hydrolysis of cellobiose and sucrose. 95°C. Column packed with 2 g of solid superacids. Flow rate 0.6 ml/h. No sodium chloride added.

Both starch and cellobiose hydrolysates from the above experiments were then fermented to ethanol with *Saccharomyces cerevisiae*. Theoretical yields were obtained, indicating that the hydrolysis step using solid superacids had not resulted in any toxic by-products that could inhibit subsequent fermentation with yeast.

In conclusion, solid superacids may be utilized in batch as well as for continuous hydrolysis of soluble carbohydrates. The catalysts can be recovered and are not subject to product inhibition. A technical-economical evaluation comparing this process with enzymatic and acid hydrolysis of carbohydrates remains to be made.

REFERENCES

1. LADISCH, M. R., K. W. LIN, M. VOLOCH & G. T. TSAO. 1983. Process considerations in the enzymatic hydrolysis of biomass. Enzyme Microb. Technol. **5:** 82–102.
2. OLAH, G. A., J. KASPI & J. BUKALA. 1977. Heterogeneous catalysis by solid superacids. 3. Alkylation of benzene and transalkylation of alkylbenzenes over graphite-intercalted Lewis acid halide and perfluorinated resin-sulfonic acid (Nafion-H) catalysts. J. Org. Chem. **42(26):** 4187–4191.
3. OLAH, G. A., T. KAUMI & D. MEIDAR. 1978. Synthetic methods and reactions: 51. A convenient and improved method for esterification over Nafion-H, a superacid perfluorinated resinsulfonic acid catalyst. Synthesis. December: 929–930.
4. TRINDER, P. 1969. Determination of glucose in blood using glucose oxidase with an alternative oxygen acceptor. Ann. Clin. Biochem. **6:** 24–27.

Kinetics of Acid-catalyzed Cellulose Hydrolysis under Low Water Conditions

S. K. SONG & Y. Y. LEE[a]

Department of Chemical Engineering
Auburn University
Auburn, Alabama 36849

Acid-catalyzed hydrolysis of cellulose is one of the major process schemes currently being developed as a part of integral biomass conversion processes. The technology as it stands is confronting a number of technical problems, which include low yield, low concentration, and poor fermentability of the sugar product. Substantial research efforts have been put forward in the past with regard to yield improvement.[1-6] In terms of economics and energy efficiency, however, the concentration of the sugar product is also vitally important. In this study, an acid-catalyzed cellulose hydrolysis process aimed at high sugar concentration as well as high yield was investigated.

In order to attain high sugar concentration one must reduce the amount of water in the reactor as much as possible while assuring that solid feed is uniformly in contact with liquid. From considerations of the microstructure of woody biomass and the results of preliminary tests, it was found feasible to carry out the reaction with a solid/liquid ratio of 1 to 1.6.

This would mean that the liquid (acid solution) fills only the pores within the wood structure leaving the external reactor voidage essentially free of water. Since the water content is at such a low level, one can afford to use a much higher level of acid than is used in dilute sulfuric acid processes. A kinetic study was therefore undertaken to investigate the range of 4 ~ 12% sulfuric acid concentrations. This also reflects the expectation that a higher kinetic yield would be obtained using a higher acid level, as indicated in earlier studies on dilute acid hydrolysis.[1-3] In terms of reaction temperature, the range of 170 ~ 190°C was chosen, the upper level being limited by experimental constraints.

It has been shown that the acid hydrolysis of cellulose can be represented by the following sequential reaction in which each reaction is of first order:

$$\text{cellulose} \xrightarrow{1} \text{glucose} \xrightarrow{2} \text{decomposed products}$$

The kinetic investigation was carried out by repeated batch experimentation using hemicellulose pre-extracted hardwood particles. The sample batch reaction data are shown in FIGURES 1 and 2. Several findings were noteworthy. Clearly, the glucose concentration in the product increases with reaction temperature. It was also confirmed that high sulfuric acid concentration favors high glucose concentration, and thus the yield. Particularly significant however, is that glucose concentrations up to 17% and corresponding yields of 55% are achievable in the low water hydrolysis scheme, thus meeting our primary objective.

The experimental results were further analyzed to verify the kinetic model. To account for the effects of temperature and acid concentration, the rate constants were

[a]To whom all correspondence should be addressed.

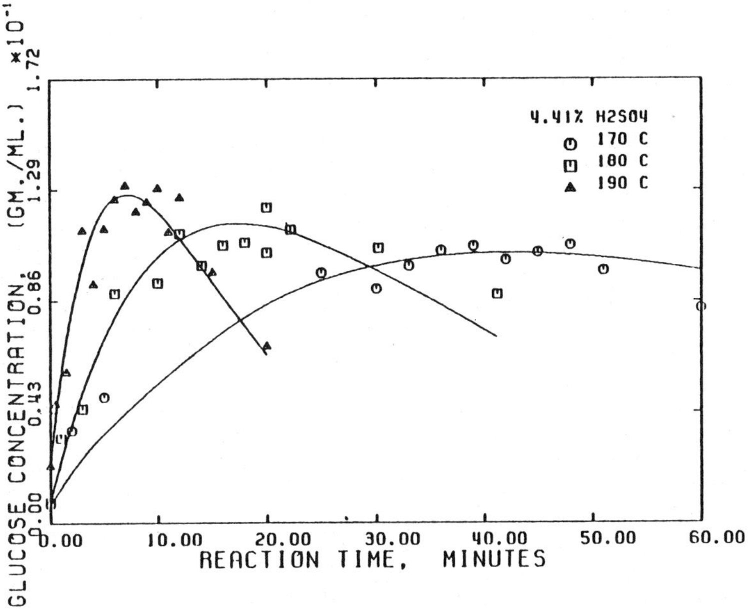

FIGURE 1. Comparison of experimental data and model prediction in batch hydrolysis of cellulose at 4.41% H_2SO_4 (--: model value).

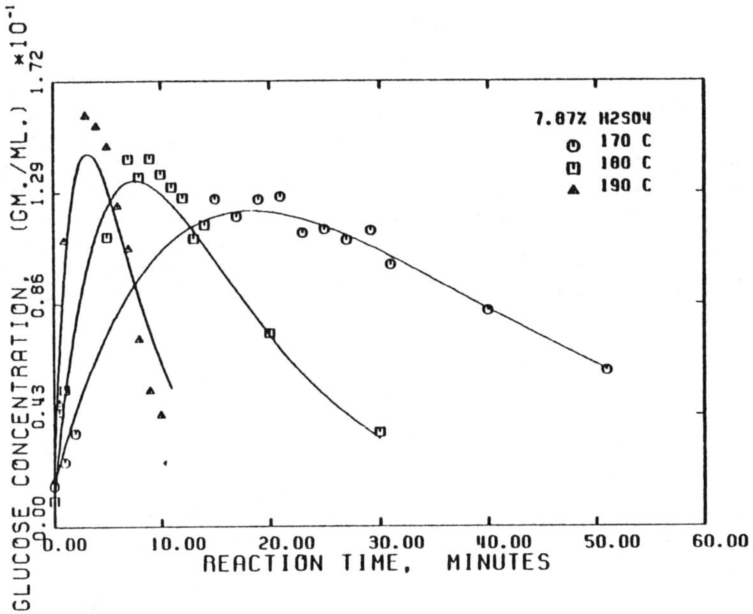

FIGURE 2. Comparison of experimental data and model prediction in batch hydrolysis of cellulose at 7.87% H_2SO_4 (--: model value).

expressed by an Arrhenius-type equation with an acid term such that:

$$k_i = k_{io}(s)^{ni} \exp\left(-\frac{E_i}{RT}\right)$$

where, s = sulfuric acid concentration in wt %; k_{io} = frequency factor; E_i = activation energy; n_i = acid concentration exponent; and i = reaction index, 1 for hydrolysis, 2 for decomposition. The kinetic parameters in this reaction model were simultaneously determined employing data from nine experimental runs. A computer program for nonlinear regression analysis was set up for this purpose.[7,8] The resulting kinetic parameters are shown in TABLE 1. The tendency of increasing yield with reaction temperature and acid concentration is reconfirmed here by the fact that the activation energy and acid exponent for the hydrolysis reaction are substantially higher than those for the decomposition reaction. The kinetic parameters are generally in accordance with those previously reported for dilute acid hydrolysis,[1-3] indicating that there is no significant change in reaction mechanism under low water and high acid conditions. As shown in FIGURES 1 and 2, the agreement between the experimental data and the model prediction was satisfactory. The kinetic model and associated parameters provided in this study should thus be a useful tool in studies involving acid hydrolysis of cellulose.

TABLE 1. Kinetic Parameters in Hydrolysis of Hardwood Cellulose[a]

i	k_{io} (min)$^{-1}$ (% acid)$^{-n_i}$	n_i	E_i (cal/g-mole)
1	6.6×10^{16}	1.64	39,500
2	6.4×10^{12}	1.10	30,800

[a]Data coverage: 170–190°C; 4.41–12.19% H_2SO_4.

In the foregoing kinetic investigation, it was proved that the low water hydrolysis scheme is quite satisfactory in providing high glucose concentration and yield. In order to make it a workable system, however, an efficient leaching mechanism must be incorporated to recover the sugar product without significant dilution. To this end, use of the counter-current leaching process was investigated. The actual experiments were performed using a multi-stage cascade extractor known as the shanks system.[9] The results have proved that the sugar trapped in biomass can be recovered at about 15% concentration. Further analysis of the data indicated that 90% of sugar recovery and 85% of concentration retainment could be achieved using this system.

REFERENCES

1. THOMPSON, D. R. & H. E. GRETHLEIN. 1979. Evaluation of a plug-flow reactor for acid hydrolysis of cellulose. Ind. Eng. Chem. Prod. Res. Dev. **18:** 166–169.
2. CHURCH, J. A. & D. WOODRIDGE. 1981. Continuous high-solids acid hydrolysis of biomass in a 1½-inch plug-flow reactor. Ind. Eng. Chem. Prod. Res. Dev. **20:** 371–378.
3. SAEMAN, J. F. 1945. Kinetics of wood saccharification hydrolysis of cellulose and decomposition of sugars in dilute acid at high temperature. Ind. Eng. Chem. **37:** 43–52.
4. SONG, S. K. & Y. Y. LEE. 1982. Counter-current reactor in acid-catalyzed cellulose hydrolysis. Chem. Eng. Commun. **17:** 23–30.
5. FAGAN, R. D., H. E. GRETHLEIN, A. O. CONVERSE & A. PORTEOUS. 1971. Kinetics of the acid hydrolysis of cellulose found in paper refuse. Environ. Sci. Technol. **5:** 545–547.

6. Rugg, B. 1981. Optimization of twin screw reactor. SERI Contractors Conference. Golden, Colorado.
7. Helwig, J. T. & K. A. Council. 1979. SAS User's Guide. SAS Institute. Cary, N.C. p. 317.
8. Ralston, M. L. & R. I. Jennrich. 1978. Dud, a derivative-free algorithm for nonlinear least squares. Technometrics **20:** 7–14.
9. King, C. J. 1980. Multistage separation processes. *In* Separation Processes. McGraw-Hill. New York. p. 172.

Effects of Compaction on the Effectiveness Factor of a Gel Particle Containing Enzyme

K. UEYAMA AND S. FURUSAKI

Department of Chemical Engineering
University of Tokyo
Tokyo, Japan 113

The packed column with gel particles that contain enzyme is one of the useful industrial immobilized enzyme reactors. Often the columns are subjected to compaction of gel particles due to their own weight or pressure drop through the column. This compaction reduces the total surface area of gel particles in contact with fluid. This effect directly decreases the overall reaction rate inside the reactor. In this work, the effects of compaction on the effectiveness factor of an enzyme-containing gel particle were investigated experimentally.

The effects of compaction on the effectiveness factor were experimentally investigated by using the hydrolysis reaction of sucrose into glucose and fructose by invertase immobilized in κ-carrageenan gel particles. Although this reaction is essentially a Michaelis-Menten-type reaction, the values of the concentration of sucrose were kept sufficiently small so that the reaction kinetics could be approximated by a first-order reaction.

TABLE 1. Characteristics of Gel Particles

Run	D_p (m)	k_1 (sec^{-1})	Φ_1	η_1
1	1.03×10^{-3}	2.33×10^{-3}	1.2	0.92
2	1.04×10^{-3}	6.02×10^{-3}	1.9	0.82
3	1.00×10^{-3}	10.2×10^{-3}	2.6	0.72

The rate constant for each gel particle was varied by changing the concentration of invertase immobilized in the gel particle. The values of the average diameter of gel particles D_p, the rate constant k_1, the Thiele modulus Φ_1, and the effectiveness factor assuming no compaction η_1 are tabulated in TABLE 1. The Thiele modulus Φ_1 is defined by Equation 1:

$$\Phi_1 = R\sqrt{k_1/D} \quad (1)$$

where R is the radius of the gel particle and D is the effective diffusion coefficient in the particle.

The value of the effectiveness factor of the gel particle under compaction η_1' was determined by using Equation 2.

$$\ln(1 - X) = -\frac{\eta_1' k_1 V_p}{F} \quad (2)$$

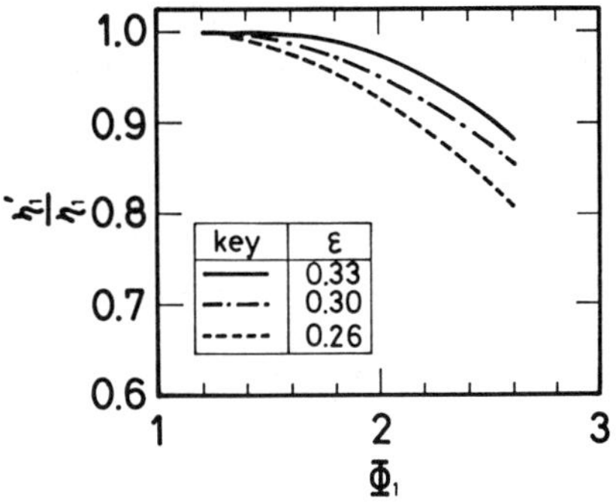

FIGURE 1. Experimental results of the effects of compaction.

where X, V_p, and F are the conversion, the volume of the particle, and the flow rate, respectively.

The values of the void fraction of a packed column were varied by changing the load imposed to the packed layer. Experiments were carried out for constant Φ_1 in TABLE 1 and different values of ϵ. Experimental values of η_1'/η_1 for $\epsilon = 0.33$, 0.30, and 0.26 were estimated by interpolation.

Experimental results thus obtained are plotted in FIGURE 1, which shows the effects of compaction on the effectiveness factor, using ϵ as a parameter.

Theoretically, the effects of compaction on the effectiveness factor were predicted as shown in FIGURE 2 by assuming that a gel particle has m flat circular surfaces contacting to adjacent particles and that each contact surface has the same size.[1] According to this theoretical result, the most important parameter was γ, which is the ratio of total contact areas to the original surface area of a gel particle before compaction.

By comparing FIGURE 1 with FIGURE 2, several interesting features can be seen.

(1) Both experimental and theoretical results show that the effects of compaction become significant for the Thiele moduli larger than unity.

(2) For $1.2 \leq \Phi_1 \leq 2.6$, FIGURE 1 and FIGURE 2 show quite similar behavior.

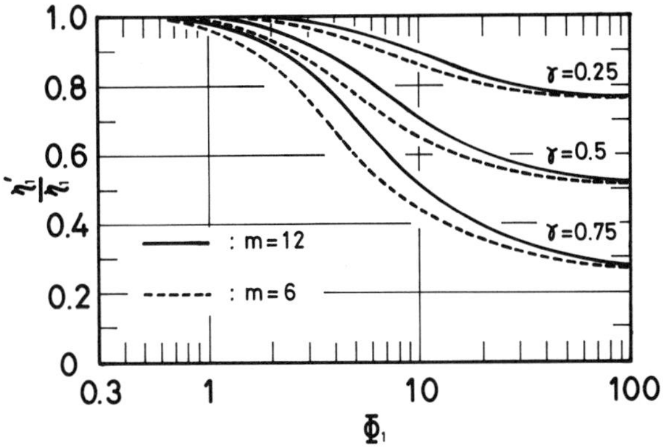

FIGURE 2. Theoretical results for change of the effectiveness factor by compaction.[1]

(3) In order to discuss the agreement between the theory and the experiments, knowledge about the relations among γ, m, and ϵ is necessary. The experimental results of the effects of compaction on the effectiveness factor agree very well with the theoretical result for $1 \leq \Phi_1 \leq 2.6$, if we assume that the values of ϵ of 0.33, 0.30, and 0.26 correspond to the values of γ of 0.5, 0.7, and 0.75, respectively.

REFERENCE

1. K. UEYAMA & S. FURUSAKI. A theoretical study on the effect of compaction on the effectiveness factor of gel particles containing immobilized enzymes. Submitted to Chemical Engineering Communications.

NOMENCLATURE

D = effective diffusion coefficient in a gel particle, m^2sec^{-1}
D_p = average diameter of gel particles, m
F = flow rate, m^3sec^{-1}
k_1 = rate constant of first-order reaction, sec^{-1}
m = number of contact surfaces on a particle
R = radius of a particle, m
V_p = volume of a particle, m^3
X = conversion
γ = ratio of the total contact area on a particle to the original surface area of a particle before compaction
ϵ = void fraction of a packed layer
η_1 = effectiveness factor of a particle without compaction
η_1' = effectiveness factor of a particle with compaction
Φ_1 = Thiele modulus for first-order reaction

PART III. ENZYMES IN PREPARATIVE CHEMISTRY

Biohydrogenation and Electromicrobial and Electroenzymatic Reduction Methods for the Preparation of Chiral Compounds[a]

H. SIMON, J. BADER, H. GÜNTHER,
S. NEUMANN, AND J. THANOS

Lehrstuhl für Organische Chemie und Biochemie
Technische Universität München
D-8046 Garching, Federal Republic of Germany

Starting from suitably substituted prochiral compounds, stereospecific reductions can lead to many chiral substrates (Eq. 1).

$$\begin{matrix} R^2 \\ R^1 \end{matrix}\!\!> = X + 2\,[H] \rightarrow H\!\!\begin{matrix} R^2 \\ | \\ R^1 \end{matrix}\!\!-XH \qquad (1)$$

There are about 190 reductases known catalyzing the reaction with X = 0 and NAD(P)H as the electron donor. Besides these pyridine nucleotide-dependent enzymes, there are only about 20 oxidoreductases operating with cosubstrates or prosthetic groups different from pyridine nucleotides, cytochromes, or oxygen. But even those are partially not reductases. About 35 pyridine nucleotide-dependent reductases are known that catalyze the reduction of C-C double bonds. For the enzymes of the nomenclature subgroup 1.3.99, it is often not clear whether they are able to reduce C-C double bonds. Looking at these facts, it is clear that pyridine nucleotide-dependent reductases play the most important role in applying biocatalysts for the synthesis of chiral compounds via reductions.

There are a huge number of interesting and also some effective methods described in the literature for the preparation of compounds of high optical purity by bioreductions. One can differentiate the use of whole cells[1,2] or that of isolated enzymes.[3–6] However, there is a discrepancy, if one looks at the apparently enormous possibilities of bioreductions on the one hand and at the low degree of general application in laboratories or of technological realization on the other hand. What are the reasons? Besides the complexity of oxidoreductases and their limited activity and stability compared to hydrolases or isomerases there is mainly the problem of NAD(P)H regeneration. Also the limited chemical stability of the pyridine nucleotides and especially their enzymatic degradation, which takes place if one does not work with pure oxidoreductases, can be serious drawbacks. So far no suitable chemical or electrochemical means seems to be known for the regeneration of NAD(P)H.

There are two main methods for NAD(P)H regeneration. The first procedure is to use isolated enzymes. Even though highly sophisticated systems are known,[5,7,8] it is a matter of fact that two enzymes, two substrates, and two products besides the pyridine

[a]This work was supported by Sonderforschungsbereich 145 of Deutsche Forschungsgemeinschaft and Fonds der Chemischen Industrie.

nucleotide in its oxidized and reduced form are involved. Such systems are already rather complex and usually reveal rather complicated kinetic behavior. The second procedure is an implicit NAD(P)H regeneration in the form of cofermentations.[1,9,10] This means that a microorganism, such as a yeast, converts an electron donor, for example glucose, in such a way that endogenous NAD(P) is reduced to NAD(P)H as an intermediate. Two problems often prevent a satisfying productivity. In the first problem, the electron donor glucose forms, for every molecule of NADH, one molecule of an electron acceptor. This acceptor, for instance, could be acetaldehyde, and/or under aerobic conditions, the NADH may be consumed in the respiratory chain. In the second problem, if resting cells are used, the endogenous pyridine nucleotide is enzymatically degraded. According to measurements, which we conducted with crude extracts of *Candida utilis* and *Saccharomyces cerevisiae,* NAD is degraded with a rate of about 0.5–1 nmol/min × mg protein.[11] In order to compare different procedures performed with cells, we suggested the use of a productivity number, PN,[12] where

$$PN = \frac{\text{mmol product}}{\text{catalyst(kg dry weight)xh}} \qquad (2)$$

Most PN values from the literature are in the region of 5–200. From these considerations, one should postulate the use of *non*pyridine nucleotide-dependent reductases and/or of electron donors that do not have the intrinsic disadvantage of forming obligatorily one mol of a competitive electron acceptor per mol of NAD(P)H.

BIOHYDROGENATIONS

A possibility for meeting one or both of the aforementioned postulates is to use hydrogen gas as an electron donor, according to Equation 3.

$$\rangle\!\!=\!X + H_2 \xrightarrow{\text{microorganism}} H\text{—}X\text{—}H \qquad (3)$$

Of course the prerequisite for the hydrogenation is the flow of electrons from hydrogen to the substrate. In clostridia, this electron transport takes place as shown in FIGURE 1.

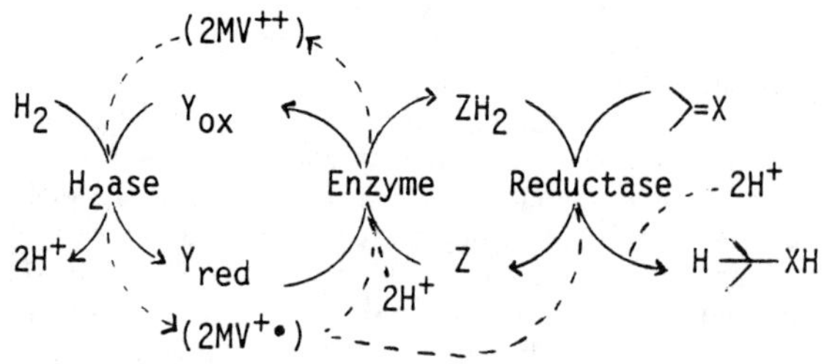

FIGURE 1. The flow of electrons from molecular hydrogen to an unsaturated compound $\rangle = X$. In clostridia, hydrogenase (H_2ase) catalyzes the reduction of ferredoxin (Y) or methylviologen (MV^{2+} = 1.1′-dimethyl-4,4′-bipyridinium dication). From Y_{red} or $MV^{+\cdot}$, electrons are transferred to NAD or other natural acceptors or directly to a suitable enzyme, which in turn delivers electrons to the unsaturated substrate.

TABLE 1. Examples of Enoates Hydrogenated to Chiral Carboxylates by *Clostridium* sp. La 1[a] in the Presence of $MV^{+\cdot}$ with Productivity Numbers from 200–3000[b,c]

Number	R^1	R^2	R^3	Enantiomeric excess (%)	Configuration
1	Me	Me	H	>96	R
2	Me	H	Me	>96	R
3	Me	Me	Me		
4	Me	C_6H_5	H		
5	OMe	(p)BrC_6H_4	H		
6	SMe	(p)BrC_6H_4	H		
7	F	C_6H_5	H	>96	R
8	Cl	(p)ClC_6H_4	H		
9	Br	C_6H_5	H		
10	NHCHO	C_6H_5	H	>96	R
11	NHCHO	C_6H_5	Me		
12	CN	OEt	H		
13	H	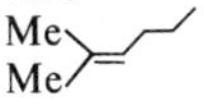	Me	>96(R)	R
14	H	Me	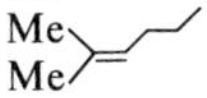	>96(S)	S
15	Me	$HOCH_2$	H	>96	R
16	H	$HOCH_2$	Me	>96	R
17	Et	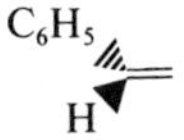		>96	R(Z)
18	Et	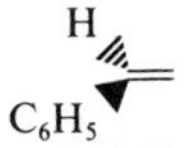		>96	R(E)

[a]*C.* sp. La 1 can be obtained from Deutsche Sammlung für Mikroorganismen under Nr. DSM 1460.

[b]See Equation 2.

[c]The general reaction sequence is:

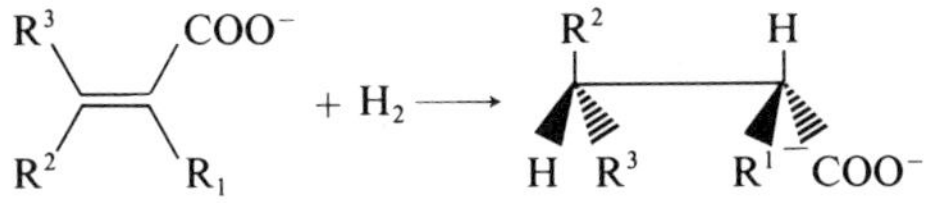

The addition of catalytic amounts of methylviologen, a rather cheap and stable compound, stabilizes and often accelerates such systems.

TABLE 1 shows a selection from about 60–70 different α, β-unsaturated carboxylates (enoates) that have been hydrogenated. R^1 can be rather different. However, it should not be too bulky. The acetylamino, instead of the formylamino, group or the ethylthio, instead of the methylthio, group seem to be too large since the α-acetylamino cinnamate or α-ethylthio cinnamate were not hydrogenated, whereas the α-formylamino and α-methylthio cinnamate are suitable substrates. R^2 can be even more different than R^1. Besides the residues indicated, many others have been successfully tested.[13,14] The substituent R^3 should not be too large and rigid. Whereas E-cinnamate is a good substrate, this is not the case for Z-cinnamate. However, a 2-methyl-2-hexenyl residue (Z-geraniate) is accepted. The steric course of the hydrogen addition is always determined by the C-C double bond and the carboxylate group, and not by R^2 or R^3. The attack to C-3 is always *trans* to that of C-2. Depending on the substituents of C-3, that can lead to a (3R) or (3S) compound. This has the consequence that substrate No. 13 (TABLE 1) leads to stereochemically pure R-citronellic acid and No. 14 to the S-citronellic acid, respectively. Also, allene carboxylates are substrates for the hydrogenation with *Clostridium* sp. La 1. Here one observes two specificities: (1) only the α, β double bond is hydrogenated or (2) the configuration at C-2 of the product is R regardless of whether the (R)- or the (S)-2-ethyl-4-phenyl-allene carboxylic acid is the substrate. The former (No. 17) leads to the Z-(2R)-2-ethyl-4-phenyl-3-butenoate and the latter to the E-configuration.[15]

As can be seen from FIGURE 1, the hydrogenation of the unsaturated substrate is a consecutive sequence of reactions. In *Clostridium* sp. La 1, the hydrogenase is usually rate-limiting when a good substrate for the final reductase (enoate reductase, *see below*) is applied. In such cases, the solution does not show the blue color of $MV^{+\cdot}$

TABLE 2. Examples for the Preparation of Stereospecifically Deuterated Compounds

COOH | ^{2}HCH | ^{2}HCH | R	Method of Preparation	References
R = C_6H_5; $4(HO)C_6H_4$; $2(HO)C_6H_4$; $2,5(HO)_2C_6H_3$; R = COOMe	hydrogenation in ^2H_2O buffer with *C.* sp. La 1	Bartl *et al.*[17] Leinberger *et al.*[18] Parker[19]
	electroenzymatic in ^2H_2O buffer with eonate reductase	Battersby *et al.*[20]
COOH | ^{2}HCH | ^{2}HC ‖ Ph Me	hydrogenation in ^2H_2O buffer with *C.* sp. La 1	unpublished
COOH | HC^{2}H | HC^{2}H | R R = C_6H_5; Me>CH<Me	hydrogenation in ^2H_2O buffer with *Peptostreptococcus anaerobius*	Parker[19] Gelsel *et al.*[21]
(4-S) [4-^{2}H] NADH	electromicrobial in ^2H_2O buffer with *C. ghoni* (PN = 20 000); *C. sordellii* (PN = 8050)	unpublished
(4-S) [4-^{2}H] NADPH	electromicrobial in ^2H_2O buffer with *C. kluyveri* (PN = 3500)	unpublished

TABLE 3. Hydrogenation of Carbonyl Group Containing Compounds to Chiral Alcohols and the Productivity Numbers for the System H_2 Gas, $MV^{+\cdot}$, Microorganism[a]

No.	Product	Productivity Number (PN)	Remarks[b]
1	(S)-3-Hydroxybutyrate[c]	4250	+18.6°; enzymatically pure
2	(2R,3S)-3-Hydroxy-2-methylbutyrate[c,d]		Starting material was (R,S)-2-methyl-3-oxo-butyrylethyl ester
3	(R)-Lactate	10000	enzymatically pure
4	(R)-2-Hydroxybutyrate	7500	enzymatically pure
5	(R)-2-Hydroxypentanate	7500	+16.5°
6	(R)-2-Hydroxy-4-methyl-pentanate	7500	+15.3°
7	(R)-2-Hydroxy-3-phenyl-propionate	5000	+31.7°
8	(R)-2-Hydroxy-3-(methylthio)-butyrate		+72.5°
9	(R)-Mandelate[e]	6000	+149°
10	(R)-3.3-Dimethyl-2.4-dihydroxybutyrate	3000	ee <98% (NMR)

[a]If not stated otherwise, the microorganism was *Proteus mirabilis* DSM30 115.

[b]Solvent and concentration are usually those for which optical rotation values are found in the literature. Molar optical rotations are given for 589 nm and room temperature.

[c]Hydrogenation of the ethyl ester with *C. kluyveri* (DSM 555).

[d]After quantitative hydrogenation, this product was found with 81%, the (2S,3R)-derivative was present with 13%.

[e]Prepared with reduced benzylviologen (*see below*).

when the hydrogenation proceeds. Using a suitable mixture of *C.* sp. La 1 and *C. kluyveri,* the rate of hydrogenation can be increased owing to the fact that *C. kluyveri* usually possesses a much higher hydrogenase activity than *C.* sp. La 1.[16] An example is the hydrogenation of substrate No. 15 (TABLE 1). With *C.* sp. La 1 alone, the productivity number 1250 was measured, while with a mixture of *C.* sp. La 1 to *C. kluyveri* at a ratio of 1.5 : 2.5, a value of 3100 could be observed.

The hydrogenation of enoates in ^2H_2O leads to stereospecifically deuterated acids. Examples are shown in TABLE 2. The electroenzymatic and electromicrobial reduction methods mentioned there will be described later. As can be seen from the cited publications, the products are optically pure within the limits of experimental error. Compared with other methods for the preparation of stereospecifically deuterated products, the procedures described here seem to be convenient and efficient.

Especially in immobilized form, these clostridia are rather stable. Cells stored at −18°C for several months and then immobilized according to Fukui *et al.*[22] have a biological half-life of 12–14 days.[23]

A second group of examples of the hydrogenation of unsaturated compounds is shown in TABLE 3. As can be seen, the productivity numbers are rather high. The bioreduction of 3-oxobutyrylesters to chiral 3-hydroxybutyrates has been often studied.[9,10] With yeasts, the optical purity of the product depends on the concentration of the substrate and productivity numbers of 300–400 have been reached.[10] (Wipf *et al.*[10] give a detailed description, in contrast to many other papers.) An exact calculation of the PN is usually not possible since the dry weight of the yeast and its growth during the reaction period is not known. In the case of the reduction of alkyl-4-chloro-3-oxo-butanoate, the stereochemical course depends heavily on the length of the alkyl residue of the alcohol component.[9] With *C. kluyveri,* a high productivity number and complete optical purity within the limits of experimental error have been observed for 3-hydroxybutyrates. The relatively high diastereomeric purity obtained after the quantitative hydrogenation of (R,S)-2-methyl-3-oxobutyrylethylester may be explained by the preferential reduction of the S enantiomer and relative fast racemization of the R-enantiomer due to the high acidity of the proton at C-2. The efficient hydrogenation of the 2-oxo-carboxylates is based on the activity of a methylviologen-dependent new reductase (*see below*). The enantiomeric purity of compounds mentioned in TABLES 1–3 have been checked by us and in other laboratories by very sensitive methods. In the limits of these methods, the compounds are optically pure.[17–21]

The advantages of microbial hydrogenations can be seen in the following facts. Relatively high productivity numbers are obtained. For the preparation of chiral carboxylic acids, they are often in the range of 700–3000 and for carbonyl compounds, several thousand. The reaction rate is easily observable through the consumption of hydrogen. The low amount of applied biocatalyst compared to cofermentations with yeasts or other microorganisms simplifies the product isolation. Hydrogen gas as electron donor does not lead to byproducts from which the wanted product must be separated.

TABLE 4. Some Kinetic Data of Enoate Reductase from *Clostridium* sp. La 1[a]

	Enzyme K Value with Reagents Shown			
K	NADH	$MV^{+\cdot}$	Cinnamate	(E)-2-Methyl-2-butenoate
K_m (m*M*)	0.013	0.4	0.01	1.5
	NAD	MV^{++}	Phenylpropionate	Aliphatic carboxylates
K_i (m*M*)	0.84	>>10	~30	>500

[a]Ratios of V_{max} for Equations 4 : 5 : 6 = 1.0 : 1.5 : 0.9

ENOATE REDUCTASE, 2-OXO-ACID REDUCTASE, ELECTROENZYMATIC REDUCTIONS

Enoate reductase, which we discovered a couple of years ago,[24] is a rather versatile enzyme. It catalyzes the following reactions[24–26] (Eq. 4–6):

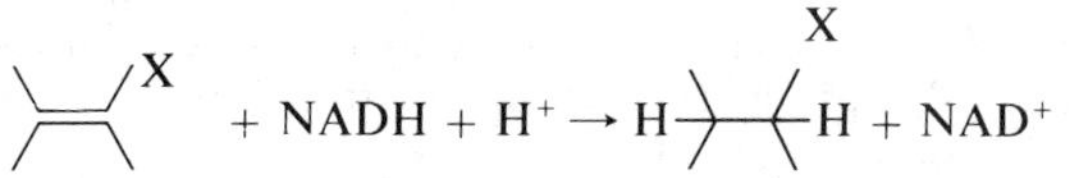

(4)

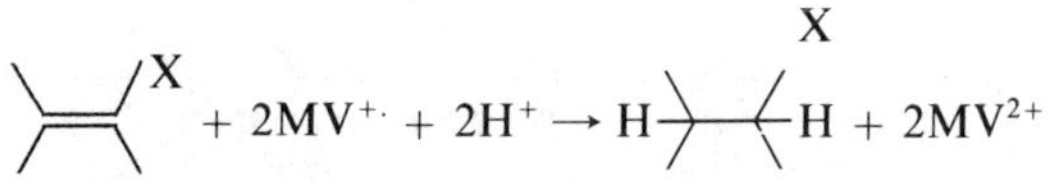

(5)

(X = COO^- or CHO)

$$NAD^+ + 2MV^{+\cdot} + H^+ \rightarrow NADH + 2MV^{2+} \qquad (6)$$

From a practical point of view, it is especially interesting that enoate reductase can be reduced by $MV^{+\cdot}$. Therefore it is applicable for *electroenzymatic* reductions. That means reductions can be conducted in a preparative scale with one enzyme and the rather stable and cheap $MV^{+\cdot}$ as an electrochemically regenerable mediator.[27] The

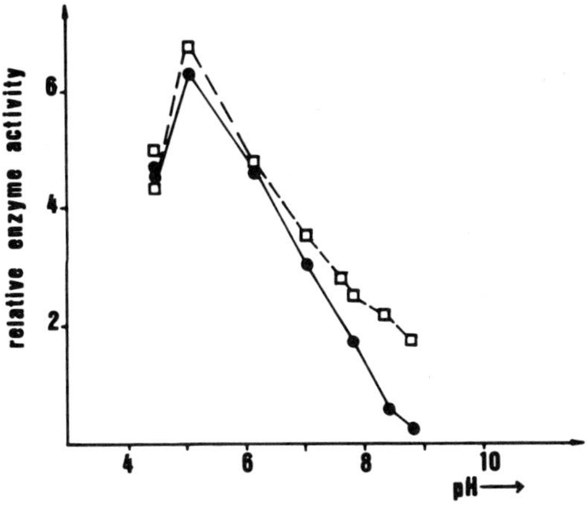

FIGURE 2. pH dependence of the initial rate of reaction 5 studied with two substrates at 30°C. A solution of 0.6 ml in a 0.2-cm cuvette contained 100 m*M* Tris-acetate, 0.5 m*M* $MV^{+\cdot}$, 1.0 m*M* cinnamate (□) or 18 m*M* (E)-2-methyl-2-butenoate (●), and depending on the expected rate 5–10 μl of an enzyme solution. Measurements at pH 6.1 and 6.95 in 100 m*M* phosphate buffer showed no differences to that in Tris-acetate buffer.

rate of reduction of enoates with reduced methylviologen is about 1.5 times faster than that with saturating concentrations of NADH. Some kinetic data can be seen from TABLE 4. The pH dependence of Reaction 5 can be seen in FIGURE 2. That for Reaction 4 is given elsewhere.[24] As shown in TABLE 5, enoate reductase from *C.* sp. La 1 can be purified in a rather simple way with satisfying yields.[26] Electroenzymatic or the later described electromicrobial reductions can be conducted in different electrochemical cells. FIGURE 3 shows rather simple and more sophisticated ones. In cell A, 1 mmol of product can be produced per hour. We assume that the volume productivity can be enlarged by a factor of 3 by rather simple means. In cell B, 50–100 μmol product/h can be produced.

Besides enoate reductase, we detected a further versatile enzyme for stereospecific reductions. This so far unknown 2-oxo-acid reductase from *Proteus mirabilis* as well as *Proteus vulgaris* catalyzes in a *non*pyridine nucleotide-dependent reaction the reduction of 2-oxo-carboxylates (Eq. 7).

$$R - COCOO^- + 2MV^{+\cdot} \text{ (or } BV^{+\cdot}\text{)}$$
$$+ 2H^+ \rightarrow (2R) - RCHOHCOO^- + 2MV^{2+} \text{ (or } BV^{2+}\text{)} \quad (7)$$

where $BV^{+\cdot}$ = reduced benzylviologen. Actually this enzyme does not react with NAD(P). Nevertheless it reduces a C-O double bond. It has an astonishingly broad substrate specificity and an excellent stereospecificity. Its activity in crude extracts of

TABLE 5. Purification of Enoate Reductase from *Clostridium* sp. La 1.

Procedure	Volume (ml)	Protein Concentration (mg/ml)	Activity (IU)	Specific Activity (IU/mg)	Yield (%)
Crude extract	162	52.1	850	0.1	100
Combined chromatography on DEAE-Sepharose and spherical hydroxylapatite	36	1.5	756	13.8[a]	89

[a]This material usually contains an impurity of about 5% which could only be detected by analytical ultracentrifugation.

Proteus mirabilis or *Proteus vulgaris* is in the range of 1–10 IU/mg protein if tested with phenylpyruvate. The membrane-bound enzyme is electrophoretically pure after about 300-fold enrichment. TABLE 6 reveals the relative rates of reduction of rather different 2-oxo carboxylates. Not only 2-oxo-monocarboxylates are substrates but also 2-oxo-dicarboxylates. Branching in the 3 position of a substrate leads to lower reduction rates. This can be seen from the comparison of 2-oxo-4-methylpentanoate with 2-oxo-3-methylpentanoate and 2-oxo-3-dimethyl-4-hydroxybutanoate. Phenyl-

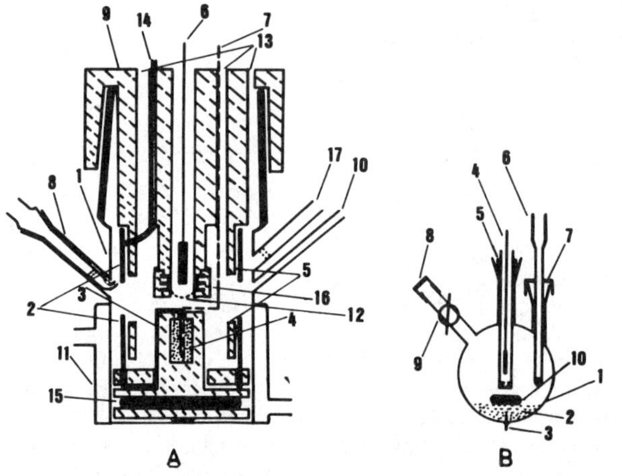

FIGURE 3. Electrochemical cells used for electroenzymatic or electromicrobial reductions.

A: With a rotating cylindrical cathode, catholyte volume of about 150 ml. 1 = Glass cell; 2 = 60 mm (height) × 50 mm (diameter) lead cathode with circular perforations midway; 3 = insulated steel wire, contacted to the cathode with its end dipped into the mercury pool 4; 4 = mercury contact; 5 = cylindrical perspex separator with perforations for suppressing vortex formation by the rotating cathode; 6 = counter electrode; 7 = contact to the working electrode; 8 = reference electrode; 9 = perspex cell cover; 10 = tube for electrolyte sampling; 11 = thermostating coat; 12 = Nafion membrane, separating catholyte from anolyte, tightly clamped between the mainframe of the cell cover and the Teflon nut 16; 13 = openings for thermometer, gas inlet and outlet; 14 = flexible Luggin capillary for measuring potential variations along the surface of the cathode; 15 = Teflon rotor with magnet and rings firmly fixing the cylindrical electrode onto the rotor; 16 = Teflon nut; and 17 = tube with sintered glass for bubbling inert gas.

B: With mercury pool cathode, catholyte volume 7–20 ml. 1 = glass cell; 2 = mercury cathode; 3 = platinum wire for electrical contact; 4 = anode; 5 = anode compartment closed at one end by a Vycor^{R} tip; 6 = reference electrode; 7 and 8 = rubber caps as oxygen-tight seals for reference electrode, gas inlet and outlet, and for withdrawing of samples through syringe.

glyoxylate fits this rule, too. For the products marked in TABLE 6 by footnote *c*, the enantiomeric purity has been determined. Again, in the limits of the methods applied, the products have been optically pure.

Some kinetic data for 2-oxo-acid reductase can be seen from TABLE 7. Reduced methylviologen exhibits a substrate inhibition. FIGURE 4 demonstrates an electroenzymatic preparation of a chiral compound with this 2-oxo-acid reductase. There was no decrease in enzyme activity during about 20 h of operational activity and an additional 9 hours standing at room temperature.

TABLE 6. Relative Rates of the Reduction of 2-Oxo-Acids to (2R)-Hydroxy-Acids with Partially Purified 2-Oxo-Acid Reductase from *Proteus mirabilis* as well as from *P. vulgaris* and $MV^{+\cdot}$ or $BV^{+\cdot}$

	Relative Activity (%)	
Substrate	*P. mirabilis*	*P. vulgaris*
Phenylpyruvate[c]	100[a]	100[b]
Pyruvate[c]	92	85
Indolylpyruvate	35	67
5-Benzyloxyindolylpyruvate	30	—
3-Fluoropyruvate	22	20
2-Oxo-4-methylpentanoate	81	—
(S)-2-Oxo-3-methylpentanoate	28	25
(R,S)-2-Oxo-3-methylpentanoate	—	46
3,3-Dimethyl-4-hydroxy-2-oxobutyrate[c]	7	4
Phenylglyoxylate[c]	16	5
2-Oxo-nonanoate	83	—
Oxalacetate	73	50
2-Oxoglutarate	78	62
2-Oxadipate	70	—
2-Oxo-(4-hydroxy-methylphosphinyl)-butanoate	24	—
3-Oxoglutarate	0	—
Hydroxyacetone	0	—

[a]Crude extracts show activities of ~1.5 IU/mg.
[b]Crude extracts show activities of 3–10 IU/mg.
[c]The products of these substrates have been checked for optical purity. Within the limits of experimental error, they turned out to be pure.

REDUCTION OF NAD AND NADP BY WHOLE CELLS OR CRUDE EXTRACTS OF DIFFERENT ORGANISMS AND ELECTROMICROBIAL REDUCTIONS

During our studies on the regeneration of reduced pyridine nucleotides, we observed that quite different anaerobic as well as aerobic cells or the crude extracts thereof catalyze the reaction (Eq. 8):

$$2MV^{+\cdot} + H^+ + NAD(P)^+ \rightarrow 2MV^{2+} + NAD(P)H \qquad (8)$$

It turned out that the rate of this reaction is often much higher than the rates of reductions of unsaturated compounds by systems made from glucose and yeast or other

TABLE 7. Some Kinetic Data for 2-Oxo-Acid Reductase from *Proteus mirabilis*

	K_m[a] (m*M*)	Influence of Products
Phenylpyruvate	0.15	0.75 m*M* MV^{2+} accelerates 3.4 times compared to experiments with ~0.01 m*M* MV^{2+}
2-Oxo-4-methylpentanoate	2.1	100 m*M* (R)- as well as (S)-phenyllactate and (S) lactate inhibit in the presence of 5 m*M* substrate the initial rate 31–35% based on the initial rate without product.
2-Oxo-3.3-dimethyl-4-hydroxy-butyrate	7.5	
2-Oxoglutarate	2.5	

[a]Initial rate kinetics show for $MV^{+\cdot}$ no Michaelis-Menten behavior. There is an increase in rates going from 0.5 m*M* $MV^{+\cdot}$ to 0.05 m*M*.

microorganisms. The rate of the reaction is therefore often sufficient for microbial reductions.

The capability of different organisms to catalyze the formation of NAD(P)H is shown in TABLE 8. The activity for the NADH formation by some yeasts depends on the carbon source of the growth medium. *Candida utilis* does not show a big difference when grown on glucose or glycerol. However, *Candida boidinii* as well as *Candida valida* grown on the glucose-containing medium possess a 50- or 100-fold higher activity than cells grown on the glycerol medium. In the case of *Rhodotorula glutinis,* glycerol-grown cells are superior by a factor of 50. Especially effective in the NAD(P)H formation are *Alcaligenes eutrophus* H 16 and the different clostridia. For *Candida utilis,* only the highest growth times were reported in TABLE 8. All the other organisms were tested without trying to optimize the activity.

In several cases, not only the reoxidation of reduced methylviologen has been measured but also the NADH as well as NADPH formation by the glutamate dehydrogenase reaction. About 80% of the expected NAD(P)H has usually been

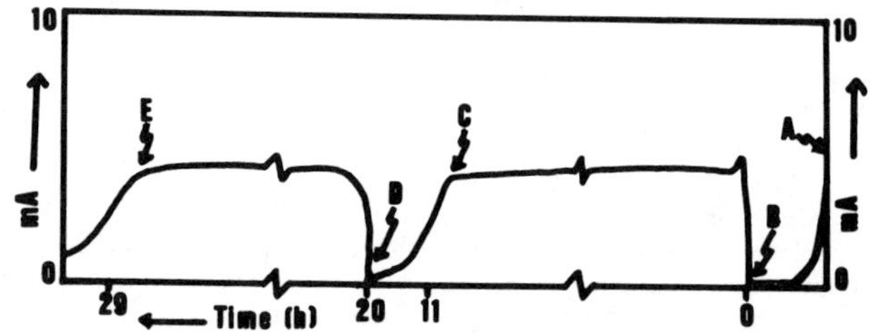

FIGURE 4. Current during an electroenzymatic reduction of 2-oxo-4-methylpentanoate by partially purified 2-oxo-acid reductase. A total volume of 25 ml contained 100 m*M* Tris-acetate pH 7.0, 0.5 mg protein, and 4 m*M* methylviologen. A: Reduction of MV^{2+} to $MV^{+\cdot}$; B: Addition of 40 m*M* substrate and setting of the voltage by a potentiostat at −700 mV against a standard calomel electrode; C: substrate becomes limiting; D: after standing overnight, 40 m*M* substrate was added again; E: substrate again almost consumed.

found. There are several uncertainties which have in common a small value for this variable. The reduction of NAD is completely or partially stereospecific as can be shown by reactions conducted in ^2H_2O-buffer. As seen by ^{1}H-NMR techniques,[28] (4S)[^{2}H]-NADH of an enantiomeric excess higher than 95% is produced by *C. sordellii* and *C. ghoni*. The product obtained with *C. kluyveri* has only an optical purity of about 80%. However, the (4S)[4-^{2}H] NADPH formed by the catalysis of *C. kluyveri* is stereochemically pure. The NADH formation by *Candida utilis* depends on the pH, drastically on the buffer, and on the ionic strength. FIGURE 5 shows the effects of pH and buffer on Reaction 8, measured in an electrochemical cell on the basis of the maximal current. For a two-electron reduction, 3.2 mA corresponds to the formation of 1 μmol of product per minute. In citrate buffer, the pH range in which the reaction is catalyzed is not so broad, and the observed activity is lower. The addition of hydroxyacetone to the system yeast, NAD, MV^{2+} in an electrochemical cell increases the maximum current, with propandiol formed as the product. Under these conditions, the reactions according to Equations 9–11 proceed[29] and can be summed up by the total reaction Equation 12.

TABLE 8. Specific Activities (Initial Rates) of the Reoxidation of $MV^{+\cdot}$ and NAD(P)H Formation in the Presence of NAD or NADP Catalyzed by Crude Extracts of Various Microorganisms

		μmol/min × mg Protein (2e transfer)				μmol/min × mg Protein (2e transfer)	
Microorganisms[a]	Carbon Source	NADH	NADPH	Microorganisms	Carbon Source	NADH	NADP
Candida utilis	glycerol	0.19	0.025	*Micrococcus luteus*	glucose	0.25	0.021
Candida utilis	glucose	0.11	0.018	*Pseudomonas aeruginosa*	glucose	0.092	0.010
Candida boidinii	glycerol	0.001	0.007	*Pseudomonas aeruginosa*	pepton + nitrate	0.25	0.021
Candida boidinii	glucose	0.071	0.014	*Alcaligenes eutrophus* H16[b]	CO_2	0.86	0.028
Candida valida	glycerol	0.0007	0.0035	*Clostridium sporogenes*	peptone	4.98	0.65
Candida valida	glucose	0.045	0.012	*Clostridium* sp. La 1	crotonate	1.64	0.78
Rhodotorula glutinis	glycerol	0.11	0.007	*Clostridium sordellii*	peptone	2.13	1.57
Rhodotorula glutinis	glucose	0.021	0.0028	*Clostridium ghoni*	peptone	3.00	0.50
Escherichia coli	glucose	0.085	0.043	*Clostridium tyrobutyricum*	peptone	3.56	0.57
Escherichia coli	pepton + nitrate	0.064	0.11	*Clostridium kluyveri*	ethanol-acetate	17.8	4.98

[a]The following organisms showed less than 0.06 IU in the presence of NAD: commercial bakers' yeast, *Bacillus subtilis, Bacillus macerans* DSM 24, *Bacillus cereus* DSM 31.

[b]Cells stored for 4 years at −10°C.

$$2MV^{2+} + 2e \longrightarrow 2MV^{+\cdot} \quad (9)$$

$$2MV^{+\cdot} + H^+ + NAD^+ \xrightarrow{\text{yeast}} NADH + 2MV^{2+} \quad (10)$$

$$NADH + H^+ + CH_3COCH_2OH \xrightarrow{\text{yeast}} NAD^+ + CH_3CHOHCH_2OH \quad (11)$$

$$\text{Total: } CH_3COCH_2OH + 2H^+ + 2e \longrightarrow CH_3CHOHCH_2OH \quad (12)$$

In such experiments, the product of current and time corresponds within a few percent to the formed product. The increase of the maximum current (FIG. 5) in the presence of the electron acceptor hydroxyacetone may be explained by the fact that in the consecutive reaction sequence, Equations 9–11, the steady-state concentration of NAD is higher in the presence of hydroxyacetone than at the time at which the maximum current is reached in the presence of NAD only. The average productivity number averaged over 18 hours was 2080. This corresponds under steady-state conditions to an average specific activity for Equation 10 of 0.07 IU/mg protein. The yeast used in this experiment was different from that mentioned in TABLE 8. This productivity number is much higher than that for conventional cofermentations. That means that even with a specific activity of 0.01 IU/mg × min for the NADH formation, efficiencies can be obtained that are comparable or higher than those usually observed in cofermentations.

In the preparation of a product from an unsaturated substrate by means of electromicrobial reduction, two situations can be differentiated. One, the electrochemically reduced mediator delivers electrons directly to a reductase, or two, it delivers them to enzymes that are capable of reducing NAD or NADP. In the former case, it is obvious that it is actually not necessary to isolate the enzyme, for instance, enoate reductase or 2-oxo acid reductase, from the corresponding microorganisms. It is

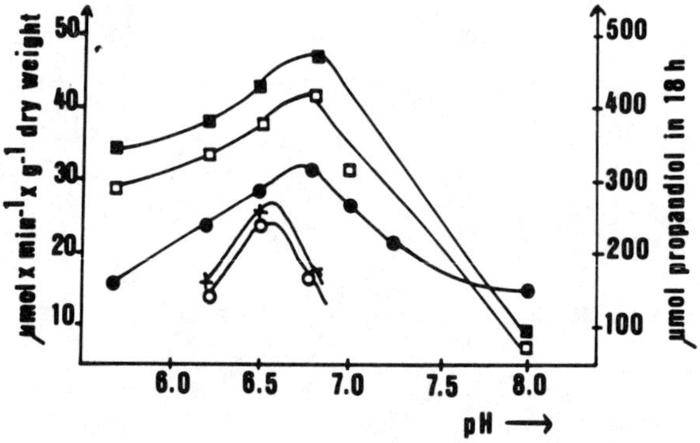

FIGURE 5. Specific activity of NADH and (R)-1.2-propandiol formation in an electrochemical cell (left-hand ordinate) and overall propandiol formation in 18 h (right-hand ordinate). A total volume of 20 ml contained 100 m*M* buffer, 4 m*M* methylviologen, 0.8 m*M* NAD and 12.6 mg (dry weight) *Candida utilis*. In experiments with hydroxyacetone, the starting concentration was 50 m*M*. (O) NADH formation in citrate buffer, (●) NADH formation in Tris-acetate, (x) propandiol formation in citrate buffer, (□) propandiol formation in Tris-acetate buffer, (■) overall propandiol formation during 18 h in Tris-acetate.

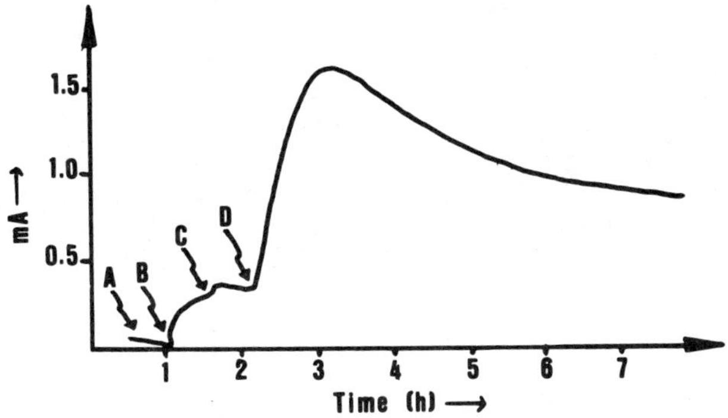

FIGURE 6. Current during an electromicrobial reduction of hydroxyacetone by the combination of crude extracts of *Candida utilis* and *Alcaligenes eutrophus* H 16. The final solution of 40 ml contained 100 m*M* Tris-acetate buffer, pH 7.0, 4 m*M* methylviologen, 0.8 m*M* NAD, 10 m*M* hydroxyacetone, 1.4 mg *Candida utilis* and 0.65 mg *Alcaligenes eutrophus*. After the reduction of the methylviologen, the components of the system were added as follows: A, crude extract of the yeast; B, NAD; C, hydroxyacetone; and D, *A. eutrophus*.

possible to use whole cells or crude extracts in an electrochemical cell. In such cases, it is advisable to use cells with a low hydrogenase activity. Otherwise, half or more of the electrons may be used up for hydrogen production. Again, two situations can be distinguished in the second case of electromicrobial reductions (Equations 9–11): (1) the NAD(P)H formation is rate limiting, or (2) the reaction that catalyzes the electron transfer from NAD(P)H to the substrate is rate limiting. We found the former case held for *Candida utilis,* in which such cells possessing a rather high NADH-dependent hydroxyacetone reductase (Eq. 11) were poor in NADH formation (Eq. 10). Such a situation can be improved by the combination of cells or crude extracts. An example is described by FIGURE 6 in which *Alcaligenes eutrophus* H 16 is combined with *Candida utilis*. For about 5 hours, an average productivity number of about 11000 can be observed.

We envisage the following advantages of electromicrobial reductions: (1) The reaction rate and product formation can be monitored continuously. (2) The rate of reductions depending on the regeneration of NAD(P)H are not limited by that of glycolysis. (3) Side reactions between substrate or product and intermediates of the glycolysis can not occur. (4) Higher productivity numbers mean less biocatalyst and therefore easier isolation of the product.

PREREQUISITES AND LIMITATIONS OF BIOHYDROGENATIONS, ELECTROENZYMATIC AND ELECTROMICROBIAL REDUCTIONS

The stereospecificity of the reduction is ruined if the mediator reacts spontaneously with the substrate. Actually this situation is rather rare. One example is phenylglyoxylate. This substrate for the formation of (R)-mandelic acid reacts spontaneously with reduced methylviologen. However, this is not the case with benzylviologen. Its use instead of methylviologen leads to optically pure mandelic acid. The reason may be that the redox potential of reduced benzylviologen is about 100 mV less negative than that of reduced methylviologen. Another prerequisite of the methods described here is the complete exclusion of oxygen. Oxygen reacts almost in a diffusion-limited rate with reduced viologens.

ACKNOWLEDGMENTS

We are grateful to Mrs. C. Frank, Mrs. C. Stuber, Mr. H. Leichmann, and Mr. L. Riesinger for skillful technical assistance. The *Alcaligenes eutrophus* cells have been a gift from Prof. H. G. Schlegel, Göttingen and several enoates from Dr. H.G.W. Leuenberger, Basel.

REFERENCES

1. KIESLICH, K. 1976. Microbial Transformations. G. Thieme Publishers. Stuttgart.
2. IIZUKA, H. & A. NAITO. 1981. Microbial Conversion of Steroids and Alkaloids. University of Tokyo Press. Springer Verlag.
3. JONES, J. B. & J. F. BECK. 1976. Asymmetric synthesis and resolutions using enzymes. *In* Applications of Biochemical Systems in Organic Chemistry. J. B. Jones, C. J. Sih & D. Perlman, Eds. Vol X Techniques of Organic Chemistry, Part I. J. Wiley & Sons. New York. pp. 107–401.
4. HUMMEL, W., H. SCHÜTTE & M.-R. KULA. 1984. New enzymes for the synthesis of chiral compounds. Ann. N.Y. Acad. Sci. **434:** This volume.
5. WHITESIDES, G. M. & C. H. WONG. 1983. Enzymes as catalysts in organic synthesis. Aldrichimica Acta **16:** 27–34.
6. HASLEGRAVE, J. A. & J. B. JONES. 1982. Enzymes in organic synthesis. 25. Heterocyclic ketones as substrates of horse liver alcohol dehydrogenase. Highly stereoselective reductions of 2-substituted tetrahydropyran-4-ones. J. Am. Chem. Soc. **104:** 4666–4671.
7. WICHMANN, R., C. WANDREY, A. BÜCKMANN & M.-R. KULA. 1981. Continuous enzymatic transformation in an enzyme membrane reactor with simultaneous NAD(H) regeneration. Biotechnol. Bioeng. **23:** 2789–2802.
8. LEUCHTENBERGER, W., M. KARRENBAUER & U. PLÖCKER. 1984. Scale-up of an enzyme membrane reactor process for the manufacture of L-enantiomeric compounds. Ann. N.Y. Acad. Sci. **434:** This volume.
9. ZHOU, B., A. S. GOPALAN, F. VANMIDDLESWORTH, W. R. SIEH & C. J. SIH. 1983. Stereochemical control of yeast reductions. 1. Asymmetric synthesis of L-carnitine. J. Am. Chem. Soc. **105:** 5925–5926 and references cited therein.
10. WIPF, B., E. KUPFER, R. BERTAZZI & H. G. W. LEUENBERGER. 1983. Production of (+)-(S)-ethyl-3-hydroxybutyrate and (−)-(R)-ethyl-3-hydroxybutyrate by microbial reduction of ethylacetoacetate. Helv. Chim. Acta **66:** 485–488.
11. SCHUETZ, H. -J. & H. SIMON. Manuscript in preparation.
12. SIMON, H. & H. GÜNTHER. 1983. Chiral synthons by biohydrogenation or electroenzymatic reductions. *In* Biomimetic Chemistry. Z. Yoshida & N. Ise, Eds. Kodansha Ltd. Tokyo. Elsevier Scientific Publishing Company. Amsterdam.
13. RAMBECK, B. 1974. Untersuchungen über die Anwendung von *Clostridium kluyveri* als stereospezifischen Austausch- und Hydrierkatalysator. Dr.rer.nat. Thesis. Technical University Munich.
14. LECOMTE, J. P. & H. SIMON. Manuscript in preparation.
15. RAMBECK, B. & H. SIMON. 1974. Stereochemical hydrogenation of (R)- or (S)-2-ethyl-4-phenylallenecarboxylic acid to *cis*- or *trans*-2-ethyl-4-phenyl-3-butenecarboxylic acid by means of *Clostridium kluyveri*. Angew. Chem. Int. Ed. Engl. **13:** 609.
16. BADER, J. & H. SIMON. 1980. The activities of hydrogenase and enoate reductase in two *Clostridium* species, their interrelationship and dependence on growth conditions. Arch. Microbiol. **127:** 279–287.
17. BARTL, K., C. CAVALAR, T. KREBS, E. RIPP, J. RÉTEY, W. E. HULL, H. GÜNTHER & H. SIMON. 1977. Synthesis of stereospecifically deuterated phenylalanines and determination of their configuration. Eur. J. Biochem. **72:** 247–250.

18. LEINBERGER, R., W. E. HULL, H. SIMON & J. RÉTEY. 1981. Steric course of the NIH shift in the enzymatic formation of homogentisic acid. Eur. J. Biochem. **117:** 311–318.
19. PARKER, D. 1983. ^{1}H and ^{2}H Nuclear magnetic resonance determination of the enantiomeric purity and absolute configuration of α-deuterated primary carboxylic acids, alcohols, and amines, J. Chem. Soc. Perkin Trans. **II:** 83–88.
20. BATTERSBY, A. R., A. L. GUTMAN, C. J. R. FOOKES, H. GÜNTHER & H. SIMON. 1981. Stereochemistry of formation of methyl and ethyl groups in bacteriochlorophyll a. J. Chem. Soc. Chem. Commun. pp. 645–647.
21. GIESEL, H., G. MACHACEK, J. BAYERL & H. SIMON. 1981. On the formation of 3-phenylpropionate and the different stereochemical course of the reduction of cinnamate by *Clostridium sporogenes* and *Peptostreptococcus anaerobius.* FEBS Lett. **123:** 107–110.
22. FUKUI, S., K. SONOMOTO, N. ITOH & A. TANAKA. 1980. Several novel methods for immobilization of enzymes, microbial cells, and organelles. Biochimie **62:** 381–386.
23. EGERER, P. & H. SIMON. 1982. Hydrogenation with entrapped *Clostridium* sp. La 1 and observation on its stability. Biotechnol. Lett. **4:** 501–506.
24. TISCHER, W., J. BADER & H. SIMON. 1979. Purification and some properties of a hitherto-unknown enzyme reducing the carbon-carbon double bond of α,β-unsaturated carboxylate anions. Eur. J. Biochem. **97:** 103–112.
25. BÜHLER, M. & H. SIMON. 1982. On the kinetics and mechanism of enoate reductase. Hoppe Seyler's Z. Physiol. Chem. **363:** 609–625.
26. KUNO, S., A. BACHER & H. SIMON. 1984. Structure of enoate reductase from *Clostridium* sp. La 1. Eur. J. Biochem. Submitted.
27. SIMON, H., H. GÜNTHER, J. BADER & W. TISCHER. 1981. Electroenzymatic and electromicrobial stereospecific reductions. Angew. Chem. Int. Ed. Engl. **20:** 861–863.
28. ARNOLD, L. J., K. YOU, W. S. ALLISON & N. O. KAPLAN. 1976. Determination of the hydride transfer stereospecificity of nicotinamide adenine dinucleotide-linked oxidoreductases by proton magnetic resonance. Biochemistry **15:** 4844–4849.
29. GÜNTHER, H., C. FRANK, H. -J. SCHUETZ, J. BADER & H. SIMON. 1983. Electromicrobial reduction using yeasts. Angew. Chem. Int. Ed. Engl. **22:** 322–323.

Enantioselective Reductions of β-keto-Esters by Bakers' Yeast

CHARLES J. SIH, BING-NAN ZHOU, ARAVAMUDAN S. GOPALAN, WOAN-RU SHIEH, CHING-SHIH CHEN, GARY GIRDAUKAS, AND FRANK VANMIDDLESWORTH

School of Pharmacy
University of Wisconsin
Madison, Wisconsin 53706

The asymmetric reduction of carbonyl compounds by resting cells of bakers' yeast[1] (*Saccharomyces cerevisiae*), a method outside the traditional arena of chemical synthesis, is central to the modern synthetic methodology for the preparation of chiral alcohols. Because bakers' yeast contains a large number of oxido-reductases,[2] it can reduce a variety of artificial ketonic substances to produce chiral carbinolic products of moderate to high enantiomeric excess (*ee*).[3] Although this phytochemical reduction has been widely used for many synthetic applications,[4–7] the enzymes responsible for these catalytic reactions have not yet been identified. Hence, it has not been possible to predict or interpret with confidence the stereochemical results of these reductions. We herein report our investigations on the enantioselective reductions of a series of β-keto esters by bakers' yeast.[8] We will show that the stereochemical course of this reduction may be altered by modifying substituents at the carboxyl end of the molecule. This observation has allowed us to prepare (R)-4-chloro-3-hydroxybutanoate, a key intermediate for L-carnitine (*1*) synthesis.

The importance of L-(−)-carnitine (*1*) in the mediation of transport of long chain fatty acids across the inner mitochondrial membrane is well known.[9] As a result of improved diagnostic procedures, an increasing number of human L-carnitine deficiency cases have been recognized and successfully treated by L-carnitine administration.[10] Because L-carnitine is currently prepared by a tedious chemical resolution procedure of the racemate, (±)-carnitinamide,[11] it was thus of interest to us to devise an efficient asymmetric synthesis of this biologically important acyl carrier:

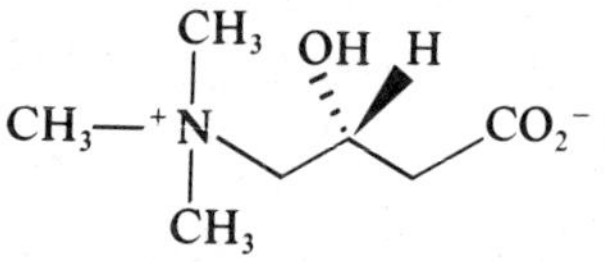

1

A number of laboratories have exposed ethyl acetoacetate (*2*) to bakers' yeast and obtained ethyl(S)(+)-3-hydroxybutanoate (*3*) of high optical purity and in reasonable yield.[12] Therefore, by analogy, we surmised that ethyl-4-chloro-acetoacetate (*4*) might be reduced to give ethyl-(R)-4-chloro-3-hydroxybutanoate, which could then be transformed into L-carnitine by known methods.[13] Unfortunately, it was found that *4* was converted predominantly to the undesired ethyl-(S)-4-chloro-3-hydroxybutanoate

(*5*), $[\alpha]_D^{23}$ −11.7° (c, 5.75, $CHCl_3$) (60% yield; *ee* = 0.55).[a] The absolute configuration of the chiral center in *4* was established by the transformation of *4* into D-carnitine chloride. This observation suggested that substitution of a chloro atom at the C-4 position of the acetoacetic ester molecule altered the stereochemical course of bakers' yeast reduction, that is, from L to D.

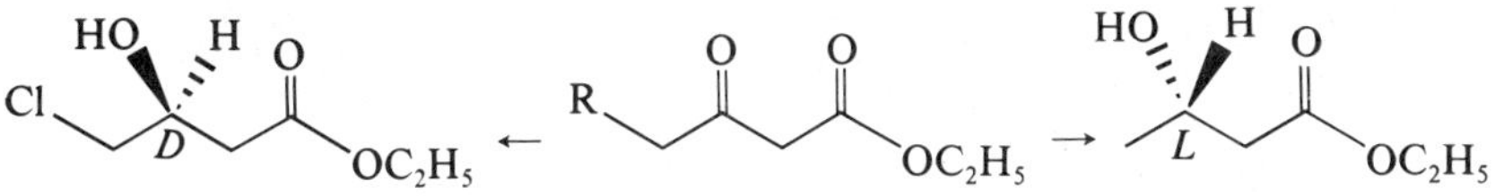

5

2 R = H
4 R = Cl

3

We then examined the reduction of *4* by a variety of yeasts and fungi[14] with a view

TABLE 1. Reduction of Ethyl-4-chloroacetoacetate (*4*) into Ethyl-(S)-(−)-4-chloro-3-hydroxybutanoate (*5*) by Microorganisms

Microorganism	Yield (%)	$[\alpha]_D^{23}$ ($CHCl_3$)
Pichia alcoholophilia	69	−21.16°
Mycoderma cerevisiae	67	−20.59°
Candida lipolytica	65	−20.98°
Candida tropicalis	57	−15.56°
Hansenula anomala	64	−18.29°
Aspergillus amstelodami	35	−20.41°
Aspergillus giganteus	68	−18.79°
Cunninghamella elegans	68	−20.44°
Cunninghamella echinulata	15	−17.28°
Cladosarum olivaceum	55	−17.46°
Mucor genevensis	59	−12.87°
Gliocladium roseum	56	−19.23°
Absidia blakesleeana	45	−21.86°

to obtaining the desired R-(+)-isomer. But again, all of the microorganisms that we examined likewise gave the undesired (S)-(−)-isomer, *5*, of varying optical purities (TABLE 1). These unexpected findings forced us to examine more closely the parameters that could influence the stereochemical course of bakers' yeast reductions.

For acyclic ketones, it is generally assumed that the stereoselectivity of bakers'

[a]The optically active 4-chloro-3-hydroxybutanoates were reacted with (+)-α-methoxy-α-(trifluoromethyl)phenylacetyl chloride in pyridine. The resulting MTPA esters were analyzed by HPLC using an Alltech μporasil (10 micron) column (4.5 mm ID × 50 cm). The column was eluted with hexane-ether (10 : 1 for C_1; 20 : 1 for C_2 and C_3; 22.5 : 1 for C_4–C_6; and 27.5 : 1 for C_7–C_{12} esters) at a flow rate of 2.6 ml per min and the absorbance at 254 nm was monitored. The enantiomeric excess (*ee*) was calculated by quantitatively measuring the peak areas of the diastereomers.

yeast reductions may be predicted by the Prelog rule,[15] where Ⓢ and Ⓛ stand for sterically smaller and larger groups, respectively. The configurations of the resulting carbinols are invariabley S, for the hydride equivalent is delivered to the Re- face of the carbonyl group. However, the applicability

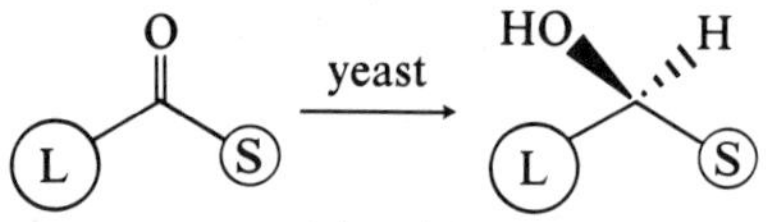

of the Prelog rule to bakers' yeast reduction of β-keto carbonyl derivatives has not been rigorously examined. TABLE 2 shows that while ethyl acetoacetate[9] and acetoacetic acid[16] were reduced mainly to their corresponding S isomers, ethyl β-ketovalerate,[17] caprioc,[16] caprylic,[16] and β-keto-6-heptenoic[4a] acids were all preferentially converted into their respective R isomers. These results suggested that by enlarging the size of the hydrocarbon end of the γ-keto carbonyl molecule, the stereochemical course of yeast reduction may be altered from S to R. The reduction of ethyl-γ-chloroacetoacetate to ethyl-(S)-(−)-4-chloro-3-hydroxybutanoate is also in accord with this supposition. Hence, we envisaged that it may be possible to alter the stereochemical course of yeast reduction of γ-chloroacetoacetic esters from S to R by manipulating the size of the ester grouping. For example, by making the ester group, Ⓛ, of the molecule considerably larger than the hydrocarbon end, Ⓢ, the hydride equivalent is delivered from the other face of the γ-chloroacetoacetate molecule. Hence, it may then be possible to prepare the requisite intermediate, (R)-γ-chloro-β-hydroxybutyrate, for L-carnitine synthesis.

To test the validity of this hypothesis, we prepared a homologous series of γ-chloroacetoacetic esters[18] ranging from C_1 to C_{12} and exposed them to bakers' yeast (FIG. 1). Although there was no significant difference in the rates of yeast reduction of γ-chloroacetoacetic esters containing one to eight carbons (n = 1 to 8), there was a

TABLE 2. Enantioselective Reduction of β-Keto-carbonyl Derivatives by Bakers' Yeast

Substrate	Product	R/S ratio
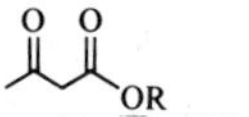 (R = H or Et)	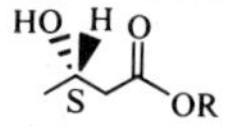	>2/98
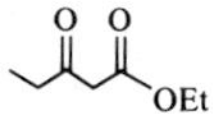	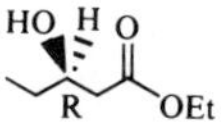	70/30
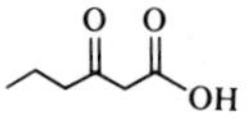	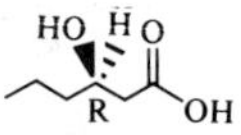	Predominantly R
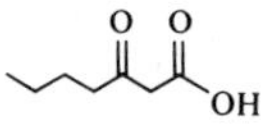	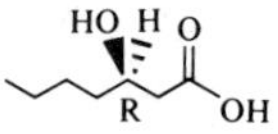	Predominantly R
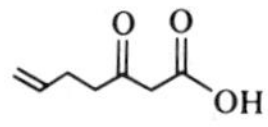	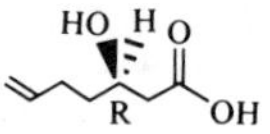	>99/1
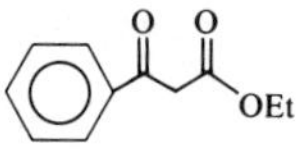	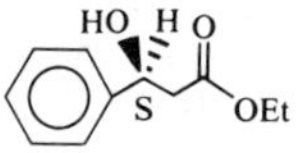	Predominantly S

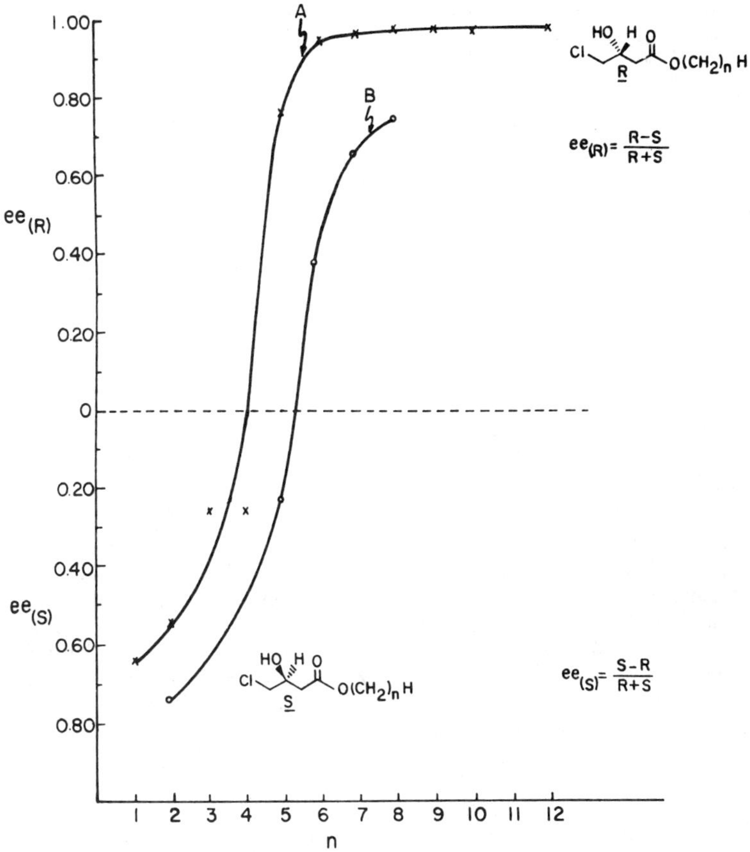

FIGURE 1. Plot of enantiomeric excess (ee) versus the size of ester grouping. Curve A: Red Star Baker yeast (4 g), tap water (20 ml), 23°C; γ-chloroacetoacetic esters (0.91 mmol). Curve B: Red Star Baker yeast (40 g), tap water (100 ml), 23°C; γ-chloroacetoacetic esters (9.0 mmol). Usual workup after 48 hours.

marked decrease in the reduction rate for the C_{12} ester. More importantly however, there was a dramatic shift in the stereochemistry of the carbinolic products formed as the size of the ester grouping was enlarged (FIG. 1). Curve A depicts that while the β-keto group of the methyl ester was reduced to its corresponding S enantiomer with an $ee_{(S)}$ of 0.65, the β-keto function of the octyl ester was reduced almost exclusively to the desired R enantiomer, $[\alpha]_D^{23}$ +15.1° (c, 4.66, $CHCl_3$) (70%), with an $ee_{(R)}$ of 0.97. Note the marked difference in the stereochemistry of the carbinols derived from the butyl and pentyl esters.

The ready availability of (R)-γ-chloro-β-hydroxybutyrate of high optical purity thus allows us to complete an asymmetric synthesis of L-carnitine. This was accomplished by reaction of octyl-(R)-γ-chloro-β-hydroxybutyrate with an excess of trimethylamine in ethanol at 80°C for two hours. The resulting crude mixture was then hydrolyzed with 3 *N* HCl for two hours. Direct crystallization from ethanol-acetone afforded L-carnitine chloride in 45% overall yield, mp 142° (dec); $[\alpha]_D^{23}$ −22.9° (c, 4.0, H_2O); lit.[19] $[\alpha]_D^{21}$ −23.7°.

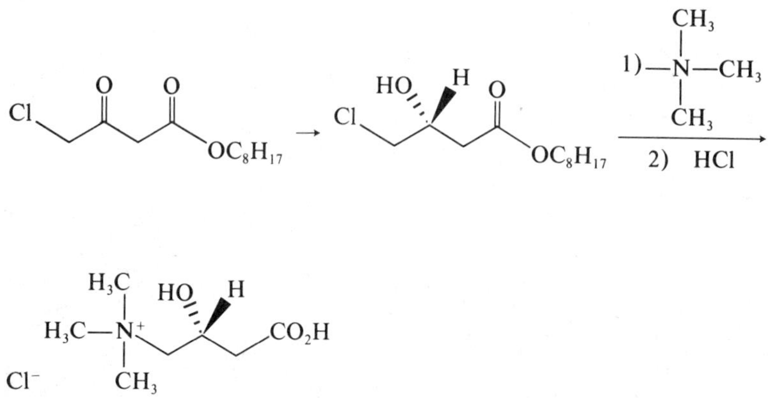

These results clearly demonstrate that the stereochemistry of yeast reduction of β-keto esters may be influenced by substituents at both ends of the molecule, which may be interpreted in two ways. If these β-keto esters are reduced by a single oxido-reductase, this enzyme is then able to interact with both faces of the carbonyl group to form two competing R and S transition states, one of which is more favored than the other. However, there exists another possibility in that bakers' yeast contains more than one oxido-reductase that can attack γ-chloroacetoacetic esters generating carbinolic products of opposite configurations at different rates (depending on V and K, which denote maximal velocity and Michaelis constant, respectively), leading to the production of unequal quantities of enantiomeric carbinols.

The first of these two possibilities may be examined only after the relevant oxido-reductase has been identified and isolated to a highly purified state. To examine the second possibility, we conducted the yeast reduction of various γ-chloroacetoacetic esters under a different set of experimental conditions (higher substrate concentration with glucose). If the reduction is catalyzed by a single enzyme, the *ee* of the carbinolic products should not be affected by changes in substrate concentrations or by the addition of a carbon source.[20] However, the results of curve B (FIG. 1) clearly show that the $ee_{(R)}$ decreased when the substrate concentration was raised tenfold in the presence of glucose, that is, the $ee_{(R)}$ for the pentyl ester at 0.045 *M* was 0.76 (curve A), but at 0.090 *M* in the presence of glucose, the S enantiomer was preferentially formed with an $ee_{(S)}$ of 0.23 (curve B). These results strongly suggest that bakers' yeast contains at least two oxido-reductases of opposite chirality producing γ-chloro-β-hydroxybutanoates of opposite configuration at different rates. This view is also consistent with the observation that there are moderate differences in the optical purity of ethyl-(S)-(+)-3-hydroxybutanoate obtained in different laboratories.[12] This variance may be attributed to the differences in the ethyl acetoacetate concentrations[21] used for the incubation as well as experimental conditions, such as the use of sucrose instead of glucose.

There are many oxido-reductases in bakers' yeast that could be responsible for catalyzing the reduction of these acetoacetic esters. Some of the oxido-reductases known to be present in *Saccharomyces cerevisiae* include: alcohol dehydrogenase

(ADH) I, II, III, and IV[22]; α-glycerophosphate dehydrogenase; D-α-hydroxy acid dehydrogenase; L-lactic dehydrogenase (L-LDH); D(−)-β-hydroxybutyryl and L(+)-β-hydroxybutyryl dehydrogenases.[23] However, crystalline yeast ADH I (EC 1·1·1·1) and purified yeast L-LDH (EC 1·1·1·27) were unable to reduce either ethyl-γ-chloroacetoacetate or octyl-γ-chloroacetoacetate. On the other hand, β-hydroxyacyl CoA dehydrogenase from porcine heart[24] (EC 1·1·1·35), an enzyme involved in the degradation of fatty acids, reduced both ethyl-γ-chloroacetoacetate and octyl-γ-chloroacetoacetate in the presence of NADH to their respective γ-chloro-3(R)-hydroxybutyrates with an *ee* of 0.98. Furthermore, we have recently found that purified fatty acid synthetase of bakers' yeast[25] readily reduced both ethyl-γ-chloroacetoacetate and octyl-γ-chloroacetoacetate in the presence of NADPH. The optical purities of the resulting γ-chloro-3-hydroxybutyrates as well as the purification of the β-hydroxyacyl CoA dehydrogenase from bakers' yeast are currently under investigation. These two enzymes are probably the major ones responsible for the reduction of these unnatural acetoacetic esters, for the normal physiological substrates of β-hydroxyacyl CoA dehydrogenase (a mitochondrial enzyme) and the β-keto-reductase component of fatty acid synthetase are the structurally analogous aceto-acetic thio esters. Furthermore, the former generates carbinolic products of the L-configuration whereas the latter generates those of the D-configuration. It is interesting to note that mammalian fatty acid synthetase[26] appears to possess a very relaxed substrate specificity, because the keto reductase component of this complex reduced cyclic ketones also.

The presence of competing enzyme systems acting on a single substrate yielding two products in living systems is a common occurrence. Some examples include: the enzymes engaged in branching pathways of amino acid biosynthesis, the enzymes involved in inactivation of aminoglycoside antibiotics, and various steroid hydroxylases competing for the steroid nucleus. The presence of two competing oxido-reductases for a ketonic substrate to yield two enantiomeric products in yeast cells is a special case of this general phenomenon. In order to quantitatively follow the stereochemical course of yeast reduction, we have derived a quantitative expression to relate the kinetic parameters of K_R and K_S (Michaelis constants of the R and S enzymes, respectively) and $y = V_R/V_S$ (maximum velocities of R and S enzymes) of the two competing enzymes to the observed stereochemical variables, *ee*, and C at fixed values of A_o.

$$ee = 1 - 2\,\frac{A_0C + (K_R - x)\ln\dfrac{A_0 + x}{A_0(1 - C) + x}}{(1 + y)\,A_0C}$$

$$\text{where } x = \frac{yK_S + K_R}{1 + y} \text{ and } \frac{y(K_R - K_S)}{1 + y} = (K_R - x)$$

When the apparent kinetic constants, K_R, and K_S and y are defined for the intact yeast cells, it is then possible to not only quantitatively follow the course of the enantioselective reduction of ketonic substrates, but also to predict *ee* for any values of C and A_o. Further, structural changes of substrates may be correlated to changes in the relative ratios of V/K and V of the competing enzyme systems. For a detailed discussion, the reader is referred to the paper by Chen *et al.*[27]

In conclusion, we have demonstrated the feasibility of directing the stereochemical course of yeast reductions of β-keto esters, for the enantioselective reduction of β-keto esters is influenced by substituents at both the hydrocarbon and the carboxyl ends of the molecule. Thus, by designing suitable substrates with differences in V and/or K values for the competing enzymes, it should be possible to prepare β-hydroxybutyrates of either S or R configurations. This approach has allowed us to obtain (R)-γ-chloro-β-hydroxybutyrate of high optical purity, and to complete an asymmetric synthesis of L-carnitine. Although not yet fully established, it is likely that the major competing enzyme systems responsible for the reduction of β-keto esters are the L-β-hydroxyacyl CoA dehydrogenase and the β-keto-reductase component of the fatty acid synthetase complex of bakers' yeast. We are currently purifying and characterizing these oxido-reductases as well as correlating their substrate stereoselectivities with those of intact bakers' yeast.

REFERENCES

1. NEUBERG, C. 1949. Biochemical reductions at the expense of sugars. Adv. Carbohydr. Chem. **4:** 75–117.
2. MACLEOD, R., H. PROSSER, L. FIKENTSCHER, J. LANYI & H. S. MOSHER. 1964. Asymmetric Reductions. XII. Stereoselective ketone reductions by fermenting yeast. Biochemistry **3:** 838–846.
3. DEOL, B. S., D. D. RIDLEY & G. W. SIMPSON. 1976. Asymmetric reductions of carbonyl compounds by yeast. II. Preparation of optically active α and β-hydroxy carboxylic acid derivatives. Aust. J. Chem. **29:** 2459–2467.
4. HIRAMA, M. & M. UEI. 1982. Chiral total synthesis of compactin. J. Am. Chem. Soc. **104:** 4251–4253.
5. LE DRIAN, C. & A. E. GREENE. 1982. Efficient stereocontrolled total syntheses of racemic and natural Brefeldin A. J. Am. Chem. Soc. **104:** 5473–5483.
6. BROOKS, D. W., P. G. GROTHAUS & W. L. IRWIN. 1982. Chiral cyclopentanoic synthetic intermediates via asymmetric microbial reduction of prochiral 2,2-disubstituted cyclopentanediones. J. Org. Chem. **47:** 2820–2821.
7. KIESLICH, K. 1969. Präparativ anivendbare mikrobiologische Reaktionen. Synthesis pp. 147–157.
8. ZHOU, B. N., A. S. GOPALAN, V. VANMIDDLESWORTH, W. R. SHIEH & C. J. SIH. 1983. Stereochemical control of yeast reductions. 1. Asymmetric synthesis of L-carnitine. J. Am. Chem. Soc. **105:** 5925–5926. (A preliminary account of this work.)
9. BREMER, J. 1977. Carnitine and its role in fatty acid metabolism. Trends Biochem. Sci. **2:** 207–209.
10. ENGEL, A. G. 1980. Carnitine Biosynthesis, Metabolism, and Function. R. A. Frenkel & J. D. McGarry, Eds. Academic Press. New York. pp. 271–285.
11. FRAENKEL, G. & S. FRIEDMAN. 1957. Carnitine. Vitam. Horm. N.Y. **16:** 73–118.
12. MORI, K. 1981. A simple synthesis of (S)-(+)Sulctol. The pheromone of *Gnathotrichus retusus,* employing bakers' yeast for asymmetric reduction. Tetrahedron **37:** 1341–1342. (See also references cited therein.)
13. OHARA, M. K. YAMAMOTO, T. KAMIYA & T. FUJISAWA. 1963. Butyric acid derivatives. Chem. Abst. **59:** 2654c.
14. HSU, C. T., N. Y. WANG, L. H. LATIMER & C. J. SIH. 1983. Total synthesis of the hypocholesterolemic agent compactin. J. Am. Chem. Soc. **105:** 593–601.
15. PRELOG, V. 1964. Specification of the stereospecificity of some oxidoreductases by diamond lattice sections. Pure Appl. Chem. **9:** 119–130.
16. LEMIEUX, R. U. & J. GIGUERE. 1951. Biochemistry of the ustilaginales. IV. The configurations of some β-hydroxy acids and the bioreduction of β-keto acids. Can. J. Chem. **29:** 678–690.
17. FRATER, G. 1979. Stereospezifische Synthese von (+)-(3R,4R)-4-Methyl-3-Heptanol, das Enantiomere eines Pheronions des kleinen Ulonensplintkäfers (*Scolytus multistriatus*). Helv. Chim. Acta **62:** 2829–2832.
18. HAMEL, J. F. 1921. Esters of γ-chloroacetoacetic acid. Bull. Soc. Chem. **29:** 390–402.

19. STRACK, E. & I. LORENZ. 1960. Die Darstellung von L-Carnitin und seiner Isomeren. Z. Physiol. Chem. **318:** 129–137.
20. CHEN, C. S., Y. FUJIMOTO, G. GIRDAUKAS & C. J. SIH. 1982. Quantitative analyses of biochemical kinetic resolutions of enantiomers. J. Am. Chem. Soc. **104:** 7294–7298.
21. WIPF, B., E. KUPFER, B. BERTAZZI & H. G. W. LEUENBERGER. 1983. Production of (+)-(S)-ethyl-3-hydroxybutyrate and (−)-(R)-ethyl-3-hydroxybutyrate by microbial reduction of ethyl acetoacetate. Helv. Chim. Acta **66:** 485–488.
22. LUTSTORF, U. & R. MEGNET. 1983. Multiple forms of alcohol dehydrogenase in *Saccharomyces cerevisiae*. Arch. Biochem. Biophys. **126:** 933–944.
23. LYNEN, F. 1961. Biosynthesis of saturated fatty acids. Fed. Proc. **20:** 941–951.
24. BRADSHAW, R. A. & B. E. NOYES. 1975. *In* Methods in Enzymology. J. M. Lowenstein, Ed. Vol. 35. Academic Press. New York. pp. 122–128.
25. SHIEH, W. R. & C. J. SIH. Unpublished data.
26. DUTLER, H., A. KULL & R. MISLIN. 1971. Fatty acid synthetase from pig liver. 2. Characterization of the enzyme complex with oxidoreductase activity for alicyclic ketones as a fatty acid synthetase. Eur. J. Biochem. **22:** 203–212.
27. CHEN, C. S., B. N. ZHOU, G. GIRDAUKAS, W. R. SHIEH, F. VANMIDDLESWORTH, A. S. GOPALAN & C. J. SIH. 1984. Bioorgan. Chem. **12:** 98–117.

New Enzymes for the Synthesis of Chiral Compounds

W. HUMMEL, H. SCHÜTTE, AND M.-R. KULA

Gesellschaft für Biotechnologische Forschung mbH
Mascheroder Weg 1
D-3300 Braunschweig, Federal Republic of Germany

The stereoselective chemical synthesis of chiral compounds is difficult to achieve and in general less efficient than enzyme-catalyzed synthesis. Nature utilizes a large number of chiral compounds, for example, amino acids, hydroxy acids, sugars, and so forth, which are derived by metabolism from simple, prochiral precursurs or by selective modification of chiral feedstocks. Chirality is the most uniform concept for encoding biological activity, recognition, and response. Therefore chiral compounds find important applications in many fields from nutrition and medicine to agriculture.

Some enzyme-catalyzed synthesis reactions have been developed already on an industrial scale, for example, production of L-aspartate and L-malic acid from fumarate.[1] However, a large number of interesting reactions require complex coenzymes together with the enzyme for catalysis. The first coenzyme-dependent process to be industrialized was carried out recently by Chibata *et al.* for the production of L-alanine from L-aspartate. The reaction is catalyzed by aspartate β-decarboxylase, which needs pyridoxal phosphate for proper functioning. In this case, the pyridoxal phosphate is regenerated during the catalytic cycle, so that only catalytic amounts are required to drive the process. In contrast, the most common coenzymes, NADH and ATP, are converted to inactive or even inhibitory forms, NAD^+ and ADP or AMP, respectively, during a catalytic cycle; thus a second reaction is needed to convert these products back to the starting material. NADH or ATP have to be supplied either in stoichiometric amounts or must be effectively regenerated simultaneously in order to achieve high conversions.

We favor the simultaneous enzyme-catalyzed regeneration of coenzymes. In the case of NADH regeneration and utilization, the development is quite advanced. Our concept is illustrated in the top part of FIGURE 1. Formate dehydrogenase can be used for the regeneration reaction; this enzyme can be produced in large amounts from *Candida boidinii.* The isolation of the enzyme has been discussed previously.[2,3] By employing a fast regeneration cycle, the stoichiometry is shifted from NADH to the consumption of the cosubstrate formate. This cosubstrate is a rather cheap and unobnoxious chemical. The equilibrium of the reaction lies far on the side of CO_2 and therefore on NADH formation, which simplifies the reaction scheme and its overall mathematical treatment. NAD(H) has to be shuffled back and forth between the consuming and regenerating enzymes. In order to minimize mass transfer resistances, we would prefer to carry out the reactions in homogeneous solutions. This can be performed technically in an enzyme membrane reactor.[4] Here the problem arises of how to keep the coenzyme in the reactor. This can be achieved by modification of the coenzyme, such as by covalently binding it to a water-soluble polymer like polyethylene glycol.[5] The modified NAD(H) is utilized efficiently by a number of dehydrogenases tested; in fact, the reaction rate of formate dehydrogenase is increased about 50% compared to the native form of the coenzyme. A typical enzyme membrane reactor, shown schematically in FIGURE 2, has been operated for extended periods of time[6,7]

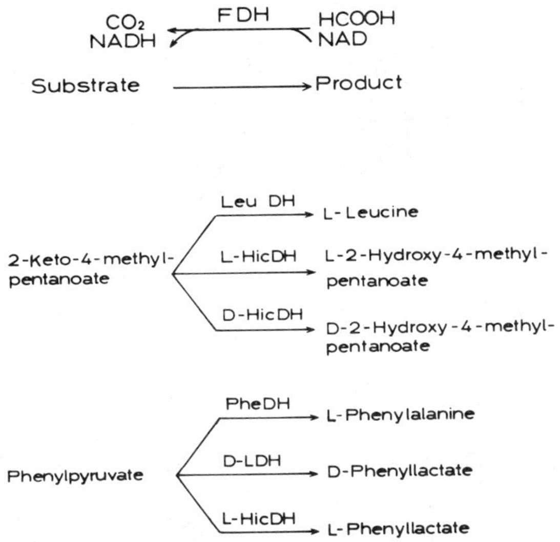

FIGURE 1. Concept for the enzyme-catalyzed production of chiral compounds emphasizing NADH regeneration via formate dehydrogenase (FDH). Several substrate-to-product examples are shown in the two lower reactions.

with space-time yields of 1 kg/l · day for L-leucine production.[8] Between 20,000 and 70,000 moles of product have been obtained per mol of NADH lost in the process. Further scale-up for this NADH-dependent process is in progress.

NADH-DEPENDENT L-AMINO ACID PRODUCTION

The confidence gained in the successful bench-scale operation of an NADH-regenerating system prompted us to look for further applications. NADH is utilized as

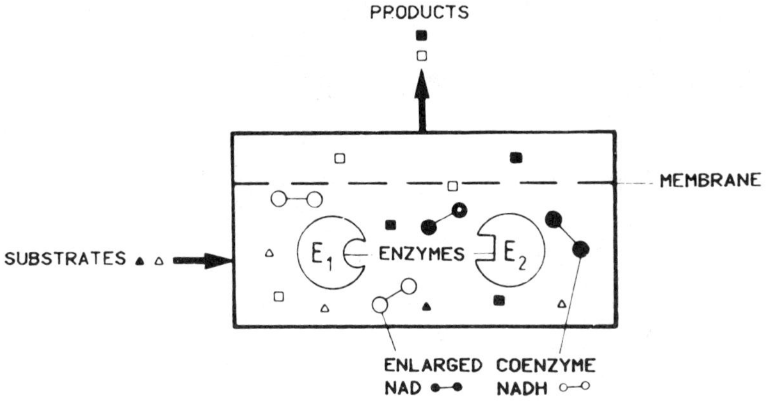

FIGURE 2. General scheme of an enzyme membrane reactor for the continuous product formation catalyzed by multienzyme systems.

TABLE 1. Comparison of Amino Acid Dehydrogenases

	Leu DH	Phe DH	Ala DH	Glu DH
Organism	*Bacillus* sp.	*Brevibacterium* sp.	*Bacillus* sp.	Widely Distributed (Microorganisms, Plants, Animals)
pH-optimum				
reductive amination	9.0	8.5[a]	8.5[b]	7.9[c]
oxidative desamination	10.5	10.5[a]	10.0[b]	9.0[c]
Optimum conc. (*M*) of NH_4^+	0.7	0.7[a]	0.3[b]	0.4[c]
Coenzyme	only NAD^+	only NAD^+	only NAD^+	NAD^+ and/or $NADP^+$
Substrates	aliphatic L-aminoacids for example: leucine, valine, isoleucine	aromatic L-amino acids for example: phenylalanine, tyrosine	short-chain amino acids for example: alanine, α-aminobutyrate	mainly glutamate

[a]Measured in crude extract.
[b]Cited from Yoshida & Freese.[14]
[c]Enzyme from *E. coli*.

a coenzyme for the reductive amination of α-keto acids to L-amino acids or the stereospecific reduction of keto- groups to hydroxyl groups. The reductive amination of 2-ketoisocaproate to L-leucine, catalyzed by leucine dehydrogenase, is discussed above. The same enzyme may be used to produce L-valine and L-isoleucine from the corresponding keto acids or to produce the nonprotein amino acid L-2-amino-3,3-dimethyl butyric acid. Besides leucine dehydrogenase, alanine dehydrogenase and glutamate dehydrogenase are well-known enzymes of this class, which play an important role in the nitrogen metabolism of cells. However, for the production of L-alanine and L-glutamic acid other methods that do not use these enzymes are well established.

We have been interested in new amino acid dehydrogenases and recently discovered and described a L-phenylalanine dehydrogenase (TABLE 1). Various microorganisms able to grow on phenylalanine as carbon and nitrogen sources were examined for the ability to convert phenylpyruvate to L-phenylalanine in the presence of NADH and

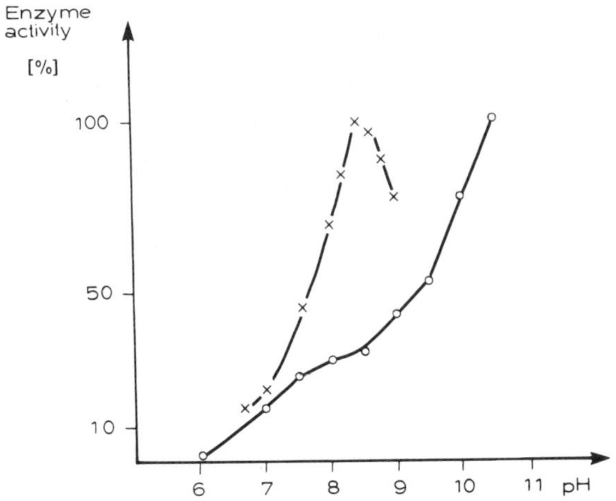

FIGURE 3. Influence of pH on activity of phenylalanine dehydrogenase (crude extract). Reductive amination (x) was measured in NH_4Cl/NH_4OH and oxidative desamination (O) in glycine buffers.

ammonia ions. One aerobic *Brevibacterium* species was found to possess the desired enzymatic activity.[9] Since among the genus *Brevibacterium* and related genera are well-known amino acid producers, we checked 25 strains from a culture collection (DSM) but failed to detect the enzyme in the crude extract of these strains. When *Brevibacterium sp.* is grown on carbohydrates, the new enzyme is inducible not only by L-phenylalanine but also by racemic D,L-phenylalanine, D-phenylalanine, and L-histidine. The enzyme is completely released from the cell during disintegration in a glass bead mill, indicating that the activity is not membrane bound. It has been partially purified to a specific activity of 30–40 IU/mg. As shown in FIGURE 3, the reductive amination reached maximal velocity around pH 8.5; while the rate of the NAD^+-dependent oxidation increased up to pH 10.5, the highest value studied. These results are comparable to other amino acid dehydrogenases, for example, leucine dehydrogenase or alanine dehydrogenase (TABLE 1). Phenylalanine dehydrogenase is also a valuable analytical tool. L-Phenylalanine can be determined, as shown in FIGURE 4, by the NAD^+-dependent oxidation reaction. This property might be developed further and exploited for process control. We examined the substrate specificity and found that the enzyme acts on several aromatic 2-ketoacids to yield L-phenylalanine, L-tyrosine, and L-tryptophane. Benzoylformic acid, however, was not a suitable substrate.

The time course of continuous L-phenylalanine production in a membrane reactor is shown in FIGURE 5. The average conversion of phenylpyruvate during the 110-hour operation was 70%. The product was isolated by ion exchange chromatography on Dowex AG 50WX8. The purity of the final product was analyzed by amino acid analysis yielding a single peak in the chromatogram corresponding to phenylalanine. The optical purity was checked in a polarimeter and by reaction with D-amino acid oxidase. Both methods indicated that the phenylalanine produced consisted of at least 99.9% L-enantiomer. The FIGURE 5 results ruled out the possibility that an efficient transamination reaction coupled with glutamate dehydrogenase was observed with the assay system. Glutamate and ketoglutarate present in the crude extract would be washed out of the reactor after 3–5 residence times; this corresponds to 15–25 hours, when the reaction was still progressing to high yields.

D-LACTATE DEHYDROGENASE

Besides L-amino acids, chiral 2-hydroxycarboxylic acids also can be obtained from 2-ketocarboxylic acids by NADH-dependent, enzyme-catalyzed processes. We devel-

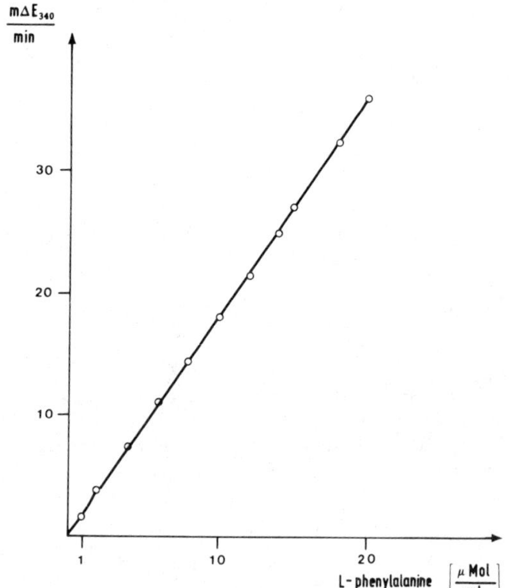

FIGURE 4. Calibration curve for the determination of L-phenylalanine using phenylalanine dehydrogenase.

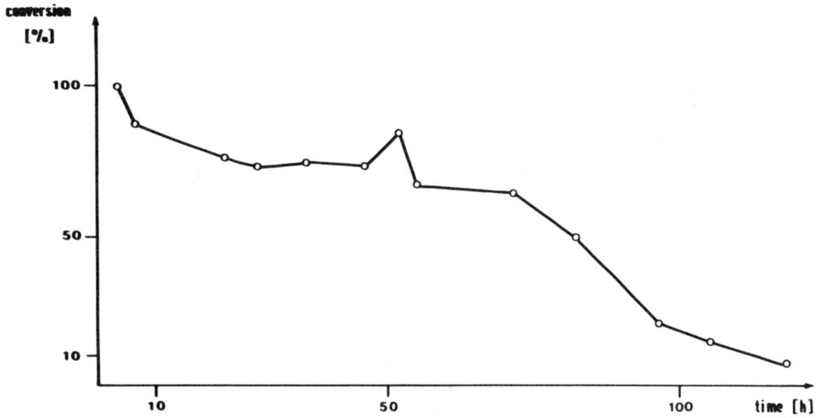

FIGURE 5. Continuous production of L-phenylalanine in an enzyme membrane reactor by reductive amination of 20 m*M* phenylpyruvate.

oped recently the large-scale production and isolation of a D-lactate dehydrogenase from *Lactobacillus confusus*.[10] This strain was selected because the D-lactate dehydrogenase was able to reduce both pyruvate and also the substrate analogue phenylpyruvate. The K_m-values for these substrates are 0.68 m*M* and 3.0 m*M*, respectively, while NADH has a favorable low K_m-value of 0.04 m*M*. The apparent V_{max} values of the highly purified enzyme are 2,144 μmol/min per mg protein at 30°C with pyruvate and 253 μmol/min with phenylpyruvate as substrate. The pH-optimum for the reduction of pyruvate was in the range 6–6.5. The reverse reaction, the NAD^+-dependent dehydrogenation, proceeded best at pH 9–11. The reaction rate of the dehydrogenation reached only 0.3% of the stereospecific reduction reaction, measured under the respective optimal conditions. Buchanan and Gibbons[11] reported that *Lactobacillus confusus* produces D-lactate during fermentation. We confirmed that the lactate dehydrogenase isolated from *Lactobacillus confusus* (DSM 20 196) reduces both substrates to the D-enantiomere.

ENZYME MEMBRANE REACTOR AS A TOOL FOR DETERMINATION OF THE STEREOSPECIFICITY OF ENZYMES

For reasons of sensitivity and convenience, the assay for D-lactate dehydrogenase as well as other dehydrogenases is frequently carried out by following the decrease in the NADH concentration during the reduction of the ketocarboxylic acid. The optical assay in this way does not allow a positive identification of the chirality of the reaction product. Many 2-hydroxycarboxylic acids exhibit a rather small optical rotation; therefore, the chirality of the product formed is not easily determined. A direct measurement of the reaction mixture is not possible due to the presence of other chiral compounds, such as NAD^+ and NADH. The membrane reactor shown schematically in FIGURE 6 allows a rather fast determination of the optical rotation of the product, even for enzymes in crude extracts. This system has been employed as part of a screening procedure for the enzymes described below. The reactor is filled with a solution of (1) substrate in 0.3 *M* sodium formate buffer pH 7.0, (2) 0.3 μmol/ml

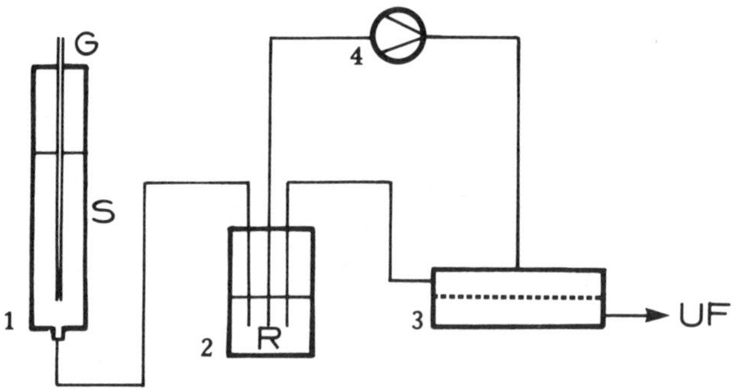

FIGURE 6. Enzyme membrane reactor for the determination of the stereospecificity of enzymes. G is a glass tube, S is substrate, and UF is ultrafiltrate. 1 is reservoir with buffered substrate solution; 2 is thermostated reaction vessel; 3 is ultrafiltration membrane reactor. (a CEC-1 unit can be conveniently employed); and 4 is circulation pump.

molecular-weight-enlarged NADH, (3) 3 IU/ml of formate dehydrogenase for coenzyme regeneration, and (4) an appropriate amount of the dehydrogenase to be tested. In the case of D-lactate dehydrogenase, 8 IU/ml were used. The coenzyme is kept in the reactor loop by means of an YM5 membrane. The reaction mixture is pumped along the surface of the ultrafiltration membrane with sufficient speed (approximately 30 ml/h) to create a small pressure differential that leads to an ultrafiltration rate of 3 ml/h. The volume in the reactor loop is kept constant, by controlling the liquid level in the thermostated mixing vessel. The total volume in the reactor loop was 10 ml. This corresponded to a residence time of 3.3 hours for an ultrafiltration rate of 3.0 ml/h. From mass balance consideration, it follows that small chiral compounds introduced with the crude extract will be washed out through the membrane in an exponential fashion. At the same time, the reaction will reach steady-state levels.[12] After 10 hours of operation, that is, three residence times, 96% of the initially present chiral compounds will be replaced by the product. After 15 hours, the reading in the ultrafiltrate should be steady.

In the experiments with lactate dehydrogenase and 6 m*M* substrate, the optical rotation in the ultrafiltrate measured at 436 nm and 25°C reached −0.016 for pyruvate and −0.097 for phenylpyruvate as substrate, respectively. Calibration carried out with D-lactate and D-phenyllactate in 0.3 *M* sodium formate buffer, pH 7.0, gave 94% and 97% conversion to the respective D-hydroxy acids. This confirmed the stereoselectivity of the enzyme-yielding D-enantiomers. A 90% conversion is sufficient to evaluate enzymes in a screening procedure. In order to determine the stereospecificity of the reaction, more highly purified enzymes and optimized reaction conditions should be employed.

2-HYDROXYISOCAPROATE DEHYDROGENASES

During the isolation of the D-lactate dehydrogenase from *Lactobacillus confusus,* we observed the selective removal of an enzymatic activity towards 2-ketoisocaproate during the second partition step. These results were analyzed in detail and led to the isolation and description of a new enzyme, the L-2-hydroxyisocaproate dehydrogenase.[13] The high resolving power of partition in aqueous phase systems is illustrated in

FIGURE 7. We assume that hydrophobic areas on the surface of the new enzyme, possibly involved in the binding of substrates with extended hydrophobic side chains, contribute to the unusually high partition coefficient. The enzyme clearly prefers the more hydrophobic polyethylene glycol-rich top phase in a PEG/salt system under conditions when most other proteins from *Lactobacillus confusus,* including D-lactate dehydrogenase, are extracted into the bottom phase. This fact was utilized in an improved isolation procedure of L-2-hydroxyisocaproate dehydrogenase from *Lactobacillus confusus.* The first extraction separating the cell debris from the crude enzyme could be performed using PEG 6000 instead of PEG 1500. The specific activity of the crude enzyme (0.125 IU/mg) was increased to 2.05 IU/mg. The purification factor increased 10-fold compared to the phase system with PEG 1500. The influence of PEG of different molecular weights and pH on the partition coefficient and specific activity of L-2-hydroxyisocaproate dehydrogenase is shown in FIGURES 8 and 9. With a second aqueous phase system, the enzyme can be shifted from the polyethylene glycol-rich top phase to the salt-rich bottom phase by adding potassium phosphate salt and sodium chloride to the top phase of the first extraction system. The partition coefficient of the enzyme decreases significantly with increasing sodium chloride concentration, as can be seen in FIGURE 10. The two partition steps described above improve the specific activity of the enzyme at least 25-fold. Further purification can be carried out as described by Schütte *et al.*[13] The stereoselectivity of the enzyme was checked as described above and found to yield the L-enantiomer.

A limited screening among 45 strains of *Lactobacillus* and *Leuconostoc* was carried out. This was done in order to find a NADH-dependent dehydrogenase, yielding the corresponding D-2-hydroxyisocaproate and to find another producer of the L-2-hydroxyisocaproate dehydrogenase. Besides *Lactobacillus confusus,* 6 strains

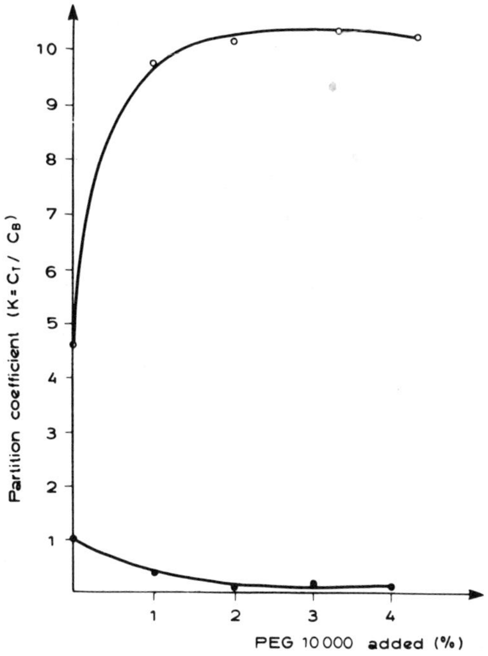

FIGURE 7. Partition of L-2-hydroxyisocaproate dehydrogenase (○) and D-lactate dehydrogenase (●) as a function of the concentration of PEG 10,000 (50% upper phase I; 10% wt/vol potassium phosphate, pH 6.1, and 200 m*M* NaCl). C_T is concentration in top and C_B in bottom phase.

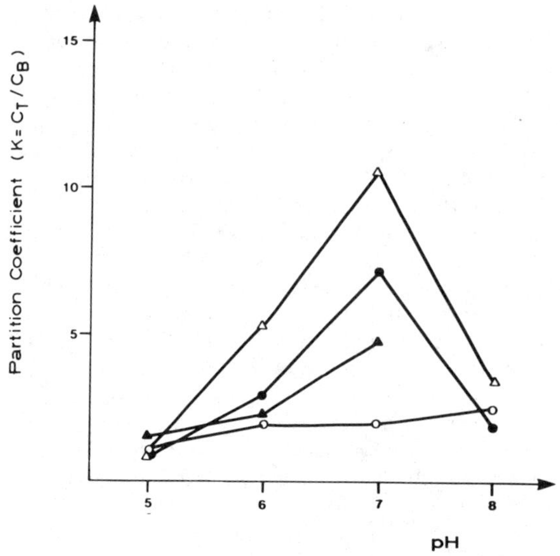

FIGURE 8. Dependence of the partition coefficient of L-2-hydroxyisocaproate dehydrogenase on pH for 20% wt/wt crude extract, 18% (wt/wt) PEG, 7% (wt/wt) potassium phosphate. Results for (○) PEG 1540; (●) PEG 4000; (△) PEG 6000; and (▲) PEG 10,000.

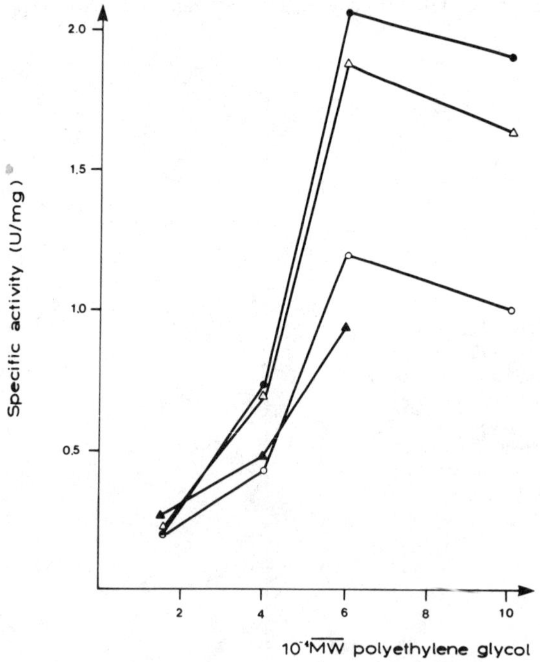

FIGURE 9. Influence of the average molecular weight of PEG on the specific activity of L-2-hydroxyisocaproate dehydrogenase extracted in the top phase under the same conditions as for FIGURE 8. Results at (○) pH 5; (●) pH 6; (△) pH 7; and (▲) pH 8.

were found to possess an NADH-dependent dehydrogenase activity towards 2-ketoisocaproate. All of the newly identified strains however reduced 2-ketoisocaproate to the D-2-hydroxycarboxylic acid. The D-2-hydroxyisocaproate dehydrogenase was purified from *Lactobacillus casei spp. pseudoplantarum* (DSM 20 008). The substrate specificity of both enzymes was investigated and is summarized in TABLE 2. There are some definite but small differences between the enzymes with regard to substrate recognition and reaction rates. Both enzymes act on various straight and branched-chain aliphatic 2-ketocarboxylic acids. The hydrophobic binding site, however, also accommodates aromatic side chains, and the D-2-hydroxyisocaproate

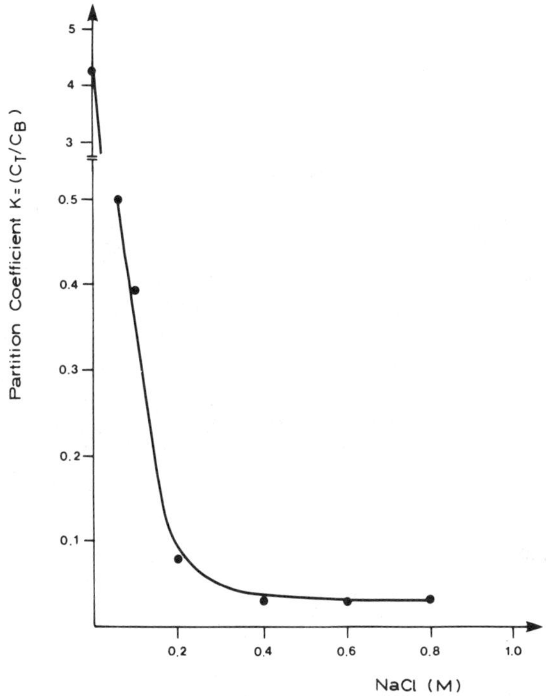

FIGURE 10. Effect of sodium chloride on the partition of L-2-hydroxyisocaproate dehydrogenase (50% top phase I; 10% (wt/vol) potassium phosphate, pH 6.0, and variable amounts of sodium chloride.)

dehydrogenase from *L. casei* is after all a better catalyst than D-lactate dehydrogenase from *L. confusus* for the stereospecific reduction of phenylpyruvate to D-phenyllactate. The screening procedure already ascertained that the dehydrogenases will accept molecular-weight-enlarged NADH. The K_m-values for the coenzymes are low enough to reach enzyme saturation in the membrane reactor in the concentration range 0.3–0.5 m*M*. TABLE 2 indicates that various L- or D-2-hydroxycarboxylic acids can be produced utilizing the new enzymes as specific catalysts. Wichmann *et al.* report on the continuous synthesis of L- and D-2-hydroxy isocaproate, respectively, elsewhere in this volume.

TABLE 2. Substrate Specifity of L- and D-2-Hydroxyisocaproate Dehydrogenase and D-Lactate Dehydrogenase for the Hydrogenation Reaction[a]

Substrate	Structure	K_m value (mM)		
		L-HicDH	D-HicDH	D-LDH
Pyruvate	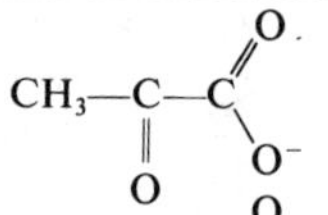	4,5	n.d.[b]	0.68
Phenylpyruvate	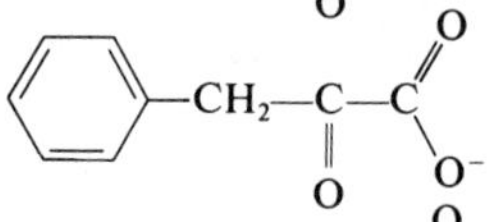	0,12	0,15	3,0
2-Ketobutyrate	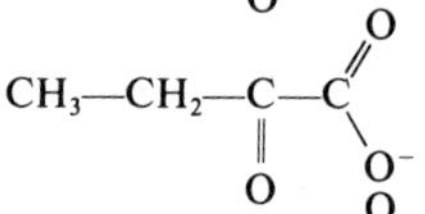	0,45	1,7	7,2
2-Ketovalerate	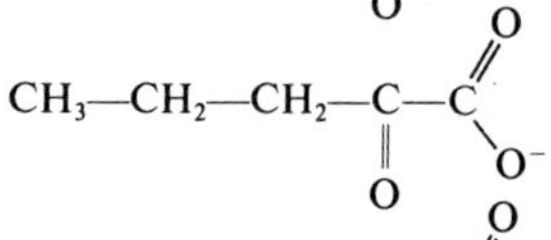	0,1	0,11	n.a.[c]
2-Ketocaproate	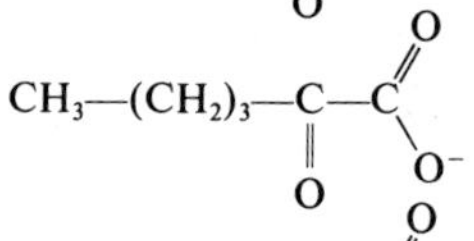	0,12	0,11	n.a.
2-Ketooctanoate	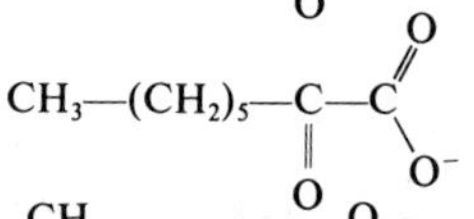	0,17	0,31	n.a.
2-Ketoisovalerate	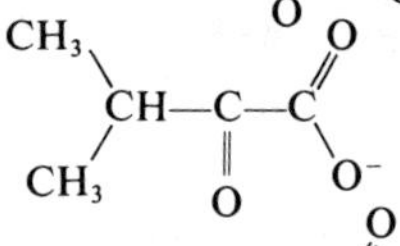	0,065	4,81	n.a.
2-Keto-3-methylvalerate	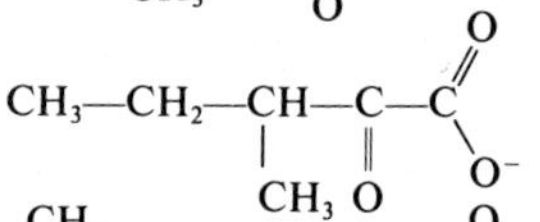	0,5	2,2	n.a.
2-Ketoisocaproate	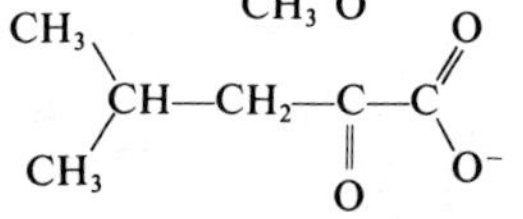	0,06	0,06	n.a.

[a]The K_m values for the substrates indicated were determined for the D-lactate dehydrogenase as described by Hummel *et al.*[10] and for the L-2-hydroxyisocaproate dehydrogenase as described by Schütte *et al.*[13] K_m-values for the D-hydroxyisocaproate dehydrogenase were determined in a mixture containing 100 mM potassium phosphate, pH 7.0; 0.1 mM NADH and rising amounts of the substrate.
[b]n.d. = not determined.
[c]n.a. = no activity.

REFERENCES

1. TOSA, T. T. SATO, T. MORI, Y. MATUO & I. CHIBATA. 1973. Continuous production of L-aspartic acid by immobilized aspartase. Biotechnol. Bioeng. **15:** 69–84.
2. KULA, M.-R., K. H. KRONER & H. HUSTEDT. 1982. Scale-up of protein purification by liquid-liquid extraction. *In* Enzyme Engineering. I. Chibata, S. Fukui, L. B. Wingard, Jr., Eds. Vol. **6:** 69–74. Plenum Press. New York.

3. KRONER, K. H., H. SCHÜTTE, W. STACH & M.-R. KULA. 1982. Scale-up of formate dehydrogenase by partition. J. Chem. Tech. Biotechnol. **32:** 130–137.
4. WANDREY, C., R. WICHMANN, A. F. BÜCKMANN & M.-R. KULA. 1978. Immobilization of biocatalysts using ultrafiltration techniques. DECHEMA Preprints. First Eur. Congr. Biotechnol. Part 1. pp. 44–47.
5. BÜCKMANN, A. F., M.-R. KULA, R. WICHMANN & C. WANDREY. 1981. An efficient synthesis of high-molecular-weight NAD(H). J. Appl. Biochem. **3:** 301–315.
6. WICHMANN, R., C. WANDREY, A. F. BÜCHMANN & M.-R. KULA. 1981. Continuous enzymatic transformation in an enzyme membrane reactor with simultaneous NAD(H) regeneration. Biotechnol. Bioeng. **23:** 2789–2802.
7. WICHMANN, R., W. HUMMEL, H. SCHÜTTE, A. F. BÜCKMANN, C. WANDREY & M.-R. KULA. 1984. Enzymatic production of optically active 2-hydroxy acids. Ann. N. Y. Acad. Sci. **434:** This volume.
8. WANDREY, C., R. WICHMANN, A. F. BÜCKMANN & M.-R. KULA. Unpublished results.
9. HUMMEL, W., N. WEISS & M.-R. KULA. 1984. Isolation and characterization of a bacterium possessing L-phenylalanine dehydrogenase activity. Arch. Microbiol. **137:** 47–52.
10. HUMMEL, W., H. SCHÜTTE & M.-R. KULA. 1983. Large-scale production of D-lactate dehydrogenase for the stereospecific reduction of pyruvate and phenylpyruvate. Eur. J. Appl. Microbiol. Biotechnol. **18:** 75–85.
11. BUCHANAN, R. E. & N. E. GIBBONS. 1974. *In* Bergey's Manual of Determinative Bacteriology. 8th Ed. Williams Wilkins. Baltimore.
12. FLASCHEL, E., C. WANDREY & M.-R. KULA. 1983. Ultrafiltration for the separation of biocatalysts. *In* Advances in Biochemical Engineering. A. Fiechter, Ed. Vol. **26:** 73–142. Springer Verlag. Berlin.
13. SCHÜTTE, H., W. HUMMEL & M.-R. KULA. 1984. L-2-hydroxyisocaproate dehydrogenase—a new enzyme from *Lactobacillus confusus* for the stereospecific reduction of 2-ketocarboxylic acids. Eur. J. Appl. Microbiol. Biotechnol. **19:** 167–176.
14. YOSHIDA, A. & E. FREESE. 1965. Enzymic properties of alanine dehydrogenase of *Bacillus subtilis*. Biochim. Biophys. Acta **96:** 248–262.

Enzymatic Synthesis of Optically Pure 1-Carbacephem Compounds

Y. HASHIMOTO, S. TAKASAWA, T. OGASA, H. SAITO, AND T. HIRATA

Tokyo Research Laboratories
Kyowa Hakko Kogyo Co., Ltd.
Tokyo, Japan

K. KIMURA

Safety Research Laboratories
Kyowa Hakko Kogyo Co., Ltd.
Yamaguchi, Japan

1-Carbacephem compounds were first synthesyzed by B. G. Christensen *et al.*[1] These compounds have an (un)substituted methyl group at C-3 (FIG. 1a) and showed slightly less potent antimicrobial activity than corresponding cephalosporin derivatives. T. Hirata *et al.*[2] reported the synthesis of a 1-carbacephem compound with no substituent at C-3 (FIG. 1b). Upon acylation of the 7-NH_2 group, this nucleus showed quite potent antimicrobial activity. But this compound was still a racemic mixture of the (6R,7S) and (6S,7R) forms (FIG. 1c,d), and optical resolution was attempted to obtain one enantiomer with presumed (6R,7S) absolute configuration that corresponds to the (6R,7R) configuration of the cephem nucleus of natural origin.

Penicillin acylase (benzylpenicillin amidohydrolase E.C.3.5.1.11) catalyzes hydrolytic deacylation of penicillin G to 6-aminopenicillanic acid (6-APA), which is an important intermediate for semisynthetic penicillins and has been used commercially in the penicillin industry. This enzyme can also catalyze N-acylation of β-lactam nuclei with organic acid esters to produce the corresponding β-lactam antibiotics.[3] There has been much work on the substrate specificity of this enzyme. Serizawa *et al.* reported that this enzyme could produce optically active 7-thienylacetyl-3-trifluoromethylcephalosporin from methyl thienyl-acetate and (±)-7-amino-3-trifluoromethyl-3-cephem-4-carboxylic acid.[4] However, the enantioselectivity of this enzymatic reaction has not been presented in detail. We applied this enzyme in the enantiospecific synthesis of D-phenylglycyl (or *p*-hydroxy-D-phenylglycyl) 1-carbacephem compounds from D-phenylglycine methyl ester and racemic mixture of 1-carbacephem nuclei, and also in enantioselective hydrolysis of (±)phenylacetyl-1-carbacephem compounds to produce optically pure 1-carbacephem nuclei from which useful antimicrobial agents should be derived by chemical acylation.

A racemic mixture of 3-H nucleus (Nor),D-phenylglycine methyl ester (D-PGM) and intact cells of *Pseudomonas melanogenum* KY-8541 (a β-lactamase-deficient mutant from KY-3937[5]) were mixed and incubated at 30°C in weakly acidic phosphate buffer at pH 6.5 for 30 min. The reaction mixture was chromatographed on reverse-phase HPLC. In FIGURE 2, a typical chromatogram of the reaction mixture is shown. Conditions of this chromatography are as follows: octadecylsilica as reverse-phase column packing, 7% methanol with 0.2 *M* phosphate as the mobile phase and flow rate at 3 ml/min. Under these conditions the diastereomeric D-phenylglycyl-Nor (PG-Nor) that was produced by chemical acylation of (±)Nor with phenylglycine derivative could be separated by this chromatography. Retention times of these

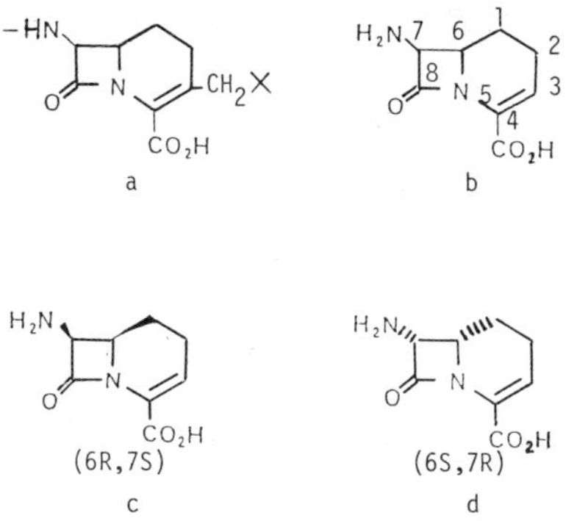

FIGURE 1. Structures of 1-carbacephem nuclei.

compounds are 6.1 min and 13.2 min, which presumably correspond to diastereoisomers with configuration of (6S,7R) and (6R,7S), respectively. It is clear in this chart that only one enantiomer of the nucleus [presumably (6R,7S)] was phenylglycylated, leaving the other enantiomer unacylated. Now it was demonstrated that penicillin acylase of *P. melanogenum* could recognize the unnatural 1-carbacephem nucleus like other β-lactam nuclei of natural origin and moreover could acylate this nucleus enantioselectively.

Many other bacteria known to produce penicillin acylase were also found to be able to synthesize PG-Nor enantioselectively (TABLE 1).

Phenylacetyl-Nor(PA-Nor) was hydrolyzed enantioselectively by *Kluyvera citro-*

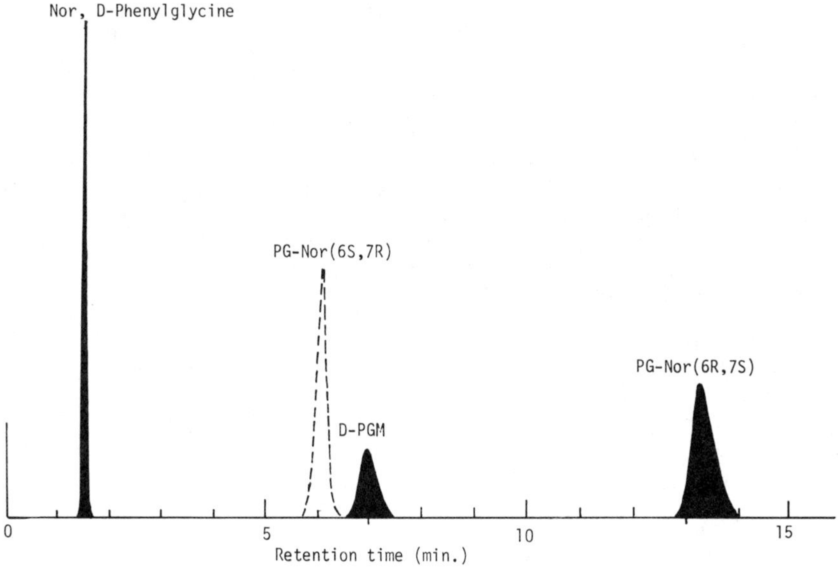

FIGURE 2. HPLC of reaction mixture, Nor(±) and D-PGM with *P. melanogenum* KY-8541. Dotted line, which corresponds to that of PG-Nor(6S,7R), was not actually detected in this chromatogram.

TABLE 1. Enantioselective Synthesis of PG-Nor (6R,7S) by Various Bacteria[a]

Bacterial Strain	Accumulation of PG-Nor (mg/ml) (6R,7S)	(6S,7R)
Acetobacter sp. KY 3035	1.02	0
Beneckea hiperoptica KY 3132	0.54	0
Gluconobacter dioxyacetonicus KY 3623	0.20	0
Mycoplana dimorpha KY 3813	0.98	0
Protaminobacter alboflavus KY 4074	1.21	0
Xanthomonas cucurbitae KY 4127	0.81	0
Kluyvera citrophila KY 7844	0.40	0
Escherichia coli KY 8219	0.06	0
Pseudomonas melanogenum KY 8541	1.06	0

[a]Nor(±) 3 mg/ml, D-PGM 10 mg/ml, and washed cells of each bacterium (10 mg/ml as dry cell weight) were mixed and incubated at 30°C, at pH 6.5 for 1 hour.

phila KY-7844, which has been known to produce penicillin acylase with broad substrate specificity.[6] Conversion yield of this hydrolysis was almost theoretical and antipode PA-Nor with presumable (6S,7R) configuration was the sole residue. Optimum conditions of this reaction were almost the same as those for penicillin G. Antipode originally contained in the substrate did not interfere with this reaction at all. The optically active nucleus thus obtained was then acylated chemically to produce a compound, KT-3767, whose antimicrobial activity is comparable to those of so-called third-generation cephalosporins. The chemical structure of KT-3767 is as follows:

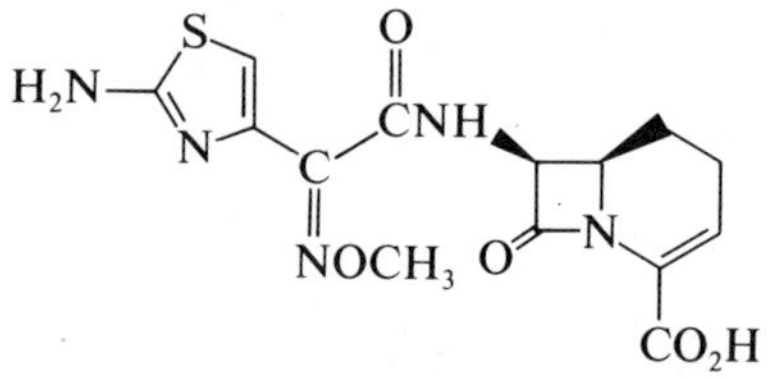

MIC values of KT-3767 are shown in TABLE 2 comparing with the racemate KT-3715, which was derived chemically from the optically inactive Nor nucleus. MIC values of KT-3767 are less than half of KT-3715; that is, the antimicrobial activity of KT-3767 is more than twice that of KT-3715. The reason why the activity of KT-3767 is more than twice that of the racemate is unknown.

TABLE 2. Antimicrobial Activity of 1-Carbacephem Derivatives[a]

Strain	MIC (μg/ml) KT-3715	KT-3767
Staphylococcus aureus 209P	12.5	6.25
Escherichia coli NIHJ JC-2	0.05	≤0.01
Klebsiella pneumoniae Y-60	0.05	≤0.01
Proteus vulgaris 6897	0.01	≤0.01
Pseudomonas aeruginosa #1	25	6.25

[a]Assayed by agar dilution method.

REFERENCES

1. R. N. GUTHIKONDA, L. D. CAMA & B. G. CHRISTENSEN. 1974. Total synthesis of β-lactam antibiotics VII. Total synthesis of (+) 1-oxacephalothin. J. Am. Chem. Soc. **96**(24): 7582–7585.
2. T. HIRATA, T. OGASA, H. SAITO, S. KOBAYASHI, A. SATO, Y. ONO, Y. HASHIMOTO, S. TAKASAWA, K. SATO & K. MINEURA. 1981. Antimicrobial activity of carbacephem compounds, novel cephalosporin-like β-lactams. 21st Interscience Conference on Antimicrobial Agents and Chemotherapy. Abstract 557.
3. A. PLASKIE, E. ROETS & H. VANDERHAEGHE. 1978. Substrate specificity of penicillin acylase of *E. coli.* J. Antibiot. **31**(8): 783–788.
4. N. SERIZAWA, K. NAKAGAWA, S. KAMIMURA, T. MIYADERA & M. ARAI. 1979. Stereospecific synthesis of 3-trifluoromethylcephalosporin derivative by microbial acylase. J. Antibiot. **32**(10): 1016–1018.
5. R. OKACHI, Y. HASHIMOTO, M. KAWAMORI, R. KATSUMATA & T. NARA. 1982. Enzymatic synthesis of penicillins and cephalosporins by penicillin acylase. *In* Enzyme Engineering. I. Chibata, S. Fukui & B. Wingard, Jr., Eds. Vol. **6:** 81–90. Plenum Press. New York.
6. M. SHIMIZU, R. OKACHI, K. KIMURA & T. NARA. 1975. Purification and properties of penicillin acylase from *Kluyvera citrophila.* **39**(8): 1655–1661.

Synthesis of Phenoxymethylpenicillin Using α-Acylamino-β-lactam Acylhydrolase from *Erwinia aroideae*

D. H. NAM

The Korea Advanced Institute of Science
Seoul, Korea

DEWEY D. Y. RYU

University of California
Davis, California 95616

During the last decade, significant advances have been made for enzymatic synthesis of semisynthetic β-lactam antibiotics. The antibiotics include: ampicillin,[1,2] amoxicillin,[3] cephalexin,[4–6] cephaloglycin,[7] cefadroxil,[8] and cephacetril.[9] Enzymatic preparation of semisynthetic antibiotics has some advantages, such as high conversion, transformation of only specific compounds, one-step reaction, and use of mild conditions without having to use toxic solvents.

It has been found that enzymatic deacylation of β-lactam antibiotics appears to be irreversible, especially in the cases of benzylpenicillin,[10] cephalexin,[4] cephaloglycin,[7] and cephacetril.[9] This means that β-lactam antibiotics can be hydrolyzed to the acid forms of their side-chain groups and their nuclei, but can be synthesized from their nuclei and the ester or amide forms of their side-chain moieties (FIG. 1).

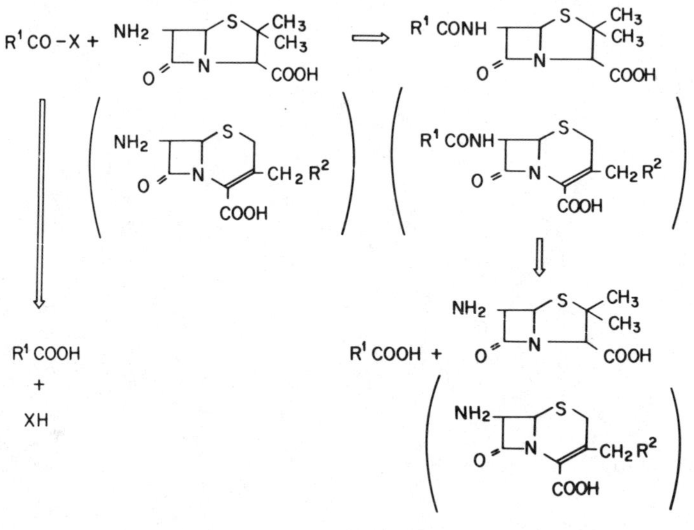

FIGURE 1. Reactions catalyzed by α-acylamino-β-lactam acylhydrolase during enzymatic synthesis of β-lactam antibiotics.

TABLE 1. Effect of Kinds and Concentrations of Acyl Donors on PNV Production at 37°C and pH 7.0[a]

Acyl Donor	Concentration (m*M*)	Overall Conversion (%)
POM	10	3.6
	20	4.8
	30	8.0
	40	11.0
	50	11.2
	60	11.8
	80	12.2
POG	30	0.6
POA	30	0

[a] 10 m*M* of 6-APA was employed as an acyl acceptor.

In this study, enzymatic synthesis of phenoxymethylpenicillin from 6-aminopenicillanic acid (6-APA) and phenoxyacetic acid methyl ester (POM) was attempted by using a partially purified α-acylamino-β-lactam acylhydrolase (ALAHase I) from *Erwinia aroideae* NRRL B-138,[11] which has been named penicillin-V acylase.

The reaction rates were carefully followed by determination of 6-APA, phenoxymethylpenicillin (PNV), phenoxyacetic acid (POA), POM, and phenoxyacetylglycin (POG) using high-pressure liquid chromatography.

Among the substrates tested, POM was preferable as a phenoxyacetyl donor to 6-APA, giving nearly 12% conversion of 6-APA to PNV (TABLE 1). When POG was employed instead of POM, only 0.6% overall conversion was achieved. It was also observed that the overall conversion increased linearly with an increase in molar ratio of POM to 6-APA up to 4:1; no more significant increase in conversion was observed beyond that ratio. During the enzymatic synthesis of PNV, the amount of PNV produced increased at a high rate initially but decreased gradually after 90 min, while the amount of POA continued to increase due to the hydrolysis of POM (FIG. 2).

In order to increase the overall conversion of 6-APA to PNV, the effect of some organic solvents was investigated by adding them to the reaction mixture (10% by volume per volume) (TABLE 2). Some improvement of PNV yield was observed in the

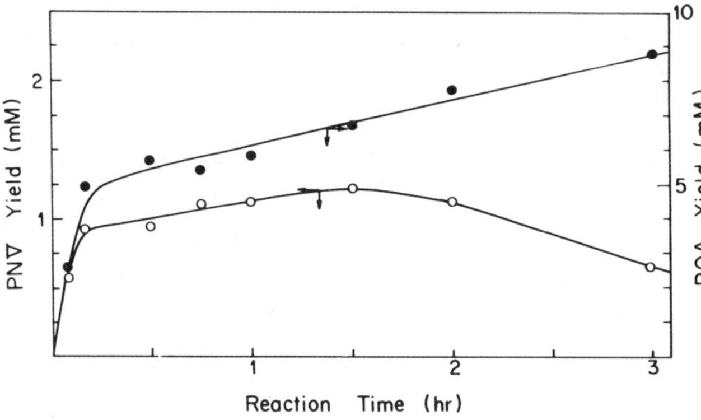

FIGURE 2. Conversion profile of 6-APA to PNV. Reaction conditions: pH 6.8, 30 m*M* of POM, 10 m*M* of 6-APA, 6.9 IU/ml of enzyme was loaded.

presence of ethanol, 2-propanol, and acetone. On the other hand, the addition of 1,4-dioxane, acetonitrile, and 2-butanol caused lowering of the yield almost to nothing.

The α-acylamino-β-lactam acylhydrolase (ALAHase I) was found to carry out three reactions simultaneously: transfer of an acyl group to an acyl acceptor to form a semisynthetic β-lactam antibiotic, hydrolysis of an acyl donor at the amide or ester bond, and hydrolysis of the semisynthetic β-lactam antibiotic that was produced by the same enzyme.

It was also observed that the hydrolysis reactions of POM and PNV were irreversible in this reaction system. The optimal pH for the three reactions was different; 9.0 for POM hydrolysis, 6.8 for the transfer of a phenoxyacetyl group to 6-APA, and 6.0 for the PNV hydrolysis. The K_m values determined for POM, 6-APA, and PNV were 33, 25, and 31 m*M*, respectively.

The overall conversion of 6-APA to PNV achieved was only 12% on a molar basis, which is lower than those achieved by penicillin-G acylase from *E. coli* (25% for benzylpenicillin)[10] and ampicillin acylase from *X. citri* (90% for cephalexin,[4] 23% for amoxicillin,[3] and 20% for cephaloglycin[7]). This could be attributed mainly to the characteristics of POM, which is slightly soluble in water, and the enzyme source itself.

TABLE 2. Effect of Different Solvents on PNV Production

Solvent (10% vol/vol)	Overall Conversion (%)
None	10.5
Methanol	9.1
Ethanol	14.5
1-Propanol	7.7
2-Propanol	14.0
1-Butanol	4.7
2-Butanol	0.5
Acetone	14.0
Dimethylsulfoxide	7.7
N,N-Dimethylformamide	10.5
1,4-Dioxane	0.0
Acetonitrile	1.1

The ampicillin acylase (ALAHase III) and penicillin-G acylase (ALAHase II) catalyze the three hydrolysis reactions in just the same way as does the ALAHase I from *E. aroideae,* which, however, showed somewhat different characteristics in its kinetic behavior.

An inhibition study indicated the participation of a phenolic group of tyrosine and a metal. This meant that the ALAHase I enzyme from *E. aroideae* was different from the ALAHase II produced from *E. coli* or *B. megaterium,* which are known to be similar to the serine proteases.

Although the results reported here are encouraging, more work will be required for complete understanding of the reaction mechanisms and kinetics of the ALAHase I enzyme and for the yield improvement before this enzymatic synthesis of semisynthetic β-lactam antibiotics could become practical.

REFERENCES

1. OKACHI, R. 1979. Enzymatic synthesis of penicillins and cephalosporins. Nippon Nogei Kagaku Kaishi **53:** R169–177.

2. NARA, T., R. OKACHI & M. MISAWA. 1971. Enzymatic synthesis of D(−)-α-aminobenzylpenicillin by *Kluyvera citrophila*. J. Antibiot. **24:** 321–323.
3. KATO, K., K. KAWAHARA, T. TAKAHASHI & S. IGARASI. 1980. Enzymatic synthesis of amoxicillin by the cell-bound α-amino acid ester hydrolase of *Xanthomonas citri*. Agric. Biol. Chem. **44:** 821–825.
4. KATO, K. 1981. Characteristics of α-amino acid ester hydrolase and its application to enzymatic synthesis of cephalosporins and penicillins. Takeda Kenkyusho Ho **40:** 253–302.
5. RHEE, D. K., S. B. LEE, J. S. RHEE, D. D. Y. RYU & J. HOSPODKA. 1980. Enzymatic biosynthesis of cephalexin. Biotechnol. Bioeng. **22:** 1237–1247.
6. FUJII, T., K. MATSUMOTO & T. WATANABE. 1976. Enzymatic synthesis of cephalexin. Process Biochem. (Oct.): 21–24.
7. NAM, D. H., H. S. SOHN & D. D. Y. RYU. 1983. Enzymatic synthesis of cephaloglycin. Bull. Korean Chem. Soc. **4:** 72–76.
8. OKACHI, R., Y. HASHIMOTO, M. KAWAMORI, R. KATSUMATA, K. TAKAYAMA & T. NARA. 1982. Enzymatic synthesis of penicillins and cephalosporins by penicillin acylase. Enzyme Eng. **6:** 81–90.
9. KONECNY, J., A. SCHNEIDER & M. SIEBER. 1983. Kinetics and mechanism of acyl transfer by penicillin acylase. Biotechnol. Bioeng. **25:** 451–467.
10. SVEDAS, V. K., A. L. MARGOLIN, I. L. BORISON & I. V. BEREZIN. 1980. Kinetics of the enzymatic synthesis of benzylpenicillin. Enzyme Microb. Technol. **2:** 313–317.
11. VANDAMME, E. J., J. P. VOETS & A. DHASE. 1971. Etude de la penicillin acylase prodnite par *Erwinia aroidene*. Ann. Inst. Pasteur **121:** 435–446.

Hydroxylation by Hemoglobin-containing Systems: Activities and Regioselectivities

DIDIER GUILLOCHON, LAURENT ESCLADE, BERNARD CAMBOU, AND DANIEL THOMAS

Laboratoire de Technologie Enzymatique
E.R.A. n° 338 du C.N.R.S.
Université de Technologie de Compiègne
B.P. 233, 60206 Compiègne Cedex, France

Hydroxylations by systems containing cytochrome P-450 are widespread in mammals. They are responsible for the oxidation of fatty acids and steroids, as well as a wide variety of xenobiotics. These enzyme systems consist of an electron donor [NAD(P)H], an electron transfer flavoprotein (cytochrome P-450 reductase), and cytochrome P-450 as terminal oxidase. The utilization of these systems in enzyme technology is doubtful because of the cost of their components and their instability.

It has been shown that hemoglobin can catalyze the hydroxylation of aniline, a typical reaction of the monooxygenase function of cytochrome P-450. This is accomplished by substituting hemoglobin for the cytochrome P-450 terminal oxidase in a NADPH-cytochrome-P-450-reductase-hemoglobin system[1] or in a NADH-riboflavin-hemoglobin system.[2] The last system is very attractive because hemoglobin is an abundant and readily available protein. Hemoglobin is cheaper than cytochrome P-450, and the flavoprotein reductase is not necessary. Nevertheless, its utilization in enzyme technology depends on the regeneration or the substitution of the electron donor by simpler principles and also on its specificity and regioselectivity. We have previously studied the activity of the NADH-riboflavin-hemoglobin system in which hemoglobin is immobilized by crosslinking with glutaraldehyde to form insoluble particles.[3] The present study represents an attempt to form a hydroxylating hemoglobin system while avoiding cofactor regeneration. The hydroxylase activity of hemoglobin, either native or crosslinked with glutaraldehyde, was studied. NADH was replaced by cheaper electron donors, such as ascorbic acid (AH_2) and dihydroxyfumaric acid (DHF). The activities with the different electron donors were compared. The regioselectivity of aniline hydroxylation was studied with the AH_2-hemoglobin system.

MATERIALS AND METHODS

Human oxyhemoglobin was prepared from freshly drawn blood as described previously.[4] Methemoglobin was prepared using potassium ferricyanide as the oxidizing agent. Excessive potassium ferrocyanide and potassium ferricyanide were removed using a Biogel P_6 column. To produce insoluble particles of oxyhemoglobin or methemoglobin, 2.4 ml of 0.05 *M* phosphate buffer, pH 6.8, containing 100 mg · ml^{-1} hemoglobin and 3.3 mg · ml^{-1} glutaraldehyde was prepared. The solution was frozen at −30°C for two hours and then warmed to 4°C. An insoluble foam structure was obtained, which was then ground. It was thoroughly rinsed with a glycine solution and 0.1 *M* phosphate buffer pH 6.8. Aniline hydroxylations were carried out at 37°C in oxygenated 0.1 *M* phosphate buffer, pH 6.8, containing 0.1 m*M* native hemoglobin or 0.26 g insoluble particles of hemoglobin, aniline, 0.1 m*M* riboflavin (sometimes omitted), and 2 m*M* NADH, AH_2, or DHF. The *p*-aminophenol concentration was spectrophotometrically determined by the method of Brodie and Axelrod.[5]

For the regioselectivity studies, the aminophenols were separated and quantitated on a reverse-phase HPLC column.

RESULTS

p-Hydroxylation of aniline was tested by two kinds of hydroxylating systems: electron-donor-riboflavin-hemoglobin and electron-donor-hemoglobin (TABLE 1). When NADH was the electron donor, riboflavin was required. On the other hand, riboflavin was not necessary when the electron donor was AH_2 or DHF. The activity of the oxyhemoglobin system was always about 2–3 times smaller than that of the methemoglobin system. The most efficient hydroxylating systems were AH_2 or DHF-methemoglobin. These were about 3–4 times more active than the NADH-

TABLE 1. Rates of Hydroxylation of Aniline by Various Hemoglobin Systems[a]

	Electron Donors					
	NADH		AH_2		DHF	
Hemoglobin States	---	RF	---	RF	---	RF
Methemoglobin	0.025	0.68	2.13	2.12	2.26	1.98
Oxyhemoglobin	0.012	0.25	1.13	1.35	0.66	0.68
Heated MetHb (a)	n.d.	0.92	1.08	---	1.42	---
Heated OxyHb (a)	n.d.	1.38	0.45	---	2.40	---
	n.d.	0.20	n.d.	0.30	n.d.	0.83

[a] *p*-Aminophenol formation rates expressed in $\mu M \cdot min^{-1}$. Kinetics were followed for 30 min. (a): An aliquot of the stock solution preheated at 90°C and then cooled was transferred in place of native oxy- or methemoglobin in the reaction mixture. n.d., not detectable; RF riboflavin.

riboflavin-methemoglobin system. We have also run controls for aniline hydroxylation using electron-donor-riboflavin-denatured hemoglobin, electron-donor-denatured hemoglobin, electron donor riboflavin, and electron donor alone. All the controls exhibited measurable hydroxylation reaction over short (30 min) or over long periods (5 hours).

Kinetic data for the NADH-riboflavin-hemoglobin, AH_2- or DHF-hemoglobin systems were compared for native and immobilized hemoglobin[3] (TABLE 2). In each case, the dependence of the rate of formation of *p*-aminophenol on the aniline concentration followed apparent Michaelis-Menten kinetics. For all the systems the $K_{m\,app}$ values for aniline were not significantly different. Activity yields after immobilization of about 30–50% were obtained in the different immobilized hemoglobin systems.

The regioselectivity of aniline hydroxylation was studied for the most efficient systems, that is, AH_2-methemoglobin and AH_2-oxyhemoglobin (FIGS. 1 and 2). Hydroxylation by the AH_2-hemoglobin systems was shown to be selective for the *para* position. A *para/ortho* ratio of 5 was obtained for both oxy- and methemoglobin. No hydroxylation in the *meta* position was detected. The detection limit for *meta*-

TABLE 2. Kinetic Data for Aniline Hydroxylations by Native Hemoglobin Systems and Immobilized Hemoglobin Systems[a]

	Native Hemoglobin		Immobilized Hemoglobin		
Systems	$K_{m\,app}$ ($M \times 100$)	Turnover Number ($min^{-1} \times 10^3$)	$K_{m\,app}$ ($M \times 100$)	Turnover Number ($min^{-1} \times 10^3$)	Activity Yield (%)
MetHb-RF-NADH	6.6	10	3.0	4.9	49
OxyHb-RF-NADH	1.8	2.5	7.8	5.0	200
MetHb-AH_2	2.5	19	1.3	4.9	26
OxyHb-AH_2	3.7	14	2.1	4.1	30
MetHb-DHF	2.4	21	1.8	5.4	26
OxyHb-DHF	2.0	10	1.6	5.2	52

[a] 3.3 to 80 mM aniline. *p*-Aminophenol formation rates expressed in $\mu M \cdot min^{-1}$. Kinetics were followed for 30 min.

aminophenol was 1.5 μM with the HPLC method used. For the control, (ascorbic acid alone) production of aminophenols is far less important, with hydroxylation in the *ortho* position preferentially obtained (*para*/*ortho* ratio = 0.3).

DISCUSSION

We have shown previously[4,5] that the hydroxylase activity of the electron donor hemoglobin systems, with or without riboflavin, can be explained by a peroxidatic

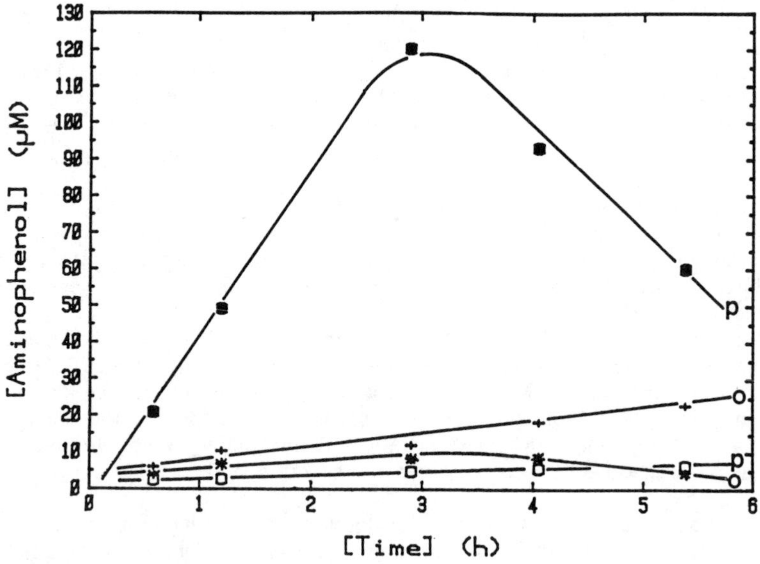

FIGURE 1. Regioselectivity of the hydroxylating system AH_2-methemoglobin. *p*-Aminophenol (■) and *o*-aminophenol (∗) formation versus time. *p*-Aminophenol (□) and *o*-aminophenol (+) formed without methemoglobin. System contained 0.1 mM methemoglobin, 30 mM aniline, 1 mM AH_2, 0.1 M phosphate buffer, pH 7.4, at 37°C.

mechanism in which H_2O_2 is produced. The presence of the electron donor NADH, AH_2, or DHF is required for the accumulation of aminophenols. The electron donors act as H_2O_2-generating molecules and have a protective effect against further oxidation of the formed aminophenols. The aniline hydroxylations obtained with denatured hemoglobins must be compared with the increase in peroxidase activity observed for methemoglobin and catalase when the hemoproteins are denatured by heating.[6] Hydroxylations obtained with electron-donor-riboflavin systems proceed by another mechanism; this involves the isoalloxazine nucleus participation as a chemical model of flavin-dependent monooxygenases.[7] The very weak hydroxylation reaction obtained with electron donors alone probably originates from reactive oxygen species besides H_2O_2 produced by reaction with oxygen. Hydroxylations by the AH_2-hemoglobin systems have a regioselectivity in the *para* position comparable to the

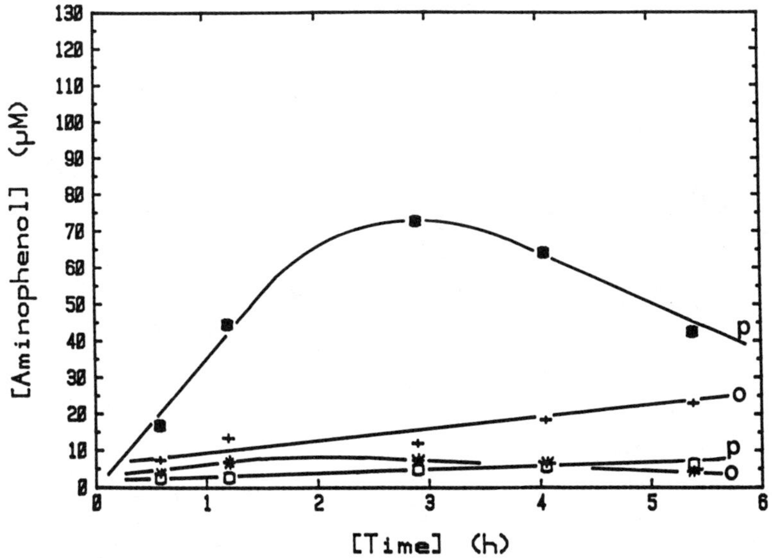

FIGURE 2. Regioselectivity of the hydroxylating system AH_2-oxyhemoglobin. *p*-Aminophenol (■) and *o*-aminophenol (∗) formation versus time. *p*-Aminophenol (□) and *o*-aminophenol (+) formed without oxyhemoglobin. Contents same as for FIGURE 1.

NADPH-cytochrome-P-450-reductase-hemoglobin system and to cytochrome P-450 monooxygenases.[1] On the other hand, ascorbic acid alone has very little hydroxylation selectivity.

In conclusion, the ascorbic acid or dihydroxyfumaric-hemoglobin systems seem very attractive for enzyme technology because (1) they can give rise to aniline hydroxylase activities similar to those of cytochrome P-450 monooxygenases and to a great regioselectivity; (2) hemoglobin is an abundant protein, available without expensive purification, (3) substrate conversion processes can be carried out without the cofactor regeneration problems; and (4) immobilization of hemoglobin to form insoluble particles can be carried out with acceptable activity yields. However, further research is necessary to characterize the specificity and regioselectivity of these hydroxylating systems with other more interesting substrates.

REFERENCES

1. Mieyal, J. J., R. S. Ackerman, J. L. Blumer & L. J. Freeman. 1976. Characterization of enzyme-like activity of human hemoglobin. J. Biol. Chem. **251**(11): 3436–3441.
2. Akhrem, A. A., S. Yu. German & D. I. Metelitsa. 1978. New model of the hydroxylating enzyme system of liver microsomes; NADH-riboflavin-hemoglobin-O_2. React. Kinet. Catal. Lett. **8**(3): 339–345.
3. Guillochon, D., J. M. Ludot, L. Esclade, B. Cambou & D. Thomas. 1982. Hydroxylase activity of immobilized hemoglobin. Enzyme Microb. Technol. **4:** 96–98.
4. Cambou, B., D. Guillochon & D. Thomas. 1984. Aniline hydroxylase activities of hemoglobin. Kinetics and mechanisms. Enzyme Microb. Technol. **6:** 11–17.
5. Guillochon, D., B. Cambou, L. Esclade & D. Thomas. 1984. Ascorbic acid and dihydroxyfumaric acid-dependent hydroxylase activity of immobilized hemoglobin. Enzyme Microb. Technol. **6:** 161–164.
6. Inada, Y., T. Kurozumi & K. Shibata. 1961. Peroxidase activity of hemoproteins. I. Generation of activity by acid or alkali denaturation of methemoglobin and catalase. Arch. Biochem. Biophys. **93:** 30–36.
7. Frost, J. W. & W. H. Rastetter. 1981. Flavoprotein monooxygenases: A chemical model. J. Am. Chem. Soc. **103**(17): 5242–5245.

Unusual Catalytic Properties of Usual Enzymes

BERNARD CAMBOU AND ALEXANDER M. KLIBANOV

Laboratory of Applied Biochemistry
Department of Nutrition and Food Science
Massachusetts Institute of Technology
Cambridge, Massachusetts 02139

In nearly all commercial applications and exploratory studies thus far, enzymes have been used to catalyze their "normal" reactions reflected in their names. It turns out that in addition to those, many enzymes can also catalyze other, sometimes quite different reactions. The latter, although unimportant for the enzyme-producing organism, can be very valuable in biotechnology, in particular in preparative organic chemistry. Some of the "unnatural" enzymatic reactions we have explored over the last few years include the following:

1. Glucose oxidase-catalyzed reduction of quinones to hydroquinones with glucose.[1]
2. Galactose oxidase-catalyzed stereospecific oxidation of prochiral and chiral aliphatic alcohols.[2]
3. Peroxidase-catalyzed selective hydroxylation of aromatic compounds.[3]
4. Sulfatase-catalyzed separation of isomeric phenols.[4]
5. Xanthine oxidase-catalyzed separation of geometric isomers of arylacroleins[5] and of positional isomers of aromatic aldehydes.[6]
6. Separation of *cis* and *trans* isomers of alicyclic alcohols catalyzed by phosphatases[7] and of unsaturated aldehydes catalyzed by alcohol dehydrogenases.[8]
7. Cholinesterase-catalyzed production of L-carnitine.[9]

In all of the above examples, the corresponding enzymes exhibited their unusual catalytic properties when exposed to unusual substrates. However, in some cases in order to display unusual catalytic functions, enzymes must also be placed in unusual environments. This is well illustrated by our recent study on esterase-catalyzed stereospecific transesterifications[10] described below.

Carboxyl esterases catalyze the following natural reaction of hydrolysis:

$$R_1COOR_2 + H_2O \rightarrow R_1COOH + R_2OH$$

In addition, most carboxyl esterases can also catalyze the reaction of transesterification in which water as a nucleophile is replaced with an alcohol:

$$R_1COOR_2 + R_3OH \rightarrow R_1COOR_3 + R_2OH$$

In this reaction, the enzyme catalyzes the transfer of the acyl group (R_1CO-) from one alcohol to another.

One can envisage the use of the above enzymatic transesterification for the resolution of racemic alcohols, $(\pm)R_3OH$. In order to accomplish that, two conditions have to be met: (1) the hydrolysis reaction must be completely suppressed by transesterification, and (2) the esterase employed must exhibit absolute stereospecific-

$$CH_3CH_2\overset{O}{\overset{\|}{C}}OCH_3 + HOCH_2CH_2R \longrightarrow CH_3CH_2\overset{O}{\overset{\|}{C}}OCH_2CH_2R + CH_3OH$$

1 R = $CH(OCH_3)CH_3$ (3-methoxy-1-butanol)

2 R = $CH(CH_3)CH_2CH_3$ (3-methyl-1-pentanol)

3 R = $CH(CH_3)CH_2CH_2CH_2CH(CH_3)_2$ (3,7-dimethyl-1-octanol)

4 R = $CH(CH_3)CH_2CH_2CH{=}C(CH_3)_2$ (citronellol)

FIGURE 1. Hog liver carboxyl esterase-catalyzed transesterification.

ity with respect to the nucleophile, that is, react only with one optical isomer of R_3OH.

We have developed a novel general approach to preparative enzymatic transesterifications. It involves the use of biphasic aqueous-organic mixtures where the substrates—the ester and the alcohol (the nucleophile)—constitute the organic phase and the enzyme is dissolved in the aqueous phase. Since proteins are insoluble in water-immiscible organic solvents, the enzyme will remain in the aqueous phase. The substrates will freely diffuse into the latter, undergo the enzymatic conversion, and then the products will diffuse back into the organic phase. The fraction of water in such a biphasic system can be made extremely low, and hence transesterification will be greatly favored over hydrolysis. Conceptually, the aforedescribed system represents an emulsion of an aqueous solution of an enzyme in a mixture of substrates. However, from the experimental standpoint, it is advantageous instead to use porous supports whose pores are filled with an aqueous solution of an enzyme: first, such beads are more

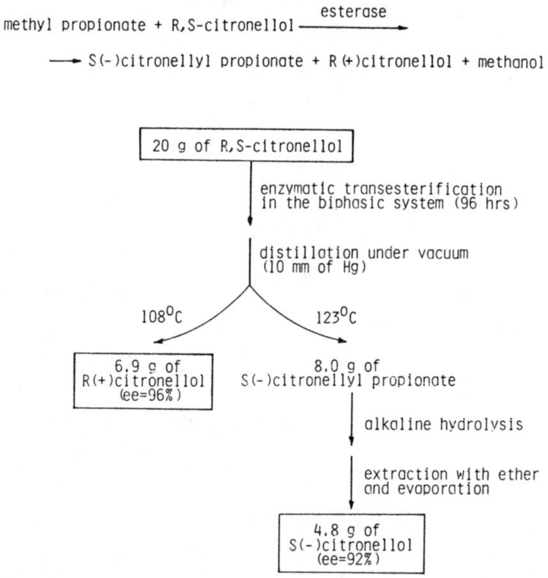

FIGURE 2. Schematic illustration of the preparative hog liver carboxyl esterase-catalyzed resolution of citronellol.

mechanically robust than water droplets; second, they can be readily separated from products at the end of the enzymatic reaction; third, they can be used repeatedly. The organic (substrate) phase should be presaturated with an aqueous buffer (of the pH optimal for the enzyme action) to avoid loss of water from the beads due to its partitioning.

We have identified two different commercially available carboxyl esterases that exhibit absolute stereospecificity in the transesterification reaction. They have been used by us for the production of gram quantities of optically active esters and alcohols via the transesterification scheme:

$$\text{matrix ester} + \text{racemic alcohol} \xrightarrow{\text{esterase}} \text{new, optically active ester} + \text{remaining, optically active alcohol}$$

Carboxyl esterase from hog liver (EC 3.1.1.1) is one of the two enzymes employed. The experimental system consisted of methyl propionate used as a matrix ester and various primary alcohols as nucleophiles. The reaction and the alcohols that we have successfully resolved are presented in FIGURE 1.

$$\text{TRIBUTYRIN} + HOCHR_1R_2 \longrightarrow CH_3CH_2CH_2\overset{O}{\overset{\|}{C}}OCHR_1R_2 + \text{DIBUTYRIN}$$

5 $R_1 = CH_3$ $R_2 = CH_2CH_3$ (2-BUTANOL)

6 $R_1 = CH_3$ $R_2 = (CH_2)_5CH_3$ (2-OCTANOL)

7 $R_1 = CH_3$ $R_2 = C_6H_5$ (SEC-PHENETHYL ALCOHOL)

8 $R_1 = CH_3$ $R_2 = CH_2CH_2CH{=}C(CH_3)_2$ (6-METHYL-5-HEPTEN-2-OL)

9 $R_1 = CH_3$ $R_2 = CH_2Cl$ (1-CHLORO-2-PROPANOL)

10 $R_1 = H$ $R_2 = CH(Cl)CH_2Cl$ (2,3-DICHLOROPROPANOL)

11 $R_1 = H$ $R_2 = CH(OH)CH_2CH_3$ (1,2-BUTANEDIOL)

FIGURE 3. Yeast lipase-catalyzed transesterification.

A typical example is shown in FIGURE 2. It deals with the preparative resolution of citronellol, both of whose isomers are widely used in perfumery. The enzyme exhibits absolute stereospecificity with respect to the nucleophile in the transesterification reaction that results in high optical purities of the R- and S-citronellols obtained. Similar data have been obtained for the other alcohols listed in FIGURE 1.

From a practical standpoint, two major requirements that must be fulfilled by the enzyme catalyst used in our transesterificaion system are:

(1) It must have a broad nucleophile specificity in order to make the approach applicable to a wide variety of chiral alcohols.
(2) It must exhibit absolute stereospecificity in the reaction with a given chiral alcohol.

Hog liver esterase meets the second but not the first of these requirements. We have found that only primary alcohols having no substituents in the first two methylene groups adjacent to the hydroxyl react at an appreciable rate. Therefore, although the enzyme exhibits absolute stereospecificity, its rather narrow substrate specificity is a major drawback for preparative purposes. To overcome this obstacle, we have endeavored to search for another commercially available esterase that will be both

TRIBUTYRIN + R,S-SEC-BUTANOL —LIPASE→

→R(−)SEC-BUTYL BUTYRATE + S(+)SEC-BUTANOL + DIBUTYRIN

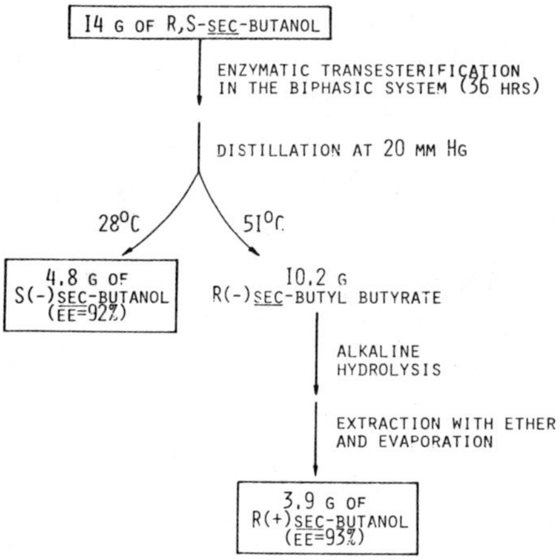

FIGURE 4. Schematic illustration of the preparative resolution of *sec*-butanol catalyzed by yeast lipase.

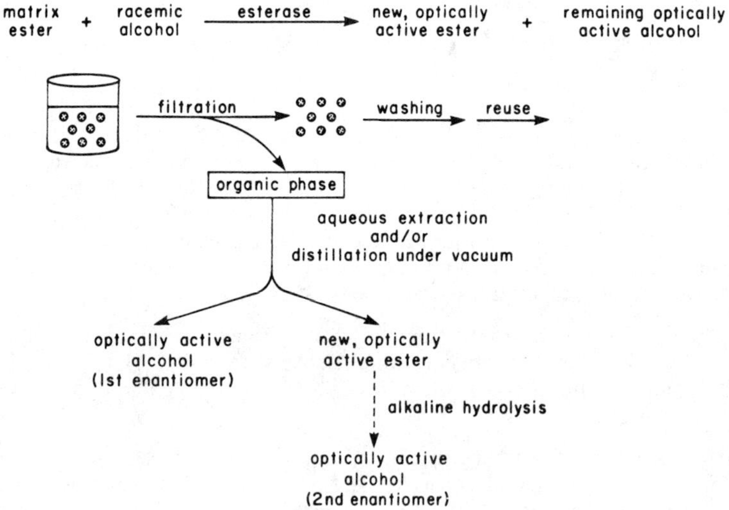

FIGURE 5. Schematic illustration of the resolution of racemic alcohols using esterase-catalyzed transesterification.

stereoselective and nonspecific in the transesterification reaction. Such an esterase was indeed found: lipase from *Candida cylindracea* (EC 3.1.1.3) was the second asymmetric enzyme catalyst we used in this study. Tributyrin was employed as a matrix ester, and several different alcohols (mainly secondary chiral alcohols that were unreactive towards hog liver carboxyl esterase) have been enzymatically resolved (FIG. 3).

As a typical example, the enzymatic resolution of *sec*-butanol is presented in FIGURE 4.

In closing, we have used the "unnatural" reaction of carboxyl esterases—transesterification—for the production of a number of optically active esters and alcohols. The general scheme of esterase-catalyzed asymmetric transesterifications in biphasic systems is shown in FIGURE 5. It is worth mentioning that the overall concentration of water in the aforedescribed systems was only a few per cent. This greatly facilitates the product recovery and also leads to a new, promising direction in enzyme technology: enzymatic processes in organic solvents.[11] In such systems, enzymes not only catalyze unnatural reactions, but they do it under unnatural conditions.

REFERENCES

1. ALBERTI, B. N. & A. M. KLIBANOV. 1982. Preparative production of hydroquinone from benzoquinone catalyzed by immobilized glucose oxidase. Enzyme Microb. Technol. **4:** 47–49.
2. KLIBANOV, A. M., B. N. ALBERTI & M. A. MARLETTA. 1982. Stereospecific oxidation of aliphatic alcohols catalyzed by galactose oxidase. Biochem. Biophys. Res. Commun. **108:** 804–808.
3. KLIBANOV, A. M., Z. BERMAN & B. N. ALBERTI. 1981. Preparative hydroxylation of aromatic compounds catalyzed by peroxidase. J. Am. Chem. Soc. **103:** 6263–6264.
4. PELSY, G. & A. M. KLIBANOV. 1983. Preparative separation of α- and β-naphthols catalyzed by immobilized sulfatase. Biotechnol. Bioeng. **25:** 919–928.
5. KLIBANOV, A. M. & P. P. GIANNOUSIS. 1982. Geometrically specific oxidation of β-arylacroleins: The preparative potential. Biotechnol. Lett. **4:** 57–60.
6. PELSY, G. & A. M. KLIBANOV. 1983. Remarkable positional specificity of xanthine oxidase and some dehydrogenases in the reactions with substituted benzaldehydes. **742:** 352–357.
7. KLIBANOV, A. M., A. CHENDRASEKHAR & B. N. ALBERTI. 1983. Enzymatic separation of *cis* and *trans* isomers of alicyclic alcohols. Enzyme Microb. Technol. **5:** 265–268.
8. KLIBANOV, A. M. & P. P. GIANNOUSIS. 1982. Geometric specificity of alcohol dehydrogenases and its potential for separation of *trans* and *cis* isomers of unsaturated aldehydes. Proc. Natl. Acad. Sci. USA **79:** 3462–3465.
9. DROPSY, E. P. & A. M. KLIBANOV. 1984. Cholinesterase-catalyzed resolution of D,L-carnitine. Biotechnol. Bioeng. **26:** No. 8.
10. CAMBOU, B. & A. M. KLIBANOV. 1984. Preparative production of optically active esters and alcohols using esterase-catalyzed stereospecific transesterification in organic media. J. Am. Chem. Soc. **106:** 2687–2692.
11. ZAKS, A. & A. M. KLIBANOV. 1984. Enzymatic calalysis in organic media at 100°C. Science **224:** 1249–1251.

Rotating Ring-Disc Electrode: Its Use in Studying Mass Transport Resistances through Immobilized Enzyme Support Matrices[a]

LEMUEL B. WINGARD, JR., JAMES F. CASTNER,[b] OSATO MIYAWAKI, AND CARL MARRESE

Department of Pharmacology
School of Medicine
University of Pittsburgh
Pittsburgh, Pennsylvania 15261

One of the continuing experimental problems in the characterization of immobilized enzymes is how to measure the rates of mass transfer of substrates and products through the immobilization matrix and also to determine the inherent activity of the immobilized enzyme. If mass transfer resistances within the immobilization matrix are very large, then attempts to measure the enzyme activity usually end up simply determining the rate of the slowest step. This may be the rate of mass transfer or a combination of the rates of mass transfer and enzyme-catalyzed reaction. The purpose of this paper is to describe how an electrochemical technique, based on a set of rotating ring and disc electrodes (RRDE), can be employed for measurement of the rates of mass transfer of an electrochemically active compound through the enzyme immobilization matrix. The method can also be applied to the study of the rates of combined reaction-mass transfer; however, this more complex situation is mentioned only briefly in this paper and is discussed more fully elsewhere.[1-3]

DESCRIPTION OF ROTATING RING-DISC ELECTRODE (RRDE)

A schematic diagram of the RRDE assembly used in these studies is shown in FIGURE 1. The enzyme immobilization matrix is attached to the surface of the platinum disc. The methods of attachment are described in a later section of this paper. The RRDE assembly, a counter electrode, and a reference electrode are placed in a solution of the enzyme substrate or the diffusate. The RRDE assembly is mounted mechnically and connected electrically so that the RRDE can be rotated at a constant speed while at the same time the potential or current at the ring and disc electrodes can be controlled or measured. For example, the potential applied to the disc or ring electrode can be selected so as to oxidize or reduce the species of interest in the solution. Under proper conditions, the resulting current can be measured and related to the local concentration or to the rate at which the species of interest is being delivered to the vicinity of the electrode surface.

[a]This work was supported by funds from NIH (AM26370), NSF (CPE8107715), and ARO (DAAG2982K0064).

[b]Present Address: E. I. duPont de Nemours Co., Biomedical Products, Wilmington, DE 19898.

THEORY AND BACKGROUND

The velocity profile for the flow of fluid across the end of the RRDE is shown in FIGURE 2. The velocity normal to the disc decreases from the limiting value at large y to zero at the disc surface (where $y = 0$). Adjacent to the disc surface is a stagnant boundary layer of fluid that is dragged along as the disc is rotated. If a reduced electroactive species is placed in the solution, then the species will undergo electrochemical oxidation at the disc electrode surface. If the potential that is applied to the disc is selected so that the oxidation occurs very rapidly, then the concentration of

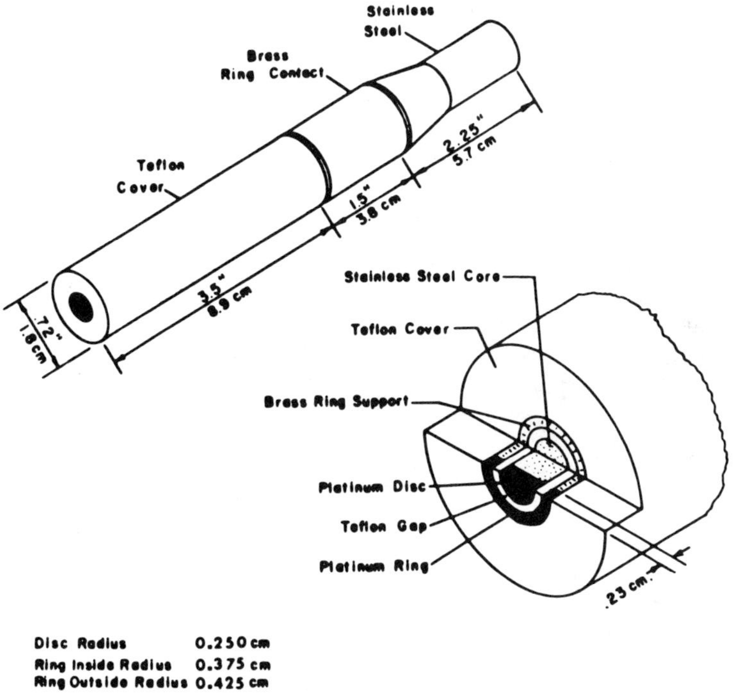

FIGURE 1. Schematic diagram of RRDE assembly. Electrical contacts with the external circuit are made using metal strips that press against the brass and stainless steel core extensions as the electrode assembly is rotated. The enzyme is immobilized on the platinum disc.

reduced species at the disc surface becomes essentially zero. This condition produces the maximum concentration gradient between the bulk solution and the surface of the disc. Therefore, if still greater potentials are applied to the disc, the concentration gradient will not change. Under such conditions, the rate of mass transport of diffusate to the disc surface is governed by the rate of transport through the stagnant fluid layer. At a constant rate of electrode rotation and with a sufficiently high applied potential, the concentrations of diffusate at different points within the stagnant layer will achieve quasi-steady-state values. The current measured at the disc under these steady-state, mass transfer rate-limiting conditions is known as the limiting current, i_1. The limiting current is defined further by Equation 1.

$$i_l = n\,F\,A\,D_r \left(\frac{\partial C_r}{\partial y} \right)_{\text{at } y\,=\,0} \tag{1}$$

where n is the number of equivalents/mole; F is 96,485 coulombs/equivalent; A is the electrode surface area; D_r is the diffusion coefficient of the reduced species within the stagnant layer; y is distance normal to the surface of the disc; and C_r is the concentration of the reduced species as a function of position within the stagnant layer. The concentration gradient at the disc surface can be obtained from the steady-state convection-diffusion equation for transport across the stagnant layer,[4] as shown in Equation 2.

$$\left(\frac{\partial C_r}{\partial y} \right)_{\text{at y}\,=\,0} = 0.62\, D_r^{-1/3}\, w^{1/2}\, \mu^{-1/6}\, C_r^* \tag{2}$$

where w is the rate of rotation, μ is the kinemetic viscosity, and C_r^* is the concentration

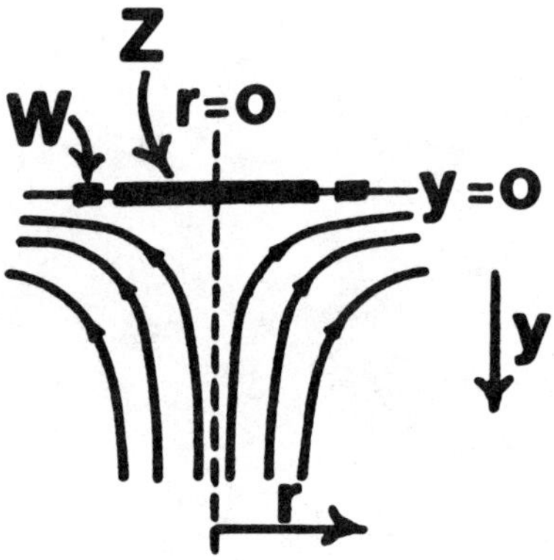

FIGURE 2. Streamlines for flow of fluid around the disc (Z) and ring (W) of the RRDE.

of reduced species in the bulk solution. Combining Equations 1 and 2 gives the Levich Equation for the limiting current, shown here as Equation 3.

$$i_l = 0.62\, n\, F\, A\, D_r^{2/3}\, w^{1/2}\, \mu^{-1/6}\, C_r^* \tag{3}$$

When a layer of immobilized enzyme is attached to the surface of the disc, the resistance to mass transport of electroactive species between the bulk solution and the disc surface is modified. If the enzyme layer is very thin, so as to be within the thickness of the boundary layer, and the porosity of the matrix is relatively open, then the system can be treated as above. It can still be considered a case of external mass transfer control, where the thickness of the stagnant layer is dependent on the fluid dynamics of the solution near the disc. When the enzyme matrix becomes relatively thick or the porosity of the matrix becomes low, the resistance to mass transfer within the matrix becomes the rate-limiting step. The system then is under internal mass transfer control; and the rate of mass transfer is no longer dependent on the solution fluid dynamics. With internal mass transfer control, Equation 3 does not apply to measurements made at the rotating disc. However, the RRDE can still be used to study the rates of reaction-diffusion within the matrix, as described briefly below.

Mell and Maloy[5] developed a mathematical model for simulating the amperometric response for a stationary disc electrode having an enzyme matrix attached to the surface of the disc. Their simulation was based on combined internal mass transfer resistance and enzyme-catalyzed reaction. The simulations were used to show how the current would be influenced by the matrix and enzyme parameters. When they tested their simulation method experimentally, they stirred the solution to minimize external mass transfer resistance. However, in their paper it is very difficult to distinguish which results were obtained experimentally and which ones came from the simulations. Several years later, Wilson and co-workers[1,6] applied the Mell-Maloy simulation approach to a rotating ring-disc electrode. Wilson treated his system as a steady-state convection-reaction case, with external mass transfer resistance and the enzyme-catalyzed reaction both occurring within the layer of stagnant fluid. He omitted any consideration of internal mass transfer resistance, even though some of his immobilized enzyme preparations were made up of a thick layer of carbon paste-enzyme matrix that filled the cavity occupied by the platinum disc in the FIGURE 1 schematic. Wilson's mathematical simulation expression contained empirical factors that apparently were included to improve the agreement between the experimental and the predicted results, but which appear to have little theoretical justification. In work reported elsewhere,[3] we have examined both external and internal mass transfer plus an enzyme-catalyzed reaction within the immobilization matrix, using a RRDE with glucose oxidase attached to the platinum disc. The reader is referred to that paper for further discussion of the reaction-mass transfer case.

For many systems, useful estimates of the effective diffusion coefficient for transport through the enzyme matrix can be made using Equation 3. For a fixed value of C_r^* and a given enzyme RRDE preparation, a plot of the limiting current versus the square root of the rotation speed should be linear in order for Equation 3 to be used. The following experimental section describes the measurement of the diffusion and shielding coefficients for the transport of the electroactive material, potassium ferrocyanide, through matrices of immobilized glucose oxidase on the platinum disc of the RRDE.

EXPERIMENTAL

In the present study, glucose oxidase was attached to the platinum disc by three methods: alkylamine silane, allylamine, and albumin glutaraldehyde. The RRDE was a Pine Instrument Model DT136 (FIGURE 1) with a removable platinum disc. Before enzyme attachment, the disc was removed, polished, and treated electrochemically to control the state of oxidation of the metal surface. The details of the procedure are given elsewhere.[3] In the alkylamine silane method, aminopropyltriethoxy silane was attached to the platinum oxide surface layer in dry benzene.[7] The amino group was activated with glutaraldehyde followed by the addition of a solution of 2.5 mg/ml glucose oxidase. The allylamine coupling procedure was similar to that for the silane since the double bond of allylamine adsorbs strongly to the platinum surface. The details of the preparation of the crosslinked albumin, glutaraldehyde, glucose oxidase matrix have already been published.[8] After attachment of the enzyme matrix to the platinum, the derivatized disc was washed and then reseated in the RRDE assembly. The reassembled RRDE was rotated in buffer for serveral hours to remove any loosely attached enzyme. The RRDE could not be refrigerated, owing to differences of thermal expansion between the materials of construction, so the experiments were done the same day. Each of the immobilization methods also was carried out with ^{125}I-labeled glucose oxidase. This was prepared by the lactoperoxidase, hydrogen peroxide, Na^{125}I procedure to give labeled enzyme of specific activity 2.6 μCi/μg and concentration of 39 μg/ml (4.2 units of glucose oxidase activity/ml).[8] The ^{125}I-labeled enzyme was used to determine the loading of enzyme onto the platinum disc.

The electrical measurements were made with a Pine Instrument Model RDE-3 four-electrode potentiostat and an X-Y recorder. A platinum screen served as the counter electrode. All potentials were measured with respect the a Ag/AgCl (saturated KCl) reference electrode. Only the disc was used for the diffusion coefficient measurements. Both the ring and disc were involved in the shielding coefficient determination.

RESULTS AND DISCUSSION

Glucose is not very electroactive, in that its rate of electrochemical oxidation or reduction proceeds slowly in comparison to that for highly electroactive materials like potassium ferrocyanide. In order for the measurements to be valid, the rate of oxidation or reduction of the indicator compound, that is the diffusate, must be very rapid compared to the rate of the mass transfer or enzyme-catalyzed reaction being measured. Ideally, one might wish to measure the diffusion coefficients for electroactive compounds having a range of molecular weights and polarities in characterizing the mass transfer resistances of the enzyme matrix. In this study, we examined only potassium ferrocyanide.

For the limiting current measurements at the disc, the electrode compartment was charged with 1.95 *M* potassium ferrocyanide and 1 *M* KCl, equilibrated to 25°C, and deoxygenated by bubbling nitrogen through the solution. The potential applied to the

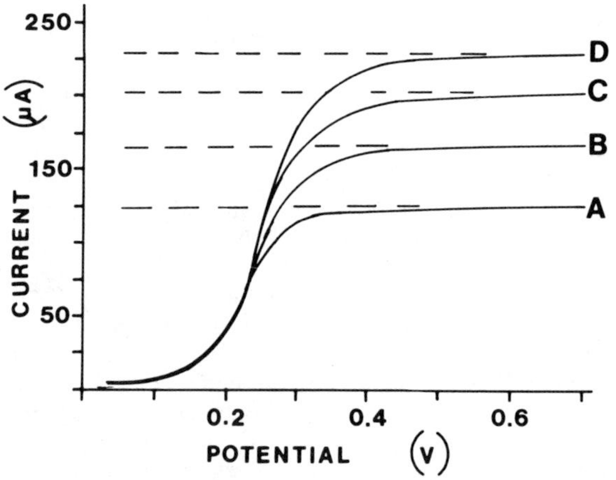

FIGURE 3. Anodic (oxidation) current at the disc as the potential applied to the disc is increased. The bulk solution contained 1.95 *M* potassium ferrocyanide and 1 *M* KCl. The enzyme was coupled to the disc using the silane-glutaralydehyde method. RRDE rotation speeds indicated as: A, 400 rpm; B, 800 rpm; C, 1200 rpm; D, 1600 rpm. Limiting currents shown by dashed lines.

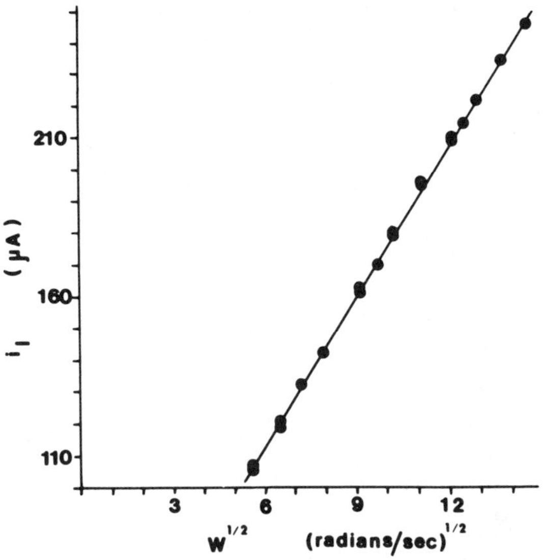

FIGURE 4. Limiting currents at different rotation rates for the enzyme coupled to the disc by the silane-glutaralydehyde method. Data obtained under conditions used for FIGURE 3 results.

disc was varied linearly from −100 mV to +700 mV at a constant rate of 200 mV/min. The half-cell potential for the conversion of ferricyanide to ferrocyanide occurs at about +160 mV with respect to Ag/AgCl. Thus, when the applied potential exceeds about +500 mV, then the oxidation of the ferrocyanide at the disc surface proceeds rapidly. The limiting current is indicated by the dashed lines in the FIGURE 3 results for alkylamine-glutaraldehyde-glucose oxidase attached to the disc. In order to test the results for adherence to the Levich Equation, the limiting current values were plotted against the square root of the rotation speed. The relationship is shown in FIGURE 4, again for the silane-glutaraldehyde method of enzyme attachment. The results in FIGURE 4 give excellent linearity, with a correlation coefficient of 0.999. From FIGURE 4, it is concluded that the results are suitable for use of Equation 3 for calculation of the diffusion coefficient. The following values were used in evaluating Equation 3 for D_r: $n = 1$ electron equivalent/mole, $A = 0.196$ cm^2, $C_r^* = 1.95$ M or mmoles/ml, and $\mu = 0.01$ cm^2/sec for water. Values for the diffusion coefficient are listed in TABLE 1 for the silane and allylamine methods of enzyme attachment. Literature results are also included in TABLE 1. The silane had essentially no effect on the ferrocyanide diffusion coefficient, whereas the silane-enzyme and allylamine-enzyme preparations both resulted in small but measurable reductions in the diffusion coefficient. This would be expected because of the much larger molecular weight of the enzyme (about 150k) as compared to the silane (less than 500). It is estimated that the silane and allylamine enzyme matrices were much thinner than the boundary layer. The immobilized enzyme was estimated to extend about 200 Å from the platinum surface; while the boundary layer thickness was calculated to be about 18 μm. No value is given in TABLE 1 for the diffusion coefficient for transport through the albumin enzyme matrix because the data did not follow the Levich Equation. The albumin enzyme matrix was about 0.3 cm thick; and the porosity of the matrix was probably quite low, based on earlier mass transfer measurements with this type of immobiliza-

TABLE 1. Parameter Values for Mass Transfer Resistance for Potassium Ferrocyanide Movement through Enzyme Matrix Immobilized on Disc of RRDE

Material Immobilized on Disc	D_r (cm^2/sec)	Shielding Coefficient	Enzyme Loading (moles/cm^2)
Control[a]	0.61×10^{-5}	0.85	—
Silane	0.66×10^{-5}	0.87	—
Silane-enzyme	0.57×10^{-5}	0.87	2.4×10^{-11}
Allylamine-enzyme	0.46×10^{-5}	0.87	9.2×10^{-11}
Albumin-enzyme	—	0.99	1.8×10^{-11}
Literature[9]	0.63×10^{-5}		

[a] Nothing attached to disc.

tion matrix.[8] A typical result for the current versus applied potential with the albumin enzyme matrix is shown in FIGURE 5. As the applied potential exceeded the value needed for oxidation of ferrocyanide, the ferrocyanide in the vicinity of the platinum disc surface was used up rapidly. The diffusional resistance of the enzyme matrix was so great that additional ferrocyanide was not supplied to the disc surface. The zone of analyte depletion gradually moved further and further into the matrix until a quasi-steady state was reached. However, the diffusional resistances were independent of the rotation speed; so that Equation 3 could not be applied to calculate D_r. This served to demonstrate the nonapplicability of Equation 3 when internal diffusional resistances become the rate-controlling factor.

The resistance to mass transport through the immobilized enzyme matrix also can

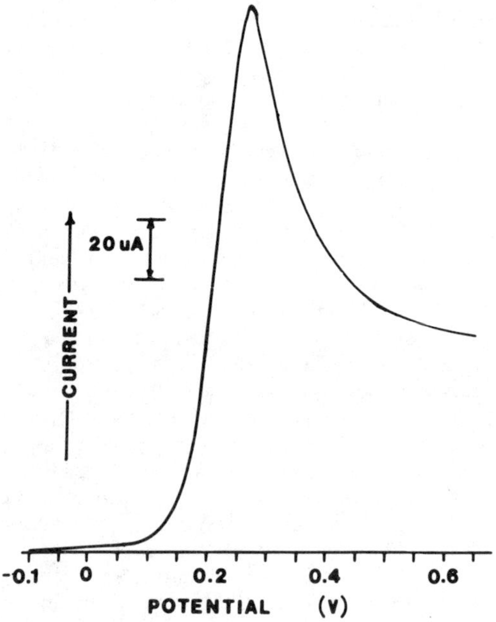

FIGURE 5. Oxidation current at the disc as a function of the applied potential for the albumin-glutaraldehyde-enzyme electrode. Conditions similar to those used for FIGURE 3 results. Rotation speed 1600 rpm.

be described in terms of the shielding coefficient. For this method, the enzyme again is immobilized on the disc; and the applied disc potential is scanned at a rate of 200 mV/min from −100 mV to +800 mV. At the same time, a fixed potential of +800 mV is applied to the ring. At +800 mV, any ferrocyanide reaching the ring will be oxidized rapidly. Thus, the ring current will provide a measure of the concentration of ferrocyanide in the vicinity of the ring. If the enzyme matrix has a low mass transfer resistance, then some of the ferrocyanide will pass through the matrix and be oxidized at the disc. The resulting ring current will be less than when the mass transfer resistance of the matrix is high. The shielding coefficient defines the loss in ring current caused by electrochemical reaction of the diffusate as it moves past the disc. Values for the coefficient also are shown in TABLE 1. The high shielding coefficient for the albumin enzyme matrix indicates that very little of the ferrocyanide reached the disc, as the flow streamlines moved the analyte across the enzyme matrix and onto the ring. This result is consistent with the high mass transfer resistance observed in attempting to apply Equation 3 to the albumin enzyme matrix.

The quantity of glucose oxidase attached to the disc by each of the three immobilization techniques is included in TABLE 1. The loadings are all fairly similar, indicating that the differences in mass transfer resistances are caused by the immobilization technique rather than by the amount of enzyme attached.

Additional electrochemical techniques based on a potential step applied to the electrode also have been used for the measurement of mass transfer resistances. The results will be presented in a separate paper.

REFERENCES

1. SHU, F. R. & G. S. WILSON. 1976. Rotating ring-disc enzyme electrode for surface catalysis studies. Anal. Chem. **48:** 1679–1686.
2. BOURDILLON, C., V. THOMAS & D. THOMAS. 1982. Electrochemical study of D-glucose oxidase autoinactivation. Enzyme Microb. Technol. **4:** 175–180.
3. CASTNER, J. F. & L. B. WINGARD, JR. 1984. Mass transport and reaction kinetic parameters determined electrochemically for immobilized glucose oxidase. Biochemistry. **23:** 2203–2210.
4. BARD, A. J. & L. R. FAULKNER. 1980. Electrochemical Methods. John Wiley & Sons, Inc. New York. pp. 286–288.
5. MELL, L. D. & J. T. MALOY. 1975. A model for the amperometric enzyme electrode obtained through digital simulation and applied to the immobilized glucose oxidase system. Anal. Chem. **47:** 299–307.
6. KAMIN, R. A. & G. S. WILSON. 1980. Rotating ring-disc enzyme electrode for biocatalytic kinetic studies and characterization of the immobilized enzyme layer. Anal. Chem. **52:** 1198–1205.
7. WINGARD JR., L. B., D. ELLIS, S. J. YAO, J. G. SCHILLER, C. C. LIU, S. K. WOLFSON JR & A. L. DRASH. 1979. Direct coupling of glucose oxidase to platinum and possible use for *in vivo* glucose determination. J. Solid Phase Biochem. **4:** 253–262.
8. WINGARD JR., L. B., L. A. CANTIN & J. F. CASTNER. 1983. Effect of enzyme-matrix composition on potentiometric response to glucose using glucose oxidase immobilized on platinum. Biochim. Biophys. Acta **748:** 21–27.
9. ADAMS, R. N. 1969. Electrochemistry at Solid Surfaces. Marcel Dekker. New York. pp. 220.

Studies of Protein Function by Various Mutagenic Strategies: β-Lactamase[a]

G. DALBADIE-MCFARLAND, J. NEITZEL, A. D. RIGGS,[b]
AND J. H. RICHARDS

Division of Chemistry and Chemical Engineering
California Institute of Technology
Pasadena, California 91125

[b]*Beckman Research Institute of the City of Hope*
Duarte, California 91010

Structure unambiguously determines the function of proteins; the linear amino acid sequence in turn dictates three-dimensional structure. Important insights into the molecular origins of the functional behavior of a particular protein become possible if one can alter in prescribed ways the structure of that protein, thereby allowing one to undertake structure-function studies of the type that have a long history of success in organic and pharmaceutical chemistry. Changes in the structure of a protein can be most expeditiously accomplished by appropriate changes in the base sequence of the DNA encoding that protein.

IN VITRO MUTAGENESIS

The techniques of *in vitro* mutagenesis provide a basis for such changes and have already been applied to studies of β-lactamase,[1,2] tyrosyl t-RNA transferase,[3,4] and dihydrofolate reductase.[5] In essence, this technique involves the hybridization of a synthetic oligonucleotide to a single-stranded template to form a double-stranded region. The specificity of the base-pairing complementarity and the rarity of a particular sequence of bases is so high that a synthetic oligonucleotide 15 bases long will likely find only a single site on a gene where all, or most all, of its bases can form Watson-Crick A = T and G ≡ C base pairs. The double-stranded region that results from this hybridization now serves as an initiation site for enzymes, particularly the Klenow fragment of DNA polymerase I, that can complete the synthesis of a strand of new DNA that is complementary to the template.

The introduction of a mutation by the synthetic oligonucleotide requires that, while most of its bases are indeed complementary to those on the template strand at the site where the oligonucleotide hybridizes, *some are not*. These will, nevertheless, be incorporated into the newly synthesized strand, which will accordingly be an exact complement of the parent everywhere except at those sites where the synthetic oligonucleotide had, by design, noncomplementary bases. There results consequently a double-stranded DNA, one of whose strands, the parent template, contains the message for the parent protein. The other strand carries the message of the mutant gene whose DNA sequence reflects that of the synthetic oligonucleotide.

[a]This work was supported by the following grants: NIH GM 16424 and ONR N00014-83-K-0487.

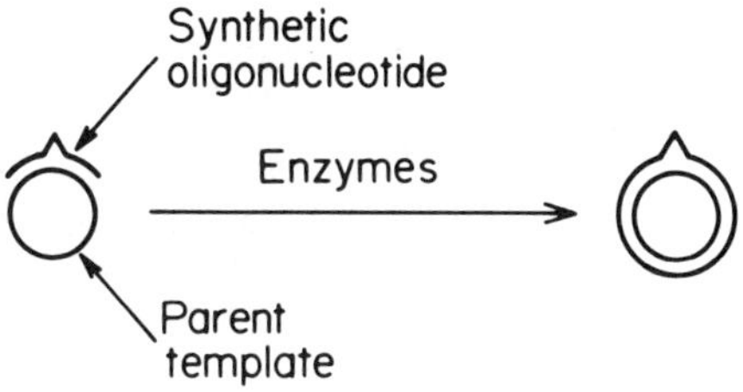

FIGURE 1. After completion of the new strand (as above), a heteroduplex plasmid will have been formed, one strand of which contains the information for the parental protein; the other strand carries the information for the new, mutant protein. The site of the mutation is indicated as —^—.

These mutations can be conveniently created in structural genes that are constituents of circular, double-stranded, extrachromosomal plasmid DNA,[1] or they can be introduced into DNA carried in a single-stranded vector such as M13.[3]

After introduction of the heteroduplex plasmid DNA into a suitable bacterial host, such as *Escherichia coli,* the normal semiconservative replication of the heteroduplex leads to two homoduplex descendents; one with the DNA sequence encoding the parental protein, the other with a DNA sequence that contains those base changes mandated by the mutagenizing synthetic oligonucleotide and that encodes the specifically designed mutant protein. FIGURES 1 and 2 outline these events.

To screen for the colonies that contain mutant as contrasted to parental genetic information, one takes advantage of the melting characteristics of DNA hybrids. Melting is a cooperative phenomenon and the melting temperature depends with great sensitivity on the number of complementary base pair interactions between the two strands. Thus, a hybrid formed between a 15-base oligonucleotide and a template with which all 15 bases form Watson-Crick base pairs will melt at a discernibly higher temperature than a similar hybrid in which one of the 15 bases does *not* complement its partner in the other strand. Thus, the synthetic oligonucleotide will hybridize perfectly, with *no* base mismatches, to the DNA of the mutant clones, whereas the synthetic oligonucleotide will hybridize to the DNA of parental clones with one or more base mismatches (at those sites where the synthetic oligonucleotide purposely contained nonparental bases). In hybridizing the mutagenizing oligonucleotide to the parental template to achieve mutagenesis, conditions of relatively low stringency are chosen such that extensive hybridization occurs even with 1 or 2 mismatches out of a total of 15 possible base pairings. For screening purposes, much more stringent conditions for hybridization are chosen that allow rapid and unambiguous identification of those clones that contain mutant DNA as contrasted to those that contain only parental DNA (see FIG. 3).

We have applied these approaches to studies of the role of the residues at the catalytic site of β-lactamase, an enzyme that catalyzes the hydrolysis of the amide bond of the lactam ring of penicillins and related antibiotics; the catalytic pathway includes an acyl-enzyme intermediate.[6] A serine residue probably participates in catalysis (7–12) and is part of a conserved Ser-Thr dyad.[12] (These residues have been

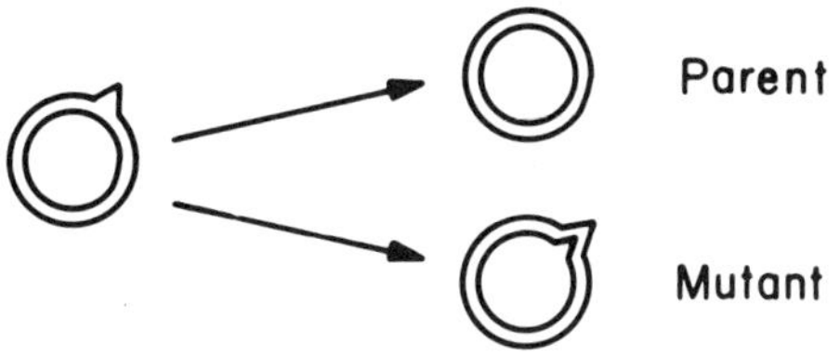

FIGURE 2. Normal semiconservative replication of the heteroduplex plasmid synthesized *in vitro* as described leads to two types of descendants, parental and mutant.

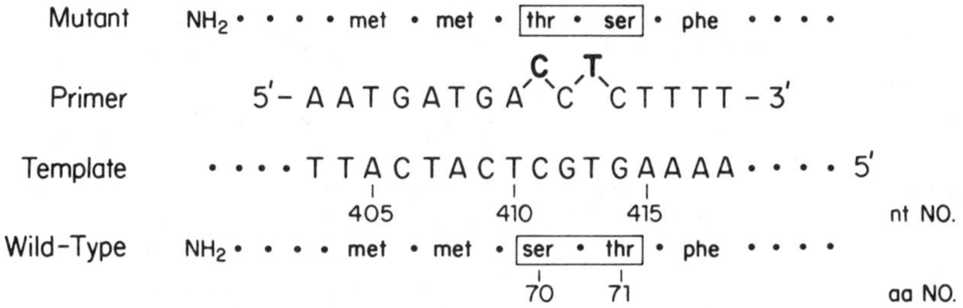

FIGURE 3. Nucleotide sequence of the synthetic oligodeoxynucleotide used as a specific mutagen and of the complementary wild-type strand. The amino acids encoded by both mutant and wild-type sequences are shown. nt NO is the nucleotide number in pBR322.[14] aa NO is the amino-acid number in β-lactamase.[12]

numbered Ser-70 and Thr-71.[12]) Presumably the hydroxyl group of Ser-70 adds nucleophilically to the carbonyl group of the β-lactam ring in a mechanism somewhat analogous to that of serine proteases. The role of Thr-71 is less clear, but it seems essential for catalytic activity because a mutant, probably with isoleucine at this position, shows no catalytic activity.[12]

FIGURE 4 outlines the use of a 16-base synthetic oligonucleotide with two deliberate base pair mismatches to prime *in vitro* DNA synthesis on pBR322 DNA containing single-stranded regions that had been generated by exonuclease treatment of nicked, double-stranded circles. The two mismatches in the primer d(A-A-T-G-A-T-G-A-C*-C-T*-C-T-T-T-T) at the bases denoted (*) will accomplish the double mutation Ser-70 → Thr and Thr-71 → Ser. We used hybridization with a ^{32}P-labeled oligonucleotide to select mutants. Conveniently, the same oligonucleotide used as a primer in the *in vitro* synthesis can be labeled with ^{32}P and also used as a hybridization probe. FIGURE 5 shows the results of such screening. Isolation of the plasmid and sequencing the DNA confirmed that the expected changes, (5′-AG*C A*CT-3′ → 5′-AC*C T*CT-3′: Ser 70 → Thr, Thr-71 → Ser), had, in fact, been introduced. A

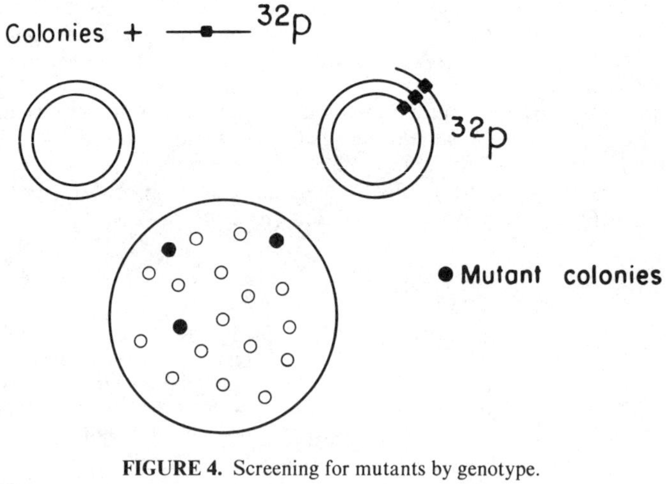

FIGURE 4. Screening for mutants by genotype.

similar procedure was used to produce the single mutant, 5′-A$\overset{*}{G}$C-3′ → 5′-A$\overset{*}{C}$C-3′: Ser 70 → Thr.

RANDOM MUTAGENESIS—REVERTANTS (THR-71 → SER)

Site-directed mutagenesis as just outlined provides a technique for generating proteins with predesigned structures and then determining function. In an alternative,

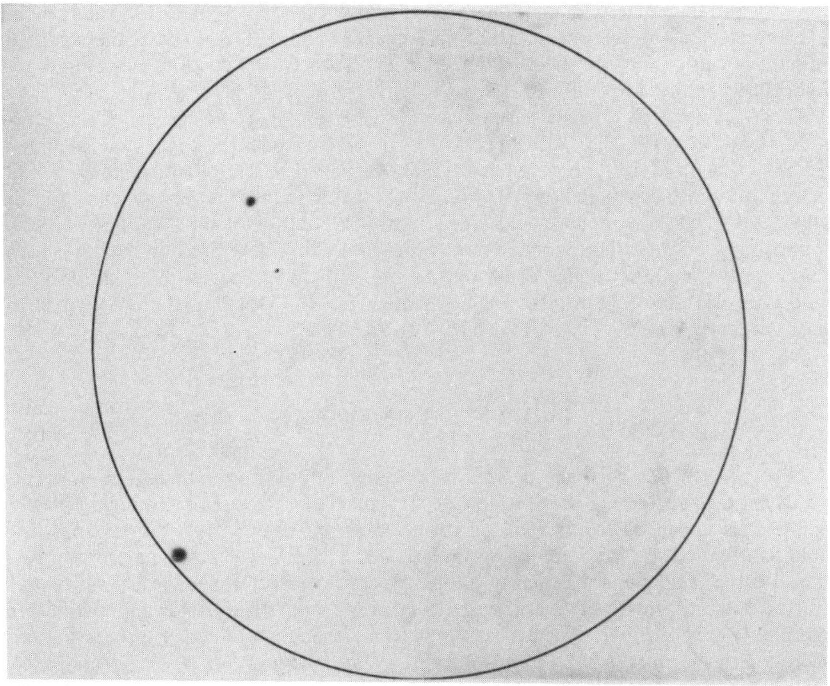

FIGURE 5. Under stringent conditions, the synthetic oligonucleotide, labeled with radioactive ^{32}P, binds preferentially to those colonies containing mutant plasmid. When the plate is exposed to film, mutant colonies, having bound the radioactive oligonucleotide, appear as dark spots. The photograph represents a real experiment in which two mutant colonies were identified by this screening technique against a background of about 2000 colonies containing plasmid for parental β-lactamase.

complementary approach, one specifies a particular function for the protein—as, for example, catalytic activity—and then determines what amino-acid sequences and structures possess this specified function. The methods for accomplishing this objective involve various types of random mutagenesis, perhaps directed to a particular region of the structural gene, followed by phenotypic screening of the many resulting mutant proteins for any that possess the specified property.

In asking questions of the possible amino-acid sequences that might be compatible with at least some degree of catalytic activity, one would like to screen phenotypically for such activity, without having any catalytically active mutants obscured by a high background of colonies that produce wild-type enzyme. To this end, we employed a mutant that has two base changes relative to the wild type and allowed random mutagenesis to occur at the normal rate characteristic of replication of plasmids in bacteria: a frequency of about 1 in 10^6 for any given base pair. Thus, we can expect a reasonable probability of examining all random mutants that differ by a single base change from the starting gene, but a very much lower probability of recovering wild-type enzyme from such a double mutant as this would require two base changes.

To produce such potentially active revertants, *E. coli* carrying the doubly mutant plasmid that encodes an inactive protein (Ser-70 → Thr; Thr-71 → Ser) was cultured in broth and then plated on agar containing 10 mg/l of ampicillin. Seven colonies that had reverted to ampicillin resistance were isolated and, in six of these, the antibiotic resistance was carried on the plasmid. Plasmid DNA from four of these colonies was hybridized against two different oligonucleotide probes: 5′-A ATG ATG ACC TCT TTT-3′ which matches the base sequence in the region of the active site of the double mutant, and 5′-ATG ATG AGC TCT TTT-3′ which encodes the single mutant Thr-71 → Ser (i.e., an active-site sequence of Ser-70, Ser-71). The oligonucleotide for the double mutant binds specifically to the DNA of the double mutant under conditions where it dissociates from both wild-type and from all four of the putative Thr-71 → Ser mutants; the oligonucleotide that encodes the single Thr-71 → Ser mutant binds to the four putative mutants under conditions where it dissociates from both the DNA of wild-type and double mutant colonies. Sequencing the DNA further confirmed the base sequence to be 5′-ATG ATG AGC TCT TTT-3′.

PROPERTIES OF MUTANTS

Proteins encoded by the β-lactamase gene and its mutants can be efficiently produced by transferring the appropriate structural gene to an overproducing plasmid where it is under the control of a hybrid *trp-lac* promoter.[13] Wild-type enzyme is abundantly produced by *E. coli* growing at 37°C with this overproducing plasmid and the wild-type enzyme is reasonably stable in a crude extract under these conditions.

In contrast, growth of *E. coli* with an overproducing plasmid for any of the three mutants described above yields little protein, though studies of processing and secretion in *Salmonella typhimurum* indicate that these events occur normally with the mutants. In fact, the mutants are thermally unstable relative to the wild-type protein; the mutants are moreover significantly less resistant to proteolysis. By maintaining cultures near 30°C, however, the mutant proteins can be prepared and isolated by standard procedures. All three mutants crossreact with rabbit antisera specific for β-lactamase. The two mutants with threonine at position 70 (Ser-70 → Thr and the double mutant Ser-70 → Thr, Thr-71 → Ser) are devoid of any detectable catalytic activity. The revertant with the net mutation Thr-71 → Ser is catalytically active with a Michaelis constant for benzylpenicillin near, and a k_{cat} of 10–30%, that of wild-type β-lactamase.

SECOND-SITE REVERTANTS

As an example of an approach to determine whether changes outside the region of the active site could restore catalytic activity to a protein that had threonine at position 70, we hybridized the synthetic oligonucleotide 5′-A ATG ATG ACC ACT TTT-3′ to the mutant plasmid that had long, single-stranded regions produced by treatment with a restriction enzyme followed by exonuclease. The binding of the synthetic oligonucleotide will protect the region of the active site from mutation during subsequent *in vitro* replications that were carried out in the presence, in turn, of a very reduced amount of each of the four deoxynucleotide triphosphates to increase the frequency of mutagenesis at sites outside the region of residue 70. No active second-site revertants have been isolated by this procedure.

MECHANISM OF CATALYSIS BY β-LACTAMASE

The only catalytically active structures yet found for β-lactamase all possess a primary nucleophile at position 70, either the hydroxyl group of serine in the wild-type enzyme and in the mutant Thr-71 → Ser or the sulfhydryl of cysteine in the mutant Ser-70 → Cys.[2] Though some caution is appropriate in extending these observations to an absolute conclusion, because one can never be absolutely certain that every possible protein structure with threonine at position 70 was examined in the search for second-site revertants, an attractive possibility is that a primary nucleophile, either CH_2OH as in wild-type enzyme, or CH_2SH as in thiolactamase, is an essential requirement for lactamase activity.

The steric hindrance in a secondary alcohol such as threonine, $-CHCH_3OH$, might effectively prevent the nucleophilic attack on the substrate that is the first step in the mechanism and thereby abolish any catalytic activity.

CONCLUSION

Applications of these techniques allow a truly systematic approach to the central question of the relation between structure and function in proteins and allow us to address such intriguing questions as: What are the essential structural features for a protein to function effectively as a catalyst, hormone, or receptor? From these studies will come striking new insights into the mechanisms by which proteins successfully carry out their myriad functions as well as the ability to design proteins with specific, novel, and useful properties.

REFERENCES

1. DALBADIE-MCFARLAND, G., L. W. COHEN, A. D. RIGGS, C. MORIN, K. ITAKURA & J. H. RICHARDS. 1982. Oligonucleotide-directed mutagenesis as a general and powerful method for studies of protein function. Proc. Natl. Acad. Sci. USA **79:** 6409–6413.
2. SIGAL, I. S., B. G. HARWOOD & R. ARENTZEN. 1982 Thiol-β-lactamase: Replacement of the active-site serine of RTEM β-lactamase by a cysteine residue. Proc. Natl. Acad. Sci. USA **79:** 7157–7160.
3. WINTER, G., A. R. FERSHT, A. J. WILKINSON, M. ZOLLER & M. SMITH. 1982. Redesigning enzyme structure by site-directed mutagenesis: Tyrosyl tRNA synthetase and ATP binding. Nature (London) **299:** 756–758.
4. WILKINSON, A. J., A. R. FERSHT, D. M. BLOW & G. WINTER. 1983. Site-directed mutagenesis as a probe of enzyme structure and catalysis tyrosyl-tRNA synthetase cysteine-35 to glysine-35 mutation. Biochemistry **22:** 3581–3586.
5. VILLAFRANCA, J. E., E. E. HOWELL, D. H. VOET, M. S. STROBEL, R. C. OGDEN, J. N. ABELSON & J. KRAUT. 1983. Directed mutagenesis of dihydrofolate reductase. Science **222:** 782–788.

6. FISHER, J., J. G. BELASCO, S. KHOSLA & J. R. KNOWLES. 1980. β-lactamase proceeds via an acylenzyme intermediate. Interaction of the *Escherichia coli* RTEM enzyme with cefoxitin. Biochemistry **19:** 2895–2901.
7. PRATT, R. F. & M. J. LOOSEMORE. 1978. 6-β-bromopenicillanic acid, a potent β-lactamase inhibitor. Proc. Natl. Acad. Sci. USA **75:** 4145–4149.
8. KNOTT-HUNZIKER, V., S. G. WALEY, B. S. ORLEK & P. G. SAMMES. 1979. Penicillinase active sites: Labelling of serine-44 in β-lactamase I by 6 β-bromopenicillanic acid. FEBS Lett. **99:** 59–61.
9. KNOTT-HUNZIKER, V., B. S. ORLEK, P. G. SAMMES & S. G. WALEY. 1979. 6 β-bromopenicillanic acid inactivates β-lactamase I. Biochem. J. **177:** 365–367.
10. LOOSEMORE, M. J., S. A. COHEN & R. F. PRATT. 1980. Inactivation of *Bacillus cereus* β-lactamase I by 6 β-bromopenicillanic acid: Kinetics. Biochemistry **19:** 3990–3995.
11. COHEN, S. A. & R. F. PRATT. 1980. Inactivation of *Bacillus cereus* β-lactamase I by 6 β-bromopenicillanic acid: Mechanism. Biochemistry **19:** 3996–4003.
12. AMBLER, R. P. 1980. The structure of β-lactamases. Philos. Trans. R. Soc. London Ser. B **289:** 321–331.
13. DE BOER, H. A., L. J. COMSTOCK & M. VASSER. 1983. The *lac* promoter: A functional hybrid derived from the *trp* and *lac* promoters. Proc. Natl. Acad. Sci. USA **80:** 21–25.
14. SUTCLIFFE, J. G. 1978. Nucleotide sequence of the ampicillin resistance gene of *Escherichia coli* plasmid pBR322. Proc. Natl. Acad. Sci. USA **75:** 3737–3741.

New Immobilization Techniques and Examples of Their Applications

KLAUS MOSBACH

Pure and Applied Biochemistry
Chemical Center, University of Lund
P. O. Box 740
S-220 07 Lund, Sweden

In the following, I would like to discuss work from my group, especially the development of two new immobilization techniques, one involving a subtle procedure suitable for cell/organelle entrapment, and the other especially suitable for the binding of affinity ligands leading to stable covalent linkages between support and ligand. The article will also describe the use of different binding techniques that allow the formation of enzyme complexes with their active sites juxtaposed to one another as well as preparation of specific enzyme-coenzyme complexes; these represent an approach that may be called "biomolecular engineering" (or "synthetic biochemistry" which, however, is not based on recombinant DNA techniques).

CELL ENTRAPMENT

A number of excellent entrapment procedures for cell immobilization are available, as amply demonstrated in these proceedings. Ideally, for instance, to obtain good flow rate, these preparations should be in a beaded form. As an example, to obtain beads of polyacrylamide, suspension polymerization in organic solvents such as toluene-chloroform is usually employed.[1] However, although this procedure has proved valuable for enzyme entrapment, the organic phase employed usually leads to loss of viability when using living cells. We later found, however, in part as a spin-off from our attempts to obtain beads of small size (2–3 μm) injectable into the blood stream for magnetic drug targeting,[2-4] that various oils such as soy or paraffin oil make excellent suspension media. Also, when employing other support material like low-gelling-temperature agarose or alginate, oil could be used in a convenient way as a suspension medium for bead formation.[5] Following this procedure, a number of different cells including microbial, plant (P. Brodelius, this volume) and animal cells were successfully entrapped leading to beads of desired diameter and with retained viability of the cells (see TABLE 1).

In the following, I would like to discuss briefly the properties of some of these preparations.

When mouse-mouse hybridoma cells that produce antibodies against *Herpes simplex* type 2 glycoprotein were entrapped in low-gelling-temperature agarose, these preparations continued to produce and export into the medium monoclonal antibodies for a long period.[6] Similarly, entrapped lymphoblastoid cells that produce interleukin 2 were productive over the two-week time period tested. More recent data obtained using entrapped Chinese hamster ovary cells showed unimpaired production levels of γ-interferon (to be published). We feel these preparations to be superior to those previously described, in which animal cells were entrapped in alginate beads, in which the cells were either embedded throughout the network of the bead[7] or kept within microcapsules, characterized by a shell of alginate-polyethyleneimine.[8] A possible drawback of the latter technique, when applied to animal cells, is the requirement for positive counterions to keep the polymeric network intact; this may adversely affect viability and growth of animal cells and/or the fragile nature of such preparations. Furthermore, larger biomolecules, for example, monoclonal antibodies, cannot easily diffuse out through the polymeric network, thus preventing their use for continuous processes. (On the other hand, harvesting of antibodies might be easier if they remain within the beads.)

In the light of the growing interest in the use of animal cell cultures for production of immunochemicals and hormones, the potential of such stable preparations producing and exporting these products for extended periods of time appears great.

In a parallel study, *Bacillus subtilis* cells, producing proinsulin, were entrapped following the above procedure.[9] It could be shown (a) that the entrapped cells did

TABLE 1. Production of "Biochemicals" by Various Cells Entrapped in Oil Phase to Obtain Beaded Agarose Particles (See esp. ref. 5)

Species	Product
Bacteria	
Mycobacterium	$\Delta^{1,4}$-androstadiene-3,17-dione
Gluconobacter oxydans	2-ketogulonic acid
Pseudomas putida	2-ketogulonic acid
Yeast	
Saccharomyces cerevisiae[a]	ethanol
Algae	
Chlorella vulgaris	oxygen
Plant cells	
Catharantus roseus[b]	ajmalicine
Animal cells	
MLA 144 (lymphoblastoid)	interleukin 2[6]
LSP 21 (hybridoma)	monoclonal antibody
Genetically engineered cells	
Bacillus subtilis	proinsulin[9]

[a]Polyacrylamide and carrageenan were also employed as supports.
[b]Carrageenan was also used.

continue proinsulin formation in the immobilized state and (b) that addition of antibiotics such as novobiocin (5 μg/ml), which inhibit DNA replication and thus cell growth, did not adversely affect formation of proinsulin by these microorganisms that had been modified with recombinant DNA technique. Novobiocin and other similarly acting antibiotics were added to prevent cell growth as too much growth away from the beads might lead to undesired extensive loss of cells into the medium. It may be that intermittent addition of those compounds will be one solution, along with others such as omission of essential growth nutrients, towards the establishment of bioreactors containing immobilized cells that can operate for long periods of time. It appears that this combination of genetic engineering and enzyme/cell technology should be pursued further in the light of the growing number of successfully cloned products into various organisms.

COVALENT BINDING OF BIOMOLECULES

A great number of immobilization methods leading to covalent attachment of enzymes to supports are currently available (e.g., *Immobilized Enzymes,* Volume 44, K. Mosbach, Ed. and the follow-up volume *Immobilized Enzymes and Cells,* K. Mosbach, Ed. in the *Methods of Enzymology* series (in preparation), or Volume 34 in the same series, *Affinity Chromatography,* W. B. Jakoby and M. Wilchek, Eds. and the follow-up volume, *Affinity Chromatography,* Volume 104, W. B. Jakoby, Ed.). The most widely used method at present involves activation of supports with cyanogen bromide. A drawback with the technique is the labile nature of the linkage formed. We found that treatment of hydroxyl group-carrying supports with sulfonyl halides is both a convenient procedure and one that overcomes these problems.[10] The reaction proceeds as outlined below.

Activation:

$$\text{Support—CH}_2\text{OH} + \text{R—SO}_2\text{Cl} \rightarrow \text{support—CH}_2\text{OSO}_2\text{—R}$$

Coupling:

$$\text{Support—CH}_2\text{OSO}_2\text{—R} + \text{H}_2\text{N—ligand} \rightarrow \text{support—CH}_2\text{—NH—ligand} + \text{HOSO}_2\text{—R}$$

$$\text{Support—CH}_2\text{OSO}_2\text{—R} + \text{HS—ligand} \rightarrow \text{support—CH}_2\text{—S—ligand} + \text{HOSO}_2\text{—R}$$

$$(\text{R} = \cdot\, \text{CF}_3\text{CH}_2 \quad \text{or} \quad \text{CH}_3 \cdot \text{C}_6\text{H}_4)$$

Enzymes and affinity ligands have been coupled to a large number of various supports following this procedure (TABLE 2), in most cases using tresyl chloride, $\cdot CF_3CH_2 \cdot SO_2Cl$,[11] and to a lesser degree, tosyl chloride, $CH_3 \cdot C_6H_4 \cdot SO_2Cl$.[12] Subsequently, an interesting alternative involving chromophoric sulfonyl halides was described by Scouten and collaborators (in this volume).

The above method should be particularly useful in applications requiring stable linkages, such as in purifications employing expensive monoclonal antibodies or in the treatment of patients using immunoaffinity techniques in extracorporeal shunt devices. In addition to polysaccharides such as agarose or cellulose, synthetic polymers such as hydroxyethylmethacrylate gels,[11] polyethylene glycol for affinity partition chromatography,[10] or silica beads used in HPLAC (high-performance liquid affinity chromatography)[13,14] are conveniently activated followed by coupling of enzymes or affinity ligands. Even surface binding of ligands such as concanavalin A to small chips used in ellipsometry for subsequent detection of affinity-bound cells is possible.[15]

Tresyl chloride can be used with advantage for coupling of pH-sensitive ligands or proteins with high yields, even at neutral pH. In the pH range 9 to 10.5, tosylated supports, similar to epoxy gels but easier to prepare, are effective for coupling ligands that are stable in this range. The degree of activation is easy to predict and to reproduce, and the activated gel does not lose coupling capacity when stored for several weeks in aqueous acidic media. The tresyl chloride method is ideally suited for the activation of sorbents applicable to high-performance liquid affinity chromatography and for activation of polyethylene glycol. In addition to amino groups, thiol groups

TABLE 2. Coupling of Proteins and Affinity Ligands to Different Supports Activated with Tresyl Chloride (Not Always under Optimized Conditions)

Support	Tresyl Groups on Activated Support (mmol/g dry support)	Ligand	Time for Coupling (h)	pH	Ligand Bound (mg/g dry support)	Coupling Yield (%)	Specific Activity Relative to Soluble Enzyme
Sepharose 4B	0.15	protein A	15	8.5	70	90	
Sepharose 4B	1.30	concanavalin A	15	7.5	360	90	
Sepharose 4B	1.30	STI[a]	15	7.5	195	68	
Sepharose 4B	1.30	STI	1.5	8.5	200	70	
Sepharose CL-6B	1.30	nucleotide[b]	15	8.2	87	32	
Diol-silica 100[c]	0.45	STI	20	8.0	250	83	
Diol-silica 100	0.45	nucleotide[b]	20	7.5	13	80	
Separon HEMA	0.14	protein A	12	8.0	10	100	
Sepharose CL-6B		hexokinase			94	53	26
Sepharose CL-6B		trypsin			97	55	50
Cellulose		trypsin			61	87	56
Diol-silica 1000		horse liver alcohol dehydrogense			21	100	95

[a]STI, soybean trypsin inhibitor.
[b]N^6-(6-Aminohexyl)-5′-AMP.
[c]Glyceryl-propyl-silica (10 μm).

react efficiently with sulfonate esters, enabling coupling of HS-carrying affinity ligands and proteins.

ENZYME-COENZYME COMPLEXES

The use of various immobilization methods for the binding of several enzymes acting in consecutive order has been pursued in my laboratory for a number of years[16,17]

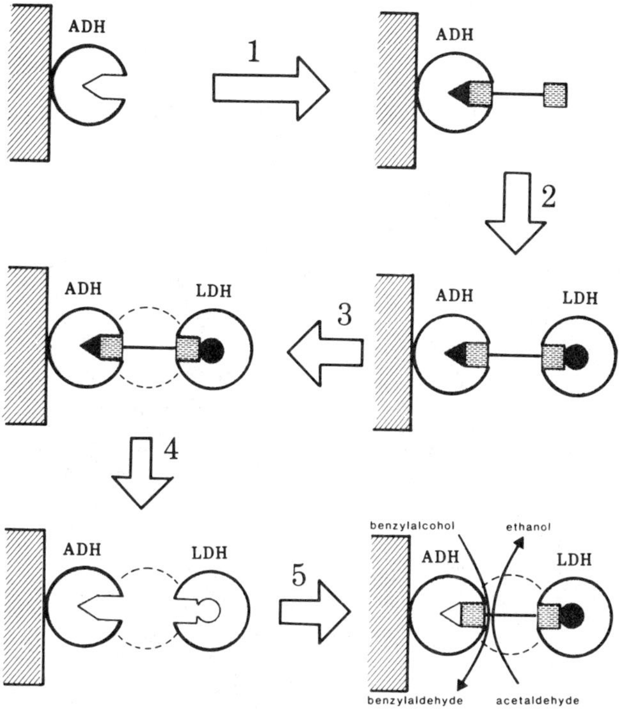

FIGURE 1. Preparation of site-to-site directed alcohol dehydrogenase (ADH)-lactate dehydrogenase (LDH) complex. Steps: 1, affinity binding of *bis*-NAD to alcohol dehydrogenase immobilized on Sepharose in the presence of pyrazole; 2, affinity binding of lactate dehydrogenase to *bis*-NAD in the presence of oxalate; 3, crosslinking with glutaraldehyde; 4, removal of pyrazole, oxalate, and *bis*-NAD by washing; 5, test for site-to-site immobilization by addition of *bis*-NAD in the presence of oxalate, giving affinity binding to lactate dehydrogenase and activity in a coupled substrate assay for alcohol dehydrogenase. The oligomeric nature of the enzymes has not been taken into account. ▤-▤, *bis*-NAD; ◀, pyrazole; ●, oxalate.

including the coupling of the members of a complete metabolic cycle, the urea cycle, to beaded agarose.[18] More recently, we attempted to bind an enzyme couple either working in consecutive order or recycling a coenzyme in such a configuration that their active sites are juxtaposed to one another.[19] This was carried out in order to obtain model systems for naturally occurring multienzyme systems, at the same time

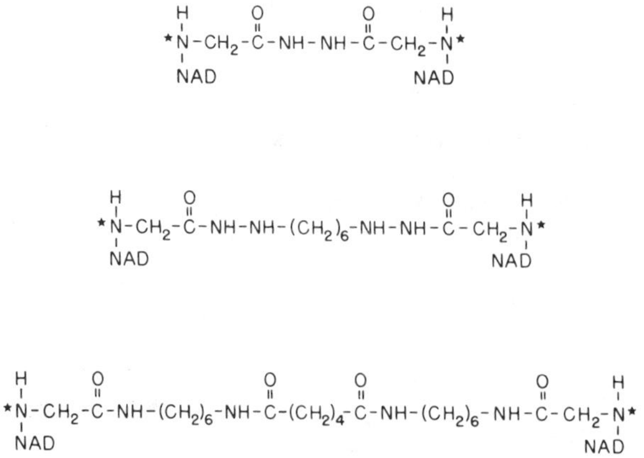

$$\overset{H}{\underset{\underset{NAD}{|}}{*N}}-CH_2-\overset{O}{\overset{\|}{C}}-NH-NH-\overset{O}{\overset{\|}{C}}-CH_2-\overset{H}{\underset{\underset{NAD}{|}}{N*}}$$

$$\overset{H}{\underset{\underset{NAD}{|}}{*N}}-CH_2-\overset{O}{\overset{\|}{C}}-NH-NH-(CH_2)_6-NH-NH-\overset{O}{\overset{\|}{C}}-CH_2-\overset{H}{\underset{\underset{NAD}{|}}{N*}}$$

$$\overset{H}{\underset{\underset{NAD}{|}}{*N}}-CH_2-\overset{O}{\overset{\|}{C}}-NH-(CH_2)_6-NH-\overset{O}{\overset{\|}{C}}-(CH_2)_4-\overset{O}{\overset{\|}{C}}-NH-(CH_2)_6-NH-\overset{O}{\overset{\|}{C}}-CH_2-\overset{H}{\underset{\underset{NAD}{|}}{N*}}$$

FIGURE 2. Various *bis*-NAD analogues used.

FIGURE 3. Schematic picture of a precipitated lactate dehydrogenase-*bis*-NAD-oxalate complex (in reality all active sites of the enzyme are occupied by *bis*-NAD). ■-■, *bis*-NAD; ●, oxalate; ⊃, lactate dehydrogenase (tetrameric).

providing some technical know-how for the coupling of enzymes in a defined, oriented fashion.

The model system studied primarily consisted of liver alcohol dehydrogenase-lactate dehydrogenase. The route followed to obtain such a preparation is depicted in FIGURE 1.

An essential component in the procedure was the use of *bis*-NAD analogues (FIG. 2), compounds recently synthesized to be used in the purification of enzymes via a technique called affinity precipitation.[20] As an example of the latter technique, precipitation of lactate dehydrogenase will be described briefly here. If this tetrameric enzyme is present together with *bis*-NAD in a 1:1 ratio, that is, enzyme subunit to NAD moiety, the enzyme will precipitate out provided a third component such as oxalate, leading to formation of a strong ternary complex, is present (FIG. 3). If one instead wishes to isolate glutamate dehydrogenase by this technique, oxalate is simply replaced with glutarate.[20] Affinity precipitation can also conveniently be employed on a preparative scale as demonstrated in the quick and specific isolation of lactate dehydrogenase from crude ox heart extract.[21]

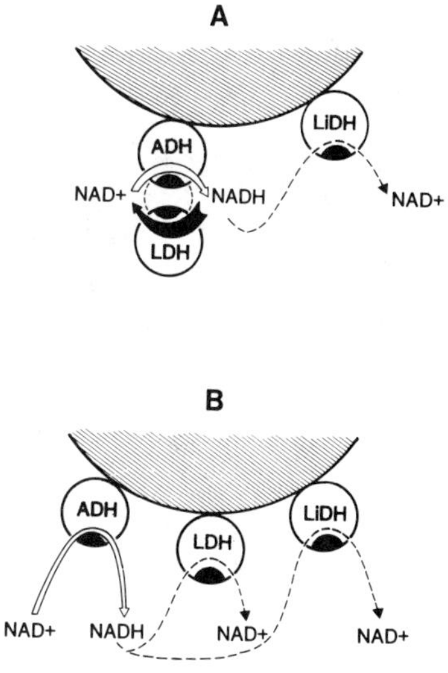

FIGURE 4. Schematic representation of the immobilized three-enzyme system. (A) Only alcohol dehydrogenase (ADH) and lipoamide dehydrogenase (LiDH) were coupled directly to tresyl chloride-activated Sepharose 4B; lactate dehydrogenase (LDH) was subsequently coupled with the site-to-site directing aid of *bis*-NAD. (B) All three enzymes were simultaneously immobilized. The oligomeric nature of the enzymes has not been taken into account.

The immobilized preparations obtained with the *bis*-NAD analogue were then tested for their enzymic activity as outlined in FIGURE 4. It was found that in the preparations with the two enzymes kept close and oriented towards one another, most of the NADH formed was reoxidized by lactate dehydrogenase positioned close to alcohol dehydrogenase and not by the competing lipoamide dehydrogenase. This contrasts with the enzymes coupled *at random* where no such channeling of formed "NADH" was observed.

Other examples of this kind of "biomolecular engineering" involve preparations that are designed for coenzyme recycling. The approaches taken and that will be discussed here briefly may appear somewhat far fetched. They are far away from the developments on amino-acid formation using enzymes and enlarged coenzymes in membrane reactors that seem close to large-scale production and that are discussed in

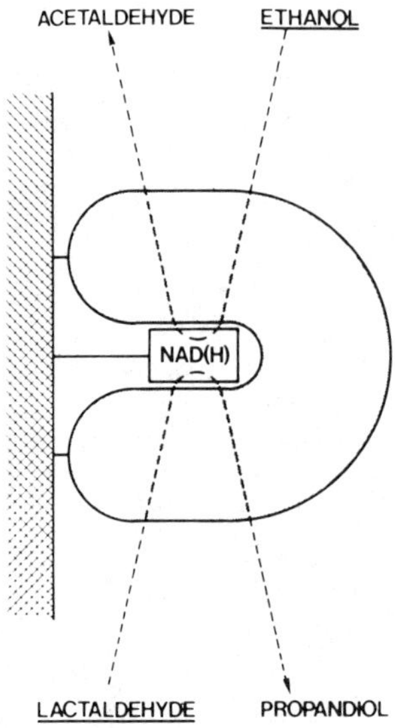

FIGURE 5. *Highly* schematic picture of a NAD-alcohol dehydrogenase complex coupled covalently to agarose and tested in a coupled substrate assay.

these proceedings by Kula and Wandrey. (Yet one should not forget that also in the latter area, the first reports describing mini-reactors were highly preliminary.)[22,23]

In one such approach, a binary NAD-alcohol dehydrogenase complex was allowed to form that was subsequently coupled to CNBr-activated agarose beads leading to preparations schematically depicted in FIGURE 5. These preparations were shown to be highly active in the absence of exogenous NAD(H) and when tested in coupled substrate assays.[24]

Alternatively, an NAD analogue with a long spacer was coupled covalently directly

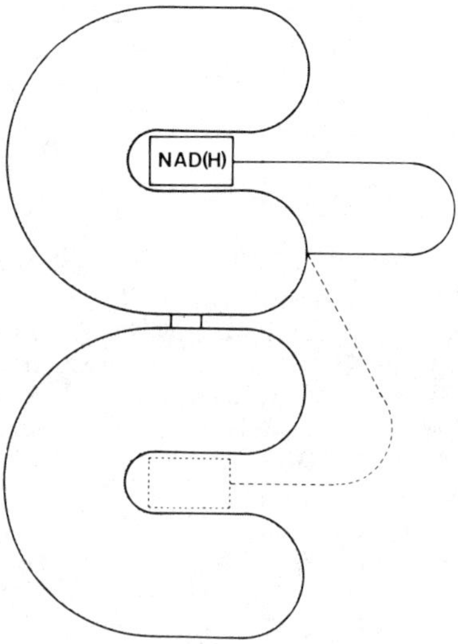

FIGURE 6. *Highly* schematic picture of covalently fixed NAD to alcohol dehydrogenase, to which complex lactate dehydrogenase has been coupled. The oligomeric nature of the enzymes has not been taken into account.

onto alcohol dehydrogenase,[25] leading to enzyme-coenzyme complexes not requiring exogenous coenzyme for activity. Such complexes could be made to interact with a second enzyme, for example lactate dehydrogenase, for NAD(H) regeneration (FIG. 6).[26] Alternatively, when the enzyme-coenzyme complex was coupled directly onto an electrode, the reduced coenzyme formed could "swing" to the electrode to become electrochemically oxidized.[27] The above examples are still of a preliminary nature, and their use may be restricted to special cases. The approach involving a permanent fixation as prosthetic group of a normally dissociable coenzyme may find application, for example, in medicine (for treatment of enzyme deficiency diseases using extracorporeal shunts). Certainly there is some attraction to such systems where the same coenzyme, after interaction with the first enzyme (that of practical interest), can "swing" to the next enzyme for regeneration.

In summary, it may very well be that attempts of the kind of "biomolecular engineering" on a non-DNA level, discussed in this section, will provide both useful model systems for the understanding of multienzyme systems as well as lead to practically useful enzyme-enzyme or enzyme-coenzyme complexes.

REFERENCES

1. NILSSON, H., R. MOSBACH & K. MOSBACH. 1972. The use of bead polymerization of acrylic monomers for immobilization of enzymes. Biochim. Biophys. Acta **268:** 253–256.
2. MOSBACH, K. & U. SCHRÖDER. 1979. Preparation and application of magnetic polymers for targeting of drugs. FEBS Lett. **102:** 112–116.
3. MOSBACH, K. Magnetic polymer particles. US patent 4.335094.
4. SCHRÖDER, U. & K. MOSBACH. Intravascularly administrable, magnetically responsive nanosphere or nanoparticle, a process for the production thereof, and the use thereof. Patent pending.
5. NILSSON, K., S. BIRNBAUM, S. FLYGARE, L. LINSE, U. SCHRÖDER, U. JEPPSSON, P.-O. LARSSON, K. MOSBACH & P. BRODELIUS. 1983. A general method for the immobilization of cells with preserved viability. Eur. J. Appl. Microbiol. Biotechnol. **17:** 319–326.
6. NILSSON, K., W. SCHEIRER, O. W. MERTEN, L. ÖSTBERG, E. LIEHL, H. W. D. KATINGER & K. MOSBACH. 1983. Entrapment of animal cells for production of monoclonal antibodies and other biomolecules. Nature **302:** 629–630.
7. NILSSON, K. & K. MOSBACH. 1980. Preparation of immobilized animal cells. FEBS Lett. **118:** 145–150.
8. LIM, F. & A. M. SUN. 1980. Microencapsulated islets as bioartificial endocrine pancreas. Science **210:** 908–910.
9. MOSBACH, K., S. BIRNBAUM, K. HARDY, J. DAVIES & L. BUELOW. 1983. Formation of proinsulin by immobilized *Bacillus subtilis*. Nature **302:** 543–545.
10. NILSSON, K. & K. MOSBACH. 1984. Immobilization of ligands with organic sulfonyl chlorides. *In* Methods in Enzymology. W. B. Jakoby, Ed. Vol. **104:** 56–69. Academic Press. New York.
11. NILSSON, K. & K. MOSBACH. 1981. Immobilization of enzymes and affinity ligands to various hydroxyl-group-carrying supports using highly reactive sulfonyl chlorides. Biochem. Biophys. Res. Commun. **102:** 449–457.
12. NILSSON, K. & K. MOSBACH. 1980. *p*-Toluenesulfonyl chloride as an activating agent of agarose for the preparation of immobilized affinity ligands and proteins. Eur. J. Biochem. **112:** 397–402.
13. LARSSON, P.-O., M. GLAD, L. HANSSON, M.-O. MÅNSSON, S. OHLSON & K. MOSBACH. 1983. High-performance liquid affinity chromatography. *In* Advances in Chromatography. J. C. Giddings, E. Grushka, J. Cazes & P. R. Brown, Eds. Vol. **21:** 41–85. Marcel Dekker, Inc. New York.
14. NILSSON, K. & P.-O. LARSSON. 1983. High-performance liquid affinity chromatography on silica-bound alcohol dehydrogenase. Anal. Biochem. **134:** 60–72.

15. MANDENIUS, C. F., S. WELIN, B. DANIELSSON, I. LUNDSTRÖM & K. MOSBACH. 1983. The interaction of proteins and cells with affinity ligands covalently coupled to silicon surfaces as monitored by ellipsometry. Anal. Biochem. **137:** 106–114.
16. MOSBACH, K. & B. MATTIASSON. 1970. Matrix-bound enzymes. Part II. Studies of a matrix-bound, two enzyme-system. Acta Chem. Scand. **24:** 2093–2100.
17. MOSBACH, K. & B. MATTIASSON. 1978. Immobilized model systems of enzyme sequences. *In* Current Topics in Cellular Regulation. B. L. Horecker & E. R. Stadtman, Eds. Vol. **14:** 197–241. Academic Press. New York.
18. SIEGBAHN, N. & K. MOSBACH. 1982. An immobilized cyclic multi-step enzyme system—the urea cycle. FEBS Lett. **137:** 6–10.
19. MÅNSSON, M.-O., N. SIEGBAHN & K. MOSBACH. 1983. Site-to-site directed immobilization of enzymes with *bis*-NAD analogues. Proc. Natl. Acad. Sci. USA **80:** 1487–1491.
20. LARSSON, P.-O. & K. MOSBACH. 1979. Affinity precipitation of enzymes. FEBS Lett. **98:** 333–338.
21. FLYGARE, S., T. GRIFFIN, P.-O. LARSSON & K. MOSBACH. 1983. Affinity precipitation of dehydrogenases. Anal. Biochem. **133:** 409–416.
22. DAVIES, P. & K. MOSBACH. 1974. The application of immobilized NAD^+ in an enzyme electrode and in model enzyme reactors. Biochim. Biophys. Acta **370:** 329–338.
23. MOSBACH, K. 1978. Immobilized coenzymes in general ligand affinity chromatography and their use as active coenzymes. *In* Advances in Enzymology. A. Meister, Ed. Vol. **46:** 205–278. John Wiley & Sons. New York.
24. GESTRELIUS, S., M.-O. MÅNSSON & K. MOSBACH. 1975. Preparation of an alcohol dehydrogenase-NAD(H)-sepharose complex showing no requirement of soluble coenzyme for its activity. Eur. J. Biochem. **57:** 529–535.
25. MÅNSSON, M.-O., P.-O. LARSSON & K. MOSBACH. 1978. Covalent binding of an NAD analogue to liver alcohol dehydrogenase resulting in an enzyme-coenzyme complex not requiring exogenous coenzyme for activity. Eur. J. Biochem. **86:** 455–463.
26. MÅNSSON, M.-O., P.-O. LARSSON & K. MOSBACH. 1979. Recycling by a second enzyme of NAD covalently bound to alcohol dehydrogenase. FEBS Lett. **98:** 309–313.
27. TORSTENSSON, A., G. JOHANSSON, M.-O. MÅNSSON, P. -O. LARSSON & K. MOSBACH. 1980. Electrochemical regeneration of NAD covalently bound to liver alcohol dehydrogenase. Anal. Lett. **13:** 837–849.

Chromophoric Sulfonyl Chloride Agarose for Immobilizing Bioligands

WILLIAM H. SCOUTEN AND WILL VAN DER TWEEL

Chemistry Department
Bucknell University
Lewisburg, Pennsylvania 17837

Nilsson and Mosbach recently described a method for the immobilization of enzymes and affinity ligands to various hydroxyl support materials using highly reactive sulfonyl chlorides.[1,2] Some years previously, Gribnau had reported with less success a similar activation procedure.[3]

We have improved the utility of the sulfonyl chloride activation procedure considerably by synthesizing a number of colored and fluorescent sulfonyl chlorides, which we have used as activating agents for agarose.[4,5] The underivatized agarose beads are clear; while the dabsyl-, dipsyl-, and dansyl-activated beads are purple, yellow-green, and light blue, respectively. These leaving groups were employed to immobilize various compounds, including sulfhydryl and amine-containing ligands and enzymes. As the coupling proceeds, the colored sulfonyl group is removed from the matrix and replaced by the nucleophilic bioligand. Consequently, the solution in which the coupling is being performed becomes colored. A quick wash of the beads allows the investigator qualitatively to assess the degree and efficiency of coupling; while a spectrum of the reacted beads (suspended in 50% glycerol) will permit quantitative determination of the number of chromophoric sulfonyl groups left. Similarly, spectral analysis of the coupling buffer will allow quantitative determination of the number of sulfonyl groups replaced by the nucleophile.

We have used this method to investigate and improve the sulfonyl chloride activation procedure in general. From this we conclude that the activation should be limited to the primary hydroxyl of each disaccharide subunit of agarose. This can be done by activating at low temperatures with minimal concentration of the activating

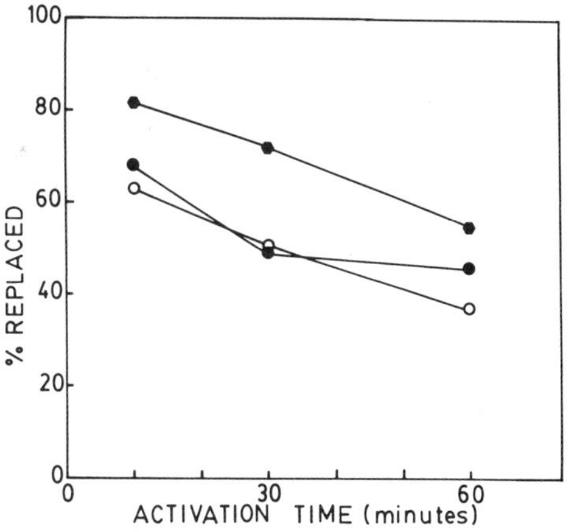

FIGURE 1. Coupling of mercaptoethanol to agarose beads activated with various amounts of dipsyl chloride for various times. Coupling was done for 4 hours at 55° with 1 *M* 2-mercaptoethanol, pH 9.0. The amount of dipsyl functions removed was determined spectrophotometrically. The mmol of dipsyl chloride/g dry gel was: (●) 3.1; (●) 15.5; and (○) 31.0.

agent. Alternatively, sulfonyl fluorides or other more selective activated sulfonyl groups could be used in place of sulfonyl chlorides. Activation of only the primary hydroxyls in necessary, as nucleophilic substitution of primary sulfonates is facile but that of secondary sulfonates is extremely difficult, often with elimination rather than substitution occurring.

As is seen in FIGURE 1, the *percentage* of replaceable sulfonates is markedly increased when activation is done for a short time under mild conditions. The sulfonyl chlorides we have most frequently employed are dansyl chloride and dabsyl chloride, both currently used in peptide NH_2-terminal analysis, and dipsyl chloride [*N*(2,4-dinitrophenyl)-4-aminobenzene sulfonyl chloride]. The latter is easily synthesized from sulfanilic acid and 2,4-dinitrofluorobenzene and is described in detail in Scouten and van der Tweel.[4] Dabsyl chloride, unfortunately, is both displaced by sulfhydryl agents and reduced by them to colorless materials. Dipsyl chloride, however, is equally usable with sulfhydryl and amine compounds.

Urease was readily coupled to dipsyl-agarose (TABLE 1), to yield a very active immobilized enzyme preparation with considerable protein coupling (410 mg g^{-1} dry gel). Thus the chromophoric sulfonyl agarose is a very promising material for affinity chromatography, immobilized enzymes, and other aspects of solid-phase biochemistry.

TABLE 1. Coupling of Urease to Dipsyl-activated Beads

Urease Employed (mg g^{-1} dry gel)	Urease Bound (mg g^{-1} dry gel)	Coupling (%)	Specific Activity[a] after Coupling	Specific Activity Retained (%)[b]
300	160	53	51	109
867	317	37	46	98
1500	410	27	35	74

[a]Activity: μmol NH_4^+ formed min^{-1} mg^{-1} urease.
[b]Activity of free urease in solution was: 47 μmol NH_4^+ formed min^{-1} mg^{-1} urease.

REFERENCES

1. NILSSON, K. & K. MOSBACH. 1981. Immobilization of enzymes and affinity ligands to various hydroxyl group carrying supports using highly reactive sulfonyl chlorides. Biochem. Biophys. Res. Commun. **102:** 449–457.
2. NILSSON, K., O. NORRLOW & K. MOSBACH. 1981. *p*-Toluenesulfonyl chloride as an activating agent for the preparation of immobilized affinity ligands and proteins: Optimization of conditions for activation and coupling. Acta Chem. Scand. **35:** 19–27.
3. GRIBNAU, T. C. J. 1977. Coupling of effector molecules to solid supports. Ph.D. Thesis, University of Nijmegen, the Netherlands.
4. SCOUTEN, W. H. & W. VAN DER TWEEL. 1984. Colored sulfonyl chlorides as activating agents for hydroxylic matrices. Proceedings, Affinity Chromatography Symposium. June, 1983. Annapolis, MD. In press.
5. SCOUTEN, W. H. 1983. U.S. Patent applied for.

New Synthetic Carriers for Enzymes

O. MAUZ, S. NOETZEL, AND K. SAUBER
Hoechst AG
Frankfurt am Main, Federal Republic of Germany

Copolymerization of vinylenecarbonate, *N*-vinylpyrrolidone, and methylenebis-acrylamide or butane dioldivinylether led us to develop a new group of carriers. The interior of the bead-shaped porous structures (FIG. 1) is accessible to enzymes of different sizes up to at least 450,000 daltons (urease). The physical properties can be modified widely by using different crosslinkers and proportions of the monomers (TABLE 1). The polymerization details are described elsewhere.[1,2] The carriers are suitable for binding of enzymes via the amino groups. This leads to stable urethane bonds as shown in FIGURE 2. The urethane bonds are preferable to the isourea bonds, which result from coupling via BrCN-activated polysaccharides.[3,4]

Coupling of enzymes proceeds at pH 8.0 with 1*M* potassium phosphate buffer at room temperature for 16 hours for the carbonate-type carrier (A) and 2½ days for the oxirane-type carrier (C).[1,2] Typical results are shown in TABLE 2. Enzyme activity was determined with an autotitrator as follows: (1) pencillin acylase, pH 7.8, 37°C, with penicillin G (2% wt/vd) as substrate;(2) trypsin, 37°C, pH 8.1, with *N*-benzoyl-L-argininethylester or soluble casein; (3) chymotrypsin, as trypsin but with N-acetyl-L-tyrosinethylester; (4) urease, with urea at pH 6.1, 30°C; and (5) cephalospo-

TABLE 1. Physical Properties of Carriers

Property	Different Types of Carriers			
	324A	249A[a]	385C	525C
Diameter (μm)	50–200	6–25	50–200	100–200
Specific surface area (m^2/g)	72	3.1	11	86
Pore volume (cm^3/g)	0.33	0.48	0.75	1.0
Medium pore diameter (nm)	23	249	159	42

[a]Not spherical.

TABLE 2. Enzyme Coupling

Enzyme	Substrate	Activity (IU/g wet wt)	Activity (IU/g dry wt)	Coupling Yield (%)	Carrier Type
Penicillin acylase	penicillin G	400	1200	70	N385C
Trypsin	BAEE	280	1100	60	M249A
		280	1100	50	M324C
		200	1050	55	N385C
Chymotrypsin	ATEE	450	1650	60	M249A
		400	1300	50	M324T
Urease	urea	50	130	80	N385C
		50	200	90	M324AT
Cephalosporinase	cephalosporin C	1300	5000	80	M249A

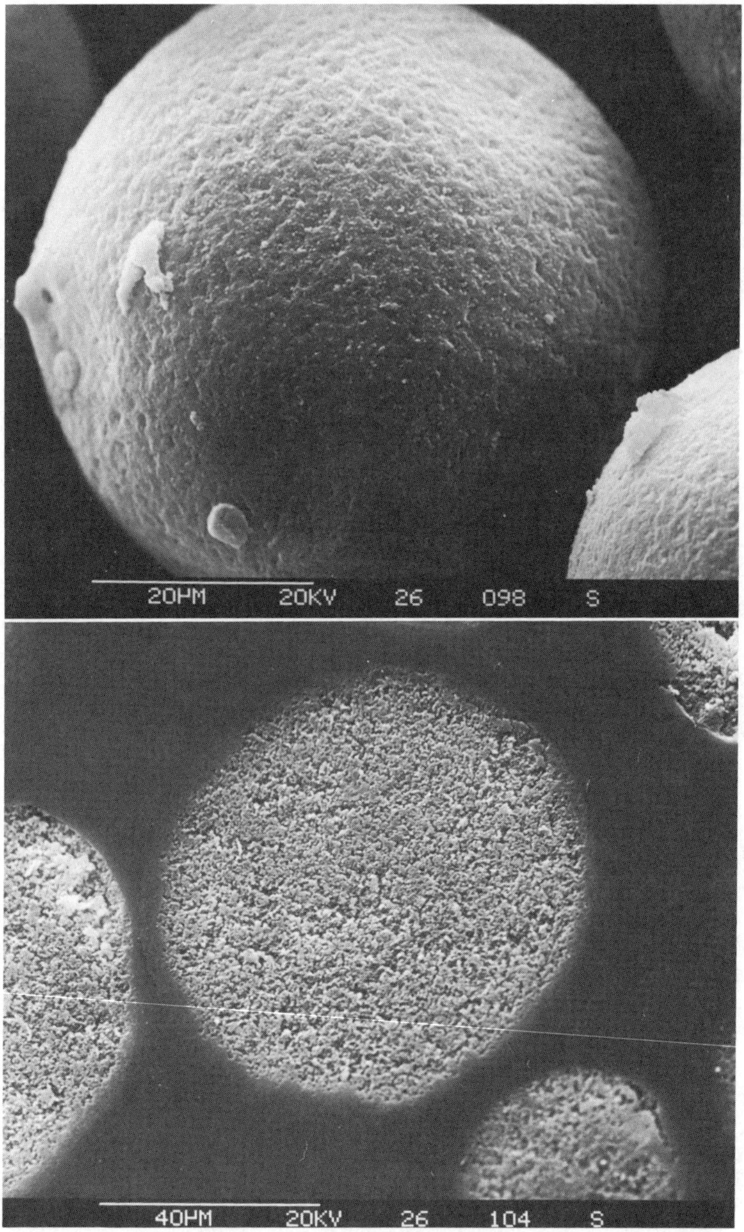

FIGURE 1. Porous particles of new carriers. *Top:* whole particle; *bottom:* sectioned particle.

rinase, with cephalosporin C as substrate at pH 7.8 37°C (1 μmol of base added in an autotitrator corresponds to μmol of lactam ring opened).

The storage and operational stabilities of the immobilized enzymes in TABLE 2 are excellent. Penicillin acylase showed no detectable inactivation during storage for 2 months and only slight inactivation (~0.2%) in repeated 2-hr batch runs at pH 7.8, 37°C, and with 6% penicillin G. With macromolecular substrates, for example casein, activities of about 25 IU/g are observed; this compares to 200–300 IU with BAEE.

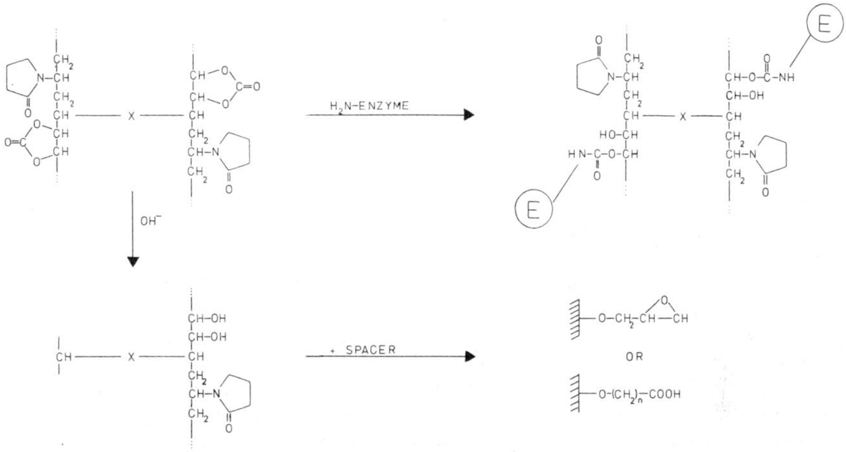

FIGURE 2. Coupling of enzyme to new carriers.

REFERENCES

1. MAUZ, O. *et al.*, 1982. German patent application P 32 43 591.
2. MAUZ, O. *et al.*, 1983. German patent application P 33 19 506.
3. M. WILCHEK, 1982. Enzyme Microb. Technol. **4:** 165.
4. M. WILCHEK, 1982. Biochem. Biophys. Res. Commun. **107:** 878.

New Methods for Activation of Polysaccharides for Protein Immobilization and Affinity Chromatography

MEIR WILCHEK, TALIA MIRON, AND JOACHIM KOHN

Department of Biophysics
The Weizmann Institute of Science
Rehovot, Israel

With the increased use of immobilized biochemicals in science and industry, it has become clear that new and improved carriers and coupling methods are of great importance.

The cyanogen bromide method that is widely used as an activating agent[1,2] is very convenient except for the extremely hazardous nature of the reagent. After we determined the mechanism of activation of polysaccharides with CNBr,[3] we were able to develop improved, milder, and nonhazardous activating agents that yielded active resins, identical, for all practical purposes, with the original CNBr-activated resins. The activation of Sepharose with CNBr in the presence of triethylamine required smaller amounts of CNBr, and the reaction could be performed at neutral pH.[4] The reaction of CNBr with tertiary amines and phenols enabled the synthesis and isolation and a variety of cyanylating reagents, including *N*-cyanothriethylammonium tetrafluoroborate, 1-cyano-4-dimethylaminopyridinium tetrafluoroborate,[5] and *p*-nitrophenylcyanate.[6] These cyano derivatives are stable and effective cyanylating agents and are also highly efficient activating agents for polysaccharides. Activation yields up to 50%, as compared to 1–2% by the regular CNBr method, were obtained. The resulting activated resins contained largely cyanate esters (Resin—O—C≡N). Activation with these reagents could be performed safely without a hood.

The product of the reaction of these activated cyanate esters with amines is the *N*-substituted isourea,[3] therefore the activation of polysaccharides with these new cyanylating agents do not overcome the inherent problems (such as leakage and charge) that accompany the CNBr immobilization procedure. In order to eliminate these problems, we developed alternative and improved methods for activation and immobilization of ligands and proteins to polysaccharides that gave chemically stable resins. The reagents used were *p*-nitrophenyl-chloroformate, *N*-hydroxysuccinimide chloroformate, and trichlorophenylchloroformate (FIG. 1). These reagents form reactive carbonates, which upon reaction with amines, give stable and uncharged carbamates (urethanes).[7] The activation and coupling can be followed spectrophotometrically and the yield of coupling is very high.

The mild reaction conditions used for activation of polysaccharides with aromatic cyanates would make this method very attractive if the imidocarbonate moieties would be converted to active carbonates. When agarose were reacted with *p*-(methylsulfide)phenylcyanate (MSPC), a resin was obtained containing *p*-(methylsulfide)phenylimidocarbonate as the active moiety. The imidocarbonates could easily be hydrolyzed to the corresponding carbonate[8] by brief exposure of the resin to ice-cold dilute sulfuric acid. The MSP moiety is a weak leaving group and therefore reacts only

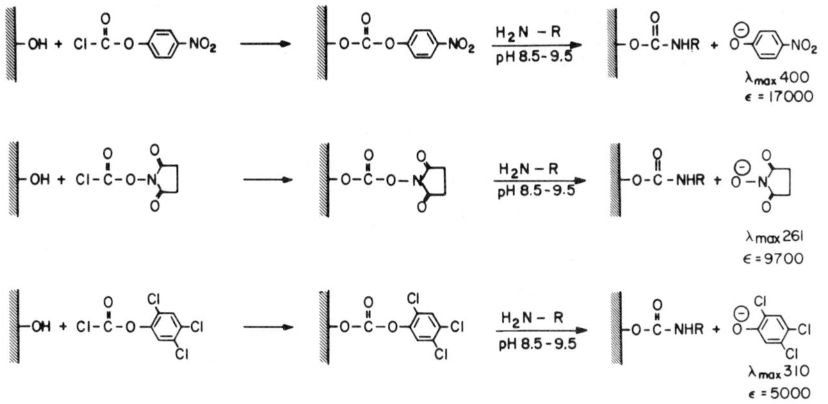

FIGURE 1. Schematic representation of the preparation of Sepharose containing active carbonates and their use for coupling amino-containing ligands.

slightly with ligands containing amino groups. The MSP resins can be further activated by oxidizing the resin-bound sulfides to the corresponding sulfoxides with diluted solutions of hydrogen peroxide for 5 minutes. The resin-bound sulfoxide is a good leaving group and could be used for immobilization of ligands with about 50% yield. Similar resins could also be obtained by reacting the polysaccharides with *p*-(methylsulfide)phenylchloroformate followed by oxidation (unpublished results).

In recent years, we described the preparation and use of aldehyde-containing resins for immobilization purposes via Schiff-base formation.[9,10] We have recently developed two additional aldehyde-containing resins. One involves the coupling of N^{ω}-amino group to any one of the above-described activated resins. Upon treatment with ninhydrin, colorless aldehyde-containing resins are obtained (FIG. 2). The other method is the oxidation of galactose residues on the agarose with the enzyme galactose oxidase. Only low amounts of aldehyde groups were introduced using this procedure. The oxidation was shown to be enzyme-mediated, since the aldehyde groups produced could be reduced with tritiated borohydride to give back the corresponding radioactive galactose. Upon further treatment with galactose oxidase, the radioactivity was released to the aqueous environment. The coupling capacity of the enzymatic prepared columns could be increased either by coupling polyacryhydrazide to the low levels of aldehyde groups formed or by direct coupling of additional galactose derivatives to the agarose before, the galactose oxidase reaction.

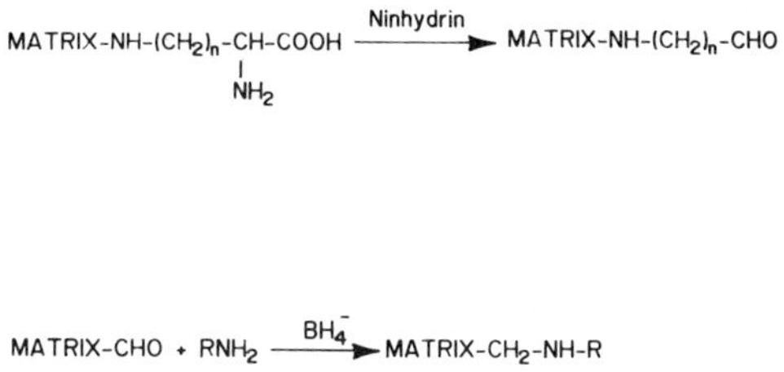

FIGURE 2. Preparation of aldehyde-containing Sepharose.

All the above-described activated resins have been used for the immobilization of amino acids, peptides, antibodies, and enzymes via their amino groups, and used for affinity chromatography and catalysis.

REFERENCES

1. AXEN, R., J. POROTH & C. ERNBACK. 1967. Chemical coupling of peptides and proteins to polysaccharides by means of cyanogen halides. Nature **214:** 1302–1303.
2. CUATRECASAS, P., M. WILCHEK & C. B. ANFINSEN. 1968. Selective enzyme purification by affinity chromatography. Proc. Natl. Acad. Sci. USA **61:** 636–643.
3. KOHN, J. & M. WILCHEK. 1982. Mechanism of activation of Sepharose and Sephadex by cyanogen bromide. Enzyme Microb. Technol. **4:** 161–163.
4. KOHN, J. & M. WILCHEK. 1982. A new approach (cyano-transfer) for cyanogen bromide activation of Sepharose at neutral pH, which yields activated resins, free of interfering nitrogen derivatives. Biochem. Biophys. Res. Commun. **107:** 878–884.
5. KOHN, J. & M. WILCHEK. 1983. 1-Cyano-4-dimethylamino pyridinium tetrafluoroborate as a cyanylating agent for covalent attachment of ligands to polysaccharide resins. FEBS Lett. **154:** 209-210.
6. KOHN, J., R. LENGER & M. WILCHEK. 1983. *p*-Nitrophenylcyanate—An efficient, convenient, and nonhazardous substitute of cyanogen bromide as an activating agent for Sepharose. Appl. Biochem. Biotech. **8:** 227–235.
7. WILCHEK, M. & T. MIRON. 1982. Immobilization of enzymes and affinity ligands onto agarose via stable and uncharged carbamate linkages. Biochem. Inter. **4:** 629–635.
8. KOHN, J. & M. WILCHEK. 1981. Procedures for the analysis of cyanogen bromide-activated Sepharose or Sephadex by quantitative determination of cyanate esters and imidocarbonates. Anal. Biochem. **115:** 375–382.
9. PITTNER, F., T. MIRON., G. P. PITTNER & M. WILCHEK. 1980. Pyridine-containing polymers: New matrices for protein immobilization. J. Am. Chem. Soc. **102:** 2451.
10. MIRON, T. & M. WILCHEK. 1981. Polyacrylhydrazido-agarose: Preparation via periodate oxidation and use for enzyme immobilization and affinity chromatography. J. Chromatogr **215:** 55–63.

Continuous Regeneration of ATP for Enzymatic Syntheses

W. BERKE, M. MORR,[a] C. WANDREY, AND M.-R. KULA[a]

Institut für Biotechnologie
Kernforschungsanlage Jülich GmbH
D-5170 Jülich, Federal Republic of Germany

[a]*Gesellschaft für Biotechnologische Forschung mbH*
D-3300 Braunschweig-Stoeckheim, Federal Republic of Germany

A process is presented that enables coenzyme ATP-dependent enzymatic synthesis pathways to be used. Generally there are two types of application. First, a phosphorylation catalyzed by a phosphotransferase (E.C. 2.7.), and second, the synthesis of compounds accompanied by cleavage of phosphate groups from ATP catalyzed by synthetases (E.C. 6.). The cofactor is effectively regenerated from ADP using a phosphate donor and the appropriate phosphotransferase.

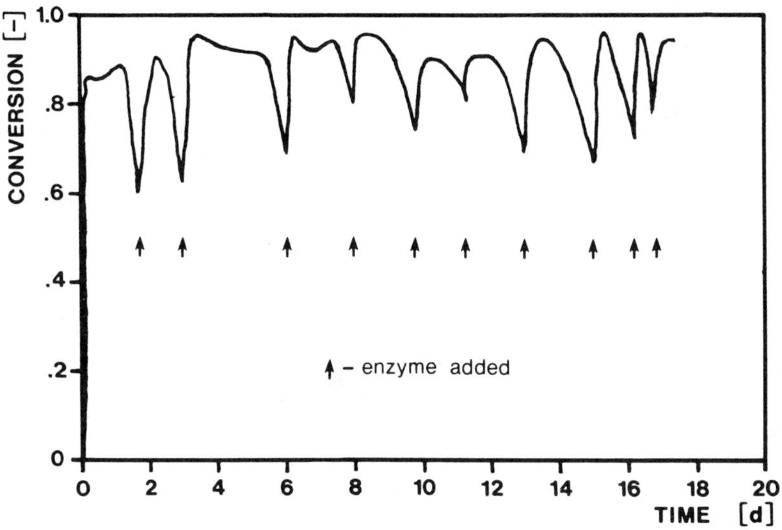

FIGURE 1. Scheme for glucose-6-phosphate production.

A suitable principle, already used for NAD regeneration, is that of the enzyme-membrane reactor.[1] In this system, the coenzyme is immobilized on a water-soluble polymer. These high-molecular-weight derivates are used entrapped within the ultrafiltration membranes together with the enzymes for production and recycling. Thus, various polymer derivatives of ATP have been prepared. The polymer is bound with various spacer structures at the N6-amino group in the adenine ring. With one exception (polyethyleneimine), the ATP is attached to polyethylenglycol with a molecular weight of 20,000.

The enzymatic activities of the ATP derivates were studied with several enzymes for both ATP-requiring and ATP-regenerating reactions. Because some of the enzymes were inactive with all of the tested derivatives, the number of possible applications was decreased. However, NAD kinase and gluthatione synthetase showed sufficient activity to obtain attractive products. Among candidates for phosphotransferases for ATP regeneration, only acetate kinase had enough activity with the N6-aminohexyl-ATP derivative for study. Depending on the type of bond to the polymer, considerable differences in enzymatic activity may result, as compared to the native coenzyme. Plots of initial reaction rates versus ATP concentration show that the structure and properties of the spacer strongly affect the native Michaelis-Menten kinetics. In the case of hexokinase with the N6-aminohexyl-ATP derivative even higher velocities can be observed. For this reason, PEG-N6-aminohexyl-ATP was selected as the appropriate coenzyme for operating the membrane reactor. Because of the high reaction velocity with hexokinase, the phosphorylation of glucose serves as an initial example (FIG. 1).

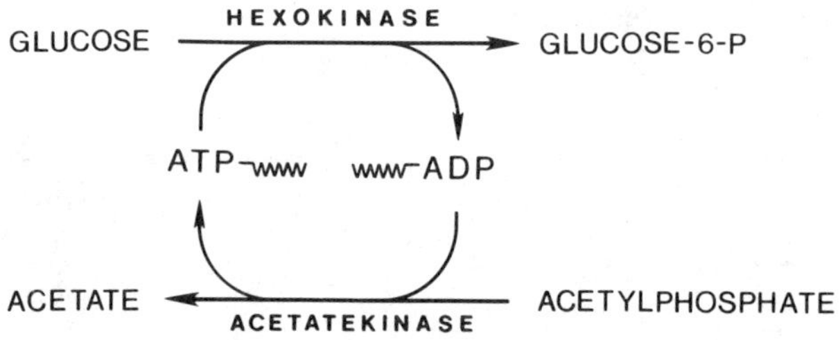

FIGURE 2. Continuous conversion of glucose to glucose-6-phosphate.

The continuous experiments were carried out in a thin-channel membrane reactor. Hexokinase, acetatekinase and PEG-ATP were continuously recirculated over the membrane surface to prevent membrane polarization. The membrane of choice was of cellulose acetate with a cut-off of 5,000 daltons. For on-line monitoring of the conversion, an autoanalyzer unit was used. With a continuous-flow technique, the outlet concentration was diluted with buffer; and the substrate or product concentration was determined enzymatically. In case of further products, the enzymatic assay could be easily changed. FIGURE 2 shows the time course of a continuous glucose-6-phosphate production experiment. Enzymes were added several times to counteract deactivation. At present, a recycling number of 12,000 has been reached.

In the immediate future, compounds more interesting than glucose-6-phosphate are going to be produced. This also will demonstrate the economic utilization of the system.

REFERENCES

1. WICHMANN, R., C. WANDREY, A. F. BUECKMANN & M. -R. KULA. 1981. Continuous enzymatic transformation in an enzyme membrane reactor with simultaneous NAD(H) regeneration. Biotechnol. Bioeng. **23:** 1341–1354.

Study and Use of Spontaneous Cofactor Regeneration in an Immobilized Enzyme-Coenzyme System

PHILIPPE MINARD, MARIE-DOMINIQUE LEGOY, AND DANIEL THOMAS

Laboratoire de Technologie Enzymatique
E.R.A. n° 338 du C.N.R.S.
Université de Technologie de Compiègne
60206 Compiègne, France

The use of immobilized dehydrogenases in preparative reactors is strongly limited by the coenzyme stoichiometric consumption. This drawback also limits some analytical applications but to a lesser extent. The simultaneous coimmobilization of dehydrogenase and coenzyme molecules, in such a way that a functional molecule of NAD was artificially maintained at a molecular distance from the active site, was tested by Gestrelius *et al.*[1] This kind of study is of great interest not only from a practical point of view for cofactor regeneration, but also for basic studies dealing with immobilized enzyme-cofactor complexes.[2] However, a prerequisite for such studies is a method that allows the quantitative transformation of reduced or oxidized forms of bound NAD. Chemical mediators, associated with a pO_2 sensor, can be used, but it seems more convenient for a study dealing with enzyme-coenzyme interactions to use strictly enzymatic methods. Nicolas *et al.*[3] used an NAD amplification system constituted by a bifunctional dehydrogenase: 3β, 17β hydroxysteroid-dehydrogenase (HSDH) of *Pseudomonas testosteroni* (E.C. 1.1.1.51). This system has been chosen as a model for a study of NAD immobilization. The sequence of reactions used by Nicolas is irreversible and gives rise to final products absorbing in the UV with high ϵ_M. Furthermore, this sequence requires only small quantities of enzymes, and continuous measurement of the cycling activity is possible.

3β, 17β HSDH and Δ5-3-ketosteroid-isomerase were a kind gift from Sempa-Chimie (France). The two enzymes were purified according to Battais *et al.*,[4] and the activities were determined as they have described.[4] Dehydroepiandrosterone (DHEA), testosterone, Δ5 androstenediol, and androstenedione were purchased from Sigma. ^{3}H DHEA was obtained from Amersham. The bienzyme system was immobilized within an artificial proteinic cross-linked structure.[5] The solid particles were studied in a continuous stirred tank reactor (CSTR). The progress of the enzymic reaction was followed by the absorbance at 248 nm of the output solution and by thin-layer chromatographic analysis and labeled compound counting.

The reaction sequence proposed by Nicolas[3] was investigated. With soluble enzyme, the labeled substrate and product concentrations were measured as a function of time. (FIG. 1). The consumption of substrate gave rise to accumulation of three products: testosterone and androstenedione, which exhibited an absorption at 248 nm linked to the conjugated Δ 4-3 ketone, and androstenediol, which did not absorb at 248 nm. Androstenediol and androstenedione were produced at exactly the same rate. Three reactions were involved in the production of one molecule of testosterone or of androstenedione: one reduction of NAD^+, one oxidation of NADH, and one isomerization of the double bond of a Δ5-4 ketone intermediate (TABLE 1). Isomerization acted as a driving force. One redox cycle for an NAD molecule could be obtained by (a) reduction and oxidation of only *one molecule* of DHEA (in this case, one molecule of testosterone is produced) and (b) reduction and oxidation of *two different molecules* of DHEA (in this case, one molecule of androstenedione and one molecule of androstenediol are obtained). Each chromophonic molecule is the result of a cyclic transformation of the coenzyme. Owing to the identity of the testosterone and androstenedione ϵ_M at 248 nm (17,000 $M^{-1} \cdot cm^{-1}$), a direct evaluation of the cofactor cycling rate was possible.

Activities of the co-immobilized enzyme-coenzyme systems were studied in a continuous stirred tank reactor. The activities were first measured without NAD in the

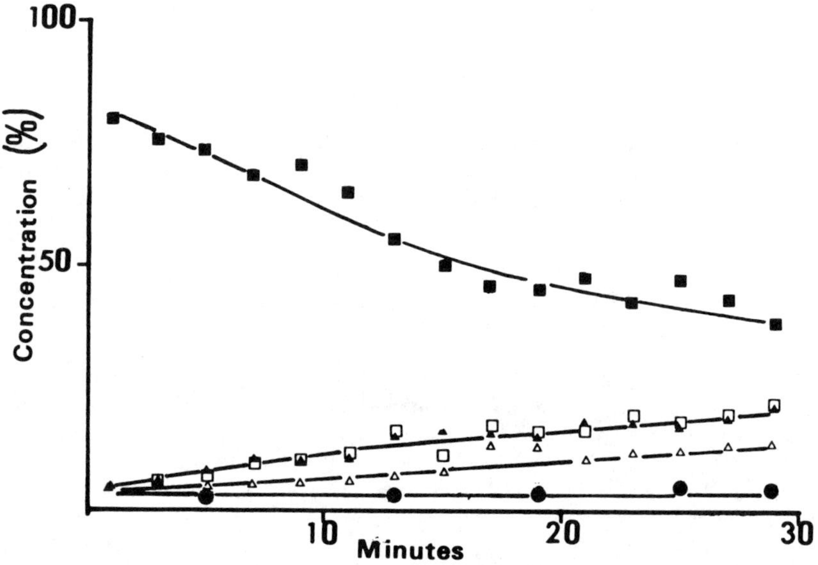

FIGURE 1. Concentrations of substrate and products as a function of time with soluble enzymes. Substrate is ^{3}H DHEA. After extraction and concentration, each sample was analyzed by thin-layer chromatography. Relative concentrations of substrate and products were determined by scintillation counting. (■) dehydroepiandrosterone (DHEA); (□) androstenedione; (▲) Δ_5 androstenediol; (△) testosterone; (●) remaining of the chromatogram.

reactor input solution. Under this condition, only immobilized NAD molecules could give activity. Diffusible NAD molecules, entrapped but not linked to the structure, should be eluted by the flow leaving the reactor. It appears that unmodified NAD^+ was able to be retained by this method of immobilization (FIG. 2, curve a). To eliminate the possibility of a stable interaction (e.g. adsorption) of NAD^+ with the cross-linked structure, the exact quantity of NAD^+ used in preparing immobilized NAD^+ was added to the NAD-free immobilized enzyme matrix. The activity was measured after the achievement of the diffusion equilibrium. Such a preparation (FIG. 2, curve b) shows decreasing activity; this is a result of the elution of unimmobilized NAD molecules. In the two cases, (FIG. 2, a and b) maximum activity can be obtained by adding NAD^+ at 2 μM concentration to the reactor input flow.

TABLE 1.

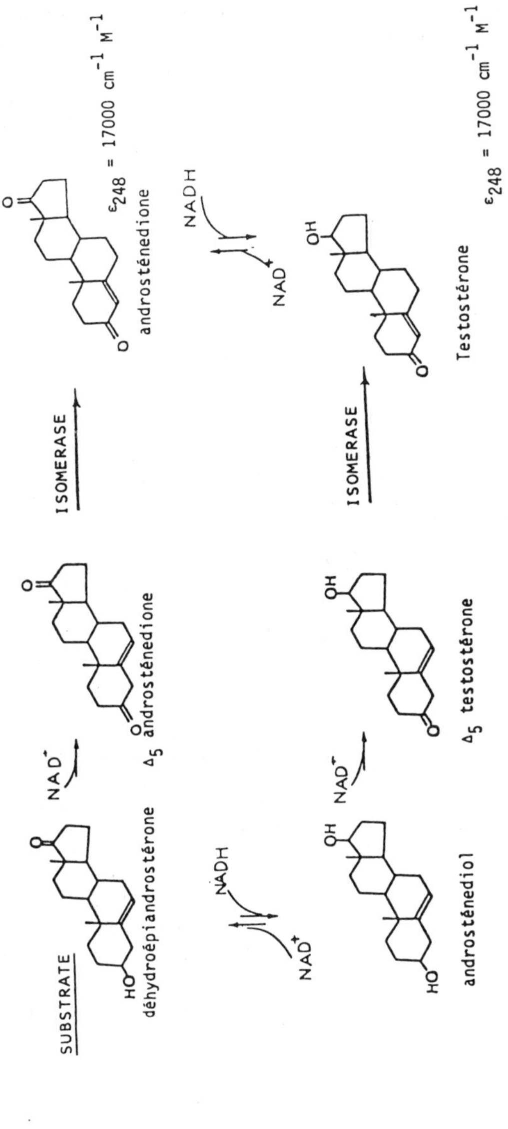
SUBSTRATE
déhydroépiandrostérone
NAD+
Δ5 androsténedione
ISOMERASE
androsténedione
ε248 = 17000 cm-1 M-1
NADH
NAD+
NADH
NAD+
androsténediol
NAD+
Δ5 testostérone
ISOMERASE
Testostérone
ε248 = 17000 cm-1 M-1

Initially the copolymerization was carried out with a coenzyme concentration 100-fold of the value of K_D. So, the copolymerization occurred in the presence of the binary complex NAD-dehydrogenase. The intrinsic activity was only 18% of the total activity obtained with 2 μM exogeneous NAD^+. This suggests that the linkage process was not able to "freeze" the molecular fitting between the coenzyme and the active site. In the proteinic structure obtained by coreticulation, the linkage sites are randomly distributed on the bulk protein and enzyme structures, and intermolecular interactions are probable events. The nature of the linkage between NAD and the cross-linked structure is not known. Glutaraldehyde is known to react in its polymeric forms with primary amino group to give as predominant reaction products conjugated Schiff bases. The spectral characteristics of these products are not observed with NAD or with adenine. A better knowledge of the chemical reactivity of adenylic compounds

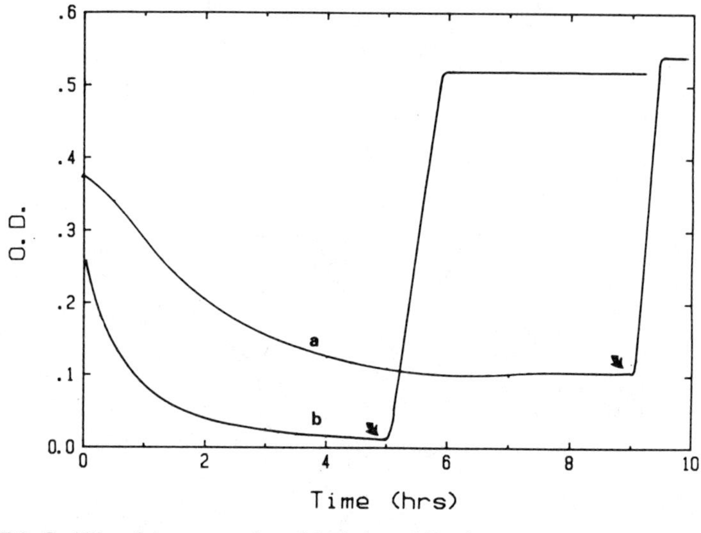

FIGURE 2. Stability of the coenzymic activity in immobilized system tested in a CSTR. Reactor feed solution does not contain NAD. At the arrow, NAD^+ at 2 μM concentration was continuously added to the entering reactor feed solution. Curve a: preparation with immobilized NAD; Curve b: preparation without immobilized NAD. The immobilized enzyme matrix was saturated after immobilization with soluble NAD (see text).

with glutaraldehyde seems necessary in order to understand our results. The possibility for direct immobilization of unmodified NAD with the dehydrogenase in a protein cross-linked structure was first suggested by Legoy.[6] The same hypothesis was formed by Pittner.[7] This work shows that it is a possible and simple way for simultaneous dehydrogenase-NAD immobilization.

REFERENCES

1. GESTRELIUS, S., M. O. MANSSON & K. MOSBACH. 1975. Preparation of an alcohol-dehydrogenase-NAD(H) sepharose complex showing no requirement of soluble coenzyme for its activity. Eur. J. Biochem. **57:** 529–535.

2. MANSSON, M. O., P. O. LARSSON & K. MOSBACH. 1978. Covalent binding of an NAD analogue to liver alcohol dehydrogenase resulting in an enzyme-coenzyme complex not requiring exogenous coenzyme for activity. Eur. J. Biochem. **86:** 455–463.
3. NICOLAS, J. C., J. CHAINTREUIL, B. DESCOMPS & A. CRASTES DE PAULET. 1980. Enzymatic microassay of serum bile acids: Increased sensitivity with an enzyme amplification technique. Anal. Biochem. **103:** 170–175.
4. BATTAIS, E., B. TERROUANNE, J. C. NICOLAS, B. DESCOMPS & A. CRASTES DE PAULET. 1977. Characterization of an associate 17β hydroxysteroid dehydrogenase activity and affinity labelling of the 3α hydroxysteroid dehydrogenase. Biochimie **59:** 909–917.
5. GUILLOCHON, D., J. M. LUDOT, L. ESCLADE, B. CAMBOU & D. THOMAS. 1982. Hydroxylase activity of immobilized haemoglobin. Enzyme Microb. Technol. **4:** 96–98.
6. LEGOY, M. D., J. M. LE MOULLEC & D. THOMAS. 1978. Chemical grafting of functional NAD in the active site of a dehydrogenase. Regeneration *in situ.* FEBS Lett. **94:** 335–337.
7. PITTNER, F. 1981. Immobilization of myo-inositol-1-phosphate synthase-containing active, self-regenerating coenzyme (NAD^+) on the same matrix. Appl. Biochem. Biotechnol. **6:** 85–90.

Immobilization of NAD Kinase

XINSONG JI, HUIXIAN LI, ZHONGYI YUAN,
AND SHUHUANG LIU

Shanghai Institute of Biochemistry
Academia Sinica
Shanghai, China

NAD Kinase catalyzes the phosphorylation of NAD to NADP in the presence of ATP. Early in 1954, T. P. Wang *et al.*[1] worked on the preparation and properties of NAD kinase from pigeon liver. Later, many researchers obtained NAD kinase from different living sources, including rat liver,[2] spinach,[3] sea urchin,[4] and various bacteria.[5–7] We have immobilized NAD kinase from chicken liver on Sepharose 4B and investigated its properties.

Immobilized NAD kinase was prepared according to Axen's method,[8] one gram Sepharose 4B and 0.4 g CNBr were reacted in 1.25 *M* phosphate buffer (pH 11.5) for 15 min, then washed with distilled water, and CNBr-activated Sepharose 4B was obtained. After 3.5 ml 0.1 *M* phosphate buffer (pH 7.5) was mixed with each 2 g CNBr-activated Sepharose 4B, 0.5 ml NAD kinase solution was added. The coupling was allowed to continue overnight at 4°C with gentle agitation. After being washed by 1 *M* NaCl, 0.1 *M* phosphate buffer (pH 7.5), the immobilized enzyme was suspended in the same buffer and stored at 4°C.

The activities of the immobilized enzyme increased with the amount of added enzyme. When the amount of enzyme was 0.5 ml, the activities were highest. When we changed the proportion of Sepharose 4B to CNBr, the highest activity was in the proportion of 1 : 0.4. For the NAD Kinase immobilization, NAD and ATP were used as protective reagents. The results are given in TABLE 1. The best results were obtained under conditions of 1 g Sepharose 4B gel (wet weight), 0.4 g CNBr, 0.5 ml enzyme solution (34 IU/ml), 6 m*M* NAD, and 8 m*M* ATP in 0.1 *M* phosphate buffer (pH 7.5). The immobilized preparation contains 3 units NAD kinase/g gel (wet weight) with an activity recovery of 20–25%.

TABLE 1. Effect of Protective Reagents on Immobilization

	H_2O	NAD (6 m*M*)	ATP (8 m*M*)	NAD (6 m*M*) + ATP (8 m*M*)
Activity of immobilized enzyme (IU/g)	3.33	3.44	3.79	4.25
Recovery of immobilized enzyme (%)	21.3	22.0	24.3	27.2

The properties of the immobilized NAD kinase also were studied. The enzyme reaction was carried out at different temperatures. The optimum temperature of the native enzyme was 50°C, but for the immobilized enzyme, it was 55°C. The enzyme reaction was carried out in 0.1 *M* phosphate buffer (pH range from 5.5 to 8.0) and 0.1 *M* Tris-HCl buffer (pH range from 7.0 to 9.0). The optimum pH of the immobilized enzyme was 6.5 (phosphate buffer) and 7.5 (Tris-HCl buffer). Only one pH optimum (pH 7.0) was observed for the native enzyme in both phosphate and Tris-HCl buffer.

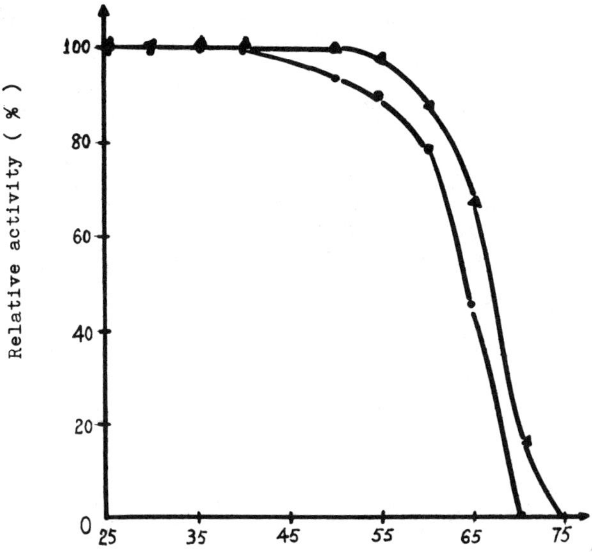

FIGURE 1. Thermobility for immobilized enzyme (▲) and native enzyme (●).

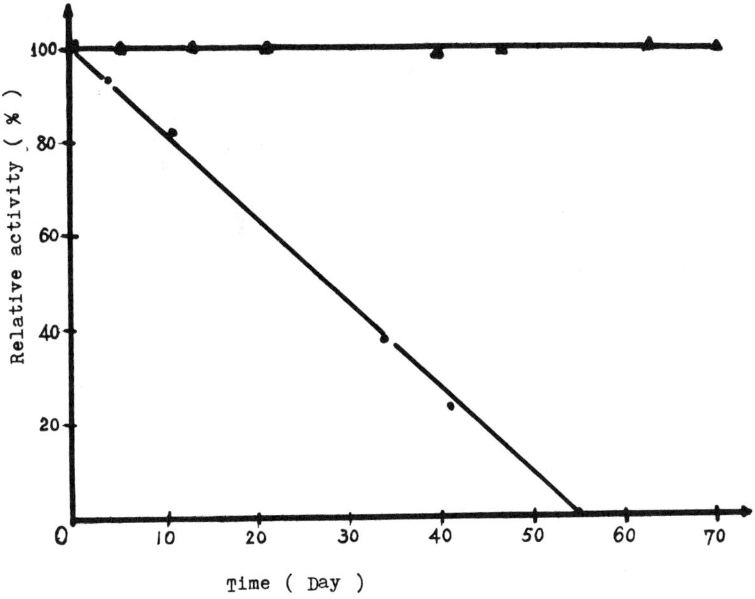

FIGURE 2. Storage stability for immobilized enzyme (▲) and native enzyme (●).

The enzyme stability was enhanced by immobilization. The native and immobilized enzymes were incubated in 0.25 *M* Tris-HCl buffer, pH 7.5, for 2 hours at different temperatures before their activities were determined. As shown in FIGURE 1, both the native and the immobilized enzyme were stable below 40°C. The immobilized enzyme was more stable than the native enzyme above 55°C. The native and the immobilized enzymes were preincubated at 37°C for 2 hours at different pH values (pH 5.5–7.0 phosphate buffer, 7.5–9.0 Tris-HCl buffer). After incubation, both the native and the immobilized enzyme were adjusted to pH 7.5, then their residual activities were determined. Both the native and the immobilized enzymes were stable. Both the native and the immobilized enzymes were stored in a refrigerator (4°C) in 0.1 *M* phosphate buffer, pH 7.5. The activity of the native enzyme was entirely lost after 55 days, but the immobilized enzyme still retained its activity without any detectable loss after 70 days (FIG. 2).

The K_m value was not significantly changed after immobilization. The values were 1.5 m*M* for NAD and 4.5 m*M* for ATP. For the native enzyme the K_m values were 1.2 m*M* and 2.5 m*M* for NAD and ATP, respectively.

These results indicate that the immobilized enzyme was more stable against heat than the native enzyme and could be stored for a longer time.

REFERENCES

1. WANG, T. P. & N. O. KAPLAN. 1954. Kinases for the synthesis of coenzyme A and triphosphopyridine nucleotide. J. Biol. Chem. **206**(1): 311–325.
2. OKA, H. & J. B. FIELD. 1968. Inhibition of rat liver nicotinamide adenine dinucleotide kinase by reduced nicotinamide adenine dinucleotide phosphate. J. Biol. Chem. **243**(4): 815–819.
3. YAMAMOTO, Y. 1966. NAD kinase in higher plants. Plant Physiol. **41**(3): 523–528.
4. BLOMQUIST, C. H. 1973. Partial purification and characterization of nicotiamide adenine dinucleotide kinase from sea urchin egg. J. Biol. Chem. **248**(20): 7044–7048.
5. KUWAHARA, M., T. TACHIKI & T. TOCHIKURA. 1972. Distribution and properties of NAD phosphorylating reaction. Agr. Biol. Chem. **36**(5): 745–754.
6. BERNOFSKY, C. & M. F. UTTER. 1968. Interconversions of mitochondrial pyridine nucleotides. Science **159**(3821): 1362–1363.
7. ORRINGER, B. P. & A. E. CHUNG. 1971. Nicotinamide adenine dinucleotide kinase from *Azotobacter vinelandii* cells. Biochim. Biophys. Acta **250**(1): 86–91.
8. AXEN, R., J. PORATH & S. ERNBACK. 1967. Chemical coupling of peptides and proteins to polysaccharides by means of cyanogen halides. Nature **214**(5095): 1302–1304.

Production and Purification of Dextransucrase from *Leuconostoc mesenteroides*, NRRL B 512 (F)

F. PAUL

Elf-Bio-Recherches
La Grande Borde Labege
31320 Castanet Tolosan, France

D. AURIOL, E. ORIOL, AND P. MONSAN

Departement de Genie Biochimique et Alimentaire
INSA, Avenue de Rangueil
31077 Toulouse Cedex, France

The enzyme dextransucrase (E.C. : 2.4.1.5) is a glycosyltransferase that catalyses the synthesis of dextran following the reaction: sucrose → dextran + fructose. Hydrolysis of sucrose provides the energy required for the condensation of D-glucopyranosyl units. Thus, dextran synthesis can be carried out *in vitro,* without energetic cofactors, using only purified enzyme and sucrose.

Dextran produced by dextransucrase from *L. mesenteroides* is a linear polyglucan containing 95% of $\alpha(1 \rightarrow 6)$ linear bonds and 5% of $\alpha(1 \rightarrow 3)$ branching bonds, which is used in several industrial applications: clinical dextran (MW 70,000), iron-dextran, dextran sulfate (MW 500,000), chemically modified dextrans, and so forth. At the present time, dextransucrase is produced from batch cultures of *L. mesenteroides* and culture broth is used directly, with or without cell elimination, to achieve the enzymatic synthesis of uncontrolled molecular weight dextran (MW $> 10^6$).

Alcoholic precipitation of the polymer, acid hydrolysis, and finally alcoholic fractional precipitation steps are used to obtain the desired molecular weight dextran.[1] This process has the following disadvantages: low enzyme productivity by cells, relatively low yield throughout the whole process of dextran synthesis, and high cost of the alcoholic precipitation. Owing to these problems, it would appear more efficient to completely dissociate enzyme production using cultures of *L. mesenteroides* and dextran synthesis *in vitro* catalyzed by purified enzyme.

This paper describes improvements in enzyme production using continuous culture of *L. mesenteroides.* After centrifugation of the cells, exocellular dextransucrase is then highly purified using a process that consists of a phase partition step between dextran and polyethylenglycol.

Dextransucrase is an inducible enzyme. Its substrate, sucrose, is the only known inducer of enzyme synthesis[2] and is also used for cell growth as, of course, for dextran synthesis as soon as dextransucrase is produced in the culture medium. Temperature, pH, and sucrose concentration are the most important factors affecting enzyme productivity.

As may be seen, enzyme productivity in continuous culture is 3 times greater than in fed-batch culture (FIGS. 1 and 2). Dextransucrase is produced during the exponential growth phase of the cells. In a continuous culture, enzyme productivity depends directly on the dilution rate (FIG. 2); optimum enzyme productivity (70 DSU $\cdot$ ml^{-1} $\cdot$ h^{-1}) is achieved at a 0.4 h^{-1} dilution rate.

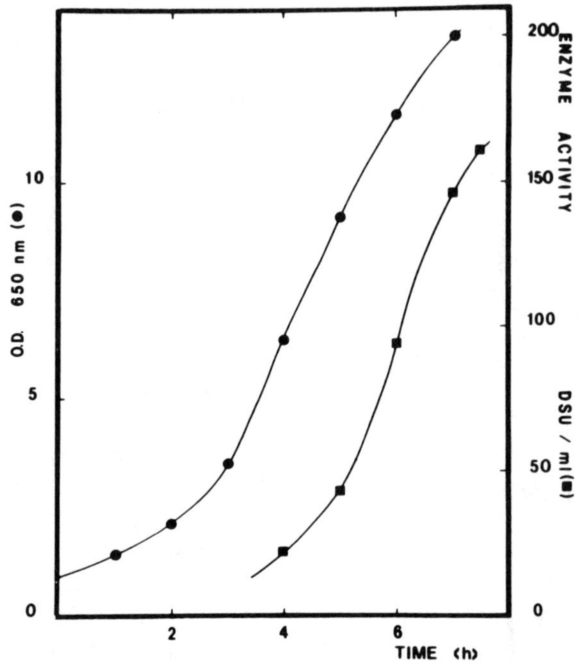

FIGURE 1. Dextransucrase activity and optical density measurement versus time in fed-batch culture of *L. mesenteroides.* Initial culture conditions: Yeast extract 20 g · 1^{-1}, K_2HPO_4 20 g · 1^{-1}, sucrose 20 g · 1^{-1}, $MnSO_4$ 0.01 g · $^{-1}$, $MgSO_4$ 0.2 g · $^{-1}$, NaCl 0.01 g · 1^{-1}, $FeSO_4$ 0.01 g · 1^{-1}, $CaCl_2$ 0.02 g · 1^{-1}, pH 6.9 at 27°C. Sucrose was added at a 20 g · $1^{-1}h^{-1}$ rate during exponential growth phase. Dextransucrase activity was measured as described by Monsan and Lopez.[5]

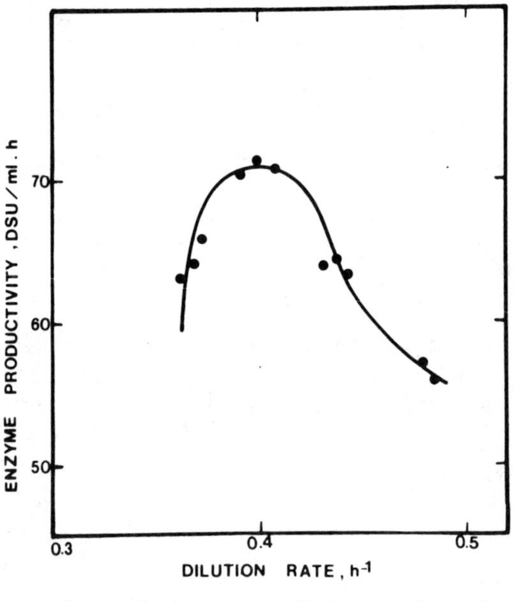

FIGURE 2. Dextransucrase productivity versus dilution rate in continuous culture of *L. mesenteroides.* Initial culture conditions were the same as in FIGURE 1.

TABLE 1. Comparison of Purification Methods of *L. mesenteroides* Dextransucrase

Purification Method	Enzyme Recovery (%)	Specific Activity (DSU · mg^{-1} protein)
Culture broth	100	30
Ultrafiltration	75	860
Gel permeation	78	2035
Phase partition between dextran and polyethylenglycol	95	>3500

In previous work, Lawford *et al.*[3] reported a maximum enzyme productivity of 9.3 DSU · ml · h^{-1} at a 0.53 h^{-1} dilution rate under conditions of energy-limited growth.

Dextransucrase from *L. mesenteroides* has already been purified by several techniques. The best results have been obtained by using ultrafiltration and gel filtration;[4,5] specific activities of 2,500 DSU/mg protein are obtained.[5] Phase partition is a general technique widely used to purify cells and organelles[6] and some endocellular enzymes.[7]

Repeated phase partitions result in a very highly purified dextransucrase preparation (TABLE 1). Specific activities above 3,500 DSU/mg protein may thus be obtained,[8] which is much greater than the highest specific activities previously reported.[4,5] Moreover, such purification conditions allow dextransucrase concentration and stabilization. This technique can be very easily scaled up and coupled with the continuous culture of *L. mesenteroides* to achieve the continuous production of large amounts of purified dextransucrase.

The presence in the culture supernatant of sucrose-hydrolyzing enzymes other than dextransucrase was assayed by specific measurement of the two reducing sugars (glucose and fructose) produced by enzymatic sucrose hydrolysis during the time course of reaction. Moreover, levansucrase activity was also determined by levan measurement. TABLE 2 shows the ratio of sucrose-hydrolyzing enzymes: dextransucrase, levansucrase, and an enzyme invertase-like activity, both in the culture supernatant, and after dextransucrase purification by phase partition. Levansucrase activity is completely eliminated, whereas the residual enzyme invertase-like activity can be attributed to dextransucrase, which can transfer glucosyl units onto H_2O.[4]

TABLE 2. Characterization of Sucrose-hydrolyzing Enzyme Activities in the Culture Supernatant and after Purification of Dextransucrase by Phase Partition

	Total Sucrose-hydrolyzing Activity (%)	Dextransucrase Activity[a] (%)	Levansucrase Activity[b] (%)	Invertase-like Activity[c] (%)
Culture supernatant	100	91.5	6	2.5
After purification by phase partition	100	96.1	0	3.9

[a]Measured as fructose appearing during the time course of reaction using a spectrophotometric method (Hexokinase/ATP, glucose-6-phosphate dehydrogenase/$NADP^+$, phosphoglucose isomerase).

[b]Measured as levan produced during the time course of reaction using a specific method of levan hydrolysis (H_2SO_4 0.4 *N*, 20 min, 55°C); released fructose is measured by the same spectrophotometric method as in *a* above.

[c]Measured as the difference between total glucose appearing during the time course of reaction and glucose-equivalent fructose incorporated in levan (levansucrase activity).

In conclusion, the continuous culture of *L. mesenteroides* allows a 3-fold increase in dextransucrase productivity when compared with usual fed-batch culture conditions. Furthermore, the enzyme can be very easily purified from the culture supernatant by a phase partition technique to obtain highly purified dextransucrase preparations, from which all the contaminating enzyme activities have been eliminated. Such preparations may be efficiently used for the synthesis of controlled molecular weight dextrans.

REFERENCES

1. JEANES, A. 1968. Dextran. *In* Encyclopedia of Polymer Science and Technology. Vol. **4:** 805–829. Interscience Publishers. John Wiley and Sons. New York.
2. NEELY, W. B. & J. NOTT. 1962. Dextransucrase, an induced enzyme from *Leuconostoc mesenteroides*. Biochemistry **6:** 1136–1140.
3. LAWFORD, G. R., A. KLIGERMAN, T. WILLIAMS & H. G. LAWFORD. 1979. Dextran biosynthesis and dextransucrase production by continuous culture of *Leuconostoc mesenteroides*. Biotechnol. Bioeng. **21:** 1121–1131.
4. ROBYT, J. F. & T. F. WALSETH. 1979. Production, purification and properties of dextransucrase from *Leuconostoc mesenteroides* NRRL B 512 F. Carbohydr. Res. **68:** 95–111.
5. MONSAN, P. & A. LOPEZ. 1981. On the production of dextran by free and immobilized dextransucrase. Biotechnol. Bioeng. **23:** 2027–2037.
6. ALBERTSSON, P. A. 1971. Partition of Cell Particles and Macromolecules. John Wiley and Sons. New York.
7. KULA, M.-R. 1979. Extraction and purification of enzymes using aqueous two-phase systems. *In* Applied Biochemistry and Bioengineering. L. B. Wingard, E. Katchalski-Katzir & L. Goldstein, Eds. Vol. **2:** 71–95. Academic Press. New York.
8. PAUL, F., P. MONSAN & D. AURIOL. 1983. Procédé de purifcation de la dextrane-saccharase. French Patent 8307650.

Characterization of a New Class of Thermophilic Pullulanases from *Bacillus acidopullulyticus*

MARTIN SCHÜLEIN AND BIRGITTE HØJER-PEDERSEN

NOVO Industri A/S
DK 2880 Bagsværd, Denmark

A novel class of pullulanase (E.C. 3.2.1.41) has recently been isolated from a new *Bacillus* species. This novel taxonomic group, *Bacillus acidopullulyticus* is characterized by production of pullulanases with temperature optima above 60°C and pH optima around 5. Two of these enzymes have been purified to homogeneity. These two enzymes are compared with the purified pullulanase from *Aerobacter aerogenes*. The enzyme activity and enzyme kinetics are also compared.

The pullulanase A was produced by a *Bacillus* strain deposited under the number NCIB 11647. Pullulanase B was produced by a *Bacillus* strain deposited under the number NCIB 11777. Pullulanase C was from *Aerobacter aerogenes*. The soluble extracellular enzymes were purified by successive chromatography through CM-Sepharose and DEAE-Sepharose. The purified enzymes were analyzed by sieve chromatography (Sephacryl S200).

Enzyme activity was determined by incubation with pullulan followed by determination of reducing power by the method of Somogyi and Nelson. One pullulanase unit NOVO (PUN) is defined as the amount of enzyme that under the given standard conditions hydrolyzes pullulan-liberating reducing carbohydrate with a reducing power equivalent to 1 μmol glucose per minute. The purified enzymes were more than 90% pure and the specific activity could not be increased by further purification.

The purified enzymes were analyzed by SDS-PAGE electrophoresis. The molecular weight of A was ~100,000, of B ~90,000, and of C about 140,000, which is in agreement with Eisele *et al.*[2] The two new purified enzymes were hydrolyzed and the amino-acid composition analyzed on a Beckman analyzer. The results are presented in TABLE 1 and compared with the amino acid composition of C from Eisele *et al.*[2] The isoelectric point of the purified enzyme was 5.0 for pullulanase A, 5.4 for pullulanase B, and 4.0 for pullulanase C. The pI was determined on a LKB Multiphor unit with LKB Anholine PAG plates. The immunological method described by Weeke[3] was used for crossed immunoelectrophoresis.

The enzymes were characterized by crossed immunoelectrophoresis (CIE). Sera against the two purified proteins did not react against pullulanase C or culture filtrate from pullulanase-producing *Bacillus* strains *megaterium* (ATCC 6459) or *mycoides* (ATCC 311027). Duplicates of the CIE plates were incubated overnight with pullulan followed by precipitation of the pullulan with acetone. A clearing zone indicates pullulanase activity. Both enzymes reacted with their own serum and with the other serum, although 3 to 10 times less antigen was necessary for obtaining the same area of immunoprecipitate with the heterologous antisera, indicating partial immunological identity.

The purified enzymes were incubated with pullulan at different temperatures and different pH values. The results are presented in FIGURES 1 and 2. It can be seen that

TABLE 1. Amino Acid Compositions

Amino Acid	Number of Amino Acids per Molecule of Purified Enzyme		
	Pullulanase A	Pullulanase B	Pullulanase C[a]
Asp	141	116	168
Thr	76	58	83
Ser	61	48	102
Glu	83	78	112
Pro	45	42	52
Gly	71	63	94
Ala	64	51	119
Val	72	63	93
Met	19	17	23
Ile	41	40	50
Leu	66	52	99
Tyr	40	32	43
Phe	31	24	37
Lys	45	43	51
His	24	19	23
Thr	16	14	25
Arg	20	22	48
Cys	2	4	7
Molecular Weight:	101,200	86,900	143,000

[a]Eisele *et al.*[2]

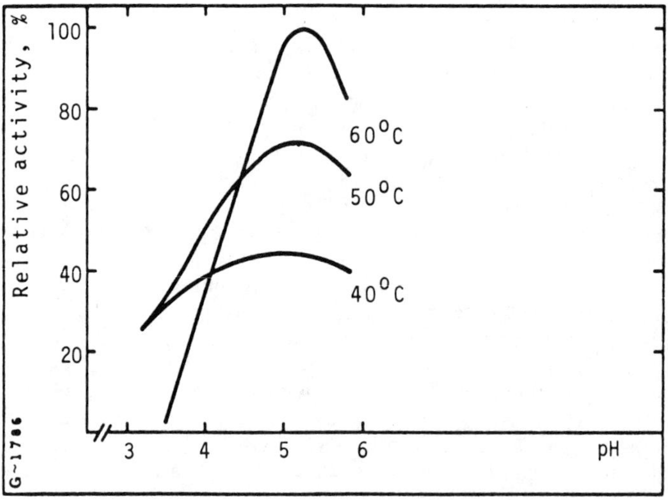

FIGURE 1. The effect of pH on enzyme activity with pullulanase A from NC1B 11647 substrate pullulan.

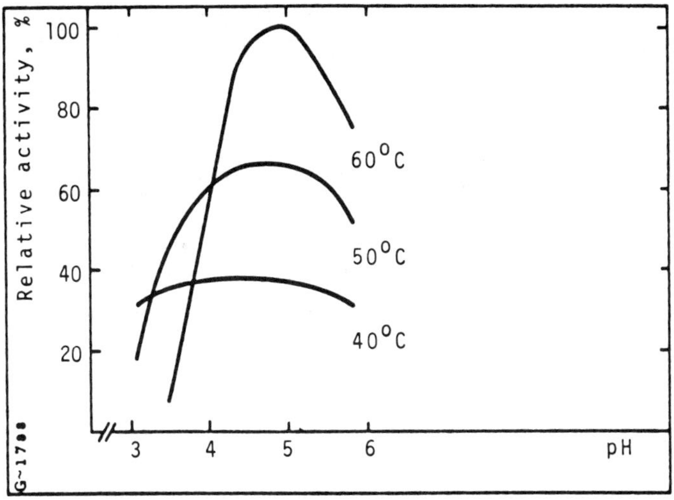

FIGURE 2. The effect of pH on enzyme activity on pullulanase B from NC1B 11777 substrate pullulan.

this new class of enzyme is very thermostable with a broad acid pH range. Pullulanase C has a lower thermostability and a pH optimum around 6.

Pullulanase from *Bacillus cereus* var. *mycoides* has a pH optimum around 6 and a temperature optimum at 50°C.[1]

The purified enzymes were incubated with both pullulan and amylopectin. The kinetics were followed by measuring the reducing power. The K_m and V_{max} were calculated from a nonlinear regression analysis, from a best-fit computer program developed at NOVO. The results are presented in TABLE 2. Pullulanase A has the highest specific activity and pullulanase B (the smallest) has the lowest K_m.

The new pullulanase enzymes A and B have minor differences. They have different amino-acid compositions and molecular weights. Pullulanase A can be inactivated with PMB and reactivated with a reducing agent such as β-mercaptoethanol. The two enzymes show partial immunochemical identity. The enzyme kinetics and activity are more similar. They are both very temperature stable and the pH activity profile shows

TABLE 2. Enzyme Characterization

	Pullulan	
	PUN/mg	K_m g/l
Pullulanase A	100	0.028
Pullulanase B	50	0.019
Pullulanase C	20	0.040
	Amylopectin	
	PUN/mg	K_m g/l
Pullulanase A	18.3	0.37
Pullulanase B	11.3	0.30
Pullulanase C	7.5	0.48

an optimum around 5. This makes them fully compatible with *A. niger* glucoamylase and opens up possibilities for processes with reduced glycoamylase requirements.

REFERENCES

1. TAKASAKI, Y. 1976. Purifications and enzymatic properties of β-amylase and pullulanase from *Bacillus cereus* var. *mycoides*. Agric. Biol. Chem. **40**(8): 1523–1530.
2. EISELE, B., I. R. RASCHED & V. WALLENFELS. 1972. Molecular characterization of pullulanase from *Aerobacter aerogenes*. Eur. J. Biochem. **26:** 62–67.
3. WEEKE, B. 1973. Crossed immunoelectrophoresis. Scand. J. Immunol. **2**(suppl. 1): 47–56.

Selection of a β-Amylase-producing Strain and Its Fermentation Conditions

HE BINGWANG, GUO JUNJUN, JIANG ZHAOYUAN, AND ZHANG SHUZHENG

Institute of Microbiology
Academia Sinica
Beijing, China

Plant β-amylases are well known and have been studied for many years. Recently, many reports on microbial β-amylase have been published. Shinke *et al.* described a mutant strain of *Bacillus cereus* with a high level of β-amylase productivity (3,600 units per ml).[1] In this paper, we will describe the selection of a high β-amylase-producing strain and its fermentation conditions.[2]

A mutant strain, M-3, was derived from *Bacillus cereus* AS 1.447 by repeated treatment with ultraviolet light, *N*-methyl-*N*-nitro-*N*-nitrosoguanidine and rifampicin. The mutant M-3 was capable of producing 5,000–7,000 U/ml of β-amylase, which was about 100 times that of the parent strain. The mutant M-3 showed filamentous forms (FIG. 1b) whereas the parent showed short rods in chains (FIG. 1a). Colonies of M-3 on beef-infusion agar plates were smaller (4–5 mm) than those of the parent (10–12 mm) and with a bigger clear zone of starch-I_2 stain (FIG. 2).

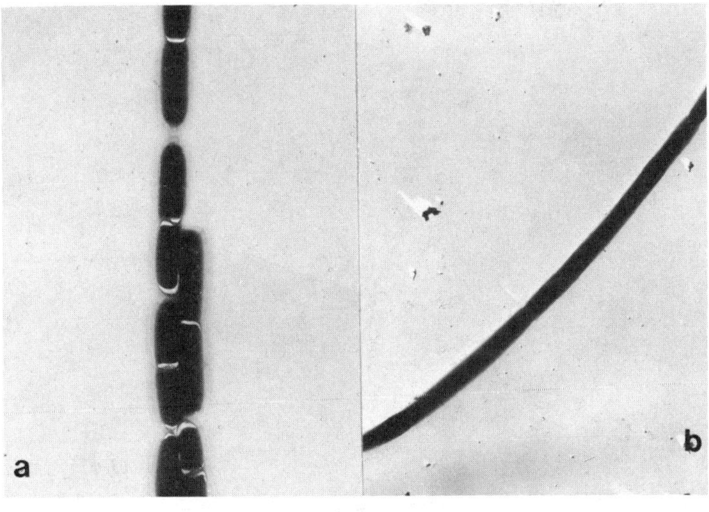

FIGURE 1. Electron microphotograph (2400×) of *Bacillus cereus* AS 1.447 (a) and mutant strain M-3 (b).

Suitable fermentation conditions have been established. The medium consisted of 100 ml beef infusion, 1 g peptone, 1 g soluble starch, 0.5 g yeast extract, and 0.5 g

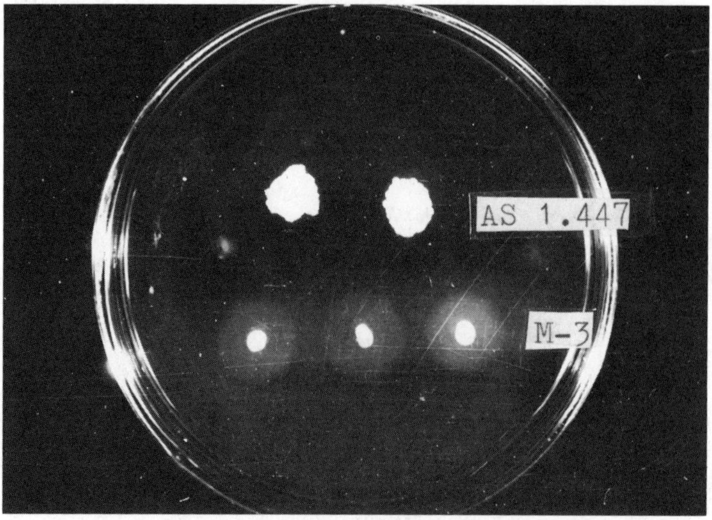

FIGURE 2. Colonies of strain AS 1.447 and strain M-3.

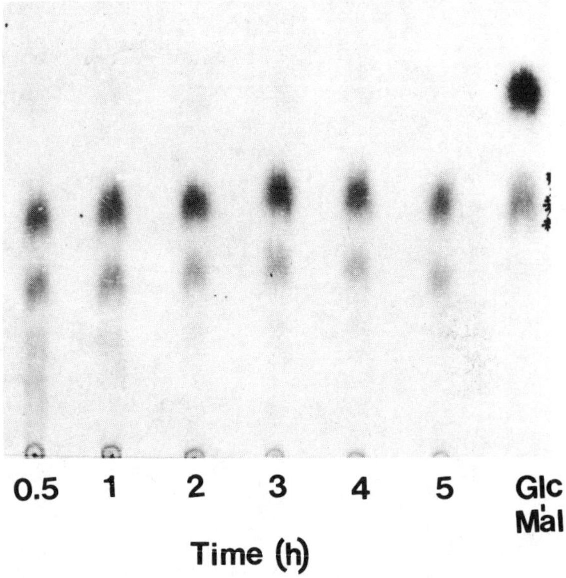

FIGURE 3. Paper chromatogram of starch hydrolysate.

TABLE 1. Activity of Various Enzymes in Culture Broth

	Enzyme (U/ml)				
Batch	β-Amylase	α-Amylase	Isoamylase	Neutral Protease	Alkaline Protease
1	5793	6	4.5	37	49
2	6070	6	9.0	37	64
3	5913	4.4	—	5	38

NaCl (pH 6.5). Maximum activity was achieved after growth at 30°C for 48 h in 250-ml flasks (50 ml medium) on a rotary shaker (200 rpm).

The optimal conditions for enzyme action were 40°C and pH 7.0. The crude enzyme was stable in the range of pH 6–9. The thermostability was not good; after treatment at 50°C for 1 h, the remaining activity was only 16.6%. The crude enzyme converted soluble starch to maltose with a conversion rate of over 85%. A paper chromatogram of a starch hydrolysate is shown in FIGURE 3. The main product was maltose, with a little maltotriose. Activities of various enzymes in the culture broth were analyzed and are shown in TABLE 1.

Pilot plant experiments have been done in 500-1 and 3,000-1 fermenters using cheaper media. In the case of 500-1 tanks, enzyme activity reached 5,000–6,000 U/ml after growth at 30°C for 52–56 h with a final pH of 7.1–7.5. In the case of 3,000-1 tanks, enzyme activity reached 6,000 U/ml within the shorter period of 30 h with a final pH about 7.1. A liquid enzyme preparation was obtained by vacuum evaporation, and solid preparations were obtained by precipitation with ammonium sulfate or alcohol. The enzyme is to be used for making maltose syrup or beer as a partial substitute for malt.

REFERENCES

1. SHINKE, R., K. AOKI, H. NISHIRA & S. YUKI. 1979. Isolation of rifampin-resistant, asporogenous mutant from *Bacillus cereus* and its high β-amylase productivity. J. Ferment. Technol. **57**(1): 53–55.
2. HE, B.-W., J.-J. GUO, Z.-Y. JIANG & S.-Z. ZHANG. 1983. Selection of a high β-amylase-producing strain and its fermentation conditions. Wei Sheng Wu Hsueh Pao **23**(1): 75–83.

Improvement of β-Amylase Production and Characterization of Polypepton Fraction Effective for β-Amylase Production[a]

R. SHINKE, H. YAMANAKA, T. NANMORI, K. AOKI,
H. NISHIRA, AND S. YUKI[b]

Faculty of Agriculture
and
[b]College of Liberal Arts
Kobe University, Rokkodaicho
Nadaku, Kobe 657, Japan

Rifampin-resistant and asporogenous mutants (*Bacillus cereus* BQ10-S1 Spo I and BQ10-S1 Spo II) were isolated from a sporogenous β-amylase producer (*B. cereus* BQ10-S1). The BQ10-S1 Spo II enzyme was more stable in the culture broth than the BQ10-S1 Spo I enzyme. Both enzymes were physicochemically and immunochemically identical and they both hydrolyzed raw starch.

Since these mutants showed high β-amylase production only in the media containing natural nitrogen sources, we attempted to isolate factors effective for enzyme production from Polypepton. Among four fractions (A, B, C, and D in order of elution on a Sephadex G-25 column), Fraction B was the most effective for β-amylase production. Fraction D, ineffective as a single addition to the basal medium (Maassen's medium), enhanced enzyme production in the presence of Fraction B. Fraction B was found to consist of basic peptidyl compounds and free amino acids. Fraction C, ineffective both as single and combined additions, consisted of tyrosine and phenylalanine. Fraction A, composed of peptidyl compounds and free amino acids, is under investigation together with Fraction B.

An asporogenous and rifampin-resistant mutant, *B. cereus* BQ10-S1, showed about seven times as high β-amylase productivity as the parent strain, BQ10-S1. However, the enzyme was unstable and the activity was reduced to about half the maximum during the 24-h shaking culture.[1] Therefore, further attempts were made to isolate more industrially promising β-amylase producers. A spore suspension of BQ10-S1 was placed on a Polypepton agar plate containing 5 μg/ml of rifampin and was UV-irradiated with a germicidal lamp (10 watt) at a distance of 50 cm for 60 to 80 sec. The UV-irradiated plates were incubated at 37°C overnight. The germination rate of the spores was 45.4%. Among one hundred rifampin-resistant strains isolated, six were found to be asporogenous by microscopic and sporulation tests. The examination of β-amylase productivity showed that one strain, designated as BQ10-S1 Spo II, was capable of producing about four times as much β-amylase as BQ10-S1. As already reported,[2] β-amylase activity was measured with 3,5-dinitrosalicylic acid reagent and one unit in the present paper is expressed as mg maltose produced at 40°C for 60 min. The morphological and physiological characteristics of BQ10-S1 Spo II were found to be almost the same as those of BQ10-S1 Spo I.[3]

[a]This work was supported in part by a Grant-in-Aid from the Ministry of Japan. The authors wish to thank the Daiwa Kasei Co. Ltd. for their financial support in carrying out the experiments.

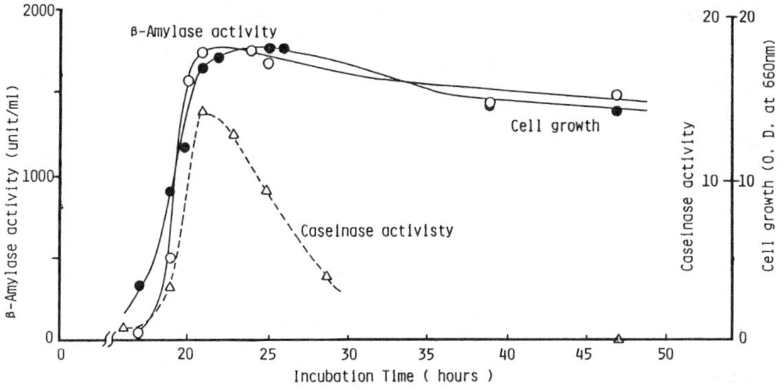

FIGURE 1. Time course of β-amylase production by shaking culture of *Bacillus cereus* BQ10-S1 Spo II.

As shown in FIGURE 1, BQ10-S1 Spo II showed maximum enzyme activity (1740 units/ml) in the 21-h culture. The enzyme activity remained nearly at the same level over 40 hours of culture. Furthermore, the enzyme activity was enhanced by repeated feeding of a new Polypepton medium and reached 3240 units/ml. This activity was about twice that of the batch culture and remained stable in the culture broth for a longer term than that of BQ10-S1 Spo I. Therefore, the enzyme yield in the case of BQ10-S1 Spo II was higher than that of BQ10-S1 Spo I.

β-Amylase produced by BQ10-S1 Spo II was purified by salting out, gel filtration, and ion-exchange chromatography. The enzyme, purified about 550-fold, turned out to be electrophoretically and ultracentrifugically homogenous. TABLE 1 shows a comparison of some characteristics of β-amylases from microbial and plant sources. The β-amylases from BQ10-S1, BQ10-S1 Spo I, and BQ10-S1 Spo II were found to be identical physicochemically, enzymatically, and immunochemically.[3] Though optimum pH's of microbial β-amylases are higher than those of plant β-amylases, optimum temperatures are not so different. The molecular weight of the β-amylases from BQ10-S1 Spo II was estimated to be 60,000. Microbial β-amylases seem to have a molecular weight of about 60,000 when measured by methods other than gel filtration on Sephadex.[3] β-Amylases so far reported contain at least one sulfhydryl

TABLE 1. Characteristics of β-Amylases from Various Sources

	Soybean	Barley	*Bacillus cereus*[a]	*Bacillus megaterium*	*Bacillus polymyxa*	*Bacillus circulans*
Optimum pH	6.0	5.0–6.0	6.0–7.0	6.0–7.0	6.5–7.5	6.5–7.5
Optimum temperature °C	—	30–40	45–50	55–65	35–45	60
Isoelectric point	5.93	5.75	8.2	9.1	8.35	4.6
Molecular weight	57,000	56,000	60,000[b]	36,000–38,000	44,000–45,000	58,000[b]
Sh content/mol	5.0	2.8	1.0	+	+	+
Hydrolysis of raw starch	+	+	+++++	+++++	+++++	

[a]β-Amylases from *Bacillus cereus* BQ10-S1, BQ10-S1 Spo I and Spo II.
[b]Values measured by methods other than gel filtration.

TABLE 2. Polypepton Fractions Effective for β-Amylase Production

Fraction	Composition	Effectiveness
Fraction A	Under experiment (peptide, etc.)	Effective
Fraction B	Peptide and free amino acids	Very effective
Fraction C	Phenylalanine and tyrosine	Little effectiveness (with Fract. B)
Fraction D	Tryptophan	Very effective (with Fract. B)

group in the molecule. However, it should be studied in detail whether or not a sulfhydryl group is involved in the active center.[3] It is also noted that microbial β-amylases have hydrolytic activity against raw starch whereas plant β-amylases were reported to have little or no activity.[4] Therefore, we examined raw starch digestion by both microbial and plant β-amylases using wheat, corn, potato, and sweet potato starches. Wheat starch was the most digestible among the starches used. Among the enzymes tested, β-amylases from BQ10-S1 Spo II was the most active against wheat starch too.[5]

Since these mutants showed high β-amylase production only in the media containing natural nitrogen sources like peptone, meat extract, and so forth,[5] we attempted to isolate factors effective for enzyme production from Polypepton. Among four fractions (A, B, C, and D in order of elution on a Sephadex G-25 column),[5] fraction B was found to be the most effective for enzyme production. Fraction C was composed of tyrosine and phenylalanine in a free state and found to have little effectiveness for enzyme production. Fraction D, though ineffective as a single addition to the basal medium (Maassen's medium), was found to increase the enzyme production in the presence of fraction B. Paper chromatography, amino-acid analysis, UV, and IR spectral analyses of fraction D revealed that fraction D was composed only of tryptophan. Fractions A and B were composed of free amino acids and peptidyl compounds that were effective for the enzyme production. The chemical composition of fraction A is currently under study. TABLE 2 shows the effectiveness of each fraction obtained on the Sephadex G-25 column.

Since fraction B was found to be the most effective among the fractions obtained by gel filtration, further fractionation of fraction B was carried out with Dowex 50W-X2 columns. As shown in FIGURE 2, sixteen fractions were separated. Each fraction was collected and freeze-dried. However, in some of the subsequent experiments, combined fractions, namely fractions B-I to B-VII were used (FIG. 2). The amino acid

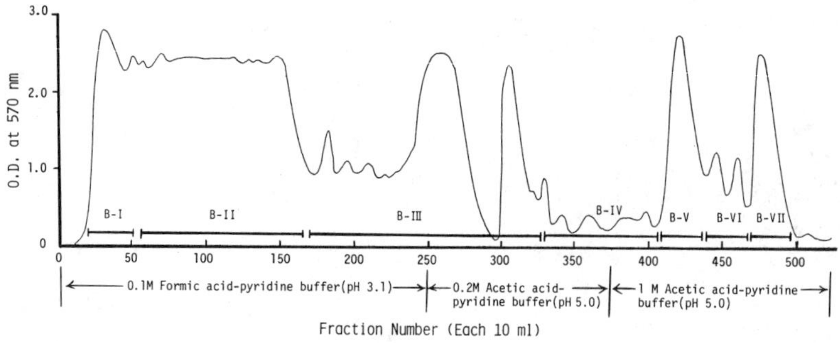

FIGURE 2. Chromatography of fraction B on Dowex 50W-X2. Column: 2.5 × 55 cm; flow rate: 30 ml/h; temperature 40°C.

composition of each fraction was analyzed in the conventional way. Fraction B-I was found to consist of 11 amino acids and the main components were Asp, Glu, and Val. Fraction B-II consisted of 14 amino acids and the main components were Asp, Glu, Leu, and Ile. Fraction B-III was found to consist of 13 amino acids, among which Asp, Glu, Val, and Pro were main components, and Fraction B-IV, of 15 amino acids, among which Glu, Val, and Pro were main components. The above fractions were all composed of free amino acids and contained no peptidyl compounds.

Fraction B-V was composed of 9 amino acids and free Lys was the main component. Fraction B-VI was found to consist of 16 amino acids and the main components were Lys, His, and Glu. Fraction B-VI contained peptidyl compounds that were effective for enzyme protection. Fraction B-VII consisted of 14 amino acids and the main component was Arg in a free state. Fraction B-VII also contained peptidyl compounds effective for enzyme production. As shown in TABLE 3, the mixture of Lys, His, and Arg, which were the main components of fractions B-VI and B-VII, were added to the basal medium and examined for the effect on enzyme production. The results, however, showed that the amount of β-amylase produced was about half that of the control. On the contrary, the addition of fractions B-VI and B-VII showed almost the same effect as that of the control. Therefore, basic peptidyl compounds were

TABLE 3. Effects of Basic Amino Acid Mixture on β-Amylase Production

	pH	Cell Growth (O.D. at 660nm)	β-Amylase Activity (units/ml)	Relative Activity (%)
Control	6.7	1.04	40.95	100
Fraction mixture	7.7	1.07	6.75	16
B-VI 0.02%, B-VII 0.02% added	7.3	1.26	45.45	111
Lys 0.01%, His 0.01%, Arg 0.02% added	7.4	1.00	22.05	54

thought to be necessary for stimulating β-amylase production under the culture conditions used. The chemical structures of these peptidyl compounds are under study. Natural nitrogen sources other than Polypepton are also under study. The comparison of chemical structures of these factors effective for enzyme production and the elucidation of the relationship between these factors and enzyme production by the mutants are necessary future studies.

REFERENCES

1. SHINKE, R., K. AOKI, H. NISHIRA & S. YUKI. 1979. Isolation of a rifampin-resistant, asporogenous mutant from *Bacillus cereus* and its high β-amylase productivity. J. Ferment. Technol. **57:** 53–55.
2. SHINKE, R., Y. KUNIMI & H. NISHIRA. 1975. Isolation and characterization of β-amylase-producing microorganisms. J. Ferment. Technol. **53:** 687–692.
3. NANMORI, T., R. SHINKE, K. AOKI & H. NISHIRA. 1983. Purification and characterization of β-amylase from *Bacillus cereus* BQ10-S1 Spo II. Agric. Biol. Chem. **47:** 941–947.
4. SHINKE, R., T. NANMORI, K. AOKI, H. NISHIRA, K. NISHIKAWA, K. YAMANE & K. NISHIDA. 1984. Sci. Rep. Fac. Agric. Kobe Univ. **16:** 309–316.
5. SHINKE, R. 1979. Studies on β-amylase production by microorganisms. Hakko Kogaku **57:** 102–113.

Microbial Glycosidases: Research and Application

ZHANG SHUZHENG, GE SUGUO, AND YANG SHOUJUN

Institute of Microbiology
Academia Sinica
Beijing, China

In our laboratory, some research groups have been working on microbial glycosidases for industrial applications as well as for basic studies.

Glucoamylase from *Monascus* sp. had been used for glucose production.[1] A mutant strain of *Aspergillus niger* AS3.4309 with much higher glucoamylase activity has been widely used for the production of glucose, starch syrup, alcohol, alcoholic beverages, and so forth.[2–4] Immobilized glucoamylase has been used to convert starch to glucose solution which was used as a raw material for glutamic acid fermentation.[5] β-Amylase of a mutant strain of *Bacillus cereus* M-3 has been produced in pilot-plant scale and tested for making maltose and beer.[6] Cellulase of *Trichoderma koningii* was used to convert furfural industrial waste to glucose, which was used to produce *Candida* sp. yeast.[7] A thermostable cellulase of *Humicola insolens* was found to be a potential carboxy methyl cellulose (CMC) hydrolyzer, which may be used to liquefy CMC fracture fluid in petroleum reservoirs.[8]

Some basic research has been done on multiple forms of glucoamylase from *Monascus rubiginosus,* such as purification, crystallization, electron microscopy of the crystals, physical and chemical properties, substrate specificity, catalytic kinetics, chemical modifications, conformational studies with UV difference, CD spectroscopy, peptide mapping, and sugar peptide bounding.[9,10] Cellulases from *Trichoderma koningii* including C_1 and multiple forms of C_x have been purified and characterized.[11,12] α-Glucosidase from *M. ruber,*[13] and pullulanase from *Aerobacter aerogenes*[14,15] have been purified and characterized, too; the latter will be described.

The crude enzyme of *Aerobacter aerogenes* was a gift from the Jiangsi Scientific Research Institute of Food and Fermentation Industry[16] and from which pullulanase

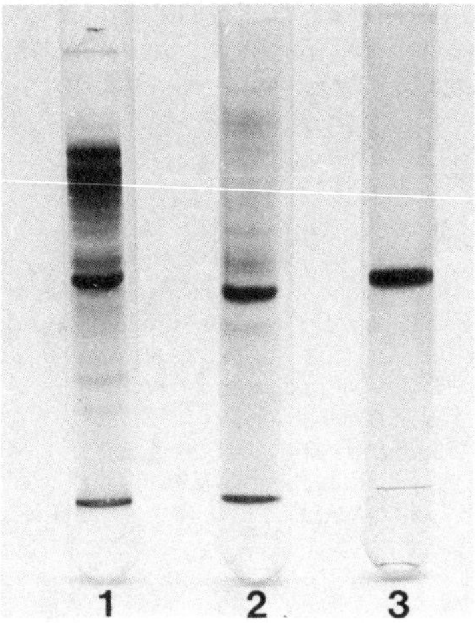

FIGURE 1. PAG electrophoresis of pullulanase. (1) Crude enzyme; (2) purified by gel filtration on Sephadex G-100 column; (3) purified by slab gel electrophoresis.

TABLE 1. Properties of Pullulanase

Optimal pH	5.3–5.8
Optimal temperature	50°C
pH stability	4.3–8.6
K_m for glutinous rice starch	2.0×10^{-2} g/ml
Isoelectric point	3.8 (IEF on PAG)
Molecular weight	51,000–52,000 (SDS PAGE)
Inhibitors (10^{-3} *M*) (relative activity)	Hg^{2+} (0), Cu^{2+} (5), Al^{3+} (7), Fe^{3+} (11)
Chemicals (10^{-3} *M*)	NEM (98), WSCCD (93), NAI (84), NBS (10^{-4} *M*, 0; 10^{-5} *M*, 9)
Carbohydrates (10^{-3} *M*)	gluconic acid-δ-lactone (93)
	maltotriose (88)
	α-cyclodextrin (46)
	β-cyclodextrin (0)
	γ-cyclodextrin (52)
K_i for β-cyclodextrin	0.55×10^{-5} *M*
Denaturants	8 *M* urea (0), 1 *M* guanidine HCl (0), 0.5% SDS (0)

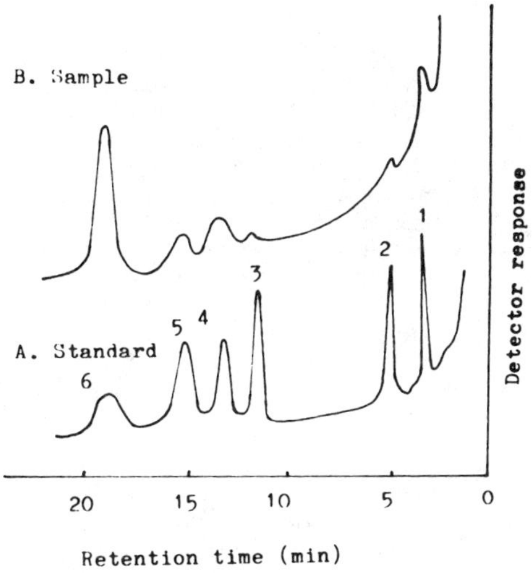

FIGURE 2. Gas chromatogram of the sugars in pullulanase: (1) rhamnose; (2) arabinose; (3) mannose; (4) galactose; (5) glucose; (6) inositol.

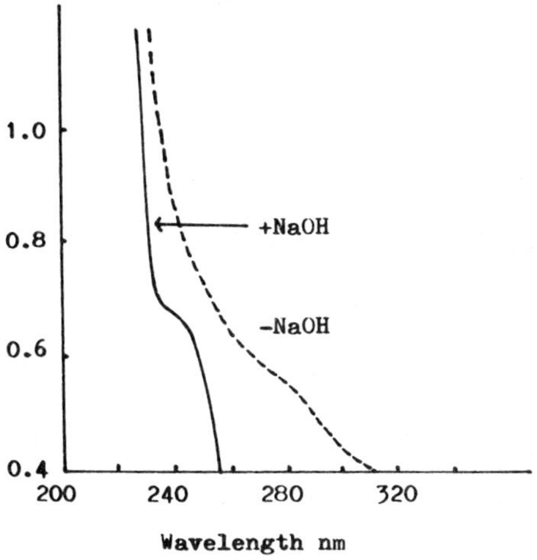

FIGURE 3. Alkali-catalyzed β-elimination of pullulanase.

(E.C.3.2.1.41) was purified by gel filtration on a Sephadex G-100 column and by preparative polyacrylamide slab gel electrophoresis. It showed a single band on disc electrophoresis (FIG. 1).

Some properties of the enzyme are summarized in TABLE 1.

Purified pullulanase contained 6% sugar, and GLC analysis (FIG. 2) showed that the sugar composition in residue/mole was galactose, 8; glucose, 6; mannose, 2; rhamnose, 2; arabinose, 1; total, 19. When the enzyme was treated with 0.2 *M* NaOH and 1 *M* $NaBH_4$ at 45°C for 24 hours, β-elimination led to a loss of 4 serine (35–31) and 3 threonine (29–26) residues, and an increase of 2 glycine (36–38) and 4 alanine (48–52) residues. The unsaturated amino acids formed during the alkali β-elimination (without $NaBH_4$) of this enzyme led to an increase in absorbance at 240 nm (FIG. 3). These results suggest that the sugars are linked to serine and threonine through 0-glycosidic linkages.

REFERENCES

1. HO, B. -W., J. -J. GUO, Y. -C. FANG & S. -C CHANG. 1973. Studies on the screening of glucoamylase-producing strains of *Monascus* and their fermentation conditions. Wei Sheng Wu Hsueh Pao **13**(2): 142–150.
2. RESEARCH GROUP ON GLUCOAMYLASE. Institute of Microbiology. Academia Sinica. 1974. Studies on the induced mutation for glucoamylase production in *Aspergillus niger*. Wei Sheng Wu Hsueh Pao **14**(1): 77–82.
3. RESEARCH GROUP ON GLUCOAMYLASE. Institute of Microbiology. Academia Sinica. 1979. Production of glucoamylase by *Aspergillus niger* A.S.3.4309. Wei Sheng Wu Hsueh Tong Pao **6**(5): 19–22.
4. RESEARCH GROUP ON GLUCOAMYLASE. Institute of Microbiology. Academia Sinica. 1980. Production of glucoamylase by *Aspergillus niger* A.S.3.4309. Wei Sheng Wu Hsueh Tong Pao **7**(4): 153–155.
5. LI, G.-S., J.-Y. HUANG, X.-F. KOU & S.-Z. ZHANG. 1983. Glucoamylase covalently coupled to porous glass. Appl. Biochem. Biotechnol. **7**(5): 325–341.
6. HO, B.-W., J.-J. GUO, Z.-Y. JIANG & S.-Z. ZHANG. 1983. Selection of high β-amylase-producing strain and its fermentation conditions. Wei Sheng Wu Hsueh Pao **23**(1): 75–83.
7. CELLULASE RESEARCH GROUP. 1977. Yeast production from cellulase-hydrolyzed furfural industrial waste. Wei Sheng Wu Hsueh Pao **17**(2): 137–142; **17**(3): 231–238.
8. CUI, F. -M., J. -H. MA, A. NA & S. -Z. ZHANG. 1983. Studies on cellulase from *Humicola insolens*. Chen Chün Hsueh Pao (Acta Mycologica Sinica) **2**(2): 119–126.
9. ZHANG, S.-Z. 1980. Comparative studies on two forms of glucoamylase from *Monascus rubiginosus Sato*. *In* Nucleic Acids and Proteins. Z. -W. Shen, Ed. Science Press. Beijing, China. pp. 402–412.
10. ZHANG, S.-Z., S.-J. YANG, J.-W. SUN, Z.-X. HE, Z.-Z. YAN, Q. LI, T.-X. XU, J.-J. GUO, Y.-S. WANG, S.-G. GE, B.-W. HE & Y.-X. WANG. 1982. Studies on glucoamylase from *Monascus rubiginosus* sato. *In* Proteins in Biology and Medicine. R. A. Bradshaw *et al.*, Eds. Academic Press. pp. 201–225.
11. NA, A., F.-M. CUI, J.-H. MA & S.-Z. ZHANG. 1982. Some properties of C_1 enzyme from *Trichoderma koningii*. Wei Sheng Wu Hsueh Pao **22**(4): 333–338.
12. NA, A., J.-H. MA, F.-M. CUI & S.-Z. ZHANG. 1983. Purification and some properties of multiple molecular forms of C_x enzymes from *Trichoderma koningii*. Chen Chün Hsueh Pao **2**(1): 50–58.
13. ZENG, Y.-C. & S.-Z. ZHANG. 1984. Studies on α-glucosidase from *Monascus ruber*. In press.
14. GE, S.-G., S.-J. YANG & S.-Z. ZHANG. 1980. Studies on pullulanase from *Aerobacter aerogenes* I. Wei Sheng Wu Hsueh Pao **20**(4): 415–420.
15. YANY, S.-J., S.-G. GE & S.-Z. ZHANG. 1981. Studies on pullulanase from *Aerobacter aerogenes* II. Wei Sheng Wu Hsueh Pao **21**(1): 68-72.
16. THE JIANGSI SCIENTIFIC RESEARCH INSTITUTE OF FOOD AND FERMENTATION INDUSTRY. 1966. Studies on the isoamylase of *Aerobacter aerogenes* 10016. Wei Sheng Wu Hsueh Pao **16**(4): 282–290.

Enzyme Recovery by Adsorption from Unclarified Microbial Cell Homogenates

PER HEDMAN AND JAN-GUNNAR GUSTAFSSON

Pharmacia Fine Chemicals
S-751 82 Uppsala, Sweden

Extraction of intracellular products from microorganisms has become an increasingly important operation. Recombinant DNA techniques are used to construct bacterial or yeast strains with the ability to produce enzymes such as coagulation factors, urokinase, tissue plasminogen activator, chymosin, and so forth. For the efficient recovery of these enzymes from the intracellular space, centrifugation and filtration are costly and time consuming. Other alternatives have been investigated, for example, extraction in aqueous two-phase systems.[1] In order to reduce the rate of proteolytic breakdown, it is preferable to adsorb the product to a solid, thereby significantly decreasing the probability of contact between proteases and the product. The earliest possible adsorption is achieved when the adsorbent is added to the homogenate immediately after the cell disruption. Then the product-adsorbent complex must be separated from the cell fragments. For the latter (solids-solids) separation, an aqueous two-phase extraction may be used if the adsorbent is designed with suitable partitioning characteristics.

The volumes of the two phases can be rather small, as the phases only need to contain the adsorbent and the cell fragments. When two-phase extraction is used without adsorbents, the phase-to-phase volume ratio affects the product yield. With adsorbents, on the contrary, the yield only depends on the amount of adsorbent, its binding strength and its capacity.

Inexpensive phase-forming chemicals such as polyethyleneglycol (PEG) + phosphates or sulfates, rather than PEG + dextran are preferred. As the phase volumes may be kept small, these inexpensive chemicals will contribute little to the total costs. The cost contribution from the adsorbents depends on the efficient use of washing buffers to increase the reusability of the adsorbent.

Cell fragments from *Saccaromyces cereviseae* and *Esherichia coli* obtained by disruption in a beadmill accumulate in the more polar salt-rich phase. Consequently, the adsorbent should be designed such that it is partitioned to the less polar PEG-rich phase. The desired partitioning behavior is achieved by derivatizing chromatography gel media with polyethyleneglycol. For the protein-binding ability, a second derivatization is necessary.

EXTRACTION OF PROTEIN A

To demonstrate that it is possible to synthesize adsorbents such as described above, we designed an adsorbent for staphylococcal Protein A.

Sepharose CL-6B was derivatized with monomethoxypolyethyleneglycol 3000 and then activated with CNBr. IgG-PEG-Sepharose was mixed with Protein A dissolved in 0.1 *M* sodium phosphate pH 7.2. Then PEG 4000 and potassium phosphate were added to 19.5% wt/wt and 11% wt/wt, respectively. From the top PEG-rich phase, the

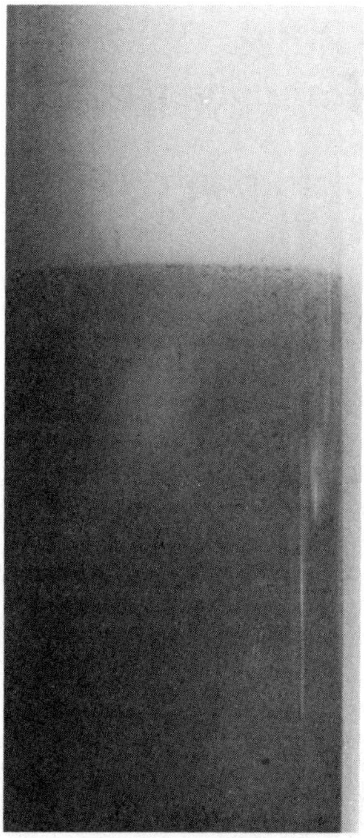

FIGURE 1. Bakers' yeast homogenate mixed with Cibacron Blue-PEG-Sepharose, PEG 4000, and potassium phosphate. Cell fragments accumulate in the bottom phase, whereas the adsorbent is partitioned to the top phase.

adsorbent was recovered and poured into a chromatography column. Finally Protein A was desorbed in 0.1 *M* glycine buffer, pH 3. The yield was 72%.

This experiment demonstrates that it is possible to bind a protein to a "carrier adsorbent" and extract the complex to a PEG-rich phase in a two-phase system.

EXTRACTION OF ALCOHOL DEHYDROGENASE (ADH) AND HEXOKINASE (HK) FROM BAKER'S YEAST

Nucleotide cofactor-dependent enzymes like ADH and HK are known to bind to Cibacron Blue F3GA.[2] In an attempt to demonstrate enzyme extraction from fresh

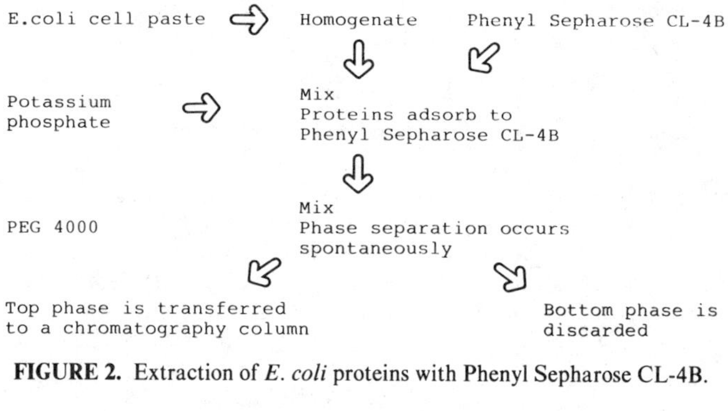

FIGURE 2. Extraction of *E. coli* proteins with Phenyl Sepharose CL-4B.

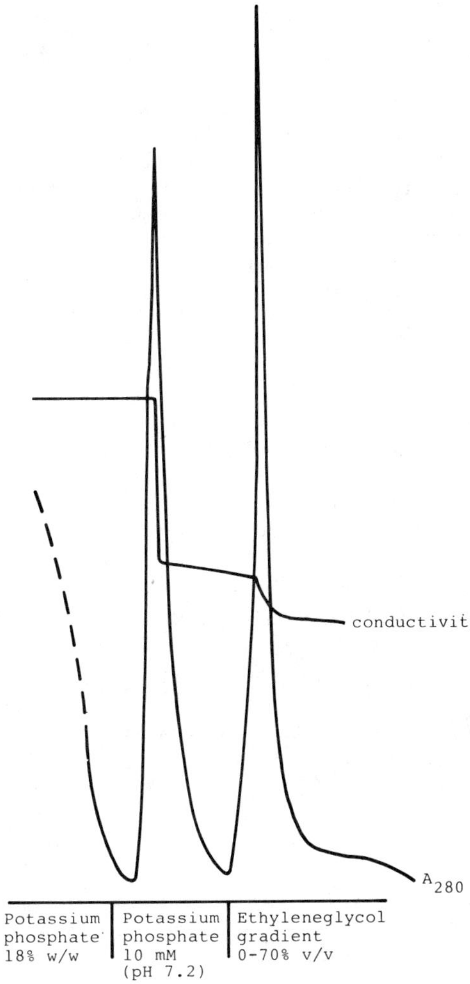

FIGURE 3. Elution of proteins from Phenyl Sepharose CL-4B, which was separated from *E. Coli* cell fragments by partitioning in an aqueous two-phase system.

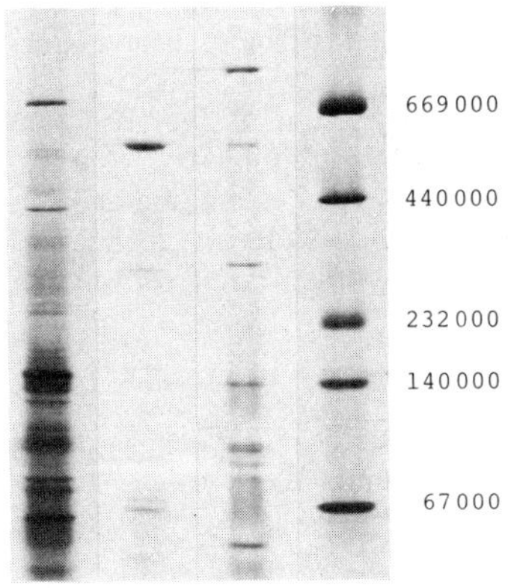

FIGURE 4. Electrophoresis of *E. coli* proteins in precast polyacrylamide gels (PAGE 4/30 Pharmacia Fine Chemicals).

Lane 1: crude homogenate;

Lane 2: first peak from elution of Phenyl Sepharose CL-4B;

Lane 3: second peak from elution of Phenyl Sepharose CL-4B;

Lane 4: Reference proteins (HMW Calibration kit, Pharmacia Fine Chemicals).

homogenized yeast, we used Cibacron Blue F3GA-substituted methoxy-PEG-Sepharose.

One-hundred grams wet yeast was homogenized in a glass beadmill and mixed with 20 ml sedimented adsorbent. PEG 4000 and potassium phosphate were added to a final concentration of 10% wt/wt and 12% wt/wt, respectively. The two phases so formed separated spontaneously. In separate experiments, the two phases were completely separated by centrifugation at 1000 × *g* in 1 minute. The upper PEG-rich phase containing the adsorbent was transferred to a column; the bottom phase containing the cell fragments was discarded.

The adsorbent was washed free from PEG in the column before enzyme elution started with 5 m*M* NAD at pH 6.4. ADH was identified by its enzymatic activity and collected in a 12-ml fraction. Similarly hexokinase was eluted at pH 8.6. The yield of these enzymes was low (less than 1 to 12% of the activity in the homogenate). The low yield could be explained by insufficient binding capacity of the adsorbent or by the high phosphate concentration in the phase system. The binding of ADH and HK to Cibacron Blue is known to be reduced at increasing salt concentrations (see FIG. 1).

After 10 consecutive adsorption-phase extraction-desorption cycles as described above, the albumin binding capacity of the adsorbent was determined. In 0.1 *M* phosphate buffer, the albumin binding capacity did not decrease after 10 cycles.

EXTRACTION OF *E. COLI* PROTEINS WITH PHENYL SEPHAROSE CL-4B

In phase systems with high salt concentrations, hydrophobic interactions are increased. To demonstrate this, Phenyl Sepharose CL-4B was used to extract hydrophobic proteins from an unclarified *E. coli* homogenate as illustrated by FIGURES 2, 3, and 4.

CONCLUSIONS

As long as PEG-salt phase systems are the only really inexpensive alternatives, protein-binding ligands sensitive to interference from high salt concentrations cannot be used. General hydrophobic ligands such as phenyl groups may be used although the anticipated degree of purification is inferior to biospecific adsorption ligands. Still, volume reduction and removal of solids without filtration or centrifugation is possible when "carrier adsorbents" are used in the recovery of proteins from cell homogenates.

The cost contribution of the adsorbent will not be prohibitive as the adsorbent retains its protein-binding capacity and partitioning behavior quite well. Further investigations are needed to determine the cost of adsorbent replacement due to losses in a large-scale process. The "carrier adsorbents" may provide an economical alternative to centrifugation or filtration as the first recovery step after cell disruption.

REFERENCES

1. K. H. KRONER, H. SCHÜTTE, W. STACH & M.-R. KULA. 1982. Scale-up of formate dehydrogenase by partition. J. Chem. Technol. Biotechnol. **321**: 130–137.
2. DEAN, P. D. G. & D. H. WATSON. 1979. J. Chromatogr. **165:** 301–319.

Immobilized Thrombolytic Enzymes Possessing Increased Affinity toward Substrate

V. P. TORCHILIN, A. V. MAKSIMENKO, E. G. TISCHENKO,
G. V. IGNASHENKOVA, AND G. A. ERMOLIN

USSR Cardiology Research Center
Institute of Experimental Cardiology
Moscow, USSR

One of the main considerations in thrombolytic enzyme therapy with plasmin, streptokinase, and urokinase is the creation of a high local enzyme concentration in the vicinity of the thrombus. Different approaches have been suggested to accomplish this. Angiographic methods are used, making it possible to infuse the enzyme solution directly to the thrombus.[1] Targeting of enzyme to the thrombus using microcontainers, such as microcapsules, liposomes, erythrocytes, or other cells, also can be useful, but requires the creation of a special system for drug release in the target zone. On the other hand, "polymeric affinity drugs," developed principally in the late 1970s by H. Ringsdorf, E. Goldberg, and others, can be of practical importance.[2,3] These affinity drugs consist of a polymeric carrier with coimmobilized enzyme and a vector molecule capable of recognizing and binding to a target site in the organism. Normally, specific antibodies are used as vector molecules; but some other compounds also can be used. This affinity approach also can be of practical use in the case of thrombolytic (fibrinolytic) enzymes.

In *in vitro* experiments, we have studied the fibrinolytic activity of a model system toward fibrin clots. The model system consists of a dextran carrier, activated by periodate oxidation, and a proteolytic enzyme, α-chymotrypsin, and a polyclonal antibody towards fibrinogen coimmobilized on the carrier (FIG. 1). These antibodies participate in immunochemical reactions with fibrinogen, fibrin, and their degradation products and bind with common antigenic determinants.

Protein binding studies with oxidized dextran (molecular weight 35,000–50,000) were performed as described in Torchilin *et al.*[4] Both the enzyme and antibody immobilization reactions can be performed simultaneously, using an enzyme-to-antibody mixture in a molar ratio of 10:1. Or, the immobilization can be done in two stages. In the first stage the enzyme is bound, and in the second stage the antibody is bound via the remaining aldehyde groups. In our work, the total number of aldehyde groups was 20–24 per 100 glycoside units. Unbound proteins were separated by gel chromatography. As a result, macromolecular conjugates were obtained with molecular weights of about 250,000 and possessing both enzymatic and immunological activity. The enzymatic activity was measured by following the initial hydrolysis rate of a low-molecular-weight specific substrate, *N*-acetyl-L-tyrosine ethyl ester, on a pH-stat. It was demonstrated that the high-molecular-weight conjugates contained ~10% of the initial enzymatic activity introduced into the immobilization reaction. This corresponded to about 50 mg of active enzyme per gram of carrier.

The following preparations were used in experiments on fibrin clot lysis: native α-chymotrypsin, dextran-chymotrypsin conjugate, and dextran-chymotrypsin-antibody conjugate. Fibrin clots were prepared by mixing 0.5 ml of 10 mg/ml fibrinogen

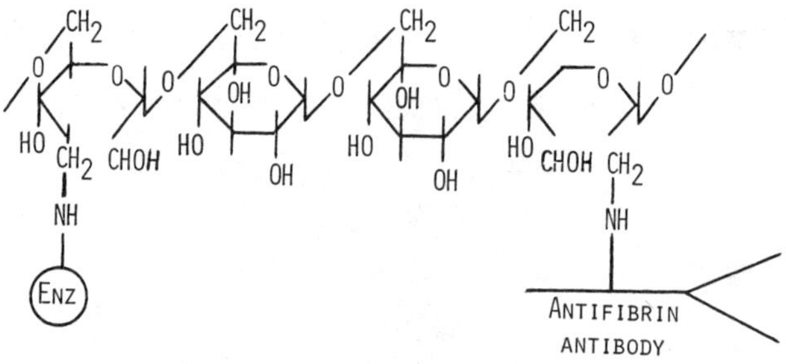

FIGURE 1. The principal structure of the ternary enzyme-dextran-antibody conjugate.

and 0.2 ml of 4 mg/ml thrombin. Each clot was placed into a plastic tube. The bottom of the tube was made of nylon netting, impermeable to the clot but easily permeable to soluble enzyme derivatives and to the products of fibrin degradation. The lower part of each tube was dipped into 0.1 *M* phosphate buffer, pH 7.4, containing appropriate enzyme preparations in equal quantities. The concentration of active centers was 8.75×10^{-8} *M*. Continuous magnetic stirring was used. To determine the relative affinity of different enzyme derivatives towards the fibrin clot, the tubes were incubated in each solution for 10 min and then were transferred into similar solutions containing no enzyme in any form. In this way the clot lysis was due only to the action of enzyme adsorbed on the clot surface during the incubation. The higher the affinity of the enzyme preparation toward fibrin, the higher was the amount of active enzyme

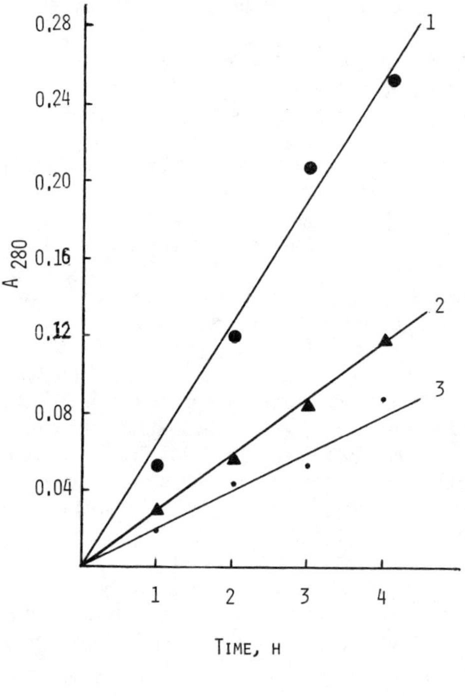

FIGURE 2. The rate of fibrin clot lysis shown as the increase in soluble product content of the solution (measured as optical density at 280 nm) under the action of adsorbed enzyme, for (•) native α-chymotrypsin, (▲) dextran-chymotrypsin, and (•) dextran-chymotrypsin-antibody.

bound to the clot and the higher would be the rate of clot dissolution. The data presented in FIGURE 2 show that the highest rate of fibrin clot lysis was observed with the enzyme-dextran-antibody conjugate. This is taken to mean that this preparation had the maximal affinity toward the substrate. The increased affinity was conditioned by the antibody binding with fibrin. At the same time, native α-chymotrypsin and dextran-chymotrypsin conjugate were only slightly adsorbed on fibrin (curves 2 and 3 in FIG. 2).

Thus, covalent binding of the enzyme and antibody to a polymeric carrier gave a preparation possessing increased affinity toward the substrate. The enzyme and the antibody preserved their activity in the ternary complex. In addition, stabilization by the soluble polymeric carrier also may result.

REFERENCES

1. CHAZOV, E. I., L. S. MATVEEVA, A. V. MAZAEV, K. E. SARGIN, G. V. SADOVSKAYA & M. YA. RUDA. 1976. Intracoronary administration of fibrinolysin for the treatment of acute myocardial infarction. Ter. Arkh. **48**(4): 8–13.
2. RINGSDORF, H. 1975. Structure and properties of pharmacologically active polymers. J. Polym. Sci. Polym. Symp. **51:** 135–153.
3. GOLDBERG, E. P. 1978. Polymeric affinity drugs for cardiovascular, cancer, and urolithiasis therapy. *In* Polymeric Drugs. L. G. Donamura & O. Vogl, Eds. Academic Press. New York, San Francisco, London. pp. 239–262.
4. TORCHILIN, V. P., I. L. REIZER, V. N. SMIRNOV & E. I. CHAZOV. 1976. Enzyme immobilization on biocompatible carriers. III. α-Chymotrypsin immobilization on soluble dextran. Bioorg. Khimia **2:** 1252–1258.

A New Viologen Mediator for Hydrogenase-catalyzed Reactions

E. ZIOMEK, W. G. MARTIN, AND R. E. WILLIAMS

Division of Biological Sciences
National Research Council of Canada
Ottawa, Ontario, Canada K1A OR6

Hydrogenase enzymes have been isolated from many different sources.[1] The enzyme activity has been used both *in vivo* and *in vitro* in the photosynthetic production of hydrogen as well as for the production of reduced pyridine nucleotides and various steroids.[2] Hydrogenases catalyze the reversible interconversion of protons and hydrogen according to the reaction:

$$2H^+ + 2M^+ \rightleftharpoons H_2 + 2M^{2+}$$

where M = a natural or synthetic mediator. The reaction can be mediated by both natural (i.e., cytochrome C_3, ferredoxin) and synthetic mediators. Among the synthetic mediators is a class of low redox potential compounds, that is, the viologens. The most widely used example is methyl viologen (see below; R = R′ = methyl).

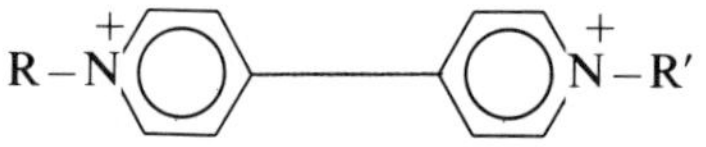

Viologen Structure

This communication describes the results of a survey of side-chain variants of the basic viologen structure. Viologens with neutral (alkyl or aryl) and charged (positive or negative) side chains (R's above) were tested for their ability to mediate hydrogen production and hydrogen uptake reactions catalyzed by the periplasmic hydrogenase of *Desulfovibrio desulfuricans,* a sulfate-reducing bacterium.

Purified hydrogenase was isolated from *D. desulfuricans,* and hydrogen production activity was assayed polarographically as previously described.[3] Hydrogen uptake activity was assayed spectrophotometrically by minor modifications of a previously described procedure.[4]

N,N′-diaminopropyl-4,4′-dipyridine (DAPV) (dibromo salt) was prepared as described by Simon and Moore.[5] Redox properties were measured under nitrogen in 0.01 *M* sodium phosphate, pH 7.0, using a differential pulse polarographic analyzer (PAR model 174A) equipped with a dropping mercury electrode. Spectrophotometric properties of the reduced viologen were determined in citrate-phosphate buffers containing a 30-fold molar excess of sodium dithionite.

One compound among the many surveyed, *N,N′*-diaminopropyl-4,4′-dipyridine (DAPV), (R = R′ = $+NH_3CH_2CH_2CH_2-$) had significantly higher activity than methyl viologen (MeV) in both the hydrogen production and hydrogen uptake reactions catalyzed by the *D. desulfuricans* hydrogenase. In comparison with MeV, DAPV was 1.6 times more reactive in the hydrogen production reaction and 15 times more reactive in the hydrogen uptake reaction (TABLE 1). The redox properties of

DAPV were also compared with MeV. The one- and two-electron reduction potentials of DAPV were located at −610 and −932 mV, respectively, versus the saturated calomel electrode in comparison with those of MeV at −675 and −1050 mV. The spectrophotometric properties of the one-electron-reduced DAPV and MeV differed. For MeV, the blue one-electron-reduced product had an $\epsilon_{600} = 12,800\ M^{-1}\ cm^{-1}$ at pH 8.8 that fell to 2,000 at pH 5.5 while the extinction coefficient of DAPV remained nearly constant over the same range (ϵ_{600} (pH 8.8) = 12,000; ϵ_{600} (pH 5.5) = 10,000).

Enzyme kinetic parameters K_m(app), k_{cat} and k_{cat}/K_m(app) of DAPV and MeV were estimated for both the hydrogen production and hydrogen uptake reactions (TABLE 2). The reactions were run under conditions of apparent saturation by the second substrate in the reaction (H^+ for production, H_2 for uptake). The K_m(app) and k_{cat} estimates for both reactions show that the Michaelis K_m(app) seemed to exert control over the reaction while k_{cat} remained relatively unchanged in the switch from MeV to DAPV. Increased binding between DAPV and the enzyme could be the cause

TABLE 1. Hydrogen Production and Uptake Mediated by Methyl Viologen and Diaminopropyl Viologen

	Reaction	
Compound	Hydrogen Production[a] (μmoles H_2/min/mg protein)	Hydrogen Uptake[b] (μmoles H_2/min/mg protein)
Methyl Viologen	10,570	1,320
Diaminopropyl Viologen	16,960	20,420

[a]Conditions: Hydrogen production activity was determined polarographically. Reaction mixtures, maintained at 37°C, contained 0.1 *M* sodium acetate, pH 5.5, and 10 m*M* sodium dithionite and enzyme (approx. 20 ng). Reaction was initiated by addition of viologen to yield a final concentration of 3.3 m*M*.

[b]Conditions: Hydrogen uptake activity was determined spectrophotometrically in sealed cuvettes. Extinction coefficients used were those given in the text. Reaction mixtures, maintained at 37°C, contained 0.1 *M* Tris-HCl, pH 7.5 and 3.3 m*M* viologen. The reaction mixture was saturated with hydrogen and the reaction initiated by addition of an aliquot of prereduced enzyme. Prereduced enzyme was prepared under nitrogen in 0.1 *M* Tris-HCl, pH 7.5, 0.1% wt/vol bovine serum albumin, 0.1 m*M* sodium dithionite.

of the observed change in K_m(app). However, K_m(app) is a complex constant composed of not only binding but also rate and redox components. Further work will be required to determine the influence that each has on the reaction.

The quantity k_{cat}/K_m(app) has been used as a general measure of the relative activity of various substrates, as well as giving an indication of the maximum turnover rate of the reaction.[6] This constant is devoid of the complexities of the two others and can serve as a more appropriate measure of the relative activity of two compounds. Both MeV and DAPV have k_{cat}/K_m(app) values of about 10^8–10^9, indicative of the fact that the reaction is diffusion-controlled. Indeed, the value of the constant of DAPV is substantially higher than those observed for catalase and carbonic anhydrase (10^7–10^8). Both reactions, production and uptake, have equally high values indicating that the initial formation of the enzyme-substrate complex and not some further stage of the mechanism is important in determining the rate of the reaction.

TABLE 2. Enzyme Kinetic Parameters for Hydrogen Production and Uptake Reactions

	Reaction					
	Hydrogen Production[a]			Hydrogen Uptake[b]		
Compound	k_{cat}^{c} (sec^{-1})	K_m(app) (mM)	k_{cat}/K_m(app) (sec^{-1} M^{-1})	k_{cat}^{c} (sec^{-1})	K_m(app) (mM)	k_{cat}/K_m(app) (sec^{-1} M^{-1})
Methyl Viologen	2.28×10^5	0.68	3.35×10^8	3.04×10^5	1.5	2.03×10^8
Diamino Propyl Viologen	2.62×10^5	0.24	1.09×10^9	5.2×10^5	0.16	3.25×10^9

[a] Measurements were made by a polarographic procedure in citrate-phosphate buffer 0.1 M, pH 6.0, containing 0.5% wt/vol bovine serum albumin. Variations in the concentration of the viologen substrate were made and the concentration of the reduced viologen was assumed to be equal to the amount of viologen added.

[b] Measurements were made by a spectrophotometric procedure with prereduced enzyme (described in TABLE 1). Variations of the viologen concentration were made in the buffer used (Tris-HCl, 0.1 M, pH 8.0, bovine serum albumin (0.5% wt/vol)).

[c] Calculated from V_{max}(app) (μmoles hydrogen evolved or utilized per 1 μmole enzyme per sec) using the expression $V_{max}(app) = k_{cat} \times [E]$. Maximum specific activity of hydrogenase used for calculation of enzyme concentration [E] was taken as 45,000 μmoles H_2/min/mg protein under standardized conditions.[7]

ACKNOWLEDGMENTS

The authors wish to thank C. J. Dicaire and J. Giroux for their technical assistance in preparing the enzyme and Dr. R. Renaud for measuring the redox potentials.

REFERENCES

1. ADAMS, M. W. W., L. E. MORTENSON & J.-S. CHEN. 1981. Hydrogenase. Biochim. Biophys. Acta **594:** 105–176.
2. KLIBANOV, M. 1983. Biotechnological potential of the enzyme hydrogenase. Process Biochem. **23:** 13–16, 23.
3. GLICK, B. R., W. G. MARTIN & S. M. MARTIN. 1980. Purification and properties of the periplasmic hydrogenase from *Desulfovibrio desulfuricans.* Can J. Microbiol. **26:** 1214–1223.
4. ERBES, D. L. & R. H. BURRIS. 1978. The kinetics of methyl viologen oxidation and reduction by the hydrogenase from *Clostridium pasteurianum.* Biochim. Biophys. Acta **525:** 45–54.
5. SIMON, M. S. & P. T. MOORE. 1975. Novel polyviologens: Photochromic redox polymers with film-forming properties. J. Polym. Sci. Polym. Chem. Ed. **13:** 1–16.
6. SUDI, J. & B. H. HAVSTEEN. 1976. On the catalytic activity of chemically modified enzymes involving two or more substrates and products. Int. J. Peptide Protein Res. **8:** 519–531.
7. ZIOMEK, E., W. G. MARTIN & R. E. WILLIAMS. 1984. Immunological and enzymatic properties of *Desulfovibric desulfuricans* hydrogenase. Can. J. Microbiol. In press.

Expression of Hydrogenase Activity in Cereals

V. TORRES,[a,b] A. BALLESTEROS,[a,c] V. M. FERNÁNDEZ,[a] AND M. NÚÑEZ[b]

[a]*Instituto de Catálisis*
C.S.I.C.
Madrid 6, Spain

[b]*Departamento de Bioquímica y Microbiología*
I.N.I.A.
Madrid 35, Spain

Several authors have reported on hydrogenase activity in plants[1-3] and animal tissues.[4] However, the data are not sufficiently convincing to exclude the possibility that such activity comes from contaminant bacteria, the most accepted view being that only prokaryots and some eukaryotic algae possess hydrogenase activity.[5] We report here some experimental evidence indicating that several *Gramineae* plants express hydrogenase activity as a response to anaerobic conditions.

Wheat (*Triticum aestivum* L., cv. Anza and *Triticum durum* L., cv. Camacho); barley (*Hordeum vulgare* L., cv., Logra); and corn (*Zea mays* L., inbreed line 2087) were used. Germination of cereal seeds was carried out on water-wet filter paper at room temperature in the dark for seven days in axenic conditions. Previously, the seeds were sanitized by soaking them for 2 minutes in 70% ethanol, then for 5 minutes in 0.2% $HgCl_2$ and exhaustively rinsed with sterile water. To estimate the bacterial contamination, appropriate dilutions from the cereal samples were anaerobically incubated in Reinforced Clostridial Medium (Oxoid CM149) at 30°C for 48 hours, and the colonies that grew were counted.

For determination of the hydrogenase activity, one gram of plant tissue was incubated for 15 min at 40°C in 50 m*M* Tris-HCl, pH 8.5, containing 1 m*M* methylviologen and 15 m*M* dithionite. The hydrogen produced was quantitatively determined by gas chromatography. To obtain the dry weight, cereal samples were dried at 110°C overnight.

We observed first that when seedlings of several cereals were confined in an anaerobic atmosphere, a considerable amount of hydrogen appeared. In order to demonstrate the effect of oxygen on the evolution of hydrogen, we next confined the seedlings in a closed aerobic atmosphere and periodically determined by gas chromatography the content of oxygen and hydrogen in the closed vial. The results in FIGURE 1 show that the phenomenon of hydrogen evolution is concomitant with the consumption of oxygen by the seedlings. This indicates that the hydrogenase activity starts to appear when the oxygen content of the surrounding atmosphere decreases from that present in air, and that the activity is higher for lower oxygen contents.

In order to differentiate the hydrogen produced by the different organs of the germinated cereal, hand-disected roots, hypocotyls or leaves were maintained under argon in a rubber-stoppered vial for 24 hours. Then the vial gas phase was analyzed for

[c]Correspondence should be sent to A. Ballesteros, Instituto de Catalisis, C.S.I.C., Serrano 119, Madrid 6, Spain.

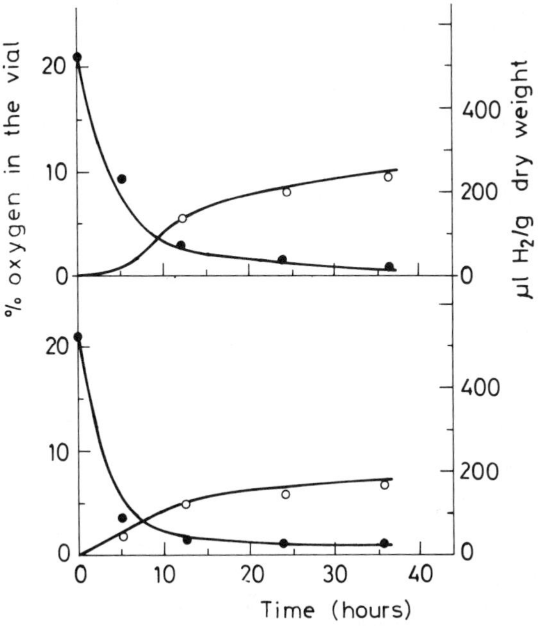

FIGURE 1. Oxygen consumption (●) and hydrogen evolution (O) by cereal seedlings enclosed in an aerobic atmosphere. Top: *T. durum;* Bottom: *Z. mays.*

weight for *T. aestivum, T. durum,* and *Z. mays*) and hypocotyls evolve (50–85 µl Hg/g dry weight) hydrogen, which suggests that hydrogenase activity is differentially expressed in those organs that can be exposed to low levels of oxygen during the plant development.

In order to prove that the observed hydrogen evolution was due to the plant tissues and not to contaminant microorganisms, seeds of three cereal species were mixed, soaked overnight in water, and germinated in crowded conditions over filter paper wetted with the soaking water. Then, each type of seedling was separated manually from the others, made anaerobic for 24 hours, and tested for hydrogenase activity. If the hydrogen evolved by each of the three cereal species comes from contaminating microorganism(s), then the organ corresponding to mixed germination would show higher hydrogenase activity due to infection with the microorganism(s) present in the

TABLE 1. Homogenization of Bacterial Background

	Hydrogenase Activity (µl H_2/min/g dry wt.)			
	Root		Hypocotyl	
Species	Unmixed	Mixed	Unmixed	Mixed
H. vulgare	123	150	74	62
T. aestivum	21	18	6.0	8.0
Z. mays	not tested	not tested	14	12
T. durum	31	29	14	10

other species. (We assume that the microorganism(s) infecting one species can infect and grow in the others as well). As can be seen in TABLE 1, the differences in the hydrogenase activity of every cereal species remained similar in seedlings germinated separately or after being mixed.

In another set of experiments, we measured the activities of organs obtained from sanitized seeds and compared them to those obtained with untreated seeds. TABLE 2 presents the data corresponding to roots. Although the number of bacteria in disinfected and untreated samples differed by more than three orders of magnitude, no significant differences in hydrogenase activity were observed. We assumed that the medium we used for anaerobic enrichment of contaminant bacteria did not make a negative selection for hydrogenase-containing microorganisms. Similar results were obtained in the case of hypocotyls.

Taken together, these data indicate that the evolution of hydrogen is accomplished by plant tissues, that is, that cereals possess the necessary genetic information to express hydrogenase as a response to anaerobic conditions. Barley seems to be the cereal most interesting to study further because of its much higher hydrogen evolution (see TABLE 1). We are presently carrying out experiments with axenic barley tissue culture. Probably, these new data will provide more conclusive proof supporting our view that the hydrogen metabolism is due to plant tissues and not to contaminant microorganisms.

TABLE 2. Hydrogenase Activity and Number of Bacteria in Sanitized (A) and Unsanitized (B) Roots

	Hydrogenase Activity (μl H_2/min/g dry weight)		No. of Bacteria/g Fresh Root	
Species	A	B	A	B
T. aestivum	15	13	8×10^4	$\gg 3 \times 10^8$
Z. mays	17	10	8×10^5	$\gg 3 \times 10^8$
T. durum	23	21	3×10^5	$\gg 3 \times 10^8$

ACKNOWLEDGMENTS

We thank J. M. Malpica for very useful discussions, and L. González Tejuca for critical reading of the manuscript. This work has been sponsored by the Program of Cultural Cooperation between the United States and Spain, and by the Spanish Comisión Asesora de Investigación Científica y Técnica.

REFERENCES

1. BOICHENKO, E. A. 1947. Hydrogenase of isolated chloroplasts. Biokhimiya **12:** 153–162.
2. RENWICK, G. M., C. GIUMARRO & S. M. SIEGEL. 1964. Hydrogen metabolism in higher plants. Plant Physiol. **39:** 303–306.
3. TOYODA, K. 1967. Hydrogen evolution in seeds of *Nelumbo nucifera* and other angiosperms. Bot. Mag. (Tokyo) **80:** 118–122.
4. KURATA, Y. 1962. On the appearance of hydrogenase, nitrate reductase and aspartase during the ontogeny of the frog. Exp. Cell Res. **28:** 424–429.
5. ADAMS, M. W. W., L. E. MORTENSON & J. S. CHEN. 1981. Hydrogenase. Biochim. Biophys. Acta **594:** 105–176.

A Novel Method of Affinity Isolation of Acetylcholine Receptor with Nylon Affinity Tubes

P. V. SUNDARAM[a]

Abteilung Klinische Chemie
Poliklinik der Universität Göttingen
D-34 Göttingen, Federal Republic of Germany

Ach receptor had been isolated earlier using agarose as an affinity support[1] in conjunction with a second column of hydroxylapatite. However, the use of the receptor as a ligand for the purification of receptor-binding proteins has not been very successful owing to a large loss of specificity and binding capacity after linkage of the receptor to the affinity matrix. Thus, though adsorbents carrying toxin as a ligand have been successfully used, receptor-linked affinity adsorbents tend to inactivate rapidly. Thus, there exists a need for efficient methods of isolation of compounds that interact specifically with the Ach receptor, such as receptor-specific antibodies that may be used as markers in the early detection of myasthenia gravis. Other unknown compounds of importance may also exist. This prompted us to look for an affinity support that has superior chemical and physical properties in terms of repeated usability and ease of application with biological fluids.

Nylon tubes (i.d. 1 mm) were activated on their inner surface to serve as an affinity matrix for use in the isolation of the electric eel acetylcholine receptor (Ach receptor) with the aid of biospecific ligands such as α-cobratoxin. The isolated receptor was in turn used as a ligand to trap specific counter-molecules such as the cobratoxin. Other proteins such as BSA were used to standardize the coupling capacities of the affinity matrix and also for immunosorption studies for trapping specific immunoglobulins. The isolated pure Ach receptor when used as a ligand retained its specificity and activity during prolonged and repeated use.

METHODS

In all the procedures, nylon tube was first partially hydrolyzed with HCl according to the method of Sundaram and Hornby[2] and the ligands were either crosslinked to the released NH_2 groups with glutaraldehyde or attached to the COOH groups on the nylon activated by soluble carbodiimide (EDAC). The toxin concentration attached to the tube was measured with pulse-labeled [^{125}I]-toxin, BSA (used as a control) by difference in UV absorption and the Ach receptor by the DEAE filter disc assay that uses ^{125}I-labeled toxin.

RESULTS AND DISCUSSION

Solubilized Ach receptor is least affected by crosslinking with EDAC, whereas glutaraldehyde and bisimidates impaired the native properties of the receptor. Thus

[a]Present address: 160 Eldams Road, Madras-600018, India.

EDAC and glutaraldehyde were chosen for coupling the ligands to the hydrolyzed nylon.

BSA that was used as a standard to compare coupling efficiencies by the two methods yielded a tube that had 420 pmol/cm (set at 100%) whereas cobratoxin coupled 41 pmol/cm (10%) and the receptor 137 pmol/cm (33%) when the EDAC method was used.

The glutaraldehyde method coupled 121 pmol/cm of BSA (set at 100%) and 15 pmol/cm of the toxin (12%). When the toxin was diluted 25-fold with BSA, the coupling yield dropped to an expected 0.6 pmol/cm or 0.5%. This shows that the coupling procedure run under controlled conditions can yield affinity tubes with known amounts of the ligands.

TABLE 1 shows the results of the performance of "affinity tubes" bearing α-cobratoxin and Ach receptor as ligands in affinity trapping and reversible binding of the corresponding counter-molecules, namely the receptor and the toxin. A noteworthy feature of the results is the fact that the higher the concentration of the receptor in the affinity tube, the less efficient it is in trapping or reversible binding of the toxin. The reason for this, as also seen in the case of the older method where agarose was used as an affinity adsorbent, is that increased interaction between the receptor and the toxin leads to a very tight binding, making elution very difficult. Thus the efficiency of the affinity tubes in the present method depended only on the density of affinity ligands rather than the total concentration of ligands linked to the support.

TABLE 1. Affinity Tube Performance Data[a]

		Capacities of:			
		Trapping		Reversible Binding	
Coupling Agent	Bound Ligand	pmol/cm	% Conc. of Linked Ligand	pmol/cm	% of Trapped Ligand
EDAC	Cobratoxin	36	26	32	89
Glutaraldehyde	Ach receptor	0.42	70	0.35	83
Glutaraldehyde	Ach receptor	3.8	26	0.4	11
EDAC	Ach receptor	7	17	0.3	4

[a]Trapping capacity of Ach receptor-linked tubes was determined with ^{125}I-labeled α-cobratoxin. Reversible binding refers to the amount of toxin that is detached by perfusion of tubes with buffer containing 0.1 *M* hexamethonium.

In a typical experiment, a receptor-bearing affinity tube when perfused with a solution containing 320 pmol of [^{3}H]-pyridoxamine-labeled cobratoxin (sp. radioactivity 4 Ci/mmol) trapped 22.3 pmol or 21% of the molar concentration of the nylon-linked ligand. The trapped toxin was eluted with a buffer containing Tween 80 (0.05%) and 0.1 *M* hexamethonium, the competing molecule against the toxin showed half-lives of elution of τ = 1.7 h whereas spontaneous elution in buffer with detergent alone had half-lives of τ = 110 h. Elution of the toxin with hexamethonium in the former experiment showed a linear relationship conforming to a first-order rate law when a semi-log plot of the maximum elutable counts minus the counts eluted at any time *t* was plotted as a function of time *t*.

CONCLUSIONS

The predictable behavior of the affinity tubes in this novel approach to affinity chromatography indicate that these preparations may be used in other forms of new applications since it has been shown that (1) high densities of ligands can be coupled, (2) the biospecific properties of the affinity ligands are preserved, and (3) the affinity tubes may be re-used and nylon appears to be of particular advantage when solubilized membrane proteins are used as affinity ligands.

REFERENCES

1. MAELICKE, A., B. W. FULPIUS, R. P. KLETT & E. REICH. 1977. Acetylcholine receptor-responses to drug binding. J. Biol. Chem. **252:** 4811–4830.
2. SUNDARAM, P. V. & W. E. HORNBY. 1970. Preparation and properties of urease chemically attached to nylon tube. FEBS Lett. **10**(5): 325–327.
3. YANG, B.-H., P. V. SUNDARAM & A. MAELICKE. 1981. Affinity chromatography and immunosorption with acetylcholine receptor attached to nylon tubes. Biochem. J. **199**(1): 317–322.

Synthetic Polymers with Enzyme-like Activities

IRVING M. KLOTZ

Department of Chemistry
Northwestern University
Evanston, Illinois 60201

For much of this century, biochemists have used chemical models as a basis for elucidating the behavior of naturally occurring macromolecules. In the last few decades, structural biochemistry has made such remarkable strides in interpreting function that it should now be possible to inaugurate a complementary program: biological molecules ought to serve as models for the design of new chemical entities. For example, chemists are intensively engaged in searches for new catalysts. The preeminent catalysts are enzymes. The molecular structure and behavior of many enzymes is known in great detail. Should it not be feasible to use this information to construct catalysts from nonliving sources?

STRATEGY

At the inception of such a program, one can plan some beginning steps on the basis of rudimentary features of enzyme kinetics. First we recognize that all enzymes are macromolecules; hence we shall assume that a polymer framework should provide a promising foundation for a synthetic catalytic entity. Secondly, we turn attention to a very elementary description of the activated-state theory of reaction rates. For a single reactant, S, the rate of transformation depends on the concentration of the activated, transition-state species, $S^{\ddagger}$, in the equilibrium

$$S = S^{\ddagger} \tag{1}$$

As indicated in FIGURE 1, the relative concentration of $S\ddagger$ is small compared to S. What can an added macromolecule, M_1 achieve? Nothing, unless it can first bind S. But even if a complex, $M \cdot S$, is formed, there will be no increase[1] in rate of S if the equilibrium constant for

$$M \cdot S = \langle M \cdot S \rangle^{\ddagger} \tag{2}$$

is the same (FIG. 1, column a) as that for nonbound substrate, (Eq. 1). In other words, if the activation free energy for the reactant is unchanged in the complex $M \cdot S$, the rate of reaction will not be modified. However, the environment in the polymer matrix, M, can be modified by chemical manipulations. If changes in macromolecular character favor $\langle M \cdot S \rangle^{\ddagger}$ (FIG. 1, column b), then the rate of reaction will be increased. We shall give an illustration of such a situation later in this paper.

When two reactants are involved in a transformation, the potential of a macromolecule is broader. For example (FIG. 2), if the second reactant, N, a potential catalytic entity, is covalently attached to the polymer to create M—N, then the concentration of the transition-state species $\langle (M—N) \cdot S \rangle^{\ddagger}$ can be raised by increasing the equilibrium constant for

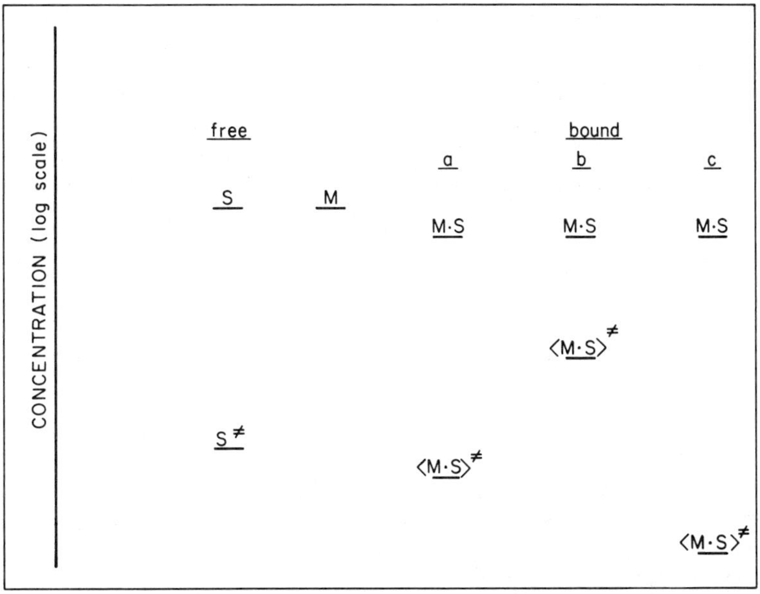

FIGURE 1. Concentration profile[1] presenting concentrations of a single reactant substrate, S, in ground state and in transition state, when free in solution or when partially bound to a macromolecule.

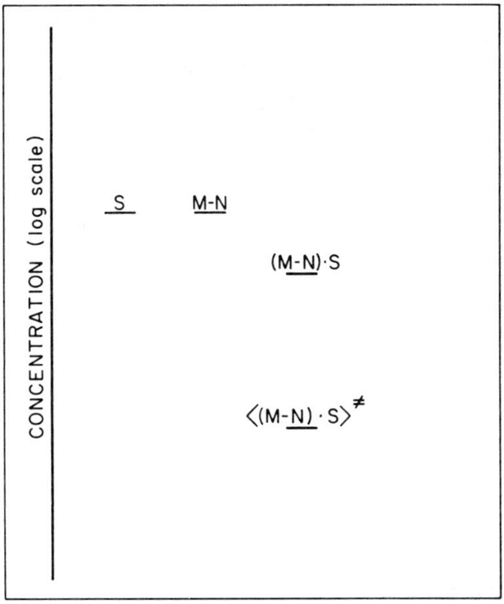

FIGURE 2. Concentration profile presenting concentrations of reacting species under various circumstances. The primary reactant, represented by S, is partially bound to M; the second participating reactant N is presumed to be covalently attached to the macromolecule M.

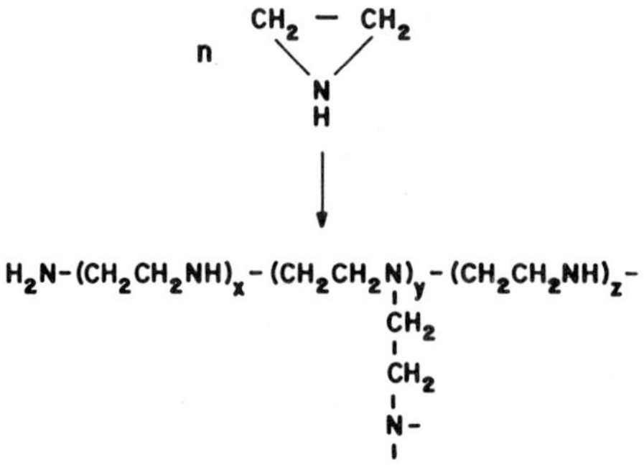

FIGURE 3. A segment of polyethylenimine.

$$M—N \cdot S = \langle (M—N) \cdot S \rangle^{\ddagger} \quad (3)$$

or for

$$S + M—N = (M—N) \cdot S \quad (4)$$

In other words, one can speed up the reaction either by lowering the free energy of activation or by increasing the affinity of M—N for the substrate S. Some illustrations of this strategy will also be presented.

With any strategy for producing a synthetic polymer catalyst, one is faced with the necessity of obtaining a macromolecule with good binding properties. Once again we can use a biological phenomenon as a model. The serum albumins have long been recognized as proteins with ubiquitous affinities for small molecules of widely different structure, especially anions.[2] Extensive studies have revealed that both apolar and electrostatic interactions are involved in binding by albumin. Thus one might expect flexible, water-soluble, synthetic polymers with suitable hydrophobic side chains to show strong affinities for small molecules. Among the many we have examined, derivatives of poly(ethylenimine) have proved the most versatile.

This highly branched polymer, which can be prepared by polymerization of ethylenimine (FIG. 3), has many amine nitrogens that provide loci for covalent attachment of a wide variety of side chains, as well as cationic sites for electrostatic interactions. We have prepared a number of derivatives of polyethylenimine that show remarkable binding properties. These have served as the basis for creating different environments to favor the activated transition state so that a variety of reactions might be speeded up.

DECARBOXYLATION REACTIONS

A particular interesting model reaction of this type is the decarboxylation of benzisoxazolecarboxylic acids, which has been thoroughly examined by Kemp and Paul.[3,4] With the nitro derivative, the reaction

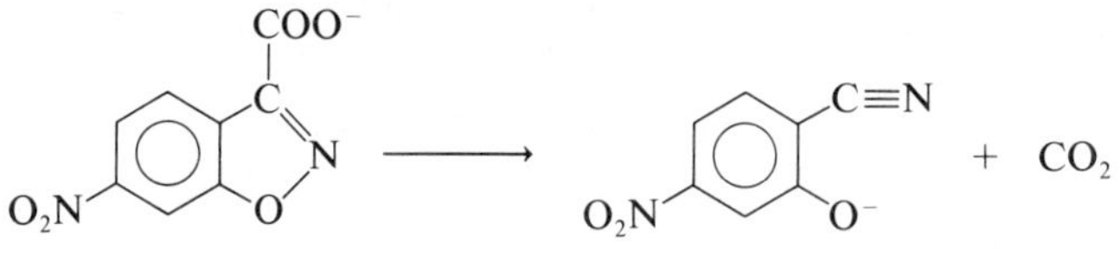

(5)

(I) (II)

can be followed readily spectroscopically by the appearance of the 2-cyano-5-nitrophenol. The rates of decarboxylation are markedly increased as one changes from water to apolar, aprotic solvents[3,4] (see TABLE 1).

As we have mentioned, modified polyethylenimines have very strong affinities for organic anions. Since these polymers with hydrophobic side chains attached can also provide apolar domains, they should be effective catalysts of decarboxylation of benzisoxazolecarboxylates dissolved in a fully aqueous solvent. Indeed, we have found[5] that in the presence of polyethylenimine with a large content of dodecyl and ethyl side chains, the turnover (first-order) rate constant in the polymer environment is about 1300 times faster than the spontaneous reaction in water (TABLE 2).

Such an accentuation in rate is comparable to that observed (TABLE 1) in an apolar solvent such as benzene. Thus it is apparent that the clustering of side chains in the polymer[6] creates microdomains with local characteristics similar to those of an apolar, aprotic solvent.

It becomes a challenge, therefore, to see if we can attain the even more striking accentuations discovered by Kemp and Paul in dipolar, aprotic solvents (see TABLE 1). In this direction, we have prepared modified polyethylenimines with

$$CH_3\overset{\overset{\displaystyle O}{\|}}{S}CH_2CO\text{—} \quad \text{and} \quad CH_3\underset{\underset{\displaystyle O}{\|}}{\overset{\overset{\displaystyle O}{\|}}{S}}\text{—}$$

groups, to compare with the amide, sulfoxide and sulfone solvents. Rates of the decarboxylation reaction, Equation (5), in the presence of some of these polymers are summarized in TABLE 2.

It is evident that increases of about 10^3-fold have again been obtained. Thus these polymers continue to provide an environment that favors formation of the activated, transition-state structure for this reaction, shown as (III).

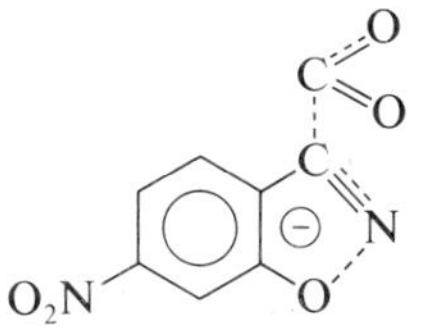

(III)

Nevertheless, the increments of 10^6–10^7-fold seen with dipolar, aprotic solvents (TABLE 1) have not been approached. Evidently the amide and oxysulfur groups attached to the polyethylenimine framework do not associate into a cluster to create a dipolar aprotic domain. This deficiency must be a reflection of the structure of the polymer. At the conclusion of this paper, we shall probe into this problem and consider ways to circumvent the constraints.

OXIDATION-REDUCTION

The oxidation of dihydronicotinamide dinucleotide (NADH) by flavoenzymes is an attractive example for our purposes, because it is a reaction that occurs in living cells. This transformation is a central step in the catalytic cycle of many reductase and

TABLE 1. Rates of Decarboxylation of 6-Nitrobenzisooxazole-3-carboxylate in Different Types of Solvent[a]

Solvent	k (sec^{-1})
H_2O	7.4×10^{-6}
Apolar, aprotic	
$CHCl_3$	8.0×10^{-4}
CCl_4	1.5×10^{-3}
C_6H_6	4.8×10^{-3}
$CH_2CH_2O—CH_2CH_2O$ (cyclic)	4.0×10^{-2}
Polar, protic, and aprotic	
$HCONH_2$	7.4×10^{-4}
$HCONHCH_3$	8.1×10^{-3}
$HCON(CH_3)_2$	$3.7 \times 10^{+1}$
$CH_3CON(CH_3)_2$	$1.6 \times 10^{+2}$
$(CH_3)_2SO$	$1.0 \times 10^{+1}$
$CH_2CH_2(SO_2)CH_2CH_2$ (cyclic)	$6.4 \times 10^{+1}$

[a]Data from Kemp and Paul.[3,4]

oxygenase enzymes. For all of these enzymes, the reduced form of the flavin moiety is generated by reaction with NADH. The reduced flavoprotein can then interact directly with potential substrates, even dissolved oxygen,[7] and the oxidized flavin is regenerated. Thus the flavin functions to mediate the transfer of reducing equivalents from dihydronicotinamides to substrates. Under aerobic conditions, the reaction scheme[8–10] can be represented by Equation 6.

$$\begin{array}{c} Fl_{ox} + HNic_{red} \underset{k_{-1}}{\overset{k_1}{\rightleftharpoons}} (Fl_{ox} \cdot HNic_{red}) \xrightarrow[H^+]{k_2} H_2Fl_{red} + Nic^+_{ox} \\ \qquad\qquad\qquad\qquad\qquad\qquad\qquad\qquad \downarrow O_2 \, k_3 \\ \qquad\qquad\qquad\qquad\qquad\qquad\qquad\qquad Fl_{ox} + H_2O_2 \end{array} \tag{6}$$

We have found[11] α-bromo-tetraacetylriboflavin, (IV),

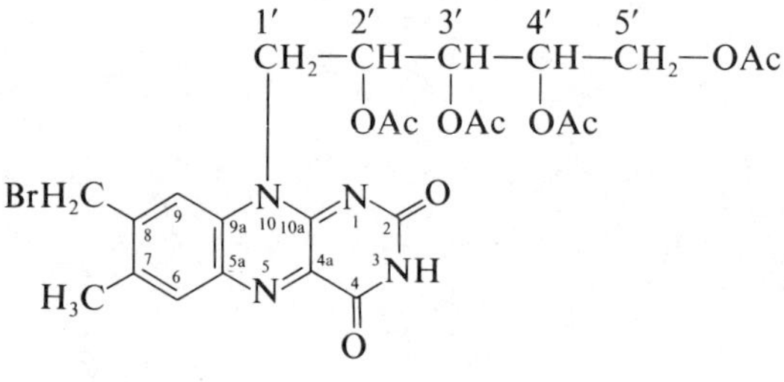

(IV)

a convenient reagent with which to prepare riboflavin derivatives of polyethylenimine. For the polymer, we have used[11] a laurylated, partially quaternized derivative to which IV was attached, and the acetyl groups removed from the ribityl side chain.

TABLE 2. Comparison of Catalytic Effectiveness of Modified Polyethylenimines in the Decarboxylation of 6-Nitrobenzisoxazole-3-carboxylate

Polymer	Rate Constant[a] in Polymer Environment × 10^3 (sec^{-1})	Second-order Rate Constant[a] × 10^{-3} ($sec^{-1}\ M^{-1}$)
None (Buffer, pH 7.4, 25°)	0.003	—
$(C_2H_4N)_m(C_{12}H_{25})_{0.25m}$	0.65[b]	0.53[b]
$(C_2H_4N)_m(C_{12}H_{25})_{0.25m}(CH_3)_{1.75m}$	1.3[b]	1.28[b]
$(C_2H_4N)_m(C_{12}H_{25})_{0.25m}(C_2H_5)_{1.75m}$	3.9[b]	5.0[b]
$(C_2H_4N)_m(C_{12}H_{25})_{0.21m}(CH_3\overset{O}{\overset{\Vert}{C}})_{0.54m}(CH_3)_{0.29m}$	2.9	1.8
$(C_2H_4N)_m(C_{12}H_{25})_{0.21m}(CH_3\overset{O}{\overset{\Vert}{C}})_{0.56m}(C_2H_5)_{0.24m}$	5.1	5.4
$(C_2H_4N)_m(C_{12}H_{25})_{0.21m}(CH_3\overset{O}{\overset{\Vert}{C}})_{0.56}(C_6H_5{-}CH_2)_{0.18m}$	12.6	7.9
$(C_2H_4N)_m(C_{12}H_{25})_{0.12m}(CH_3\overset{O}{\overset{\Vert}{C}})_{0.56m}(CH_3)_{0.48m}$	2.0	0.39
$(C_2H_4N)_m(C_{12}H_{25})_{0.12m}(CH_3\overset{O}{\overset{\Vert}{C}})_{0.56m}(C_2H_5)_{0.17m}$	4.4	1.12
$(C_2H_4N)_m(C_{12}H_{25})_{0.21m}(CH_3\underset{O}{\underset{\Vert}{\overset{O}{\overset{\Vert}{S}}}})_{0.22m}$	1.11	0.22
$(C_2H_4N)_m(C_{12}H_{25})_{0.21m}(CH_3\overset{O}{\overset{\Vert}{S}}CH_2CO)_{0.23m}$	1.67	0.92

[a]D. Mirejovsky and I. M. Klotz, unpublished experiments.
[b]From Suh *et al.*[5]

When solutions of NADH and flavopolymer are mixed in the presence of air, Equation 6 can be followed from spectroscopic changes in the region of 340 nm (FIG. 4), the position of the characteristic absorption peak of NADH ($HNic_{red}$) in solution. Thus the disappearance of this reactant can be followed quantitatively with time. It should also be noted that the absorbance of the peak at 455 nm remains constant. Since this band arises from the flavin, its constancy indicates that there is no change in concentration of oxidized flavopolymer during the aerobic oxidation of NADH. Clearly under aerobic conditions, the reaction of reduced flavin with dissolved oxygen k_3 in Equation 6 is very rapid in comparison with the oxidation of NADH by the flavopolymer. The flavopolymer is continuously recycled and acts as a bona fide catalyst. Fiftyfold excess of NADH over total flavin concentration has been found to react to completion in aerobic solution.

To make quantitative comparisons, we have analyzed the kinetic behavior in terms of the mechanistic scheme of Equation 6. From this, one can obtain a first-order rate constant that reveals the rate when the NADH substrate is totally imbedded in the flavopolymer environment. One can also compute a second-order rate constant which, in essence, is a measure of the transformation efficiency of the collision process between reactants under different circumstances. Values of these parameters are listed in TABLE 3.

The rate of oxidation of NADH by flavopolymer is substantially larger than that of the small-molecule riboflavin model system. From TABLE 3, we see that the second-order constant for flavopolymer is over 100 times greater than that for free riboflavin.

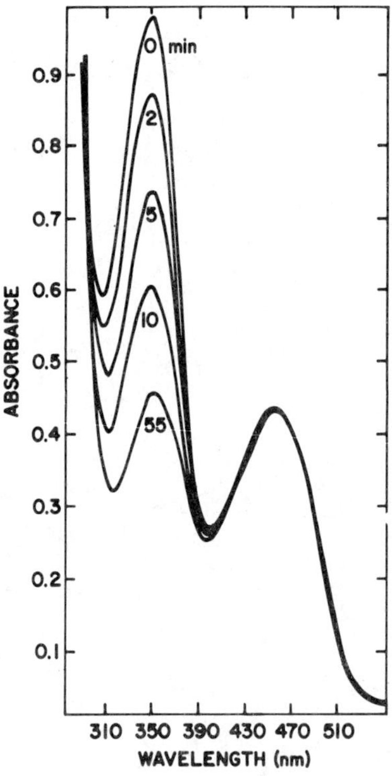

FIGURE 4. Visible spectrum of a mixture of (7×10^{-5} *M*) NADH with flavopolymer (4×10^{-5} *M* in flavin) as a function of time, in the presence of air.

TABLE 3. Comparison of Catalytic Effectiveness in the Oxidation of Dihydronicotinamide Adenine Dinucleotide (NADH)

Catalyst	Rate Constant in Polymer Environment (min^{-1})	Second-order Rate Constant $min^{-1}M^{-1}$
None; $(O_2) = 2 \times 10^{-4}$ *M*	1.9×10^{-4}	
Flavopolymer $[(C_2H_4N)_m(C_{12}H_{25})_{0.25m}$-$(CH_3)_m(C_{17}H_{19}N_4O_6)_{0.05m}]$	5.9	5.4×10^3
Riboflavin (Suelter and Metzler[8])		4.5×10^1
Cytochrome b_5 reductase (Strittmatter[12])	1.4	
Triphosphopyridinenucleotide-cytochrome c reductase (Masters *et al.*[13])	1–2	
Melilotate hydroxylase (Strickland and Massey[14])	1.4	
Imidazole acetate monooxygenase (Maki *et al.*[15])	2	

The first-order constant for flavopolymer can also be compared with the pseudo-first-order constant for the spontaneous reaction between NADH and dissolved oxygen. In this case, the enhancement by flavopolymer is some 10^4-fold.

TABLE 3 also lists the reported diaphorase activities of several flavoenzymes.[12-16] Each of these catalyzes the oxidation of NADH in the presence of O_2 without my additional substrate. In the absence of their natural substrates, the enzymatic activity is sluggish. The rate constants for these nonspecific enzymatic flavin-catalyzed oxidations approximate that of flavopolymer catalysis. It should be noted, however, that under natural circumstances, the flavoenzymes are much more effective catalysts; for example, second-order rate constants are in the range[17-19] of 10^3–10^8 $M^{-1}s^{-1}$.

Other artificial systems containing flavins have also been described in the literature. These include micelles,[20] polysoaps,[21] and other synthetic polymers.[22] An interesting hybrid creation is the flavin derivative of the hydrolytic enzyme papain.[23] All of these preparations show modest accelerations in rate of oxidation of reduced nicotinamide derivations, somewhat smaller than that seen with the flavopolyethylenimine described here.

In all of these systems, it is evident that binding of NADH to the flavin moiety is crucial to effective catalysis. Blankenhorn[10] has discussed the role of substrate binding to flavin catalysis. Strength of binding reflects the stability of the charge-transfer complex between NADH and flavin. Typical dissociation constants in the small-molecule system for this charge-transfer complex are of the order of 0.1 *M*. Binding by our flavopolymer is two orders of magnitude greater, the dissociation constant being near 10^{-3} *M*. The cationic polymeric matrix may serve to concentrate the anionic NADH substrate in the vicinity of the flavin and also to stabilize the charge-transfer complex.

The rate of NADH oxidation by flavin is also directly related to the potential difference between the flavin and the dihydronicotinamide.[10] The flavopolymer may function to enlarge this potential difference in two ways. The flavin potential could be increased by the highly cationic matrix of the polymer, these cationic loci providing an environment very favorable for electron donation. Also, as has been pointed out by Kassner for heme derivatives,[24] a hydrophobic environment dramatically increases reduction potential. The hydrophobic domains of the flavopolymer may have a similar influence on the reduction potential of attached flavin.

It is thus apparent that a suitably modified polyethylenimine derivative can supply a favorable environment for accelerating oxidation-reduction reactions, as well as the decarboxylations described above.

HYDROLYSIS

From the viewpoint of potential practical application, these reactions are the most enticing, and we have focused most attention on them. Our major effort has been directed toward moieties that can function as nucleophiles, but these might also act as catalysts through general acid–general base mechanisms. Some attention has also been devoted to specific hydroxyl ion catalysis.

(V)

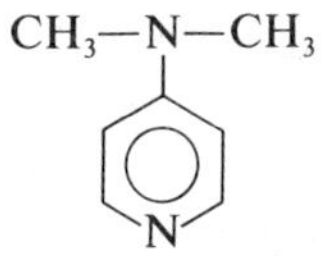

(VI)

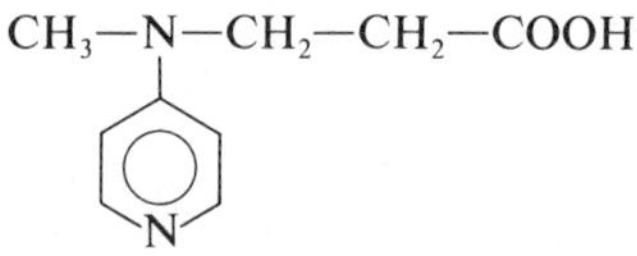

(VII)

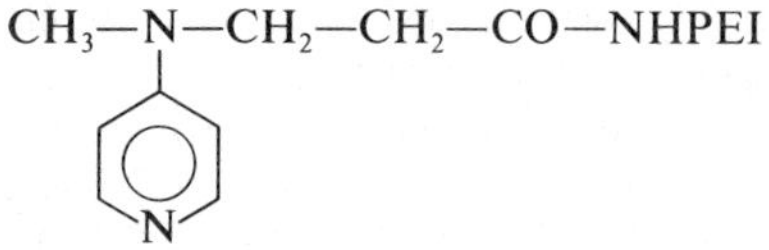

(VIII)

Among the many nucleophiles (imidazole, mercaptan, hydroxamate, etc.) that we have attached to polyethylenimines, by far the most promising are the aminopyridines. In bulk solution, pyridine (V) itself has been shown to catalyze the hydrolysis of esters.[25] Nevertheless, we have detected no catalysis of cleavage of activated esters attached to polyethylenimines. Evidently, in the polymer environment, the pK_a and nucleophilicity of the pyridine nitrogen are reduced substantially. On the other hand, 4-N,N-dialkylaminopyridines (VI) have been reported[26,27] to be acylation catalysts far superior to pyridine. Furthermore, acylation reactions with these dialkylaminopyridines are faster in apolar than in polar solvents.[28] Hence a compound of the type of (VI) seemed to be a promising moiety for attachment to modified polyethylenimines with apolar substituents to create a superior macromolecular nucleophilic catalyst.

In practice, this goal required the preparation of entities such as (VII) that contain one arm with a —COOH group for linkage to the polyethylenimine (PEI). In this way, materials such as (VIII) have been prepared,[29,30] with a variety of different substituents on the amino nitrogen. A selection of these and some of their properties are listed in TABLE 4.

A comparison of observed second-order rate constants for the small-molecule nucleophiles and for polymer-bound nucleophiles is given in TABLE 4. In both situations, the rate increases markedly with pH in consequence of the numerical value of the pK_a of the pyridine nitrogen (FIG. 5). Nevertheless, at all pH values, the nucleophile in the polymer environment is a more effective catalyst.

Attached to the polymer framework, the alkylaminopyridines manifest differences in behavior depending on their structure (TABLE 4). Although the added three CH_2 groups in (X) as compared to (IX) provide an additional spacer to increase the linkage length to the macromolecule, no effect on activity is evident. In contrast, change in the molecular structure of the bridge introduced by the ring compounds of (XI) and (XII) lead to a substantial change in catalytic effects. Since the nonbound nucleophiles show little difference in intrinsic nucleophilicity (TABLE 4), the distinctions that arise in the polymer environment may be a result of the rigidity of the bridge to the framework when ring structures (XI) and (XII) are used. This limitation of flexibility could facilitate the approach to a transition-state structure. Also listed in TABLE 4 is a nucleophile (XIII) with a very long tail, $CH_3(CH_2)_9$-, in place of the CH_3 group of (IX). Our initial thought was that this hydrophobic chain would draw the pyridine into the apolar cluster provided by the large number of lauryl groups originally placed in the polymer. The lack of any increased catalysis is puzzling but may be rationalized in terms of conformational factors to be described later.

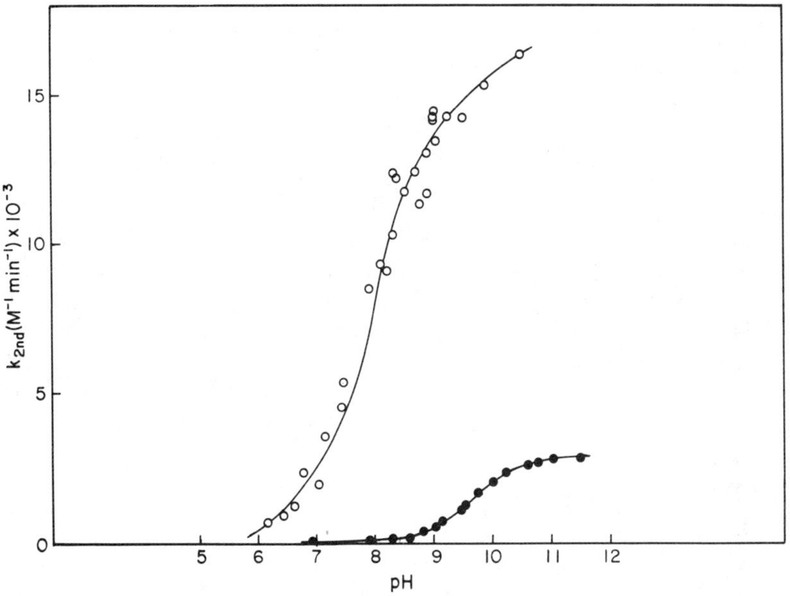

FIGURE 5. Rate-pH profiles for (●) dimethylaminopyridine (VI) in aqueous solutions, and (○) for $(C_2H_4N)_m(C_{12}H_{25})_{0.12m}(C_9H_{11}N_2O)_{0.10m}$, a 60,000 molecular-weight polyethylenimine with 10% of the nitrogens covalently attached (by an amide bond) to 3-[*N*-methyl-*N*-(4-pyridyl)amino]propionic acid and 12% linked to lauryl groups.

Another puzzling feature of the behavior of these nucleophiles appears in a comparison of preparations (XIa) and XIb). Chemically these are essentially the same but (XIb) contains substantially less nucleophile. On a normalized basis, that is, with catalytic activities expressed in terms of unit concentration of nucleophile, the polymer derivative with the lower content of aminopyridine, (XIb), is much more effective than that, (XIa), with the higher amount. A possible origin of this enigma will also become apparent in an examination of conformation models of the polymer.

The mode of catalytic action of the alkylaminopyridines is definitely nucleophilic in character. Such a mechanism can be represented by Equation 7.

TABLE 4. Properties of Polyethylenimines with Covalently Linked Aminopyridines

Polymer[a]	pK_a of Pyridine		Second-order Rate Constant Observed (M^{-1} min^{-1}), pH 9.2	
	Attached to Polymer	Free in Solution	Attached to Polymer	Free in Solution
$(C_2H_4N)_m(C_{12}H_{25})_{0.117m}^-$ (IX) 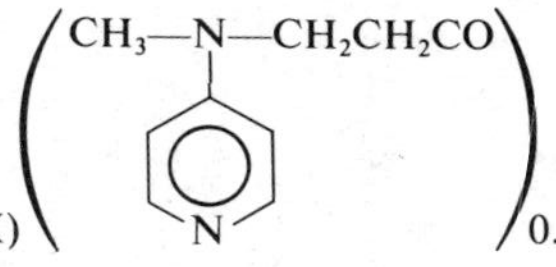 0.09m	7.31	9.78	34,000	544
$(C_2H_4N)_m(C_{12}H_{25})_{0.117m}^-$ (X) 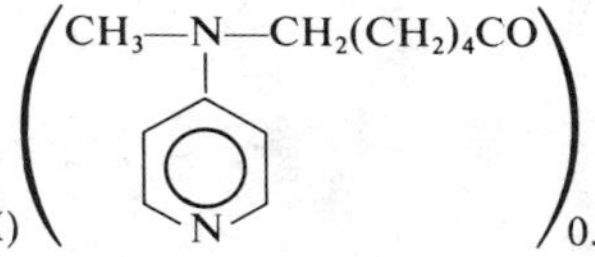 0.08m	7.54	9.96	34,000	617
$(C_2H_4N)_m(C_{12}H_{25})_{0.117m}^-$ (XIa) 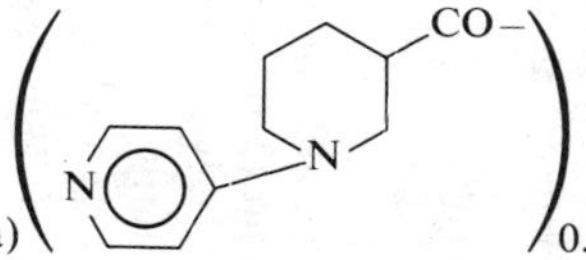 0.03m	6.64	9.59	53,000	616

$(C_2H_4N)_m(C_{12}H_{25})_{0.117m}$– (XIb) 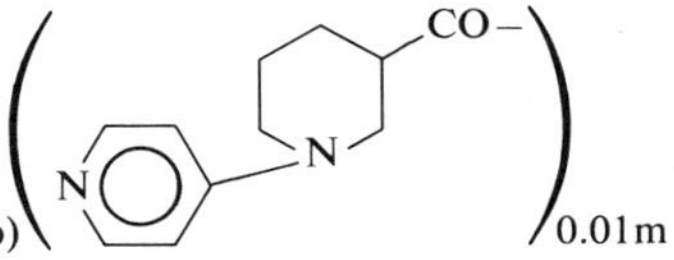	6.68	9.59	164,000	616
$(C_2H_4N)_m(C_{12}H_{25})_{0.1117m}$– (XII) 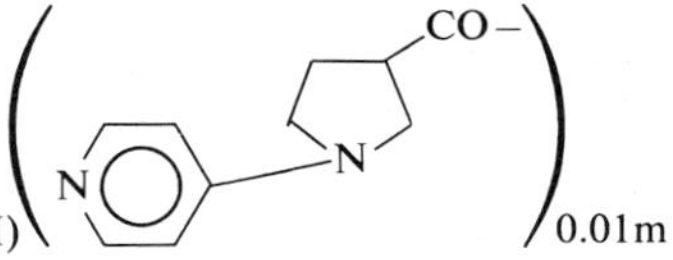	7.51	9.94	163,000	662
$(C_2H_4N)_m(C_{12}H_{25})_{0.117m}$– (XIII) 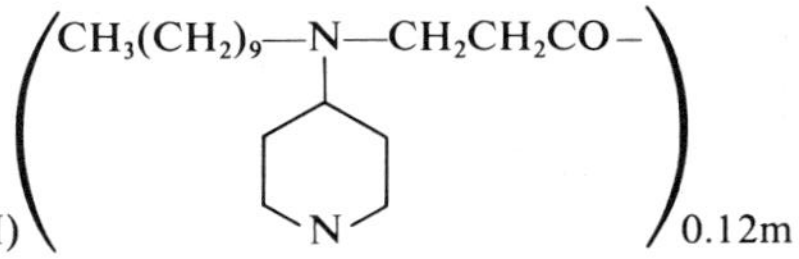	5.82	9.90	35,000	748

[a] A linear form of the polymer can also be obtained by hydrolysis of polyethyloxazolines (H. B. Collier, Dow Chemical Co.)

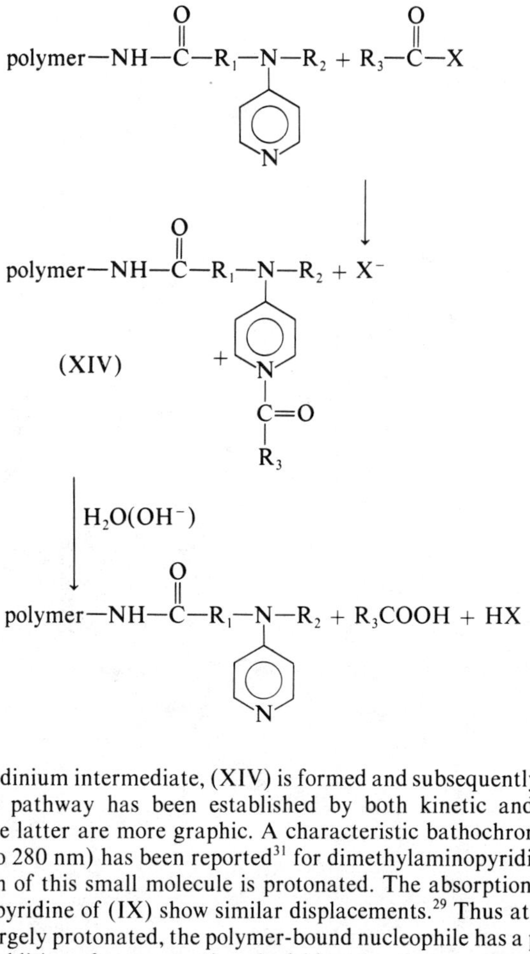

$$\text{polymer}-\text{NH}-\overset{\overset{O}{\|}}{C}-R_1-\underset{\underset{\underset{N}{\bigcirc}}{|}}{N}-R_2 + R_3-\overset{\overset{O}{\|}}{C}-X$$

$$\downarrow$$

$$\text{polymer}-\text{NH}-\overset{\overset{O}{\|}}{C}-R_1-\underset{\underset{\underset{\underset{R_3}{|}}{\underset{C=O}{|}}}{\underset{\overset{+}{N}}{\bigcirc}}}{N}-R_2 + X^-$$

(XIV)   (7)

$$\downarrow H_2O(OH^-)$$

$$\text{polymer}-\text{NH}-\overset{\overset{O}{\|}}{C}-R_1-\underset{\underset{\underset{N}{\bigcirc}}{|}}{N}-R_2 + R_3\text{COOH} + \text{HX}$$

That an acylpyridinium intermediate, (XIV) is formed and subsequently hydrolyzed in the mechanistic pathway has been established by both kinetic and spectroscopic experiments. The latter are more graphic. A characteristic bathochromic shift (from λ_{max} of 259 nm to 280 nm) has been reported[31] for dimethylaminopyridine, (VI), when the ring nitrogen of this small molecule is protonated. The absorption spectra of the polymer-linked pyridine of (IX) show similar displacements.[29] Thus at pH 7.3, where the pyridine is largely protonated, the polymer-bound nucleophile has a peak at 280 nm (FIG. 6). Upon addition of excess acetic anhydride, an entirely new peak at 315 nm is generated (FIG. 6), which disappears rapidly. Clearly the 315-nm peak reveals the accumulation and disappearance of an acylpyridinium intermediate, (XIV).

Let us turn now to a specific OH^- ion-catalyzed hydrolysis. For this purpose, we have studied the kinetics of hydrolysis of 4-nitro-3-carboxytrifluoroacetanilide, (XV),

$$CF_3-\overset{\overset{O}{\|}}{C}-NH-\underset{\underset{COOH}{}}{\bigcirc}-NO_2$$

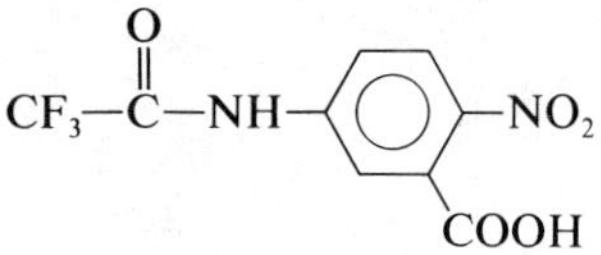

(XV)

as a model for amide hydrolysis.

The hydrolysis of trifluoroacetanilides is known[32–39] to proceed according to the scheme shown in Equation 8.

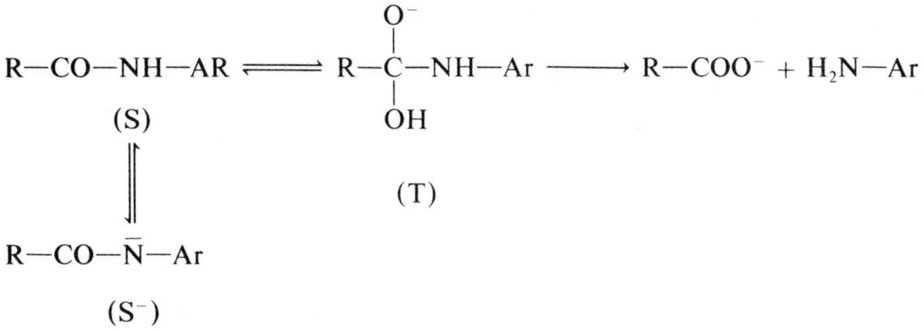

(8)

Only the hydroxide ion and the water molecule (in a general base-assisted reaction) have been found to make a nucleophilic attack on the anilide substrate (S) to give the tetrahedral intermediate (T). The subsequent breakdown of the tetrahedral intermediate is catalyzed by hydroxide ion, general acids and general bases. In the presence of polyethylenimine, additional complications are introduced[40] since the species (S), (S^-), and (T) can be changed when they are bound to the polymer. Nevertheless, the kinetics of the hydrolysis can be analyzed as a function of pH and the kinetic constants evaluated.[40]

Experimental observations of the spontaneous reaction show that a plateau is reached at high pH (above 10). This behavior reflects the increasing conversion of (S) to unreactive (S^-) as the pH is raised, which balances the rise in OH^- concentration. When pseudo-first-order rates are measured under conditions of excess polymer over substrate, one can extract values of the catalytic constant, k_{cat}, in the polymer environment. Values of k_{cat} as a function of pH are shown in FIGURE 7. At high pH, a plateau is also reached, again due to formation of unreactive S^-. At pH values near 7, the rate of hydrolysis is accelerated over a hundredfold in the presence of the polymer, as is apparent in FIGURE 7. Quantitative evaluations of such enhancements for a variety of modified polyethylenimines[40] give numbers from 6–180.

CONFORMATION OF MODIFIED POLYETHYLENIMINES

Some insights into the three-dimensional features of polyethylenimines with added hydrophobic and functional groups have been obtained with ^{19}F nuclear magnetic resonance probes[41,42] and from studies of excimer fluorescence spectra[43] of attached pyrene derivatives. The disposition of groups deduced from these investigations has also been confirmed in recent spin-label experiments.[44] All three probes show that pendant apolar groups associate into a hydrophobic cluster (FIG. 8), which then must be flanked by

$$-CH_2CH_2-\overset{|}{N}-$$

residues of the polymer framework. The polyamine chain is sufficiently long to reach around a cluster but not so long or bulky to cause excessive crowding at the surface. The number of apolar groups in a cluster is unknown but the features of importance are

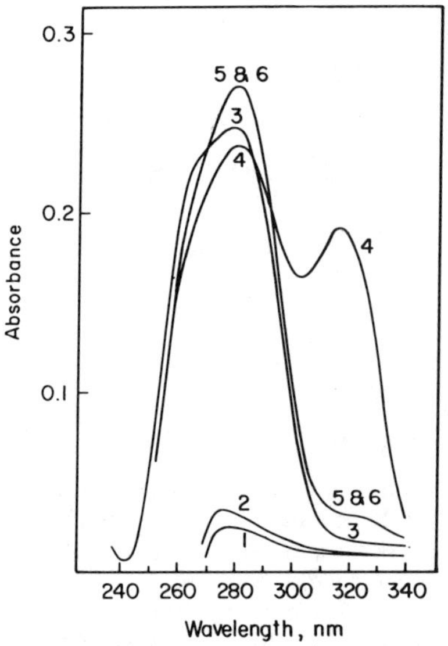

FIGURE 6. Ultraviolet spectra of modified polyethylenimine, $[(C_2H_4N)_m(C_{12}H_{25})_{0.12m}(C_9H_{11}N_2O)_{0.10m}]$, upon reaction with anhydride in 0.05 *M* Bistris buffer, pH 7.30: 1, buffer only; 2, buffer and acetic anhydride only; 3, buffer and polymer; 4, buffer, polymer, and acetic anhydride at time $t = 0$; 5, buffer, polymer, and acetic anhydride at $t = 3.5$ min after mixing; 6, buffer, polymer, and acetic anhydride at time $t = 7$ min after mixing. Concentration of $C_9H_{11}N_2O$ pyridine moiety, 4.0×10^{-5} residue molar, of acetic anhydride 7.5×10^{-4} *M*.

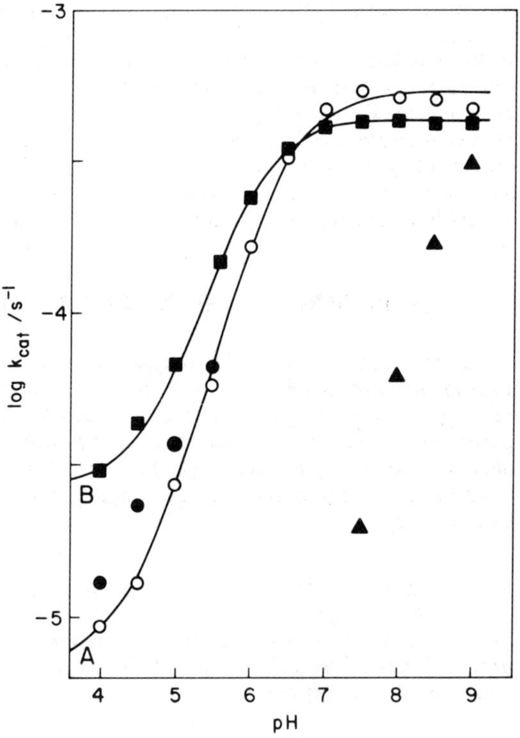

FIGURE 7. The pH dependence of k_{cat} for the hydrolysis of anilide (XV) catalyzed by lauryl PEI, $[(C_2H_4N)_m(C_{12}H_{25})_{0.12m}]$; curve A is for chloride as anion, curve B, acetate. Spontaneous rates (in 0.02 *M* buffer) are also shown as ▲; above pH 10 these approach a plateau.

not qualitatively changed by variations in aggregate size. In the limit of very large size, the clusters would have to assume a nonspherical shape. Coincident with this, the area per apolar group at the surface of the cluster would be somewhat reduced eventually to such an extent that the polyamine framework would become crowded. However, before such a situation could be attained, electrostatic and solvation effects would presumably stop the further growth of any one individual cluster. Apolar groups at a distance from the initial aggregate but on the same polyamine matrix could associate into an additional cluster so that several such domains may be connected in any single macromolecule.

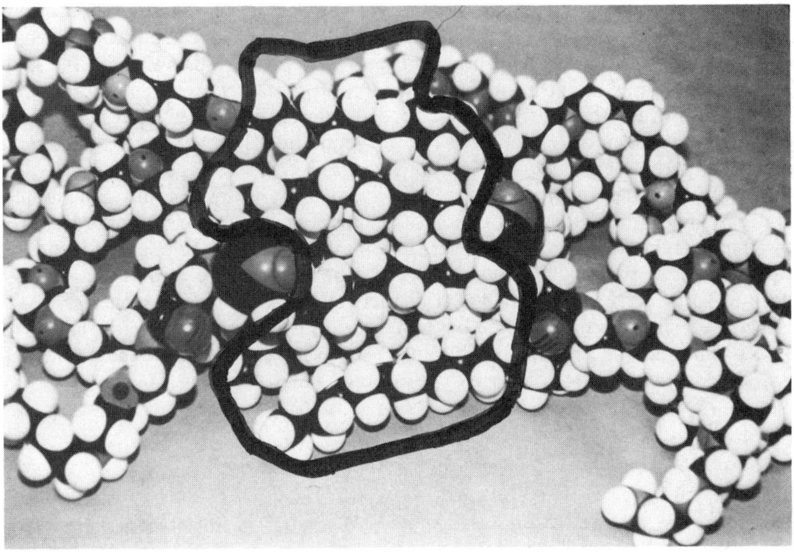

FIGURE 8. Photograph of space-filling molecular model of a segment of modified polyethylenimine constituted of 100 ethylenimine residues, 10 lauryl groups, and 2 aminopyridines. In the central area, delineated by a black thick border, is the hydrophobic cluster of apolar lauryl residues showing only black and white balls (carbon and hydrogen atoms). Visible as indentations in the border of this cluster on the left and right are two aminopyridine moieties. Flanking the apolar domain are the ethylenimine residues of the polymer framework which can be distinguished by the frequent appearance of gray balls (nitrogen atoms) amidst the blacks and whites (carbon and hydrogen atoms of —CH_2— groups).

Added moieties, such as the

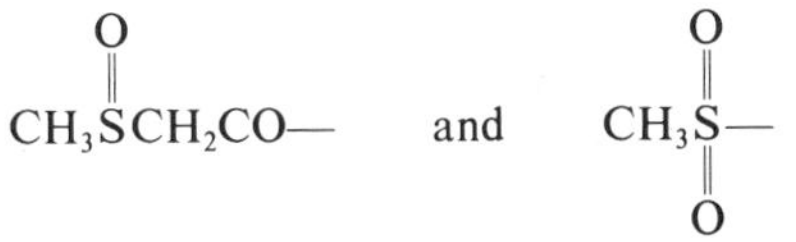

groups introduced to accentuate decarboxylation reactions, are attached covalently to amine nitrogens and hence must be situated at the periphery of the hydrophobic region (FIG. 8). Thus it is evident that they are not able to self-associate into a dipolar aprotic domain, and hence we can see why this modification has only minor effects on the rate of the reaction of Equation 5. The conformational model of FIGURE 8 suggests that the amide, sulfoxide, or sulfone groups should be incorporated within the apolar domain, where the bound substrate (I) must reside. To achieve that goal, it will be necessary to develop some new chemical procedures for placing the dipolar aprotic groups within the hydrocarbon chains *per se*.

The space-filling model (FIG. 8) also provides an explanation for the decreasing effectiveness of increasing amounts of covalently attached nucleophile (compare XIa and XIb in TABLE 4). It should be recognized that the attachment of the first aminopyridine nucleophile will take place at sites that provide the strongest attractive interactions, presumably nitrogens of the polymer framework that are in the vicinity of the hydrophobic domain and are sterically most accessible. These will be a small fraction of the total potential loci. With successive further increments of nucleophile adduct, covalent bonds will be formed at sites increasingly distant from the hydrophobic cluster. Thus the latter moieties will be at progressively larger distances from the domain in which the substrate is embedded. Hence they are essentially ineffective. Again, then, to increase effectiveness of nucleophiles, they too must be embedded within the hydrophobic cluster.

It is thus apparent that a suitably modified polyethylenimine can supply a favorable environment for decarboxylation, oxidation-reduction, or hydrolysis reactions. This polymer is a remarkably versatile matrix for the attachment of a variety of different types of functional groups. However, the virtues of its structure are also the source of some constraints. Molecular models of the polymer, nevertheless, suggest strategies for circumventing some of the constraining limitations. With imagination and determination, we should be able to create even more effective macromolecular catalysts, some of which may prove to be of practical utility.

REFERENCES

1. KLOTZ, I. M. 1976. Free energy diagrams and concentration profiles for enzyme-catalyzed reactions, J. Chem. Ed. **53:** 159–160.
2. KLOTZ, I. M. 1949. The nature of some ion-protein complexes. Cold Spring Harbor Symp. Quant. Biol. **14:** 97–112.
3. KEMP, D. S. & K. PAUL. 1970. Decarboxylation of benzisoxazole-3-carboxylic acids. J. Am. Chem. Soc. **92:** 2553–2554.
4. KEMP, D. S. & K. PAUL. 1975. The physical organic chemistry of benzisoxazoles. J. Am. Chem. Soc. **97:** 7305–7312.
5. SUH, J., I. S. SCARPA & I. M. KLOTZ. 1976. Catalysis of decarboxylation of nitrobenzisoxazolecarboxylic acid and of cyanophenylacetic acid by modified polyethylenimines. J. Am. Chem. Soc. **98:** 7060–7064.
6. KLOTZ, I. M. 1978. Synzymes: synthetic polymers with enzymelike catalytic activities. Adv. Chem. Phys. **39:** 109–176.
7. SINGER, T. P. & E. B. KEARNEY. 1950. The nonenzymatic reduction of cytochrome c by pyridine nucleotides and its catalysis by various flavins. J. Biol. Chem. **183:** 409–429.
8. SUELTER, C. H. & D. E. METZLER. 1960. The oxidation of a reduced pyridine nucleotide analog by flavins. Biochim. Biophys. Acta **44:** 23–33.
9. PORTER, D. J. T., G. BLANKENHORN & L. L. INGRAHAM. 1973. The kinetics of lumiflavin reduction by *N*-methyl-1,4-dihydronicotinamide. Biochem. Biophys. Res. Commun. **52:** 447–452.
10. BLANKENHORN, G. 1975. Intermolecular complexes between *N*-methyl-1,4-dihydronicotinamide and flavins. Biochemistry **14:** 3172–3176.
11. SPETNAGEL, W. J. & I. M. KLOTZ 1978. Oxidation of dihydronicotinamide adenine dinucleotide by a flavin derivative of polyethylenimine. Biopolymers **17:** 1657–1668.

12. STRITTMATTER, P. 1966. NADH-cytochrome b_5 reductase. *In* Flavins and Flavoproteins. E. C. SLATER, Ed. Elsevier. New York. pp. 325–340.
13. MASTERS, B. S. S., H. KAMIN, Q. H. GIBSON & C. H. WILLIAMS JR. 1965. Studies on the mechanism of microsomal triphosphopyridine nucleotide-cytochrome c reductase. J. Biol. Chem. **240:** 921–931.
14. STRICKLAND, S. & V. MASSEY. 1973. The purification and properties of the flavoprotein melilotate hydroxylase. J. Biol. Chem. **248:** 2944–2952.
15. MAKI, Y., S. YAMAMOTO, M. NOZAKI & O. HAYAISHI. 1969. Studies on monooxygenases. J. Biol. Chem. **244:** 2942–2950.
16. KATAGIRI, M., H. MAENO, S. YAMAMOTO, O. HAYAISHI, T. KITAO & S. OAE. 1965. Salicylate hydroxylase, a monooxygenase-requiring flavin adenine dinucleotide. J. Biol. Chem. **240:** 3414–3417.
17. JABLONSKI, F. & M. DE LUCA. 1977. Purification and properties of the NADH and NADPH-specific FMN oxidoreductases from *Beneckea harveyi*. Biochemistry **16:** 2932–2936.
18. HONMA, T. & Y. OGURA. 1977. Kinetic studies of the old yellow enzyme. Biochim. Biophys. Acta **484:** 9–23.
19. SINGER, T. P. 1968: The respiratory chain-linked dehydrogenases. *In* Biological Oxidations. T. P. SINGER, Ed. Interscience. New York. pp. 339–377.
20. SHINKAI, S. & T. KUNITAKE. 1977. Coenzyme models. Bull. Chem. Soc. Jpn. **50:** 2400–2405.
21. SHINKAI, S., S. YAMADA & T. KUNITAKE. 1978. Coenzyme models. 10. Rapid oxidation of NADH by a flavin immobilized in cationic polyelectrolytes. Macromolecules **11:** 65–68.
22. KONDO, L., T. FUJITA & K. TAKEMOTO. 1973. Functional monomers and polymers. Makromol. Chem. **174:** 7–13.
23. LEVINE, H. L. & E. T. KAISER. 1978. Oxidation of dihydronicotinamides by flavopapain. J Am. Chem. Soc. **100:** 7670–7677.
24. KASSNER, R. J. 1972. Effects of nonpolar environments on the redox potentials of heme complexes. Proc. Natl. Acad. Sci. USA **69:** 2263–2267.
25. SHEINKMAN, A., S. I. SUMINOV & A. N. KOST. 1973. *N*-Acylpyridinium salts and corresponding fused benzo-derivatives. Russ. Chem. Rev. **42**(8): 642–661. (Also, references therein.)
26. STEGLICH, W. & G. HÖFLE. 1969. *N,N*-dimethyl-4-pyridinamine, a very effective acylation catalyst. Angew. Chem. Int. Ed. Engl. **8:** 981.
27. HÖFLE, G., W. STEGLICH & H. VORBRÜGGEN. 1978. 4-dialkylaminopyridines as highly active acylation catalysts. Angew. Chem. Int. Ed. Eng. **17:** 569–583.
28. HASSNER, A., L. R. KREPSKI & V. ALEXANIAN. 1978. Aminopyridines as acylation catalysts for tertiary alcohols. Tetrahedron **34:** 2069–2076.
29. HIERL, M. A., E. P. GAMSON & I. M. KLOTZ. 1979. Nucleophilic catalysis by polyethylenimines with covalently attached 4-dialkylaminopyridine. J. Am. Chem. Soc. **101:** 6020–6022.
30. DELANEY, E. J., L. E. WOOD & I. M. KLOTZ. 1982. Polyethylenimines with alternative alkylaminopyridines as nucleophilic catalysts. J. Am. Chem. Soc. **104:** 799–807.
31. PENTIMALLI, L. 1964. Ossidazione di 2-amino-pyridine con peracidi organici. Gazz. Chim. Ital. **94:** 458–466.
32. ERIKSSON, S. O. & C. HOLST. 1966. Hydrolysis of Anilides II. Hydrolysis of trifluoro- and trichloroacetanilide by hydroxyl ions and by some other catalysts. Acta Chem. Scand. **20:** 1892–1906.
33. ERIKSSON, S. O. & L. BRATT. 1967. Hydrolysis of anilides II. aminolysis and general acid-catalyzed alkaline hydrolysis of trifluoroacetanilide. Acta Chem. Scand. **21:** 1812–1822.
34. ERIKSSON, S. O. 1968. Hydrolysis of anilides IV. Hydroxylaminolysis, hydrazinolysis, and general acid-catalyzed alkaline hydrolysis of trifluoroacetanilide. Acta Chem. Scand. **22:** 892–906.
35. STAUFFER, C. E. 1972. Hydrolysis of substituted trifluoroacetanilides. Some implications for the mechanism of action of serine proteases. J. Am. Chem. Soc. **94:** 7887–7891.
36. POLLACK, R. M. & T. C. DUMSHA. 1973. Specific base-general acid catalysis of the alkaline hydrolysis of *p*-nitrotrifluoroacetanilide. J. Am. Chem. Soc. **95:** 4463–4465.

37. Pollack, R. M. & T. C. Dumsha. 1975. Imidazole-catalyzed hydrolysis of anilides. Nucleophilic catalysis or proton-transfer catalysis? J. Am. Chem. Soc. **97:** 377–380.
38. Komiyama, M. & M. L. Bender. 1978. Effect of phenyl substituents on the hydrolyses of trifluoroacetanilides catalyzed by imidazolyl cation and hydroxide ion. J. Am. Chem. Soc. **100:** 5977–5978.
39. Huffman, R. W. 1980. Amine catalysis of the hydrolysis of trifluoroacetanilide. J. Org. Chem. **45:** 2675–2680.
40. Suh, J. & I. M. Klotz. 1984. Catalysis of anilide hydrolysis by polyethylenimine derivatives. J. Am. Chem. Soc. **106:** 2373–2378.
41. Johnson, T. W. & I. M. Klotz. 1971. Fluorine nuclear magnetic resonance study of the conformation of a polyethylenimine derivative, J. Phys. Chem. **75:** 4061–4064.
42. Johnson, T. W. & I. M. Klotz. 1974. Fluorine magnetic resonance studies of conformation of poly(ethylenimine)s. Macromolecules **7:** 618–623.
43. Pranis, R. A. & I. M. Klotz. 1977. Conformational behavior of polyethylenimine derivatives revealed by fluorescence spectroscopy. Biopolymers **16:** 299–316.
44. Sisido, M., K. Akiyama, Y. Imanishi & I. M. Klotz. 1984. Dynamics and hydrophobic binding of lauryl-quaternized polyethylenimine in aqueous solution. Macromolecules **17:** 198–204.

Semisynthetic Enzymes[a]

E. T. KAISER

Laboratory of Bioorganic Chemistry and Biochemistry
The Rockefeller University
New York, New York 10021

The design of an enzyme from its constituent amino acids is a major goal of research in our laboratory. The problem of creating a new enzyme can be divided into two aspects, the designing of a structure that includes a binding site and the preparation of a suitable catalytic site. The prediction of tertiary structure from primary amino acid sequence is not yet at a stage where the design of a long polypeptide chain to fold up into an enzyme structure would be undertaken with confidence. In a more modest approach that we hope will lead eventually to the construction of tertiary structures, we have succeeded in building biologically active peptides and polypeptides in cases where, as a first approximation, tertiary structural considerations can be neglected.[1] Specifically, we have found that for peptides or proteins that bind along membrane surfaces, amphiphilic secondary structures are frequently encountered, and we have been able to model such systems successfully. A typical structure that we have encountered is the amphiphilic α-helix, functioning as an important constituent of surface-active peptides ranging from apolipoprotein A-I[2–4] through the toxin melittin[5,6] to peptide hormones such as calcitonin and β-endorphin.[8–10] Recent work in our laboratory has shown that we can readily design also amphiphilic β-strands[11] and that such peptides can act as excellent mimics of the corresponding natural peptide systems. In all of these studies in which amphiphilic secondary structures have been prepared, we have attempted to construct sequences that have minimum homology to the corresponding naturally occurring amino-acid region.[3] This can be done as long as the secondary structural region can be delineated as a separate entity from the active site. Thus, in the case of the peptide hormone calcitonin, for example, we have found that the structure can be considered to consist of three parts.[7] At the NH_2-terminus there is a disulfide loop involving seven amino acids. This loop is a sort of "active site" for the hormone. Residues 8–22 correspond to an amphiphilic α-helix and the remainder of the peptide reaching to the COOH-terminus is rather hydrophilic with a structure that we do not as yet completely understand. For calcitonin, it has proven possible to take the 8–22 region of a model peptide and to design it as an amphiphilic α-helix, minimizing the homology to the natural amino acid sequence, and to generate an extremely active model peptide with calcitonin-like activity.[7] Indeed, in our most recent studies, we have succeeded in preparing a peptide with a highly altered 8–22 region that is as active as the most active calcitonin known, namely the salmon species.[12]

In all of our work on modeling secondary structure, we have focused on regions outside the "active site." When one turns to designing an enzyme system, however, it is clear that we have a considerable way to go, despite our success in preparing secondary structural units. In particular, we have not as yet, even on an empirical basis, designed a tertiary structure by the construction of an appropriate sequence and, also, in our secondary structural modeling in those systems that have "active sites," we have held

[a]The support of National Science Foundation grants CHE-8114418 and CHE-8218637 is gratefully acknowledged.

the latter regions constant. As a result of these features of our secondary structural modeling, we have decided, additionally, to approach the question of designing enzymes from the opposite end of the spectrum, as it were. In our work on what we term the "chemical mutation" of enzyme active sites, we have taken a defined tertiary structure for which the x-ray crystallographic information and solution studies are well in hand, and we have attempted to introduce a new active site on such a structure.[13] In a sense, what we have done here is almost the converse of the secondary structural modeling, since in the semisynthetic work, we have held the tertiary structure constant and focused on changing the active site.

THE "CHEMICAL MUTATION" APPROACH TO ACTIVE-SITE DESIGN

In searching for appropriate systems where a new active site might be introduced, we first considered earlier studies. Probably, the most pertinent investigation involved the conversion of subtilisin to thiol subtilisin achieved independently by Bender[14,15] and Koshland[16,17] using chemical modification. In essence, what was found was that the conversion of the serine hydroxyl of subtilisin to the cysteine sulfhydryl of thiol subtilisin resulted in an enzyme that had considerably lower activity than the native species. The modified enzyme was capable of acting as a catalyst typically for the hydrolysis of highly reactive substrates such as *p*-nitrophenyl esters, but it was no longer a proteinase. The reasons why the thiol subtilisin was not a very effective enzyme remain somewhat obscure. Among the various possibilities that may be considered is that the mechanism of action of a proteolytic enzyme that acts in acyl transfer requires the formation and breakdown of tetrahedral intermediates and that the various steps thus involved must proceed at optimal rates for overall catalysis to occur successfully. Conceivably, the conversion of the hydroxyl group of subtilisin to the thiol form might result in a retardation of an important proton transfer step affecting the formation and/or breakdown of the tetrahedral intermediates involved in acyl transfer or, possibly, might result in a steric effect that might destabilize the formation of the tetrahedral intermediate. In any case, it was our feeling that, despite the apparent simplicity of converting one enzyme active-site nucleophile to another, the probability that a highly active species might be generated by such conversion was not great. It was our philosophy that, if we could find a system that did not require rapid proton transfers to occur for the catalytic act to proceed, we might have better success in designing a new catalytic center in a preexisting enzyme active-site milieu.[13]

STUDIES ON SEMISYNTHETIC ENZYMES

For this reason, we elected to look at the possibility that we could modify the active site of a proteolytic enzyme with a suitable coenzyme analogue. The proteolytic enzyme with which we have done the most work is papain. This has proven to be a good choice because not only are its x-ray structure and those of various of its derivatives available, but also a great deal is known about the solution chemistry of the enzyme. Intelligent guesses can be made as to the ways in which potential substrates might fit into the binding site if the active-site Cys-25 of papain were modified with an appropriate coenzyme analogue. Additionally, the groove that exists near the active-site Cys-25 residue is about 25-Å long. Model building (Lab Quip) based on the x-ray structure of the enzyme allowed us to see that if the sulfhydryl were modified with a coenzyme analogue, it would be possible to bind suitable substrates within the binding pocket. As the class of coenzyme analogues with which we began work, we chose the flavins. The reason for this choice was that we felt that it might be possible to produce a flavopapain species that could be an effective catalyst without involving the occurrence of rapid proton transfers requiring the assistance of appropriately placed protein functional groups. In other words, we felt that it was quite likely that rapid catalysis of a reaction such as the oxidation of a dihydronicotinamide might be achieved simply by holding the substrate in close proximity to the flavin in an appropriate binding-site environment. This hypothesis has been amply justified by the experiments that we have performed.

In considering what flavin molecules we might attach to the active-site sulfhydryl, it was crucial for us to concern ourselves not only with the question of whether or not the binding site would be still available for a substrate but also to establish the likelihood that the flavin would remain in the binding site rather than flipping out onto the protein surface. If the latter phenomenon occurred, this could pose a very serious problem since the objective of our work was to hold the flavin within the binding-site cavity of papain in close proximity to an appropriate substrate molecule. In examining the active-site region of papain, it became clear to us that if we linked the flavin molecule to papain's Cys-25 residue via a benzyl position as in the 8α-substituted system shown in structure 1 below, the flavin might not have a strong reason for remaining in the binding site rather than flipping out onto the surface of the protein.[13,18] Of course, we considered that the nature of the group on the N^{10}-position might be important in this regard. Our expectation that flavins corresponding to structure 1 would not remain in the binding site was borne out by the experimental results of our modification studies and the kinetics done with the modified enzymes. Thus, when we modified papain with 8α-bromotetraacetylriboflavin or with 8α-bromolumiflavin, in either case the flavopapains obtained showed no evidence for saturation kinetics using a range of dihydronicotinamides, including N^1-benzyl-1,4-dihydronicotinamide and N^1-propyl-1,4-dihydronicotinamide, at concentrations of up to approximately 10^{-3} M.[19] Furthermore, the second-order rate constants calculated for the oxidation of the dihydronicotinamides by these flavopapains showed no significant rate accelerations relative to the same reactions catalyzed by the corresponding model flavins.

In contrast, flavopapains 2[18,19] and 3[20] derived from alkylation of the active-site sulfhydryl by 7α-bromoacetyl flavin 4 and by 8α-bromoacetyl flavin 5 showed saturation kinetics at relatively low dihydronicotinamide concentrations, especially in the case of the 8α-acetyl flavin-derived species 3. For the case of flavopapain 2, the k_{cat}/K_m value measured at pH 7.5 for the oxidation of N_1-benzyl-1,4-dihydronicotinamide was about twenty times larger than the second-order rate constant for the oxidation reaction catalyzed by the corresponding model flavin.[19] Typically, the apparent K_m values for the oxidation of alkyl-1,4-dihydronicotinamides were in the vicinity of 10^{-4} M. The most rapidly oxidized dihydronicotinamide was the N^1-hexyl species that underwent reaction with flavopapain 2 with a k_{cat}/K_m value of 10,500 M^{-1} sec^{-1} at pH 7.5 and 25.0°C.

Flavopapain 3 was a considerably more effective catalyst than flavopapain 2 for the oxidation of alkyl-1,4-dihydronicotinamides. Indeed, with N^1-hexyl-1,4-dihydronicotinamide as a substrate, flavopapain, 3 exhibited a k_{cat}/K_m value of 570,000 M^{-1} sec^{-1} at pH 7.5 and 25°C.[20] This means that flavopapain 3 shows catalytic activity comparable to that displayed by all but the most efficient naturally occurring flavin-containing oxidoreductases known.

In addition to flavopapains 2 and 3, we have studied flavopapain 6 derived from the alkylation of the active-site Cys-25 residue with 6α-bromoacetyl-10-methylisoalloxa-

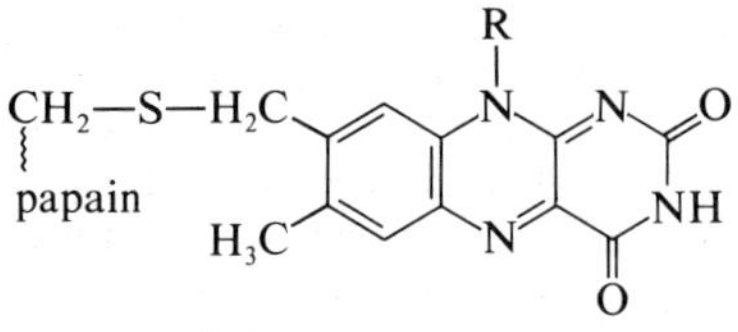

R = tetraacetylribose
or R = methyl

(1)

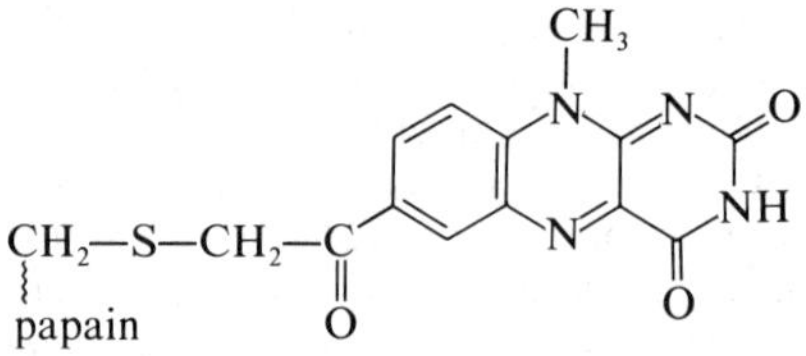

(2)

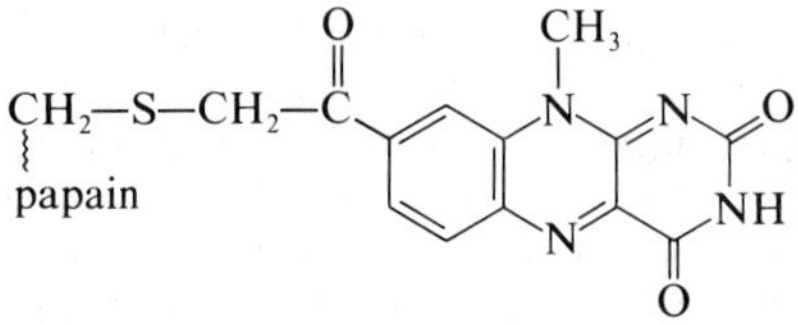

(3)

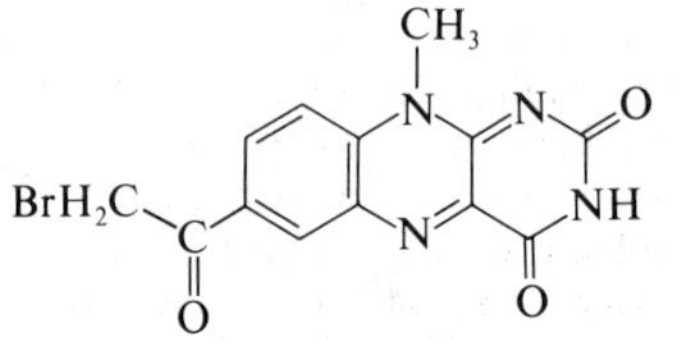

(4)

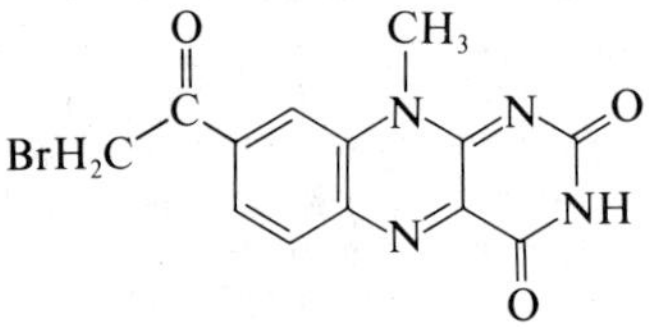

(5)

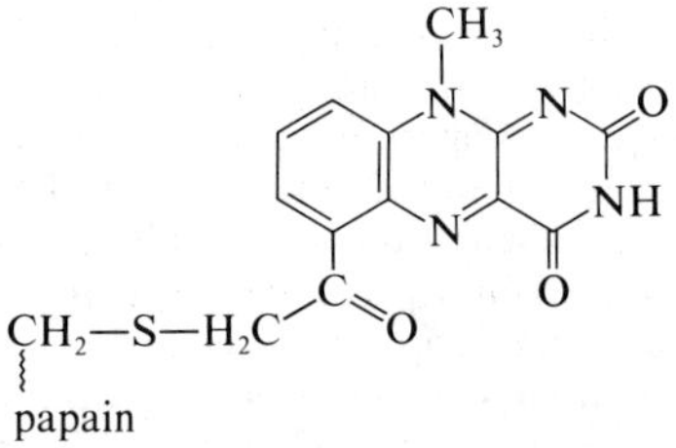

(6)

zine. Flavopapain 6 is a very poor catalyst for the oxidation of N^1-alkyl-1,4-dihydronicotinamide. For example, the k_{cat}/K_m value seen for the oxidation of N^1-benzyl-1,4-dihydronicotinamide by flavopapain 6 is approximately equal to the second-order rate constant observed for the corresponding model reaction using 6-acetyl-10-methylisoalloxazine as the catalyst.

Drenth *et al.* have published x-ray diffraction studies of covalent papain-inhibitor complexes obtained from the reaction of chloromethyl ketone peptide substrate analogues with the sulfhydryl of Cys-25.[21] In the various structures examined, the carbonyl group of what had formerly been the chloromethyl ketone group was found in close proximity to two potential hydrogen-bond donating groups, the backbone NH of Cys-25 and a side-chain NH of Gln-19. The design of the flavin moieties of flavopapains 2, 3, and 6 was based on the idea that building in the possibility of a similar interaction through the presence of the carbonyl moiety of the acetyl groups attached to the flavin ring system might help to constrain the covalently bound flavin to the interior of the enzyme close to the substrate binding site. Indeed, as has been seen, flavopapain 3 is an excellent catalyst and flavopapain 2 is a moderately effective one. The finding that flavopapain 6 is a poor catalyst shows how exquisitely sensitive the catalytic efficiency of the semisynthetic enzyme produced by "chemical mutation" of the active site of an enzyme can be to the exact design of the new catalytic group introduced.

CONCLUSION

Although a major step forward in the design of enzyme active sites has been made, particularly in view of the high catalytic activity of flavopapain 3, much remains to be done. Most of the "chemical mutation" experiments performed have started with the papain tertiary structure. We are now extending our work on semisynthetic enzymes to other templates including lysozyme and glyceraldehyde-3-phosphate dehydrogenase. Additionally, flavins containing strongly electron-withdrawing groups and having the potential to act as enzyme-modifying agents have been prepared with the expectation that the semisynthetic enzymes obtained from their use may be more powerful oxidoreductases than the flavopapains studied to date. Another route we are taking is to try to prepare semisynthetic enzymes containing other coenzyme analogues such as thiazolium species so that a greater diversity of reactions can be studied. Finally, we have initiated site-specific mutagenesis experiments utilizing recombinant DNA methodology that may allow us to position groups, such as a cysteine sulfhydryl, suitable for "chemical mutation" at almost any location we choose in a protein's binding site.

ACKNOWLEDGMENTS

This research was carried out by my coworkers whose articles are cited in the text.

REFERENCES

1. KAISER, E. T. & F. J. KEZDY. 1983. Secondary structures of proteins and peptides in amphiphilic environments (a review). Proc. Natl. Acad. Sci. USA **80:** 1137–1143.
2. KROON, D. J., J. P. KUPFERBERG, E. T. KAISER & F. J. KÉZDY. 1978. Mechanism of lipid-protein interaction in lipoproteins. A synthetic peptide-lecithin vesicle model. J. Am. Chem. Soc. **100:** 5975–5977.
3. FUKUSHIMA, D., J. P. KUPFERBERG, S. YOKOYAMA, D. J. KROON, E. T. KAISER & F. J. KÉZDY. 1979. A synthetic amphiphilic helical docosapeptide with the surface properties of plasma apolipoprotein A-I. J. Am. Chem. Soc. **101:** 3703–3704.
4. YOKOYAMA, S., D. FUKUSHIMA, F. J. KÉZDY & E. T. KAISER. 1980. The mechanism of activation of lecithin: cholesterol acyltransferase by apolipoprotein A-I and an amphiphilic peptide. J. Biol. Chem. **255:** 7333–7339.
5. DEGRADO, W. F., F. J. KÉZDY & E. T. KAISER. 1981. Design, synthesis and characterization of a cytotoxic peptide with melittin-like activity. J. Am. Chem. Soc. **103:** 679–681.
6. DEGRADO, W. F., G. F. MUSSO, M. LIEBER, E. T. KAISER & F. J. KÉZDY. 1982. Kinetics and mechanism of hemolysis induced by melittin and by a synthetic melittin analogue. Biophys. J. **35:** 353–361.
7. MOE, G. R., R. J. MILLER & E. T. KAISER. 1983. Design of a peptide hormone: Synthesis and characterization of a model peptide with calcitonin-like activity. J. Am. Chem. Soc. **105:** 4100–4102.
8. TAYLOR, J. W., R. J. MILLER & E. T. KAISER. 1981. Design and synthesis of a model peptide with β-endorphin-like properties. J. Am. Chem. Soc. **103:** 6965–6966.
9. TAYLOR, J. W., R. J. MILLER & E. T. KAISER. 1982. Structural characterization of β-endorphin through the design, synthesis, and study of model peptides. Mol. Pharmacol. **22:** 657–666.
10. BLANC, J. P., J. W. TAYLOR, R. J. MILLER & E. T. KAISER. 1983. Examination of the

requirement for an amphiphilic helical structure in β-endorphin through the design, synthesis, and study of model peptides. J. Biol. Chem. **258:** 8277–8284.

11. OSTERMAN, D. & D. KENDALL. Unpublished results.
12. MOE, G. R. & E. T. KAISER. Manuscript in preparation.
13. LEVINE, H. L., Y. NAKAGAWA & E. T. KAISER. 1977. Flavopapain: synthesis and properties of semisynthetic enzymes. Biochem. Biophys. Res. Commun. **76:** 64–70.
14. POLGAR, L. & M. L. BENDER. 1966. A new enzyme containing a synthetically formed active site. Thiol-subtilisin. J. Am. Chem. Soc. **88:** 3153–3154.
15. POLGAR, L. & M. L. BENDER. 1970. Simulated mutation at the active site of biologically active proteins. Adv. Enzymol. **33:** 381–400.
16. NEET, K. E. & D. E. KOSHLAND. 1966. Conversion of serine at the active site of subtilisin to cysteine-chemical mutation. Proc. Natl. Acad. Sci. USA **56:** 1606–1611.
17. NEET, K. E., A. NANCI & D. E. KOSHLAND. 1968. Properties of thiol-subtilisin. The consequences of converting the active serine residue to cysteine in a serine-protease. J. Biol. Chem. **243:** 6392–6401.
18. LEVINE, H. L. & E. T. KAISER. 1978. Oxidation of dihydronicotinamides by flavopapain. J. Am. Chem. Soc. **100:** 7670–7677.
19. KAISER, E. T., H. L. LEVINE, T. OTSUKI, H. E. FRIED & R. M. DUPEYRE. 1980. Studies on the mechanism of action and stereochemical behavior of semisynthetic model enzymes. Adv. Chem. Ser. **191:** 35–48.
20. SLAMA, J. T., S. R. ORUGANTI & E. T. KAISER. 1981. Semisynthetic enzymes: Synthesis of a new flavopapain with high catalytic efficiency. J. Am. Chem. Soc. **103:** 6211–6213.
21. DRENTH, J., K. H. KALK & H. M. SWEN. 1976. Binding of chloromethyl ketone substrate analogues to crystalline papain. Biochemistry **15:** 3731–3738.

Molecular Imprinting[a]

GÜNTER WULFF

Institute of Organic Chemistry II
University of Düsseldorf
D-4000 Düsseldorf, Federal Republic of Germany

INTRODUCTION

Enzymes can be used as valuable models for the development of catalysts used in chemical synthesis. One characteristic of enzyme catalysis is the binding of the reacting substrate in a perfectly fitting cavity or hole that contains, in the correct stereochemistry, functional groups for binding, catalysis, and group transfer (active center). It is exactly this feature that has been very difficult to mimic in synthetic catalysts. During the last few years, remarkable progress has been achieved in the design of organic molecules based on molecular recognition. Thus it was possible to use cyclodextrins as the binding site carrying catalytically active groups attached in a controlled stereochemistry to individual hydroxyl groups of the molecule.[1,2] Another approach to construct tailor-made cavities for binding and catalysis uses crown ethers, cryptates, cyclophanes, and similar ring systems.[3–6]

In the above-mentioned examples, substances of low molecular weight were used as carriers. We tried to use synthetic polymers and studied in detail the possibilities of producing microcavities or holes of specific shape within the polymeric matrix by imprinting procedures.[7–11] Basically, the use of polymers makes the system even more complicated than with substances of low molecular weight, but in principle, the typical properties of the enzymes can be better mimicked, since enzymes are polymers as well.

It is not the aim of our research to mimic an individual enzyme and acquire information on the mechanism of enzyme catalysis as it is the main purpose of some other research groups. We are more interested in the use of these substances for preparatory work. We are trying to use the principles of enzyme catalysis to design catalysts working in a similar way to enzymes, but to utilize functional groups for binding and catalysis, substrates, and reactions other than those occurring in nature.

THE PREPARATION OF CAVITIES IN POLYMERS BY MOLECULAR IMPRINTING

As a first step in the design of organic catalysts acting similarly to enzymes, the construction of just the binding site was attempted. For our imprinting procedure, we bound suitable binding groups in the form of polymerizable vinyl derivatives to a template molecule (see FIG. 1).[7–10] This monomer was then copolymerized in the presence of a high amount of cross-linking agent in order to obtain a rigid polymer with chains in a fixed arrangement. Subsequently, the templates were split off and the remaining cavities possessed a shape and an arrangement of functional groups that

[a]Financial support was provided by Fonds der Chemischen Industrie, the Minister für Wissenschaft und Forschung des Landes Nordrhein-Westfalen, and Deutsche Forschungsgemeinschaft.

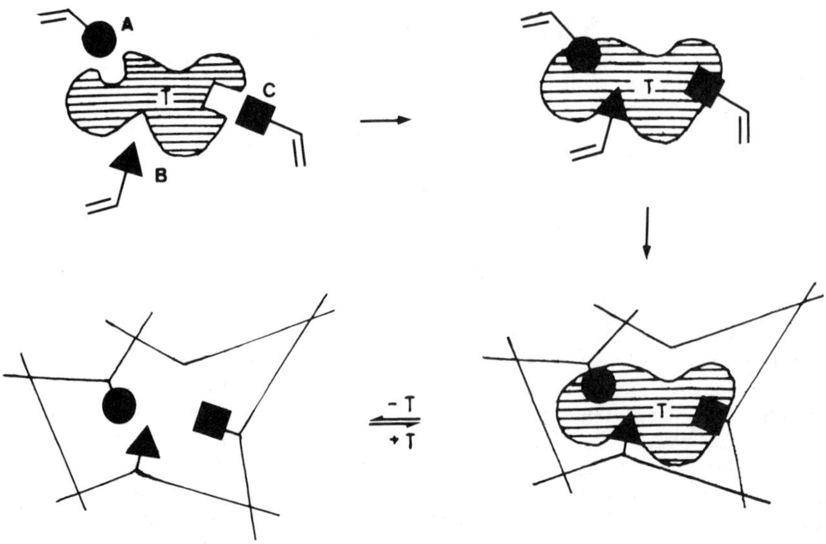

FIGURE 1. The binding of suitable binding groups in the form of polymerizable vinyl derivatives to a template molecule.

corresponded to that of the template. The functional groups in this polymer were located at quite different points of the polymer chain; they are held in spatial relationship by the cross-linking of the polymer.

This approach is in contrast to those already mentioned where a low-molecular-weight part carries the stereochemical information. The method resembles, however, to some extent that of Dickey[12,13] who, some 30 years ago, precipitated silicic acid in the presence of certain dyes and found a small but detectable preference for adsorption for these substances to the resulting silica.

Several different templates and binding groups have been used by our group.[7–11,14–19] Most of the optimization of the procedure has been done with phenyl-α-D-mannopyranoside as the template, to which two molecules of 4-vinylphenylboronic acid have been bound by ester linkages to each of two hydroxyl groups (see FIG. 2). Here, compared to FIGURE 1, a more simple case with only *two identical* binding groups was realized. Boronic acid was chosen as the binding group since it was known to undergo an easily reversible interaction with diol groupings. The α-D-phenylmannoside was a chiral template and therefore the accuracy of the steric arrangement in the cavity could be tested by the ability of the polymer to resolve the racemate of the template after splitting off the original template.

Monomer 1 (FIG. 1) was incorporated in a macroporous polymer by radical copolymerization. The best results were obtained[11] by a copolymerization of 5% of monomer 1 with 95% of glycoldimethacrylate in the presence of the same volume of benzene/acetonitrile (1:1) as an inert solvent. In this case, a polymer with an inner surface area of 382 m^2/g was obtained. By cleavage with methanol/water in a continuous mode, 82% of the template could be split off. The rest of the templates were apparently entrapped in highly cross-linked parts of the polymer and could not be reached by water/methanol. On equilibration of this polymer with the racemate of the template in a batch procedure, the same enantiomer was preferably taken up as had previously been used for the preparation of the polymer.

The specificity for racemic resolution was expressed as the separation factor α known from chromatography. It is the ratio of the distribution coefficients between the solution and polymer of the D-and L-form ($\alpha = K_D/K_L$). An α-value of 3.66 was obtained in this case. Thus in the simple batch procedure, an enrichment of the L-form in the filtrate of 12.8% and of the D-form at the polymer of 40.4% was obtained. These results clearly show that it was indeed possible by this new approach to obtain polymers with tailor-made cavities having a high specificity for one enantiomer of a racemate.

EXAMPLES FOR IMPRINTING PROCEDURES FROM OTHER RESEARCH GROUPS

The described approach for preparing selective binding sites by molecular imprinting during polymerization has been used by other groups as well. There are quite a number of examples of selective memory of a synthesized polymer for its origin. For selective binding of metal ions, polymers were prepared from twofold polymerizable bidentate metal complexes. After splitting off the metal ion, out of a mixture of different metal ions, the original one was preferably bound, as the groups of Kabanov,[20] Tsuchida,[22] Neckers,[23] and Kuchen[24] could show. The preparation of cross-linked polymers in the presence of certain dyes also resulted in selective binding ability for this dye (Klotz *et al.*,[26] Mosbach *et al.*[27]). Stereoselective reactions with substrates bound in specific cavities of the polymer were observed by the groups of Shea[28,29] and Neckers.[30,31] Belokon *et al.*[32] found that stabilization of the initial conformation and configuration of a copolymerized metal complex was possible by the polymer matrix formed around the complex during cross-linking. By cross-linking water soluble starch in the presence of a template, Shinkai *et al.*[33] obtained a resin with memory for its origin. They claim to have a tailor-made cyclodextrin. Keyes[34] described the preparation of a semisynthetic enzyme by denaturating a native protein and subsequent cross-linking in the presence of a template.

THE ORIGIN OF THE MEMORY EFFECT

Now the question arises as to the way the polymers are imprinted by the template and how this imprinting is stabilized after the cleavage of the template. The main stabilization should be realized by cross-linking. FIGURE 3 shows the dependence of

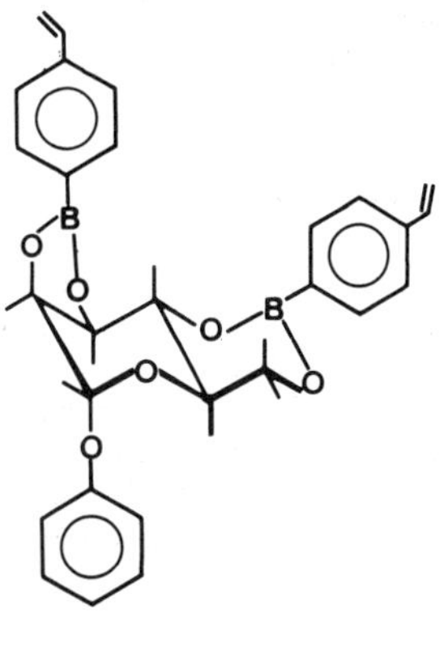

FIGURE 2. Two molecules of 4-vinylphenylboronic acid bound by ester linkages to each of two hydroxyl groups with phenyl-α-D mannopyranoside as the template.

selectivity for racemic resolution on the amount of cross-linking.[11] It can be seen that at low cross-linking (<10%), no specificity at all is obtained. Up to 50% cross-linking, the α-value increases slowly to a value of 1.50. A dramatic increase in specificity is observed on increasing the cross-linking from 50 to 65% (α-value of 3.04). This is a fourfold increase in selectivity in this small range of cross-linking. At higher cross-linking, there is some further increase to 3.66 (95% cross-linking). These experiments clearly show the predominant role of cross-linking for the accuracy of the imprinting procedure.

It is interesting to note that polymers with high cross-linking and high selectivity freed from the templates possess a swellability of more than 60% in the solvent used for

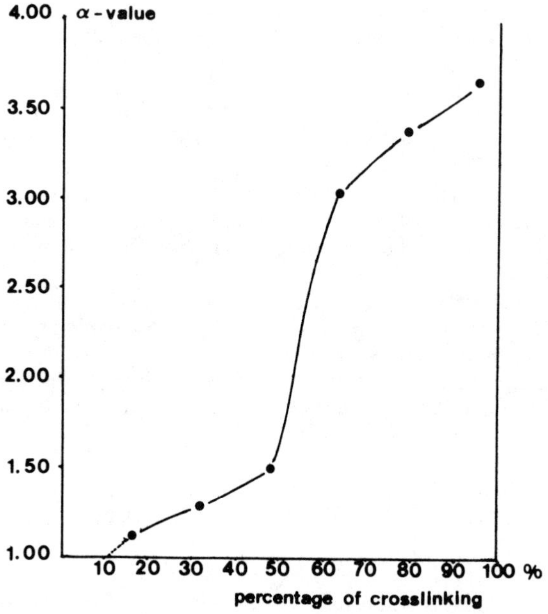

FIGURE 3. Dependence of the specificity of the polymers on the amount of cross-linking agent. In all cases, glycoldimethacrylate in the given percentage. 5% of template monomer 1 and up to 100% methylmethacrylate were copolymerized.[11] 1 ml acetonitrile/benzene 1:1 per 1 g monomeric mixture was used as inert solvent. The separation factor α was determined in the batch procedure.[7-10]

equilibration in the batch procedure. On addition of the template, the original volume is restored by deswelling.[35] Assuming that the swelling occurs for preference in the microcavities by the solvation of the functional groups, the microcavities apparently restore their original form on binding the substrate as the reduced swellability and the substantial specificity indicate (an analogy to "induced fit"?).

Another possible reason for the observed racemic resolution could be the existence of the optically active templates that could not be split off. Careful investigations show[7-10] that racemic resolution is not brought about by the interaction of the racemate with the remaining templates. It could, however, have been shown that in certain cases an asymmetric copolymerization occurred giving rise to asymmetric linear parts of the poly-vinyl chains.[11,36,37] The interaction between the asymmetric polymer chains and the racemate contributed to some extent to racemic resolution. Belokon attributed the conformational stabilization in his case to cyclopolymerization.[32] Cyclopolymerization is the reason for the asymmetric copolymerization as well.[11]

THE FUNCTION OF THE BINDING GROUPS

With our method of preparing synthetic receptor sites, the binding groups have a twofold function. First, during polymerization a strong interaction between template and binding groups should be present, so that the template molecule can fix the binding groups at the growing polymer chains in a defined stereochemistry. Second, after splitting off the template, the binding groups should be able to undergo an easily reversible binding interaction with the template or similar substates. It is generally said that binding in enzymes is a complex and highly cooperative overlap of hydrophobic, electrostatic, charge-transfer, and dipol-dipol interaction. In addition, hydrogen bonding and covalent bonds play an important role. There is good evidence that in most enzymes at least reactive intermediates during catalysis are covalently bound (double displacement reaction).[38]

We have chosen covalent binding in most cases since it is preferable during polymerization. For a fast and reversible binding reaction, however, the activation energy in most cases is too high. It could be lowered by the addition of suitable catalysts (e.g., piperidine in the case of boronic acids[39]) or by the introduction of appropriate neighboring groups.[40] In the case of the binding reaction of diols by polymeric phenyl boronic acids, the introduction of nitrogen-containing neighboring groups in the *ortho* position resulted in an enhancement of the kinetics of the esterification and hydrolysis of the esters by 6–8 orders of magnitude.[41] In this way, functional group-specific binding could be developed. In addition to the use of these groups in imprinting techniques, they can also be used for a type of chemoselective affinity chromatography.[42]

REFERENCES

1. TABUSHI, I., T. NABESHIMA, H. KITAGUCHI & K. YAMAMURA. 1982. Unsymmetrical introduction of two functional groups into cyclodextrin. Combination specificity by use of *N*-benzyl-*N*-methylaniline-*N*-oxide caps. J. Am. Chem. Soc. **104:** 2017–2019.
2. BRESLOW, R. & A. W. CZARNIK. 1983. Transamination by pyridoxamine selectively attached at C-3 in β-cyclodextrin. J. Am. Chem. Soc. **105:** 1390–1391.
3. CRAM, D. J. & H. E. KATZ. 1983. An incremental approach to hosts that mimic serine proteases. J. Am. Chem. Soc. **105:** 135–137.
4. BEHR, J.-P., J.-M. LEHN & P. VIERLING. 1982. Molecular receptors. Structural effects and substrate recognition in binding of organic and biogenic ammonium ions by chiral polyfunctional macrocyclic polyethers bearing amino-acid and other side chains. Helv. Chim. Acta **65:** 1853–1867.
5. CANCEILL, J., A. COLLET, J. GABARD, F. KOTZBYA-HIBERT & J.-M. LEHN. 1982. Speleands. Macrocyclic receptor cages based on binding and shaping subunits. Synthesis and properties of macrocycle-cyclotriveratrylene combinations. Helv. Chim. Acta **65:** 1894–1897.
6. TABUSHI, I. & K. YAMAMURA. 1983. Water soluble cyclophanes as hosts and catalysts. Top. Curr. Chem. **113:** 145–182.
7. WULFF, G. & A. SARHAN. 1972. On the use of enzyme-analogue-built polymers for racemic resolution. Angew. Chem. Int. Ed. Engl. **11:** 341–342.

8. WULFF, G., A. SARHAN & K. ZABROCKI. 1973. Enzyme-analogue-built polymers and their use for the resolution of racemates. Tetrahedron Lett. **1973:** 4329–4332.
9. WULFF, G., W. VESPER, R. GROBE-EINSLER & A. SARHAN. 1977. Enzyme-analogue-built polymers, IV. On the synthesis of polymers containing chiral cavities and their use for the resolution of racemates. Makromol. Chem. **178:** 2799–2816.
10. WULFF, G. & A. SARHAN. 1982. Models of the receptor sites of enzymes. *In* Chemical Approaches to Understanding Enzyme Catalysis. B. S. Green, Y. Ashani & D. Chipman, Eds. Elsevier. Amsterdam. pp. 106–118.
11. WULFF, G., R. KEMMERER, J. VIETMEIER & H.-G. POLL. 1982. Chirality of vinyl polymers. The preparation of chiral cavities in synthetic polymers. Nouveau J. Chim. **6:** 681–687.
12. DICKEY, F. H. 1949. Preparation of specific adsorbents. Proc. Natl. Acad. Sci. USA **35:** 227–229.
13. DICKEY, F. H. 1955. Specific adsorption. J. Phys. Chem. **59:** 695–707.
14. WULFF, G. & I. SCHULTZE. 1978. Enzyme-analogue-built polymers, IX. Polymers with mercapto groups of definite cooperativity. Isr. J. Chem. **17:** 291–297.
15. WULFF, G., M. LAUER & B. DISSE. 1979. Über enzymanalog gebaute Polymere, X. Über die Synthese von Monomeren zur Einführung von Aminogruppen in Polymere in definiertem Abstand. Chem. Ber. **112:** 2854–2865.
16. WULFF, G., I. SCHULTZE, K. ZABROCKI & W. VESPER. 1980. Über enzymanalog gebaute Polymere, XI. Bindungsstellen im Polymer mit unterschiedlicher Zahl der Haftgruppen. Makromol. Chem. **181:** 531–544.
17. WULFF, G. & E. LOHMAR. 1979. Enzyme-analogue-built polymers, XII. Specific binding effects in chiral microcavities of cross-linked polymers. Isr. J. Chem. **18:** 279–284.
18. WULFF, G. & J. GIMPEL. 1982. Über enzymanalog gebaute Polymere, XVI. Über den Einfluss der Flexibilität der Haftgruppen auf die Racemattrennungsfähigkeit. Makromol. Chem. **183:** 2469–2477.
19. WULFF, G., W. BEST & A. AKELAH. 1984. Enzyme-analogue-built polymers, XVII. Investigations on the racemic resolution of amino acids. Reactive Polymers. **2:** 167-74.
20. KABANOV, V. A., A. A. EFENDIEV & D. D. ORUDZHEV. 1976. Sorbent. U.S.S.R. Pat. 502,907. 15. Feb. 1976. Chem. Abstr. **85:** 6685 d, 1976.
21. EFENDIEV, A. A. & V. A. KABANOV. 1982. Selective polymer complexons prearranged for metal ions sorption. Pure Appl. Chem. **54:** 2077–2092.
22. NISHIDE, H., J. DEGUCHI & E. TSUCHIDA, 1976. Selective adsorption of metal ions on cross-linked poly (vinylpyridine) resin prepared with a metal ion as a template. Chem. Lett. **1976:** 169–174.
23. GUPTA, S. N. & D. C. NECKERS. 1982. Template effects in chelating polymers. J. Polym. Sci. Polym. Chem. Ed. **20:** 1609–1622.
24. KUCHEN, W. & U. BRAUN. Unpublished results.
25. BRAUN, U. 1983. Ionenselektive Austauscherharze durch vernetzende Copolymerisation vinylsubstituierter Dithiophosphinatometallkomplexe. Dissertation, University of Düsseldorf.
26. TAKAGISHI, T. & I. M. KLOTZ. 1972. Macromolecule-small molecule interactions; Introduction of additional binding sites in polyethyleneimine by disulfide cross-linkages. Biopolymers. **11:** 483–491.
27. ARSHADY, R. & K. MOSBACH. 1981. Synthesis of substrate-selective polymers by host-guest polymerization. Makromol. Chem. **182:** 687–692.
28. SHEA, K. J. & E. A. THOMPSON. 1978. Template synthesis of macromolecules. Selective functionalization of an organic polymer. J. Org. Chem. **43:** 4253–4255.
29. SHEA, K. J., E. A. THOMPSON, S. D. PAUDEY & P. S. BEAUCHAMP. 1980. Template synthesis of macromolecules. Synthesis and chemistry of functionalized macroporous polydivinylbenzene. J. Am. Chem. Soc. **102:** 3149–3155.
30. DAMEN, J. & D. C. NECKERS. 1980. Memory of synthesized vinyl polymers for their origins. J. Org. Chem. **45:** 1382–1387.
31. DAMEN, J. & D. C. NECKERS. 1980. Stereoselective synthesis via a photochemical template effect. J. Am. Chem. Soc. **102:** 3265–3267.
32. BELOKON, Y. N., V. I. TARAROV, T. F. SAVEL'EVA, M. M. VOROB'EV, S. V. VITT, V. F. SIZOY, N. A. SUKHACHEVA, G. V. VASIL'EV & V. M. BELIKOV. 1983. Memory of a

polymeric matrix, stabilizing the initial conformation of potassium bis [*N*-(5-methacryloylamino)salicylidene-(S)-norvalinato]-cobaltate III in the deuterium exchange. Makromol. Chem. **184:** 2213–2223 and earlier papers in the same journal.
33. Shinkai, S., M. Yamada, T. Sone & O. Manabe. 1983. Template synthesis from starch as an approach to tailor-made "cyclodextrin." Tetrahedron Lett. **24:** 3501–3054.
34. Keyes, M. H. 1982. Verfahren zur Herstellung halbsynthetischer Enzyme. Deutsche Offenlegungsschrift 3147947, 8.7.1982.
35. Sarhan, A. & G. Wulff. 1982. Über enzymanalog gebaute Polymere, XIV. Stereospezifische Haftungen über Amidbindungen oder elektrostatische Wechselwirkung. Makromol. Chem. **183:** 1603–1614.
36. Wulff, G., K. Zabrocki J. Hohn. 1978. Optically active polyvinyl compounds with chirality in the main chain. Angew. Chem. Int. Ed. Engl. **17:** 535–537.
37. Wulff, G. & J. Hohn. 1982. Chirality of polyvinyl compounds. 2. An asymmetric copolymerization. Macromolecules **15:** 1255–1261.
38. Spector, L. B. 1982. Covalent Catalysis by Enzymes. Springer Verlag. New York.
39. Wulff, G., W. Dederichs, R. Grotstollen & C. Jupe. 1982. Specific binding of substances to polymers by fast and reversible covalent interactions. *In* Affinity Chromatography and Related Techniques. T. C. J. Gribnau, J. Visser & R. J. F. Nivard, Eds. Elsevier. Amsterdam. pp. 207–216.
40. Lauer, M. & G. Wulff. 1983. Arylboronic acids with intramolecular B-N interaction. Convenient synthesis through *ortho*-lithiation of substituted benzylamines. J. Organomet. Chem. **256:** 1–10.
41. Lauer, M., H. Böhnke, R. Grotstollen, M. Salehnia & G. Wulff. 1984. Zur Chemie von Haftgruppen, IV. Über eine außerordentliche Erhöhung der Reaktivität von Arylboronsäuren durch Nachbargruppen. Chem. Ber. In press.
42. Wulff, G. 1982. Selective binding to polymers *via* covalent bonds. The construction of chiral cavities as specific receptor sites. Pure Appl. Chem. **54:** 2093–2102.

Synzymes: Synthetic Hydrogenation Catalysts

G. P. ROYER

Biotechnology Division
Corporate Research Department
Standard Oil Company (Indiana)
Naperville, Illinois 60566

K. S. HATTON AND W.-S. CHOW

Department of Biochemistry
Ohio State University
Columbus, Ohio 43210

There is growing interest in synthetic catalysts that mimic enzymes.[1] Bioorganic chemists study enzyme models with the hope that results of their investigations will aid in the understanding of enzyme mechanism. With the development of industrial catalysis to the point where improvement in selectivity is a central theme in research, many believe that abiotic catalysts constructed with enzymes in mind could be of value in the chemical process industry. Also, it should be said that enzymatic processes (even those based on immobilized enzymes) are limited in their applicability despite the fact that enzymes have a preeminent position in terms of catalytic power and specificity. Limitations of enzymes as industrial catalysts are listed in TABLE 1. Cost and instability are irrevocably linked. An enzyme that demonstrates very good stability could command a high price; a very inexpensive enzyme could be viewed as a disposable reagent. It is unfortunately still true that enzymes (even those produced from recombinant organisms) are often too expensive for industrial application.

Instability at high temperature and in nonaqueous phases is a serious limitation. The use of heat as a thermodynamic driving force is common in chemical processing even when very effective catalysts are available. Biochemical reactions occur at low temperature; they are driven with coupled reactions involving reduced coenzymes and nucleoside triphosphates. Many compounds of industrial interest are not soluble in water. Organic solvents in general are not compatible with enzymes.

Another fundamental limitation of enzyme use in chemical processing is based on the fact that enzymes have evolved to be excellent catalysts at low reactant concentration, which means that they have very high affinities for their substrates. When the product and substrate are structurally similar, product inhibition occurs. In nature this effect is offset by consumption of the product by sequential reactions such as those in a metabolic pathway. In an industrial reactor it is often appropriate to have a high feed concentration and total conversion. These conditions would be detrimental to many enzymes. Substrate inhibition (FIG. 1) would occur at low conversion and product inhibition would occur at high conversion. Although alternative reactor designs could be employed to partially offset these drawbacks, substrate inhibition and product inhibition remain as fundamental limitations of enzyme use in industry.

Consideration of the development of "synzymes" leads immediately to the question of feasibility at reasonable cost in a reasonable time period. One obvious question is how faithfully must the structure of the enzyme active center be duplicated? In the case of the binding site, only a rough approximation would be required; very strong

TABLE 1. Advantages and Limitations of Biocatalysts

Advantages
• Great rate enhancements under mild reaction conditions
• Specificity, especially regiospecificity and stereospecificity
Limitations
• Cost
• Instability
• Phase incompatibility
• Product inhibition
• Substrate inhibition

binding can be detrimental as pointed out above. In the case of the catalytic site, again there is much room for error; enzymes produce rate enhancements of 10^6–10^{14}, so even a near miss could result in a very effective catalyst. As pointed out elsewhere, (Maugh,[1] Royer,[2] and Kaiser, this volume) impressive rate enhancement and selectivity have been demonstrated with abiotic catalysts inspired by enzymes.

Klotz has pointed out the advantages of using globular, water soluble polymers as the starting point for preparation of catalysts that mimic enzyme action. A number of examples of synthetic catalysts that show Michaelis-Menten kinetics have been produced. Presumably, attachment of binding groups to the polymer in the vicinity of catalytic groups is the basis of the observed kinetic behavior. We have been interested in the preparation of solid-phase catalysts that permit chemical manipulation of the environment of the catalytic center. Since palladium is a widely used metal in catalysis of organic reactions, we felt that hydrogenation catalysts prepared from supported polyethylenimine (PEI) and palladium salts would be of academic interest and, at least potentially, of industrial interest.

PEI has been developed by Dow Chemical in two forms. Polymerization of aziridine yields branched PEI. Polymerization of 2-ethyl-2-oxazoline followed by hydrolysis yields linear polyethylenimine (L-PEI, FIG. 2). We have prepared supported PEI in several forms. Attachment of PEI to silica gel results in a material that is useful in chromatography.[3] We have produced PEI "ghosts" that are useful as a catalyst support.[4–6] PEI and L-PEI have been incorporated into the same support.

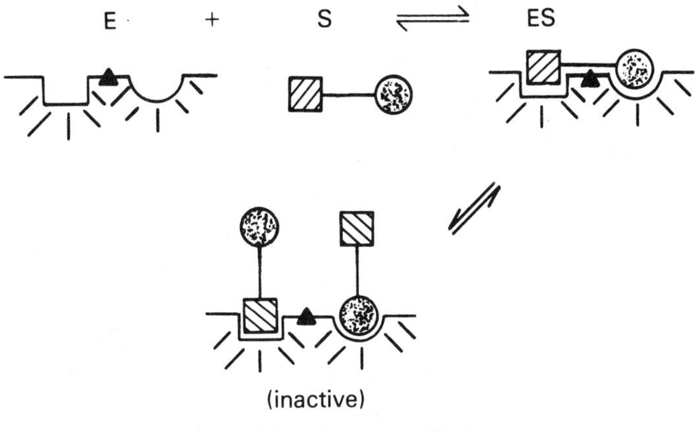

FIGURE 1. Substrate inhibition.

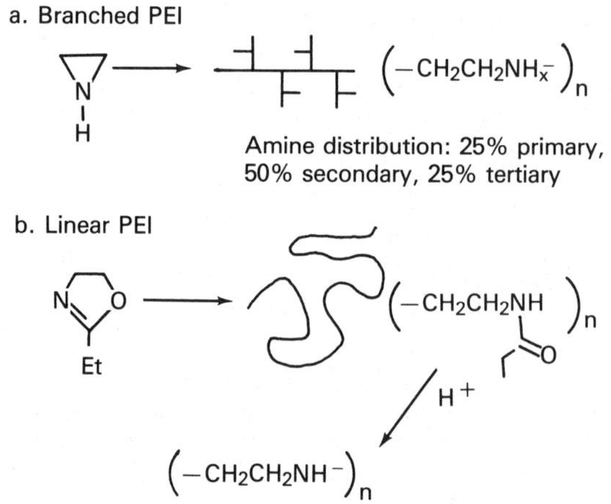

FIGURE 2. Preparation of branched and linear PEI.

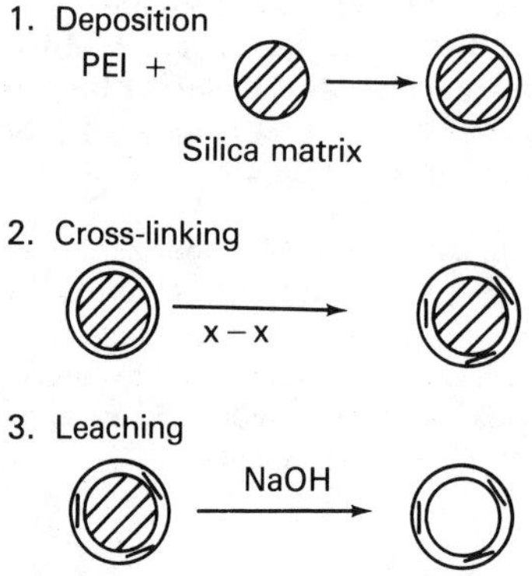

FIGURE 3. Preparation of PEI "ghosts."

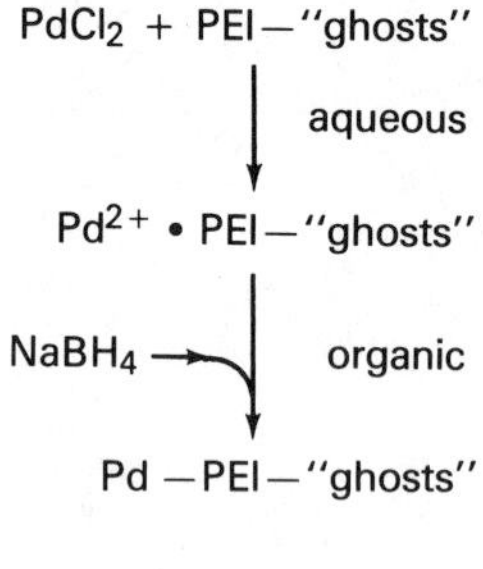

FIGURE 4. Preparation of Pd-PEI "ghosts."

The most interesting catalyst that we have found is the adduct of palladium with PEI "ghosts" (Pd/PEI-G). PEI "ghosts" are made in a three-step process, PEI is adsorbed onto an inorganic matrix. The PEI layer is then cross-linked. Part or all of the inorganic matrix is dissolved away to leave a material of high surface area and high capacity for binding metal ions (FIG. 3). Pd/PEI-G is prepared as shown in FIGURE 4. In aqueous solution, the PEI support binds palladium very well. The mixing of the two is followed by a rapid decolorizing of the supernatant solution. The coordinated palladium is then reduced with sodium borohydride to give the active catalyst. Formic acid (or ammonium formate) is an especially good hydrogen donor for Pd/PEI-G. In the reaction shown below, the rate of the reaction catalyzed by Pd/PEI-G is very much faster than the reaction catalyzed by palladium on carbon.

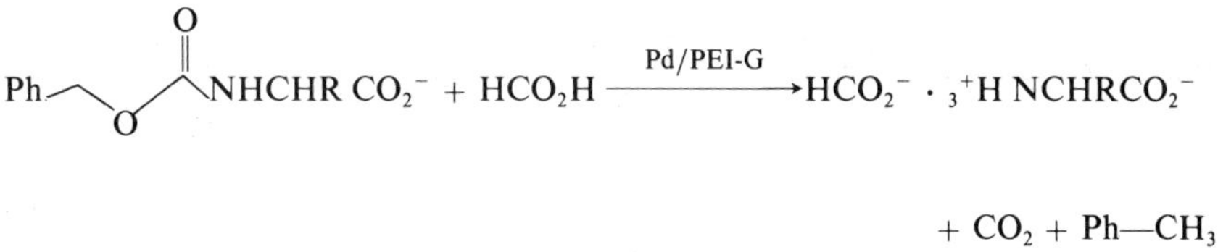

These results are shown in FIGURE 5. It has been pointed out that this system has advantages for use in peptide synthesis.[5]

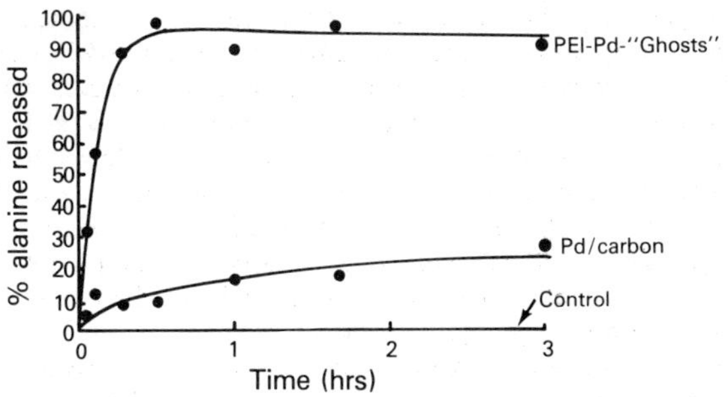

FIGURE 5. Comparison of rate of the reaction catalyzed by Pd/PEI-G and the rate of the reaction catalyzed by palladium on carbon.

The catalyst described above (Pd/PEI-G bead) was prepared with porous silica beads (40 mesh, Corning) as the starting material. A second version of the catalyst was prepared in an analogous procedure using silica gel as the support (Grade 923, W. R. Grace). This material is a free-flowing powder that settles out readily from suspension. It can be used in a Parr shaker. Pd/PEI-G in the powder form works very well with hydrogen gas as the source of hydrogen. The reduction of aromatic nitro groups with the aid of palladium catalysts is frequently performed in the plant as well as the research laboratory. We used the reaction shown below to assess the activity and handling properties of the Pd/PEI-G powder:

$$\text{Ph-NO}_2 + \text{H}_2\text{(g)} \xrightarrow{\text{Pd/PET-G}} \text{Ph-NH}_2$$

The data shown in FIGURE 6 indicate an excellent operating stability for the catalyst.

The reactions were done in a three-neck flask fitted with an overhead stirrer. Hydrogen was introduced through a gas dispersion tube. The reactions were carried out at atmospheric pressure. No external cooling was used that led to a temperature increase during the reaction from 25°C to 56°C. The Pd/PEI powder settles rapidly from suspension, which permits isolation of the product simply by decantation. Palladium on carbon requires filtration with a filter aid. Also, both catalysts have good mechanical stability and high specific activity; they are not pyrophoric.

The activity and properties of the basic catalyst were encouraging. We then set out to improve the catalysts by modification of the matrix to which the metal was bound. The hydrophobic character of the test reactant below,

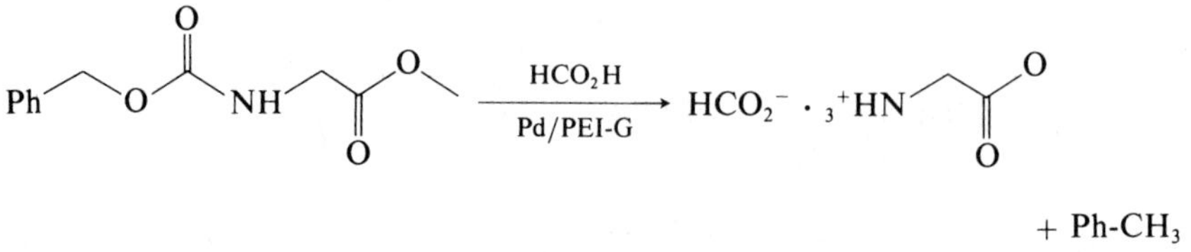

z-glycine-*tert* butyl ester, suggested that interaction of the reactant with the catalyst could be helped by introduction of hydrophobic groups into the catalyst support. Repulsion of the formate salt of the product amine was hypothesized. Octanoyl, octyl, and lauroyl derivatives of Pd/PEI-G beads were prepared using the acid chloride and the bromides, respectively. The degree of substitution was estimated in each case by the reduction in primary amine content as judged by the ninhydrin test. The presence of the hydrophobic groups does not result in rate enhancement. The catalysts are prepared by substitution of the most accessible amines with hydrophobic groups followed by the attachment of palladium. The most plausible model of the catalyst structure would show the binding groups on the outside and the metal on the inside. One possible consequence of this arrangement would be nonproductive binding of the reactant on the margin of the polymer in such a way that the palladium sites are not accessible. However, supports produced with L-PEI and PEI in combination gave catalysts with only slightly improved activity.

Despite the fact that our attempts to mimic enzymes were not successful, the catalysts that we prepared have a number of features that make them of practical value at least in benchtop operations. In TABLE 2, the advantages are listed. One important

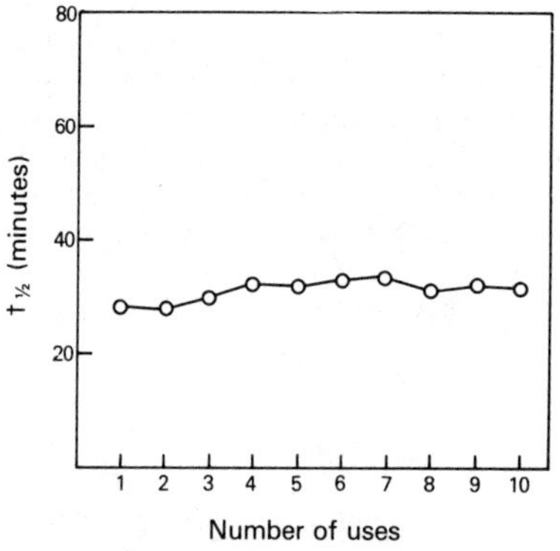

FIGURE 6. Stability of Pd/PEI-G powder in the reduction of nitrobenzene.

TABLE 2. Advantages of Pd/PEI Catalysts

• Active with formic acid and H_2 gas
• Nonpyrophorbic
• Stable
• Easily recoverable for reuse

feature of the catalysts is the low pyrophoric potential. At one per cent palladium loading, the catalysts appear to be safe even in the presence of methanol and other organic solvents. The activity and stability are also attractive. Both catalysts are easily recoverable, which permits reuse and eliminates the slow filtration process necessary for palladium/carbon.

Although the results reported here do not prove that a new "synzyme" has been prepared, we are pleased with the practical value of the catalysts and that a base line has been established.

REFERENCES

1. MAUGH, T. H., II. 1983. Catalysts that break nature's monopoly. Science **221:** 351–354.
2. ROYER, G. P. 1980. Synthetic enzyme-like catalysts (synzymes). Adv. Catal. **29:** 197–224.
3. WATANABE, K., W.-S. CHOW & G. P. ROYER. 1982. Column chromatography on polyethylenimine-silica. **127:** 155–158.
4. MEYERS, W. E. & G. P. ROYER. 1977. Catalysis of *p*-nitrotrifluoroacetanilide hydrolysis by an imidazole derivative of PEI "Ghosts." J. Am. Chem. Soc. **99:** 6141–6142.
5. COLEMAN, D. R. & G. P. ROYER. 1980. New hydrogenation catalyst: Palladium-polyethylenimine "Ghosts." J. Org. Chem. **45:** 2268–2270.
6. ROYER, G. P. U.S. Patent 4326009.

Amino Acid and NMR Analysis of Oxidized *Mucor miehei* Rennet

SVEN BRANNER, PETER EIGTVED,
MOGENS CHRISTENSEN, AND HENNING THØGERSEN

NOVO Industri A/S
DK-2880 Bagsvaerd, Denmark

Microbial rennet derived from the fungus *Mucor miehei* has been used extensively in cheese manufacture all over the world. The application was further extended after the introduction about 4 years ago of chemically modified products with reduced thermostability. Low thermal stability is an advantage in the manufacture of certain types of cheese, for example Emmental, by making it possible to inactivate residual enzyme activity in cheese whey by pasteurization. Chemical modification of *Mucor miehei* rennet with oxidizing agents such as peracetic acid, hydrogen peroxide, and sodium hypochlorite has proved useful for reduction of thermal stability.

CHEMICAL MODIFICATION

Crude preparations of *Mucor miehei* rennet from the commercial manufacture of rennilase were oxidized with sufficient oxidizing agent to give high degree of destabilization without activity loss. For amino acid and NMR-analyses, samples of purified enzyme were prepared by conventional procedures.

AMINO-ACID ANALYSIS

Both 6 *N* HCl and 4 *M* MSA (methane sulfonic acid) hydrolysis at 110°C were used.

Methionine sulfoxide, which can be expected in an oxidized protein, is known to be reduced to methionine during 6 *N* HCl hydrolysis. Methionine sulfone is stable. Methionine sulfoxide, tryptophane and cystine are better analyzed after MSA hydrolysis, due to rapid destruction in HCl. Cystine is first derivatized to S-4-pyridyl-hetyl-cysteine. Hydrolysis times were varied in order to make the best estimates of degraded/incompletely hydrolyzed amino acids: 6 *N* HCl hydrolysis for 16–144 hours, MSA for 24–72 hours. The results for methionine sulfoxide are minimum values based on 24 hours of hydrolysis. (See TABLE 1 for amino-acid composition.)

CONCLUSIONS FROM AMINO-ACID ANALYSIS

Our determination of the amino-acid composition of the native enzyme is very close to published results.[1–3] Threonine is the only residue where our determination deviates from the published results. The most pronounced effect of the oxidations is the formation of methionine sulfoxide, which is formed most effectively with peracetic acid. No other differences are significant, but there is a tendency to tyrosine degradation due to the oxidations. Further investigations are necessary to demonstrate if this change is real. In no case could methionine sulfone be detected.

NUCLEAR MAGNETIC RESONANCE SPECTROSCOPY

400-MHz ^{1}H-NMR spectra were recorded at 25°C on a Bruker WM 400 instrument. The NMR analyses were made on resolution-enhanced spectra. The spectra were expanded around 1.9–2.2 ppm where absorption from methyl groups in methionine residues is expected, and around 2.6–2.8 ppm where the absorption from methyl groups in methionine sulfoxide residues is expected. Unfortunately, a peak from an acetic acid impurity is observed in the methionine area. The exact position of this signal is pH-dependent, and it is eliminated in the schematic representation of the peak positions (FIG. 1).

TABLE 1. Amino Acid Composition

	Native Enzyme		Oxidized Enzyme			
Amino Acid	In the Literature	Our Determination	CH_3COOOH	H_2O_2 pH 3.5	H_2O_2 pH 7	NaClO
Met(O)	—	0	>3.6	>2	>3.2	<0.7
Asp	46–48	47	47	47	47	47
Thr	32	30	30	30	30	30
Ser	34–39	35–36	36	36	36	36
Glu	25–27	25	25	25	25	25
Pro	19	19	19	19	19	19
Gly	37–38	37	37	37	37	37
Ala	29	29	29	29	29	29
Cys	4	4	4	4	4	4
Val	26–28	28	28	28	28	28
Met, Met(O)	6–7	7	6–7	7	7	7
Ile	20–23	21	21	21–22	22	21
Leu	21–22	22	22	22	22	22
Tyr	21–22	21	20	21	20–21	20
Phe	22–23	23	22–23	23	23	23
Lys	10	10	10	10	10	10
His	2	2	2	2	2	2
Trp	3	3	3	3	3	3
Arg	7	7	7	7	7	7

CONCLUSIONS FROM THE NMR SPECTRA

In the spectrum of native enzyme seven peaks are observed between 1.95 and 2.12 ppm that may correspond to the seven methionine residues in the enzyme. By oxidation, two of these peaks invariably disappear (2.084 and 2.088 ppm), and the intensity of some other peaks changes, depending on the oxidant. In all cases, new peaks are found in the methionine sulfoxide area after oxidation, and the positions depend on the oxidant. The application of nuclear magnetic resonance spectroscopy to detect methionine oxidation has been demonstrated and should be a general technique, useful in protein and peptide chemistry.

SUMMARY

Chemical modification of *Mucor miehei* rennet with different oxidizing agents has proved useful for reduction of the thermal stability of this enzyme. Low thermal stability is necessary to make it possible to inactivate residual enzyme activity in cheese whey by pasteurization. Amino-acid analysis of oxidized *Mucor miehei* protease shows that methionine is the only affected amino-acid residue. One or more of the seven methionine residues in the enzyme is oxidized to methionine sulfoxide residues depending on the nature and the concentration of the oxidizing agent. However, it is not possible, based on the amino-acid analysis alone, to differentiate between the oxidative reagents that have been used for the modification. In the ^{1}H-NMR spectra at

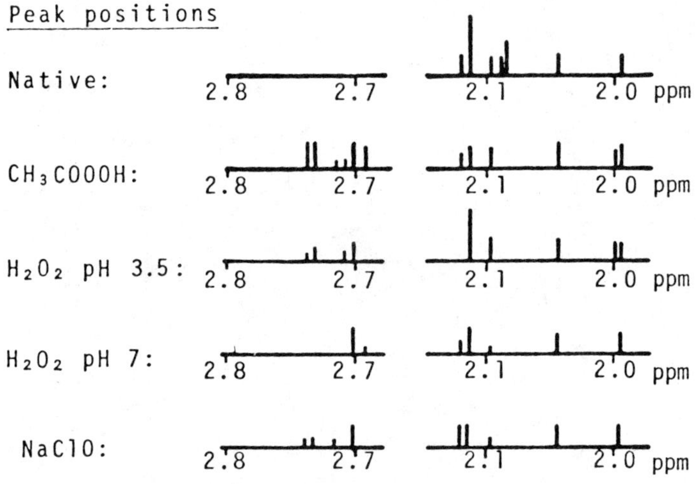

FIGURE 1. Schematic representation of peak positions.

400 MHz of native and chemically modified *Mucor miehei* protease, peaks can be assigned to methyl groups from each of the methionine and methionine sulfoxide residues in the enzyme molecules. By this technique, it may thus be possible to correlate the peak positions to the nature of the oxidation.

REFERENCES

1. OTTESEN, M. & W. RICKERT. 1970. The isolation and partial purification of an acid protease produced by *Mucor miehei*. C. R. Trav. Lab. Carlsberg **37:** 301–325.
2. RICKERT, W. S. & J. R. ELLIOTT. 1973. Acid proteases from species of *Mucor:* Molecular weight of *Mucor miehei* protease from amino-acid analysis data. Can. J. Biochem. **51:** 1638–46.
3. FAERCH, A.-M. & B. FOLTMANN. 1979. The primary structure of *Mucor miehei* milk-clotting proteinase. Poster presentation, 11th IUB-Congr. Toronto.

New Properties of Chymotrypsin Modified by Fixation with a Hydrophobic Molecule: Hexanal

M. H. REMY, C. BOURDILLON, AND D. THOMAS

Laboratoire de Technologie Enzymatique
E.R.A. n° 338 du C.N.R.S.
Université de Technologie de Compiègne
BP 233, 60206 Compiègne, France

Modified enzymes are an interesting step between the native enzyme and the insolubilized form.[1] Indeed, insolubilization causes some alterations of kinetic properties and heat stability that may be explained by (1) diffusional limitations, (2) modification of the amino acids in the active or regulatory center, and (3) locking of the tertiary structure. By elimination of the steric hindrance, the modified enzyme can become a model system.[2] The aim of our present work is to study the kinetic properties of an enzyme modified with different hydrophobic agents. α-Chymotrypsin was chosen as a model hydrophilic enzyme. The mechanism of hydrolysis involves two intermediates.

$$E + A \underset{k_{-1}}{\overset{k_{+1}}{\rightleftharpoons}} EA \xrightarrow{k_2} \underset{+X}{EA'} \xrightarrow{k_3} E + Y \tag{1}$$

$$V_m = \frac{k_2 k_3 E_o}{k_2 + k_3} \tag{2}$$

$$K_m = \frac{k_{-1} + k_2}{k_1}\left(\frac{k_3}{k_2 + k_3}\right) \tag{3}$$

For amide and anilide substrate $k_3 \gg k_2$, so $V_m = k_2 E_o$ and $K_m = (k_{-1} + k_2)/k_1$. For ester substrate $k_2 \gg k_3$; and the appropriate expressions can be developed in a similar way.

Bovine pancreatic α-chymotrypsin (E.C. 3.4.21.1.) or chymotrypsinogen at 8×10^{-4} *M* was dissolved in 5.0 ml 0.1 *M* borate buffer, pH 8, and coupled to different aldehydes (2×10^{-2} *M*). The products so formed were reduced by a mild reducing agent, 0.2 *M* sodium cyanoborohydre.[3] The reaction mixture was incubated at 4°C for 1 hour. Dialysis against 10^{-4} *M* HCl was used to separate the modified protein from unwanted components of the reaction solution. The amino groups were determined with 2, 4, 6-trinitrobenzenesulfonic acid.[4] Free and modified α-chymotrypsins were tested in 2×10^{-3} *M* glutaryl-L-phenylalanine-4-nitroanilide (GPNA) in 0.05 *M* Tris-HCl buffer, pH 8, with 0.02 *M* $CaCl_2$. The variation of nitroanilide concentration was followed by measuring the absorbance at 410 nm.[5] Measurement with casein as substrate was realized by the method of Rich.[6] Measurement with benzoyltyrosinamide (BTA) as substrate was by the method of Moore.[7] For acetyl-L-tyrosine ethyl ester (ATEE), the initial rate of hydrolysis was followed by measuring the rate of liberation of protons at constant pH.

Different aldehydes were used to modify the hydrophobicity of soluble α-chymotrypsin. The charge in the different α-chymotrypsin derivatives was verified by electrofocusing. Distributions were similar to the native enzyme. Two major bands were observed in each case. TABLE 1 shows the measured apparent Michaelis constants and the limiting rates at pH 8 with GPNA as substrate. An interesting result was obtained with α-chymotrypsin modified with hexanal. After modification, the activity was increased 1.5-fold with no change in the Michaelis constant. Some other substrates also were tested. When using the amide as substrate (a system with $k_3 \gg k_2$), the apparent rate with hexanal-chymotrypsin was also increased. When using casein as a substrate, the modified enzyme was less active. For ATEE, V_m was decreased to 83% (in this case $k_2 \gg k_3$). An initial check was carried out to see that all enzyme molecules were chemically modified. An octyl-sepharose CL-4B column with 4 M NaCl as the elutant, was used. No further purification was required. The degree of substitution of α-chymotrypsin amino groups was expressed in relative arbitrary units, 100% corresponding to all the free amino acids able to react with TNBS. After hexanal modification, 85% of the amino groups were measurable and after acetaldehyde substitution, 35%. We found that isoleucine-16 or alanine-149 were involved in the properties of modified α-chymotrypsin. First, chymotrypsinogen was modified by hexanal; 24% of the amino groups were modified when measured with TNBS (100% corresponding to native chymotrypsinogen). The chymotrypsinogen was then activated

TABLE 1. Kinetic Parameters of Native and Modified Chymotrypsin[a]

	K_m mM	V_m %
Chymotrypsin	0.57	100
Chymotrypsin + sodium cyanoborohydride	0.48	76
Ethyl-chymotrypsin	0.38	64
Butyryl-chymotrypsin	0.51	95
Hexyl-chymotrypsin	0.55	164
Benzyl-chymotrypsin	0.48	80
Trimethylacetyl-chymotrypsin	0.40	91

[a]Activity was measured with GPNA as substrate, on 0.05 M Tris-HCl buffer pH 8, 0.02 M, temperature 25°C.

by trypsin; and the Michaelis constant and maximal velocity were the same for both α-chymotrypsin and the hexyl-α-chymotrypsin. The activity, as a function of pH, gave the same bell-shaped curve between pH 6 to 9.5, but the optimum pH for the hexanal chymotrypsin was shifted by one pH unit toward the alkaline pH.

The inactivation process was tested at low enzyme concentrations to eliminate autolysis phenomena and at pH 8. The denaturation rate was determined at different temperatures for construction of Arrhenius plots. The energy, which must be brought to denature the catalytic site (activation energy), was equal for both forms of the enzyme; but the different enzyme preparations were not in the same temperature domain. The modified enzyme was less stable than the native enzyme (FIG. 1).

Kinetic experiments also were performed at different temperatures for the hexyl-chymotrypsin, ethyl-chymotrypsin, and free enzyme. Michaelis constants and the maximal velocity were determined by means of the Hanes-Woolf correlation. The same K_m value was observed no matter which form of the modified enzyme was tested. The maximal velocity was determined in the temperature range in which the thermal denaturation was negligible. From the data, Arrhenius plots were drawn for the three forms of the enzyme (FIG. 2). The increase of V_m for the hexyl-α-chymotrypsin was not

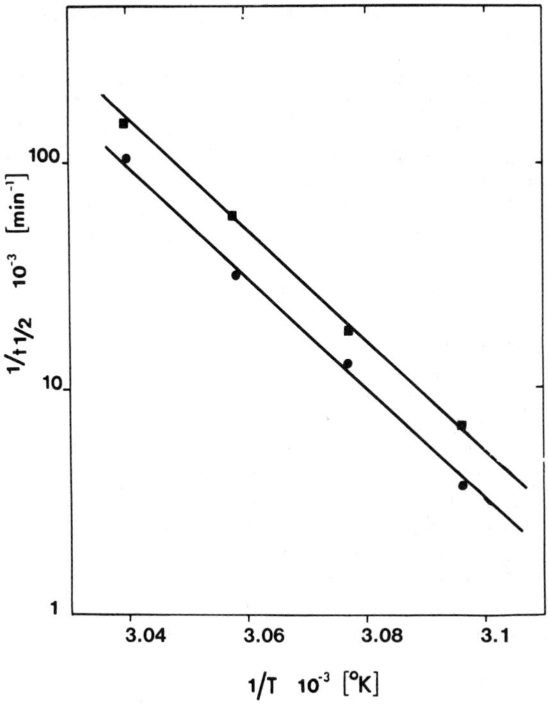

FIGURE 1. Arrhenius plot for thermal denaturation of enzyme. The t 1/2 was determined for each temperature between 50°C and 56°C for native chymotrypsin (●) and hexyl-chymotrypsin (■).

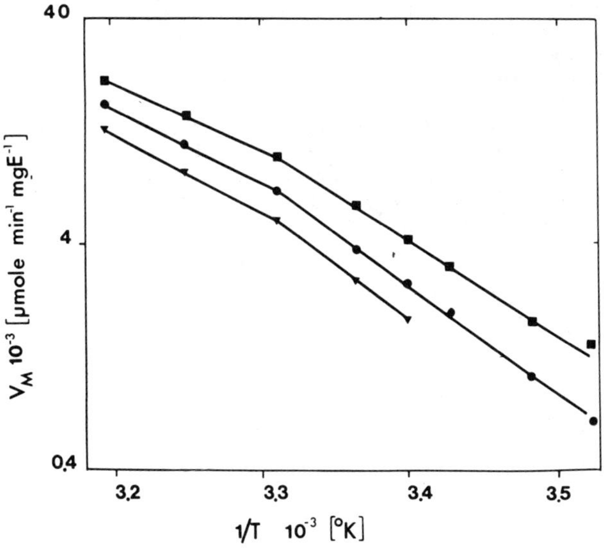

FIGURE 2. Arrhenius plot for the chymotrypsin-catalyzed hydrolysis of GPNA: native chymotrypsin (●), hexyl-chymotrypsin (■), and ethyl-chymotrypsin (▲).

due to a modification of the catalytic process itself. The energy barrier to be overcome in order to go from substrate to product was not modified. Two hypothesis can be discussed. First, the number of active sites was increased. Second, the microenvironment of the enzyme site caused an increase in the interaction of the substrate with the protein or in the solubilization of the product. The number of active sites of chymotrypsin can be measured by an assessment of *p*-nitrophenylacetate hydrolysis. For the native and hexyl enzymes, 52% and 37% of the sites were active, respectively. The increase of V_m in the case of hexyl-chymotrypsin was obviously not linked to activation of latent sites. When GPNA was used as substrate, V_m increased by 164%, and k_2 increased to 230%. For ATEE, k_3 was slightly increased to 116%.

From these results, two kinds of problems are apparent. The results must be consistent with both kinetic and thermodynamic rules. As far as the kinetics are concerned, with a substrate such as GPNA, V_m is equal to $k_2 E_o$; and the increase of V_m could be a simple increase of k_2 linked to a modification of the site or of the protein. As far as the thermodynamics are concerned, we have to explain an increase of the catalytic velocity together with a stagnation of the activation energy. This unusual phenomenon can be due to a simple and identical shift of the thermodynamic state of EA* and EA′*, when compared with EA and EA′. The activation energy is the same for both native and modified enzymes, but it is not placed at the same level in the energetics scale. The activation energies of thermal denaturation for native and modified enzyme also are the same, but the velocities of denaturation are different. A recent discussion of the interaction between the macromolecule and the active site may be of interest.[8]

REFERENCES

1. Chibata, I. 1978. Immobilized Enzymes. Hodsansha, Ed. Tokyo.
2. Remy, M. H., D. Guillochon & D. Thomas. 1981. Comparative study of native and chemically modified chymotrypsin as monomers. J. Chromatogr. **215:** 87–91.
3. Gray, G. R. 1974. The direct coupling of oligosaccharides to proteins and derivatized gels. Arch. Biochem. Biophys. **163:** 426–428.
4. Fields, R. 1971. The measurement of amino groups in protein and peptides. Biochem. J. **124:** 581–590.
5. Erlanger, B. F., F. Edal & A. G. Cooper. 1966. The action of chymotrypsin on two new chromogenic substrates. Arch. Biochem. Biophys. **115:** 206–210.
6. Rick, W. 1974. Chymotrypsin. Methods of enzymatic analysis. **2:** 1006–1009.
7. Moore, S. 1968. Amino-acid analysis. J. Biol. Chem. **243:** 6281–6284.
8. Payens, T. A. J. 1983. Why are enzymes so large? Trends Biochem. Sci. **8:** 46.

PART VI. IMMOBILIZED MICROBIAL AND PLANT CELLS AND ORGANELLES

Reproductive Immobilized Cells[a]

A. CONSTANTINIDES[b] AND G. K. CHOTANI[c]

Department of Chemical and Biochemical Engineering
Rutgers University
Piscataway, New Jersey 08854

Immobilized growing microbial systems are efficient and have potential for high-volume productivities. For example, the potential of entrapped yeast cells to produce ethanol by fermentation of solubilized sugars has been recognized by several researchers.[1-3] In almost all applications, fermenters with immobilized living cells fall into the category of three-phase (gas-liquid-solid) reactors with solid acting as a catalyst. For such heterogeneous catalyst systems, the major requirements are: active state during use, large effectiveness factor, regeneration of the catalyst, and optimum operating conditions of temperature, concentrations, and flow rates. Under the conditions of entrapment in gel particles, the yeast cells multiply in the support matrix, substrate and product diffuse, a favorable microenvironment prevails, and mainly reproductive cells are retained by the support. In a liquid-solid immobilized microbial system, gas production, for example, carbon dioxide, poses a severe problem. The diffusion of the gaseous products can be the rate-limiting step. It might accumulate in the matrix, reduce the liquid-solid contact area leading to an unsteady operation, and sometimes squash the cell-carrying particles. In order to resist internal pressure gradients due to gas production, the cell-entrapping particles might have to be made mechanically strong and elastic.

The system chosen for this study is ethanol production via anaerobic fermentation by the yeast *Saccharomyces cerevisiae* (ATCC 4126). This choice is ideal in several ways:

- *Saccharomyces cerevisiae* is one of the enzymatically stable organisms and is relatively easy to keep free of contaminants.
- The overall metabolism is simple and the only major products are ethanol and carbon dioxide when grown anaerobically.
- The organism grows well on simple sugars as the sole source of carbon and fermentation characteristics are reasonably well understood.
- The activation of the biocatalyst in the immobilized state can be easily carried out with growth nutrients and/or air.
- These yeasts are known to have no pathogenicity.
- As eukaryotes, yeasts may be able to express genes from other eukaryotes, including humans.[4]

The immobilized growing-cell reactor systems are superior to conventional free-cell systems in overcoming limitations such as low cell densities and product yields, washout, and high cost of large containers. Improved ethanol productivities have been reported by several investigators.[5-7] However, the fermenters reported so far have

[a]The authors wish to thank the National Science Foundation for supporting this work through Grants CPE 80-10865 and CPE 82-15786.

[b]To whom correspondence should be addressed.

[c]Present address: BERC, IIT Delhi, New Delhi 110016, India.

commonly been of the packed-bed type. Such reactors face several undersirable problems during their performance in terms of carbon dioxide evolution, causing dead spaces, channeling, pressure buildup, and matrix disruption. The malfunctioning of packed-bed fermenters is likely to pose uncertainties in the scale-up. In our work, we have adapted two reactor designs, based on conventionally used process equipment in the chemical industry: first, a column fermenter with beds of fluid-agitated solids; second, a shallow-bed, cross-current flow fermenter.[20]

MATERIALS AND METHODS

Medium

The growth medium contained 1% glucose, 0.5% peptone, 0.3% yeast extract, and 0.3% malt extract (pH 5.0). The fermentation medium had the composition:[1] 0.15% yeast extract, 0.25% NH_4Cl, 0.55% K_2HPO_4, 0.3% citric acid, 0.1% NaCl, 0.025% $MgSO_4 \cdot 7H_2O$ (pH 4.5), 1% KCl or 0.15% $CaCl_2 \cdot 2H_2O$. The glucose source was bakery-grade cerelose.

Organism

Saccharomyces cerevisiae ATCC 4126 was employed in these studies.

Entrapment Materials

Kappa carrageenan (Marine Colloids, NJ), sodium alginate (Stauffer Chemical Co., CT), and micron-sized silica (Glidden Pigments, MD).

Cell Immobilization

Yeast cells were cultured aerobically for twenty-four hours at 30°C in the growth medium on a rotary shaker at 200 rpm in a 500-ml indented flask. Twenty-five

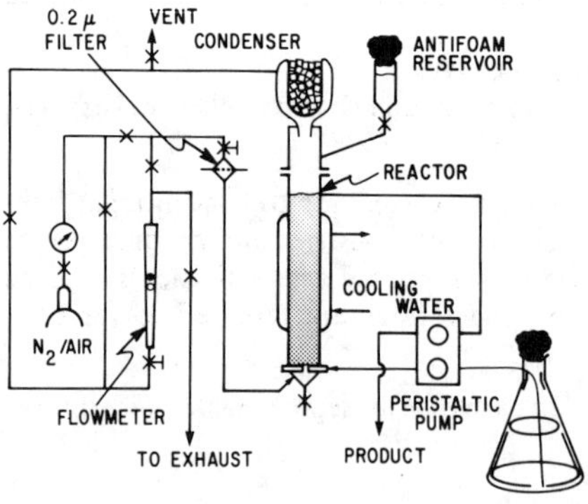

FIGURE 1. Fluid-agitated bioreactor system.

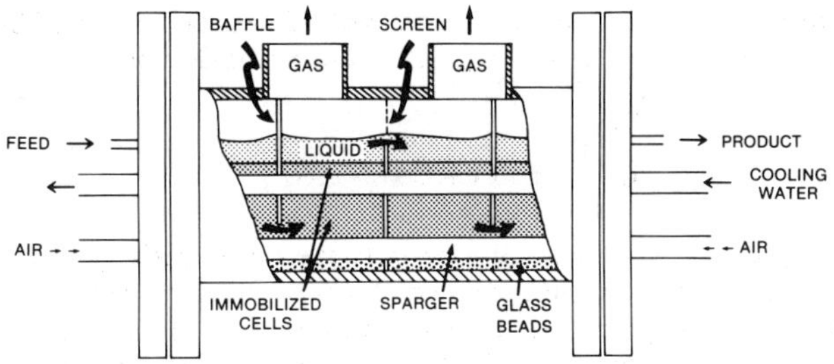

FIGURE 2. Cross-flow bioreactor system.

milliliters of the broth (7.5×10^7 cells/ml) was mixed with 75 ml of the sterile solution containing gelling agent (4 gm carrageenan or 1 gm alginate) and silica and the mixture was gelled into particles with 2% KCl or 0.65% $CaCl_2 \cdot 2H_2O$ solution, respectively, under sterile conditions.

Bioreactors

The fluid-agitated bed-type fermenter consisted of a jacketed glass column of 40 mmϕ × 210 mm with a composite liquid-gas distributor at the base. Dry nitrogen or air was used as the fluidizing medium. The detailed experimental setup is shown in FIGURE 1. Since the immobilized cell particles (1 to 4 mmϕ with carrageenan; 0.1 to 2 mmϕ with alginate) were relatively coarse, spouting[8] achieved the same purpose as fluidization does for fine solids. PALL rigimesh was used for fresh feed distribution and a sintered porous disc at the center sparged the gas and imparted momentum to the solid particles. The systematic cyclic movement of particles in a spouted bed, as against the more random motion in fluidization, may be a feature of critical value for fermentation systems. The product stream was drawn through a wire mesh candle with the help of a peristaltic pump, leaving the solids behind in the column. The product gas stream was freed of condensibles by contacting with an ice-fed condenser and its flow rate was metered with a rotameter. The fermenter was maintained at constant temperature by circulating water from a Haake temperature bath.

The cross-current flow type fermenter was similar to a shell and tube evaporator system provided with disengaging space. As shown in FIGURE 2, it consisted of a 50 mmϕ × 150 mm Lexan plastic tube with flanges at both ends. It was provided with 25% cut removable baffles mounted on two cooling tubes and separated by 45-mm-long spacers. Five baffles were used to divide the entire reactor into three sections and at the same time induce the liquid flow at right angles to the axis of the reactor. The liquid was caused to flow downward in three sections with the help of risers, to counteract buoyancy of the gas-producing solids. The end flanges carried inlet and outlet nozzles for the liquid flow. The bioreactor temperature was controlled by water circulation through the cooling tubes at a sufficiently high velocity with the help of a controlled temperature bath. The overhead gas space was connected to an ice-fed condenser to remove moisture and other volatile components for trouble-free measurements of CO_2 flow rate with the help of a rotameter. The reactor was filled with 3 mmϕ glass beads to the baffle-cut level to prevent solids from escaping along with the liquid flow. The entire reactor assembly was treated with 75% alcohol followed by a thorough wash with sterile distilled water before loading the biocatalyst particles.

EXPERIMENTAL

The entrapped cells were allowed to grow to their peak activity first in the growth medium and then in the complete medium for fermentation. There was a 100-fold increase in the cell density and total cell concentration of the pregrown biocatalyst was found to be 2×10^9 per gram with only 20% of the cells viable. The pregrown particles were loaded into the reactor and fed with mixed solutions of the basal medium and the cerelose, sterilized separately. A peristaltic pump (Pharmacia P-3) was used to match

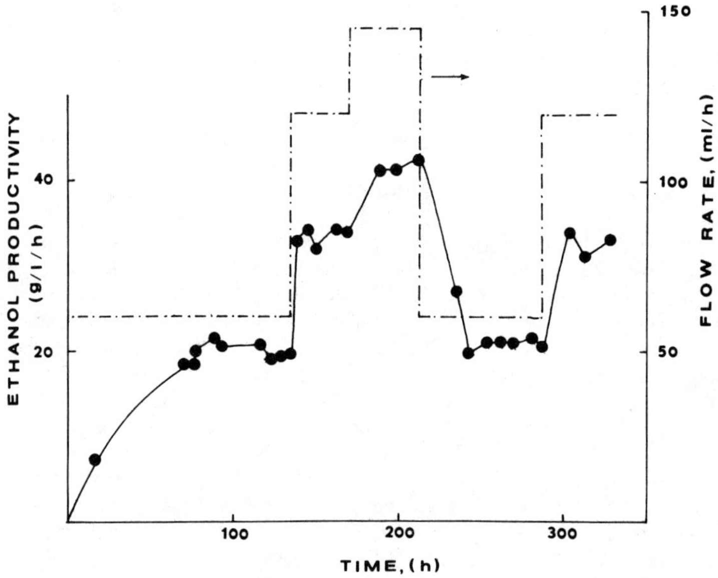

FIGURE 3. Performance of fluid-agitated bioreactor under changing feed rate.

the inflow and outflow of the liquid. The effluent samples were filtered through 0.45-micron manifold (Nalgene) diluted with 1% propanol before analyzing for ethanol and glucose.

Ethanol was assayed by gas chromtography (Hewlett Packard, Model 5880A) using propanol as internal standard. Glucose concentration was determined by an enzyme electrode method (Yellow Springs Instrument, Model 27). The immobilized cell density was determined after dissolving the particles in sterile physiological saline by shaking at 30°C for 30 min. Total and methylene blue counts[9] were obtained with a Brightline hemocytometer and an Olympus microscope. Viable cell counts were performed by spreading a 1:250,000-diluted sample on the surface of plates of agar growth medium. After four to six days of incubation at 25°C, the colonies were counted on six plates for each sample.

TABLE 1. Performance of the Fluid-Agitated Bioreactor

Gelling Agent % wt/vol	Filler % wt/vol	Particle Size mm	Reactor Volume ml	Feed Flow Rate ml/h	Ethanol in Effluent g/l	Total Cells in Effluent g/l	Fludizing Gas Rate cc/min	Gel to Liquid Ratio	Ethanol Productivity g/l gel/h
4.0[a]	8.0	4–5	200	60	40.3	0.9	80	1.5	20.2
4.0[a]	8.0	4–5	200	120	33.8	0.5	80	1.5	30.8
4.0[a]	1.0	3–4	200	120	32.9	0.6	200	0.7	49.3
1.0[b]	2.0	1–2	200	200	30.3	1.0	80	0.7	75.8
1.0[b]	1.0	0.5–1.5	60	60	31.7	0.8	None	0.7	79.2

[a]Carrageenan.
[b]Alginate.

RESULTS

Fluid-agitated Bioreactor

FIGURE 3 shows the performance of the bioreactor under varying flow rate. The notable features include fast response and negligible hysteresis of the ethanol productivity over 1.3 to 3.3 h mean residence time on a total reactor volume basis. In order to induce agitation and at the same time maintain anaerobic conditions, the system was sparged with pure nitrogen gas. At 1 vvm (gel volume basis) gas flow rate, when the height/diameter ratio was about five, the particles were in uniform motion to eliminate any particle-particle or particle-wall stationary contact. Under these operating conditions, 90% of the feed glucose could be converted, since the backmixing was limited. However, at the higher sparging rate, an excessive backmixing caused by the random motion of the particles yields lower conversion. Interestingly, no sparging gas was needed for a height-to-diameter ratio of one. In this case, carbon dioxide produced during the fermentation was sufficient to induce limited backmixing. These results have been summarized in TABLE 1. The reactor temperature was maintained at 30°C and the glucose concentration was 91.2 g/l. A substantial enhancement in ethanol productivity on gel basis was realized by the use of sodium alginate in place of carrageenan as gelling material by virtue of the reduction in mean size of the particles and hence larger surface area per unit mass of the solids. As shown in FIGURE 4, periodic aeration (for 6 h) improved the ethanol productivity by about 20% due to replenishment of the inactive cells and increase in cell density under aerobic growth.

Cross-Flow Bioreactor

Runs were made under different conditions of glucose concentration, flow rate, and temperature for a 3-section bioreactor with downward liquid flow. TABLE 2 lists the salient data. Productivities have been calculated on the basis of the solids volume. Since no sparging medium was used in this type of reactor system, the gelling solution

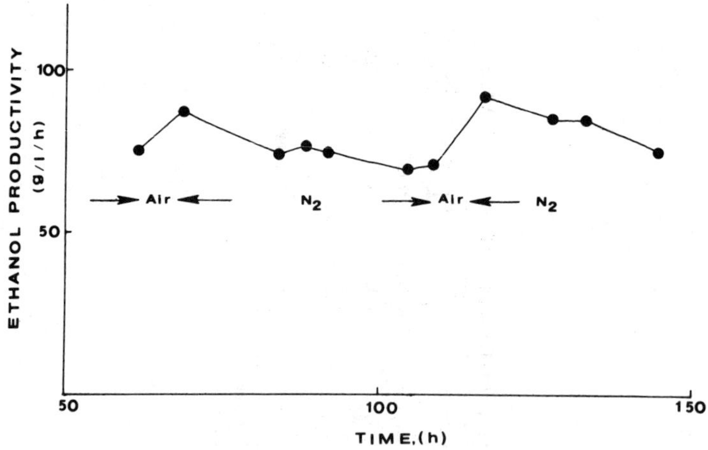

FIGURE 4. Effect of periodic aeration on ethanol productivity.

TABLE 2. Performance of the Cross-Flow Bioreactor

Glucose in Feed g/l	Temp. °C	Flow Rate ml/h	Ethanol in Effluent g/l	Cells in Effluent 10^7/ml	Cells in Particles 10^9/g	Ethanol Productivity g/l gel/h
46.5	25		10.9	2.7	3.6	20.4
	30	108.0	13.6	4.4	3.9	25.5
	35		15.8	6.2	3.6	29.6
92.3	25		20.2	4.3	4.8	17.3
	30	55.5	28.9	4.5	3.7	24.8
	35		31.8	3.8	2.9	27.2
185.0	25		56.6	4.4	2.5	14.2
	30	18.0	71.9	3.4	3.2	18.0
	35		80.5	2.4	2.4	20.1

(1% alginate) was blended with excess silica (6% wt/vol) to get denser particles. The acidic nature of the silica apparently did not enhance viscosity of the cast solution and the particles were approximately 1 mmϕ size. The conversion efficiency of glucose to ethanol was practically 100%. During the fermentation, CO_2 gas is evolved equimolarly with ethanol and its rate measurement is an excellent indicator of the reactor performance. Under steady-state conditions, cell growth is insignificant and negligible sugar is consumed for products other than ethanol. In this shallow reactor system, evolved CO_2 gas exits from the biocatalyst bed promptly and allows an intimate solid-liquid contact all the time leading to stable operating conditions. The packed fraction was ~0.4 in the liquid-catalyst layer.

Effect of Product Inhibition on Cell Growth

The growth-inhibitory effect of alcohol has been found to depend on its concentration inside the yeast cells.[10] Also, the higher the temperature, the greater is the inhibitory effect of alcohol. Furthermore, ethanol affects fermentation activity less than it does growth and, as a result, certain strains are more tolerant than others. A number of researchers have observed that reductions in growth rate, fermentation rate, and cell viability due to ethanol are all separable phenomena.[11,12] The yeast strain used in our study was tested for a simple model described by Equations 1 to 4.

$$\mu = \mu_{max}\, f(S)\, g(P)\, h(X) \tag{1}$$

$$f(S) \equiv \left(\frac{K_{S_0}}{S} + 1 + \frac{S}{K_{S_2}}\right)^{-1} \tag{2}$$

$$g(P) \equiv \left(1 - \frac{P}{P_m}\right) \tag{3}$$

$$h(X) \equiv \left(1 - \frac{X}{X_m}\right) \tag{4}$$

where Equation 1 relates specific growth rate (μ) to maximum rate of growth (μ_{max}) under the conditions of substrate inhibition, product inhibition, and cell "crowding"

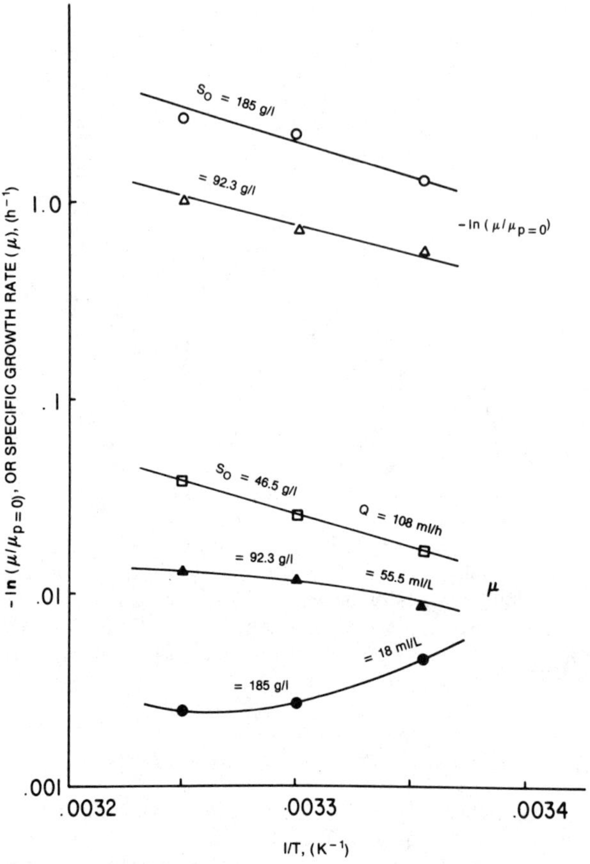

FIGURE 5. Effect of temperature on specific cell growth rate under inhibition conditions.

effect, described by Equations 2, 3, and 4, respectively. In our immobilized-cell system, the cells are more or less confined to a very small depth (about ten cell layers exist near the surface of a particle as seen under the microscope). The maximum packing density for *Saccharomyces cerevisiae* cells (7-μ spheres) has been calculated as 5.6×10^{12} cells per liter, which corresponds to 100 g/l (dry basis).[13] The maximum density of cells achievable in a particular system is determined by the growth-limiting substrate flux, the inhibitory product concentration, the maximum packing density, and the cell death rate. As a first approximation, the cell decay rate has been assumed to be negligible and the crowding effect is considered constant under all operating conditions. From the data given in TABLE 2, the shortest doubling time is 19 h at 35°C and 46.5 g/l glucose in the feed. Under the product inhibition effect, the doubling time rose to 283 h at 35°C and 185 g/l glucose level. FIGURE 5 is an Arrhenius plot for the cell growth. A distinct reversal effect of temperature can be observed for the highest glucose concentration.

Equation 3 can be considered to be one form of a more general exponential-type expression:[14]

$$g(P) \equiv \bar{e}^{k_P P} \tag{5}$$

Unlike Equation 3, the general form does not define a maximum ethanol concentration above which the cell growth ceases completely. However, it is more useful in isolating the product inhibition effect on growth. The inhibition constant (k_p) is related to temperature (Eq. 6). For simplicity, the substrate inhibition effect can be assumed to be varying negligibly with temperature.

$$k_p = k_p^o \exp(-E_p|RT) \quad (6)$$

so that

$$\mu = \mu_{P=0} \exp(-k_p^o P \exp(-E_p|RT))$$

and

$$\ell n\left(-\ell n \frac{\mu}{\mu_{P=0}}\right) = -E_P|RT + \ell n(k_p^o P) \quad (7)$$

From Equation 7 and the growth data for high glucose concentration in the feed, E_p has been estimated as 15 kcal/g mole and k_p^o is about 10^9 $(g/l)^{-1}$. The specific growth rate under no inhibition ($\mu_{P=0}$) was estimated from the lowest feed sugar level, that is, 46.5 g/l, data. Surprisingly, the activation energy of growth (E) was found to be nearly the same as that of the inhibition reactions. The constant $1/k_P$, which signifies ethanol concentration at which cell growth stops completely, was estimated from the constants E_P and k_p^o, and has the value of 50 g/l at 30°C, significantly below the tolerance limit for the fermentation activity.

Reactor Performance

In order to model product formation by the immobilized cell particles in the cross-current flow reactor, the effect of axial liquid dispersion was studied first. The tracer experimental study for the residence time distribution (RTD) was carried out by

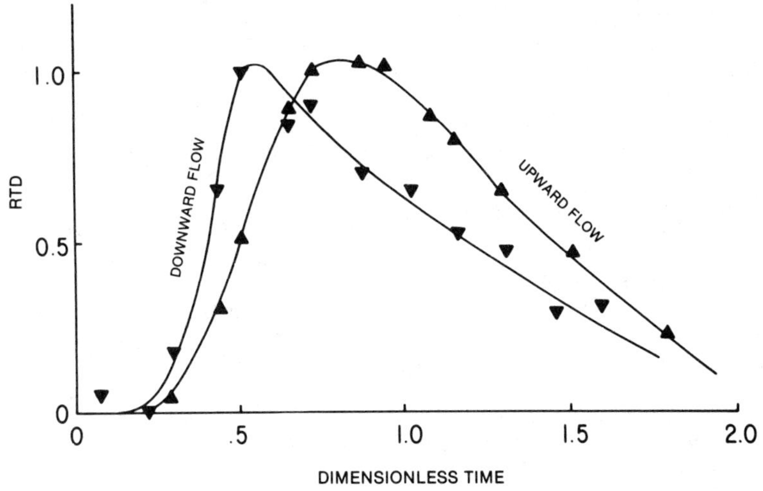

FIGURE 6. Experimentally measured RTD for the cross-flow reactor system.

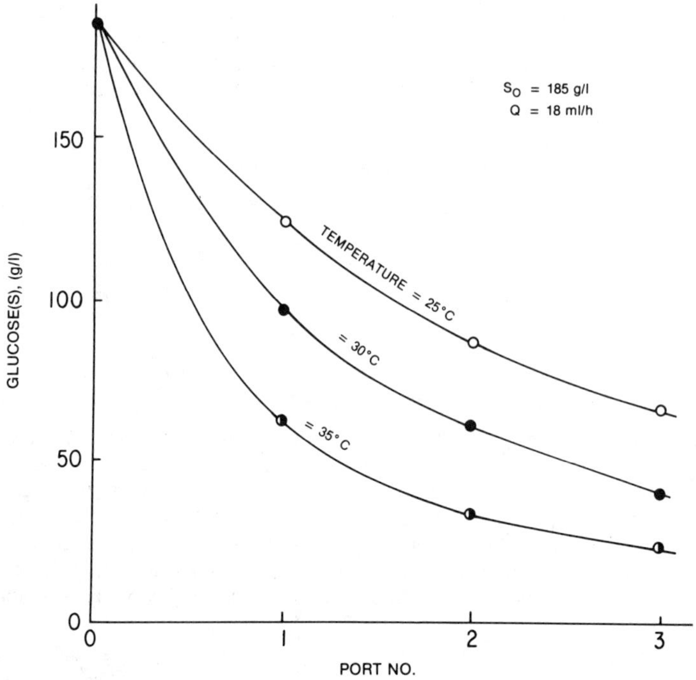

FIGURE 7. Profile of glucose concentration along the reactor axis at various temperatures.

injecting one hundred microliters of isobutanol into the feed port of the reactor. The impulse response was analyzed over the next two space times by sampling at regular time intervals. For the conditions of inlet glucose concentration 185 g/l, 30°C, 40% packing density, 7-h residence time and 180-ml operating volume distributed equally among three chambers, RTD response has been plotted in FIGURE 6. When the direction of the liquid flow through the main chambers was downward, this type of response corresponded to a train of seven stirred tanks (CSTR) in series. On reversal of the flow through the reactor, the distribution corresponded to eight CSTRs in series. Although there is no exact way to compare the tanks-in-series and dispersion models, the variances for the two models can be equated and the overall Peclet number comes out approximately as 14, showing an intermediate amount of axial dispersion.[15]

FIGURES 7 to 9 demonstrate the changes taking place between inlet and outlet ports in glucose, ethanol, and total cell concentrations. This study was done at three different temperatures: 25, 30, and 35°C. The total cell density along the length decreases due to two factors: substrate depletion and product accumulation. An increase in temperature increases the product inhibition severity and thus the cell density decreases. Therefore, the rate of fermentation varies along the length of the reactor under the accumulated effect of decrease in substrate and cell concentrations, and increase in product concentration. The effect of temperature on ethanol production rate at three different glucose concentrations: 46.5, 92.3, and 185 g/l, is shown in FIGURE 10. Based on the observation that axial dispersion, as predicted by the RTD response, is not significant, the cross-current flow reactor can be approximated as an ideal plug-flow reactor (PFR). Equations 8 to 11 describe such performance for ethanol fermentation.

$$-\frac{dS}{dt} = \frac{\eta\nu_{max}S}{K_{S_o} + S + \frac{S^2}{K_{S_2}}}\left(1 - \frac{P}{P_m}\right) \quad (8)$$

$$P = Y_{PS}(S_0 - S) \quad (9)$$

$$\eta\nu_{max}\bar{t} = \frac{P_m}{K_{S_2}}\left(\frac{S_0 - S}{Y_{PS}}\right) + \frac{\left(\frac{P_m}{Y_{PS}} - S_0\right)\left(K_{S_2} - \frac{P_m}{Y_{PS}} + S_o\right) - K_{S_o}K_{S_2}}{P_m - Y_{PS}S_o} \cdot \ell n \frac{P_m}{P_m + Y_{PS}(S - S_0)} + \frac{K_{S_o}K_{S_2}}{P_m - Y_{PS}S_o}\ell n \frac{S_o}{S}\Bigg) \quad (10)$$

where $Y_{PS} = 0.5$, $P_m = 95$ g/l, $K_{S_o} = 1$ g/l, $K_{S_2} = 450$ g/l.[16–18] The effective maximum rate ($\eta\nu_{max}$), estimated from the residual substrate concentration(s), and the residence time ($\bar{t}$) when plotted against 1/T gives overall activation energy as slope for three different glucose concentrations in feed: 46.5, 92.3, and 185 g/l. For the initial part of the reactor and at the highest feed concentration of glucose, the overall kinetics seems to be reaction controlled with activation energy of 18 kcal/g mole. However, in the

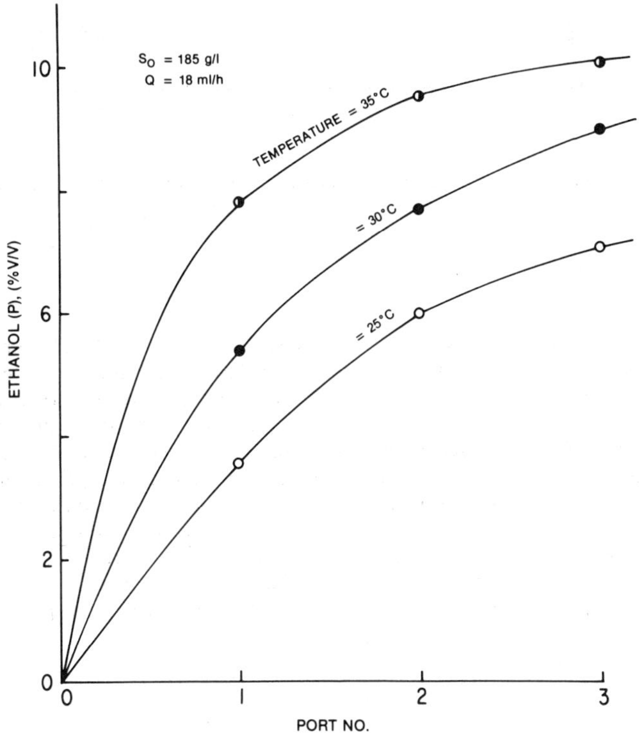

FIGURE 8. Profile of ethanol concentration along the reactor axis at various temperatures.

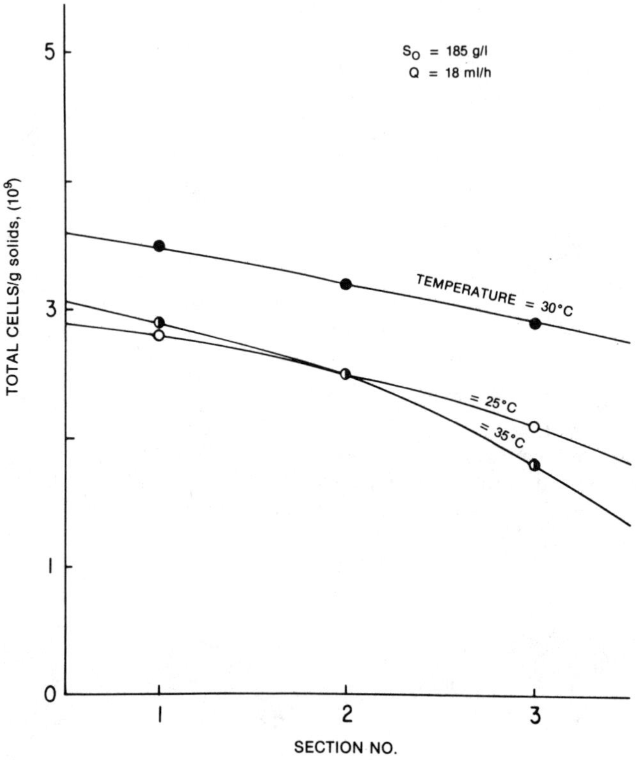

FIGURE 9. Profile of immobilized total cell density along the reactor axis at various temperatures.

latter part of the reactor, mass transfer effects play a significant role. The fermentation kinetics appears to be substrate limitation controlled at feed glucose of 46.5 g/l. At intermediate concentration, that is, 92.3 g/l, activation energies were found in the range of 5 to 16 kcal/g mole. In the above analysis, as first approximations, the ethanol tolerance limit (P_m) was assumed invariant over the studied temperature range, and the cell densities were assumed to be constant over the entire length of the reactor. However, lack of fit ($R^2 = 0.904$) results for the data obtained at extreme conditions of the temperature and the glucose concentration in the feed due to these assumptions. As shown in FIGURE 9, the steady-state cell density varies significantly with position in the reactor at 35°C operating temperature. Therefore, the maximum rate (ν_{max}) has to be set proportional to the cell growth rate (μ) in which case, the rate expression should be

$$-\frac{dS}{dt} = \eta\nu_{max}(T)\left(\frac{S(1 - P/P_m)}{K_{S_o} + S + S^2/K_{S_2}}\right)^2 \qquad \textbf{(11)}$$

Equation 11 yields, on integration, an expression similar to Equation 10, which fits the data extremely well with $R^2 = 0.999$ for $S_o = 185$ g per liter.

DISCUSSION AND CONCLUSIONS

The immobilized systems discussed in this paper offer high-productivity alternatives to the production of growth-associated products such as ethanol as exhibited by nearly theoretical yield and minimal cell synthesis. Under diffusion-limited conditions, the activity is more or less confined to the outer shell of the particles, resulting in low effectiveness factors. This disadvantage can be overcome by using either extremely fine particles in fluidized-bed reactor or a shell-structured catalyst that consists of an inert core where no active cells are present and an outer shell containing the immobilized growing cells. The desirable properties such as size and density required for fluidization can be easily manipulated by the type of gel and various grades of micron-sized silica (acidic or neutral pH, porosity, etc.). An important characteristic of the fluidized bed is the constant uniform movement of the particles, that is, the cells on the immobilizing particles move from low substrate and high product concentration near the outlet to high substrate and low product concentration towards the inlet part of the reactor. A constant biomass can be achieved by ensuring a steady state between rates of growth and cell leakage from the immobilizing matrix under fluid dynamic forces and collisions among particles. TABLE 3 makes a comparison of fluid-agitated bed and cross-current fermenters with fixed-bed type. A distinctive feature of a fluid-agitated bioreactor operating under air sparge such as a spouted bed, is the aerobic zone at the

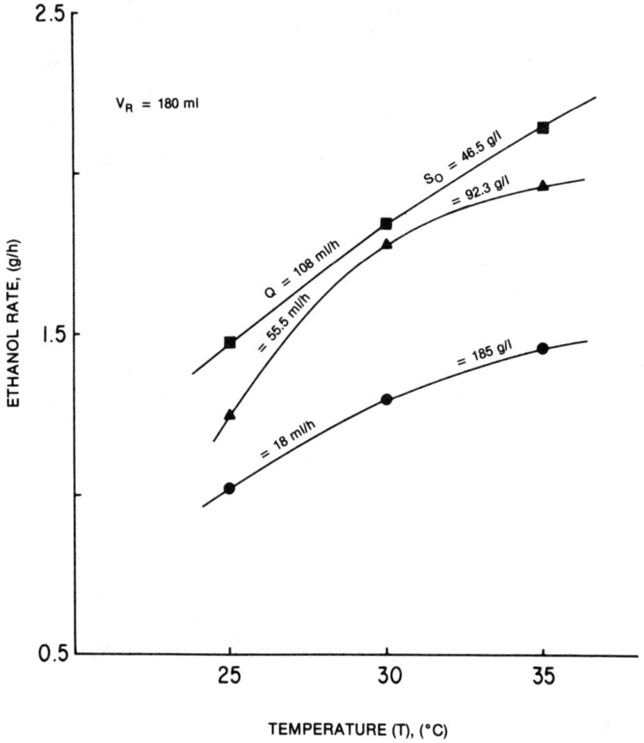

FIGURE 10. Effect of temperature on rate of ethanol production at various glucose concentrations in feed.

central core and nearly oxygen-free conditions exist in the annulus. In this system, the residence time of the biocatalytic particles in the aerobic phase is very short because of the high air velocity, and the intense turbulence gives high oxygen transfer. Therefore, very limited aerobic growth takes place leading to cyclic activation of the particles emerging from the inactiviting fermentation phase throughout the annulus. Alternatively, by drawing an analogy from the conventional petroleum cracker-regenerator using heterogeneous catalyst bed reactors, a dual-column fluidized bed system is an operationally stable design for continuous fermentation by immobilized cells in their viable state. A stream of solids can be added to and withdrawn from the bed, retention time of which can be controlled by adjusting the ratio between the feed rate and the volume of the bed. Such a system would dampen any instabilities likely to occur during continuous operation of a bioreactor. An inert gas (nitrogen) in an anaerobic system and air in an aerobic system can be used as a fluidizing medium with the liquid flow upward or downward through the bed making a very effective three-phase reactor system.

For continuous ethanol fermentation, an ideal PFR is the most effective design of the reactor. In a cross-current flow reactor, by increasing the number of cross-flows, the performance is expected to match that of a PFR. In other words, ethanol inhibition effect can be attenuated by lowering liquid back-mixing effect.[19] A cross-flow reactor system allows various ideal reactor configurations required for large-scale production of chemicals under substrate and production-inhibited fermentations. A large baffle spacing followed by smaller spacing is an example of a minimized volume setup of CSTR and PFR in series, found optimum for immobilized yeast cell fermentation with high sugar concentration feed.

TABLE 3. Comparisons of Immobilized Cell Bioreactors

Fixed Bed	Fluid-agitated Bed	Cross-Flow
No mixing	Good mixing	Intermediate mixing
Low fermentation rate	High fermentation rate	High fermentation rate
Unstable operation	Stable operation	Stable operation
Low effectiveness factor	High effectiveness factor	High effectiveness factor
Radial temperature gradients	Good temperature control	Good temperature control
Solids-free feed	Suspensions processable	Solids-free feed
Low axial dispersion	High axial dispersion	Intermediate axial dispersion

REFERENCES

1. WADA, M., J. KATO & I. CHIBATA. 1978. A new immobilization of microbial cells. Eur. J. Appl. Microbiol. Biotechnol. **8**(4): 241–247.
2. KIERTAN, M. & C. BUCKE. 1977. The immobilization of microbial cells, subcellular organelles and enzymes in calcium alginate gels. Biotechnol. Bioeng. **19:** 387–397.
3. WILLIAMS, D. & D. M. MUNNECKE. 1981. The production of ethanol by immobilized yeast *Saccharomyces cerevisiae* cells. Biotechnol. Bioeng. **23**(8): 1813–1826.
4. SPINKS, A. 1982. Targets in biotechnology. Proc. R. Soc. Lond. B Biol. Sci. **214**(1196): 289–304.

5. SITTON, O. C. & J. L. GADDY. 1980. Ethanol production in an immobilized cell reactor. Biotechnol. Bioeng. **22:** 1735–1748.
6. GHOSE, T. K. & K. K. BANDYOPADHYAY. 1980. Rapid ethanol fermentation in immobilized yeast *Saccharomyces cerevisiae* cell reactor. Biotechnol. Bioeng. **22**(7): 1489–1496.
7. MOO-YOUNG, M., J. LAMPTEY & C. W. ROBINSON. 1980. Immobilization of yeast cells on various supports for ethanol production. Biotechnol. Lett. **2**(12): 541–548.
8. MATHUR, K. B. & N. EPSTEIN. 1974. Spouted Beds. Academic Press. London.
9. THOMAS, D. C., J. A. HUSSACK & A. M. ROSE. 1978. Plasma-membrane lipid composition and ethanol tolerance in *Saccharomyces cerevisiae.* Arch. Microbiol. **117:** 239–245.
10. NAVARRO, J. M. & G. DURAND. 1978. Alcohol fermentation: Effect of temperature on ethanol accumulation within yeast cells. Ann. Microbiol. **129B:** 215–224.
11. TROYER, J. R. 1953. A relation between cell multiplication and alcohol tolerance in yeasts. Mycologia. **45:** 20–39.
12. BROWN, S. W., S. G. OLIVER, D. E. F. HARRISON & R. C. RIGHELATO. 1981. Ethanol inhibition of yeast growth and fermentation. Eur. J. Appl. Microbiol. Biotechnol. **11:** 151–155.
13. PIRT, S. J. 1975. Principles of microbe and cell cultivation. Blackwell Scientific Publications. Oxford. pp. 12–13.
14. NAGATANI, M., M. SHODA & S. AIBA. 1968. Kinetics of product inhibition in alcohol fermentation. J. Ferment. Technol. **46**(3): 241–248.
15. BAILEY, J. E. & D. F. OLLIS. 1977. Biochemical Engineering fundamentals. McGraw-Hill. New York. p. 542.
16. LEVENSPIEL, O. 1980. The monod equation: A revisit and a generalization to product inhibition situations. Biotechnol. Bioeng. **22:** 1671–1687.
17. BAZUA, C. D. & C. R. WILKE. 1976. Effect of alcohol concentration on kinetics of ethanol production by *S. cerevisiae.* First Chemical Congress of the North American Continent. Mexico City. Dec. 1–5.
18. LEE, J. M., J. F. POLLARD & G. A. COULMAN. 1983. Ethanol fermentation with cell recycling: Computer simulation. Biotechnol. Bioeng. **25:** 497–511.
19. YAMANE, T. & S. SHIMIZU. 1982. The minimum-sized ideal reactor for continuous alcohol fermentation using immobilized microorganism. Biotechnol. Bioeng. **24:** 2731–2737.
20. CHOTANI, G. K. & A. CONSTANTINIDES. 1984. Immobilized cell cross-flow reactor. Biotechnol. Bioeng. **26:** 217–220.

NOMENCLATURE

E	activation energy of cell growth (kcal g mole^{-1})
E_p	activation energy of product inhibition (kcal g mole^{-1})
k_p	product inhibition constant (lg^{-1})
k_p^o	constant defined in Arrhenius equation (6)
K_{S_0}	substrate saturation constant (g/l)
K_{S_2}	substrate inhibition constant (g/l)
P	product concentration (g/l)
P_m	product concentration limit (g/l)
Q	feed flow rate (ml/h)
S	substrate concentration (g/l)
S_o	substrate concentration in feed (g/l)
t	time (h)
$\bar{t}$	residence time (h)
T	temperature (K)
X	cell concentration (g/l)
X_m	cell concentration limit (g/l)
Y_{PS}	substrate to product yield constant

GREEK

η	effectiveness factor
μ	specific cell growth rate (h^{-1})
μ_{max}	specific cell growth rate limit (h^{-1})
$\mu_{P=O}$*	growth constant at no product inhibition (h^{-1})
ν_{max}	specific reactor productivity limit ($gl^{-1}h^{-1}$)
$\eta\nu_{max}$	effective specific reactor productivity limit (gl^{-1}/h^{-1})

Microbiological Measurements by Immobilization of Cells within Small-Volume Elements

JAMES C. WEAVER, PETER E. SEISSLER, STEVEN A. THREEFOOT, JEFFREY W. LORENZ, TIMOTHY HUIE, AND RONALD RODRIGUEZ

Harvard-M.I.T. Division of Health Sciences and Technology
Cambridge, Massachusetts 02139

ALEXANDER M. KLIBANOV

Department of Nutrition and Food Science
Massachusetts Institute of Technology
Cambridge, Massachusetts 02139

Rapid microbial measurements have long been recognized to be important in a variety of fields.[1–4] There are important reasons for wanting rapid determination of the count or enumeration, and also rapid identification of the microorganism. Here we report two versions of a general approach based on the use of small-volume elements (SVEs). The use of gel microdroplets as a general means of providing manipulable SVEs has been described by Weaver.[5]

The purposes of this paper are to describe (1) the basic concept of using small-volume elements (SVEs), (2) the possibility of a purely electrical approach, and (3) the use of gel microdroplets (GMDs) as SVEs.

BASIC CONCEPTS

Cultivation of microorganisms on or within a gel-containing growth medium is a well-established method for obtaining an enumeration or count of the microorganisms. At a suitable dilution, each cell is sufficiently separated from other cells such that distinct visible colonies are formed after many (typically 25 to 30) cell divisions.[6] Strictly speaking, this approach provides a count of the concentration of "colony forming units." Some microorganisms naturally form clusters or aggregates while others do not. Thus, colony-forming units may represent either aggregations of cells or isolated single cells. It is the concentration of these entities that is determined when obtaining a count.

A well-known attribute of conventional cultivation is its slowness. Rapidly growing organisms such as *Escherichia coli* have doubling times of 20 min under proper conditions. Even so, approximately 12 hours are required to obtain colonies that can be readily counted on a petri dish. More slowly growing organisms require a correspondingly longer time to reach colony sizes that are detectable, and can thereby form the basis of a count. Although these properties of conventional cultivation are well known, they are repeated here to emphasize an important but fundamental point: traditional methods for obtaining a count depend on growth (replication), *not* metabolism.

In contrast, more recently developed methods such as the collection and measurement of radioactive $^{14}CO_2$, or the measurement of the change in electrical properties of the medium are fundamentally based on metabolism.[7-9] However, these metabolically based methods do not provide a count. Further, as shown schematically in FIGURE 1, if stressed or injured cells are present, a macroscopic metabolic method can give a false negative. This shortcoming is particularly serious in the case of testing for sterility. A significant attribute of the SVE approach is that it is possible to obtain rapid measurements while still obtaining a count. Simply put, this possibility is based on the fact that metabolic measurements are made on individual cells or a small number of cells that originate from one single cell or colony-forming unit. Because the SVE approach is based on measurements of individual cells, the measurement time is essentially independent of the count (cell concentration in the sample).

Still other recently developed methods are based on assays of composition, and are essentially insensitive as to whether the cells are living or dead. Examples of such methods are the use of labeled antibodies to surface antigens, pyrolysis–mass spectrometry and genetic probes.

The basic idea underlying the SVE approach is illustrated in FIGURE 2. A macroscopic liquid sample is assumed to be well stirred, such that any microorganisms are randomly distributed. For purposes of illustration, we consider a hypothetical example in which the average cell concentration is 10^5 cells/ml. Then, as a purely mathematical construction, we imagine subdividing the macroscopic volume into a large number of identical microscopic small-volume elements (SVEs). For clarity, we imagine the SVEs to be small cubes. As the hypothetical SVEs are made smaller and smaller, a point is reached wherein most of the SVEs contain no cells. Those SVEs that are occupied have a high probability of containing 1 cell. The probability of initial multiple occupation (2 or more cells) is low.

A significant property of the singly occupied SVEs is that the effective cell concentration is high. For example, if the volume of each SVE is 10^{-5} ml, then the presence of 1 cell corresponds to a concentration of 10^5 cell/ml. We now imagine that a

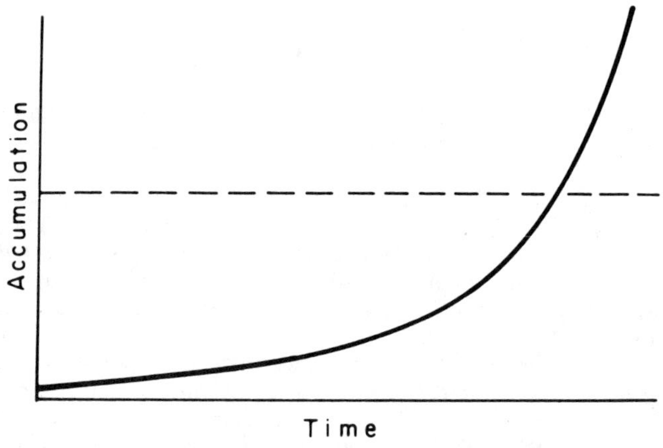

FIGURE 1. Schematic representation of accumulated metabolic signal from a total population of microorganisms. With stressed or injured cells, the signal ($^{14}CO_2$ or electrical conductance change) will cross the threshold value (dashed line) later, constituting a false negative.

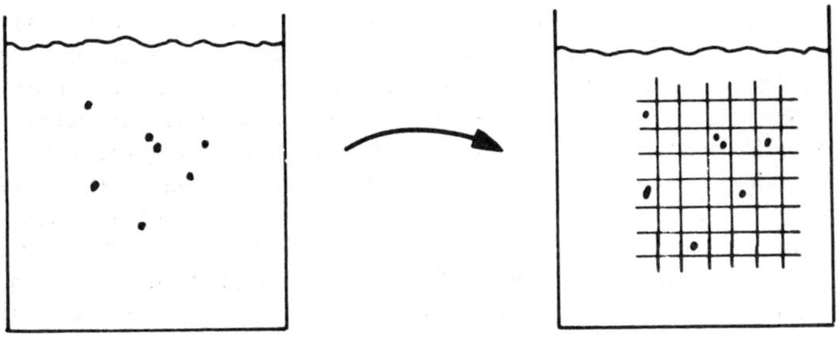

FIGURE 2. Schematic representation of a macroscopic volume containing randomly distributed microorganisms. Right portion represents a hypothetical or mathematical subdivision of the macroscopic volume into small-volume elements (SVEs), represented here as small cubes.

barrier to mass transport of metabolites and other cell products surrounds each SVE. Because metabolites often accumulate rapidly in macroscopic cell suspensions with an average concentration of 10^5 cell/ml, we expect occupied SVEs to experience a correspondingly rapid change in metabolite concentration.

General expectations for the SVE approach include both counting (of occupied and unoccupied SVEs), and quantitative determination of the metabolite concentration in each SVE. Specifically, if the number of unoccupied ($n = 0$) and occupied ($n = 1$ and $n \geq 2$) SVEs can be determined, and the volume of the GMDs can also be determined, the concentration of cells (the count) can be obtained. The sum of the individual SVE volumes yields the total volume of the measured sample. At the same time, the digital process of simply counting the number of occupied SVEs provides the total number of cells in the measured sample. The use of Poisson statistics can allow for correction for the few multiply occupied SVEs. Continuing our example based on 10^{-5} ml SVEs, the count is expected to be obtained rapidly compared to the petri dish method. Rapidly growing microorganisms that require 12 hours for a count by plating should yield a count by the SVE method in an hour or less, given that these same microorganisms cause measurable changes in metabolite concentration in an hour or less in a macroscopic sample.

Because the SVE method is based on measurements on the metabolic effects of single cells or their progeny (microcolonies), the variability of individual cells can also be investigated.

ELECTRICAL CONDUCTANCE APPROACH

Others have shown that a careful measurement of the electrical properties of a growth medium is closely related to the total metabolism of a growing suspension of microorganisms.[8,9] Here it is important to emphasize the value of obtaining a *final* output that is an electrical signal.[10] A basic attribute of the SVE method is the possibility of collecting a large amount of microscopic data from a macroscopic sample. A large microscopic data stream can be expected, because several measurements or assays can be made on the contents of each SVE. Fortunately, the widespread existence of microprocessors provides a realistic way to deal with a large data stream.[10,11]

Because computers fundamentally operate with electrical signals, the metabolic-based measurement on each SVE must eventually be converted into an electrical signal.

The electrical conductance of a medium containing viable cells generally changes with time. Microbial conversions change uncharged nutrients into dissociable species, such as low-molecular-weight acids. A number of rapidly growing organisms produce conductance changes that can be measured in a macroscopic suspension of cells. At a concentration in the range 10^5 to 10^6, cell/ml measurable conductance changes occur within 1 hour.[8,9] Not only can rapidly growing organisms be detected, slowly growing organisms have a sufficiently high level of metabolism that they also can be measured. Thus, measurement of electrical conductance is a candidate for use in the SVE approach.

The electrical conductance-based measurement of microbial metabolism is similar to the measurement of radioactive $^{14}CO_2$ release that occurs when radiolabeled nutrients are provided.[7] In both cases, stressed or injured cells can lead to false negatives.

APPLICATION OF ELECTRICAL CONDUCTANCE MEASUREMENTS TO SVEs

Microfabrication techniques developed primarily by the semiconductor industry can be adopted to make large arrays of small devices.[12–15] Thus, as shown in FIGURE 3a, it is possible to envision a sheet of insulating material perforated with a number of holes, each forming a small cylindrical volume. As shown in FIGURE 3b, platinum thin-film electrodes can be vacuum deposited at each end, thereby providing an SVE that is a miniature chamber wherein electrical conductance measurements can be made. Electrical access to each could be made by thin-film metallic electrodes, with electrical insulation provided by a passivating layer such as sputtered silicon nitride.

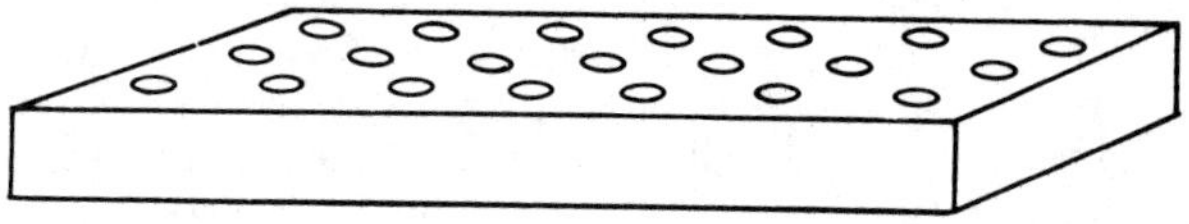

(a)

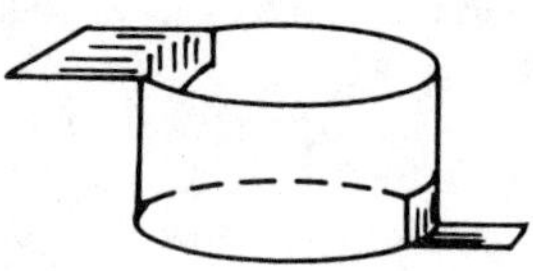

(b)

FIGURE 3. Electrically insulating sheet containing in (a) a number of small cylindrical holes, which serve as SVEs. (b) is a close-up view of one such SVE, showing possible electrical connection via vacuum-deposited thin-film inert metal electrodes.

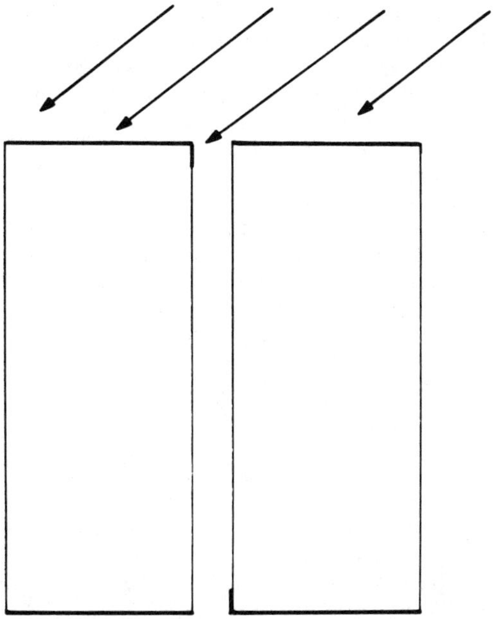

FIGURE 4. Cross-section sketch of single prototype SVE made from a segment of a precision-bore glass capillary. Thin-film metal electrodes were vacuum deposited at about a 45-degree angle, so that electrical contact between the SVE interior and an external electronic conductance-measuring apparatus could be made.

In order to explore experimentally the general possibility of the use of the SVE-based approach, several individual SVEs were constructed by cutting precision-bore glass capillaries into 3-mm lengths. Following careful cleaning, an initial thin adherence film of chromium was applied at about 45 degrees by vacuum deposition (FIG. 4). Subsequently a thin film (about 1000 Å) of platinum was similarly applied, providing an inert metal electrode at each end of the cylindrical SVE, which penetrated partly into the SVE. Macroscopic metal electrodes were pressed against the end of each SVE, allowing connection one at a time to an external impedance bridge that was operated at frequencies in the range of 10^3 to 10^5 Hz. Temperature control was passive, provided by enclosing ten such SVEs within a styrofoam container.

In spite of several modifications of this basic exploratory apparatus, a major difficulty became apparent. Sampling of a liquid could be readily accomplished by initially filling the SVEs with water, and then exploiting capillary action to draw in a sample. However, evaporative loss was fairly rapid, in spite of many precautions. Evaporative loss of water from the aqueous solution filling a SVE causes two major problems: (1) the ionic concentrations increase, changing the conductance, and (2) the temperature changes, also changing the conductance because the temperature coefficients of typical media are about 1.5%/°C.[9] Severe signal drift therefore occurred within a short time, usually a matter of minutes. Although it is possible to consider strategies such as immersion of such SVEs in oil after filling with sample, such a protocol appears awkward.

The electrical conductance approach was not pursued further when several other problems were identified. Five anticipated problems associated with direct electrical measurements on an array of SVEs are summarized here: (1) The SVE approach is attractive if large numbers of SVEs can be routinely used. An $n \times n$ array with $n = 10$ provides 10^2 electrical SVEs, requiring 2n = 20 connections to the array. However if 10^6 SVEs are to be used, the $n \times n$ array requires $2n = 2 \times 10^3$ connections. Thus, there is a problem concerning the connections to a semiconductor chip, which are a weak link. The formidable number of connections required is a deterrent. (2) Another problem arises since the electrical thin-film leads on an array must be passivated, so that the array can be readily contacted with an aqueous sample. (3) Following a sampling protocol involving an aqueous sample, a thin film of aqueous electrolyte may reside on the surface of the array. Yet the SVEs must be electrically isolated. (4) Evaporative loss is a problem because it can alter electrolyte concentration and cause temperature changes, with both phenomena altering the SVE conductance. (5) And finally, electrical conductance changes associated with microbial metabolism are general in the sense that changes in concentration of different metabolites can cause the same change in conductance. There is no biochemical specificity.

To summarize, although direct electrical measurements are attractive in that an electrical signal output is obtained, the application to routine measurements on *large* numbers of SVEs has several significant anticipated problems. For these reasons another approach was sought.

OPTICAL MEASUREMENTS AND SVEs

The combination of SVEs and optical measurements appears much more attractive. A tremendous number of biochemical assays exist that are based on optical measurements, primarily colorimetric or flourescent.[16] These assays are often based on enzymatic analysis[17,18] or on immunoassays.[19,20] By using different wavelengths, it is often possible to carry out several measurements simultaneously. Further, optical measurements are rapid and noncontacting. Thus, it should be possible to make separate measurements rapidly on large numbers of SVEs. Finally, if contact is not required, the SVEs can be mobile and therefore manipulable. This attribute is extremely important, because it provides for the possibility of cell sorting and screening based on measurements of either single cells or microcolonies.

GEL MICRODROPLETS

Gel microdroplets (GMDs) are microbially and optically compatible SVEs that can be readily produced in large numbers. GMDs consist of small beads of gel within which microorganisms can be entrapped. For rapid, metabolic-based measurements, GMDs can be provided with a barrier to mass transport of metabolites by either a coating with a low-permeability substance or by suspension in a low-solubility liquid.

One method of producing GMDs is shown in FIGURE 5. A liquid medium containing nutrients, indicator dyes, a gelable substance such as sodium alginate and microorganisms are forced through a vibrating nozzle. The stream of liquid is broken up into liquid droplets with a narrow distribution of sizes. The liquid droplets become approximately spherical because of surface tension at the interface of the droplets and fluid into which the droplets emerge. The diameter of the liquid droplets is a function of the orifice diameter, flow rate, and vibrational frequency.[21] The creation of liquid droplets by this method is the basis of ink jet printing, and of flourescence-activated cell sorting.

As an example, we have produced calcium alginate GMDs by this method. A medium containing sodium alginate is passed through a 50-micron piezoelectric axially vibrated orifice that is driven at about 2×10^4 Hz. Thus, in 50 sec, 10^6 droplets are produced. The liquid droplets can be charged electrically, so that they slightly repel each other as they travel to a stirred vessel containing calcium chloride. This repulsion prevents droplets from colliding and coalescing, particularly at the surface of the calcium chloride solution. In a gently stirred suspension, the liquid droplets gel by exchanging Ca^{2+} for Na^{+} and forming ionic crosslinks. Alginate GMDs are denser than water, and will settle within minutes if not stirred. Nevertheless, there is no significant aggregation. Alginate GMDs can be stored for months at room temperature without aggregation or apparent degradation. FIGURE 6 is a photograph of 90-micron alginate GMDs containing litmus for easy visualization.

This process generates GMDs with a fairly narrow size distribution. An important attribute of the vibrating orifice method is that it is a serial process. If needed, several orifices could be used in parallel. Recent developments in ink jet printing technology (which is based on the vibrating orifice method for producing liquid ink droplets) demonstrate that miniature droplet generators can be fabricated using the photolithographic processes of the semiconductor industry.

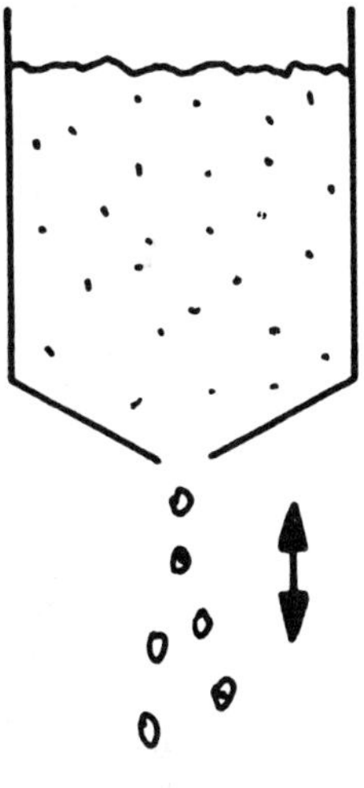

FIGURE 5. Simplified drawing of a droplet generator based on a vibrating orifice.

GMDs can also be readily produced by a simple parallel process.[22] A liquid sample containing a substance capable of undergoing a temperature-induced gelation is added to a moderately sitrred liquid hydrocarbon. Dispersion rapidly occurs, often producing a population of GMDs with a wide distribution of sizes. Similar procedures, usually employed for producing larger gel particles, have been reported by Nilsson and co-workers.[23,24]

A typical procedure of ours consists of the following: 0.3 ml of 3% agarose (Sigma type VII) solution is introduced into a 50-ml beaker containing 35 ml of 38 C mineral oil while stirring moderately. The agarose solution also contains a defined growth medium, the colorimetric indicator bromothymol blue at a concentration of 4×10^{-3} M and a test concentration of a dye-tolerant strain of *E. coli*. Another sample of the test concentration is plated to obtain a count. The defined medium contains 1×10^{-3} M $NaNH_4HPO_4$, 1×10^{-4} M KCl, 1×10^{-6} M $CaCl_2$, 1×10^{-4} M $MgSO_4$, 5×10^{-2} glucose supplemented with essential amino acids and vitamins. The initial pH is about 7.6, so that the GMDs are initially deep blue. The moderate stirring of the mineral oil disperses the aqueous liquid sample into droplets within seconds of introduction. While stirring is continued, the 50-ml beaker is surrounded by an ice bath for 15 minutes, causing gelation of the liquid droplets and thereby creating GMDs. A coarse nylon filter (typically 88-micron square mesh opening) is used to remove any larger GMDs. An aliquot of the filtered GMD suspension in mineral oil is placed in a microconcavity slide. The GMDs are placed in a 37°C environment, in which the larger GMDs settle to the bottom of the miroconcavity slide within 45 minutes. The slides are removed at intervals for examination and photography under a light microscope.

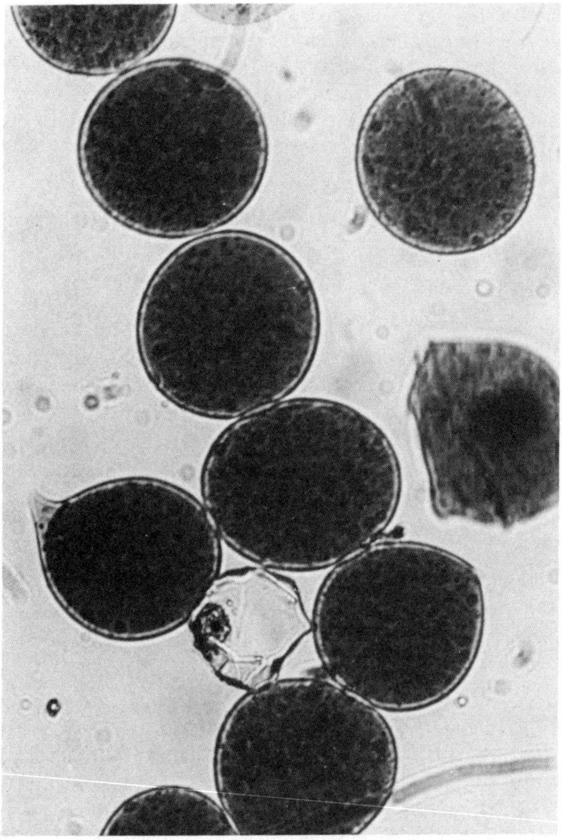

FIGURE 6. Reproduction of a photomicrograph of 90-micron-diameter calcium alginate GMDs.

In the experiments described here, a colorimetric rather than fluorescent pH indicator dye is used in order to determine whether well-known simple pH indicators can be used with GMDs to test for the presence of bacteria. As a further condition of the test, no effort was made to force anaerobic conditions (beyond partially degassing the mineral oil before use in order to minimize bubble formation upon stirring).

The following sequence of observations was typical. Initially the entire population of GMDs is deep blue. After 5¾ hours, small (about 20-micron diameter) occupied

GMDs have changed to yellow. After 9¼ hours, even the largest (about 70-micron diameter; about 40 times larger volume) occupied GMDs have also begun to change color. In another two hours, a bright yellow color is achieved in the occupied GMDs. A negative control (uninoculated sample) population of GMDs remained deep blue even after 24 hours.

A detailed analysis of the size distribution was not made. Instead, simple inspection showed that most of the sample volume was represented by GMDs with a diameter of about 70 microns. The corresponding volume was $V_{GMD} = 1.8 \times 10^{-7}$ ml. The independently determined initial cell concentration was 9×10^5 cell/ml. The estimated average number of cells per 70-micron GMD was thus 0.16. Straightforward application of Poisson statistics yields the following probabilities:

$$P(n,\bar{n}) = (\bar{n})^n e^{-\bar{n}}/n!$$

where $\bar{n} = 0.16$ and n is the actual number. We now estimate the probability of an unoccupied GMD ($n = 0$), the probability of a singly occupied GMD ($n = 1$), and the probability of a multiply occupied GMD ($n \geq 2$), with $\bar{n} = 0.16$. The estimates showed that $P(0,\bar{n}) = 0.85$; $P(1,\bar{n}) = 0.14$; and $P(\bar{n} \geq 2) = 0.01$. The experimental observations are consistent with these probabilities.

ACKNOWLEDGMENTS

We thank A. J. Sinskey, R. F. Gomez, and S. K. Burns for stimulating discussions and E. J. Costa for experimental assistance during early phases of this investigation. This work was supported partially by Cambridge Research Laboratory: Ortho Diagnostics Systems Inc.

REFERENCES

1. Hedén, C-G & T. Illéni Eds. 1975. New Approaches to the Identification of Microorganisms. Proceedings of a Symposium on Rapid Methods and Automation in Microbiology. John Wiley & Sons. New York.
2. Sharpe, A. N. & D. S. Clark, Eds. 1978. Mechanizing Microbiology. Thomas. Springfield, MA.
3. Johnston, H. H. & S. W. B. Newsom, Eds. 1977. Rapid Methods and Automation in Microbiology. Research Studies Press. Forest Grove, Oregon.
4. Tilton, R. C., Ed. 1982. Rapid Methods and Automation in Microbiology. Proceedings of the Third International Symposium on Rapid Methods and Automation in Microbiology. American Society for Microbiology. Washington, DC.
5. Weaver, J. C. 1983. Process for isolating microbiologically active material. U. S. Patent 4,339,219. Process for measuring microbiologically active material. U. S. Patent 4,401,755. Washington, DC.
6. Sharpe, A. N. 1978. Some theoretical aspects of microbiological analysis pertinent to mechanization. *In* Mechanizing Microbiology. A. N. Sharpe & D. S. Clark, Eds. Thomas. Springfield, MA. pp. 19–40.
7. Reasoner, D. J. & E. E. Geldreich. 1978. Rapid Detection of Water-Borne Fecal Coliforms by $^{14}CO_2$ Release. *In* Mechanizing Microbiology. A. N. Sharpe & D. S. Clark, Eds. 1978. Thomas. Springfield, MA. pp. 120–139.
8. Ur, A. & D. F. J. Brown. 1975. Monitoring of bacterial activity by impedance measurements. *In* New Approaches to the Identification of Microorganisms. Proceedings of a Symposium on Rapid Methods and Automation in Microbiology. C-G. Hedén & T. Illéni, Eds. John Wiley & Sons. New York.

9. Richards, J. C. S., A. C. Jason, G. Hobbs, D. M. Gibson & R. H. Christie. 1978. Electronic measurement of bacterial growth. J. Phys. E. **11:** 560–568.
10. Weaver, J. C. & S. K. Burns. 1981. Potential impacts of physics and electronics on enzyme-based analysis. *In* Analytical Applications of Immobilized Enzymes and Cells. Vol. **3**. L. B. Wingard, E. Katchalski-Katzir & L. Goldstein, Eds. Academic Press. New York. pp. 272–308.
11. Weaver, J. C. Enzyme electrodes and related technologies. *In* Clinical Biochemistry Nearer the Patient. V. Marks & K. G. M. M. Alberti, Eds. Churchill-Livingstone. Submitted.
12. Zemel, J. N. 1977. Chemically sensitive semiconductor devices. Res. Dev. April: 38–44.
13. Cheung, P. W., D. G. Fleming, M. R. Neumann & H. K. Wen, Eds. 1978. Theory, Design, and Biomedical Applications of Solid State Chemical Sensors. CRC Press. Boca Raton, FL.
14. Angell, J. B., S. C. Terry & P. W. Barth. 1983. Silicon micromechanical devices. Sci. Am. **248:** 44–55.
15. Petersen, K. E. 1982. Silicon as a mechanical material. Proc. IEEE **70:** 420–457.
16. Cantor, C. R. & P. R. Schimmel. 1980. Biophysical Chemistry. W. H. Freeman. San Francisco. pp. 349–480.
17. Bergmeyer, H. U. 1974. Methods of Enzymatic Analysis. Second English Edition. Academic Press. New York.
18. Guilbault, G. G. 1976. Handbook of Enzymatic Methods of Analysis. Marcel Dekker. New York.
19. Engvall, E. 1977. Enzyme-linked immunosorbent assay, ELISA. *In* Biomedical Applications of Immobilized Enzymes and Proteins. Vol. **2**. T. M. S. Chang, Ed. Plenum Press. New York. pp. 87–96.
20. Guesdon, J-L. & S. Avrameas. 1981. Solid phase enzyme immunoassays. *In* Analytical Applications of Immobilized Enzymes and Cells. L. B. Wingard, E. Katchalski-Katzir & L. Goldstein, Eds. Academic Press. New York. pp. 207–232.
21. Kachel, V. & E. Menke. 1979. Hydrodynamic properties of flow cytometric instruments. *In* Flow Cytometry and Sorting. E. R. Melamed, P. F. Mullaney & M. L. Mendelsohn, Eds. John Wiley & Sons. New York. pp. 41–59.
22. Fasse, P. J. 1983. Parallel production of gel microdroplets. B. S. Thesis. Massachusetts Institute of Technology. Unpublished.
23. Nilsson, K., W. Scheirer, O. W. Merten, L. Östberg, E. Liehl, H. W. D. Katlinger & K. Mosbach. 1983. Entrapment of Animal Cells for the Production of Monoclonal Antibodies and Other Biomolecules. Nature **302:** 629–630.
24. Nilsson, K., S. Birnbaum, S. Flygare, L. Linse, U. Schröder, U. Jeppsson, P-O Larsson, K. Mosback & P. Brodelius. 1983. A general method for the immobilization of cells with preserved viability. Eur. J. Appl. Microbiol. Biotechnol. **17:** 319–326.

Growth of Procaryotic Cells in Hollow-Fiber Reactors

HARVEY W. BLANCH, T. BRUCE VICKROY, AND CHARLES R. WILKE

Department of Chemical Engineering
University of California,
Berkeley, California 94720

Hollow-fiber reactors, in the form of ultrafiltration modules, have been used for immobilizing enzymes and for culturing both plant and mammalian cells. Enzymes may be immobilized in hollow-fiber reactors without chemical modification. Consequently the disadvantages of chemical immobilization, such as loss of activity, are avoided. The kinetic data obtained in such systems reflect the kinetics of the native enzyme rather than that of a modified enzyme.[1,2,3] In the case of mammalian cells, the advantages of this type of immobilization are due primarily to the low-shear environment in which the cell is placed, coupled with a ready supply of nutrients, subject only to diffusion through the cell mass. These types of reactors have been mainly operated with cells grown outside hollow fibers, so that the cell mass occupies the shell side, with nutrient supplied through the lumen of the fibers. This is of some advantage for mammalian cells as the resulting growth resembles normal tissue growth.[4] This type of operation has been shown to increase the per cell productivity of biologicals over that of cell surface culture techniques.[5]

Plant cells have also been immobilized in hollow-fiber devices and the low-shear environment is of obvious advantage in maintaining cell viability and in achieving high cell concentrations.[6,7] An alternative to immobilization of enzymes is the immobilization of whole cells which, in nonviable form, are able to catalyze the desired reaction. This has been examined in hollow-fiber systems for a number of cases, and acceptable reaction rates and stability has been obtained.[8,9]

The hollow-fiber reactor offers several advantages over more "conventional" types of cell immobilization, which primarily rely on entrapping cells in various polymer matrices. By selection of fibers with appropriate molecular weight cut-offs, cells and some high-molecular-weight products are separated in the reactor itself. Thus the usual filtration step that follows fermentation is avoided, and in principle, selective membranes could be used to provide a partial purification of product from other extracellular metabolites. Contamination of the feed stream does not contaminate the cell culture; however, the use of nonsoluble raw materials is limited to those that would pass through the ultrafiltration membrane, that is, the substrates must practically be restricted to soluble, prefiltered materials.

The growth of viable cells in hollow-fiber reactors offers advantages in that extremely high cell densities may be obtained, and consequently reactor productivities can be greatly enhanced. One of the problems associated with the high cell density is the necessity to ultimately restrict growth so that large pressures do not build up in the reactor and cause collapse of the fibers. As a model system, the present work will examine the production of lactic acid from glucose, as this reaction does not involve gas uptake or evolution. The question of transport of oxygen to a densely packed cell mass in these type of reactors has not been satisfactorily resolved, although gas-carrying fibers and gas supply and removal from the shell side of the reactor have been examined.[10]

The geometry of the hollow-fiber reactor most commonly employed to date is shown in FIGURE 1. Other configurations include cultivation of cells inside the fibers, in analogy with Rony's work with enzymes,[1] and suspensions of cells have been recirculated on either the inside or outside of the fibers.[11,12] A novel approach for the cultivation of animal cells supplies air and CO_2 through the fiber lumen and cells grow outside the fibers while nutrient medium flows around them.[5] In order to enhance mass transfer to the cell mass, the use of two sets of fibers has been proposed;[13] one set carries substrate, which then is transported convectively through the cell mass to a second set of fibers that remove the product stream.

Earlier reports on the growth of viable cells in hollow-fiber reactors include those where the product was a primary metabolite,[14,15] an enzyme (β-lactamase),[9] and lactic acid.[12] Both isotropic and anisotropic fibers have been used, although cell breakthrough in the case of anisotropic fibers has favored the use of isotropic fibers, such as Celgard (Celanese Corp.).

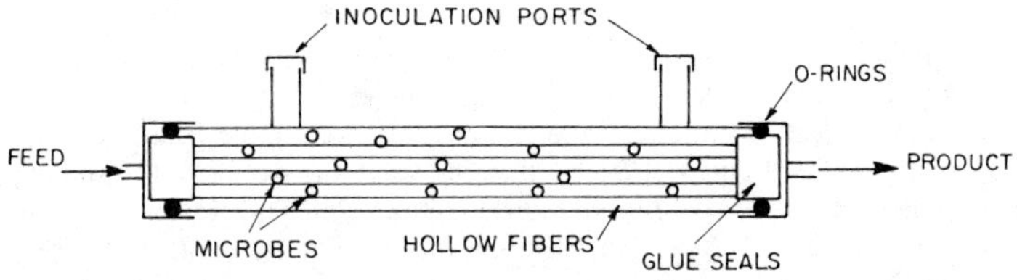

FIGURE 1. Schematic diagram of a typical hollow-fiber reactor.

LACTIC ACID PRODUCTION BY *LACTOBACILLUS DELBRUECKII*

As described above, the production of lactic acid from glucose provides an excellent model system to examine the usefulness of the hollow-fiber reactor. The homofermentative growth of *L. delbrueckii* does not involve gas transport, and the growth of the organism may potentially be uncoupled from lactic acid production. The strain used was NRRL-B445, grown on the medium of VickRoy *et al.*[14] Reactors were sterilized with ethlyene oxide and wetted with 50% ethanol before use. Fibers, composed of microporous polyproplyene (Celgard, Celanese Corp.), were type X-10, i.d. 100 microns, o.d. 150 microns. These were potted into reactors of various lengths ranging up to 61 cm. In all experiments described, temperature was controlled at 45°C. Other experimental details may be found in VickRoy *et al.*[14] The experimental apparatus is shown in FIGURE 2, for operation in a single-pass mode.

Initial experiments were conducted in a batch recycle mode; the effluent from the reactor was returned to the feed vessel, which was operated with pH control, and the time course of the conversion was followed. Typical results are shown in FIGURE 3 for the case of a 6.5-cm reactor containing 300 fibers. During the course of the experiment, the organism slowly grew to finally occupy the entire shell space, achieving extremely high final cell concentrations; from 200 to more than 500 gm dry weight/liter. It is clear from these values that the cells had a considerably reduced water content, as the concentrations are in excess of the theoretical packing densities of spheres or rods, and

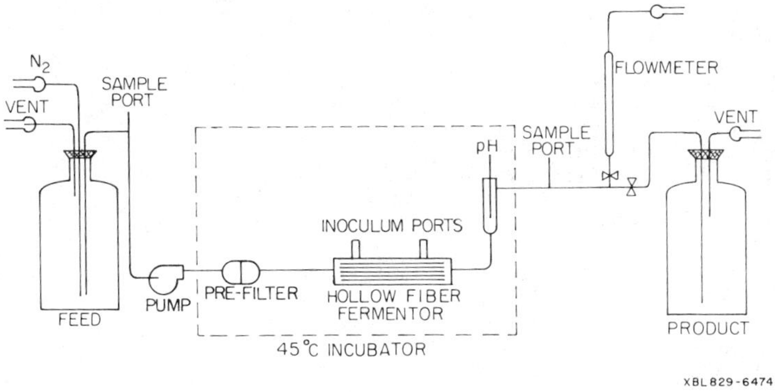

FIGURE 2. Schematic of experimental apparatus for continuous single-pass operation of the hollow-fiber reactor.

on microscopic examination, very little cellular debris was found. The results of the batch recycle experiments, shown in TABLE 1 (from VickRoy *et al.*[14]), indicate the increase in productivity with the increase in surface area of fibers in the system. Estimates of the specific productivity of the cells show a considerable decrease in the hollow-fiber case as compared to batch growth on the same medium. The large increase in volumetric productivity is clearly due to the very large increase in cell mass per unit volume.

Attempts were made to see if it were possible to limit the growth of cells in the

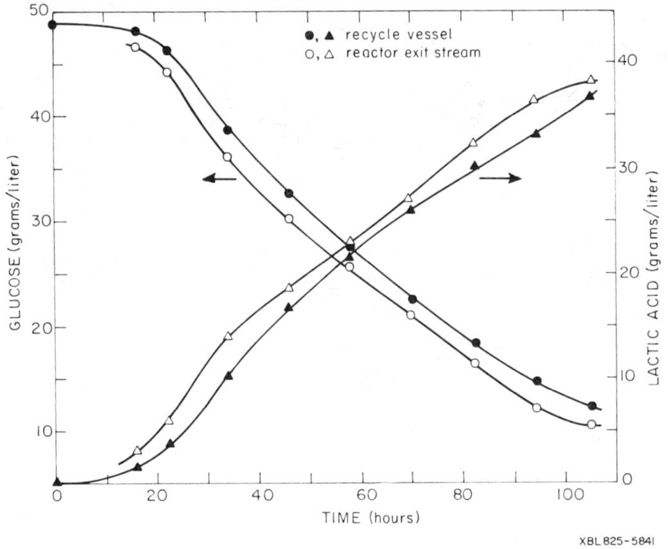

FIGURE 3. Batch recycle of the hollow-fiber reactor with 300 fibers. Initial glucose concentration 50 g/l.

TABLE 1. Comparison of Lactic Acid Fermentation Systems

Organism	System	Per Cent Conversion	Cell Conc. (g/l)	Lactic Acid Conc. (g/l)	Volumetric Productivity (g/l-h)	Reference
L. delbrueckii NRRL-B445	Cell recycle CSTR	≈100	54	35	76	VickRoy, *et al.*[16]
L. delbrueckii NRRL-B445	Batch	≈100	≈7–8[a]	45	1–2[b]	Leudeking & Piret[17]
L. delbrueckii NRRL-B445	Batch with dialysis	70[a]	≈11[a]	35	2–3[b]	Freidman & Gad-en[19]
L. delbrueckii NRRL-B445	CSTR	≈100	≈7–8[a]	38	7	Luedeking & Piret[18]
L. delbrueckii NRRL-B445	Hollow-fiber reactor (300 fibers)	4[c]	350	2	100	VickRoy *et al.*[14]
L. delbrueckii special strain	Calcium alginate gel beads	≈100	≈67[c]	≈46[a]	≈3[a]	Stenroos *et al.*[20]

[a]Estimated from reported data.
[b]Adjusted for total cycle time.
[c]For a single pass with a residence time of 3 to 4 sec.

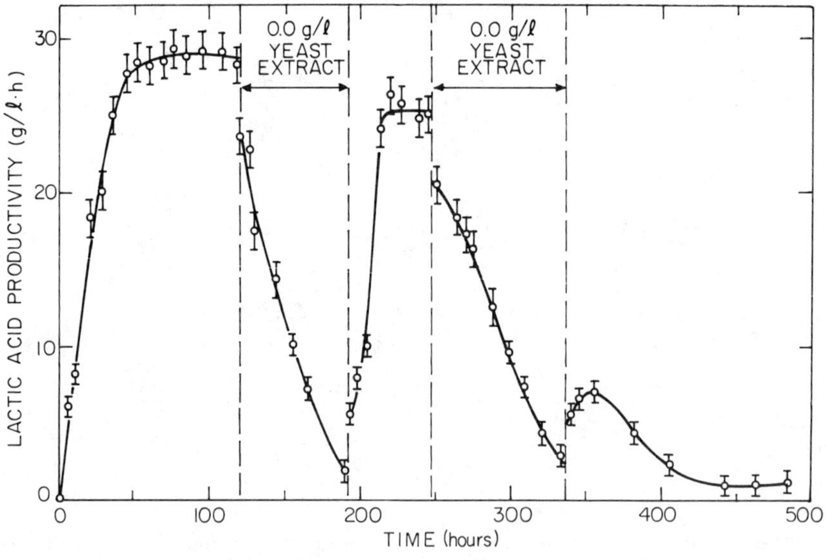

FIGURE 4. The effect of removal of nitrogen in the feed stream on lactic acid production. Continuous-flow operation of a 108-fiber reactor with a 5 g/l feed glucose concentration.

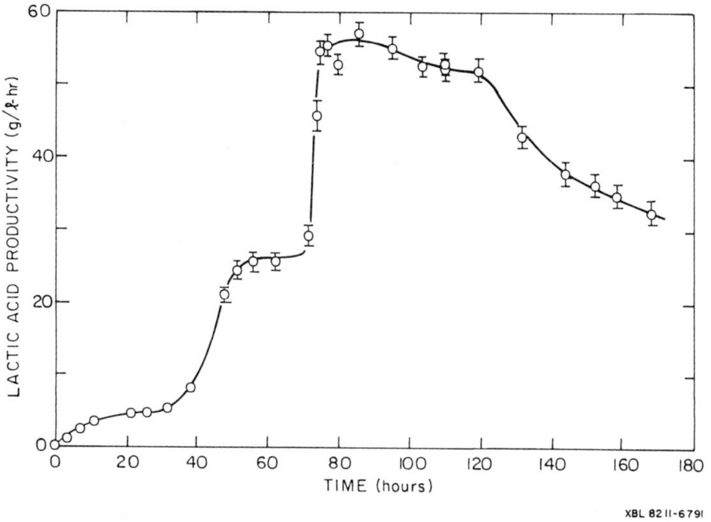

FIGURE 5. The effect of initial glucose concentration on reactor productivity. Continuous operation of a 208-fiber reactor, inlet glucose concentration 2.5 g/l for 0–75 h, 5.0 g/l for 75–170 h. Initial lag due to equipment failure.

reactor to avoid possible collapse of fibers due to these high cell densities. Specifically this was attempted by controlling the supply of nitrogen, in the form of yeast extract, in the influent medium. As can be seen from FIGURE 4, this approach was unsuccessful in establishing cryptic growth in the system. The biosynthetic ablilty of the cells slowly decayed with time. Complex nitrogen was required for cell maintenance and metabolism.

FIGURE 6. The effect of surface area on reactor productivity. Batch recycle operation (O, ●) with 50 g/l glucose and 30 g/l yeast extract. Continuous operation (Δ, ▲) with 5 g/l glucose concentration and 3.0 g/l yeast extract. Open symbols represent data taken at maximum productivity, closed symbols are steady-state values.

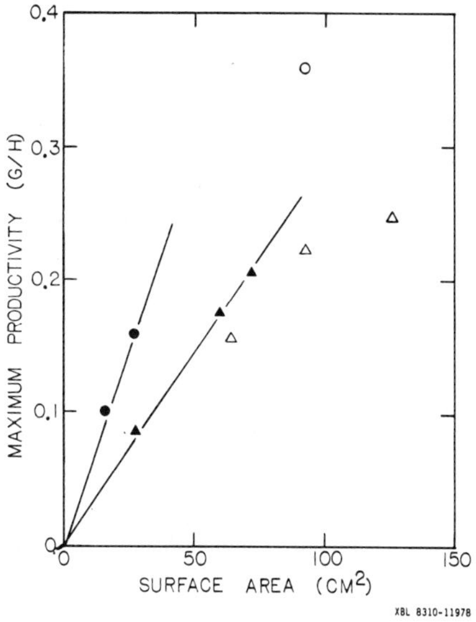

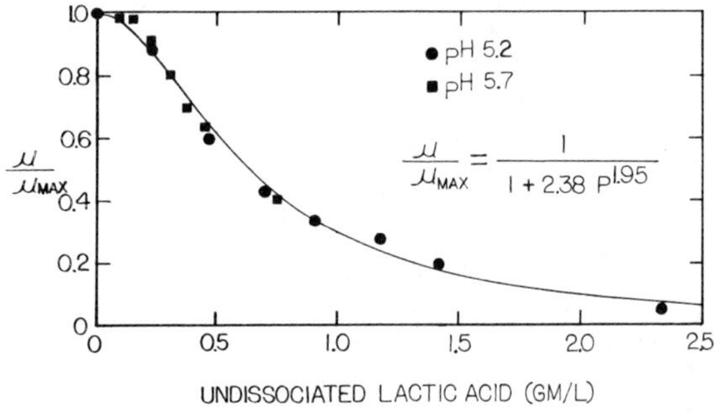

FIGURE 7. The effect of undissociated lactic acid on the growth of *L. delbrueckii.*

The results of the batch recycle experiments indicated that the productivity of the system was controlled by the surface area available for substrate diffusion, and this was confirmed at low levels of glucose in the feed as shown in FIGURE 5. An increase in concentration from 2.5 to 5.0 g/l showed a doubling of lactic acid productivity; this was not observed in the corresponding batch cultures. The effect of surface area on productivity is summarized in FIGURE 6. Both batch recycle and single-pass modes of operation are compared. The diffusion of glucose through the fiber wall and into the cell mass appears to be the rate-determining step.

When the glucose concentration is increased, the rate of production of lactic acid is decreased due to both lactic acid inhibition and pH inhibition. In operating the hollow-fiber reactor at glucose concentrations greater than 10 gm/l in the incoming feed without pH control in the reactor, these effects become significant. In the batch

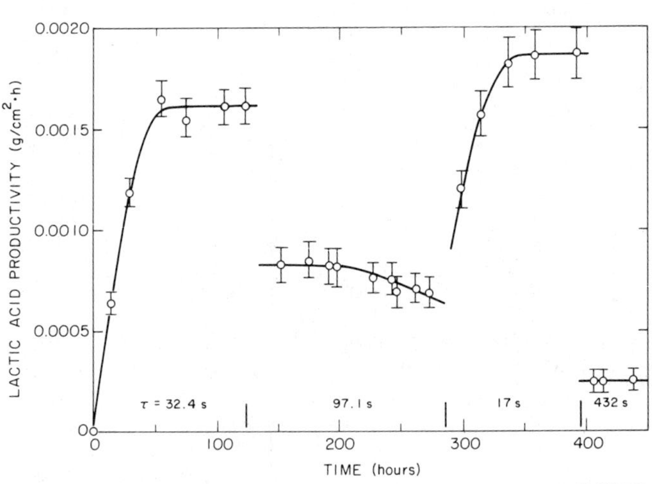

FIGURE 8. The effect of residence time on the productivity of a 61-cm hollow-fiber reactor with 600 fibers. Inlet glucose concentration 20 g/l.

recycle mode, this does not represent a problem as the per pass conversion is small and the recycle vessel is used to maintain the pH at a constant value. The effect of lactic acid inhibition appears to be due to the presence of the protonated species; lactate does not appear to be inhibitory. Batch growth experiments were conducted to quantify this effect, by varying both pH and lactic acid concentration. The results are shown in FIGURE 7. Thus two effects must be considered: the decrease in growth as the pH falls below the optimum, and the concentration of lactic acid, which depends on the pH and the total amount of lactic and lactate forms. A similar phenomenon appears to govern the kinetics of acetic-acid production.

The inhibition caused by lactic acid restricted the conversion possible when the reactor was operated in a single pass, continuous-flow mode. Short residence times result in high productivity and low conversion. The effect of residence time on productivity, expressed in terms of grams lactic acid produced per area of fiber per hour, is shown in FIGURE 8. FIGURE 9 indicates the conversion and productivity data for two concentrations of glucose in the feed stream. As can be seen, at these higher

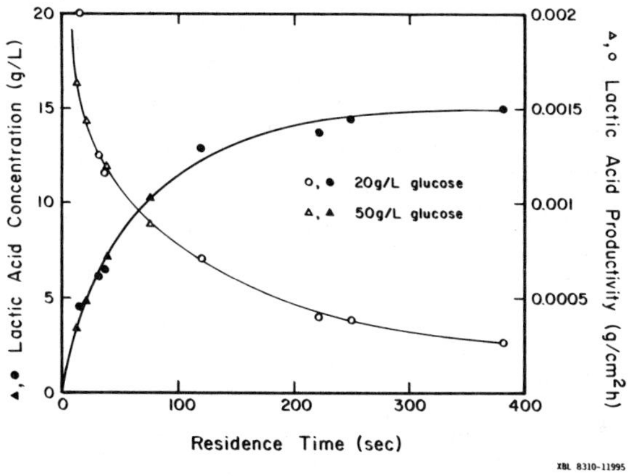

FIGURE 9. The effect of residence time on the productivity and conversion of a 16.5-cm hollow-fiber reactor. Inlet glucose concentration 20 and 50 g/l.

substrate levels, the effect of mass transfer limitation is no longer apparent, as the conversion is primarily governed by the higher concentration of lactic acid present, which moves the system from diffusion to kinetic control. Thus, doubling the feed concentration does not double the productivity. Nevertheless, conversions of up to 75% were attained at glucose concentrations of 20 g/l and residence times of 200 seconds. The productivity of these systems is around 100 times that of a conventional batch fermentation.

The results of varying the initial pH in the incoming feed stream and of varying the reactor length at several glucose concentrations is shown in FIGURE 10. The data lie essentially on the same curve except in the case where the initial pH is 5.0, a value that results in rapid inhibition by lactic acid. The results suggest the strong effect of lactic acid inhibition, which overshadows initial substrate concentration effects. The potential ability of scaling up these reactors is also apparent, as results were obtained with a tenfold change in reactor length.

CONCLUSIONS

The hollow-fiber reactor demonstrates extremely high volumetric productivities, due to the large cell concentrations that can be obtained. Although high conversions were not possible in the system examined when operated in a single-pass mode, due to the drop in pH resulting from lactic acid production, the potential for reactors with productivities far greater than batch or continuous fermentation is apparent. The further development of these systems will require means for supply of oxygen and removal of CO_2, and a means of regulation of pH in the reactor itself. An alternative approach has been reported using cross-flow filtration as a means to separate and recycle cells to a continuous fermentation.[16] Lactic acid productivities of 76 g/l · hr were obtained with complete conversion of up to 50 g/l glucose in the feed stream.

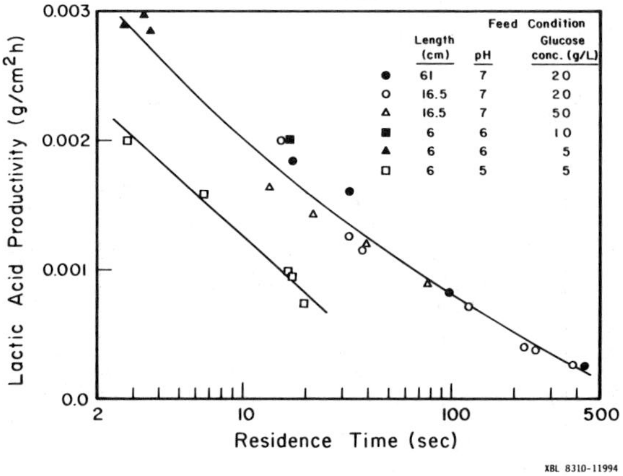

FIGURE 10. The effect of residence time on the productivity of various lengths of hollow-fiber reactors with various inlet conditions.

REFERENCES

1. RONY, P. R. 1971. Multiphase catalysis. II. Hollow-fiber catalysis. Biotechnol. Bioeng. **13**: 431–447.
2. LEWIS, W. & S. MIDDLEMAN. 1974. Conversion in a hollow-fiber/enzyme reactor. AIChE J. **20**: 1012–1014.
3. WATERLAND, L. R., C. R. ROBERTSON & A. S. MICHAELS. 1975. Enzymatic catalysis using asymmetric hollow-fiber membranes. Chem. Eng. Commun. **2**: 37–47.
4. KNAZEK, R. A., P. M. GULLINO, P. O. KOHLER & R. L. DEDRICK. 1972. Cell culture on artificial capillaries. An approach to tissue culture *in vitro*. Science **178**: 65–67.
5. KU, K., M. J. KUO, J. DELENTE, B. S. WILDI & J. FEDER. 1981. Development of a hollow-fiber system for large-scale culture of mammalian cells. Biotechnol. Bioeng. **23**: 79–95.
6. SHULER, M. L. 1981. Production of secondary metabolites from plant tissue culture. Problems and prospects. Ann. N. Y. Acad. Sci. **369**: 65–80.
7. JOSE, W., H. PEDERSEN & C. H. CHIN. 1983. Immobilization of plant cells in a hollow-fiber reactor. Ann. N. Y. Acad. Sci. In press.

8. Kan, J. K. & M. L. Shuler. 1978. Urocanic acid production using whole cells immobilized in a hollow-fiber reactor. Biotechnol. Bioeng. **20**: 217–230.
9. Inloes, D. S., D. P. Taylor, S. N. Cohen, A. S. Michaels & C. R. Robertson. 1983. Ethanol production by *S. cerevisae* immobilized in hollow-fiber membrane bioreactors. Appl. Environ. Microbiol. In press.
10. Robertson, C. R. 1983. Personal communication.
11. Charley, R. C., J. E. Fein, B. H. Lavers, H. G. Lawford & G. R. Lawford. 1983. Optimization of process design for continuous ethanol production by *Zymomonas mobilis*. Biotechnol. Bioeng. **5**: 169–174.
12. Mattiason, B., M. Ramstorp, L. Nilsson & B. Hahn-Hağerdal. 1981. Comparison of the performance of a hollow-fiber microbe reactor with a reactor containing alginate-entrapped cells. Biotechnol. Lett. **3**: 561–566.
13. VickRoy, T. B., C. R. Wilke & H. W. Blanch. 1982. Rapid production of biological products in a densely packed microbial membrane reactor. Univ. Calif. Patent Disclosure 82–267–1.
14. VickRoy, T. B., H. W. Blanch & C. R. Wilke. 1982. Lactic acid production by *Lactobacillus delbrueckii* in a hollow-fiber fermentor. Biotechnol. Lett. **4**: 483–488.
15. Inloes, D. S. 1982. Immobilization of bacterial and yeast cells in hollow-fiber reactors. Ph.D. Thesis. Stanford University.
16. VickRoy, T. B., D. K. Mandel, D. K. Dea, H. W. Blanch & C. R. Wilke. 1983. The application of cell recycle to continuous fermentative lactic acid production. Biotechnol. Lett. **5**: 665–670.
17. Luedeking, R. L. & E. Piret. 1959. A Kinetic study of the lactic acid fermentation batch process at controlled pH. Biotechnol. Bioeng. **1**: 393–412.
18. Luedeking, R. L. & E. Piret. 1959. Transient and steady states in continuous fermentation. Theory and experiment. Biotechnol. Bioeng. **1**: 431–459.
19. Friedman, M. R. & E. L. Gaden. 1970. Growth and acid production by *Lactobacillus delbrueckii* in a dialysis culture system. Biotechnol. Bioeng. **12**: 961–974.
20. Stenroos, S. L., Y. Y. Linko & P. Linko. 1982. Production of lactic acid with immobilized *Lactobacillus delbrueckii* Biotechnol. Lett. **4**: 159–164.

Immobilized Viable Plant Cells

PETER BRODELIUS

Institute of Biotechnology
Swiss Federal Institute of Technology
CH-8093 Zürich, Switzerland

Isolated plant cells from higher plants may be cultivated in submerged cultures in analogy to microorganisms. There are, however, some major differences between plant and microbial cells that make the cultivation of the former cell type more complicated. A plant cell in culture is very large compared to a microbial cell. It may be 50–100 μm in diameter and the volume can be 100–200,000 times larger than the volume of a microorganism. These large cells are held together by only a fragile wall and they are in most cases very sensitive to shear forces. In addition to this, plant cells in suspension cultures grow in aggregates containing from a few up to hundreds of cells. These characteristics of plant cells make their cultivation on a large scale technically more difficult.

Another major difference between plant and microbial cells is their growth rates. The dividing time for plant cells in culture may vary from 24 to 60 h. While one plant cell has turned into two cells after 24 h, one microbial cell has during the same time turned into 4×10^8 (assuming a dividing time of 30 min). Slow growth of plant cells is considered to be a major limitation in their utilization for the production of complex biochemicals.

The slow growth of plant cells poses some special problems for culturing in large volumes. It may take up to two weeks from inoculation to stationary phase and, in addition, a plant cell culture cannot be diluted too much (normally not more than 10–15-fold) because the cells condition their own medium. Thus, the time needed for growth from a small culture (100 ml) to a technical-scale culture (1000 l) may be as long as 2 months, very different from the time required for large-scale cultivation of a micoorganism.

Despite the limitations and problems in cultivating cells from higher plants, there is wide interest in developing plant tissue culture techniques for commercial production of natural products. These compounds include such groups as pharmaceuticals, flavors, and fragrances. A few plant products, some of which may be considered for production by plant tissue culture, and their market values, are listed in TABLE 1. The price of plant products may vary from a few dollars per kilo up to 5 million dollars per kilo (vincristine). A price of around 1000 dollars per kilo has been estimated as the lowest price of a compound worth producing by plant tissue culture at the current levels of technology.

The first chemical product from plant tissue culture was recently introduced on the market. The compound, shikonin, is a dye and pharmaceutical used in Japan for its antibacterial and anti-inflammatory properties. It is produced by cultures of *Lithospermum erythrorhizon* developed in a two-stage batch process by Mitsui Petrochemical Industries Ltd., Tokyo, as outlined in FIGURE 1. The cells are cultured for three weeks, and the yield of product is estimated to be around 5 kg per batch. The yearly consumption of pure shikonin in Japan is around 150 kg per year, so the Mitsui process can supply a major part of the shikonin consumed.

The Mitsui process may be considered a major breakthrough in plant tissue culture technology, even though the operation is carried out on a relatively small scale. It has,

TABLE 1. The Price and Estimated Markets for Some Plant Secondary Products[a]

Compound	Use	Wholesale Price (US$/kg)	Estimated Retail Market (10^6 US$)
Ajmalicine	circulatory problems	1500	5.25 (world)
Codeine	sedative	650	50 (US)
Digitoxin, digoxin	heart disorders	3000	20–55 (US)
Jasmine	fragrance	5000	0.5 (world)
Pyrethrins	insecticide	300	20 (US)
Quinine	malaria, flavor	100	5–10 (US)
Shikonin	dye, anti-inflammation	4000	0.7 (Japan)
Spearmint	flavor, fragrance	30	85–90 (world)
Vincristine, vinblastine	anticancer	5×10^6	18–20 (US)

[a]Adapted from Curtin.[20]

however, demonstrated that a secondary metabolite can be produced by plant tissue culture techniques on a commercial basis.

The success of the shikonin process is considered to be an exception that will be difficult to duplicate. Many problems remain in applying plant cell culture on a large scale, as has been indicated in this introduction. We have, for a few years, been investigating ways to overcome, or at least reduce, these problems by immobilizing the cells. Other problems are introduced by using this type of biocatalyst.

Since our first report on the immobilization of plant cells,[7] many investigations on this subject have been reported. Almost exclusively, entrapment techniques have been utilized to immobilize large, sensitive plant cells, as summarized in TABLE 2.

In our laboratory, we have used various polysaccharides to entrap plant cells under very mild conditions.[1,2] Recently, a general method for the immobilization of viable cells in spherical polysaccharide gel beads was developed.[3] The method is based on dispersion of the cell/polymer suspension in an inert hydrophobic phase (e.g. vegetable oil). Gel formation is induced after droplets of appropriate size have been formed.

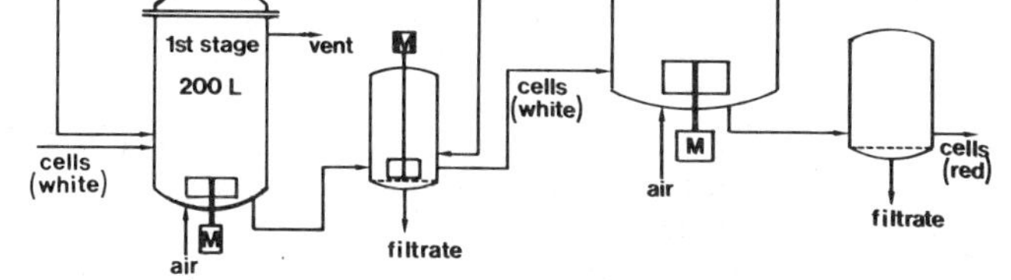

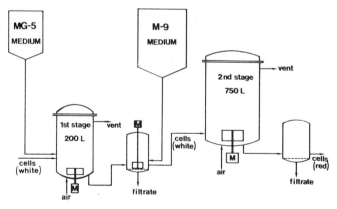

FIGURE 1. Flow sheet for the two-stage process used for production of shikonin by cultivation of *Lithospermum erythrorhizon*. Note: The inoculated cells and the cells obtained after the first stage are white, whereas the cells obtained after the second stage are red owing to the shikonin produced. (Adapted from Curtis.[20])

TABLE 2. Polymers Used for Entrapment of Whole Plant Cells[a]

Polymer	Gelling Mechanism	Viability after Immobilization
Alginate	ionotropic	+
Carrageenan	ionotropic (thermal)	+
Agarose	thermal	+
Agar	thermal	+
Gelatin	cross-linking with glutaraldehyde	−
Polyacrylamide	radical polymerization	−
Polyacrylamide + alginate or xanthan gum	radical polymerization	+
Prepolymerized polyacrylamide	cross-linking with dialdehyde	+
Hollow fibers	−	+

[a]For references see TABLE 3.

Using this procedure, we have immobilized, in various beaded gels, bacteria, fungi, yeasts, algae, plant, and animal cells, with viability and biosynthetic capacity preserved.

Our attempts to immobilize plant cells in polyacrylamide by radical polymerization of acrylamide and bisacrylamide, in analogy to the immobilization of microbial cells, resulted in nonviable preparations. An elegant approach to overcome the toxicity of the reagents has been taken by Freeman and coworkers.[4] The plant cells were entrapped in gels made of prepolymerized, linear, water-soluble polyacrylamide, partially substituted with acylhydrazide groups. Gelation was effected under mild conditions by the addition of controlled amounts of dialdehyde (e.g. glyoxal). Plant cells have also been entrapped in a mixture of polyacrylamide and polysaccharides.[5] The cells were mixed with a viscous polysaccharide solution (alginate or xanthan gum) before they were added to the toxic mixture utilized in making the polyacrylamide gel. The cells remained viable after immobilization and were clearly protected by the viscous polysaccharide solution.

Plant cells may conveniently be immobilized by entrapment in hollow-fiber reactors.[6,7] The cells are placed on the shell side of the fibers and nutrient medium is rapidly recirculated through the fibers. The plant cells appear to be unaffected by the entrapment, and this relatively simple technique may prove useful for large-scale operations.

Some factors that certainly will influence the choice of immobilization method for a large-scale operation are:

(1) ease of preparation,
(2) toxicity,
(3) cost of polymer, and
(4) mechanical strength of gel.

A gel meeting these criteria is Ca-alginate. The preparation of alginate-entrapped cells is very easy and it can be employed on a large scale without complications. In this context, it can be mentioned that a pilot plant operated (3500-l reactor volume) by Kyowa Hakko Kogyo Co., Ltd, Tokyo, for the production of ethanol by alginate-entrapped yeast cells, has shown very promising results. The alginate-entrapped cells are easily prepared in the column reactor and this technique should also be applicable to plant cells. The price of alginate is low, and it is available in unlimited amounts. The mechanical stability of alginate is acceptable and, furthermore, it is widely used as a food additive. A product from a process based on alginate-entrapped cells may therefore pass the FDA more easily.

Kappa-carrageenan is almost as suitable as alginate for the immobilization of plant cells on a large scale. However, the cost of immobilization will be higher since a special preparation (low in potassium) will be required for the immobilization of sensitive plant cells.

As mentioned above, hollow-fiber reactors may be the method chosen for a large-scale operation with immobilized plant cells.

TABLE 2 lists conditions under which plant cells remain viable after immobilization. There are many methods available to study the viability of immobilized plant cells for example, plasmolysis, respiration, dry weight increase and various staining techniques. Recently, we initiated studies on such cells by ^{31}P-NMR, which is a noninvasive method that gives direct information about phosphate uptake and metabolism, as well as the energization of the cells (ATP/ADP-ratio).[8,9] Some ^{31}P-NMR spectra of plant cells are shown in FIGURE 2.

A viable cell preparation is a requirement when large parts of the cell metabolism are to be used for synthetic purposes. While slow growth of plant cells in culture is a major limitation in general, paradoxically, growth of cells leads to cell leakage and

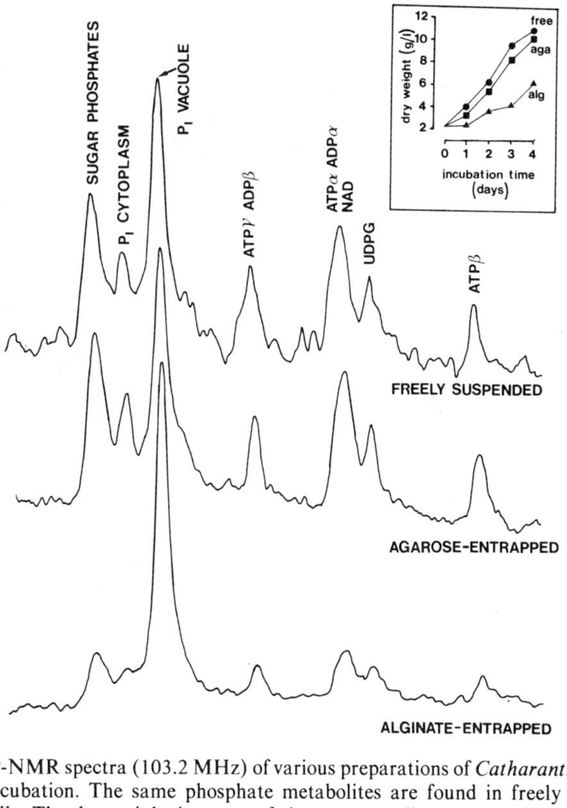

FIGURE 2. ^{31}P-NMR spectra (103.2 MHz) of various preparations of *Catharanthus roseus* cells after 4 days incubation. The same phosphate metabolites are found in freely suspended and immobilized cells. The dry weight increase of the corresponding preparations is shown in the inserted diagram. (From Vogel and Brodelius.[9])

TABLE 3. Use of Immobilized Plant Cells for the Production of Biochemicals

Species	Type of Reaction	Immobilization Method	Substrate and Product	Operation	Duration	References
Digitalis lanata	bioconversion	entrapment in alginate	digitoxin → digoxin	batch	33 days	Brodelius *et al.*[1]
				continuous	70 days	Brodelius *et al.*[10]
			methyldigitoxin → methyldigoxin	batch change	180 days	Alferman *et al.*[21]
Daucus carota	bioconversion	entrapment in alginate	digitoxigenin → periplogenin	batch change	12 days	Jones and Veliky[22]
				continuous	21 days	Veliky and Jones[16]
Catharanthus roseus	bioconversion	entrapment in agarose	cathenamine → ajmalicine isomers	batch	1 day	Felix *et al.*[13]
Mentha sp.	bioconversion	entrapment in prepolymerized polyacrylamide	(−)menthone → neomenthol	batch	1 day	Galun *et al.*[4]
Catharanthus roseus	synthesis from precursors	entrapment in alginate	tryptamine + secologanin → ajmalicine isomers	batch	4 days	Brodelius *et al.*[1]
		entrapment in alginate agarose agar carrageenan	tryptamine + secologanin → ajmalicine isomers	batch	5 days	Brodelius and Nilsson[2]
		entrapment in alginate agarose	tryptamine + secologanin → ajmalicine isomers	batch change	14 days	Brodelius and Nilsson[14]
Catharanthus roseus	*de novo* synthesis	entrapment in alginate agarose agar carrageenan	sucrose → ajmalicine	batch	14 days	Brodelius and Nilsson[2]
		entrapment in alginate	sucrose → serpentine	batch	40 days	Brodelius *et al.*[10]
		entrapment in alginate/polyacrylamide	sucrose → ajmalicine	batch change	220 days	Lambe and Rosevear[12]
		entrapment in xanthan/polyacrylamide	sucrose → serpentine	batch change	180 days	Lambe and Rosevear[12]
Morinda citrifolia	*de novo* synthesis	entrapment in alginate	sucrose → anthraquinones	batch	21	Brodelius *et al.*[11]
Glycine max	*de novo* synthesis	entrapment in hollow-fiber reactors	sucrose → phenolics	continuous	30 days	Shuler[7]
Petunia hybrida	*de novo* synthesis	entrapment in hollow-fiber reactors	sucrose → phenolics	continuous	21 days	Jose *et al.*[6]
Lavandula vera	*de novo* synthesis	entrapment in urethane prepolymers	sucrose → pigments			Tanaka *et al.*[23]
Solanum aviculare	*de novo* synthesis	covalent linkage to polyphenyleneoxide beads	sucrose → steroid glycosides	continuous	11 days	Jirku *et al.*[24]
Capsicum frutescens	*de novo* synthesis	adsorption to macroporous polyurethane foam	sucrose → capsaicin	batch	10 days	Lindsey *et al.*[25]

destruction of beads within a relatively short time. In a continuous process, such free cells or fragmented beads would interfere.

Attempts are now being made to restrict growth within the beads without affecting the viability and biosynthetic capacity of the cells. Restricted growth has been achieved by limiting a nutrient in the medium (e.g. phosphate or sulfate)[10] or by changing the hormone concentration of the medium.[10,11] The formulation of new media, which keep the immobilized cells in a stationary phase over an extended period of time, would be most valuable for the further development of this technique.

Immobilized plant cells will suffer from the same drawbacks as other immobilized biocatalysts. Diffusion limitations are a major problem of most immobilized biocatalysts, but this may be reduced, as compared to immobilized microorganisms, because of the relatively slow metabolism of plant cells. There is, however, always a risk that internal diffusion limitations will lead to death of cells, and thereby loss of biomass and biosynthetic capacity, in the interior of the beads. This type of problem may be reduced by decreasing the size of the biocatalyst particles and/or lowering the cell loading of the preparation.

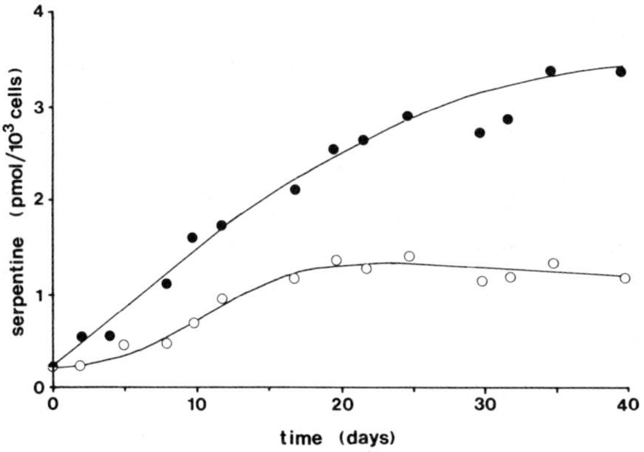

FIGURE 3. Kinetics of the *de novo* synthesis of the indole alkaloid serpentine by *Catharanthus roseus* cells under phosphate-limiting conditions (1 m*M*). Free cells (—○—); alginate-entrapped cells (—●—). (From Brodelius *et al.*[10])

Immobilized plant cells have been used in model studies for various biosynthetic purposes, as summarized in TABLE 3. The whole spectrum of biosynthetic reactions has been exemplified, that is, bioconversions, synthesis from precursors and *de novo* synthesis. It appears that any biosynthetic activity observed in freely suspended cells can also be expressed by immobilized viable cell preparations. The productivity may even be enhanced in the immobilized cells. Some examples of increased productivity with immobilized cells are illustrated in FIGURES 3 and 4.

Are any significant differences observed between freely suspended and immobilized plant cells and, if so, do they favor immobilization? At first glance, the metabolism (primary and secondary) of the cells appears not to be altered to any great extent by immobilization. For instance, phosphate uptake and metabolism by free and immobilized cells showed no significant differences when studied *in vivo* by ^{31}P-NMR (FIG. 2). Upon closer examination, however, some differences appear, and the prolonged stationary phase observed for the immobilized cells is probably the most striking difference. Freely suspended cells may be utilized for 2–3 weeks, while

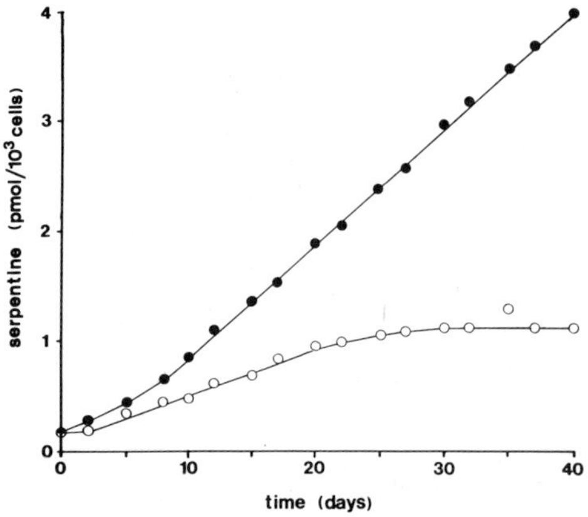

FIGURE 4. Kinetics of the *de novo* synthesis of the indole alkaloid serpentine by *Catharanthus roseus* cells under osmotic stress (0.5 *M* mannitol). Free cells (—○—); alginate-entrapped cells (—●—). (From Brodelius *et al.*[10])

immobilized cells have been used for up to 6–7 months, as indicated in TABLE 3 and FIGURE 5. The productivity per biomass unit can therefore be increased considerably. The prolonged production phase will be of great importance in further development of immobilized plant cell technology, since the cost of biomass generation is considered a major limitation in the production of chemicals by cultivated plant cells.

Plant cells in suspension culture normally grow in aggregates of various sizes, which may range from a few up to hundreds of cells per aggregate. The agitation of such a heterogeneous and, at the same time, sensitive population of cells may impose problems. A stable preparation of the biocatalyst, obtained after immobilization, can be employed readily in a continuous process.

In general, a simpler bioreactor design may be employed after immobilization of cells. This is mainly because immobilized plant cells are protected against shear forces by the polymeric backbone of the matrix. The sensitivity of freely suspended cells to shear forces is considered a major problem in scaling up plant cell cultivations. If the immobilization method employed results in mechanically stable preparations, a relatively uncomplicated reactor design may be used on a large scale. There will be no need for the development of sophisticated reactor configurations for each application since it can be expected that the same reactor may be used for different plant cell cultures that have been made similar by immobilization.

Many plant products are stored within the producing cells. For isolation of product(s), extraction of the cell mass is required. It has, however, been reported that certain preparations of immobilized cells spontaneously release, at least partly, products that normally are stored within the vacuoles of the cells.[10,12] One of the

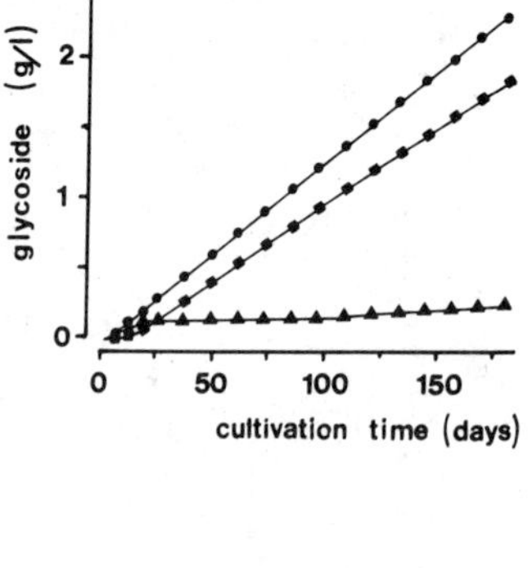

FIGURE 5. Bioconversion of β-methyldigitoxin to β-methyldigoxin by alginate-entrapped cells of *Digitalis lanata* in a batch procedure. Medium was changed every third day. After a short lag phase, the hydroxylation continued for at least 70 days at a constant rate. (—●—) methyldigitoxin added; (—◆—) methyldigoxin formed; (—▲—) methyldigitoxin unconverted. (From Alfermann *et al.*[21])

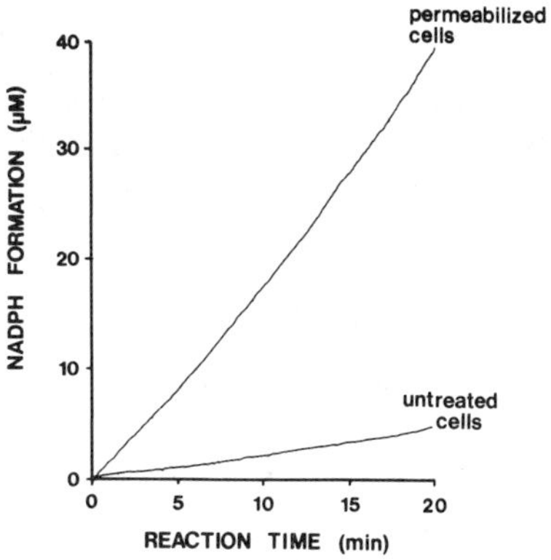

FIGURE 6. Time course of the isocitrate dehydrogenase reaction in agarose-entrapped cells of *Catharanthus roseus.* The reaction mixture (5 ml substrate solution plus 1 g of wet beads containing 20% (wt/wt) of cells) was monitored continuously at 340 nm by pumping the supernatant through a flow cell at a rate of 5 ml/min. Permeabilization treatment was carried out with 5% DMSO for 30 min. Before the enzymatic assay, the beads were washed free of DMSO. (From Brodelius and Nilsson.[14])

inherent advantages of an immobilized biocatalyst is its possible continuous operation. The spontaneous leakage of products actually makes it possible to realize this distinct advantage. The mechanism involved in the spontaneous release of product(s) from entrapped cells is not fully understood, but is not due to lysis of cells.

When no spontaneous release of product occurs, the cells may be induced to release intracellularly stored products by permeabilization of cell membranes. Since the product is most often stored within cellular vacuoles, the procedure must make both the plasma membrane and the tonoplast permeable to the compound of interest. We have recently focused our research on the possibility of carrying out such a permeabilization procedure without affecting the viability of the immobilized cells.

The permeability of the plasma membrane can be monitored by measuring the activity of various coenzyme-requiring enzymes within the cells.[13] As an example, the time course of the isocitrate dehydrogenase reaction for intact and permeabilized plant cells is shown in FIGURE 6. Various agents have been tested for the permeabilization. DMSO appears to be suitable for this operation. Permeability as a function of DMSO concentration is shown in FIGURE 7, for agarose-entrapped plant cells. The DMSO concentration required for permeabilization of the plasma membrane may vary depending on the plant species. *Catharanthus* cells treated with up to 5% DMSO for 30 min recover relatively quickly, and grow at almost the same rate as untreated cells, as shown in FIGURE 8.

^{31}P-NMR may be employed to study the permeability of the tonoplast within immobilized cells as illustrated in FIGURE 9.[9]

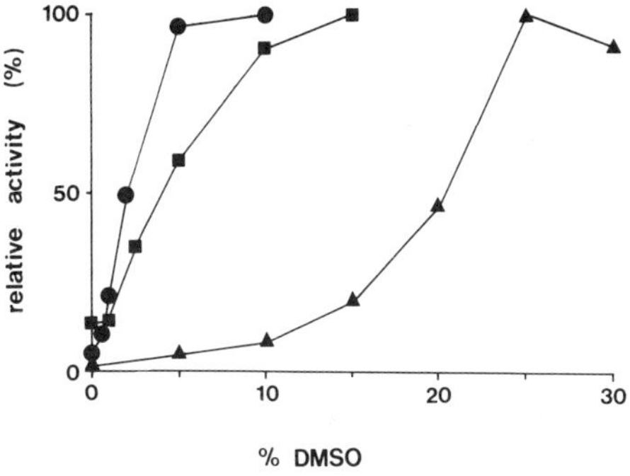

FIGURE 7. Relative isocitrate dehydrogenase activity as a function of DMSO concentration during the permeabilization procedure. 100% enzyme activity is defined as the maximal activity observed for each cell preparation. Beads (1 g wet weight containing 20% (wt/wt) of cells) were treated with the DMSO concentrations indicated in medium (10 ml) for 30 min then washed and assayed as described in the legend for FIGURE 6. (—●—) *Catharanthus roseus*; (—■—) *Daucus carota*; (—▲—) *Datura innoxia.* (From Brodelius and Nilsson.[14])

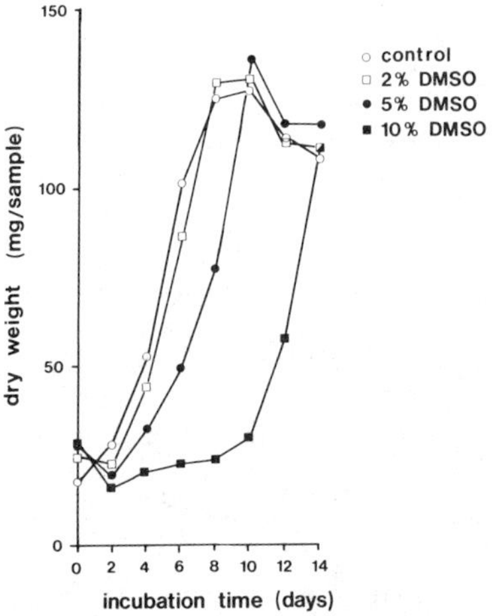

FIGURE 8. Dry weight of various preparations of freely suspended cells of *Catharanthus roseus* as a function of incubation time. The cells were treated with the DMSO concentrations indicated for 30 min before inoculation into growth medium. (From Brodelius and Nilsson.[14])

The inorganic phosphate ions stored within vacuoles of *Catharanthus* cells may be released into the surrounding medium by DMSO treatment of the cells.

DMSO appears to be suitable for the permeabilization of immobilized plant cells and, with an appropriate DMSO treatment, up to 90% of the product may be released from the cells without loss of viability.[14]

We have recently suggested a process based on immobilized plant cells with an intermittant release of product(s), as shown schematically in FIGURE 10. To test the outlined process, some model studies have been carried out.[14] Cells entrapped in alginate or agarose, producing an alkaloid, could be used at least three cycles as shown in FIGURES 11 and 12. In fact, the productivity increased for each cycle. We ascribe

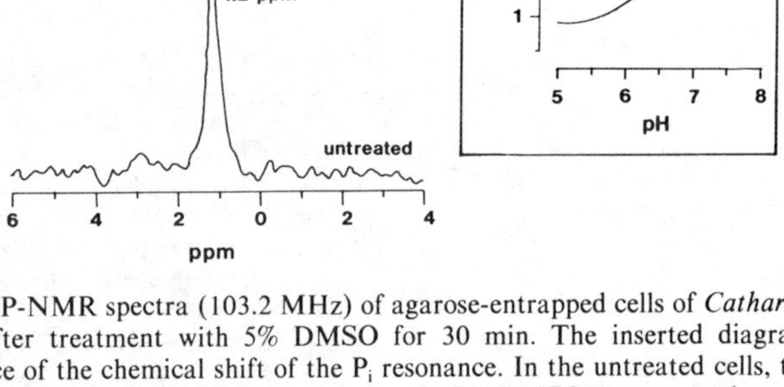

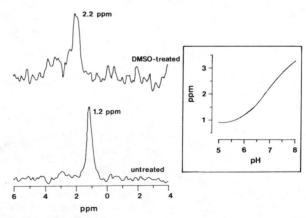

FIGURE 9. ^{31}P-NMR spectra (103.2 MHz) of agarose-entrapped cells of *Catharanthus roseus* before and after treatment with 5% DMSO for 30 min. The inserted diagram shows the pH-dependence of the chemical shift of the P_i resonance. In the untreated cells, the P_i ions are found in an environment at pH 6.0 (= vacuoles) and after DMSO-treatment, the peak has shifted its position to a chemical shift corresponding to a pH of 7.0 (= the external pH). Note also the dilution (signal to noise ratio) of the P_i ions in the DMSO-treated cells.

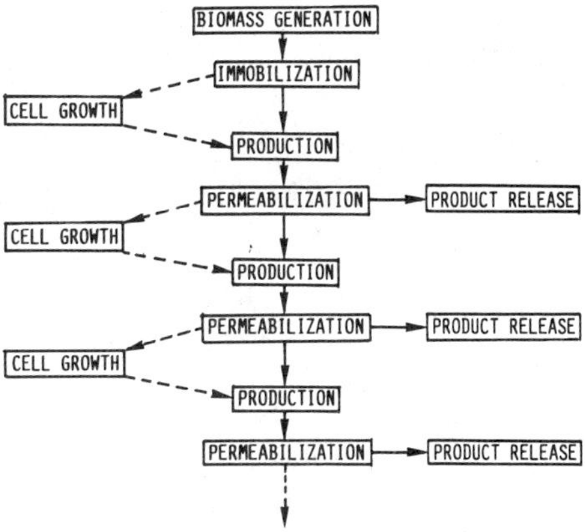

FIGURE 10. Schematic diagram of a proposed semicontinuous process for the production of intracellularly stored products.

this increased production to the fact that the biomass within the beads increases during the experiments. After optimization, it should be possible to run under steady-state conditions. These model studies have shown that it is possible to release product(s) from immobilized cells intermittantly, thereby making it possible to reutilize the biomass over an extended period of time. An additional advantage is that the product can be recovered from a relatively small volume as compared to a continuous release.

As mentioned earlier, it can be expected that immobilized plant cells may be utilized in uncomplicated reactors. Relatively little research has been carried out on the design of reactors for immobilized plant cell preparations. Most experiments have been carried out in simple batch configurations (that is, shaker flasks), but there are examples of packed-bed reactors,[1,10] bubble column reactors,[15,16] and a fluidized-bed reactor.[17] A comparison of the hydroxylation capacity of alginate-entrapped *Digitalis*

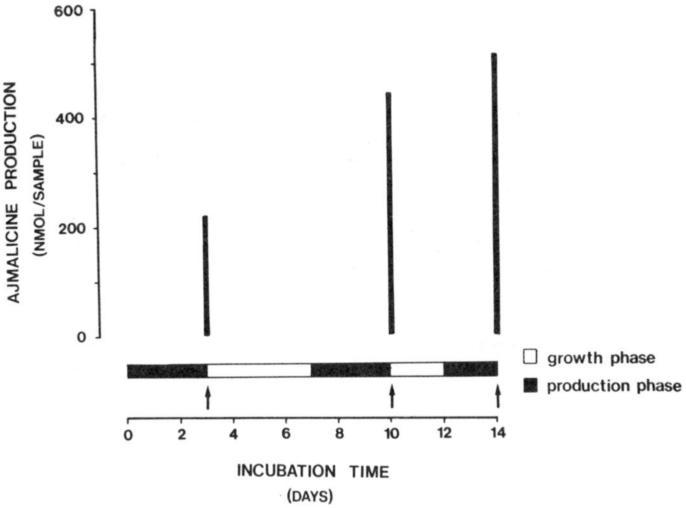

FIGURE 11. Semicontinuous production of the indole alkaloid ajmalicine by immobilized cells of *Catharanthus roseus* according to the process outlined in FIGURE 10 (agarose-entrapped cells). (From Brodelius and Nilsson.[14])

cells in shaker flasks and a bubble column reactor disclosed no significant differences. Recently tubular hollow-fiber reactors were introduced as potential immobilized plant cell reactors.[6,7] In this reactor type, the cells are entrapped in the extrafiber space within the reactor and medium is passed through the fibers. This appears to be a convenient way simultaneously to immobilize plant cells and solve the problem of reactor design. Further studies are needed to evaluate this approach fully.

A number of model studies on immobilized plant cells have been carried out (cf. TABLE 3). Some distinct differences, as compared to freely suspended cells, have been observed that make immobilization of plant cells for the production of biochemicals an attractive alternative. These differences include the following:

(1) increased productivity,
(2) prolonged production (stationary) phase,
(3) protection of the sensitive cells,
(4) in certain cases, spontaneous release of product(s), and
(5) possibility of inducing product release by permeabilization.

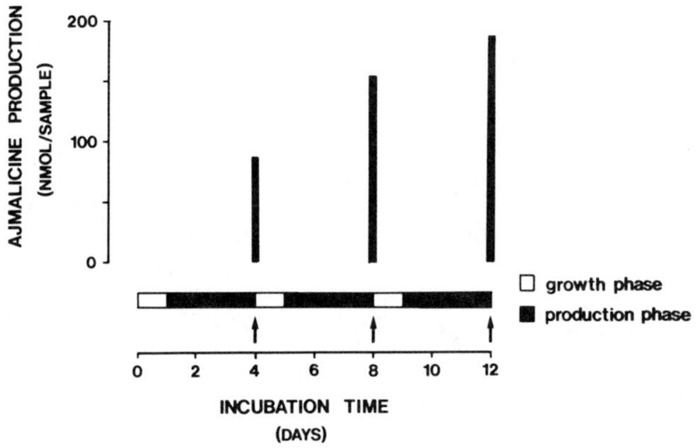

FIGURE 12. As in FIGURE 11, using alginate-entrapped cells.

In summary, new processes for the production of plant metabolites by cultivated cells would most likely separate growth from production phases. Growth could be carried out in suspension cultures while production could be carried out by immobilized cells, with some distinct advantages. In general, immobilized plant cells can be manipulated more easily than freely suspended cells. This fact may also influence the future development of plant technology.

For further information, some recent reviews on immobilized plant cells are available (see Brodelius[18] and Brodelius and Mosbach[19]).

REFERENCES

1. BRODELIUS, P., B. DEUS, K. MOSBACH & M. H. ZENK. 1979. Immobilized plant cells for the production and transformation of natural products. FEBS Lett. **103**: 93–97.
2. BRODELIUS, P. & K. NILSSON. 1980. Entrapment of plant cells in different matrices. A comparative study. FEBS Lett. **122**: 312–316.
3. NILSSON, K., S. BIRNBAUM, S. FLYGARE, L. LINSE, U. SCHRÖDER, U. JEPPSSON, P. O.

LARSSON, K. MOSBACH & P. BRODELIUS. 1983. A general method for the immobilization of cells with preserved viability. Eur. J. Appl. Microbiol. Biotechnol. **17**: 319–326.

4. GALUN, E., D. AVIV, A. DANTES & A. FREEMAN. 1983. Biotransformation by plant cells immobilized in cross-linked polyacrylamide-hydrazide. Planta Med. **49**: 9–13.
5. ROSEVEAR, A. & C. A. LAMBE. 1982. Improvements in or relating to production of chemical compounds. European Patent Application 82301571.4.
6. JOSE, W., H. PEDERSEN & C. K. CHIN. 1983. Immobilized plant cell cultures of *Petunia hybrida*. Ann. N.Y. Acad. Sci. **413**: 409–412.
7. SHULER, M. L. 1981. Production of secondary metabolites from plant tissue culture—problems and prospects. Ann. N.Y. Acad. Sci. **369**: 65–79.
8. BRODELIUS, P. & H. J. VOGEL. 1984. ^{31}P-NMR studies on phosphate uptake mechanism of *Catharanthus roseus* and *Daucus carota* cells. Submitted for publication.
9. VOGEL, H. J. & P. BRODELIUS. 1984. *In vivo* ^{31}P-NMR comparison of freely suspended and immobilized plant cells. J. Biotechnol. In press.
10. BRODELIUS, P., B. DEUS, K. MOSBACH & M. H. ZENK. 1981. Catalysts for the production and transformation of natural products having their origin in higher plants, process for production of the catalysts, and use thereof. European Patent Application 80850105.0.
11. BRODELIUS, P., B. DEUS, K. MOSBACH & M. H. ZENK. 1980. The potential use of immobilized plant cells for the production and transformation of natural products. *In* Enzyme Engineering. H. H. Weetall & G. P. Roger, Eds Vol. **5**: 373–381. Plenum Press. New York.
12. LAMBE, C. A. & A. ROSEVEAR. 1983. Review of plant and animal cell immobilization. Proc. Biotech83. London. May 4–6 1983. pp. 565–576.
13. FELIX, H., P. BRODELIUS & K. MOSBACH. 1981. Enzyme activities of the primary and secondary metabolism of simultaneously permeabilized and immobilized plant cells. Anal. Biochem. **116**: 462–470.
14. BRODELIUS, P. & K. NILSSON. 1983. Permeabilization of immobilized plant cells, resulting in release of intracellularly stored products with preserved viability. Eur. J. Appl. Microbiol. Biotechnol. **17**: 275–280.
15. MORITZ, S., I. SCHULLER, C. FIGUR, A. W. ALFERMANN & E. REINHARD. 1982. Biotransformation of cardenolides by immobilized *Digitalis* cells. *In* Proc. 5th Intl. Cong. Plant Tissue & Cell Culture. A. Fujiwara, Ed. Japanese Association for Plant Tissue Culture, Tokyo. pp. 401–402.
16. VELIKY, I. A. & A. JONES. 1981. Bioconversion of gitoxigenin by immobilized plant cells in a column bioreactor. Biotechnol. Lett. **3**: 551–554.
17. MORRIS, P., N. J. SMART, & M. W. FOWLER. 1983. A fluidized-bed vessel for the culture of immobilized plant cells and its application for the continuous production of fine cell suspension. Plant Cell Tissue Organ Culture **2**: 207–216.
18. BRODELIUS, P. 1983. Immobilized plant cells. *In* Immobilized Cells and Organelles. B. Mattiasson, Ed. Vol. **1**: 27–55. CRC Press. Boca Raton, FL.
19. BRODELIUS, P. & K. MOSBACH. 1982. Immobilized plant cells. *In* Advances in Applied Microbiology. A. Laskin, Ed. Vol **28**: 1–26. Academic Press. New York.
20. CURTIN, M. E. 1983. Harvesting profitable products from plant tissue culture. Biotechnology **1**: 649–657.
21. ALFERMANN, A. W., W. BERGMANN, C. FIGUR, U. HELMBOLD, D. SCHWANTAG, I. SHULLER & E. REINHARD. 1983. Biotransformation of β-methyldigitoxin to β-methyldigoxin by cell cultures of *Digitalis lanata*. *In* Plant Biotechnology (Society for Experimental Biology seminar series; 18) University Press. Cambridge. pp. 67–74.
22. JONES, A. & I. A. VELIKY. 1981. Examination of parameters affecting the 5β-hydroxylation of digitoxigenin by immobilized cells of *Daucus carota*. Eur. J. Appl. Microbiol. Biotechnol. **13**: 84–89.
23. TANAKA, A., K. SONOMOTO & S. FUKUI. 1983. Application of living cells immobilized by prepolymer methods. Ann. N.Y. Acad. Sci. This volume.
24. JIRKU, V., T. MACEK, T. VANEK, V. KRUMPHANZL & V. KUBANEK. 1981. Continuous production of steroid glycoalkaloids by immobilized plant cells. Biotechnol. Lett. **3**: 447–450.
25. LINDSEY, K., M. M. YEOMAN, G. M. BLACK & F. MAVITUNA. 1983. A novel method for the immobilization and culture of plant cells. FEBS Lett. **155**: 143–149.

Semicommercial Production of Ethanol Using Immobilized Microbial Cells[a]

H. SAMEJIMA

Kyowa Hakko Kogyo Co., Ltd.
Ohtemachi 1-6-1, Chiyoda-ku,
Tokyo 100, Japan

M. NAGASHIMA, M. AZUMA, S. NOGUCHI,
AND K. INUZUKA

Kyowa Hakko Kogyo Co., Ltd., Hofu Plant
Technical Research Laboratories,
Kyowa-cho 1-1, Hofu-shi,
Yamaguchi-ken 747, Japan

Following the successive oil crises of the 70s, the Research Association for Petroleum Alternatives Development (RAPAD) was organized in Japan in May 1980 under the auspices of the Ministry of International Trade and Industry of the Japanese government. Biomass conversion and utilization was one of three main projects of the association. The biomass conversion and utilization project consists of two main targets: conversion of cellulosic materials to fuel ethanol and continuous ethanol fermentation using immobilized yeast cells.

With respect to alcohol fermentation, the batchwise fermentation process is still dominant today. However, in order to produce fuel alcohol more economically, it is necessary to attain much improved productivity and reduced manufacturing cost in comparison with the conventional process. For these requirements, several processes, such as a yeast cell recycling system[1] and a flocculated cell system[2,3] have been proposed. RAPAD decided to develop continuous fermentation processes using immobilized growing yeast cells according to the recent development of immobilized cell technology.[4,5] Kyowa Hakko has chosen the calcium alginate gel beads process and the JGC group has chosen the photocross-linkable resin sheet process. This paper presents the results of studies on an advanced fermentation process developed by Kyowa Hakko.

SELECTION OF SUPPORTING MATERIALS

Chibata[6] reported that carrageenan was a suitable supporting material for immobilization of living yeast cells. Various supporting materials were examined with respect to the following selection standards. These standards are (a) higher carrier activity, (b) availability of the material in quantity, (c) low cost of immobilization, (d) ease of scaling-up of the preparation, (e) mechanical strength for long-run operation, and (f) physiological safety of the material.

For the screening of supporting materials, living cells of *Saccharomyces cerevisiae*

[a]This work was done as a part of the Biomass Utilization Project of the Research Association for Petroleum Alternatives Development in Japan.

were entrapped with various polymers, and ethanol productivity and other characteristics of such immobilized cells were examined using laboratory-scale glass reactors. Some of the experimental results are shown in TABLE 1. Use of calcium alginate gel beads resulted in the best ethanol productivity. Also, sodium alginate solution does not coagulate at room temperature until it contacts a calcium chloride solution. This characteristic makes it easier to handle alginate in large-scale operations compared with other hydrogel polymers. Considering those merits and other selection standards, calcium alginate was finally selected as the entrapping agent.

DESIGN OF REACTOR

Following bench-scale experiments using calcium alginate gel beads, three types of prototype reactors were constructed in order to decide the pilot plant design, as shown in FIGURE 1. The first is a tall, tower-type reactor with adjustable L/D ratio from 7 to 10. The second is a short, column-type reactor with adjustable ratio from 1.5 to 2. The

TABLE 1. Activity of Various Immobilized Carriers

Immobilizing Process	Alcohol-producing Activity	
	mg alc/g gel, h	mg alc/ml gel, h
Porous epoxy resin	26	15
Nylon microcapsule (mc)	2	1
Unsat. polyester mc	4	2
Acetyl butyl cellulose mc	14	8
Porous polystyrene mc	14	7
Polyacryl amide gel	36	20
Calcium alginate gel	40	22
Low methoxy pectin gel	35	19
Carrageenan gel	40	22
Agar gel	40	22
Silica sol	50	28

third consists of a number of vertical plates between which the substrate solution flows laterally in a zig-zag course from one end of the reactor to the other. As a result of these prototype experiments, tower or column-type reactors were tentatively chosen for the next-step pilot plant.

Continuous preparation of immobilized cell beads was carried out as shown in FIGURE 2. The sodium alginate solution mixed with yeast cell suspension was showered from the top nozzle of the reactor into the calcium chloride solution in the reactor. The drops of alginate solution coagulate quickly when they contact the calcium chloride solution. The preparation of cell beads was completed within several hours in the case of prototype reactors. Therefore, no special equipment was needed for the preparation of gel beads, and the preparation can be done under aseptic conditions when the system is sterilized.

After further studies on the operational characteristics, it was recognized that there was extensive turbulent flow due to the carbon dioxide gas evolved during the fermentation. It was decided to connect two columns in series to accomplish higher conversion yields. The good fluidization obtained resulted in good mixing in the reactor. The immobilized cell bed could be filled to 60% of the total volume of each

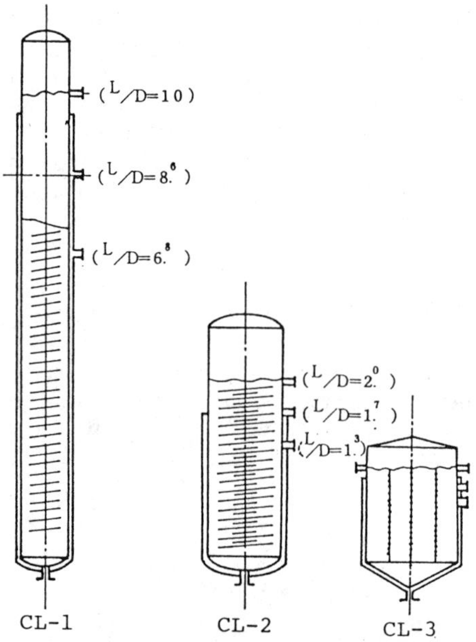

FIGURE 1. Design of prototype reactors: CL-1 column reactor, L/D = 7–10, H = 6 m, capacity, 1 kl; CL-2 column reactor, L/D = 1–2, H = 2.5 m, capacity 1 kl; and CL-3 rectangular reactor (gas-liquid separation type) H = 1 m, capacity 1 kl.

reactor. From these results, a tentative process flow was decided upon. At this stage, a medium sterilization system and a seed fermenter were still included in the total flow in order to keep the system aseptic.

PREVENTION OF CONTAMINATION

In early experiments, long-run operations sometimes suffered contamination by certain microbes. One of these was identified as *Acetobacter oxydans,* an acid producer encountered frequently. In order to get better fermentation yields and make the long-run operation more stable, it was necessary to protect the process from such contamination. Studies on the characteristics of the isolated contaminants were carried out. A very effective way to prevent contamination problems was to keep the initial pH of the inlet substrate solution at 4.0 with sulfuric acid. Addition of some bactericidal substances was also effective. By employing such measures, the contamination problem was practically eliminated; moreover the process had become operable without sterilization of the inlet medium.

FIGURE 2. Preparation of immobilized yeast cells.

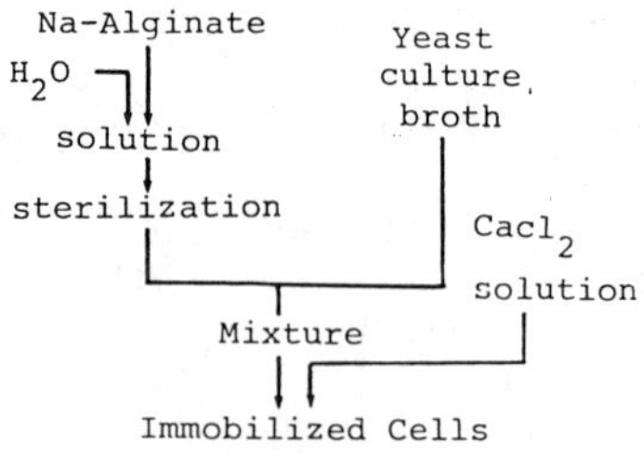

IMPROVEMENT OF YEAST VIABILITY

It was observed that the activity of the immobilized yeast cells gradually decreased during long-run operations. Therefore, maintenance of yeast cell viability was also investigated for the stabilization of the process. As shown in FIGURE 3, yeast growth inside of a gel bead is usually near the surface of bead. It was found that the cell activity was greatly promoted by the presence of dissolved oxygen and certain sterols and unsaturated fatty acids.[7] Therefore, entrapment of such lipids into gel beads was tried. Yeast growth was greatly enhanced, and a higher productivity of ethanol (30 to 50 g ethanol/l gel · h) was attained. Moreover, suitable aeration from the bottom of reactor further enhanced the operational stability of immobilized yeast cells. FIGURE 4 is a photomicrograph of a cross section of a gel bead after such improvement. The yeast layer became much thicker. FIGURE 5 shows an electron-microscopic photograph of a

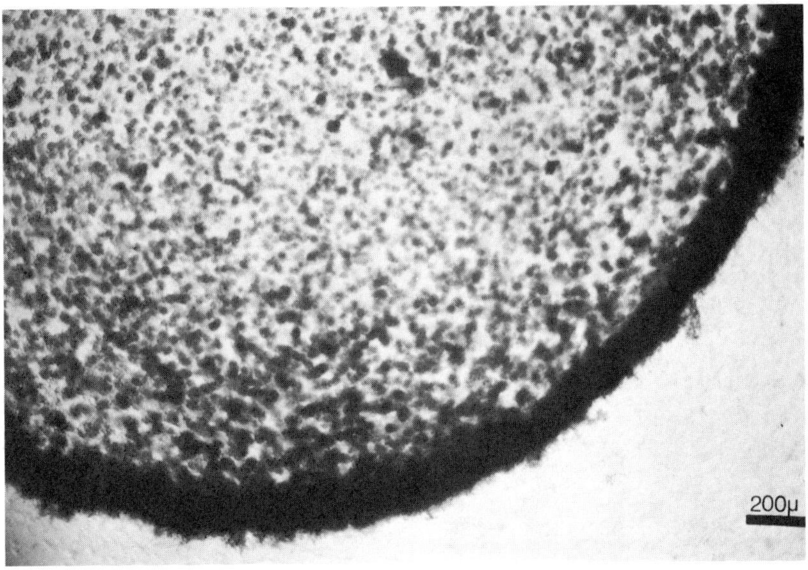

FIGURE 3. Cross section of gel bead prepared by the tentative immobilization process.

cross section of a gel bead in which the initial inoculum size of yeast cells was very small. Despite this, a more abundant growth of yeast cells was observed around the gel.

SELECTION AND IMPROVEMENT OF STRAINS

Suitable yeast strains were first selected from the culture collection of Kyowa Hakko using laboratory-scale column reactors. Those strains can produce 8 to 9% (vol/vol) of ethanol steadily for long periods at 30 to 32°C. Strain improvement was also carried out in order to get strains with greater alcohol tolerance. An example of long-run performance is shown in FIGURE 6. In this case, the strain produced more than

10% (vol/vol) of ethanol for a long time. Higher temperature strains that tolerate 35°C are also available.

PILOT PLANT OPERATIONS

FIGURE 7 shows the improved process flow. In this process, sterilization facilities and seed culture fermenters are eliminated. Even larger than 10 kl-sized reactors can be operated by this process. Our goals in the process development were to simplify the total process flow and to minimize the initial investment and operational costs. The final pilot plant was completed in March 1982. The pilot plant is composed of two channels of reactors; one channel consists of two columns (each 1 kl) in series, and the other consists of three columns (0.8 kl × 1, 0.6 kl ×2). Total column volume is 4 kl and the total productivity under standard conditions is 2.4 kl of pure ethanol per day.

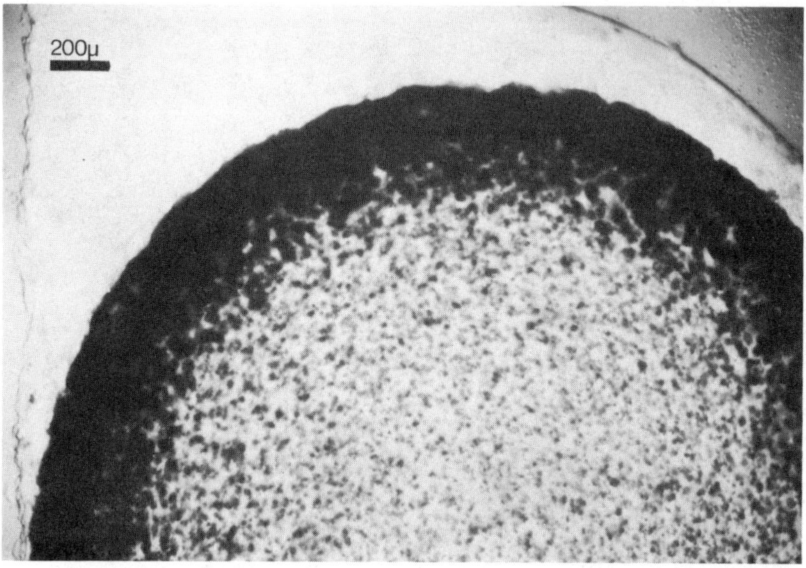

FIGURE 4. Cross section of gel bead prepared by the improved method.

Pilot plant operations have been carried out since April 1982. FIGURE 8 shows an operation that was initiated in September 1982. In this case, 8.5 to 9% (vol/vol) ethanol had been constantly produced from diluted cane molasses for more than 4000 hours (about 6 months) without any trouble. The productivity of ethanol was calculated as about 20g/l of total volume/h (33g/l of gel/h). This means that 600 liters of pure ethanol was produced per day using a 1-kl-sized reactor. FIGURE 9 shows another pilot plant run. In this case, an alcohol-tolerant strain was employed, and more than 10% (vol/vol) of ethanol was steadily produced for more than 2000 hours. Although the productivity of alcohol is lower than that of the standard operation (alcohol concentration is 8.5% (vol/vol), this process showed more than 10 times greater productivity than that of the conventional batchwise fermentation.

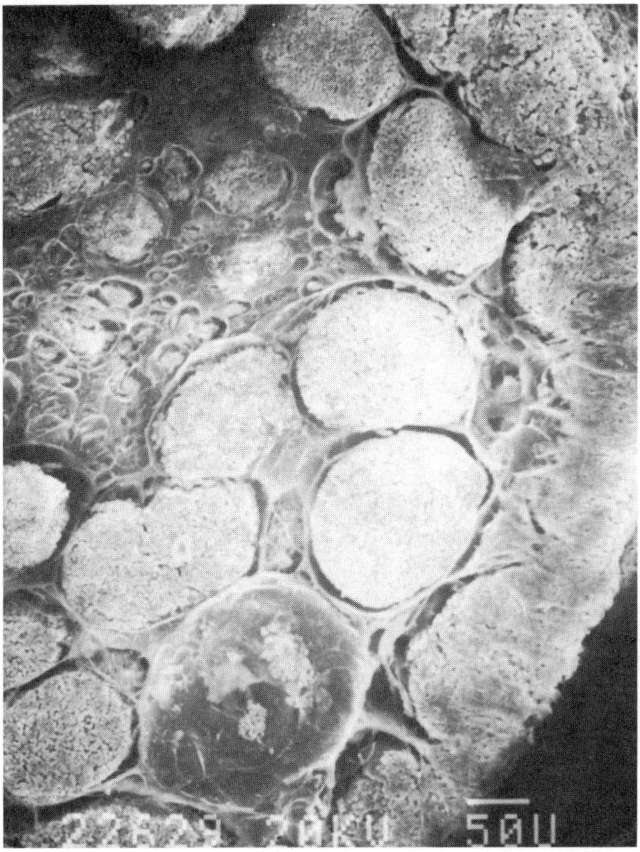

FIGURE 5. Electron photomicrograph of gel bead prepared by the further improved method.

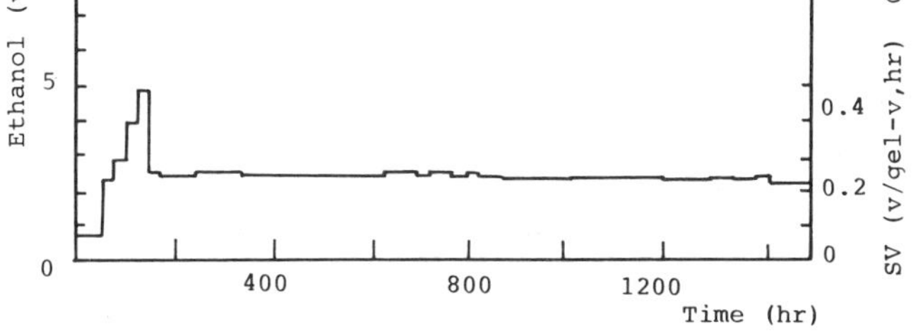

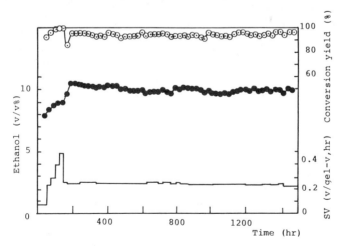

FIGURE 6. Continuous operation in laboratory for higher concentration of ethanol. (200-ml columns in series, strain: T-620) Ethanol, ●; space velocity, —; conversion yield, ⊙.

This pilot plant was operated with a computer-controlled system. FIGURE 10 shows a process diagram displayed on the CRT screen. Also, automatic analysis of ethanol and residual available sugar was carried out with computer-aided instrumentation. It enables further decreases in labor costs.

A large-scale, semicommercial plant was constructed at the Hofu Plant in April 1983. The plant consists of two 10-kl-sized reactors in series. Operation of the plant started in May 1983. Even in this large-scale plant, more than 8% (vol/vol) of ethanol has been steadily produced for more than 1000 hours, as shown in FIGURE 11. About 95% of conversion yield against theoretical value was attained.

EVALUATION OF THE PROCESS

In comparison with the conventional batchwise process, the present process can be evaluated as follows:

(1) The productivity of alcohol is more than 20 times higher than conventional batchwise fermentation under the standard conditions (alcohol concentration

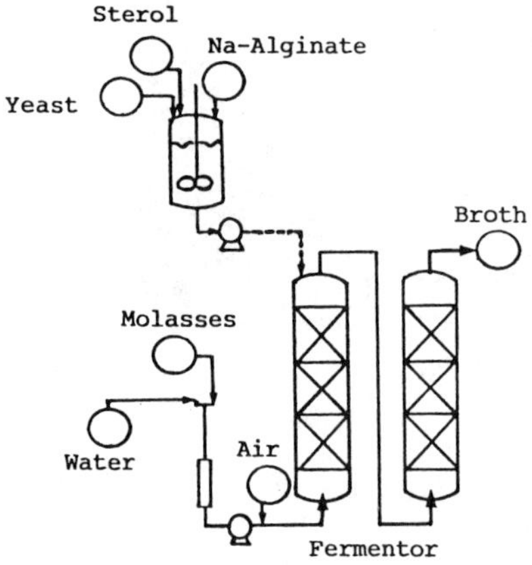

FIGURE 7. Process flow diagram of the pilot operation.

is 8 to 9% (vol/vol). This enables a significant decrease in the initial investment cost of fermenters.

(2) This process can be operated more than 4000 hours without sterilization and seed culture steps. This means a significant savings in energy and labor costs.

(3) This process does not need any special equipment for the preparation of immobilized yeast cells. This may be an advantage over other immobilized cell systems.

(4) This process can be easily operated automatically by a computer-controlled system, which enables further decreases in labor costs.

(5) Higher alcohol concentration, for example, more than 10% (vol/vol) can be attained, if a suitable strain is employed. In this case, more than 10 times higher productivity can be obtained.

(6) Conversion yield of alcohol is about 95% against theoretical value on the basis of assimilable sugar charged to the fermenter.

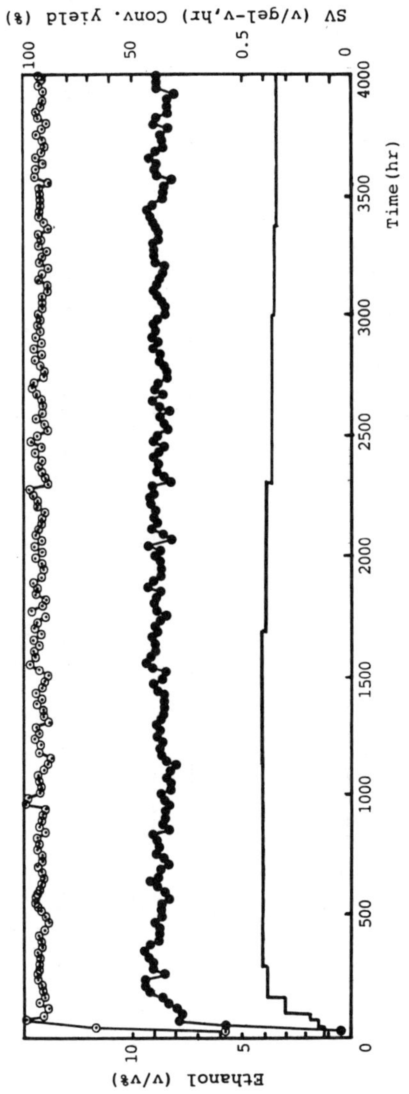

FIGURE 8. Continuous alcohol fermentation by immobilized growing yeast cells in the pilot plant (Run 7). Conversion yield, ⊙; ethanol, ●; space velocity, —.

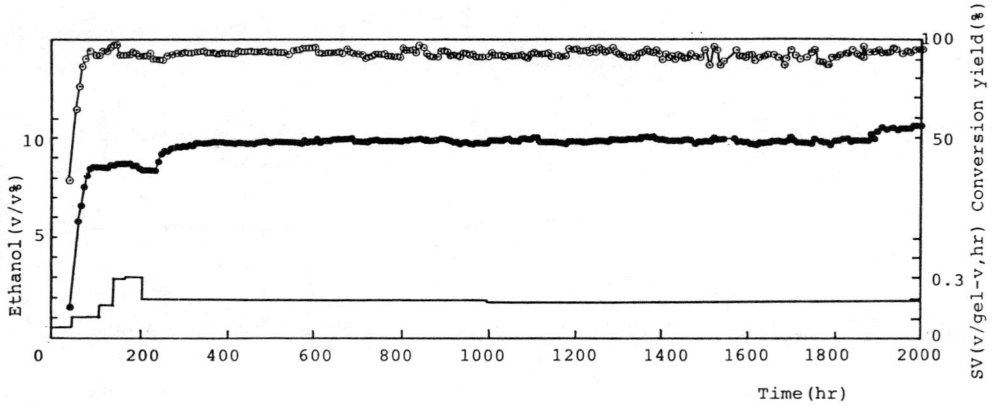

FIGURE 9. Continuous alcohol fermentation by immobilized growing yeast cells in the pilot plant (Run 8). Conversion yield, ⊙; ethanol, ●; space velocity,—.

This process represents a new trend of alcohol fermentation technology, and can economize the alcohol production process remarkably. Furthermore, this process can be improved by combining with other technologies such as flash fermentation.

SUMMARY

Growing cells of *Saccharomyces cerevisiae* immobilized in calcium alginate gel beads were employed in fluidized-bed reactors for continuous ethanol fermentation from cane molasses and other sugar sources.

Some improvements were made in order to avoid microbial contamination and to

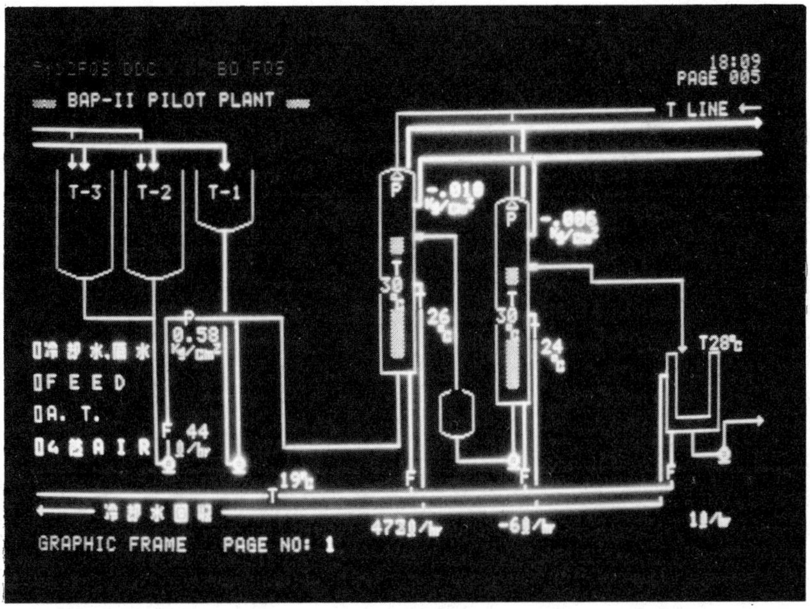

FIGURE 10. Process diagram displayed on the CRT screen of the pilot plant.

keep cell viability in order to pursue stable, long-run operations. Especially, entrapment of sterol into immobilized gel beads enhanced ethanol productivity more than 50 g ethanol/liter-gel · h and prolonged life stability for more than half a year. Cell concentration in the carrier was estimated at greater than 250 g dry cell/l gel.

A pilot plant with total 4 kl column volume was constructed and has operated since 1982. It was confirmed that 8 to 10% (vol/vol) ethanol-containing broth was continuously produced from nonsterilized, diluted cane molasses for over half a year. The productivity of ethanol was calculated as 0.6 kl ethanol/kl reactor volume per day with 95% of conversion yield against theoretical yield in the case of 8.5% (vol/vol) ethanol broth.

A large semicommercial-scale plant that consists of two 10 kl-sized reactors was constructed in April 1983 and has operated since May 1983. The results showed

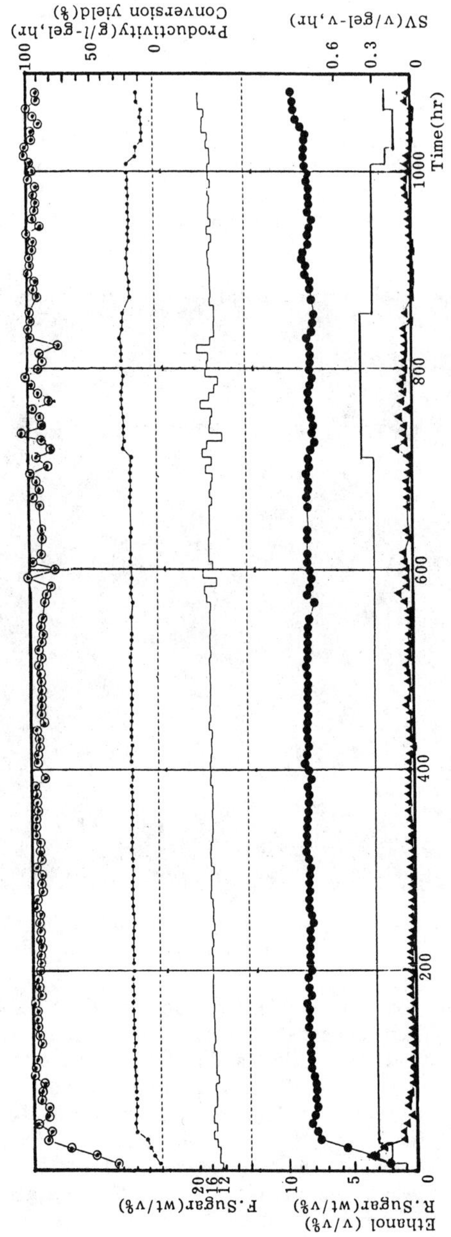

FIGURE 11. Operating result of the immobilized process in the semicommercial plant. ● ethanol, ○ conversion yield, ▲ residual sugar, — feed sugar, — feeding rate as SV, -●- productivity.

feasibility of this process for the commercial production of ethanol. Merits of the process are discussed in comparison with the conventional batchwise ethanol fermentation process.

REFERENCES

1. BOINOT, F. 1941. United States Pat. 2230318.
2. PRINCE, I. G. & J. P. BARFORD. 1982. Biotechnol. Lett. **4**(4): 261–268.
3. ARCURI, E. J., R. M. WORDEN & S. E. SHUMATE. 1980. Biotechnol. Lett. **2**(11): 499–504.
4. NAVARRO, J-M. & G. DURAND. 1976. Ind. Aliment. Agr. **93**: 695–703.
5. KIERSTAN, M. & C. BUCK. 1977. Biotechnol. Bioeng. **19**: 387–397.
6. WADA, M., J. KATO & I. CHIBATA. 1980. Eur. J. Appl. Microbiol. Biotechnol. **10**(4): 275–287.
7. ANDREASEN, A. A. & T. J. B. STIER. 1954. J. Cell. Comp. Physiol. **43**: 271–281.

Applications of Immobilized Lactic Acid Bacteria[a]

P. LINKO, S.-L. STENROOS, AND Y.-Y. LINKO

Laboratory of Biotechnology and Food Engineering
Department of Chemistry
Helsinki University of Technology
SF-02150 Espoo 15, Finland

T. KOISTINEN, M. HARJU, AND M. HEIKONEN

Research and Development Department
Division of Process Technology
Valio Finnish Co-operative Dairies' Association
Kalevankatu 56 B
SF-00180 Helsinki 18, Finland

Both lactic acid as an additive and lactic acid fermentation have a number of food applications. Furthermore, biotechnically produced lactic acid from renewable carbohydrate sources or wastes would be a potential feedstock for chemical synthesis via lactonitrile and for lactide polymers.[1] Because homofermentative lactic acid bacteria produce no carbon dioxide from sugars, substrate carbon is more efficiently utilized in lactic acid than in ethanol fermentation. Lactic acid is produced industrially both by chemical synthesis and by fermentation. When lactic acid is produced fermentatively by selection of the appropriate microorganisms, L-lactic acid, D-lactic acid, or a mixture can be obtained. Batch fermentation has been employed in industrial-scale lactic acid production. Continuous lactic acid processes have been reported,[2–9] but as far as is known, they are not used on production scale.[10] Only a few reports have been published on immobilized microorganisms for lactic acid production.[11–15] Compere and Griffith[15] grew a mixed culture of commercial lactic acid bacteria on a solid support covered by a gelatin-polyelectrolyte film in a column reactor. The lactic acid content of sour whey increased from 1.4 to 2.1% in one passage through the column. Compere and Griffith[16,17] also produced lactic acid continuously from wood molasses using a similar packed-column reactor. Divies[12] employed polyacrylamide-entrapped *Lactobacillus delbrueckii* for continuous lactic acid production from glucose, and *L. casei* for fermenting maleic acid to lactic acid. Divies[18] also used both calcium alginate- and polyacrylamide-immobilized mixed cultures of *L. bulgaricus* and *Streptococcus thermophilus* both for preferementation of milk and for yogurt production. We have earlier reported[13,14,19] that lactic acid can be produced by calcium alginate gel-immobilized *L. delbrueckii* (recently identified as *L. casei* subsp. rhamnosus) in very good yields. In our recent work, we have used several lactic acid bacteria for lactic acid production and for applications in dairy processing.

[a]Financial support for this work was received from The Ministry of Trade and Industry, the Academy of Finland, the Research Council for the Technical Sciences, and from the Orion Corporation Research Foundation.

IMMOBILIZATION OF LACTIC ACID BACTERIA

Lactobacillus bulgaricus, L. pentosus, and *L. helveticus* cells were immobilized in calcium alginate gel beads. The wet cell mass (17% d.m. of *L. bulgaricus,* 29% d.m. of *L. pentosus,* and 14% d.m. of *L. helveticus*) was suspended in 6 or 8% sodium alginate solution (the ratio of cell mass to alginate was 1:5) and extruded under pressure through 0.6-mm diameter hollow needles into 0.5 *M* calcium chloride to obtain biocatalyst beads of about 2–3 mm in diameter. The beads were allowed to harden for at least 20 min before filtering.

Immobilization of whole cells in calcium alginate is a mild and easy method.[20] Nilsson *et al.*[21] studied the viability of calcium alginate-entrapped *Pseudomonas denitrificans* by dissolving the gel in phosphate buffer. 80–85% of the cells could be recovered from the dissolved alginate gel. Garde *et al.*[22] found that *Rhodopseudomonas capsulata* cells grew well in calcium alginate beads. After immobilization, there was one bacterium per cavity and after a 24-h incubation, there were 8–12 bacteria. We found that immediately after entrapment of *L. helveticus* cells in calcium alginate, the number of living lactic acid bacteria cells was about 3.0×10^7 to 8.9×10^7 per g gel. After 20 days of intermittent prefermentation of milk with the immobilized biocatalyst, the number of lactic acid bacteria was increased to 1.2×10^8 per g gel.

Calcium-chelating agents such as phosphates and certain cations, such as K^+ or Mg^{2+}, disrupt alginate gel by solubilizing the bound Ca^{2+}.[23] To guarantee good stabilization of gel, some experiments were carried out in the presence of 0.005 *M* $CaCl_2$.

LACTIC ACID PRODUCTION FROM VARIOUS SUBSTRATES

We have used several immobilized lactic acid bacteria for lactic acid production from glucose, lactose, xylose, wood hydrolysates and from dairy industry by-products. Both batch and continuous fermentations were usually performed in packed-bed column reactors. Finely powered calcium carbonate was added to the substrate as buffer; in batch experiments pH was also automatically controlled by the addition of sodium hydroxide. In batch recycle column processes, substrate was continuously circulated through the column. In continuous experiments, substrate was fed in from the top of the column reactor, and the carbon dioxide liberated from the buffer during the fermentation was allowed to escape freely.

We have reported earlier[13,14] that with calcium alginate-entrapped *L. delbrueckii,* (*L. casei* subsp. rhamnosus), 93% lactic acid yield could be obtained in a packed-bed column reactor in continuous fermentation at a 7.9-h residence time from 4.8% (wt/vol) glucose. At a 4.9-h residence time, the yield was still about 80%. About 90–95% of the lactic acid was L-lactic acid. With the same bacterial strain in a batch recycle column reactor, about 95% lactic acid yield was achieved in about 40 h. With immobilized *L. bulgaricus,* the lactic acid yield in a batch process was somewhat lower (FIG. 1). Lactic acid yield was about 75% after a 70-h fermentation from 5.0% (wt/vol) glucose, when pH was controlled automatically. About 85% of the lactic acid formed was D-lactic acid. With calcium alginate-immobilized *L. pentosus,* a 90% lactic acid yield from 5.0% (wt/vol) glucose was reached at 40 h with 1.0% (wt/vol) yeast extract as nutrient and at 25 h with all MRS-medium[23] substituents as nutrient in a batch recycle column reactor process (FIG. 2). Immobilized *L. pentosus* produced a racemic mixture of L- and D-lactic acid. According to Lockwood[25] and Casida,[26] 93–95% lactic acid yield could be achieved in a conventional batch fermentation

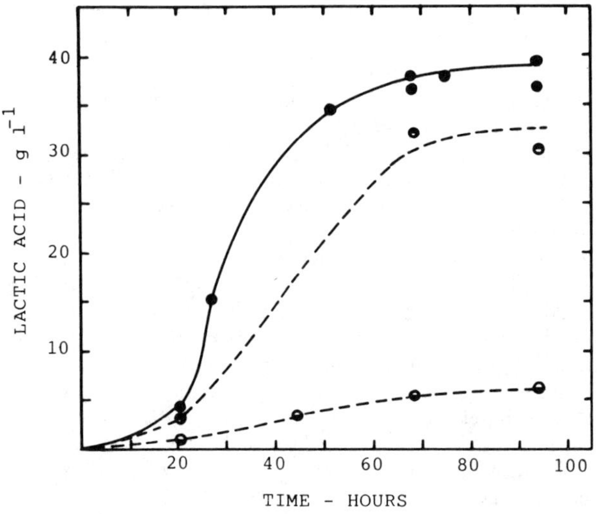

FIGURE 1. Batch production of lactic acid from glucose (5.0% wt/vol) with immobilized *L. bulgaricus* (1.0% wt/vol yeast extract, automatic pH control; 45°C, biocatalyst 50 g/200 ml). ● total lactic acid, ◓ D-lactic acid, ◒ L-lactic acid.

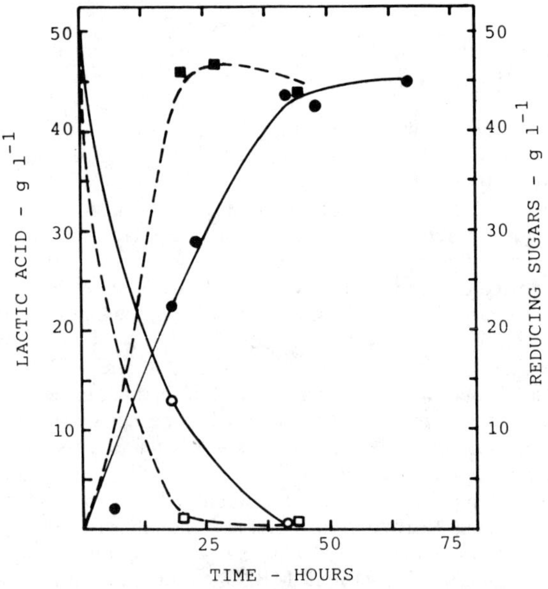

FIGURE 2. Lactic acid production from glucose (5.0% wt/vol) with immobilized *L. pentosus* in a batch recycle column reactor: (A—) with 1.0% wt/vol yeast extract as a nutrient and (B---) in MRS-medium (50 g l^{-1} $CaCO_3$; 30°C; pH 6.4; biocatalyst 20 g/80 ml). ● and ■, lactic acid, ○ and □, reducing sugars.

process in 4 to 6 days. Consequently, calcium alginate-immobilized lactic acid bacteria fermented glucose at a relatively rapid rate.

Many industrial chemicals, including lactic acid, could be produced from wastewaters from the pulp, paper, and fiberboard industries. These wastewaters contain also 5-carbon sugars, so that for lactic acid fermentation a microorganism must be chosen that also can ferment pentoses. We used *L. pentosus,* which ferments ribose to one mole of lactic and one mole of acetic acid. When pure xylose was fermented with calcium alginate-immobilized *L. pentosus* in a batch recycle column reactor, a 90% lactic acid yield was reached in 120 h from 5.0% (wt/vol) xylose in MRS-medium (FIG. 3). *L. pentosus* ferments xylose well, but much slower than glucose (FIG. 2). When a mixture of glucose and xylose (2.5% wt/vol each) was used as the carbon

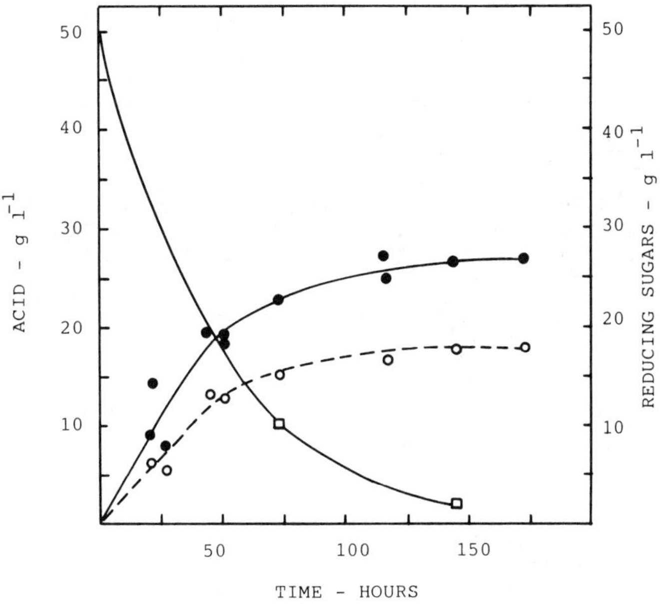

FIGURE 3. Lactic acid production from xylose (5.0% wt/vol) with immobilized *L. pentosus* in a batch recycle column reactor (MRS-medium, 50 g l^{-1} $CaCO_3$, 0.7 g l^{-1} $CaCl_2$; 30°C; biocatalyst 20 g/80 ml). ● lactic acid, ○ acetic acid, □ reducing sugars.

source, a 90% lactic acid yield was achieved in 50 h. With 5.0% (wt/vol) glucose and 5.0% (wt/vol) xylose, it took 180 h to reach 90% lactic acid yield (FIG. 4). Xylose was fermented faster to lactic and acetic acid in a glucose-xylose mixture than with either sugar alone. A lactic acid fermentation rate with higher sugar concentration could obviously be improved with adaptation.

L. pentosus also fermented well a wood hydrolysate, which contained 5.7% (wt/vol) reducing sugars (56% pentoses and 44% hexoses) (FIG. 5). In a batch process with calcium carbonate as buffer and MRS-medium substituents as nutrient, the concentration of reducing sugars decreased from 57 g l^{-1} to 7 g l^{-1} at 70 h both with free *L. pentosus* cells and with immobilized cells. Lactic acid concentration was then 40 g l^{-1}. A longer lag-phase at the beginning of the fermentation with immobilized

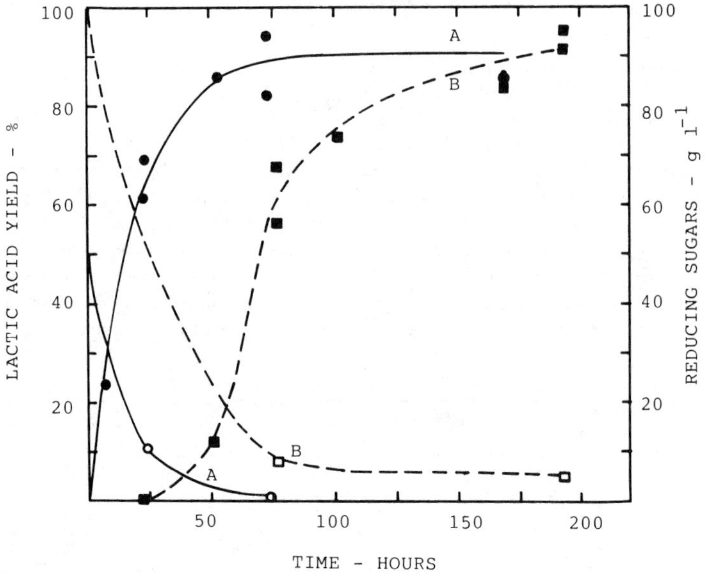

FIGURE 4. Batch production of lactic acid from glucose-xylose-mixture (A, 2.5% wt/vol both sugars, and B, 5.0% wt/vol both sugars) with immobilized *L. pentosus* (MRS medium, 50 g l^{-1} $CaCO_3$, 0.7 g l^{-1} $CaCl_2$; 30°C; biocatalyst 5 g/100 ml). ● and ■, lactic acid; ○ and □, reducing sugars.

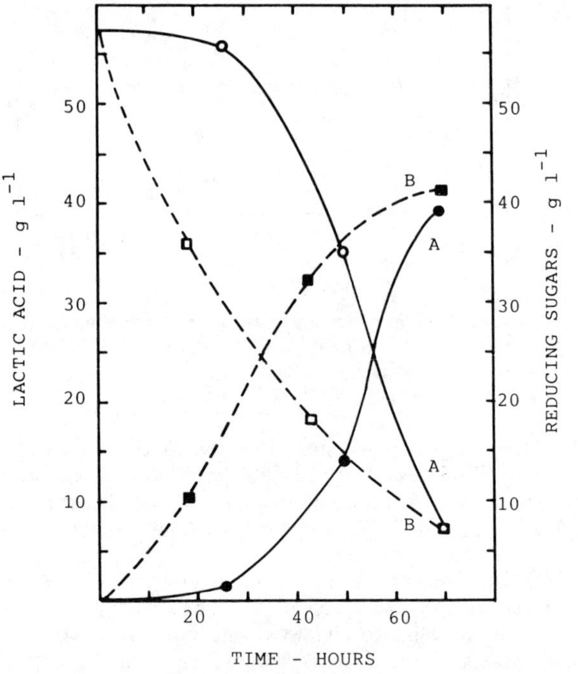

FIGURE 5. Batch production of lactic acid from wood hydrolysate with (A) immobilized *L. pentosus* cells and (B) with free cells (MRS medium, 50 g l^{-1} $CaCO_3$, 0.7 g l^{-1} $CaCl_2$; 30°C; biocatalyst 5 g/100 ml, free cells 5 ml/100 ml). ● and ■, lactic acid; ○ and □, reducing sugars.

cells was probably due to storage of biocatalyst beads after immobilization. In a batch recycle column reactor, when wood hydrolysate was used as the carbon source, the maximum lactic acid concentration was about 4.5% (wt/vol).

Leonard *et al.*[27] have produced lactic acid from sulfite waste liquor with *L. pentosus.* They reached 3.0% (wt/vol) lactic acid concentration at 48–72 h. In the fermentation, however, very little acetic acid was formed, showing that pentose sugars in sulfite liquor were fermented slowly. Griffith and Compere[17] produced lactic acid continuously from wood molasses in a fixed-film packed tower reactor. With wood molasses, which contained 8.0% (wt/vol) sugars as a substrate, the maximum lactic acid concentration that could be reached was 5.0% (wt/vol).

We have also fermented lactose-containing raw materials with *L. bulgaricus,* which is a typical lactose-fermenting microorganism used in lactic acid production. With calcium alginate-immobilized *L. bulgaricus,* about 55–60% lactic acid yield

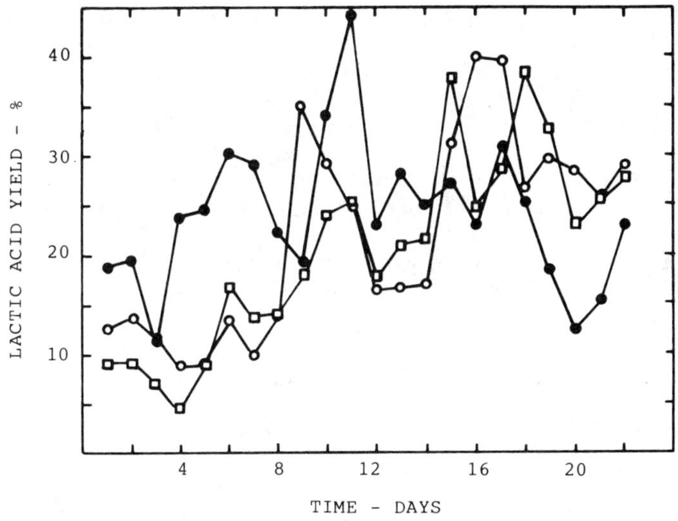

FIGURE 6. Continuous production of lactic acid with immobilized *L. bulgaricus* ($\tau \sim$ 12.5 h; 45°C; pH 6.0) from: ● 5.0% wt/vol lactose, 1.0% wt/vol yeast extract, 30 g l^{-1} $CaCO_3$; ○ 7.5% wt/vol lactose, 1.5% wt/vol yeast extract, 45 g l^{-1} $CaCO_3$; □ 10.0% wt/vol lactose, 2.0% wt/vol yeast extract, 45 g l^{-1} $CaCO_3$.

could be obtained in a batch recycle column process, when 5–9% (wt/vol) lactose was used as carbon source and yeast extract as nutrient.[14] In continuous production of lactic acid from 5–10% (wt/vol) lactose, the maximum lactic acid yields varied between 37 and 44% at 12.5 h residence time (FIG. 6). Lactic acid yield was higher at the beginning of the run with lower lactose concentration, but after 10 days, no difference was observed. In the lactic acid production with immobilized *L. bulgaricus,* when lactose was used as substrate, there was much galactose left in the solution after fermentation. The *L. bulgaricus* strain which we used in our experiments fermented well only the glucose formed in lactose hydrolysis.

Cheese whey, a by-product of the dairy industry, has earlier been used as substrate in industrial lactic acid production.[28,29] Today, in addition to the production of pure lactic acid, the production of ammonium lactase as a feed supplement from cheese whey has been investigated.[30–32] According to Kosikowski,[33] the annual world production of liquid whey in 1979 was about 72×10^9 kg and is increasing. We have used immobilized *L. bulgaricus* for lactic acid fermentation from by-products of the dairy industry. In batch production, lactic acid yield from whey and whey UF-permeate was 70% and from demineralized whey 80% at 90 h (FIG. 7). In continuous column operation with UF-permeate used as substrate and 1.0% (wt/vol) yeast extract as nutrient, the lactic acid yield was about 20% at the very short (3.2-h) residence time. With longer residence times, the lactic acid yield could not be improved. Sterilized UF-permeate with 1.0% yeast extract as nutrient was not a good medium for continuous lactic acid fermentation with our *L. bulgaricus*-strain. If UF-permeate with 1.0% (wt/vol) yeast extract as nutrient was fermented with *L. bulgaricus* without buffer, the lactic acid yield was 1.5–2.0 g l^{-1} at 11-min residence time and about 3.0 g l^{-1} at 50 min residence time. pH of the substrate fell from original 6.1 to 4.5 and 4.0, respectively (FIG. 8). The packed-bead columns were very stable during the time operated. Compere and Griffith[15] could increase the lactic acid content of sour whey from 1.4% to 2.1% with a fixed-film packed tower reactor at 10–20 h residence time. They used a mixed culture containing lactobacilli and lactose-fermenting yeasts derived from commercial kefir culture.

With immobilized lactic acid bacteria, the costs of lactic acid production could perhaps be reduced, if the problems involved in product isolation can be solved.

PREFERMENTATION OF MILK

Lactic acid bacteria are employed in the dairy industry, for example, for production of cheese, butter, and many types of cultured milk products. The rate of

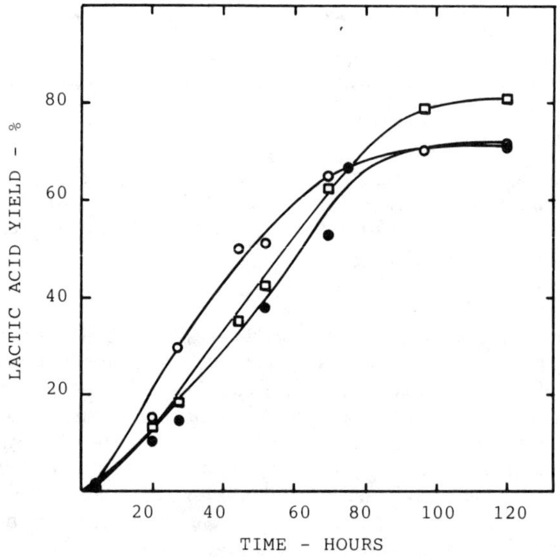

FIGURE 7. Batch production of lactic acid from ● whey, ○ whey UF-permeate, and □ demineralized whey with immobilized *L. bulgaricus* (5.0% wt/vol lactose, 1.0% yeast extract, 30 g l^{-1} $CaCO_3$; 45°C; biocatalyst 10 g/100 ml).

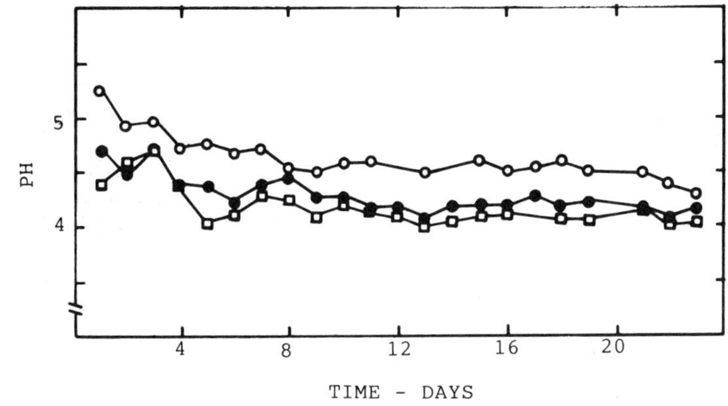

FIGURE 8. Continuous fermentation of whey UF permeate with immobilized *L. bulgaricus* (5.0% wt/vol lactose, 1.0% wt/vol yeast extract; 45°C; pH 6.1). ○ $\tau \sim$ 11 min. ● $\tau \sim$ 28 min, □ $\tau \sim$ 50 min.

lactic acid production and lactic acid concentration has an effect on the structure of the milk product. Continuous fermentations in the dairy industry have many advantages compared with conventional batch processes. They are fast, the quality of the product is unchangeable, and the process control is easy. In continuous processes it is, however, sometimes difficult to avoid contamination. Continuous fermentations have been used in laboratory scale for yogurt and sour curd cheese production,[34–36] and also at least one industrial-scale continuous yogurt process has been reported.[37]

Divies[18] has employed calcium alginate-immobilized *L. bulgaricus* and *S. thermophilus* mixed culture for prefermentation of milk. It was found that the pH of milk decreased to 5.45 at 14 min residence time (based on bead volume) in a mixed vessel with the product overflow. Divies employed polyacrylamide entrapped yogurt bacteria for continuous yogurt production. With 20 min residence time (based on bead volume), the pH of milk was 4.8 and the lactic acid concentration 8.3 g l^{-1}. The reactor was operated continuously 10 days.

We have studied the prefermentation of milk with calcium alginate-immobilized lactic acid bacteria in a packed-bead column reactor. Our purpose was to lower the pH of milk from 6.6–6.8 to 6.2. The columns were used intermittently. They were operated continuously with milk (40–42°C) for about 5 h per day, followed by circulating (21°C) 5% (wt/vol) lactose solution through the columns. In continuous prefermentation of milk with immobilized *L. helveticus,* the maximum lactic acid productivity was observed, when the residence time was 20 min (FIG. 9). By using a residence time of 10 min, the pH of milk decreased 0.4 pH units, and in a residence time of 20 min, it decreased by 0.7 pH units (FIG. 10). If longer residence times were used, the pH of milk fell more than 0.7 pH units and then some casein of milk precipitated. The casein precipitate made the operation of the biocatalyst columns difficult.

Two columns with a 15-min residence time were used for continuous prefermentation of nonfat milk for four weeks. Columns were operated 5 h per day, after which the lactose solution was circulated through the columns. The pH of the lactose solution was adjusted to 3.5 or 4.0. Low pH prevented the growth of contaminating microorganisms. During the weekends, biocatalyst beads were stored at +4°C in lactose (5%, wt/vol) solution at pH 4.0. The stabilization of the preferementation process with immobilized *L. helveticus* took about four days (FIG. 11). After that, the lactic acid productivity varied between 2.8–5.6 mg g^{-1} h^{-1} in both biocatalyst columns. The amount of

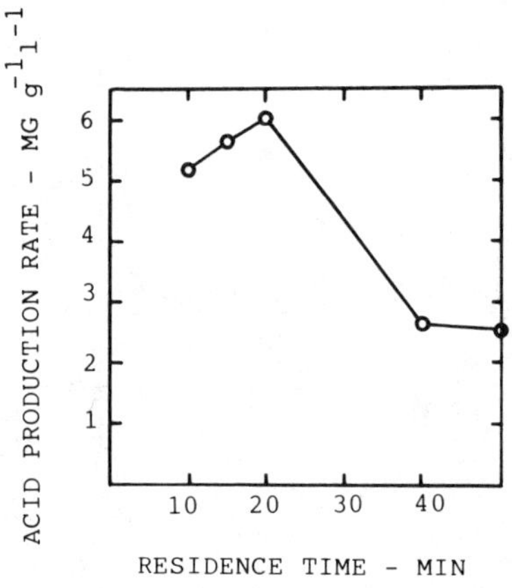

FIGURE 9. The effect of residence time on lactic acid production rate in continuous prefermentation of nonfat milk with immobilized *L. helveticus*.

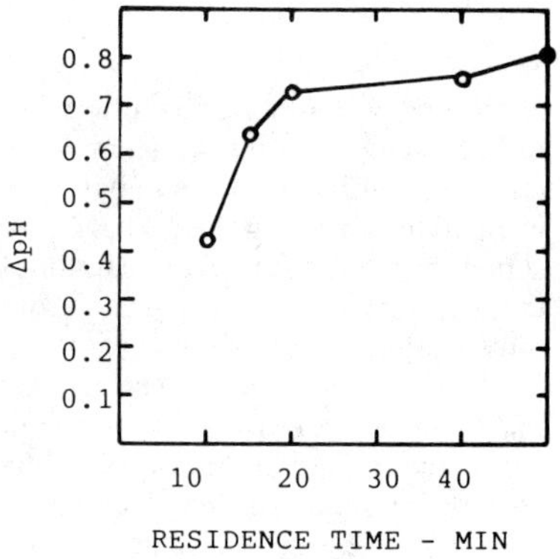

FIGURE 10. The effect of residence time on pH in continuous prefermentation of nonfat milk with immobilized *L. helveticus*.

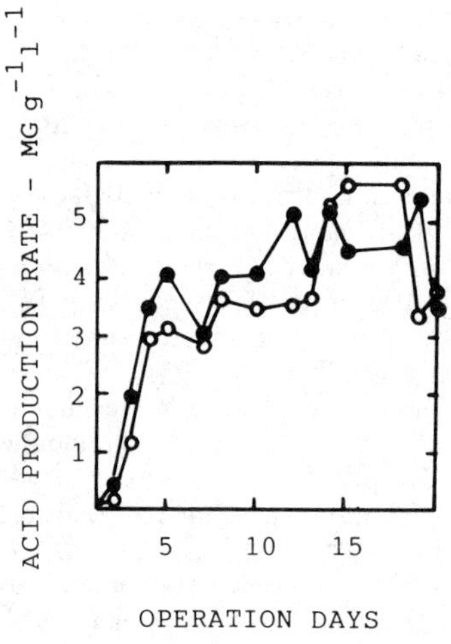

FIGURE 11. Lactic acid production rate in continuous prefermentation on nonfat milk with immobilized *L. helveticus*. At night, lactose solution was circulated through the columns, the pH of which was ○ 3.5 or ● 4.0.

lactobacilli in the beads was about 6.0×10^7 per g gel immediately after immobilization and during the four weeks operation, on the average, was 1.8×10^8 per g gel. In the product, there were 700–25,000 microorganisms per ml solution. In these experiments, no problems with contaminations existed.

We also studied the continuous prefermentation of milk without interruption during three days with calcium alginate-immobilized *L. bulgaricus*. UHT milk was fed through the packed-bead column at 30-min residence time for 20 hours. The pH of milk decreased from 6.6 to 5.9. This pH was too low, because there was casein precipitate in the column. The residence time was therefore shortened to 12 min. After 24 hours stabilization, the pH of milk was again 5.9. The residence time was then shortened to 10, 9, 7, and 6 min, and we found that a 6-min residence time was enough to lower the pH of UHT milk to 6.2 (TABLE 1).

CONCLUSIONS

It has been shown that calcium alginate-immobilized lactic acid bacteria can be applied both for lactic acid production and for applications in dairy processing.

TABLE 1. Prefermentation of Milk with Immobilized *Lactobacillus bulgaricus*[a]

Residence Time (min)	pH
30[b]	5.90
12[c]	5.90
10[d]	6.12
9[d]	6.15
7[d]	6.17
6[d]	6.20

[a]45°C; pH of UHT-milk 6.6).
[b]The time of stabilization: 20 h.
[c]The time of stabilization: 24 h.
[d]The time of stabilization: 2 hr.

Immobilized lactic acid bacteria can be used for lactic acid production from dilute waste waters, for example, by-products of the dairy industry and wood hydrolysate. Furthermore, in dairy processing, the prefermentation of milk can be performed continuously with immobilized lactic acid bacteria in a packed-bead column reactor.

REFERENCES

1. LIPINSKY, E. S. 1981. Chemicals from biomass: Petrochemical substitution options. Science **212:** 1465–1471.
2. WHITTIER, E. O. & L. A. ROGERS. 1931. Continuous fermentation in the production of lactic acid. Ind. Eng. Chem. **23:** 532–534.
3. KELLER, A. K. & P. GERHARDT. 1975. Continuous lactic acid fermentation of whey to produce a ruminant feed supplement high in crude protein. Biotechnol. Bioeng. **17:** 997–1018.
4. COX, G. C. & R. D. MACBEAN. 1977. Lactic acid production by *Lactobacillus bulgaricus* in supplemented whey ultrafiltrate. Aust. J. Dairy Technol. **32**(1): 19–22.

5. MARSHALL, K. R. 1970. Continuous production of lactic acid. *In* Proceedings, 18. International Dairy Congress. Sydney. p. 447.
6. STIEBER, R. W., G. A. COULMAN & P. GERHARDT. 1977. Dialysis continuous process for ammonium-lactate fermentation of whey: Experimental tests. Appl. Environ. Microbiol. **34:** 733–739.
7. STIEBER, R. W. & P. GERHARDT. 1981. Dialysis continuous process for ammonium lactate fermentation: Simulated prefermentor and cell-recycling systems. Biotechnol. Bioeng. **23:** 523–534.
8. MILKO, E. S., O. V. SPERELUP & I. L. RABOTNOWA. 1966. Die Milchsäuregärung von *Lactobacterium delbrückii* bei kontinuierlicher Kultivierung. Z. Allg. Mikrobiol. **6:** 297–301.
9. HANSON, T. P. & G. T. TSAO. 1972. Kinetic studies of the lactic acid fermentation in batch and continuous cultures. Biotechnol. Bioeng. **14:** 233–252.
10. MIALL, L. M. 1978. Organic acids. *In* Economic Microbiology, Vol. 2. Primary Products of Metabolism. A. H. Rose, Ed. Academic Press. London. pp. 47–119.
11. GRIFFITH, W. L. & A. L. COMPERE. 1975. A new method for coating fermentation tower packing so as to facilitate microorganism attachment. Dev. Ind. Microbiol. **17:** 241–246.
12. DIVIES, C. (Institut National de la Recherche Agronomique). 1975. Procédé enzymatique utilisant des microorganismes inclus. Fr. Demande 7 720 503.
13. STENROOS, S.-L., Y.-Y. LINKO & P. LINKO. 1982. Production of L-lactic acid with immobilized *Lactobacillus delbrueckii*. Biotechnol. Lett. **4**(3): 159–163.
14. STENROOS, S.-L., Y.-Y. LINKO, P. LINKO, M. HARJU & M. HEIKONEN. 1982. Lactic acid fermentation with immobilized *Lactobacillus* sp. *In* Enzyme Engineering. I. Chibata, S. Fukui & L. B. Wingard, Jr., Eds. Vol. **6:** 229–301. Plenum Press. New York.
15. COMPERE, A. L. & W. L. GRIFFITH. 1975. Fermentation of waste materials to produce industrial intermediates. Dev. Ind. Microbiol. **17:** 247–252.
16. COMPERE, A. L. & W. L. GRIFFITH. 1980. Industrial chemicals and chemical feedstocks from wood-pulping wastewaters. Tappi **63**(2): 101–104.
17. GRIFFITH, W. L. & A. L. COMPERE. 1977. Continuous lactic acid production using a fixed-film system. Dev. Ind. Microbiol. **18:** 723–726.
18. DIVIES, C. (Institut National de la Recherche Agronomique). 1977. Procédé enzymatique utilisant des microorganismes inclus. Fr. Demande 7 524 509.
19. LINKO, P. 1981. Immobilized live cells. *In* Advances in Biotechnology. M. Moo-Young, C. W. Robinson & C. Wezzina, Eds. Vol. **1:** 711–716. Pergamon Press. New York.
20. LINKO, Y.-Y., L. WECKSTRÖM & P. LINKO. 1980. Sucrose inversion by immobilized *Saccharomyces cerevisiae* yeast. *In* Food Process Engineering. P. Linko & J. Larinkari, Eds. Vol. **2:** 81–91. Applied Science Publishers. London.
21. NILSSON, I., S. OHLSON, L. HÄGGSTRÖM, N. MOLIN & K. MOSBACH. 1980. Denitrification of water using immobilized *Pseudomonas denitrificans* cells. Eur. J. Appl. Microbiol. Biotechnol. **10:** 261–274.
22. GARDE, V. L., B. THOMASSET & J.-N. BARBOTIN. 1981. Electron microscopic evidence of an immobilized living cell system. Enzyme Microb. Technol. **3:** 216–218.
23. CHEETHAM, P. S. J., K. W. BLUNT & C. BUCKE. 1979. Physical studies on cell immobilization using calcium alginate gels. Biotechnol. Bioeng. **21:** 2155–2168.
24. MAN, J. C., M. ROGOSA & M. E. SHARPE. 1960. A medium for the cultivation of lactobacilli. J. Appl. Bact. **23:** 130–135.
25. LOCKWOOD, L. B. 1979. Production of organic acids by fermentation. *In* Microbial Technology, Microbial Processes. H. J. Peppler & D. Perlman, Eds. Vol. **1:** 356–387. Academic Press. New York.
26. CASIDA, L. E. 1964. Industrial Microbiology. pp. 304–314. John Wiley and Sons. New York.
27. LEONARD, R. H., W. H. PETERSON & M. L. JOHNSON. 1948. Lactic acid fermentation of sulfite waste liquor. Ind. Eng. Chem. **40:** 57–67.
28. BURTON, L. V. 1937. Byproducts of milk. Food Ind. **9:** 571–575, 617–618, 634–636.
29. PRESCOTT, S. C. & C. G. DUNN. 1959. Industrial Microbiology. McGraw-Hill. New York. pp. 304–331.

30. STIEBER, R. W. & P. GERHARDT. 1979. Dialysis continuous process for ammonium-lactate fermentation: Improved mathematical model and use of deproteinized whey. Appl. Environ. Microbiol. **37:** 487–495.
31. COTON, S. G. 1980. Subject: Whey technology, the utilization of permeates from the ultrafiltration of whey and skim milk. J. Soc. Dairy Technol. **33**(3): 89–94.
32. COX, G. C. & R. D. MACBEAN. 1977. Lactic acid production by *Lactobacillus bulgaricus* in supplemented whey ultrafiltrate. Aust. J. Dairy Technol. **32**(1): 19–22.
33. KOSIKOWSKI, F. W. 1979. Whey utilization and whey products. J. Dairy Sci. **62:** 1149–1160.
34. LINKLATER, P. M. & C. J. GRIFFIN. 1971. A laboratory study of continuous fermentation of skim milk for the production of sour curd cheese. J. Dairy Res. **38:** 137–144.
35. DRIESSEN, F. M., J. UBBELS & J. STADHOUDERS. 1977. Continuous manufacture of yogurt. Biotechnol. Bioeng. **19:** 821–829, 841–851.
36. MACBEAN, R. D., R. J. HALL & P. M. LINKLATER. 1979. Analysis of pH-stat continuous cultivation and the stability of the mixed fermentation in continuous yogurt production. Biotechnol. Bioeng. **21:** 1517–1541.
37. VAN DER LOO, L. G. W. 1980. Kontinuierliche Zubereitung von Rührjoghurt in industriellem Ausmass. Dtsch. Milchwirtsch. **29:** 1199–1202.

Gel Entrapment of Whole Cells and Enzymes in Crosslinked, Prepolymerized Polyacrylamide Hydrazide

AMIHAY FREEMAN

Center for Biotechnology
The George S. Wise Faculty of Life Sciences
Tel-Aviv University
Tel-Aviv, 69978, Israel

Immobilization of whole cells offers several advantages as a tool in biotechnology: easy separation of biocatalyst from the reaction mixture, reuse of biocatalyst, flexibility in reactor choice, high concentration of biocatalyst per unit volume of reactor, and improved storage, thermal, and operational stability. Throughout the last decade, these advantages led to the applications of immobilized cells as biocatalysts on a production scale[1,2] as well as an analytical tool.[3]

Immobilization of whole cells may be achieved via several approaches—adsorption, crosslinking, chemical binding, and gel entrapment. However, gel entrapment techniques seem to be the most popular approach employed for whole cell immobilization.[4] Gel entrapment is usually based on suspending the cells to be immobilized in a solution containing water-soluble polymers or monomers, followed by the induction of gelation of the whole system via physical or chemical means. These include cooling, pH or salinity changes, initiation of polymerization, or the addition of an appropriate crosslinking agent.[4]

A process designed for the efficient entrapment of whole cells should allow for:

(1) High retention of cell *viability;* the biological activity of the entrapped cells should not be impaired by the immobilization conditions.
(2) The *porosity* of the formed gel should be uniform and controllable. Free exchange of substrates, products, cofactors, gases, and so forth is essential for efficient performance of the immobilized cells.
(3) The gel obtained should retain good mechanical, chemical, and biological *stability*. It should not be easily degraded by enzymes, solvents, pressure changes, or shearing forces.
(4) The gel should be composed of reasonably priced components and the process should be amenable to ready scaling up.

In view of these requirements, a method for whole cell immobilization via controlled chemical crosslinking of prepolymerized polyacrylamide-hydrazide was developed. In the following, the principle of this method is presented and its application for whole cell and enzyme immobilization is described.

THE PRINCIPLE

The method presented was developed on the basis of the following guidelines:

(a) Water-soluble prepolymerized synthetic polymer should be used as the backbone for the formation of the gel matrix. The cells to be immobilized are to be suspended in a solution of this prepolymer;

(b) This prepolymer should contain controlled amounts of functional groups to allow for chemical crosslinking;
(c) The crosslinking reaction, taking place in the presence of the cells, should be carried out under mild physiological conditions.

Of several potential water-soluble synthetic prepolymers, linear water-soluble polyacrylamide was found to be most suitable. This polymer may be readily synthesized, purified and stored for a long time. Moreover, several chemical approaches are available for its derivatization to generate functional groups for crosslinking.

One of the most convenient ways to introduce such groups, without significantly affecting the chemical nature of this polymer, is controlled hydrazinolysis.[5] This reaction may be carried out directly upon completion of the polymerization reaction, thus simplifying the procedure. The degree of hydrazinolysis is easily controlled by the time and temperature of the reaction, and the content of acylhydrazide groups may be readily determined by a simple colorimetric assay. Purification of the polyacrylamide-hydrazide (PAAH) thus obtained is simple, and it may be stored as a dry powder for long periods of time.[6]

For the crosslinking reaction, Schiff base formation between the acylhydrazide groups on the polymeric backbone and dialdehydes was found to be highly advantageous. This reaction takes place readily under mild conditions and over a wide pH range.[4–9] By controlling the amount of dialdehydes added on a stoichiometric ratio to the amount of acylhydrazide groups, excess of this reagent is avoided. Although Schiff base formation is considered to be a reversible reaction, it is considerably stabilized in the case of acylhydrazides combined with aldehydes. In particular, a highly stable conjugated system is formed when glyoxal forms a "bridge" between two polymeric chains by reacting with two acylhydrazides (see FIG. 1).

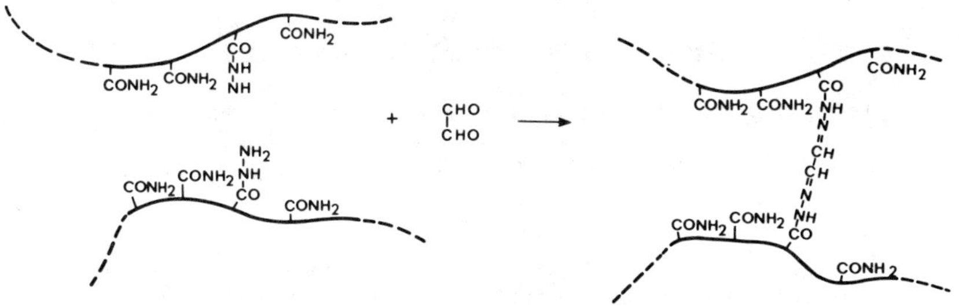

FIGURE 1. Crosslinking of polyacrylamide-hydrazide by glyoxal.

The mild and controlled conditions employed for the reaction leading to gel formation may thus be assumed to allow for high retention of immobilized biological activity. By changing the chain length and chemical nature of the crosslinking dialdehydes and monitoring the concentration of the polymeric backbone, it may be anticipated that gel porosity will be readily affected.

The procedure finally adopted for whole cell immobilization according to this method consists of the following steps:

(1) Polyacrylamide-hydrazide (acylhydrazide content—0.8 meq/g of dry polymer) can be synthesized according to the literature,[6] or purchased from commercial sources. A 3% (wt/vol) solution of this polymer is then prepared in

distilled water. Buffers or salts are added to this solution as solids or concentrated solutions, with minimal dilution effect (not more than 1:10).
(2) The cells to be immobilized, either as cell pellets or highly concentrated suspension, are then added and suspended into an ice-cold solution of the prepolymer.
(3) The crosslinking agent (preferably buffered glyoxal-hydrate trimer solution, 5% wt/vol) is then introduced with strong mixing (employing volume ratio 1:20, glyoxal to polymer solution). Gelation is effected within several seconds.
(4) The gel thus formed is allowed to undergo self-hardening over ice. After 20 minutes, it is cut into cubes of about 0.5 cm to allow free liberation of water from the spontaneously shrinking gel.
(5) Following incubation at 4°C for 1–24 hrs (preferably more than 20 hrs), the gel is fragmented by passing it through a syringe (outlet diameter 2 mm). Following washings with the appropriate buffer, the immobilized cells are ready for use.

The method developed may be applicable to the immobilization of whole cells (microbial, mammalian, and plant cells), organelles and, following an appropriate adaptation, for enzymes too. Experience with the application of this technique for the immobilization of cells and enzymes is summarized below.

IMMOBILIZATION OF MICROBIAL CELLS

Antibiotic Production by Immobilized Streptomyces clavuligerus *Cells*

As a first case study for testing the feasibility of this system, the immobilization of *Streptomyces clavuligerus* cells was attempted. In this study, carried out with Dr. Y. Aharonowitz, the two alternative routes for obtaining whole cells immobilized in crosslinked polyacrylamide were compared: polymerization of water-soluble monomers (acrylamide and *bis*-acrylamide) in the presence of the cells, and the controlled chemical crosslinking by dialdehydes of prepolymerized polyacrylamide-hydrazide. It was clearly shown that the retention of immobilized biocatalytic activity was only 55% for the former, but over 90% for the latter.[6] This finding is in accord with other reports on the toxic effects exhibited by the acrylamide monomer.[4]

During the course of this study, another experimental difficulty was encountered. In order to verify the retention of activity, in terms of *specific* immobilized activity, the exact amount of immobilized cells had to be determined. Since some of the entrapped cells may be released during gel fragmentation and washing, the initial cell input does not represent, as a rule, the amount of cells presented in the final immobilized cell system. Since the gels involved in this study are chemically crosslinked, a procedure had to be worked out for the determination of protein content of cells entrapped in chemically crosslinked synthetic gels. The development of this method (see Freeman *et al.*[7]) enabled us to correlate measured immobilized activity with the amount of entrapped cell mass.

Steroid Reduction by Immobilized Mycobacterium *sp. Cells*

Mycobacterium sp. cells, capable of reducing the Δ^1 double bond of steroids, were immobilized in polyacrylamide-hydrazide crosslinked by glyoxal. It was found that these cells retained 100% of their reduction capacity [measured as the capability to reduce 1,4-androstanediene-3,17-dione (ADD) to 4-androstane-3,17-dione (AD)]. Moreover, essentially the same reaction rate was obtained for the immobilized cells and the corresponding control of freely suspended cells.[8]

Glucose Conversion into Alcohol by Immobilized Saccharomyces cerevisiae

In parallel with the growing interest in alcohol production by fermentation, the immobilization of yeasts for the continuous conversion of carbohydrates into ethanol has been attempted by several groups.[9–12] The glycolytic activity of entrapped yeasts may serve as an interesting case study, where several fundamental aspects of whole cell immobilization may be investigated: retention of the multienzyme catalytic activity involved throughout the immobilization procedure; gel porosity, which enables free exchange of substrates and products (including free liberation of the formed CO_2) and the capability to allow controlled cell growth throughout the continuous process, without disintegration of the gel. Immobilization of *Saccharomyces cerevisiae* (bakers' yeast) in crosslinked polyacrylamide-hydrazide gel was therefore attempted.

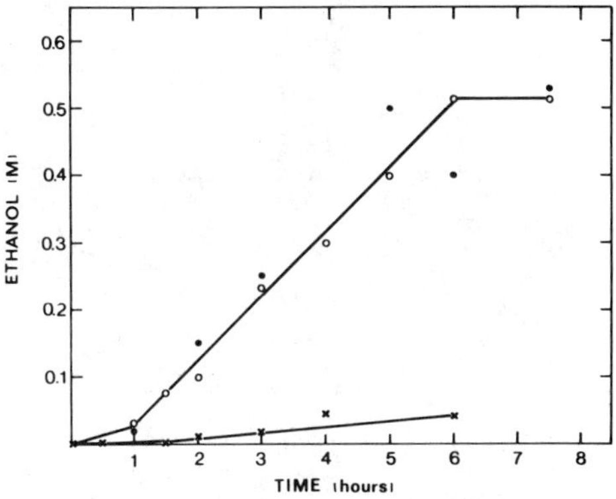

FIGURE 2. Determination of relative specific glycolytic activity of immobilized yeast. Yeasts were immobilized in polyacrylamide-hydrazide crosslinked by glyoxal, and in parallel by copolymerization of acrylamide and bisacrylamide in the presence of the cells. Gel samples, each containing the same amount of entrapped yeasts, were assayed for initial glycolytic activity. An equal amount of freely suspended yeasts served as control. Yeasts entrapped in PAAH-glyoxal gel, ● ; yeasts entrapped in common polyacrylamide gel, x ; control, O. For experimental details, see Pines and Freeman.[13]

Full retention of glycolytic activity was recorded for yeasts entrapped in this gel, while only 15% of the activity survived entrapment in gel made via polymerization of acrylamide monomers in the presence of the cells (see FIG. 2 and Pines and Freeman[13]). This sensitivity of yeasts towards the toxicity of the acrylamide monomer is in accord with the results of a detailed study reported by Siess and Divies.[11] Our results demonstrate again the superiority of employing a prepolymerized polymeric backbone for gel formation over the polymerization of monomers in the presence of the cells.

The conversion of carbohydrates into ethanol takes place at relatively high substrate concentrations. Large amounts of CO_2 are formed as a by-product and the process is accompanied by a continuous growth of cell mass. These characteristics make this process a good test case for the selection of the most appropriate crosslinking dialdehyde, which leads to the formation of a porous and stable gel structure. Our study clearly revealed that glyoxal, the shortest dialdehyde, is the most suitable crosslinking agent for polyacrylamide-hydrazide. Of the three dialdehydes tested—glyoxal, glutardialdehyde, and polyvinylalcohol with aldehyde terminal groups (obtained via periodate oxidation of polyvinylalcohol),[6] diffusion-free kinetics were recorded only for gels made with glyoxal. Moreover, the gel made by polyacrylamide hydrazide crosslinked by the oxidized polyvinyl-alcohol exhibited resistance to the free liberation of CO_2. This gel swelled considerably during the process, in a way resembling bread during baking. No deformation of gel structure, or change in reaction kinetics (as compared to the control of freely suspended cells) were observed for the gel made by crosslinking with glyoxal (see FIG. 2). Moreover, a spontaneous process of shrinking is exhibited only by gels crosslinked by glyoxal. This shrinking, down to about half of the initial gel volume, increases gel strength as well as gel porosity. Scanning electron micrographs, obtained for lyophylized gel particles containing immobilized yeasts, show that a uniform sponge-like porous structure is formed throughout this gel (see FIG. 3). The capability of this immobilized yeast system to operate continuously was demonstrated with 10% and 20% glucose solutions,

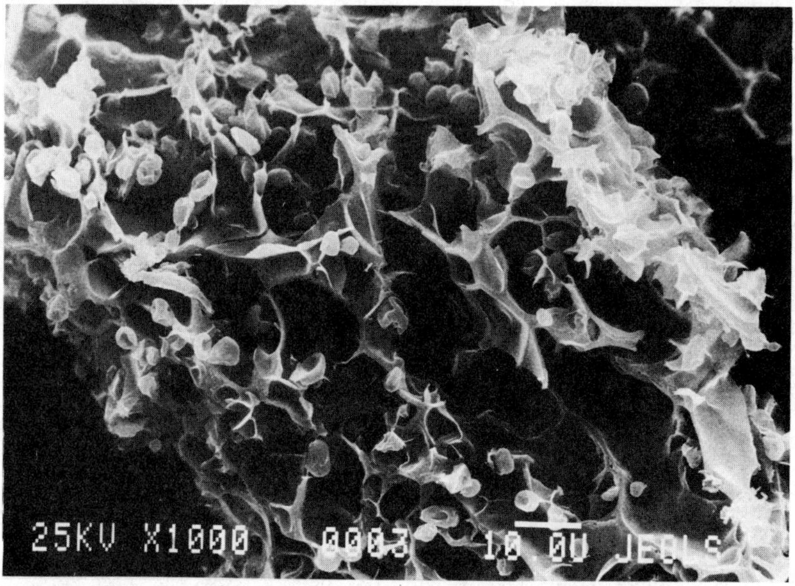

FIGURE 3. Scanning electron micrographs of the inner structure of a gel containing immobilized yeast. The gel, made from polyacrylamide-hydrazide crosslinked by glyoxal, was freeze-dried and mechanically broken. The exposed internal surface was casted with gold and the sample was observed in a scanning electron microscope (Jeol JSM 35C). The bar indicates 10 microns (magnification: × 1000, shown here reduced to 65%). Scanning electron microscopy was carried out by Mrs. E. Freeman of the Israel Institute for Biological Research.

for at least up to the period of the two weeks tested. Moreover, substantial enhancement in survival of cells in the presence of high ethanol concentrations was recorded. The entrapped yeast was able to grow inside this gel matrix throughout the continuous conversion of glucose into ethanol.

MONOTERPENE REDUCTION BY IMMOBILIZED *MENTHA* CELLS

In view of the growing interest in the immobilization of plant cells (derived from cell suspension cultures),[14] the immobilization of *Mentha* cells, as a model system, was attempted. Two cell lines, previously shown to be capable of reducing stereospecifically (−) menthone to (+) neomenthol, or (+) pulegone to (+) isomenthone[15,16] were immobilized in polyacrylamide-hydrazide crosslinked by glyoxal. This study (in collaboration with Prof. E. Galun and Dr. D. Aviv) showed that the reduction capacity was fully retained by these cells following the immobilization, and was essentially equal to that recorded for the freely suspended cell control.[17] Furthermore, the same kinetics of biotransformation was measured for the immobilized cells and the control. In view of several reports in the literature (e.g. Brodelius *et al.*[18]) describing diffusional limitations observed for plant cells entrapped in Ca-alginate gels, this result is encouraging. The potential inherent in the use of this gel system for the design of continuous biotransformation carried out by immobilized plant cells is under investigation.

IMMOBILIZATION OF LIVER MICROSOMES

Though immobilization techniques have been mostly applied for either whole cell or purified enzymes, attempts have been made by several groups to immobilize organelles (e.g. mitochondria, chloroplasts[19]). One of the intrinsic difficulties often encountered with such immobilizations is the capability of these rather unstable systems to survive the immobilization process. In view of the high retention of entrapped activity achieved by the mild, controlled crosslinking of polyacrylamide-hydrazide by glyoxal, the immobilization of liver microsomes was attempted. The monooxygenase activity (*O*-demethylation of *p*-nitroanisole) of the Cyt P-450 system of these microsomes served as an assay for the characterization of the immobilized rat liver microsomes. In this study, carried out in collaboration with Dr. A. Yawetz, Prof. A. S. Perry, and Prof. E. Katzir, high retention of immobilized monooxygenase activity was recorded (88% of the microsomal fraction input).[20] The kinetics of monooxygenase activity of the immobilized microsomes was quite similar to that of the freely suspended control (e.g. same V_{max}; similar K_m values), indicating only minor diffusional limitations for this substrate. The immobilized microsomal monooxygenase system could be operated continuously for several hours at 37°C, provided that adequate amounts of an NADPH-generating system were added periodically. Under similar conditions, a control microsomal suspension lost its enzymic activity within 90 minutes.

GEL ENTRAPMENT OF ENZYMES IN POLYACRYLAMIDE-HYDRAZIDE CROSSLINKED BY GLYOXAL

The capability to entrap whole cells and organelles with high retention of immobilized activity raised the question of whether this mild and controlled gel entrapment may also be applicable for the immobilization of purified enzymes. The first technical question was whether the mean pore size, which allowed for the retention of whole cells, would be efficient for the entrapment of enzyme molecules. A preliminary experiment carried out with acetylcholineesterase (MW ~ 350,000) showed that only 9% of the enzyme input was retained in a gel prepared according to the procedure described above for whole-cell immobilization.[21] Significant reduction in mean pore size was therefore essential. The porosity of gels made from polyacrylamide-hydrazide crosslinked by dialdehydes may be affected by changing the concentration of prepolymer, by the nature of the crosslinking agent, the degree of crosslinking, and chemical modifications of the polymeric backbone. The prepolymer

TABLE 1. Gel Entrapment of Enzymes in Polyacrylamide-hydrazide (MW ~ 100,000) Crosslinked by Glyoxal[a]

Enzyme	MW	Yield of Entrapment[b]
Trypsin	23,800	14%
Alkaline phosphatase	86,000	27%
Glucose oxidase	180,000	51%
Acetylcholine esterase	350,000	64%

[a]Enzymes were dissolved in 15% (wt/vol) PAAH and the solution injected into buffered 1% (wt/vol) ice-cold glyoxal. The gel "noodle" thus obtained was further fragmented into particles of mean diameter 0.25 mm by means of a blade homogenizer. Activity was assayed following filtration and appropriate washings. The activity of these enzymes is not impaired throughout the process, so there is no significant reduction in total input activity.

[b]$\frac{\text{entrapped activity}}{\text{input activity}} \times 100\%$

employed for whole-cell immobilization allowed for the preparation of solutions up to 5% (wt/vol). By reducing the molecular weight of this polymer to about one-half (100,000 versus 180,000), the maximal solubility could be increased to 15% (wt/vol). When this polymer was used at the concentrations of 5, 10, and 15% (wt/vol) and crosslinked by glyoxal, the retention of acetylcholine esterase activity was 15, 40, and 64%, respectively. Thus, the procedure finally adopted for the entrapment of enzymes is based on dissolving the enzyme to be immobilized in 15% (wt/vol) polyacrylamide-hydrazide solution and injecting this viscous solution into an ice-cold, 1% (wt/vol) buffered glyoxal solution. The gel "noodle" thus obtained could be further fragmented into approximately 0.25-mm gel particles by means of a blade homogenizer.

The applicability of the procedure for the gel entrapment of enzymes of various molecular weights could thus be investigated. The data in TABLE 1 show that enzymes

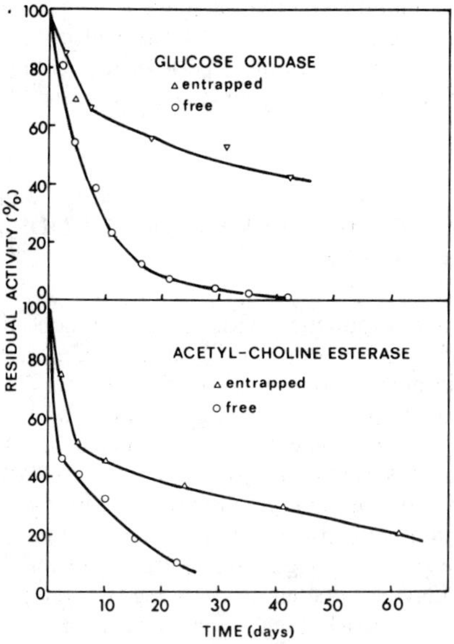

FIGURE 4. Storage stability of glucose oxidase and acetylcholine esterase at 37°C. The enzymes were entrapped in 15% PAAH crosslinked by glyoxal, as described in the legend to TABLE 1.

of molecular weight higher than 150,000 may be successfully retained by this gel. Moreover, these results indicate that the enzymes are physically retained within this gel. Characterization of the two immobilized enzymes of higher molecular weight—glucose oxidase and acetylcholine esterase—showed that only minor perturbations were recorded for the kinetics of these enzymes at the immobilized state (apparent K_m values of 2.8×10^{-4} *M* and 2.2×10^{-2} *M* vs. 2.4×10^{-4} *M* and 2.2×10^{-2} *M* for normal K_m values of acetylcholine esterase and glucose oxidase, respectively). Moreover, a significant enhancement of thermal and storage stability was recorded for the two enzymes following entrapment in this gel system (see FIG. 4).

The ability to introduce chemical modifications onto the prepolymeric polyacrylamide-hydrazide (e.g. via copolymerization) allowed for the development of a new method for the preparation of thin, uniform enzyme membranes. This method was found to be useful, in particular for the easy construction of enzyme electrodes based on pH electrodes.[22]

In conclusion, the results described above show that the method developed fulfill the requirements for an efficient method of gel entrapment, at least in terms of viability and porosity. The potential inherent in controlled chemical modifications of the synthetic prepolymerized backbone is still to be explored. It may be anticipated that controlled changes in the microenvironment of the entrapped biocatalyst may impose significant effects on the biotransformation kinetics, as well as on the stabilization of the immobilized biocatalyst. These aspects are now under investigation.

REFERENCES

1. LINKO, P. & Y. Y. LINKO. 1983. Applications of immobilized microbial cells. *In* Immobilized Microbial Cells. I. Chibata and L. B. Wingard, Eds. Academic Press. New York. pp. 54–154.
2. CHIBATA, I. & T. TOSA. 1980. Immobilized microbial cells and their applications. TIBS **5**(4): 88–90.
3. SUZUKI, S. & I. KARUBE. 1981. Bioelectrochemical sensors based on immobilized enzymes, whole cells and proteins. *In* Applied Biochemistry and Bioengineering. L. B. Wingard, E. Katchalski-Katzir & L. Goldstein, Eds. Vol. **3**: 145–174. Academic Press. New York.
4. KLEIN, J. & F. WAGNER. 1983. Methods for the immobilization of microbial cells. *In* Immobilized Microbial Cells. I. Chibata & L. B. Wingard, Eds. Academic Press, New York. pp. 12–51.
5. INMAN, J. K. & H. M. DINTZIS. 1969. The derivatization of crosslinked polyacrylamide beads. Controlled introduction of functional groups for the preparation of special-purpose, biochemical adsorbents. Biochemistry **8**: 4074–4082.
6. FREEMAN, A. & Y. AHARONOWITZ. 1981. Immobilization of microbial cells in crosslinked, prepolymerized, linear polyacrylamide gels. Antibiotic production by immobilized *Streptomyces clavuligerus* cells. Biotechnol. Bioeng. **23**: 2747–2759.
7. FREEMAN, A., T. BLANK & Y. AHARONOWITZ. 1982. Protein determination of cells immobilized in crosslinked synthetic gels. Eur. J. App. Microbiol. Biotechnol. **14**: 13–15.
8. FREEMAN, A. & Y. AHARONOWITZ. 1980. β-Lactam antibiotic biosynthesis and 17-ketosteroid reduction by bacteria immobilized in crosslinked prepolymerized linear polyacrylamide. Proc. VIth Int. Ferment. Symp. p. 121.
9. WADA, M., J. KATO & I. CHIBATA. 1981. Continuous production of ethanol in high concentration using immobilized growing yeast cells. Eur. J. Appl. Microbiol. Biotechnol. **11**: 67–71.
10. WILLIAMS, D. & D. M. MUNNECKE. 1981. The production of ethanol by immobilized yeast cells. Biotechnol. Bioeng. **23**: 1813–1825.
11. SIESS, M. H. & C. DIVIES. 1981. Behaviour of *Saccharomyces cerevisiae* cells entrapped in

polyacrylamide gel and performing alcoholic fermentation. Eur. J. Appl. Microbiol. Biotechnol. **12:** 10–15.

12. KOLOT, F. B. 1980. New trends in yeast technology—immobilized cells. Process Biochem. **15**(7): 2–8.
13. PINES, G. & A. FREEMAN. 1982. Immobilization and characterization of *Saccharomyces cerevisiae* in crosslinked, prepolymerized polyacrylamide-hydrazide. Eur. J. Appl. Microbiol. Biotechnol. **16:** 75–80.
14. BRODELIUS, P. & K. MOSBACH. 1982. Immobilized plant cells. Adv. Appl. Microbiol. **28:** 1–26.
15. AVIV, D., E. KROCHMAL, A. DANTES & E. GALUN. 1981. Biotransformation of monoterpenes by *Mentha* cell lines: Conversion of menthone to neomenthol. Planta Med. **42:** 236–243.
16. AVIV, D., A. DANTES, E. KROCHMAL & E. GALUN. 1983. Biotransformation of monoterpenes by *Mentha* cell lines: Conversion of pulegone-substituents and related unsaturated α-β ketones. Planta Med. **47:** 7–10.
17. GALUN, E., D. AVIV, A. DANTES & A. FREEMAN. 1983. Biotransformation by plant cells immobilized in crosslinked polyacrylamide-hydrazide. Monoterpene reduction by entrapped *Mentha* cells. Planta Med. **49:** 9–13.
18. BRODELIUS, P., K. NILSSON & K. MOSBACH. 1981. Production of α-keto acids. Part I. Immobilized cells of *Trigonopsis variabilis* containing D-amino acid oxidase. Appl. Biochem. Biotechnol. **6:** 293–308.
19. OCHIAI, H., A. TANAKA & S. FUKUI. 1983. Immobilized organelles. *In* Immobilized Microbial Cells. I. Chibata & L. B. Wingard, Eds. Academic Press. New York. pp. 153–187.
20. YAWETZ, A., A. S. PERRY, A. FREEMAN & E. KATCHALSKI-KATZIR. 1984. Monooxygenase activity of rat liver microsomes immobilized by entrapment in a crosslinked prepolymerized polyacrylamide-hydrazide. Biochim. Biophys. Acta **798:** 204-209.
21. FREEMAN, A. & T. BLANK. 1984. Gel entrapment of enzymes in Cross-linked prepolymerized polyacrylamide-hydrazide. Manuscript in preparation.
22. TOR, R. & A. FREEMAN. 1984. A new method for the preparation of enzyme membranes on solid surfaces: Preparation and characterization of enzyme-pH-electrodes. Manuscript in preparation.

Large-Scale Bacterial Fuel Cell Using Immobilized Photosynthetic Bacteria

ISAO KARUBE, HIDEAKI MATSUOKA, HIDEKI MURATA, KAZUHITO KAJIWARA, AND SHUICHI SUZUKI

Research Laboratory of Resources Utilization
Tokyo Institute of Technology
Nagatsuta-cho, Midori-ku
Yokohama, 227 Japan

MITSUO MAEDA

Central Research Laboratories
Mitsubishi Electric Co.
Tsukaguchihoncho, Amagasaki
Hyogo, 661 Japan

Hydrogen is attracting attention as one of the clean fuel resources. Various bacteria and algae produce hydrogen under anaerobic conditions.[1] However, because the hydrogen evolution system, especially hydrogenase or nitrogenase, in bacteria is unstable, it is difficult to use whole cells for continuous hydrogen production.[2]

Recently, immobilization techniques for enzymes and bacteria have been developed for industrial application of these biocatalysts. Hydrogen-producing bacteria, *Clostridium butyricum*, were immobilized in natural and synthetic polymers and the immobilized whole cells continuously evolved hydrogen from diluted molasses.[3-5] Hydrogen exhibits excellent reactivity at electrodes. Therefore, immobilized hydrogen-producing bacteria (3 kg wet cells) have been applied to a phosphoric acid fuel cell system.[6] Hydrogen produced (400–800 ml · min^{-1}) was supplied to the fuel cell. About 10–12 W and a current of 10–12 A were obtained for 10 hours. However, hydrogen-producing bacteria also produced organic acids such as acetic acid, lactic acid, and propionic acid. The efficiency of the hydrogen production from molasses would be improved if hydrogen were to be produced from organic acids; photosynthetic bacteria produce hydrogen from organic acids.

This paper describes a large-scale bacterial fuel cell using immobilized photosynthetic bacteria. *Rhodospirillum rubrum* was employed for the fuel cell system.

CULTIVATION AND IMMOBILIZATION OF PHOTOSYNTHETIC BACTERIA

Rhodospirillum rubrum IFO 3986 was cultivated in a medium containing 1% yeast extract, 0.05% $MgSO_4$ and 0.1% K_2HPO_4 (pH 7.0). The cultivation was performed for 3–4 days at 37°C under fluorescent light (5000 lux). The bacteria were collected by centrifugation at 6000 × *g* for 15 min. The wet cells were suspended in a Tris-HCl buffer solution (0.05 *M*, pH 7.0) containing 2% sodium alginate. The bacterial suspension was dropped into 1% calcium chloride solution through a small capillary. The small beads (diameter 2–3 mm) formed were stored at 4°C.

ANALYTICAL METHODS FOR ORGANIC ACIDS AND BIOGAS

Organic acids in a reaction mixture were determined by high-pressure liquid chromatography (Shimazu Seisakusho, Model LC-3A; column:SCR-101H). Gas components were analyzed by gas chromatography (Shimazu Seisakusho, Model GC-3BT; column:molecular sieve 5A, 60–80 mesh).

HYDROGEN PRODUCTION FROM ORGANIC ACIDS BY IMMOBILIZED *R. RUBRUM*

R. rubrum is a purple nonsulfur photosynthetic bacterium. The photosynthetic bacteria contain a single photosystem and do not evolve O_2. Reductant for the fixation of CO_2 is obtained from organic and inorganic sources rather than from water.

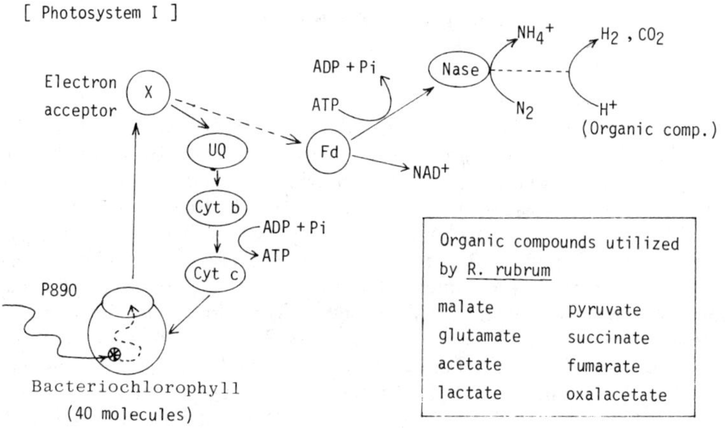

FIGURE 1. Postulated scheme of H_2 evolution by photosynthetic bacteria.

Photosynthetic cyclic electron transport is coupled to photophosphorylation. The purple photosynthetic bacteria, *Rhodospirillaceae*, have been extensively studied with regard to H_2 metabolism.[7–9] The purple nonsulfur *Rhodospirillaceae* utilize H_2, H_2S and organic substrates. The ability of photosynthetic bacteria to produce large quantities of H_2 and CO_2 during photosynthetic growth on organic compounds has been known for several decades and the presence of hydrogenase has been demonstrated. However, H_2 production occurs only in the absence of NH_4^+ and is inhibited by N_2. H_2 evolution does not occur in the dark and is inhibited by uncouplers of phosphorylation. It has become apparent that H_2 production is indeed catalyzed by nitrogenase.[10–12] Nitrogenase catalyses N_2 reduction and ATP-dependent H_2 evolution. The hydrogen-producing system in *R. rubrum* is schematically shown in FIGURE 1. *R. rubrum* utilizes malate, pyruvate, glutamate, succinate, acetate, fumarate, lactate, and oxalate.[13] Continuous hydrogen production was demonstrated in a bioreactor using photosynthetic bacteria from waste waters containing organic acids, though the productivity and stability were low.[14,15]

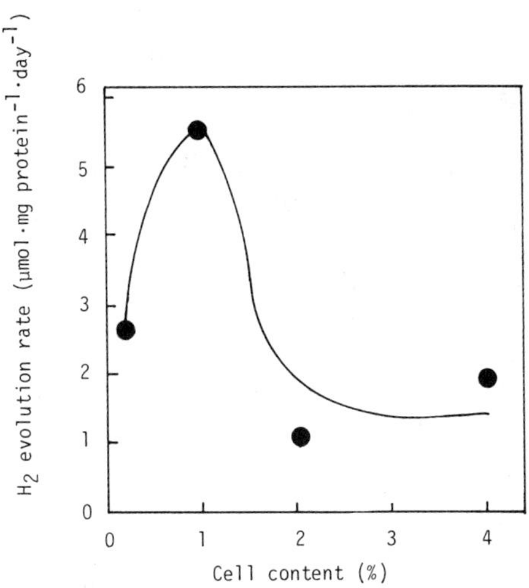

FIGURE 2. Effect of cell content on the H_2 evolution by immobilized *R. rubrum*. Gel: 5 g of 2% calcium alginate containing 50 mg cells; substrate: 10 ml of 15 m*M* acetate; temperature: 30°C, pH: 7.0; light intensity: 5000 lux.

OPTIMUM CONDITIONS FOR HYDROGEN PRODUCTION

The effect of cell content in calcium alginate beads on the H_2 evolution rate is shown in FIGURE 2. A 15-m*M* acetate solution was used as a substrate for hydrogen evolution. The highest hydrogen evolution rate was observed at a cell content of 1%. FIGURE 3 shows the effect of the alginate concentration on the H_2 evolution rate. The optimum concentration of alginate was approximately 1%. However, this gel preparation was too soft to use. Thus, 2% gel and 1% cell content were employed in subsequent experiments.

The effect of initial pH on hydrogen evolution was examined. The optimum pH of

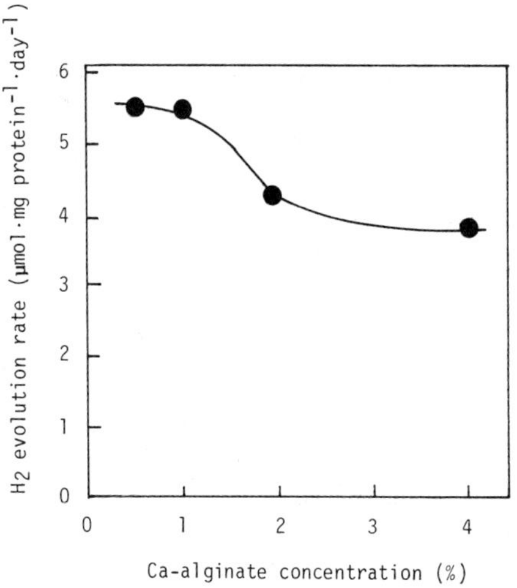

FIGURE 3. Effect of alginate concentration on H_2 evolution by immobilized *R. rubrum*. Immobilized cell content: 1%. Other conditions were the same as for FIGURE 1.

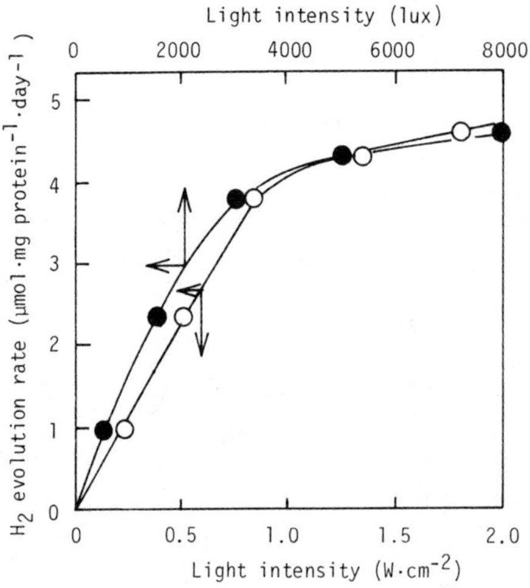

FIGURE 4. Influence of light intensity on H_2 evolution by immobilized *R. rubrum.* Immobilization: 1% calcium alginate gel. Other conditions were the same as for FIGURE 3 except for light intensity employed.

the immobilized photosynthetic bacteria was 7.0 and further increase or decrease of pH decreased the H_2 evolution rate.

FIGURE 4 shows the effect of light intensity on the H_2 evolution rate. The H_2 production rate increased with increasing light intensity up to 5000 lux. Thus, initial pH of 7.0, 30°C and light intensity of 5000 lux were employed.

HYDROGEN EVOLUTION FROM ORGANIC ACIDS

Hydrogen evolution from organic acids was performed in a batch system using immobilized photosynthetic bacteria. Fifteen m*M* of acetate, proprionate and butyrate were employed, respectively. The results are shown in FIGURE 5. Hydrogen was evolved

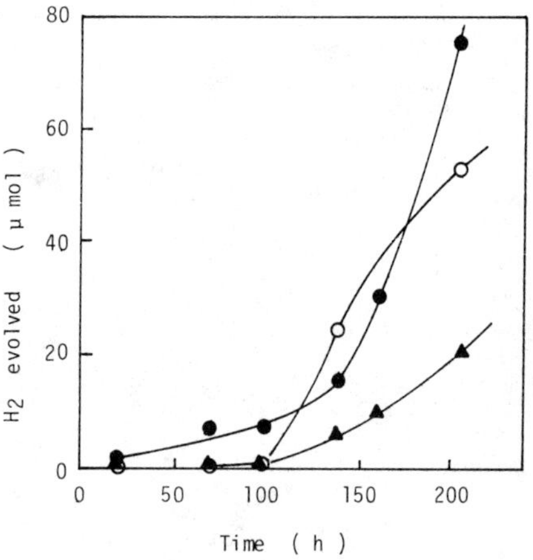

FIGURE 5. H_2 evolution from organic acids by immobilized *R. rubrum.* Immobilized cells: 5 g of 2% calcium alginate gel containing 50 mg wet cells; temperature: 30°C; pH: 7.0; light intensity: 5000 lux; substrate: 10 ml of organic acid; (○) acetate 15 m*M*; (●) propionate 15 m*M*; and (▲) butyrate 15 m*M*.

from these organic acids by immobilized *R. rubrum*. The H_2 evolution rate from butyrate was the lowest among these three organic acids. Waste waters were obtained from a bioreactor using immobilized hydrogen-producing bacteria (*Clostridium butyricum*) and diluted molasses (sugar content, 0.87%). The waste water containing 13 m*M* acetate, 6 m*M* propionate, and 30 m*M* butyrate was used as a substrate for hydrogen evolution. FIGURE 6 shows the time course of hydrogen evolution from the waste water. Hydrogen was continuously evolved for 320 h from the waste water by immobilized photosynthetic bacteria. Therefore, a bioreactor was constructed for continuous hydrogen production.

CONTINUOUS HYDROGEN PRODUCTION BY IMMOBILIZED PHOTOSYNTHETIC BACTERIA

Hydrogen production by immobilized whole cells was performed in a flow system using a 140-ml bioreactor. About 82 g of immobilized photosynthetic bacteria were

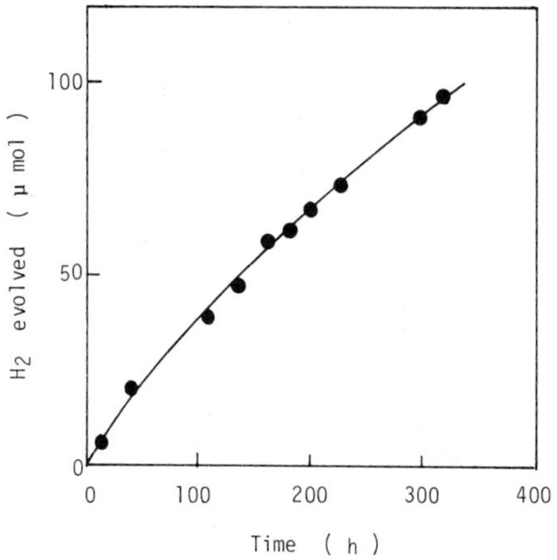

FIGURE 6. Continuous H_2 evolution from waste water containing organic acids with a batch system. Organic acids content: acetate 13 m*M*, propionate 6 m*M*, butyrate 30 m*M*. Other conditions were the same as for FIGURE 5.

employed. The waste water was supplied to a bioreactor at a flow rate of 30 ml · h^{-1}. The reaction was performed at pH 7.0, 30°C under a light intensity of 5000 lux. The hydrogen produced was determined by gas chromatography. FIGURE 7 shows the time course of hydrogen production in the bioreactor containing immobilized photosynthetic bacteria.

The rate of hydrogen production was gradually increased and reached a steady state (26 ml · day^{-1} after 3 days). Immobilized photosynthetic bacteria continuously evolved hydrogen over a 6-day period. As previously reported, the growth of bacteria in calcium alginate gel matrices during the reaction provides an obvious explanation for the increase in the activity of hydrogen production with continuous utilization of the immobilized bacteria. A steady-state hydrogen production rate was obtained after 3 days, which means that the interstitial space was filled with active bacteria at that time.

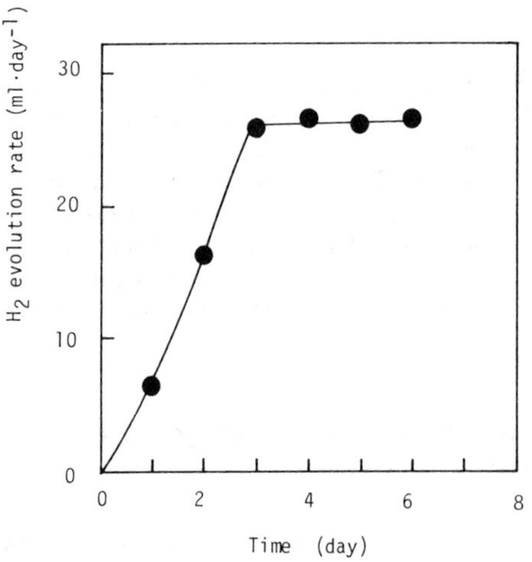

FIGURE 7. Continuous H_2 evolution from waste water with a small-scale, flow-type bioreactor. Reactor volume: 140 ml; immobilized cells: 82-g gel containing 820 mg wet cells; feed rate: 30 ml · h^{-1}, substrate: waste water containing acetate 14 m*M*, lactate 16.2 m*M*, propionate 11 m*M*. Other conditions were the same as for FIGURE 6.

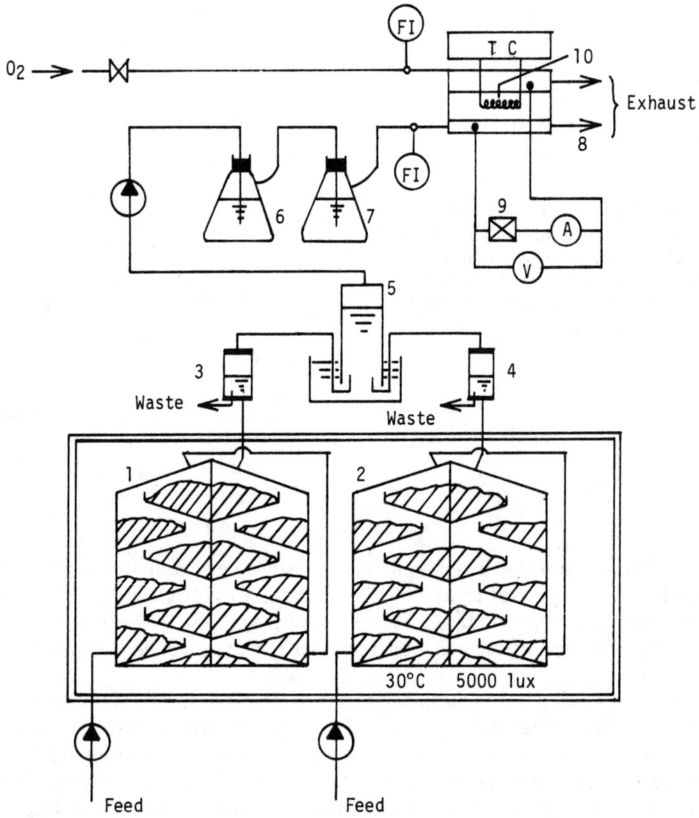

FIGURE 8. Schematic diagram of photosynthetic bacterial fuel cell system. (1, 2) Bioreactors for immobilized *R. rubrum;* (3–5) H_2 reservoirs; (6) KOH solution; (7) water; (8) phosphoric acid fuel cell; (9) variable electronic load; (10) heater; (FI) flow indicator; and (TC) temperature control.

STORAGE STABILITY OF IMMOBILIZED BACTERIA

Immobilized whole cells were stored in Tris-HCl buffer solution (0.1 *M*, pH 7.0) at 4°C under air. The hydrogen productivity of the immobilized whole cells was examined by gas chromatography. The hydrogen productivity was maintained for more than 14 days at 4°C.

PHOTOSYNTHETIC BACTERIAL FUEL CELL SYSTEM

A photosynthetic bacterial fuel cell system is shown in FIGURE 8. Two large, transparent acrylate bioreactors (20 l) were constructed for immobilized photosyn-

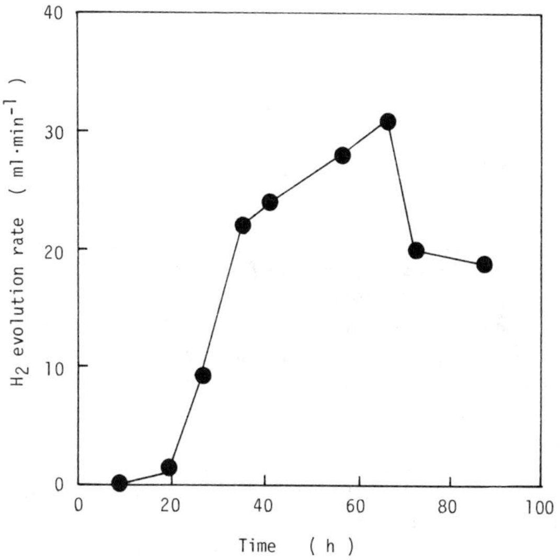

FIGURE 9. Continuous production of hydrogen in an immobilized *R. rubrum* reactor. Reactor volume: 20 liters × 2; immobilized cells: 18 = kg gel containing 180 g wet cells; feed rate: 2 l · h^{-1}. Other conditions were the same as for FIGURE 7.

thetic bacteria. About 18 kg of immobilized bacteria were placed in two reactors. The reactors were operated at pH 6.8–7.0, 30°C, and light intensity of 5000 lux. The waste water was continuously transferred to the reactors at a flow rate of 2.0 l · h^{-1}. Immobilized bacteria continuously evolved hydrogen over 90 hours, as shown in FIGURE 9. Initially, the amount of hydrogen produced increased with increasing reaction time. Hydrogen produced by the reactors was transferred to an anode of a phosphoric acid fuel cell system at 19–31 ml · min^{-1}. Oxygen was also transferred to a cathode at 0.1 l · min^{-1}.

A phosphoric acid fuel cell system consisted of two carbon plate separators, a cathode, and an anode. The carbon plate separators have gas passages for hydrogen and oxygen. The anode and cathode consisted of hydrophobic carbon fibers, platinum black catalysts and carbon pellets. The matrix containing silicon carbide and

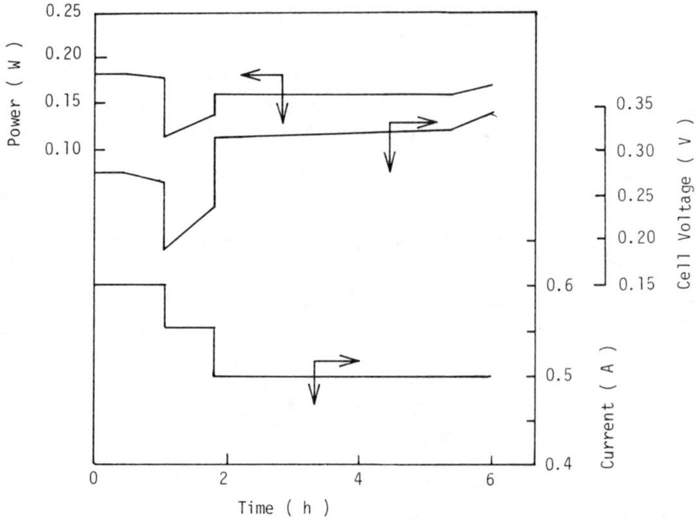

FIGURE 10. Continuous output of electric power.

phosphoric acid (100%) separated the anode and the cathode. The fuel cell system was equipped with an electronic control system for a constant current. By adjusting an electronic load manually, a steady-state current could be maintained during the operation. The current and cell voltage were measured by an ammeter and an electrometer and displayed on a recorder. The current density and the cell voltage increased as the temperature of the fuel cell was raised. In the following experiments, the fuel cell was operated at 220°C.

FIGURE 10 shows the time course of the electric current, cell voltage, and electric power. In order to remove toxic components, biogas was passed through 2 traps containing concentrated KOH solution (30%) and tap water, respectively. A stable current from 0.5 to 0.6 A was obtained for 6 h. Power of 0.16–0.18 W was observed during operation. The maximum power output of 0.4 W was obtained for one hour. During operation, the hydrogen productivity decreased gradually. This might be due to denaturation of the nitrogenase system.

TABLE 1 shows organic acids in the waste water of the inlet and the outlet of the reactor. As shown in TABLE 1, organic acids such as acetic, lactic, and propionic acids were assimilated by the immobilized photosynthetic bacteria. About 86 to 54% of the

TABLE 1. Changes in Main Organic Compounds in Waste Water during the Operation of Photosynthetic Bacterial Fuel Cell System

Compounds	Inlet ($mmol \cdot l^{-1}$)	Outlet ($mmol \cdot l^{-1}$)	Difference ($mmol \cdot l^{-1}$)
Acetate	14.0	6.4 (46%)	7.6 (54%)
Lactate	16.2	4.8 (30%)	11.4 (70%)
Propionate	11.0	1.5 (14%)	9.5 (86%)
Butyrate	0	0	0
COD	19000 ppm	13000 ppm (68%)	6000 ppm (32%)

organic acids were converted to hydrogen and carbon dioxide in the bioreactor. The chemical oxygen demand (COD) of the waste water at the inlet of the reactors was 19,000 ppm and that at the outlet was 13,000 ppm. Only 32% of the COD was removed in the bioreactors. Therefore, the waste water still contained unassimilable organic compounds.

SUMMARY

Hydrogen-producing photosynthetic bacteria, *Rhodospirillum rubrum,* were immobilized in a calcium alginate (2%) gel. The immobilized photosynthetic bacteria (180 g wet cells) were employed for continous production of hydrogen from the waste water containing organic acids. The immobilized bacteria continuously produced hydrogen (19–31 ml · min^{-1}) for more than 90 hours. Hydrogen produced was supplied to a phosphoric acid fuel cell. About 0.16–0.18 W and a current of 0.5 to 0.6 A were obtained for 6 hours.

REFERENCES

1. THAUER, R. K., K. JUNGERMANN & K. DEKKER. 1977. Energy conservation in chemotropic anaerobic bacteria. Bacteriol. Rev. **41:** 100–180.
2. GEST, H., M. D. KAMEN & H. M. BREGOFF. 1950. Studies on the metabolism of photosynthetic bacteria. V. Photoproduction of hydrogen and nitrogen fixaton by *Rhodospirillum rubrum.* J. Biol. Chem. **182:** 153–170.
3. KARUBE, I., T. MATSUNAGA, S. TSURU & S. SUZUKI. 1976. Continuous hydrogen production by immobilized whole cells of *Clostridium butyricum.* Biochim. Biophys. Acta **444:** 338–343.
4. SUZUKI, S., I. KARUBE, T. MATSUNAGA, S. KURIYAMA, N. SUZUKI, T. SHIROGAMI & T. TAKAMURA. 1980. Biochemical energy conversion using immobilized whole cells of *Clostridium butyricum.* Biochimie **62:** 353–358.
5. KARUBE, I., S. SUZUKI, T. MATSUNAGA & S. KURIYAMA. 1981. Biochemical energy conversion by immobilized whole cells. Ann. N.Y. Acad. Sci. **369:** 91–98.
6. SUZUKI, S., I. KARUBE, H. MATSUOKA, S. UEYAMA, H. KAWAKUBO, S. ISODA & T. MURAHASHI. 1983. Biochemical energy conversion by immobilized whole cells. Ann. N.Y. Acad. Sci. **413:** 133–143.
7. ADAMS, M. W. W., L. E. MORTENSON & J.-S. CHEN. 1981. Hydrogenase. Biochim. Biophys. Acta **594:** 105–176.
8. GEST, H. 1963. Metabolic aspects of bacterial photosynthesis. *In* Bacterial Photosynthesis. H. Gest, A. San Pietro & L. P. Vernon, Eds. Antioch Press. Yellow Springs, Ohio. pp. 129–150.
9. ADAMS, M. W. W. & D. O. HALL. 1979. Properties of the solubilized membrane-bound hydrogenase from the photosynthetic bacterium *Rhodospirillum rubrum.* Arch. Biochem. Biophys. **195:** 288–299.
10. SHICK, H. J. 1971. Interrelationship of nitrogen fixation, hydrogen evolution and photoreduction in *Rhodospirillum rubrum.* Arch. Microbiol. **75:** 102–109.
11. SWEET, W. J. & R. H. BURRIS. 1981. Inhibition of nitrogenase activity by NH_4^+ in *Rhodospirillum rubrum.* J. Bacteriol. **145:** 824–831.
12. YOCH, D. C. & J. W. GOTTO. 1982. Effect of light intensity and inhibitors of assimilation on NH_4^+ inhibition of nitrogenase activity in *Rhodospirillum rubrum* and *Anabaena* sp. J. Bacteriol. **151:** 800–806.
13. MITSUI, A., Y. OHTA, J. FRANK, S. KUMAZAWA, C. HILL, D. ROSNER, S. BARCIELLA, J. GREENBAUM, L. HAYNES, L. OLIVA, P. DALTON, J. RADWAY & P. GRIFFARD. 1980. Photosynthetic bacteria as alternative energy sources. Overview on hydrogen production

research. *In* Alternative Energy Sources II. Vol. 8. Hydrogen Energy. T. N. Veziroglu, Ed. Hemisphere Publishing Co. Washington. pp. 3483–3510.

14. ZÜRRER, H. & R. BACHOFEN. 1979. Hydrogen production by the photosynthetic bacterium *Rhodospirillum rubrum*. Appl. Environ. Microbiol. **37:** 789–793.
15. KIM, J. S., K. ITO & H. TAKAHASHI. 1982. Production of molecular hydrogen in outdoor batch cultures of *Rhodopseudomonas spaeroides*. Agric. Biol. Chem. **46:** 937–941.

The Implication of Reaction Kinetics and Mass Transfer on the Design of Biocatalytic Processes with Immobilized Cells[a]

J. KLEIN,[b] K.-D. VORLOP,[b] AND F. WAGNER[c]

[b]*Institute of Technical Chemistry*
[c]*Institute of Biochemistry and Biotechnology*
Technical University
3300 Braunschweig, Federal Republic of Germany

Many results have been published that are related to the use of immobilized whole-cell biocatalysts in general, but in most cases these data are based on short-time batch experiments of a semiquantitative or qualitative nature. Research directed towards large-scale application of immobilized cell systems must include three steps in the laboratory scale:

(1) Selection and development of strains with high specific activity, including the use of genetic engineering techniques.[1,2]
(2) Development of immobilization procedures that give particles having good mechanical properties and that provide a high level of catalyst efficiency—to make the best use of the high specific activity—and that should be simple and easy to scale up, and last but not least, that are inexpensive.[3,4]
(3) Detailed studies on the short-range and long-range kinetic behavior of the free and immobilized cells to develop a process model used in the selection of the most appropriate reactor configuration and process scheme, including the identification of the critical control variables.

The first two aspects have been covered in earlier contributions;[1–4] it is the intention of this paper to present data on the third aspect for three typical immobilized cell processes as a basis for scale-up calculation and possibly pilot-scale experiments.

PRODUCTION OF 6-AMINOPENCILLANIC ACID (6-APA)

6-APA can be produced by a specific enzymatic cleavage reaction from penicillin G or penicillin V, where *E. coli* cells are used in the former case[1] and *Pleurotus ostreatus* cells in the latter.[5,6] These cells have been developed to high activity levels and immobilization in a chitosan matrix has been found to be the most effective

[a]Support of this work by the Federal Ministry of Science and Technology (B.M.F.T.) under grant No. PTB 8150 is gratefully acknowledged.

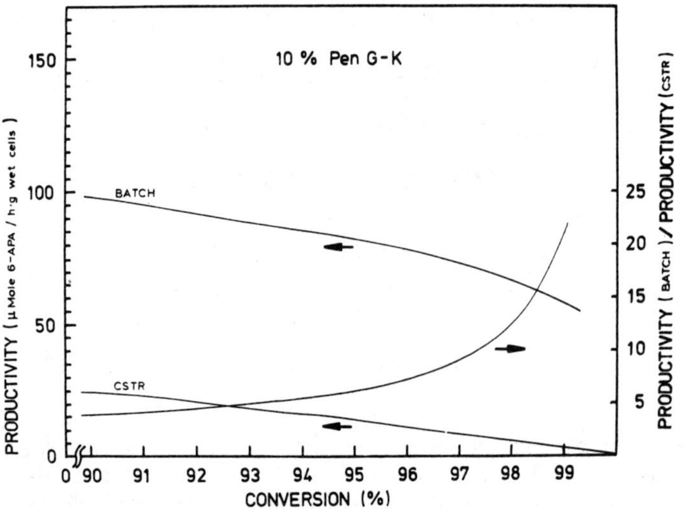

FIGURE 1. Comparison of the productivity of two reactor configurations (batch and CSTR) for the cleavage of penicillin G at high degrees of conversion.

method.[7] In accordance with a study on immobilized enzymes,[8] the following kinetic equation has been applied to describe the kinetics of the immobilized cells as well:

$$v = \frac{v_{max}}{1 + \frac{K_m^S}{c_S}\left[1 + \frac{c_A}{K_i^A}\right] + \frac{c_P}{K_i^P}\left[1 + \frac{K_m^S}{c_S}\right] + \frac{c_A \cdot c_P}{c_S} \cdot \frac{K_m^S}{K_i^A \cdot K_i^P} + \frac{c_S}{K_i^S}} \quad (1)$$

where

K_m and K_i are the Michaelis and the inhibition constants and where the indices refer to S = penicillin (substrate), A = 6-APA and P = phenylacetic acid.

Strict control of pH is of crucial importance in this process, such that a packed-bed catalyst column reactor is excluded. A general comparison of two well-mixed reactors, a batch and a CSTR system, is given in FIGURE 1, demonstrating the much higher productivity of the batch scheme with increasing conversion. Since conversion has to be close to 100% (>98%) and the residence time has to be uniform and short (less than 2 hours), the batch scheme is clearly optimal.

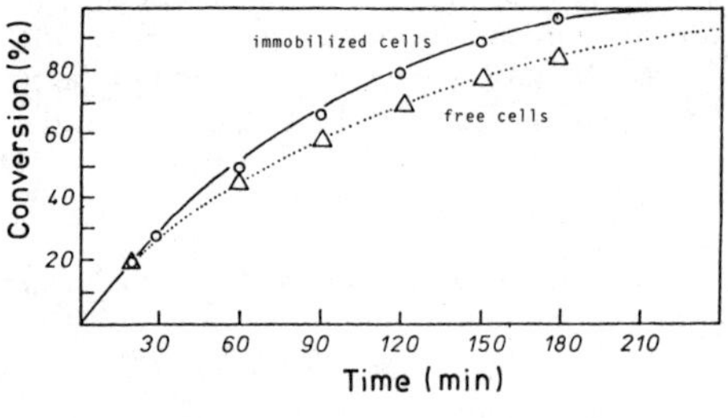

FIGURE 2. Cleavage of penicillin G with *E. coli* pHM12 cells: comparison of the conversion curves for free and immobilized cells (10% Pen G, 37°C, pH = 7.8).

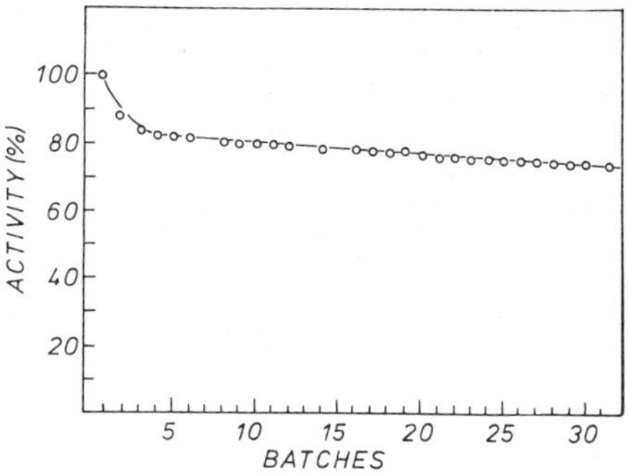

FIGURE 3. Cleavage of penicillin G with immobilized *E. coli* cells in a chitosan carrier: operational stability in repeated batch runs (10% Pen G, 37°C, pH = 7.8).

Based on the kinetic equation (Eq. 1), the conversion-time curve can be calculated, as given in FIGURE 2 for immobilized and free pHM12-cells. Comparing the free and immobilized cells under the boundary condition of identical initial activity, the total conversion time for immobilized cells is shorter. This can be understood considering the fact that the immobilized cell reaction undergoes a transition from a diffusion-controlled to a reaction-controlled regime, and therefore to adjust both systems to identical v_0 values, a somewhat higher amount of active biomass has to be fed to the reactor in the immobilized form.

Using a cell system with very high specific activity, on the order of 250 IU/g wet cells, diffusion control with the immobilized cell systems has to be expected. The quantitative evaluation of this phenomenon[1] reveals the fact that reduction of the particle size to the level of <1 mm is of greater importance compared to variations in cell loading. In this way, at a cell loading of 30% and a particle size of approximately 1 mm diameter, residual activities on the order of 40%, and thus absolute activities on the order of 30 IU/g wet catalyst can be obtained (300 IU/g dry catalyst). These preparations also show very good operational stability (see FIG. 3).

The kinetic model (Eq. 1) has also been used to develop a computer program for the calculation of the batch reaction for certain variables. One example is the dependence of the cleavage activity (units per reactor volume) as a function of substrate concentration (% penicillin G) required for 98% conversion to be reached within 2 hours. Such figures are given in TABLE 1.

As mentioned before, penicillin V is an alternative as a substrate for 6-APA production. Usually a different cell system is used[5] since the specific activity of conventional *E. coli* cells is too low, but using genetic engineering techniques and improved fermentation control, the activity levels of *E. coli* and *Pleurotus* cells for

TABLE 1. Cleavage of Penicillin G with Immobilized *E. coli* (pHM12) Cells[a]

Catalyst Activity (IU/liter)	Penicillin G Concentration (g/liter)
13,100	120
10,000	100
7,200	80
3,870	50
2,040	30

[a]Catalyst activity (IU/l) versus penicillin G concentration required for 95% conversion within 90 min.

penicillin V cleavage become comparable. Therefore the selection of the catalyst type might be reconsidered. Such a comparison is shown in FIGURES 4a and b, where the time-conversion functions—under the condition of identical initial activity v_0—are compared. In a low-conversion experiment, very comparable amounts of catalyst are required, but looking upon the practically important complete conversion case, the much shorter reaction time for the *P. ostreatus* cell catalyst becomes obvious. This is related to the much more favorable inhibition constants for the *P. ostreatus* cells compared to the *E. coli* cells, (a factor of 10 for K_i^S and a factor of 50 for K_i^A) at comparable values of v_{max} (70 versus 105) and K_m^S (7.5 versus 7.0).

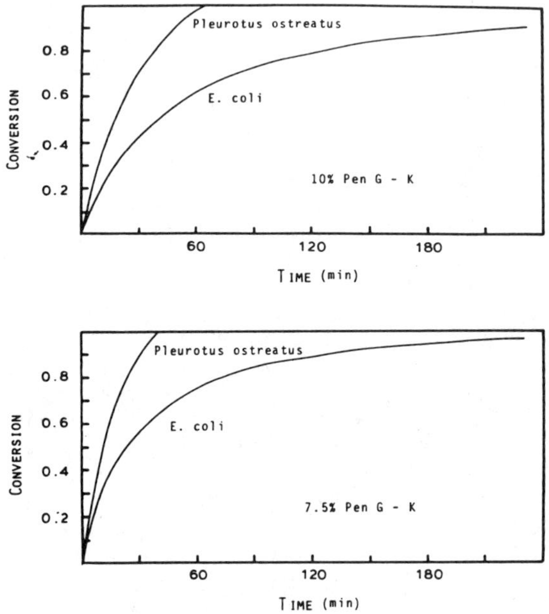

FIGURE 4. Cleavage of penicillin V with immobilized cells (*Pleurotus ostreatus* and *E. coli* pHM12 in chitosan). Comparison of conversion curves for identical v_0 values (*P. ostreatus:* v_{max} = 70 IU/g dry cells, $K_m^S = 7.5$, $K_i^S = 1.600$, $K_i^P = 240$ m*M*/1. *E coli:* v_{max} = 105 IU/dry cells, K_m^S = 7.0, $K_i^S = 175$, $K_i^P = 48$, $K_i^A = 4.5$ m*M*/1).

PRODUCTION OF L-TRYPTOPHAN

The synthesis of L-tryptophan can be achieved using indole and L-serine as precursers and the tryptophan synthetase activity of *E. coli* B10 cells. The reaction behavior of free cells and cells immobilized in a chitosan-tripolyphosphate carrier has been described.[2] While the formation of L-tryptophan is of first order with respect to L-serine, the cosubstrate, indole, has a very inhibitory effect, such that a low indole concentration in the reaction mixture is of primary importance. Continuous production using a CSTR systems could therefore be used.

An attractive alternative, however, is given by a fed-batch system, where L-serine is converted in a batch mode, while a continuous feed of indole—coupled to the indole

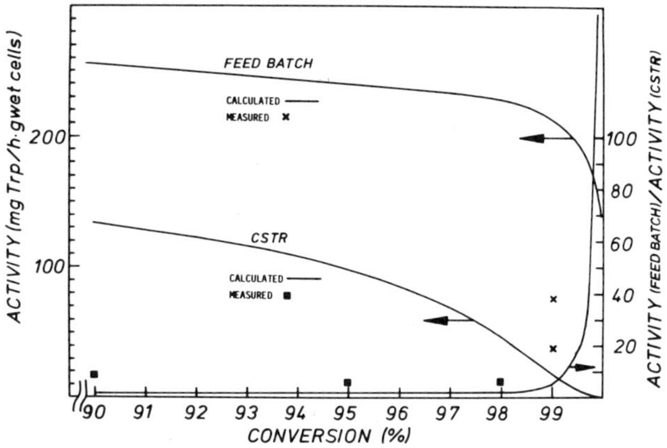

FIGURE 5. Comparison of the productivity of two reactor configurations (CSTR and fed-batch) for the production of L-tryptophan from indole and L-serine at high degrees of conversion.

consumption by a newly designed indole gas phase sensor—provides an overall optimum of the L-tryptophan productivity. The theoretical control scheme, as shown in FIGURE 5, can indeed be realized practically.

A comparison of the productivities of the CSTR and the fed-batch system is given in FIGURES 6 and 7. The ratio of reactor productivities increases drastically with increasing conversion, where again in any practical production, a complete consumption of both substrates has to be realized. Primarily due to effects of diffusional limitations, the experimental values do not yet reach the theoretical values (calculated under the assumption of negligible diffusion resistance). In continuous production runs, an average value of 15 mg L-tryptophan/g wet cells × h was obtained; the fed-batch value, obtained in a 2-l laboratory reactor, of 80 mg L-tryptophan/g wet cells × h, however, demonstrates experimentally the superiority of the latter system. Working with a catalyst volume of 300 ml/l reactor and a cell loading of 30 g wet cells/100 g wet catalyst, this corresponds to reactor volume-related productivities of 15 g/l · h and 96 g/l · h, respectively.

Experiments with particles of variable diameter clearly show the relevance of intraparticle diffusion for the performance of this heterogenous catalyst (TABLE 2). As shown before,[2] growth of the immobilized *E. coli* cells within the carrier matrix could

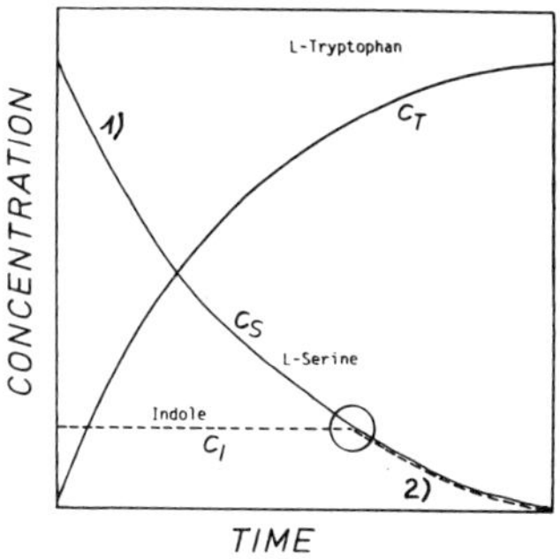

FIGURE 6. Control scheme for the production of L-tryptophan in a fed-batch system with immobilized cells: Schematic time profiles of the concentrations of L-serine (c_S), indole (c_S), and L-tryptophan (c_T).

be induced by nutrient addition. In such a way (owing to transport-limited growth conditions), cell multiplication predominantly occurs closer to the surface of the catalyst particle, leading to a nonuniform intraparticle cell density profile. A comparison of the activity of such a catalyst preparation with preparations having an initially immobilized high and uniform cell distribution demonstrates the much better performance of the former, due to the much higher catalyst efficiency values[9] (TABLE 3).

Not only the reaction rate, but also the operational stability of the cells is negatively influenced by higher indole concentrations; therefore, the fed-batch reaction scheme is also advantageous for the half-life of the immobilized cell catalyst. Furthermore, a shift of the kinetic constants as a function of time is observed such that the optimum indole concentration shifts to lower values with time. A three-dimensional representation of this effect is given in FIGURE 8 for immobilized cells, and comparable results have been found for free cells.

To understand this behavior, the fact has to be mentioned that the optimum indole

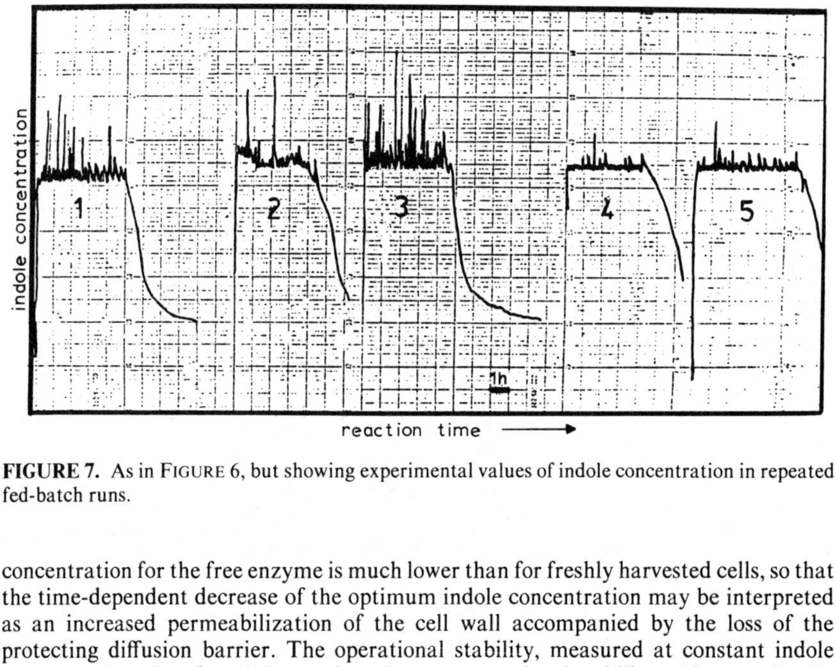

FIGURE 7. As in FIGURE 6, but showing experimental values of indole concentration in repeated fed-batch runs.

concentration for the free enzyme is much lower than for freshly harvested cells, so that the time-dependent decrease of the optimum indole concentration may be interpreted as an increased permeabilization of the cell wall accompanied by the loss of the protecting diffusion barrier. The operational stability, measured at constant indole concentration, therefore, is lower than the true operational stability with an optimally adjusted indole time profile (see FIG. 9).

PRODUCTION OF ETHANOL

Continuous ethanol production with immobilized living and multiplying cells is now a well-established example, where the cell fixation allows the uncoupling of product formation from cell growth kinetics. Using different methods of immobilization of yeast cells, high operational stabilities could be demonstrated at a semicommer-

TABLE 2. Production of L-Tryptophan from Indole and L-Serine with *E. coli* B10 Cells, Immobilized in Carrageenan[a]

Type	Cell Concentration V_B (g wet cells/ml wet cat)	Polymer Concentration V_P (g/ml wet cat)	$D_{eff} \cdot 10^4$ (cm^2/min)	Particle Diameter (mm)	Indole/ L-Serine (mg/100 ml)	Activity (mg Trp/h · g wet cells) Measured (%)	Calculated (%)
Free cells	—	—	—	—	150	205 (100)	205 (100)
Carrageenan biocatalysts	0.1	0.02	5.16	~2	150	170 (83)	197 (96)
Carrageenan biocatalysts	0.1	0.02	5.16	~2	50	113 (55)	144 (70)
Carrageenan biocatalysts	0.1	0.02	5.16	~4	150	122 (60)	168 (82)
Carrageenan biocatalysts	0.1	0.02	5.16	~4	50	68 (33)	98 (48)

[a]Catalyst efficiency as a function of particle size. $[D_{eff} = D_0 \cdot e^{-a \cdot (V_P + b \cdot V_P)}]$

TABLE 3. Production of L-Tryptophan from Indole and L-Serine with *E. coli* B10 Cells Immobilized in Chitosan[a]

	Particle Diameter (mm)	Loading (g wet cells/g wet cat) · 100	Activity (mg Try/h · g wet cat)	Theoretical (%)
Incubation	2.5	~20	~35	80–100
Immobilization	2.2	~40–42	~28	~30

[a]Catalyst activity and efficiency as a function of cell distribution in the catalyst bead.

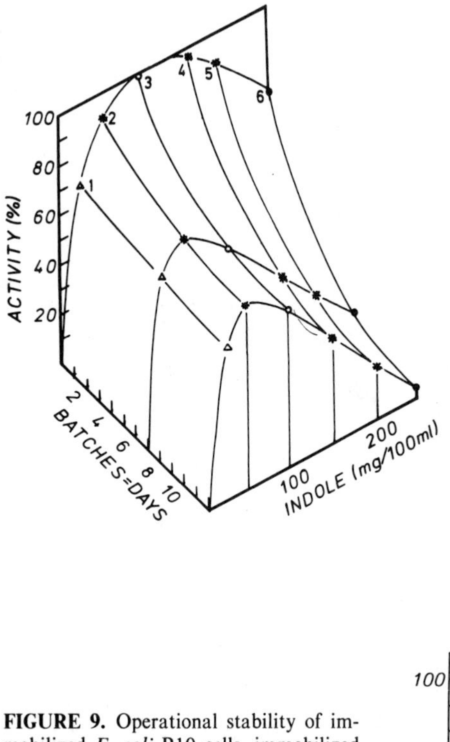

FIGURE 8. Productions of L-tryptophan from indole and L-serine with *E. coli* B10 cells in chitosan. Activity as a function of indole concentration and time of operation.

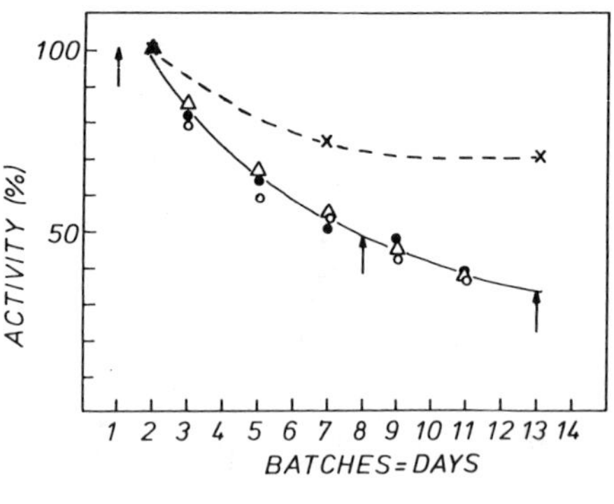

FIGURE 9. Operational stability of immobilized *E. coli* B10 cells, immobilized in chitosan, in the production of L-tryptophan from indole and L-serine.

Solid line: Activity decay at constant indole concentration.

Dotted line: Activity decay at variable indole concentrations, according to the maxima in FIGURE 7.

TABLE 4. Ethanol Production with Immobilized Yeast Cells. Comparison of Reactor Productivities, Based on Total Reactor Volume (V_T)

Reactor	Glucose Concentration (g/l)	Conversion (%)	Productivity (g EtOH/l · h)	Reference
Packed bed, horizontal	196	95	27	Shiotani & Yamane[14]
Packed bed, vertical	127	63	54	Williams & Munnecke[15]
Packed bed, vertical	120	83	62	Daugulis *et al.*[16]
Packed bed, vertical	150	98	25	Tanaka *et al.*[11]
Packed plates	170	95	11	Azuma *et al.*[10]
Packed bed, horizontal	150	98	25.8	Klein & Kressdorf[12]
Packed bed, vertical	150	100	24.6	Klein & Kressdorf[12]
Fluidized bed, 3 stages	150	98	27.2	Klein & Kressdorf[12]

TABLE 5. Ethanol Production with Immobilized *Zymomonas mobilis* Cells: Comparison of Reactor Productivities, Based on Total Reactor Volume (V_T)

Reactor	Glucose Concentration (g/l)	Conversion (%)	Productivity (g EtOH/l · h)	Reference
Packed bed, horizontal	100	97	29.1	Grote *et al.*[17]
Packed bed, vertical	100	83	31.7	Arcuri[18]
Packed bed, vertical	100	82	40	Linko & Linko[19]
Packed bed, vertical	105	64	52.8	Bland *et al.*[20]
Packed bed, vertical	100	97	32.4	Krug & Daugulis[21]
Fluidized bed, 3 stages	150	98	56.5	Klein & Kressdorf[13]

cial production scale, with productivities of 11 g/l · h[10] and 25 g/l · h,[11] respectively (based on total reactor volume).

For immobilized yeast cells (*Saccharomyces cerevisiae*), but even more so for *Zymomonas mobilis* cells with their considerably higher specific activities, the reactor configuration should meet the following requirements:

(1) High degree of local mixing to provide optimum degassing conditions for CO_2 liberation and separation without channeling and an accompanying liquid phase/solid phase contact area.
(2) Residence time distribution close to plug flow conditions to minimize the reactor volume under the conditions of very high conversion ($\mu > 98\%$).

A three-stage, fluidized-bed reactor, where mixing is provided by the continuous feed effluent stream and by CO_2 formation, was shown to be superior to a packed-bed reactor in a vertical or horizontal arrangement[12] (TABLE 4).

For *Zymomonas* cells, immobilized in small particle size Ca-alginate gels, the three-stage fluidized-bed reactor also showed the highest productivities in comparison to the available literature values[13] (TABLE 5). Compared to the specific activity of the free cells of 2.1 g EtOH/g wet cells × hour, the overall productivity of 0.130 g EtOH/g wet catalyst × hour leaves considerable room for improvement.

As shown in TABLE 6, the specific activity of the immobilized cells—in a well-stirred reactor at a very high ratio of liquid volume to catalyst volume—was determined to be 0.282 g EtOH/(g wet catalyst × hour). At a level of cell loading of 20%, this would give an estimated specific activity of 1.5 g EtOH/(g wet cells × hour) or, in other words, an internal effectiveness factor due to pore diffusion of about $\mu =$ 0.70. The difference to the lower value of 0.130 g EtOH/(g wet catalyst × hour) must therefore be attributed primarily to external mass transfer problems.

Model experiments of ethanol formation with immobilized *Zymomonas* cells under well-controlled conditions of external mixing revealed a pronounced dependence of the

TABLE 6. Ethanol Production with Immobilized *Zymomonas mobilis* Cells: Comparision of Activity of Free and Immobilized Cells

System	Activity
Free cells	2.1 g EtOH/h · g wet cells
Immobilized cells (activity test/batch)	0.282 g EtOH/h · g wet catalyst
Immobilized cells (three-stage reactor system)	0.13 g EtOH/h · g wet catalyst

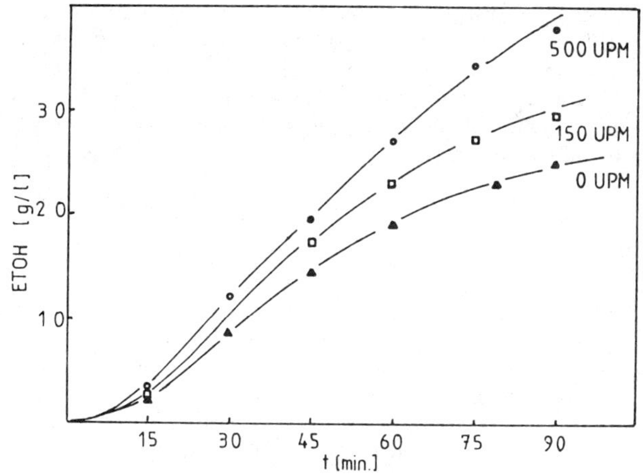

FIGURE 10. Ethanol production with immobilized *Zymomonas mobilis* cells in Ca-alginate: Dependence of reaction rate on stirrer speed in batch reaction runs.

conversion rate on the degree of mixing, expressed by increasing stirrer speed. These results are shown in FIGURE 10, which can be converted to a presentation of EtOH concentration (at 90 min) as a function of the Sherwood number, which again is proportional to the ¾ power of the stirrer speed (FIG. 11).

It is therefore evident that an increase of the mixing intensity in the staged, fluidized reactor will considerably improve the productivity of ethanol formation with immobilized cells. Such experiments are under way in our laboratory.

CONCLUSIONS

Immobilization of cells is in many cases intuitively identified with a continuously operated, packed-bed column reactor. While such processes are indeed in operation in industry,[22] various other reaction schemes will find application as well. Such different schemes have to be chosen if the conditions of high substrate conversion at controlled residence time have to be combined with the necessity for intensive mixing (pH control,

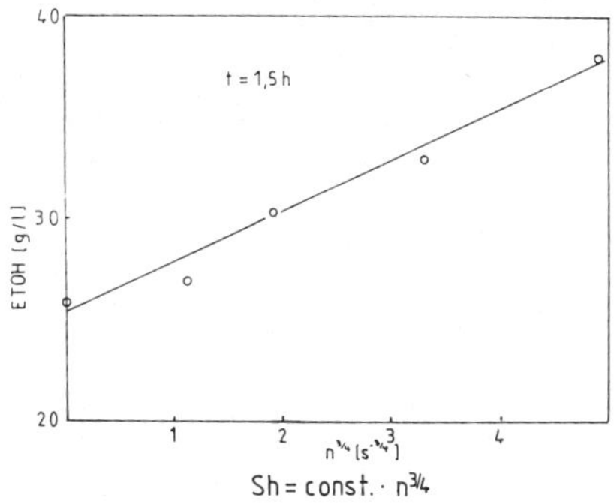

FIGURE 11. Ethanol production with immobilized *Zymomonas mobilis* cells in Ca-alginate. Dependence of ethanol formation (g/l · 90 min) on Sheerwood number.

controlled addition of cosubstrate with poor solubility, gas uptake, or gas formation). Usually with cell strains of high activity, some loss of activity due to transport limitation is encountered. But based on the high cell concentration, productivities on the order of 10–100 g/l · h are reached, which is comparable to petrochemical reactions. If good operational stabilities can be demonstrated, such immobilized-cell catalysts should find increasing interest in the production of chemicals.

SUMMARY

Based on an experimentally developed kinetic model and on information on all relevant types of mass transfer, different reactor models can be established and used in process optimization calculations. Three examples—production of 6-APA, production of L-tryptophan, and production of ethanol—are chosen to illustrate why very different process schemes (batch reactor, fed-batch reactor, and continuous, staged fluidized-bed reactor) have to be used to optimize the productivity.

REFERENCES

1. Klein, J. & F. Wagner. 1982. Immobilization of whole microbial cells for the production of 6-aminopenicillanic acid. *In* Enzyme Engineering V. H. H. Weetall & G. P. Royer, Eds. Plenum Publishing Corp. New York. pp. 335–345.
2. Wagner, F., S. Lang, W.-G. Bang, K.-D. Vorlop & J. Klein. 1982. Production of L-tryptophan with immobilized cells. *In* Enzyme Engineering VI. I. Chibata, S. Fukui & L. B. Wingard, Jr., Eds. Plenum Publishing Corp. New York. pp. 251–257.
3. Klein, J., P. Washausen, M. Kluge & H. Eng. 1980. Physical characterization of biocatalyst particles obtained from polymer entrapment of whole cells. Enzyme Engineering V. H. H. Weetall & G. P. Royer, Eds. Plenum Publishing Corp. New York. pp. 359–362.
4. Klein, J. & G. Manecke. 1982. New Developments in the preparation and characterization of polymer-bound biocatalysts. Enzyme Engineering VI. I. Chibata, S. Fukui & L. B. Wingard, Jr., Eds. Plenum Publishing Corp. New York.
5. Stoppok, E., F. Wagner & F. Zadrazil. 1981. Identification of a penicillin V acylase-processing fungus. Eur. J. Appl. Microbiol. Biotechnol. **13:** 60–61.
6. Kluge, M., J. Klein & F. Wagner. 1982. Production of 6-aminopenicillanic acid by immobilized *Pleurotus ostreatus*. Biotechnol. Lett. **4:** 293–296.
7. Vorlop, K.-D. & J. Klein. 1981. Formation of spherical chitosan biocatalysts by ionotropic gelation. Biotechnol. Lett. **3:** 9–14.
8. Warburton, D., P. Dunnill & M. D. Lilly. 1973. Conversion of benzylpenicillin to 6-aminopenicillanic acid in a batch reactor and continuous feed, stirred-tank reactor using immobilized penicillin amidase Biotechnol. Bioeng. **15:** 13–25.
9. Borchert, A. & K. Buchholz. 1979. Inhomogeneous distribution of fixed enzymes inside carriers. Biotechnol. Lett. **1:** 15–20.
10. Samejima, H., M. Nagashima, M. Azuma, S. Noguchi & K. Junzuka. 1984. Semicommercial production of ethanol using immobilized microbial cells. Ann. N.Y. Acad. Sci. **434:.** This volume.
11. Tanaka, A., K. Sonomoto & S. Fukui. 1984. Various applications of living cells immobilized by prepolymer methods. Ann. N.Y. Acad. Sci. **434:.** This volume.
12. Klein, J., B. Kressdorf, K. D. Vorlop & J. W. Becke. 1983. Production of ethanol with immobilized cells. Proc. 29th JUPAC Congr. June '83. Cologne, F.R.G.
13. Klein, J. & B. Kressdorf. 1983. Improvement of productivity and efficiency in ethanol production with Ca-alginate-immobilized *Zymomonas* cells. Biotechnol. Lett. **5:** 497–502.
14. Shiotani, T. & T. Yamane. 1981. A horizontal packed-bed bioreactor to reduce CO_2 gas

holdup in the continuous production of ethanol by immobilized yeast cells. Eur. J. Appl. Microbiol. Biotechnol. **13:** 96–101.
15. WILLIAMS, D. & D. M. MUNNECKE. 1981. The production of ethanol by immobilized yeast cells. Biotechnol. Bioeng. **23:** 1813–1825.
16. DAUGULIS, A. J., N. M. BROWN, W. R. CLUET & D. B. DUNLOP. 1981. Production of ethanol by adsorbed yeast cells. Biotechnol. Lett. **3:** 651–656.
17. GROTE, W., K. J. LEE & P. L. ROGERS. 1980. Continuous ethanol production by immobilized cells of *Zymomonas mobilis*. Biotechnol. Lett. **2:** 481–486.
18. ARCURI, J. R. 1982. Continuous ethanol production and cell growth in an immobilized-cell bioreactor employing *Zymomonas mobilis*. Biotechnol. Bioeng. **24:** 595–604.
19. LINKO, P. & Y. Y. LINKO. 1982. Continuous ethanol fermentation by immobilized biocatalysts. Enzyme Engineering VI, I. Chibata, S. Fukui & L. B. Wingard, Jr., Eds. Plenum Publishing Corp. New York. pp. 335–342.
20. BLAND, R., H. CHEN, W. JEWELL, W. BELLAMY & R. ZALL. 1982. Continuous high-rate production of ethanol by *Zymomonas mobilis* in an attached-film, expanded-bed reactor Biotechnol. Lett. **4:** 323–328.
21. KRUG, T. A. & A. J. DAUGULIS. 1983. Ethanol production using *Zymomonas mobilis* immobilized on an ion-exchange resin. Biotechnol. Lett. **5:** 159–164.
22. CHIBATA, I. & T. TOSA. 1980. Immobilized microbial cells and their applications. Trends Biochem. Sci. **5:** 88–90.
23. CHIBATA, I. 1979. Immobilized microbial cells with polyacrylamide gels and carrageenan and their industrial applications. *In* Immobilized Microbial Cells. K. Venkatsubramanian, Ed. ACS Symp. Ser. **106:** 187–202.

Continuous Production of L-Alanine: Successive Enzyme Reactions with Two Immobilized Cells

TETSUYA TOSA, SATORU TAKAMATSU, MASAKATSU FURUI, AND ICHIRO CHIBATA

Tanabe Seiyaku Co. Ltd.
Osaka, Japan

The application of immobilized microbial cells for transformation of organic compounds is a subject of increased interest. We are producing L-aspartic acid and L-malic acid industrially by using immobilized *Escherichia coli* and immobilized *Brevibacterium flavum,* respectively.

Since 1965, L-alanine has been produced industrially from L-aspartic acid by a batchwise enzymatic method using L-aspartate β-decarboxylase activity of *Pseudomonas dacunhae* cells. To develop a more efficient process for the production of L-alanine, we had studied continuous production of L-alanine from L-aspartic acid using L-aspartate β-decarboxylase activity of *P. dacunhae* cells immobilized with κ-carrageenan.[1] However, when a conventional column reactor was employed for this enzyme reaction, L-alanine was not produced efficiently owing to back-mixing of the substrate solution and deviation of pH from the optimum range due to liberation of CO_2 gas during the enzyme reaction. These problems were solved by using a closed column reactor (FIG. 1) in which the enzyme reaction was performed at high pressure. The efficiency of the closed column reactor was 50% higher than that of the conventional one.[2]

On the other hand, in 1973 we succeeded in the industrial production of L-aspartic acid from ammonium fumarate using the aspartase activity of immobilized *E. coli* cells. Thus, in order to improve the productivity of L-alanine, we investigated its production from ammonium fumarate by the two enzyme reactions according to the following equation:

$$\text{Fumaric acid} + NH_3 \underset{\text{aspartase}}{\longleftrightarrow} \text{L-Aspartic acid} \xrightarrow[\beta\text{-decarboxylase}]{\text{L-aspartate}} \text{L-Alanine} + CO_2$$

L-Alanine was found to be produced efficiently with removal of fumarase and alanine racemase activities of these two microorganisms.[3]

Besides enzyme activities of immobilized microorganisms, operational stability is also an important problem for the industrial production of L-alanine. As the operational stability of L-aspartate β-decarboxylase activity of immobilized *P. dacunhae* cells was lower than that of aspartase activity of immobilized *E. coli* cells, various methods were tested to obtain immobilized *P. dacunhae* cells having higher activity and stability. As shown in TABLE 1, pH and glutaraldehyde treatment of cells before immobilization was found to be superior to other treatments.[4]

We further investigated the L-alanine productivity of coimmobilized *E. coli-P. dacunhae* cells in order to minimize diffusion resistance of L-aspartic acid within the gel. As shown in TABLE 2, however, the productivity of the coimmobilized preparations

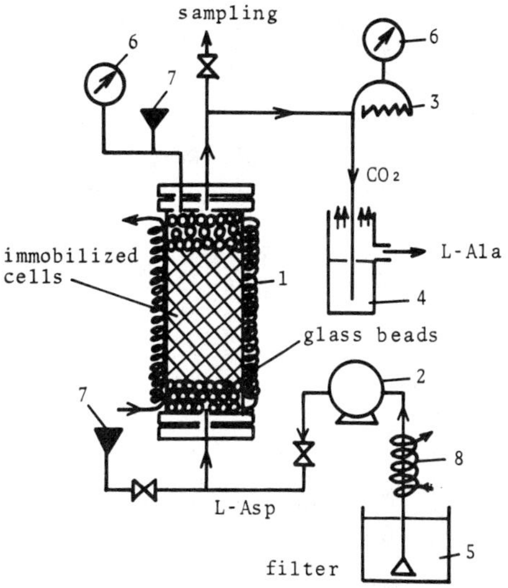

FIGURE 1. Closed column reactor for continuous production of L-alanine using immobilized microorganisms.

(1) Reactor, (2) plunger pump, (3) pressure control valve, (4) receiver, (5) substrate solution tank, (6) pressure gauge, (7) safety valve, and (8) heat exchanger.

was found to be inferior to that of the mixture of two immobilized cells.[5] One explanation for this result may be the higher substrate inhibition by L-aspartic acid of L-aspartate β-decarboxylase of the coimmobilized preparation than that of the mixture of the two types of immobilized cells.

We next investigated the most efficient conditions for these two enzyme reactions by immobilized *E. coli* cells and immobilized *P. dacunhae* cells on the basis of rate equations. The sequential reaction using the column reactor containing immobilized *E. coli* cells and the closed column reactor containing immobilized *P. dacunhae* cells was

TABLE 1. Comparison of L-Alanine Productivity of Immobilized *P. dacunhae* Cells[a]

Cells Used for Immobilization	Relative Productivity[b]
Untreated	100
Glutaraldehyde-treated[c]	222
pH-treated[d]	122
pH- and glutaraldehyde-treated[e]	722

[a]Immobilization was carried out using κ-carrageenan.[1]

[b]The productivity of immobilized cells for continuous production of L-alanine was calculated from the following equation:

$$\text{Productivity} = \int_0^{t_{1/2}} E_0 \exp(-K_d \cdot t)dt$$

where E_0 is the initial enzyme activity, K_d is the decay constant, t is the operational period and $t_{1/2}$ is the half-life of the enzyme activity. The productivity of immobilized untreated cells is taken as 100.

[c]The cells were treated with 5 m*M* glutaraldehyde at pH 6 and 10°C for 10 min before immobilization.

[d]The cells were treated at pH 4.75 and 30°C for 1 h before immobilization.

[e]The pH-treated cells were further treated with 5 m*M* glutaraldehyde at pH 6 and 10°C for 10 min before immobilization.

TABLE 2. Comparison of L-Alanine Production by Batchwise Reaction Using Mixture of Immobilized *E. coli* Cells, Immobilized *P. dacunhae* Cells, and Coimmobilized *E. coli-P. dacunhae* Cells

Immobilized Cells	Efficiency μmol/h · g of gel (at 99% conversion)
Mixture of immobilized *E. coli* and immobilized *P. dacunhae*	417
Coimmobilized *E. coli-P. dacunhae*	294

found to be most efficient for L-alanine production from ammonium fumarate. In this sequential reactor system, the pH of the effluent from the immobilized *E. coli* cell column was adjusted to pH 6.0 from pH 8.5 and charged into the immobilized *P. dacunhae* cell column. This was done because the optimum pHs of aspartase of immobilized *E. coli* cells and L-aspartate β-decarboxylase of immobilized *P. dacunhae* cells are 8.5 and 6.0, respectively.

In 1982, Tanabe Seiyaku Co. Ltd. successfully industrialized this continuous production system of L-alanine from ammonium fumarate with two immobilized

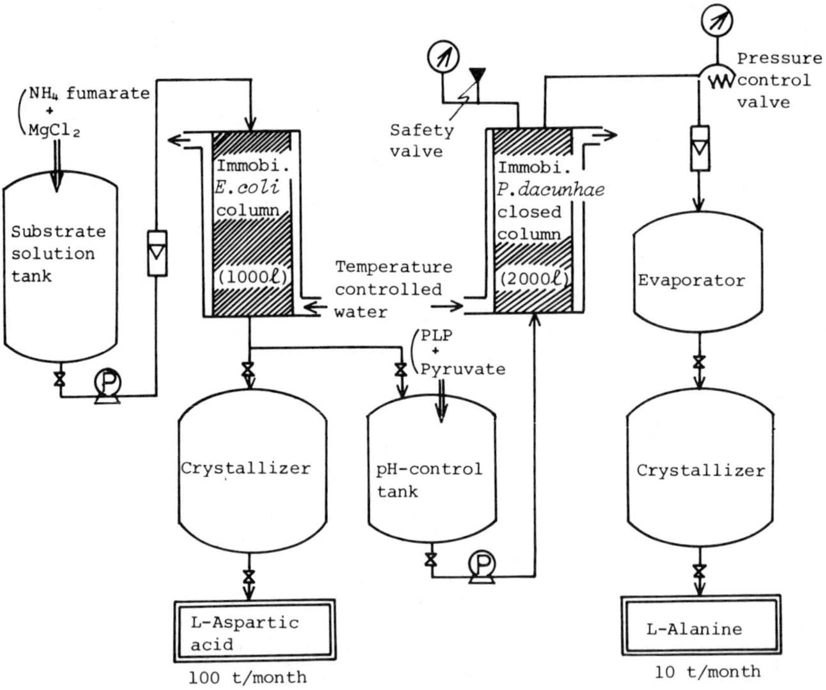

FIGURE 2. Flow diagram for continuous production of L-aspartic acid and L-alanine by immobilized microorganisms (Tanabe Seiyaku Co., Ltd.).

microorganisms (FIG. 2). This is considered to be the first industrial application of sequential enzyme reactions using immobilized cells of two species.

REFERENCES

1. YAMAMOTO, K., T. TOSA & I. CHIBATA. 1980. Continuous production of L-alanine using *Pseudomonas dacunhae* immobilized with carrageenan. Biotechnol. Bioeng. **22:** 2045–2054.
2. FURUI, M., K. YAMASHITA, A. SUMI & I. CHIBATA. 1981. Continuous production of L-alanine using immobilized *Pseudomonas dacunhae*—continuous closed reaction. Proceedings of the Autumn Meeting of the Society of Chemical Engineers. Japan. p. 159.
3. TAKAMATSU, S., I. UMEMURA, K. YAMAMOTO, T. SATO, T. TOSA & I. CHIBATA. 1982. Production of L-alanine from ammonium fumarate using two immobilized microorganisms—elimination of side reactions. Eur. J. Appl. Microbiol. Biotechnol. **15:**.
4. TAKAMATSU, S., T. TOSA & I. CHIBATA. 1983. Continuous Production of L-alanine using *Pseudomonas dacunhae* immobilized with κ-carrageenan. J. Chem. Soc. Jpn. **9:** 1369–1376.
5. SATO, T., S. TAKAMATSU, K. YAMAMOTO, I. UMEMURA, T. TOSA & I. CHIBATA. 1982. Production of L-alanine from ammonium fumarate using two types of immobilized microbial cells. *In* Enzyme Engineering VI. I. Chibata, S. Fukui & L. B. Wingard, Jr., Eds. Plenum Publishing Corporation. New York. pp. 271–272.

2,3-Butanediol Production by Immobilized Cells of *Enterobacter* sp.[a]

H. KAUTOLA, Y.-Y. LINKO, AND P. LINKO

Laboratory of Biotechnology and Food Engineering
Department of Chemistry
Helsinki University of Technology
SF-02150 Espoo 15, Finland

After the Second World War, there was little interest in production of 2,3-butanediol by fermentation until the late 1970s. 2,3-Butanediol may be used for the production of 1,3-butadiene, which can be used for making polybutadiene and styrene-butadiene gums. Butadiene can be produced from ethanol,[1] but its production directly from 2,3-butanediol is cheaper.[2] It can also be used as liquid fuel because of its great fuel value (27.2 MJ/kg).

2,3-Butanediol can be produced by many different microorganisms, but only a few species are suitable for industrial production. Typical 2,3-butanediol producers are *Enterobacter, Klebsiella, Serratia, Aeromonas* and some *Bacillus* spp. The formation of 2,3-butanediol is connected to the production of acetoin, ethanol, acetic acid, lactic acid, formic acid and succinic acid.[3] The above-mentioned organisms are able to use pentoses and pentitols as their only carbohydrate source. *Enterobacter aerogenes* resembles *Klebsiella pneumoniae,* and they were earlier regarded as the same strain. Chua *et al.*[4] produced 2,3-butanediol using κ-carrageenan bead-entrapped cells with glucose as substrate. The cells were not as sensitive to pH, temperature, and dissolved oxygen as the native cells. Furthermore, the formation of acetoin was significantly lower with immobilized cells. The production rate in a shake flask fermentation was 0.5 $g/dm^3/h$ and the yield was 60% of the theoretical value. In the continuous fermentation, the yield was 50% of the theoretical value, and 3 g/dm^3 could be produced at least for 10 days. Chambers *et al.*[5] used, for 2,3-butanediol production, a bacterium isolated from decaying wood. Immobilizing cells on 12-mm Rasching rings, they were able to convert a 100 g/dm^3 xylose solution at a rate three times as great as that for their conventional batch reactor.

We have studied 2,3-butanediol production with immobilized cells. The organism, isolated from paper mill process waters, was identified as *Enterobacter* sp., and was shown to be the best 2,3-butanediol producer together with *Klebsiella pneumoniae.*

Nutrients were supplied by Difco Labs. (USA), sodium alginate by BDH Chemicals (England), gelatin by Riedel de Haën AG (FRG), cellulose diacetate by Eastman Kodak Co. (USA), nylon net by 3M Co. (USA). Xylose was obtained from China.

Enterobacter sp. free cells were cultivated by batch fermentation (unless otherwise mentioned) in a medium containing (weight/volume) 0.2% peptone, 0.3% yeast extract, 0.25% ammonium chloride, 0.02% magnesium chloride, 0.5% dipotassium hydrogen phosphate, and 5% xylose. The pH was 5.75. The cultivation was carried out for 22 h at 32 °C. After centrifugation (5900 *g*) the dry weight of the cells was 23%. Repeated batch fermentations were carried out in 250-cm^3 Erlenmeyer flasks

[a]Financial support was received from the Academy of Finland, the Research Council for the Technical Sciences, and from the Kemira Co. Research Foundation.

containing 100 cm^3 of medium. The free cells (1.1 g) or the immobilized cells (6.5 g) were cultivated for 24 h (100 rpm, 32°C) after which time the medium was changed to a fresh one. In continuous fermentations by 6.5 g immobilized cells, in a reactor (ϕ 15 mm, height 60 mm), 2,3-butanediol was produced mainly at a residence time of 7 hours. Cells were immobilized as follows:

Alginate: Sodium alginate (8% wt/wt; autoclaved; 20 g) was mixed with cells (typically 4 g wet weight), and this mixture was extruded into 0.5 *M* calcium chloride solution and mixed for 20 minutes. The calcium alginate beads were washed with sterilized distilled water.

Gelatin: Warm 20% gelatin solution (20 g) was mixed with cells (4 g wet weight) and the solidified gelatin was cut into small cubes (3-mm sides), which were then shaken with 1% glutaraldehyde for an hour at +8°C (1 g immobilized cells + 2 cm^3 glutaraldehyde).

Cellulose diacetate (CDA): Cellulose diacetate beads were prepared in two different ways: 15% (wt/wt) CDA in dimethylsulfoxide (20 g) was mixed with 4 g (wet weight) freeze-dried cells, or 18.5% CDA in acetone and dimethylformamide (4:3)

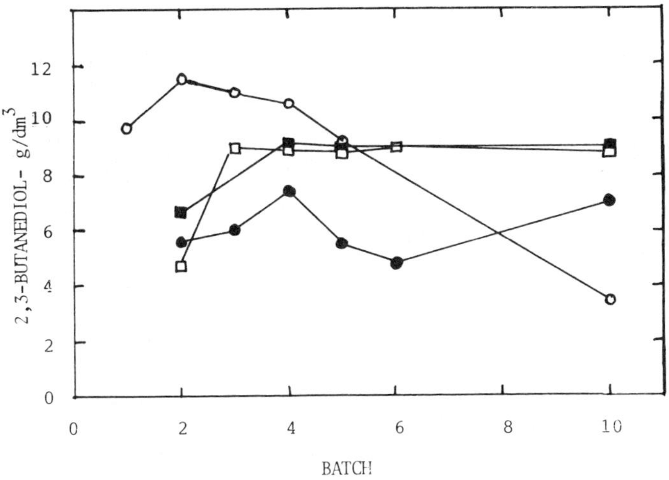

FIGURE 1. The effect of immobilization of *Enterobacter* sp. cells with various supports: calcium alginate (○), gelatin (●), CDA(DMSO) (□), CDA (DMF + acetone) (■) (repeated batch fermentation).

(20 g) was mixed with 4 g (wet weight) freeze-dried cells. Both mixtures were extruded into water and washed many times with sterilized water.

Nylon net: The cells were cultivated on the nylon net by circulating medium for 24–48 hours.

2,3-Butanediol and acetoin were determined by gas chromatography (Varian Aerograph 1520): column size 2 m, ϕ ¼ inch, glass tube; packing material, Chromosorb 101, mesh size 80/100; column temperature, 200°C; detector temperature, 235°C; injector temperature, 220°C; sample size 1 μl; flow rate of the nitrogen carrier gas 30 cm^3/min. 2,3-Butanediol was used as an internal standard. Reducing sugars were determined by the methods of Nelson[6] and Somogyi.[7]

The best 2,3-butanediol-producing microorganism isolated from paper mill process waters was identified as *Enterobacter* sp. The production of 2,3-butanediol was investigated first in batch fermentation with free cells using 2 dm^3 of medium. The maximum 2,3-butanediol yield was 64% of the theoretical value from xylose, the other products were mainly acetoin, ethanol, and acetic acid. The yield was at least as high as obtained by Chua *et al.*[4] (60% yield) from glucose with *Enterobacter aerogenes* IAM1133.

In comparing immobilization of cells on various supports, the highest 2,3-butanediol concentration (12 g/dm^3, yield 48%) was obtained with calcium alginate bead-entrapped cells and nylon net-immobilized cells. But the presence of phosphate in the medium caused softening of the alginate gel beads after 4–5 days in a repeated batch fermentation. In the continuous reactor fermentation, however, nylon net-immobilized cells produced 12 g/dm^3 (yield 48%) over 35 days when calcium carbonate was added to the medium. Without calcium carbonate, production declined

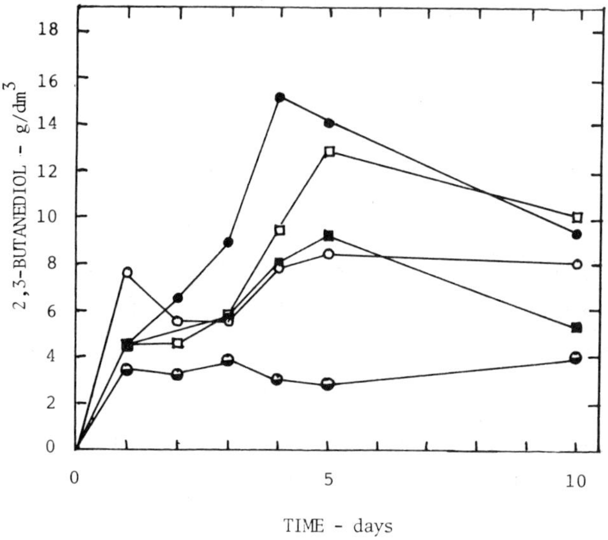

FIGURE 2. The effect of cell concentration in preparation of alginate bead-entrapped cells for production (repeated batch fermentation). 14% (◒), 29% (■), 58% (□), 86% (●), 144% (○) cells (d.m.)/alginate (d.m.).

after 5–6 days, when the maximum yield had been reached. In a repeated batch fermentation, the most stable production was obtained with cellulose diacetate bead-entrapped cells (9 g/dm^3, yield 36%). Gelatin cube-entrapped cells produced 7 g/dm^3 (yield 28%) (FIG. 1).

The optimum cell concentration for 2,3-butanediol production with calcium alginate bead-entrapped cells was 58–86% cells (d.m.)/alginate (d.m.). This optimum was reached in five days (FIG. 2). Cell concentration of 58% was used throughout this study.

The following medium composition was used: 5% xylose, 0.3% yeast extract, 0.2% peptone and no salts; 0.25% ammonium chloride; 0.25% ammonium chloride, and 0.02% magnesium chloride; 0.02% magnesium chloride. The production (8 g/dm^3) was lower than with the ordinary medium (12 g/dm^3). The most stable production was

TABLE 1. The Effect of Different Xylose Concentrations on 2,3-Butanediol Production Using Calcium Alginate Bead-entrapped Cells in Repeated Batch Fermentation

Xylose Concentration (%)	5th Batch Product (g/dm^3)	5th Batch Yield (%)	9th Batch product (g/dm^3)	9th Batch Yield (%)
2	4.3	43	5.5	53
5	9.0	36	9.8	39
7	15.0	42	10.6	33
10	17.0	34	11.4	21

reached, using alginate bead-entrapped cells, with medium of the following composition: 5% xylose, 0.3% yeast extract, 0.2% peptone and 0.02% magnesium chloride.

The effect of xylose concentrations on the production of 2,3-butanediol in a repeated batch fermentation with calcium alginate bead-entrapped cells is shown on TABLE 1. After five batches, the maximum yield of 43% was achieved with 2% xylose. Similar results were obtained with other concentrations. After 9 batches, decreases in yield were observed as the initial xylose concentration increased. Xylose concentrations of 2 to 7% could be used.

In a continuous column fermentation, with a residence time of 7 hours, 2,3-butanediol can be produced during a 40-day period by calcium alginate bead-entrapped cells with substrate consisting of 5% xylose, 0.3% yeast extract, 0.2% peptone and 0.02% magnesium chloride (FIG. 3). The yield, 32%, was low because of the lack of phosphate and ammonium in the substrate. With nylon net-immobilized

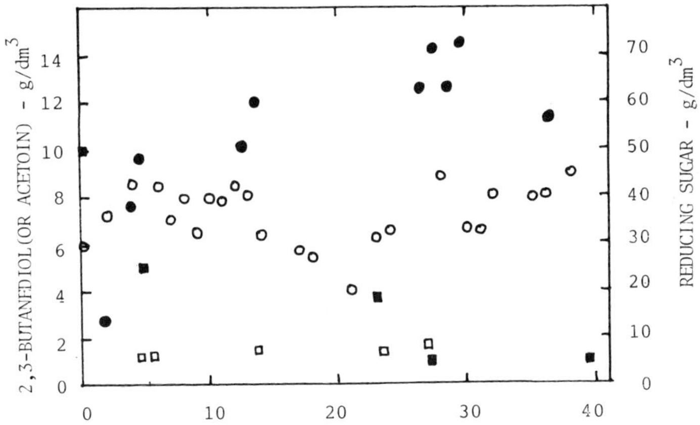

FIGURE 3. Continuous column fermentation with calcium alginate bead-entrapped *Enterobacter* sp. cells (○) (substrate: 5% xylose, 0.3% yeast extract, 0.2% peptone, 0.02% magnesium chloride, pH 5.75, 32 °C, τ = 7 h) and on nylon net-immobilized cells (●) (substrate: 5% xylose, 0.3% yeast extract, 0.2% peptone, 0.02% magnesium chloride, 0.5% dipotassium hydrogen phosphate, 0.25% ammonium chloride, 1% calcium carbonate). Reducing sugars (■) and acetoin (□) production by nylon net-immobilized cells.

cells, production was 12 g/dm^3 (yield 48%) over 35 days when calcium carbonate was added.

In continuous operation, residence times up to 2 hours increased 2,3-butanediol production linearly. Longer residence times did not increase production.

Scanning electron micrographs clearly showed that gelatin cube-entrapped cells grew more both on the surface and in the cross section after a long (27 days), continuous production of 2,3-butanediol. The same effect was obtained in a shorter time with calcium alginate bead-entrapped cells.

It was shown that immobilized cells of *Enterobacter* sp. can produce 2,3-butanediol from xylose in yields as high as reported for *Enterobacter aerogenes* from glucose in conventional fermentation. The stability of the biocatalyst was suitable for continuous column reactor operation. The low optimum of xylose concentration indicates that these methods could be suitable for production of 2,3-butanediol from wastes containing xylose.

REFERENCES

1. WADDONS, L. A. 1973. Chemicals from Petroleum. 3rd Edition. John Wiley & Sons. New York. p. 80.
2. FLICKINGER, M. C. & G. T. TSAO. 1978. Fermentation substrates from cellulosic materials: Fermentation products from cellulosic materials. *In* Annual Reports on Fermentation Processes. D. Perlman, Ed. Vol. **2:** 23–42. Academic Press. New York.
3. ROSENBERG, S. L. 1980. Fermentation of pentose sugars to ethanol and other neutral products by microorganisms. Enzyme Microb. Technol. **2:** 185–193.
4. CHUA, J. W., S. ERARSLAN & S. KINOSHITA. 1980. 2,3-Butanediol production by immobilized *Enterobacter aerogenes* IAM1133 with κ-carrageenan. J. Ferment. Technol. **58:** 123–127.
5. CHAMBERS, R. P., Y. Y. LEE & T. A. MCCASKEY. 1979. Liquid fuel and chemical production from cellulosic biomass. Hemicellulose recovery and pentose utilization in a biomass processing complex. Proc. 3rd Annual Biomass Energy Systems Conf. Solar Energy Research Institute. Technical Report. pp. 255–264.
6. NELSON, N. 1944. A photometric adaptation of the Somogyi method for the determination of glucose. J. Biol. Chem. **153:** 373–380.
7. SOMOGYI, M. 1952. Notes on sugar determination. J. Biol. Chem. **195:** 19–23.

Transformation of Steroids by Immobilized Microbial Cells

YANG LIAN-WAN AND ZHONG LI-CHAN

Institute of Microbiology
Academia Sinica
Beijing, China

The microbial transformation of steroids is very important for the production of pharmacologically useful steroid, but the conversion rate is usually low. The process of purification is complex and expensive. The transformation of steroids by immobilized cells is widely discussed, but little is reported about the practical use in industrial production.

Transformation of steroids with microbial enzymes for dehydrogenation at the C_1 and C_2 of the A ring, resulting in increase of biological activity, is of considerable commercial value in the production of pharmaceuticals.

Using the above-mentioned facts, we studied the transformation of steroids by immobilized *Arthrobacter simplex* By-2-13 cells with 3-ketosteroid-Δ^1-dehydrogenase.

The cells of *A. simplex* By-2-13 were entrapped in two kinds of gel: 10% polyacrylamide gel[1] and 5% mixed gel (mixture of 5% sodium alginate and 5% crude gelatin).

TABLE 1. Polyacrylamide Gel-entrapped *A. simplex* By-2-13 Cells

Optimal pH	7.0–8.5 (7.0–8.5)[a]
Optimal temperature	35°C (30°C)
K_m	0.417 mM (0.33 mM)
pH stability	6.0–7.0 (6.0–7.0)
Heat stability	Good (Poor)
Storage stability	−22°C, 5 months no loss of activity
Substrate concentration	0.2–0.4%
Conversion rate	Over 95%
Operational stability	30°C, 25 days no loss of activity

[a]Native cells in parentheses.

One gram of *A. simplex* By-2-13 cells immobilized in 5 g of 10% polyacrylamide gel retained about 80% of the original enzyme activity, which could be increased about 40% by addition of a cofactor, menadione. The immobilized cells were used to transform hydrocortisone to prednisolone. Some basic parameters of the immobilized cells are listed in TABLE 1.

One gram of *A. simplex* By-2-13 cells was mixed with 5 g of 5% mixed gel. The mixture was hardened with 0.1 M $CaCl_2$. The recovery of enzyme activity of the immobilized cells was about 95%. The immobilized cells were used to transform cortisone acetate to prednisone acetate. Suitable reaction conditions are listed in TABLE 2.

TABLE 2. Mixed Gel-entrapped *A. simplex* By-2-13 Cells

Reaction medium	0.5% Fish peptone
Reaction temperature	25–35°C
Reaction pH	6.0–7.2
Substrate concentration	0.5–1.0
Conversion rate	About 95%
Operational stability	34–35°C, 17 days no loss of activity

The main feature of the two kinds of immobilized microbial cells are as follows:(1) the entrapping methods are very simple; (2) the recovery of enzyme activity is high; (3) the storage and operational stability are good; and (4) the conversion rate of substrate into product is high.

Information concerning the conversion of cortisone acetate into prednisone acetate by mixed gel-entrapped microbial cells is unavailable. We developed a method using the mixed gel to entrap *A. simplex* By-2-13 cells. It was successfully applied to convert cortisone acetate into prednisone acetate with promising results.

REFERENCE

1. CHIBATA I., T. TOSA & T. SATO. 1974. Immobilized aspartase-containing microbial cells: Preparation and enzymatic properties. Appl. Microbiol. **27**(5): 878–885.

Malolactic Fermentation and Secondary Product Formation in Wine by *Leuconostoc oenos* Cells Immobilized in a Continuous-flow Reactor[a]

PAOLO SPETTOLI, MARCO P. NUTI, ANGELO DAL, BELIN PERUFFO, AND ARTURO ZAMORANI

Istituto di Chimica Agraria e Industrie Agrarie
Università di Padova
Via Gradenigo 6
35100 Padova, Italy

Bacterial fermentation of malic acid and its conversion into lactic acid have long been recognized, and are promoted where grapes ripen with excessive amounts of malic acid. Malolactic fermentation is usually considered desirable for the biological stability of Italian wines, in particular for premium quality red wines; on the other hand this process is difficult to control and the decision by the winemaker to promote it or not is quite adventurous. A good correlation between malic acid decarboxylation and relative concentration of the bacterial population in wine was shown by Rice and Mattick.[1] Since then, attempts have been made to stimulate the malolactic fermentation by inoculation with strains of *Leuconostoc oenos*; however, this inoculation method has been only partially successful,[2] and therefore the use of reactors with entrapped *L. oenos* cells has been suggested. The potentialities of immobilized microbial cells have been explored only in recent years.[3]

Some data on the use of immobilized whole cells for malic acid fermentation on grape juice have been reported by Monsan and Durand.[4] Divies[5,6] studied a continuous alcoholic and malolactic fermentation using whole cells immobilized in polyacrylamide and calcium alginate gels. Our previous studies,[7,8] focusing on the technological traits of malolactic activity from *L. oenos* immobilized in calcium alginate gels and on the immobilization of *L. oenos* in calcium alginate gels, have suggested the possibility of controlling the biochemical transformation of malic acid in model solutions as well as in wine. In this context, Totsuka and Hara[9] and Gestrelius[10] examined the possibility of performing this process by passing wine through reactors containing immobilized viable cells. However, no matter which immobilized microbial cells are used, their metabolism will affect the sensory characteristics of wine undergoing malolactic fermentation. Unfortunately, no data are available that allow a comparison between columntreated and untreated wines.

The purpose of the present study was to perform the decarboxylation of *l*-malic into *l*-lactic acid in wines by the use of entrapped cells and to investigate secondary product formation along with the evolution of nitrogenous fractions.

[a]This research was supported by the Consiglio Nazionale delle Ricerche.

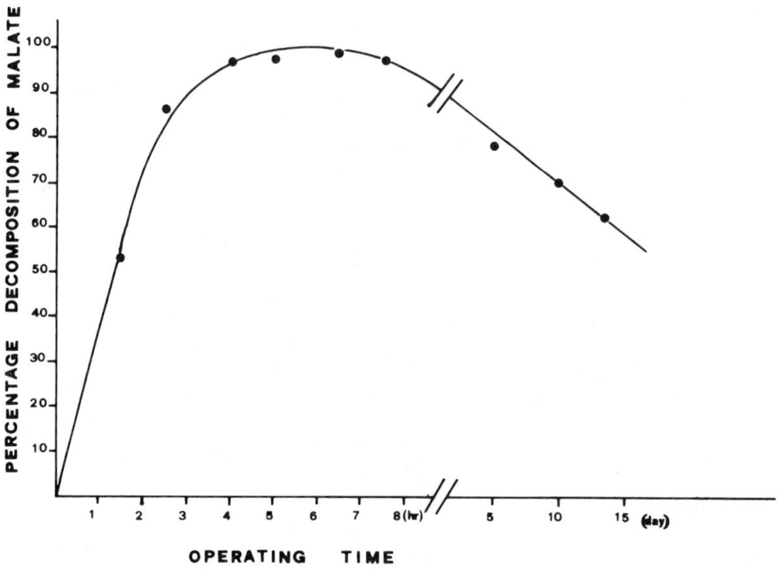

FIGURE 1. *l*-Malic acid decomposition as function of time. Red wine was continuously passed through a packed-bed column of Ca alginate-immobilized *Leuconostoc oenos* ML 34 cells at 20°C at a flow rate of 0.80 vol/h.

DESIGN OF A CONTINUOUS-FLOW BIOREACTOR

Leuconostoc oenos ML 34 (courtesy of R. E. Kunkee) was grown in tomato juice broth as described in a previous paper.[8] The cell immobilization was performed by adding one volume of wet cell pellet (up to 50% cells, wet weight) to 2 volumes of 2.5% wt/vol sodium alginate solution and the mixture was dripped into 100 m*M* $CaCl_2$.

TABLE 1. Chemicals in a Red Wine before and after Fermentation by Immobilized *L. oenos* ML 34 Cells in a Continuous Reactor System

Chemical (measure)	Before ML Fermentation	After ML Fermentation	Difference
Alcohol (% vol/vol)	10.20	—	
Free SO_2 (mg/l)	4.50	—	
Total SO_2 (mg/l)	18.13	—	
pH	3.20	3.48	+0.28
Titratable acidity (g/l tartaric acid)	10.74	8.13	−2.61
l-Lactic acid (g/l)	0.27	2.87	+2.60
l-Malic acid (g/l)	4.50	0.06	−4.44
Acetic acid (g/l)	0.12	0.44	+0.32
Citric acid (g/l)	0.57	0.12	−0.45
Glucose (g/l)	0.052	0.025	−0.027
Fructose (g/l)	0.570	0.180	−0.390
Glycerol (g/l)	6.00	5.85	−0.15
Diacetyl (mg/l)	0.07	0.07	0.00
Acetoin (mg/l)	0.21	0.05	−0.16

After 2–3 h, the bead-type gels, up to 2 mm in diameter, were harvested and incubated for 24–48 h at 25°C in grape juice (reconstituted grape juice with 30 m*M* *l*-malate, 10% ethanol, pH 4.8)[10] and thoroughly washed with wine before use.

Red wine was continuously passed through a packed column of 2.6 cm i.d. at 20°C at a flow rate of 0.80 vol/h. A nitrogen pressure of about 0.17 atm was maintained in the reactor in order to keep the produced CO_2 in solution and to prevent contact with air. The flow was controlled accurately by a peristaltic pump system. Red table wine from a local vineyard (cv. Raboso) was subjected to chemical analyses before and after flow through immobilized microorganisms according to previously described methods.[8]

The immobilized cells showed a good effectiveness in degrading *l*-malic acid as seen from FIGURE 1. The increase in *l*-lactic acid and the decrease in titratable acidity

TABLE 2. Amino Acids, Ammoniacal and Total Nitrogen Contents in a Red Wine before and after Fermentation by Immobilized *L. oenos* ML 34 Cells in a Continuous Reactor System

Free Amino Acids (mg/l)	Before Fermentation	After Fermentation	Difference %
Aspartic acid	12.06	9.17	−23.96
Threonine	7.27	4.48	−38.38
Serine	4.11	3.18	−22.63
Glutamic acid	11.30	8.56	−24.25
Proline	442.41	395.03	−10.71
Glycine	2.60	2.14	−17.70
α-Alanine	3.25	3.12	−4.00
Cystine	4.70	traces	—
Valine	traces	traces	—
Methionine	traces	traces	—
Isoleucine	traces	traces	—
Leucine	0.90	0.65	−27.78
Tyrosine	1.09	traces	—
Phenylalanine	0.70	traces	—
Histidine	3.01	2.00	−33.56
Lysine	2.48	1.87	−24.60
Arginine	3.88	3.61	−6.96
Total free amino acids	499.76	433.81	−13.20
Ammoniacal N (mg/l)	1.16	0.99	−14.66
Total N (mg/l)	106.40	98.00	−7.90

were equivalent to 58.6% and 58% of the original *l*-malic acid (TABLE 1), respectively.

The 0.28-unit increase of pH was in agreement with previous data on malolactic fermentation in wine.[11]

SECONDARY PRODUCT FORMATION

As expected, the increase in volatile acidity (+0.32 g/l acetic acid) could be related to a decrease of citric acid (−0.45 g/l). However, diacetyl and acetoin concentrations in the wine after fermentation (TABLE 1) were found constant and lower, respectively. This behavior would be commercially important because in low concentrations diacetyl and acetoin can be a quality factor.[2] The decreases in glucose and fructose were substantially in agreement with other published data,[12] while glycerol remained almost unchanged. Peynaud and Domercq[13] studied the amino-acid content in wines inoculated with pure cultures of *Leuconostoc oenos* sp. or *Lactobacillus* sp. After fermentation, the amount of amino acids, total nitrogen, and ammonium nitrogen was about 10–30% lower than in the uninoculated control wine. Our data are in agreement with these results as seen from TABLE 2; therefore the malolactic fermentation performed as described seems to provide also a better biological stability of the wine.

The results obtained so far seem to indicate the possbility of inducing and controlling malolactic fermentation by Ca alginate-immobilized *Leuconostoc oenos* ML 34 cells in a continuous-flow reactor.

ACKNOWLEDGMENT

The authors are grateful to Mrs. Maria L. Morandi for skillful technical assistance.

REFERENCES

1. RICE, A. C. & L. R. MATTICK. 1970. Natural malo-lactic fermentation in New York state wines. Am. J. Enol. Vitic. **21:** 145–152.
2. RANKINE, B. C. 1977. Developments in malo-lactic fermentation of Australian red table wines. Am. J. Enol. Vitic. **28:** 27–33.
3. CHIBATA, I. 1979. Immobilized microbial cells. Application. *In* Cellules Immobilisees. J. M. Lebeault & G. Durand, Eds. Compiègne. pp. 7–44.
4. MONSAN, P. & G. DURAND. 1976. Applications récentes des microorganismes et des enzymes dans les industries agricoles et alimentaires. Ind. Aliment. Agric. **93:** 543–551.
5. DIVIES, CH. 1975. Brevet A.N.V.A.R. n° 7524509.
6. DIVIES, CH., M. H. SIESS & M. F. JEANBLANC. 1979. Utilisation des levures incluses en fermentation alcoolique. *In* Cellules Immobilisees. J. M. Lebeault & G. Durand, Eds. Compiègne. pp. 151–172.
7. SPETTOLI, P., M. P. NUTI, A. BOTTACIN & A. ZAMORANI. 1982. Technological traits of malo-lactic activity from *Leuconostoc oenos* ML 34 immobilized in calcium alginate gels. *In* Use of Enzymes in Food Technology. P. Dupuy, Ed. Versailles. Paris. pp. 545–548.
8. SPETTOLI, P., A. BOTTACIN, M. P. NUTI & A. ZAMORANI. 1982. Immobilization of *Leuconostoc oenos* ML 34 in calcium alginate gels and its application to wine technology. Am. J. Enol. Vitic. **33:** 1–5.
9. TOTSUKA, A. & S. HARA. 1981. Decomposition of malic acid in red wine by immobilized microbial cells. Hakkokogaku **59:** 231–237.
10. GESTRELIUS, S. 1982. Potential application of immobilized viable cells in the food industry: Malolactic fermentation of wine. *In* Enzyme Engineering. I. Chibata, S. Fukui & L. B. Wingard, Jr., Eds. Vol. **6:** 245–250. Plenum Press. New York.
11. KUNKEE, R. E. 1974. Malo-lactic fermentation and winemaking. *In* Chemistry of Winemaking. A. D. Webb, Ed. Adv. Chem. Ser. **137:** 151–170. Washington, D. C.
12. MELAMED, N. 1962. Determination des sucres résiduels des vins, leur relation avec la fermentation malolactique. Ann. Technol. Agric. **11:** 5–32.
13. PEYNAUD, E. & S. DOMERCQ. 1961. Etudes sur les bactéries lactiques des vins. Ann. Technol. Agric. **10:** 43–60.

Immobilization of Microorganisms

JOHN L. GAINER AND DONALD J. KIRWAN
Department of Chemical Engineering
University of Virginia
Charlottesville, Virginia 22901

Whole cell immobilization has been the subject of numerous reviews. Most work on whole cell immobilization has involved the entrapment of the cells within gels of either polyacrylamide or carrageenan, or within crosslinked collagen matrices. Although the entrapment method appears to allow the organisms to be maintained in an environment similar to that of free cells, the gel does introduce a significant additional mass transfer resistance. To avoid this problem, it is possible to attach cells using electrostatic and covalent coupling methods, and our work has been involved with such techniques.

Since most organisms have a negative surface charge, we have studied their adsorption on positively charged ion-exchange resins. We began by studying the adsorption of a nitrogen-fixing bacterium, *Azotobacter vinelandii,* to such anionic resins, and found that the adsorption was faster with one particular resin, Cellex-E (Bio-Rad). It was noted that the cells are able to reproduce while adsorbed. The cells can be desorbed by increasing the ionic strength of the solution (FIG. 1), and we can model the adsorption process with a Freundlich isotherm. It is particularly interesting to note that it is possible, with such adsorbed cells, to construct reactors having effective cell concentrations up to 10^{11} per ml of reactor volume. This implies that it should be possible to construct much smaller reactors for a given conversion.

A. vinelandii immobilized by electrostatic adsorption to Cellex-E maintained reproductive capacity and nitrogen fixation for more than 330 hours. Although the respiration rate was comparable for both free and immobilized cells, the nitrogen fixation rate for immobilized cells was slightly different than that for free cells, with the maximum occurring at a lower dissolved oxygen concentration. Other organisms, such as *Acremonium chrysogenum* and *Penicillium digitatum* have also been adsorbed this way, and current studies are focusing on the rate of product formation after immobilization.

We have been using cyanuric chloride as a coupling agent and following an enzyme attachment procedure proposed by Kay and Lilly[1] as well as one by Smith and Lenhoff.[2] The method of Kay and Lilly appears to work better for yeasts, and the Smith and Lenhoff method appears to be better for bacteria.

We have attached a number of bacteria and yeasts with these methods and are still developing optimum conditions for the immobilization process. Some initial observations, though, are quite interesting. We have placed immobilized cell preparations in a continuous-flow stirred reactor, and, unless the dilution rate (reciprocal of hydraulic residence time) was large, the population of free cells grew quite large. The coupled cells apparently reproduce and the results suggest that their specific growth is higher than that of free cells. In order to test the activities of only the immobilized cells, the dilution rates were increased until the free cell population was negligible in the chemostat (*i.e.,* above the washout rate). We studied *Azotobacter vinelandii* and *Saccharomyces cerevisiae* in this way (see FIG. 2). It can be seen that the nitrogenase activity of the immobilized *A. vinelandii* could be maintained for over a month. The decline may or may not be related to a change in the viability of the cells, as

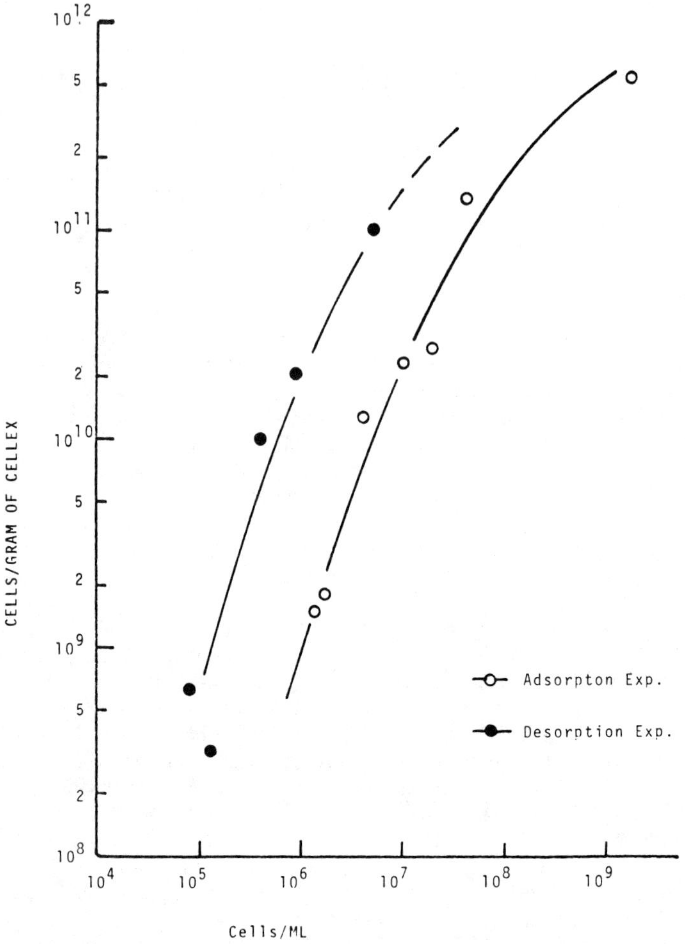

FIGURE 1. Adsorption of *A. vinelandii* (UW 590) on Cellex E anionic resin in Burk's medium.

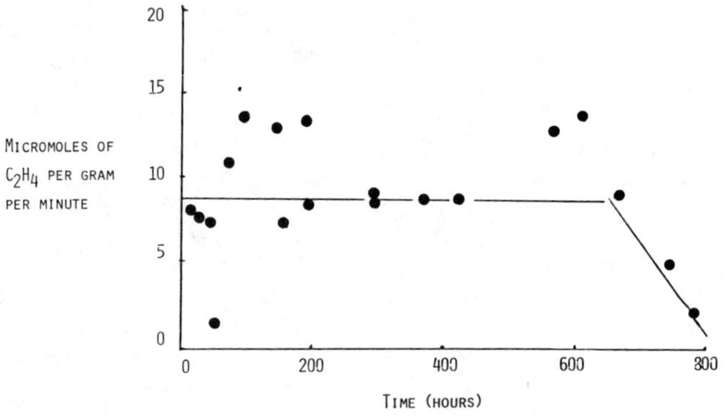

FIGURE 2. Nitrogenase activity of covalently bound *A. vinelandii.*

contamination in the fermentor occurred at the very long times. Current work involves determining optimum conditions for immobilizing different microorganisms, and testing the effect of immobilization on product formation rates.

REFERENCES

1. KAY, G. & M. D. LILLY. 1970. The chemical attachment of chymotrypsin to water-insoluble polymers using 2-amino-4, 6-dichloro-*s*-triazine. Biochim. Biophys. Acta **198:** 276–285.
2. SMITH, N. L. & H. M. LENHOFF. 1974. Covalent binding of proteins and glucose-6-phosphate dehydrogenase to cellulosic carriers activated with *s*-triazine trichloride. Anal. Biochem. **61:** 392–415.

Production, Purification, and Immobilization of *Bacillus subtilis* Levan Sucrase

PATRICE PERLOT

Societe Beghin-Say
D.R.D. 54, avenue Hoche 75360
Paris Cedex 08, France

PIERRE MONSAN

Department de Genie Biochimique
ERA-CNRS 879 INSA, avenue de Rangueil
F31077 Toulouse, France

The most common mechanism of polysaccharide synthesis (starch, glycogen, etc.) involves high-energy phosphorylated monomer intermediates (sugar nucleotides). In addition to this general mechanism, the synthesis of two polysaccharides may be achieved in a cell-free system using sucrose as substrate and either dextran sucrase (E.C. 2.4. 1.5., α=(1-6)D-glucan:D-fructose 2-glucosyl transferase) or levan sucrase (E.C. 2.4.1.10, β-(2-6)D-fructan:D-glucose 1-fructosyl transferase) as catalyst. In both cases, the energy required for polymerization is provided by the hydrolysis of the glycosidic bond of the sucrose molecule. Levan sucrase synthesizes a ramified D-fructofuranose polymer, levan, in which fructose units are linked by β-(2-6) bonds in the main chain[1] and branching results from β-(2-1) bonds.[1] Levan sucrase is an exocellular enzyme produced by several bacteria.[1] *Aerobacter levanicum* and *Bacillus subtilis* have been extensively studied as sources of this inducible enzyme. Levan sucrase has been highly purified from the culture supernatant of *Bacillus subtilis.*[2] In the present work, we have studied the production, purification, and immobilization of levan sucrase from batch cultures of *Bacillus subtilis* C4, a constitutive mutant of *Bacillus subtilis* B.S.5 var. nigra. This strain was a gift from R. Chambert (I.R.B.M., Univ. Paris VII, France).

In standard conditions, levan sucrase is obtained by growing *Bacillus subtilis* C4 in a medium containing glycerol (1% wt/vol) as carbon source, $(NH_4)_2$ SO_4 (0.33% wt/vol), K_2HPO_4 (1.42% wt/vol), KH_2PO_4 (0.41% wt/vol), $MgSO_4$ (0.006% wt/vol), $MnSO_4$ (0.0017% wt/vol), and ammonium iron III citrate (0.0022% wt/vol).

Iron is added to the culture medium to stabilize the levan sucrase.[3] FIGURE 1A shows a typical growth curve with substrate consumption and levan sucrase activity obtained by cultivating *Bacillus subtilis* C4 in the standard medium described above. After a lag phase, an exponential growth phase is obtained and the maximum specific growth rate (μ_m) is 0.6 h^{-1}. Levan sucrase production starts at the beginning of the exponential growth phase and in the middle of this phase, the activity reaches a maximum value of 1.9 LSU/ml (1 levan sucrase unit LSU is the amount of enzyme that catalyzes the synthesis of 1 μmole of glucose per minute at 37 °C with 0.33 *M* sucrose solution in a phosphate buffer 0.05 *M* pH 6.0), and then decreases. This decrease is attributed to protease excretion by the cells. At the stationary phase of the culture, no levan sucrase activity was detected in the supernatant. It was also observed

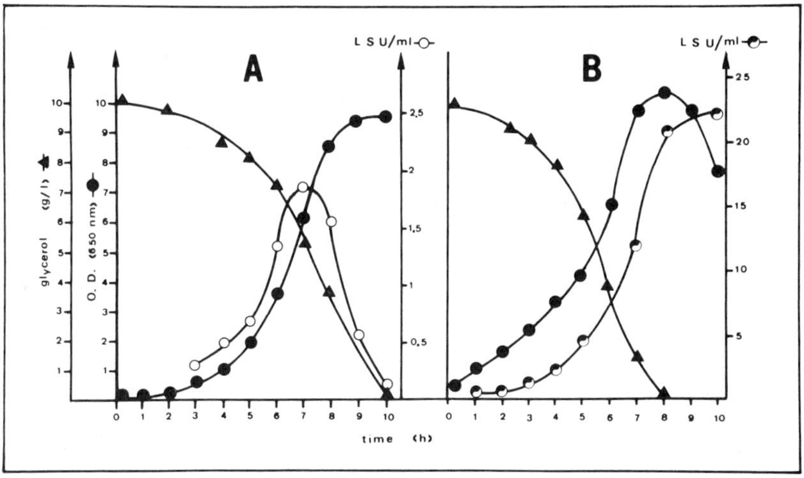

FIGURE 1. *Bacillus subtilis* C4 growth curve at 37 °C (K_la = 1300 h^{-1}). A: standard culture conditions; B: culture with addition of 1.67 g l^{-1} antifoaming agent (Rhodorsil 410).

that during levan sucrase production, enzyme activity is directly linked to biomass production.

The microorganism used in this study needs to be cultivated in very aerobic conditions, in which case extensive foaming occurs. This foaming is prevented by using an antifoaming agent, Rhodorsil 410 (Rhône Poulenc Industries), which is a silicon oil derivative. The addition of this antifoaming agent at the beginning of the culture has consequences both on the growth of *Bacillus subtilis* C4 and on levan sucrase production.

When the microorganism is cultivated in the standard medium containing the antifoaming agent at a concentration of 1.67 g l^{-1}, a linear growth phase, characteristic of a limitation, may be observed. Injection of pure oxygen during the linear growth phase induces an exponential growth phase. This result shows that addition of the antifoaming agent results in an oxygen transfer limitation. Optimum conditions of aeration and agitation for the culture of *Bacillus subtilis* C4 in a standard medium containing 1.67 g l^{-1} of antifoaming agent have been determined by cultivating the bacteria with different values of oxygen transfer coefficient (K_la). The results described in TABLE 1 show that the oxygen transfer coefficient must be greater than the critical value of 1280 h^{-1} to obtain an exponential growth phase.

Furthermore, when *Bacillus subtilis* C4 is cultivated in a medium containing

TABLE 1. Influence of Aeration and Agitation Rate on *Bacillus subtilis* C4 Growth and on Levan Sucrase Production

Oxygen transfer coefficient (h^{-1})	680	996	1280	3047
Growth rate	Linear	Linear	Expon.	Expon.
Maximum specific growth rate (h^{-1})	—	—	0.38	0.52
Maximum optical density (650 nm)	6.0	9.5	11.0	11.5
Final levan sucrase activity (LSU/ml)	3.2	10.0	21.6	21.0

Rhodorsil 410, there is no decrease in levan sucrase activity during the second part of the exponential growth phase. An example of a growth curve obtained under such conditions is given in FIGURE 1B. Levan sucrase is produced throughout the whole growth phase and enzyme activity reaches a maximum value of 21 LSU/ml at the stationary phase. Although the mechanism by which the antifoaming agent acts to give high levels of enzyme activity is unknown, two hypotheses may be made. The first hypothesis is directly derived from the usual effects of antifoaming agents on bacteria. By their physical properties, these products facilitate the excretion of enzymes either by releasing proteins attached to the bacteria or by increasing their permeation through the membrane. It should be noted that other antifoaming agents of the same type (silicone oil derivatives) do not have a similar effect and that Rhodorsil 410 has no effect on purified levan sucrase.

Levan sucrase contained in the fermentation broth of *Bacillus subtilis* C4 is purified in three steps: ultrafiltration (cutoff 10,000), nucleic acid precipitation, and gel permeation chromatography. As shown in TABLE 2, ultrafiltration of the fermentation broth does not greatly increase the specific activity of the levan sucrase solution, but this step allows the elimination of small molecules and the reduction of the volume of the solution. Nucleic acids are then eliminated by protamine sulfate precipitation. Gel permeation chromatography of concentrated levan sucrase solution on Ultrogel AcA 34 is the last step in enzyme purification. Levan sucrase is eluted in the void volume although its molecular weight is lower than the exclusion limit of the gel. Study of the protein peak containing levan sucrase activity shows that the enzyme is uniformly distributed in this peak and its specific activity is 160 LSU/mg of protein. The total yield of the purification process is 60%.

Purified levan sucrase has been immobilized by covalent coupling onto a glutaraldehyde-activated amino silica, Spherosil (Rhône Poulenc Industrie). Glutaraldehyde activation of amino silica[4] and the conditions for levan sucrase immobilization[5] have been optimized. The fact that silica is available with different specific areas but with the same particle diameter allows the influence of internal diffusion limitations to be studied. As shown in FIGURE 2A, the enzymatic activity of the carrier decreases after a few minutes' reaction. The initial reaction rate is a function of the specific area of the carrier and optimum activity (2800 LSU/g of silica) is reached at 120 m^2/g. After a few minutes' reaction, the synthesis of high-molecular-weight levan (above 10^6 MW) results in a sharp increase in intraparticular medium viscosity. In consequence, mass transfer limitations are then highly increased and the activity of the levan sucrase molecules immobilized within the porous structure becomes very low. Then, only the enzyme molecules immobilized near the silica bead surface still remain active with the result that the residual activity becomes independent of the specific area of the support

TABLE 2. Purification of Levan Sucrase from *Bacillus subtilis* C4 Fermentation Broth

Step	Specific Activity (LSU/mg Protein)	Purification Factor		Yield (%)	
		Step	Total	Step	Total
Fermentation broth	4.1	0	0	100	100
Ultrafiltration (cutoff 10,000)	5.6	1.4	1.4	73.5	73.5
Protamine sulfate precipitation	9.0	1.6	2.0	90.0	66.2
Gel permeation AcA34	161.0	18.0	39.3	90.0	59.5

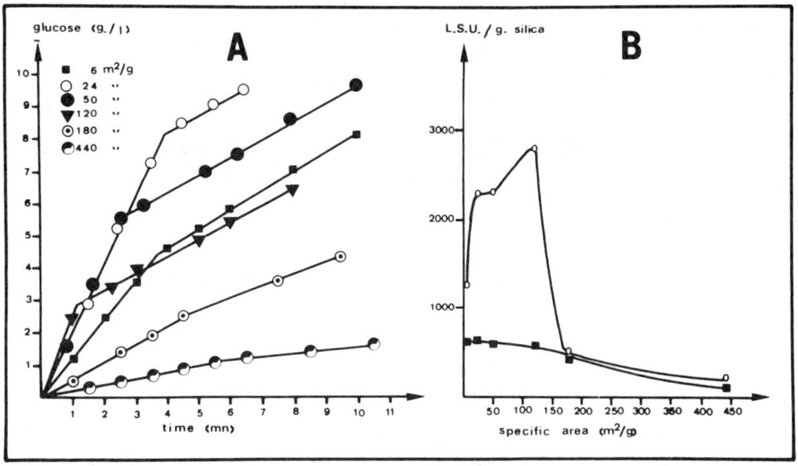

FIGURE 2. Influence of porous silica (Spherosil) specific area on immobilized levan sucrase activity. A: glucose production as a function of reaction time; B: initial activity —O— and residual activity after 5-min reaction —■—.

(FIG. 2B). The immobilization of an enzyme producing a high-molecular-weight product thus gives a good illustration of the interference of intraparticular mass transfer phenomena with immobilized enzyme kinetics.

In conclusion, the optimization of the culture conditions of *Bacillus subtilis* C4 and the specific effect of the addition of an antifoaming agent has allowed an elevenfold increase in levan sucrase activity, avoiding the decrease occurring in standard culture conditions. The purified free levan sucrase preparations can be efficiently used for producing high-molecular-weight levan polymers.

REFERENCES

1. AVIGAD, G. 1968. Levans. Encycl. Polym. Sci. Technol. **8:** 711–718.
2. DEDONDER, R., E. JOZON, G. RAPOPORT, Y. JOYEUX & A. FRISTCH. 1963. La levane sucrase, β-D-fructofuranosyle transferase de *Bacillus subtilis*. Obtention de l'enzyme pure. Bull. Soc. Chim. Biol. **45:** 477–492.
3. DEDONDER, R. 1960. La levane sucrase de *Bacillus subtilis*. Bull. Soc. Chim. Biol. **42:** 1745–1761.
4. MONSAN, P. 1977–1978. Optimization of glutaraldehyde activation of a support for enzyme immobilization. J. Mol. Catal. **3:** 371–384.
5. PERLOT, P. 1980. Production, Purification et Application de la levane saccharase de *Bacillus subtilis*. These Doct. Ing. N ° 32. INSA. Toulouse.

Oxygen Supply to Immobilized Cells Using Hydrogen Peroxide

OLLE HOLST,[a] HANS LUNDBÄCK[b] AND BO MATTIASSON[c]

[a]*Department of Applied Microbiology*
[b]*Department of Analytical Chemistry*
[c]*Department of Pure and Applied Biochemistry*
Chemical Center, University of Lund
P.O. Box 740
S-220 07 Lund, Sweden

The supply of oxygen to aerobic organisms is a crucial point in fermentation technology. This is even more accentuated in systems with immobilized cells, owing to high cell densities and diffusion limitations introduced by the immobilization matrix.[1] The decomposition of hydrogen peroxide to water and oxygen has been suggested as a way to get more oxygen into the matrix.[2] The decomposition occurs inside the matrix and is catalyzed by catalase from the immobilized cells. However, high concentrations of hydrogen peroxide will affect the stability of the immobilized cells, and thus a controlled addition of the oxygen source is necessary. In order to develop a suitable method for the determination of hydrogen peroxide in the reaction medium, an earlier method, intended for monitoring of hydrogen peroxide in pickling baths,[3] was modified.[4] The method was based on flow injection analysis (FIA) with amperometric detection. We report here the use of this FIA system in an experimental setup (FIG. 1) for computer-controlled addition of hydrogen peroxide to a reactor containing immobilized cells operated under aerobic conditions.

Gluconobacter oxydans (ATCC 621) was immobilized in Ca-alginate (0.03 g dry cells/g wet gel) for conversion of glycerol (CH_2OH-CHOH-CH_2OH) to dihydroxyacetone (CH_2OH-CO-CH_2OH). Cultivation and immobilization were accomplished as described earlier.[2] The medium consisted of 0.5 *M* glycerol, 0.05 *M* succinate-buffer, pH 5.0, and hydrogen peroxide as indicated elsewhere. $CaCl_2$ (5 m*M*) was added in order to stabilize the alginate gel. The reaction temperature was 26°C. Dihydroxyacetone was determined as reducing substance using the DNS-method.[5] Hydrogen peroxide was determined in a FIA system, using amperometric detection,[4] by oxidation on a glassy carbon electrode at +1.2 V versus SCE. The current was proportional to the concentration over a wide range. Thirty samples could be analyzed per hour. The analytical system showed high stability, and the electrochemical interferences from compounds in the reaction mixture were minimized by dilution of the sample in the FIA system. To maintain a constant concentration of peroxide in the reactor, the analytical signal was processed in a computer programed as a PID-regulator. The regulator controlled a peristaltic pump, which added the peroxide to the reactor. The concentration of dissolved oxygen was followed by means of a commercial Clark electrode.

Increased productivity in systems with immobilized whole cells using hydrogen peroxide as an oxygen source has been demonstrated earlier.[2] The earlier experiments were carried out with packed columns, which were fed with substrate solution including hydrogen peroxide. This made it difficult to interpret the effect of different hydrogen peroxide concentrations on the stability and on the survival of the immobil-

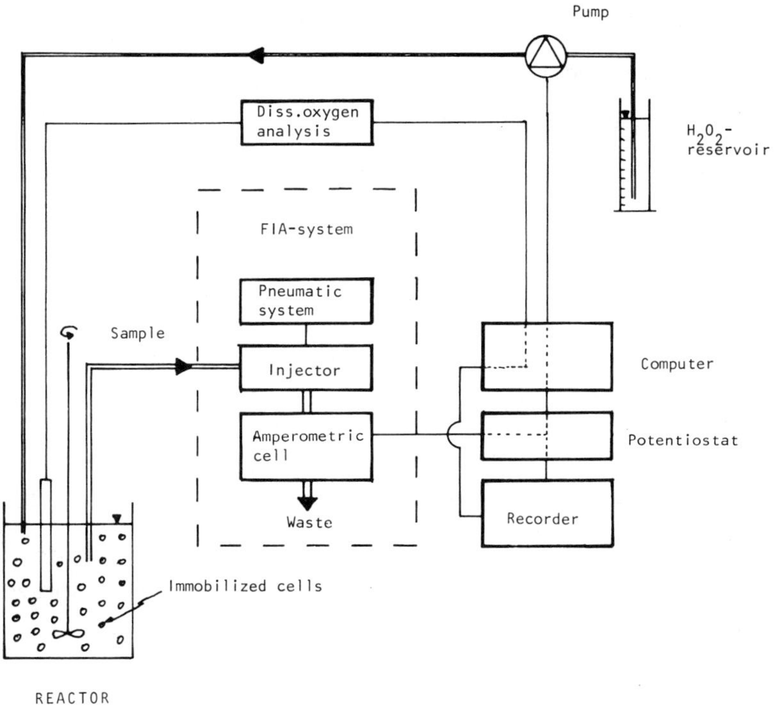

FIGURE 1. Experimental setup.

ized microorganisms. In the work reported here, batch experiments were performed with the concentration of hydrogen peroxide maintained at the desired value by continuous addition of peroxide. The rate of addition was controlled by using an FIA system connected to a computer. In FIGURE 2, the effect of different concentrations of hydrogen peroxide on the initial specific productivity is shown. A significant increase in specific productivity with increasing hydrogen peroxide concentration is noted. In order to evaluate the effect of hydrogen peroxide on the stability of the preparation, the oxygen and peroxide consumption rates were measured before and after a period of five

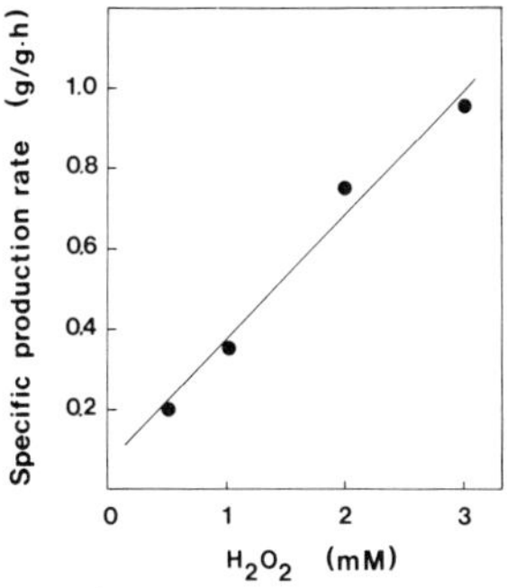

FIGURE 2. Initial specific production rate of dihydroxyacetone as a function of hydrogen peroxide concentration.

hours. A decrease in activity was noted at all peroxide concentrations used. The decrease was observed for both oxygen consumption rate and peroxide consumption rate. However, the influence of different concentrations seemed to be rather small.

From this work, the following conclusions might be drawn: (1) the use of hydrogen peroxide as an oxygen source drastically increases the productivity of immobilized aerobic cells; (2) the stability of the immobilized cells decreases with increasing concentration of hydrogen peroxide; and (3) FIA seems to be a promising method for controlling this process.

REFERENCES

1. ENFORS, S.-O. & B. MATTIASSON. 1983. Oxygenation of processes involving immobilized cells. *In* Immobilized Cells and Organelles. B. Mattiasson, Ed. Vol. **II:** 41–60. CRC-Press. Boca Raton, Florida.
2. HOLST, O., S.-O. ENFORS & B. MATTIASSON. 1982. Oxygenation of immobilized cells using hydrogen peroxide; A model study of *Gluconobacter oxydans* converting glycerol to dihydroxyacetone. Eur. J. Appl. Microbiol. Biotechnol. **14:** 64–68.
3. LUNDBÄCK, H. 1983. Amperometric determination of hydrogen peroxide in pickling baths for copper and copper alloys by flow injection analysis. Anal. Chim. Acta **145:** 189–196.
4. LUNDBÄCK, H. G. JOHANSSON & O. HOLST. Determination of hydrogen peroxide for application in aerobic cell systems oxygenated via hydrogen peroxide. Anal. Chim. Acta. In press.
5. MILLER, G. L., R. BLUM, W. E. GLENNON & A. L. BURTON. 1960. Measurements of carboxymethylcellulase activity. Anal. Biochem. **2:** 127–132.

Metabolic Behavior of Immobilized Cells—Effects of Some Microenvironmental Factors[a]

BO MATTIASSON,[b] MATS LARSSON,[b] AND
BÄRBEL HAHN-HÄGERDAL[c]

[b]*Department of Pure and Applied Biochemistry*
[c]*Department of Applied Microbiology*
Chemical Center, University of Lund
P.O. Box 740
S-220 07 Lund, Sweden

Variations in microenvironmental conditions have been demonstrated to be important on the subcellular level. Model studies on immobilized enzymes have helped to clarify the picture. However, when cells have been immobilized, changed metabolic behavior has been reported in several cases, but very little is known about the mechanisms behind these changes. The new environment created in the immobilization process differs from that in free solution, for example, by the presence of high concentrations of polymers and by impeded diffusion of reactants (nutrients). The presence of polymer can be important in the sense that these molecules can directly interact with the cells, but probably the most overlooked effect is that polymers cause a change in water activity of the medium. When setting up a system to study the effects of the microenvironment on the metabolic behavior of cells, it may be useful to simplify the system. Suspensions of cells in water solutions of polymers represent such a simplified model.

In the present study, polyethylene glycol (PEG) was used to study the effects of a decreased water activity on cell metabolism, thus simulating the behavior of immobilized cells. In order to prepare suspensions with a defined water activity, methods for calculating the reduction in water activity for buffered polymer-rich media have been developed.[1] The water activity is defined as:

$$A_w = \frac{P}{P_0}$$

where P = vapor pressure of water over the solution and P_0 = vapor pressure over pure water. When a solution becomes more concentrated, the vapor pressure decreases and A_w falls. A_w in a PEG-solution was calculated according to Norrish:[2]

$$A_w = X_1 \exp(-K X_2^2)$$

where X_1 and X_2 are molar fractions of water and PEG, respectively, and K is calculated from the equation given by Chirife and Fontan:[3]

$$K = 1.6\,(MW/100)^2$$

[a]This work was supported by The Swedish Natural Science Research Council and the Swedish Board for Technical Development.

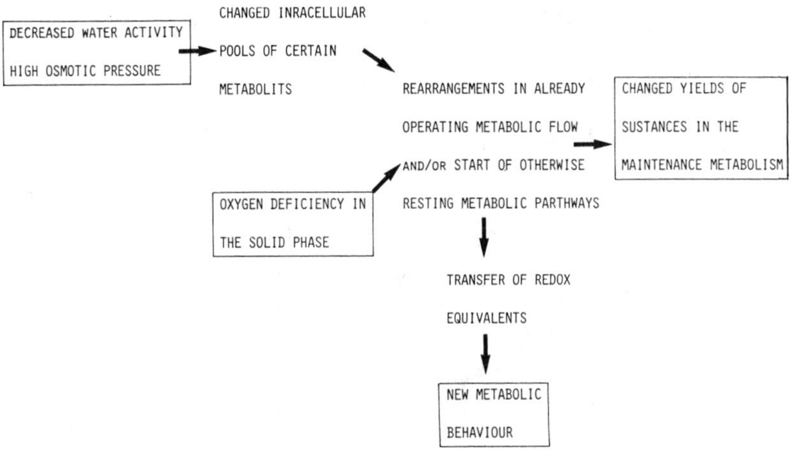

FIGURE 1. Influence of water activity and oxygen supply on metabolism of immobilized cells.

The relationship between osmolarity and water activity is given by:

$$\text{osmolarity} = 55.51/A_w - 55.51$$

Osmolarity is an additive property. It is therefore possible to calculate the osmolarity for all the individual components in a medium, add them up and calculate a water activity close to the real value by using the equation above.

Based on literature data and on some experiments of our own, a hypothesis was formulated concerning the mechanisms by which microenvironmental factors affect the cell metabolism[4] (see FIG. 1). Subsequent experiments have been made to test the validity of the hypothesis. Some of the results are presented here. The fundamental idea in the hypothesis is that the conditions offered by the microenvironment influence the physiological status of the cell. This change then affects the observed metabolic behavior of the cells. One well-known system that has been studied earlier is that of the alga *Dunaliella*. This organism can live in 4 *M* NaCl as well as in a medium of physiological ionic strength. To compensate for the high salt in the environment, these cells build up an elevated level of glycerol intracellularly. At high concentrations of the polyalcohol, a new set of enzymes are induced, which dramatically change the flow patterns in the metabolic network. Furthermore, at elevated osmotic pressures, the

TABLE 1. Ethanol Production Rate in Batch Fermentations in the Presence of Dextran T10 or PEG 1540

Additive	Conc. % (wt/wt)	A_w	Ethanol Concentration (g/l) 6 Hours	Ethanol Concentration (g/l) 50 Hours
Blank	—	0.994	6.21	26.91
Dextran T10	9.5	0.993	7.82	27.68
Dextran T10	18.7	0.992	8.06	26.70
PEG 1540	12.5	0.990	8.98	28.43
PEG 1540	21.6	0.986	9.64	27.75

maintenance metabolism requires a larger share of the total metabolic throughput of the cell. This also influences the observed metabolic behavior.

In batch experiments with *Saccharomyces cerevisiae* in a medium of glucose (60 g/l), yeast extract (2.5 g/l) and inorganic salts supplemented with different water soluble polymers, some rather drastic effects were observed.[1] In the system studied, ethanol was analyzed using gas chromatography. From TABLE 1 it is seen that a marked increase in ethanol production was obtained in the initial phase of the experiment, whereas the end values were very much the same. Since these were batch experiments, they have no practical value *per se* because in batch experiments a total conversion must be obtained. However, the results clearly point towards the possibility of using water activity as a control parameter when setting up biotechnical processes. In a similar study, it was found that a medium with decreased water activity switched its metabolism from xylitol to ethanol production for pentos fermentation with *Candida tropicalis.*[5]

In another set of experiments, the earlier-mentioned halotolerant alga *Dunaliella parva* was studied.[6] The alga was grown in a medium of inorganic salts at a

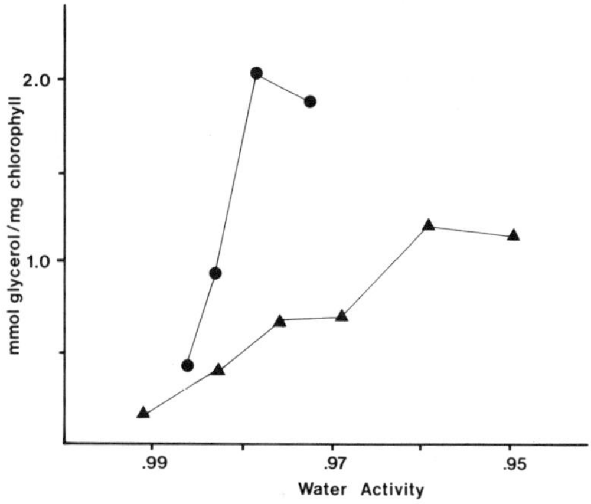

FIGURE 2. Glycerol concentration in medium with (▲) NaCl or (●) PEG 600 as water activity reducing agent.

temperature of 26°C under constant light. The glycerol content in the medium was analyzed after the cells were removed by centrifugation. Intracellular glycerol was measured after heat treatment of the cells. The glycerol assay was based on the use of glycerol dehydrogenase (E.C.1.1.1.6), which converts glycerol to dihydroxyacetone with the concomittant conversion of NAD^+ to $NADH + H^+$. The chlorophyll content was analyzed spectrophotometrically in acetone/water (8/2 vol/vol) solution. A plot of the glycerol concentration in the medium as a function of water activity after a fixed period of incubation is shown in FIGURE. 2. When either NaCl or PEG was used in order to reduce the water activity, a changed metabolic behavior was observed. To eliminate the risks of unwanted side effects of stabilizing agents that might be present in PEG, high quality PEG, intended for cell biology experiments, was used. Alternatively, the PEG was purified by repeated crystallization from ethanol before use. PEG turned out to be much more efficient than NaCl in altering the metabolism of these cells. This indicates that the effects observed are not only a result of the changes in water activity. Further investigations on this subject are necessary.

The conclusions that might be drawn so far are: (1) the effects of the aqueous polymer medium on the metabolism of cells is consistent with that reported for immobilized cells; (2) changes in the water activity in the microenvironment influence the metabolic behavior of the cells; and (3) water activity may be used as a control parameter in biotechnical processes.

REFERENCES

1. HAHN-HÄGERDAL, B., M. LARSSON & B. MATTIASSON. 1982. Shift in metabolism towards ethanol production in *Saccharomyces cerevisiae* using alterations of the physical-chemical microenvironment. Biotechnol. Bioeng. Symp. No. **12:** 199–202.
2. NORRISH, R. S. 1966. An equation for the activity coefficients and equilibrium relative humidities of water in confectionary syrup. J. Food Technol. **1:** 25–39.
3. CHIRIFE, J. & C. FERRO FONTAN. 1980. A study of the water activity-lowering behavior of polyethylene glycols in the intermediate moisture range. J. Food Sci. **45:** 1717–1719.
4. MATTIASSON, B. & B. HAHN-HÄGERDAL. 1982. Microenvironmental effects on metabolic behaviour of immobilized cells, a hypothesis. Eur. J. Appl. Microbiol. Biotechnol. **16:** 52–55.
5. LOHMEIER-VOGEL, E., B. JÖNSSON & B. HAHN-HÄGERDAL. Shifting product formation from xylitol to ethanol in pentose fermentation with *Candida tropicalis* by altering environmental parameters. Submitted.
6. RESLOW, M., A. PERSSON, M. LARSSON & B. MATTIASSON. The influence of water activity on glycerol production by *Dunaliella parva.* In preparation.

Various Applications of Living Cells Immobilized by Prepolymer Methods

ATSUO TANAKA, KENJI SONOMOTO, AND SABURO FUKUI

Department of Industrial Chemistry
Faculty of Engineering
Kyoto University
Yoshida, Sakyo-ku
Kyoto 606, Japan

Entrapment is the most attractive method of immobilizing not only single enzymes but also multiple enzymes, cellular organelles, microbial cells, plant cells, and animal cells. In these cases, physicochemical properties as well as the network structure of the gels used have significant effects on the apparent activities and stabilities of entrapped biocatalysts. We have developed new entrapping methods by using synthetic prepolymers of photocrosslinkable resins[1] and urethane resins,[2] and have applied the entrapped biocatalysts for the production of useful compounds in aqueous systems, water-organic cosolvent systems, and organic solvent systems.[3–5] The specific features of these prepolymer methods are that not only is immobilization carried out under simple and mild conditions, but also the network structure as well as physicochemical properties of the gels can be adjusted according to the desired purposes by selecting the proper prepolymers. Application of immobilized biocatalysts is now being expanded to more complicated reactions using living or growing cells, which provide self-proliferating and self-regenerating catalytic systems. In this paper, we report several examples of immobilized living or growing cells prepared by the prepolymer methods (TABLE 1). As described below, living or growing cells entrapped with photocrosslinkable resin prepolymers or urethane prepolymers have been found to be useful catalysts for multistep reactions.

HYDROXYLATION OF STEROIDS

Fungal spores, such as those of *Rhizopus stolonifer, Curvularia lunata,* and *Sepedonium ampullosporum,* were entrapped with photocrosslinkable resin prepolymers and allowed to germinate inside gel matrices. Entrapped mycelia thus prepared hydroxylated steroids at a specific position in aqueous reaction systems containing suitable water-miscible organic cosolvents (11α-hydroxylation of progesterone,[6] 11β-hydroxylation of cortexolone,[7,8] and 16α-hydroxylation of estrone). A close relationship was observed between the growth of mycelia and the activities of the gel-entrapped mycelia, both of which were markedly influenced by the network structure of the gels, which can be controlled by the chain length of the prepolymers employed (FIG. 1).

Living cells of *Corynebacterium* sp. were used for 9α-hydroxylation of 4-androstene-3,17-dione (4-AD).[9] In this case, the desired product, 9α-OH-4-AD, was an intermediate in the metabolic pathway of the substrate, and its release from gels to the external solvent before further metabolism was essential. The hydrophilicity-hydrophobicity balance of the gels has a significant effect on this phenomenon. In the

TABLE 1. Application of Living or Growing Cells Entrapped with Prepolymers

Cells (condition)	Application	Reference
Corynebacterium sp. (living)	9α-Hydroxylation of steroid	Sonomoto *et al.*[9]
Rhizopus stolonifer (living)	11α-Hydroxylation of steroid	Sonomoto *et al.*[6]
Curvularia lunata (living)	11β-Hydroxylation of steroid	Sonomoto *et al.*[7,8]
Sepedonium ampullosporum (living)	16α-Hydroxylation of steroid	—
Propionibacterium sp. (growing)	Production of vitamin B_{12}	Yongsmith *et al.*[10]
Streptomyces rimosus (growing)	Production of oxytetracycline	—
Methanogenic bacterium (growing)	Production of methane	Anasawa & Beppu[11]
Yeast (growing)	Production of ethanol	Nojima[12]
Lavandula vera (growing)	Production of pigments	—

aqueous reaction system employed, hydrophilic gels were found to be preferable. Introduction of dimethyl sulfoxide (15%) to the reaction system facilitated the extraction of 9α-OH-4-AD from gels having a hydrophobic nature as well as a hydrophilic nature. In this case, the cells entrapped in either type of gel gave similar conversion activities.

PRODUCTION OF VITAMIN B_{12}

Vitamin B_{12}, having a complicated structure, was produced *de novo* by a strain of *Propionibacterium* sp. entrapped with a suitable urethane prepolymer or photocross-linkable resin prepolymer. Addition of the optimal concentrations of cobaltous ion and 5,6-dimethyl benzimidazole to the medium enhanced the vitamin production by the entrapped cells. The polyurethane-entrapped cells could be utilized for several successive batches of fermentation.[10]

PRODUCTION OF OXYTETRACYCLINE

Streptomyces rimosus growing inside polyurethane gels accumulated oxytetracycline in the fermentation medium (FIG. 2). The entrapped cells were active for at least two months. A slow release system for a drug might be constructed by implanting such

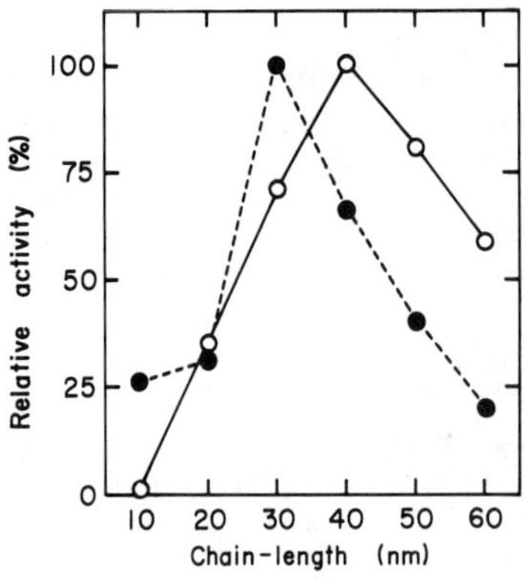

FIGURE 1. Effect of prepolymer chain length on the steroid hydroxylation activities of entrapped mycelia. (○), 11β-hydroxylation by *Curvularia lunata;* (●), 11α-hydroxylation by *Rhizopus stolonifer.* Hydrophilic photocrosslinkable resin prepolymers (ENT) used: 10 nm, ENT-1000; 20 nm, ENT-2000; 30 nm, an equi-weight mixture of ENT-2000 and ENT-4000; 40 nm, ENT-4000; 50 nm, an equi-weight mixture of ENT-4000 and ENT-6000; 60 nm, ENT-6000.

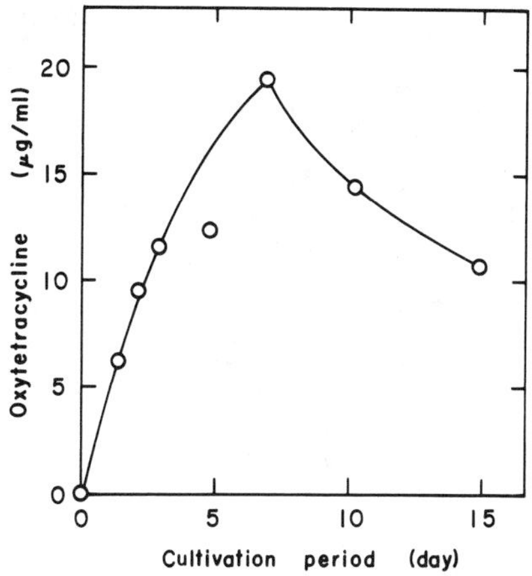

FIGURE 2. Production of oxytetracycline by *Streptomyces rimosus* cells entrapped with a urethane prepolymer.

entrapped cells in animals. The polyurethane gels were found not to be harmful for animals when implanted.

ENTRAPMENT OF PLANT CELLS

Immobilization of plant cells, using *Lavandula vera* as an example, was tried using urethane prepolymers and photocrosslinkable resin prepolymers. Of the prepolymers examined, the cells entrapped with a urethane prepolymer giving hydrophobic gels of suitable network structure grew inside gels, the growth being identified by the increase in the oxygen uptake (TABLE 2). The entrapped cells accumulated blue pigments when L-cysteine was added to the medium.

PRODUCTION OF BIOFUELS

A methanogenic bacterium entrapped with a urethane prepolymer produced methane in the presence of formate.[11] Our photocrosslinkable resin prepolymer method has been applied to the immobilization of living yeast cells for the continuous production of ethanol by the Research Association for Petroleum Alternatives Development (RAPAD) in Japan. Pilot-scale production of alcohol from molasses is being carried out using reactors equipped with photocrosslinked films containing growing yeast cells.[12]

TABLE 2. Oxygen Uptake of *Lavandula vera* Cells[a]

Cells	Q_{O_2} (μmol · h^{-1} · $system^{-1}$) Cultivated for: 0 Week	3 Weeks
Free	6.13	80.6
Ca-Alginate-entrapped	2.87	45.0
Polyurethane-entrapped	nil	56.0

[a]System, 25 ml medium.

REFERENCES

1. FUKUI, S., A. TANAKA, T. IIDA & E. HASEGAWA. 1976. FEBS Lett. **66:** 179.
2. FUKUSHIMA, S., T. NAGAI, K. FUJITA, A. TANAKA & S. FUKUI. 1978. Biotechnol. Bioeng. **20:** 1465.
3. FUKUI, S., K. SONOMOTO, N. ITOH & A. TANAKA, 1980. Biochimie **62:** 381.
4. FUKUI, S. & A. TANAKA. 1982. Enzyme Engineering **6:** 191.
5. FUKUI, S. & A. TANAKA. 1984. Adv. Biochem. Eng./Biotechnol. **29:** 1.
6. SONOMOTO, K., K. NOMURA, A. TANAKA & S. FUKUI. 1982. Eur. J. Appl. Microbiol. Biotechnol. **16:** 57.
7. SONOMOTO, K., M. M. HOQ, A. TANAKA & S. FUKUI. 1981. J. Ferment. Technol. **59:** 465.
8. SONOMOTO, K., M. M. HOQ, A. TANAKA & S. FUKUI. 1983. Appl. Environ. Microbiol. **45:** 436.
9. SONOMOTO, K., N. USUI, A. TANAKA & S. FUKUI. 1983. Eur. J. Appl. Microbiol. Biotechnol. **17:** 203.
10. YONGSMITH, B., K. SONOMOTO, A. TANAKA & S. FUKUI. 1982. Eur. J. Appl. Microbiol. Biotechnol. **16:** 70.
11. ANASAWA, H. & T. BEPPU. 1982. Abstr. Annu. Meet. Agric. Chem. Soc. Jpn. p. 411.
12. NOJIMA, S. 1983. Chem. Econ. Eng. Rev. **15**(4): 17.

Immobilization of Yeast Cells on Transition Metal-activated Pumice Stone[a]

J. M. S. CABRAL, M. M. CADETE, J. M. NOVAIS, AND J. P. CARDOSO

Laboratório de Engenharia Bioquímica
Instituto Superior Técnico
1000 Lisboa, Portugal

The immobilization of whole cells is a matter of increasing practical and research concern for biotechnological applications and provides an alternative to fermentation and intracellular enzyme immobilization. A simple process for immobilizing biocatalysts on inorganic carriers is the metal-link method and its variations.[1-4] In this communication, a study of the immobilization of *Saccharomyces cerevisiae* (bakers' yeast) cells with invertase activity onto transition metal-activated pumice stone is reported.

Pumice stone, a volcanic material (mainly 67% SiO_2 and 15% Al_2O_3) from the Azores Islands, previously pretreated with acid and having a mean particle size of 300 μm and a surface area of 5.7 m^2/g, was chosen as the cell support.[2] Yeast cells were immobilized onto pumice stone by the metal-link method and some of its variations. In the first step, the pumice stone was activated by reaction of the transition metal salt, M Cl_x, with $pumic_e$ SiOH groups for 30 hours at 45°C and dried to give the oxychloride derivative plus HCl. The coupling steps are summarized in TABLE 1.

The results of the hydrolysis of sucrose (2% wt/vol) at pH 4.5 and 30°C, using immobilized yeast cells, are shown in TABLES 2 and 3 and FIGURE 1. From these results, several conclusions may be drawn. The yeast cells chelated on hydrous metal oxide derivatives of pumice stone displayed higher initial invertase activity than on other derivatives. The introduction of an organic bridge between the cells and the metal activator led to a decrease of the initial invertase activity of the immobilized cells (TABLES 2 and 3). The operational stability and the initial activity of yeast cells immobilized on hydrous metal oxide derivatives of pumice stone seem to be dependent on the valence of the metal activator. Metal activators with high valence (+3, +4) led to higher half-lives than metals with low valence (+2), possibly due to a stronger metal interaction with the silanol groups of the carrier,[4] and thus to a more stable metal oxide layer on the pumice stone (TABLE 2). The low level of initial invertase activity and the unstable preparation of yeast cells immobilized on vanadium(III)-activated pumice stone were possibly due to inhibition of yeast invertase by vanadium(III) (TABLE 2). The ionic linkage between the carboxyl derivative and the yeast cells led to a comparatively unstable immobilized cell preparation. However, *Saccharomyces cerevisiae* cells immobilized by covalent linkage on the carbonyl derivative of titanium(IV)-activated pumice stone displayed a very stable behavior. In continuous operation the covalently coupled preparation showed only a slight decrease in invertase activity over a two-month period (TABLE 3 and FIG. 1).

[a]Financial support was received from the Instituto Nacional de Investigação Científica, Lisboa, Portugal.

TABLE 1. Derivatization of Transition Metal-activated Pumice Stone Oxychloride Derivative (—SiOM(OH)$_x$Cl$_y$) and Coupling of Whole Cells

Activator	Reaction Conditions	Derivative	Cell Coupling Mechanism
H_2O		—SiOM(OH)$_n$ (Hydrous metal oxide)	—SiOM ← CELL, (OH)$_{n-1}$ (Chelation)
H_2O + tannic acid	45°C, 1 h	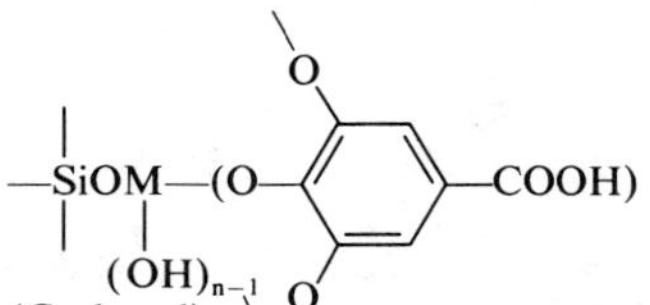 (Carboxyl)	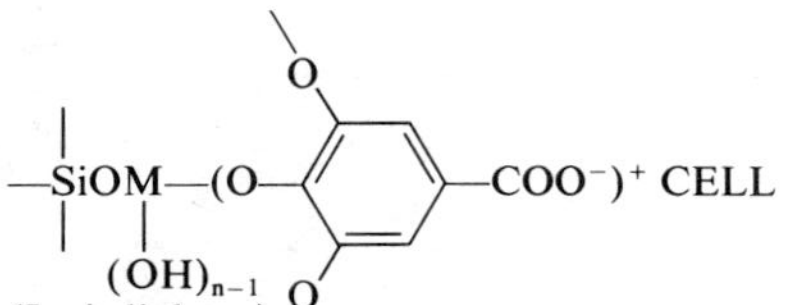 (Ionic linkage)
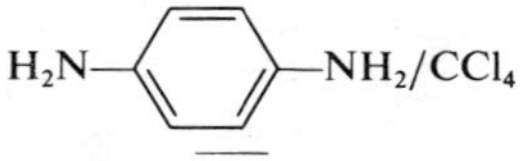	45°C, 1 h	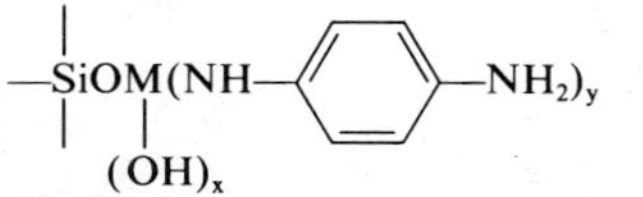 (Arylamine)	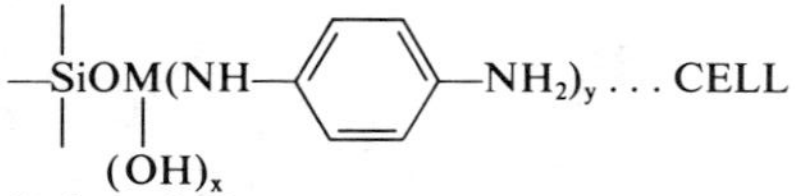 (Adsorption)
$H_2N(CH_2)_6NH_2/CCl_4$	45°C, 1 h	—SiOM(NH(CH$_2$)$_6$NH$_2$)$_y$, (OH)$_x$ (Alkylamine)	—SiOM(NH(CH$_2$)$_6$NH$_2$)$_y$ . . . CELL, (OH)$_x$ (Adsorption)
$H_2N(CH_2)_6NH_2/CCl_4$ + + OHC(CH$_2$)$_3$CHO	45°C, 1 h + 25°C, 1 h, pH 7	—SiOM(NH(CH$_2$)$_6$N=CH(CH$_2$)$_3$CHO)$_y$, (OH)$_x$ (Carbonyl)	—SiOM(NH(CH$_2$)$_6$N=CH(CH$_2$)$_3$CH=N—CELL)$_y$, (OH)$_x$ (Covalent linkage)

TABLE 2. Long-term Batch Operational Stability of *Saccharomyces cerevisiae* Cells Immobilized on Hydrous Metal Oxide Derivatives of Pumice Stone

Metal Salt	Initial Invertase Activity ($IU\ g^{-1}$ matrix)	Half-life (h)
$CoCl_2$	299	122
$FeCl_2$	241	112
$SnCl_2$	316	125
$AlCl_3$	502	108
$FeCl_3$	219	140
VCl_3	161	42
$SnCl_4$	517	158
$TiCl_4$	386	144
$ZrCl_4$	294	145

TABLE 3. Long-term Batch Operational Stability of *Saccharomyces cerevisiae* Cells Immobilized on Several Derivatives of Titanium(IV)-activated Pumice Stone

Derivative	Initial Invertase Activity ($IU\ g^{-1}$ matrix)	Half-life (h)
Alkylamine	201	115
Arylamine	104	216
Carbonyl	190	424
Carboxyl	177	79

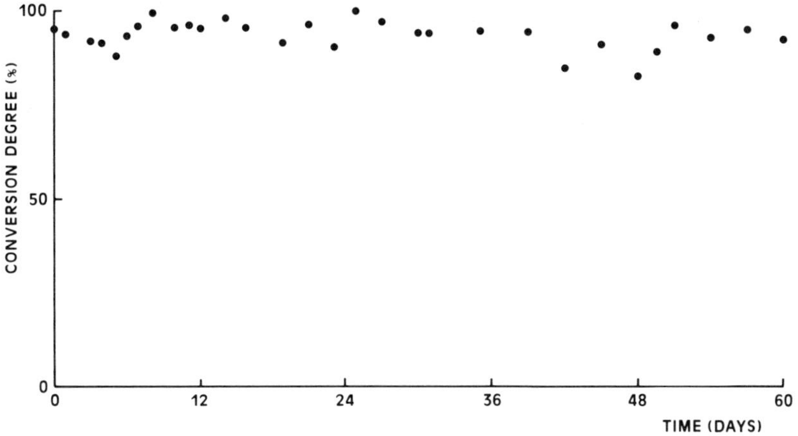

FIGURE 1. Long-term operational stability of *Saccharomyces cerevisiae* cells with invertase activity immobilized by covalent coupling on a carbonyl derivative of titanium (IV)-activated pumice stone.

REFERENCES

1. CABRAL, J. M. S., J. M. NOVAIS & J. P. CARDOSO. 1981. Immobilization of amyloglucosidase on alkylamine derivatives of metal-link-activated inorganic supports. Biotechnol. Bioeng. **29**(9): 2083–2092.
2. DIAS, S. M. M., J. M. NOVAIS & J. M. S. CABRAL. 1982. Immobilization of yeast cells on titanium-activated inorganic supports. Biotechnol. Lett. **4**(4): 203–208.
3. CABRAL, J. M. S., J. M. NOVAIS, J. F. KENNEDY & J. P. CARDOSO. 1983. Immobilization of biocatalysts on new route transition metal-activated inorganic supports. Enzyme Microb. Technol. **5**(1): 30–32.
4. CABRAL, J. M. S., J. F. KENNEDY & J. M. NOVAIS. 1982. Investigation of the binding mechanism of glucoamylase to alkylamine derivatives of titanium(IV)-activated porous inorganic supports. Enzyme Microb. Technol. **4**(4): 337–342.

Immobilization of Plant Protoplasts

L. LINSE AND P. BRODELIUS[a]

Department of Pure and Applied Biochemistry
Chemical Center, University of Lund
P.O. Box 740
S-220 07 Lund, Sweden

During recent years, many successful experiments have been carried out on the immobilization of plant cells in various polymers.[1,2] Protoplasts lacking cell walls are very sensitive to mechanical stress. However, it is possible to stabilize these fragile cells by immobilization. This has been demonstrated with *Vicia faba* protoplasts in alginate beads.[3] Here it was possible to build up turgor pressure intracellularly, using the alginate gel instead of the cell wall as the counter-pressure.

We have used cells of *Daucus carota* Ca68 (the culture was kindly supplied by Dr. I. A. Veliky, National Research Council, Canada) as a model system to study immobilization of protoplasts in polymer gels. *D. carota* Ca68 cells perform a 5-β-hydroxylation of the steroidal compound digitoxigenin to periplogenin, according to the reaction shown in FIGURE 1. The immobilization of the intact cells in alginate and the hydroxylation capacity of these immobilized cells have been studied extensively.[4,5] The hydroxylation reaction is dependent on viability of the cells.[5] It appears as if the plasma membrane has to be intact for enzyme activity to occur. We have used this enzymatic reaction to monitor the intactness and viability of the protoplasts.

The immobilization has been carried out according to a new "two-phase" method.[6] The principle is that a polymer-cell/protoplast suspension is suspended in an inert hydrophobic phase in order to form droplets. The polymer is subsequently induced to form a gel. Spherical beads with entrapped cells/protoplasts are thus readily obtained.

The stability of immobilized and freely suspended protoplasts towards mechanical stress has been compared by exposing the protoplast preparations to media of different osmolarities and by using them in repeated hydroxylation experiments.

PROTOPLAST IMMOBILIZATION

Cells from a two-day-old suspension culture of *D. carota* Ca68, cultivated in 71v medium, were collected on a nylon net (100 μm) and washed with medium A (0.5 *M* sorbitol and 50 m*M* $CaCl_2$, pH 5.6–5.8). Preplasmolysis in medium A was carried out for 30 min and then cellulase (0.5% wt/vol) and pectinase (0.5% wt/vol) were added. After approximately 16 h, the protoplast suspension was filtered through a nylon net (45 μm) and centrifuged at 100 $\times$ *g*. The protoplasts were then washed twice with medium A and finally the concentration was adjusted to 2×10^7 protoplasts/ml.

Protoplast suspension (5 ml) was mixed with carrageenan (10 g; 3% wt/vol in 0.5 *M* sorbitol) at 30°C, and the mixture was added under continuous stirring to vegetable oil (60 ml) kept at the same temperature. The emulsion was cooled to 10°C whereupon the carrageenan solidified. Spherical beads were obtained, and by varying the stirring

[a]Present address: Institute of Biotechnology, Swiss Federal Institute of Technology, Hönggerberg, CH-8093 Zürich, Switzerland.

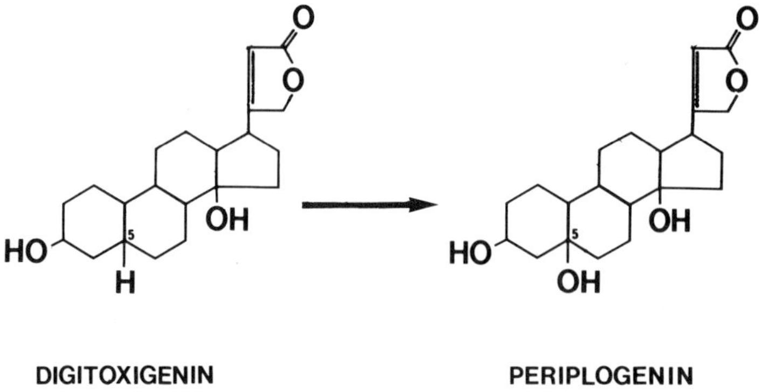

FIGURE 1. The hydroxylation reaction catalyzed by free and immobilized cells and protoplasts of *Daucus carota.*

speed during preparation the bead size may be adjusted (0.2–2 mm). The formed beads were collected, washed and incubated in 0.3 *M* KCl (in medium A) for 30 min to increase the mechanical stability of the gel. The beads were then sieved through metal screens and the fraction 0.5 to 1 mm was used in the experiments.

Immobilized protoplasts are shown in FIGURE 2. In FIGURE 2A, the edge of the bead can be seen. As indicated by fluorescein diacetate staining (FIG. 2B), the immobilized protoplasts are fully viable.

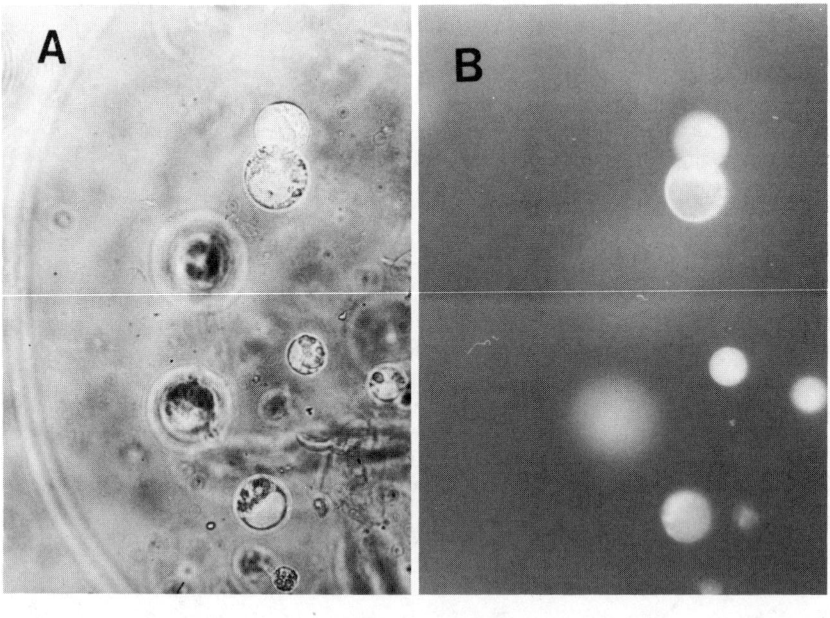

FIGURE 2. (A) Micrograph of immobilized protoplasts of *Daucus carota.* (B) The same preparation after staining with fluorescein diacetate.

Immobilization of protoplasts in agarose was carried out in the same manner with the exceptions that the polymer solution had to be heated to 35°C and that no incubation in the KCl solution was required.

The hydroxylation capacity of protoplasts was not affected by heat treatment (5 min) up to 40°C. The immobilization, however, was carried out at a lower temperature (30–35°C) and therefore the hydroxylation capacity of the immobilized protoplasts is not affected to any great extent.

LYSIS EXPERIMENTS

Protoplasts (0.25 ml suspension or 0.75 g beads) were incubated in various media (5 ml) for 15 min on a rotary shaker (125 rpm). The sorbitol concentration in these media varied between 0 and 0.5 *M* and the calcium concentration was 10, 25, or 50 m*M*. The viability of the protoplasts was then tested in a 5-β-hydroxylation experiment. The free or immobilized protoplasts were transferred to medium A (5 ml) and subsequently the hydroxylation was initiated by the addition of digitoxigenin (125 μg). After 18 hours on a rotary shaker (125 rpm, 25°C), the protoplasts were removed by

TABLE 1. Relative 5-β-Hydroxylase Activity of Free and Immobilized Protoplasts after Osmotic Shock Treatment

Sorbitol Concentration During Treatment (M)	5-β-Hydroxylase Activity (%) 10 m*M* Ca^{2+} Free	10 m*M* Ca^{2+} Immob.	25 m*M* Ca^{2+} Free	25 m*M* Ca^{2+} Immob.	50 m*M* Ca^{2+} Free	50 m*M* Ca^{2+} Immob.
0.50	100	100	70	100	80	100
0.20	56	89	100	98	86	96
0.10	0	55	47	78	94	84
0.05	0	19	19	65	100	78
0.00	0	5	0	24	80	65

centrifugation and the supernatant was analyzed by reversed-phase HPLC. The following isocratic system was used: LiChrosorb RP-18 (5 μm, 130 × 4.6 mm); mobile phase: methanol : phosphate buffer (10 m*M*, pH 7.0) (6 : 4); flow rate: 1.1 ml/min; sample size: 10 μl; detection: UV (220 nm).

The results of these lysis experiments are summarized in TABLE 1. At the highest calcium concentration (50 m*M*) the sorbitol concentration appeared to have a very small effect, but at lower calcium concentrations (10 or 25 m*M*) the viability of the protoplasts decreased with decreasing sorbitol concentration. The immobilized protoplasts showed a considerably increased tolerance to osmotic shock as compared to the corresponding free protoplasts. Lysis experiments with protoplasts immobilized in agarose have also been carried out. The results obtained resemble those observed with carrageenan.

REPEATED HYDROXYLATION EXPERIMENTS

To study the viability over an extended period of time, the protoplast suspension (0.67 ml) or the beads (2.0 g) were used in four consecutive hydroxylation experiments. The experiments were carried out as described for the lysis experiments with the exceptions that the amount of free or immobilized protoplasts used was higher and that the reaction time was 2 hours.

Between the experiments, the product and the unreacted substrate were removed by washing. From the results summarized in TABLE 2, it can be concluded that the immobilized protoplasts are stabilized against mechanical stress.

TABLE 2. Relative 5-β-Hydroxylase Activity after Repetitive Use of Free and Immobilized Protoplasts

	5-β-Hydroxylase Activity (%)	
Batch Number	Free	Immobilized
1	100	100
2	72	86
3	53	78
4	40	77

CONCLUSIONS

Using the two-phase technique, it is possible to immobilize protoplasts in spherical beads of various sizes under mild conditions. The protoplasts remain viable and appear to be stabilized by the polymeric matrix. Consequently, immobilized protoplasts can be handled more easily and used for various purposes, such as studies on cellular processes and regeneration, production of metabolites, and bioconversions.

REFERENCES

1. BRODELIUS, P. & K. MOSBACH. 1982. Immobilized plant cells. *In* Advances in Applied Microbiology. A. Laskin, Ed. Vol. **28:** 1–26. Academic Press. New York.
2. BRODELIUS, P. 1983. Immobilized plant cells. *In* Immobilized Cells and Organelles. B. Mattiasson, Ed. Vol. **1:** 27–55. CRC Press. Boca Raton, FL.
3. SCHEURICH, P., H. SCHNABL, U. ZIMMERMAN & J. KLEIN. 1980. Immobilisation and mechanical support of individual protoplasts. Biochim. Biophys. Acta **598:** 645–651.
4. JONES, A. & I. A. VELIKY. 1981. Effect of medium constituents on the viability of immobilized plant cells. Can. J. Bot. **59:** 2095–2101.
5. JONES, A. & I. A. VELIKY. 1981. Examination of parameters affecting the 5-β-hydroxylation of digitoxgenin by immobilized cells of *Daucus carota.* Eur. J. Appl. Microbiol. Biotechnol. **13:** 84–89.
6. NILSSON, K., S. BIRNBAUM, S. FLYGARE, L. LINSE, U. SCHRÖDER, U. JEPPSSON, P.-O. LARSSON, K. MOSBACH & P. BRODELIUS. 1983. A general method for the immobilization of cells with preserved viability. Eur. J. Appl. Microbiol. Biotechnol. **17:** 319–326.

Growth and Cardenolide Production by *Digitalis lanata* in Different Fermentor Types

P. MARKKANEN, T. IDMAN, AND V. KAUPPINEN

Helsinki University of Technology
Laboratory of Biochemistry and Food Technology
Kemistintie 1
02150 Espoo 15, Finland

Production of secondary metabolites with suspension cultures of plant cells is a fascinating research area, although only a few feasible processes have been developed. One group of potential products are cardenolides produced by *Digitalis* species.

The production of cardenolides has been studied by many groups. A significant hindrance to progress has been the use and interpretation of different analysis methods, although adoption of RIA (radioimmunoassay)[1] partly solved this problem. It was long believed that *Digitalis* strains could not produce cardenolides in suspension cultures. However, we now know that cardenolides can be produced by differentiating cells of *D. lanata*,[2,3] although the yields have been less than those obtained from whole plants. Zenk[4] used *D. lanata* cells for bioconversion when producing digoxin.

MATERIALS AND METHODS

The *Digitalis lanata* strain used was kindly donated by L. Björk (Royal Inst. Technol., Stockholm, Sweden). Cell mass was estimated by measurements of wet or dry weight (mean dry wt 8.2%).

Modified Murashige-Skoog medium was used in all cultivations. The sucrose concentration was 3% (wt/vol), the temperature was 24°C, and the pH was 6.0 before sterilization. Benzyladenine (0.2 mg/l) was used as growth hormone.

Both radioimmunoassay (RIA) and high-pressure liquid chromatography (HPLC) were used for cardenolide assays. Digitoxin was determined according to the method of Vogel[5] and digoxin using the reagent kit of Farmos Diagnostica (Turku, Finland). HPLC analyses were performed according to the gradient elution method of Wichtl *et al.*[6] Extracts for RIA were made as described by Weiler and Zenk[1] with 20% ethanol, and for HPLC according to the procedure of Wichtl *et al.*[6]

RESULTS

Experiments were carried out in a mechanically stirred laboratory fermentor (5 dm^3) equipped with different agitator systems (FIG. 1), in an air-lift fermentor (FIG. 2), and in a 1 dm^3 Kluyver flask (FIG. 3). Some experiments were also performed in shake flasks (100 cm^3 medium in 250-cm^3 flasks).

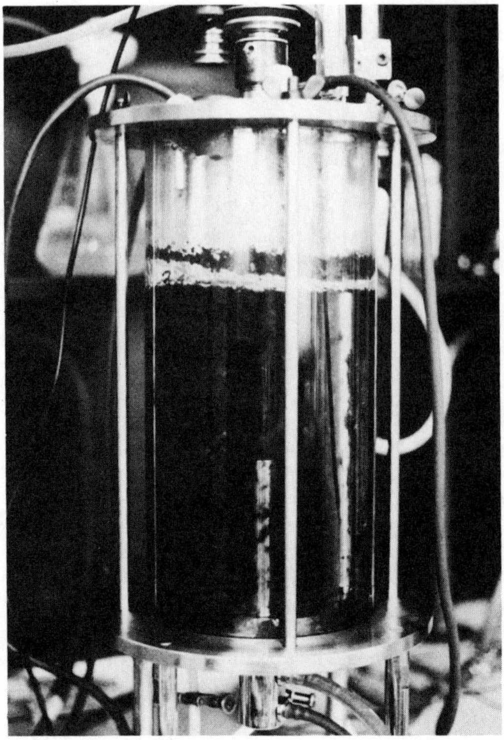

FIGURE 1. Mechanically stirred laboratory fermentor.

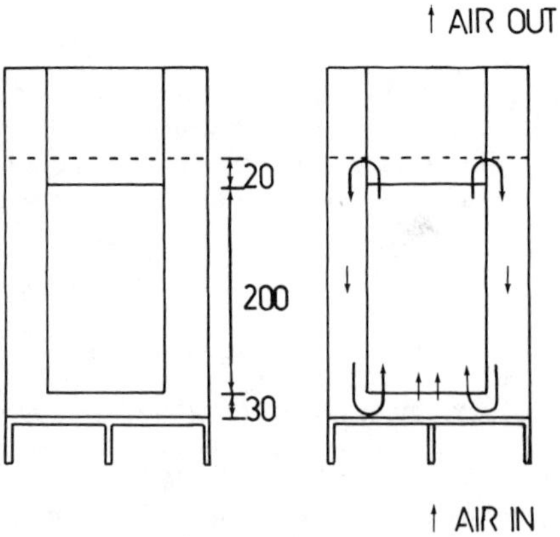

FIGURE 2. Air-lift fermentor.

FIGURE 3. Kluyver flask.

FIGURE 4. Agitation system for the mechanically stirred fermentor.

TABLE 1. Cardenolide Concentrations in Different Cultivations[a]

	HPLC Cardenolide pmol/mg (dry wt)					RIA pmol eq/mg (dry wt)	
	Digitoxin	Lanat. A	Git.	Digitoxig.	Digoxin	Digitoxin	Digoxin
1	0.68	20	22	18	—	54	—
2	0.85	35	2.3	1.7	—	174	3.3
3	0.34	2.3	0.67	1.0	—	12	0.82
4	—	4.8	1.3	0.35	—	72	—
5	—	14	7.8	4.6	—	49	—
6	—	24	—	4.1	4.0	63	—
7	3.1	19	2.0	19	—	92	—

[a](1, 2) Mechanically stirred fermentor; (3, 4) Kluyver flask (no. 4 with gibberellic acid); (5) air-lift fermentor; and (6, 7) shake flasks (no. 6, 14 d; no. 7, 21 d).

Mechanically Stirred Fermentations

The turbine-type agitator could not be used because plant cells do not tolerate high agitation speeds. We tested many agitation systems and that shown in FIGURE 4 seemed suitable, giving an upward flow and maintaining homogeneity. The agitation speed was 1.5 sec^{-1} and the aeration rate 1 vol/vol/min. Results from two cultivations are given in TABLES 1 and 2.

Cells were grown in aggregates (0.2–2 cm). Within three weeks, the culture medium turned brown and some of the cardenolides were excreted into the medium.

Air-lift Fermentations

Kluyver flask cultivations were performed using 650 ml of culture broth. The homogeneity of the culture was easy to maintain by changing the aeration rate depending on cell mass: 0.4, 0.5, and 0.6 1/min for the first, second, and third weeks, respectively. Gibberellic acid (5 mg/1) increased cell mass and doubled cardenolide production. The aggregates were bright green and seemed much better than in the mechanically stirred fermentor. Cell differentation was also more obvious: shoots were clearly visible. Cells grew in large aggregates, but were not harmed by the mixing. Results are given in TABLES 1 and 2.

Fermentations in a Draft Tube Air-lift Fermentor

A laboratory fermentor equipped with a draft tube gave rise to cell concentrations as high as in the mechanically stirred fermentor but cardenolide yields were lower

TABLE 2. Growth Parameters of *Digitalis lanata*

	Inoculum (wet wt/g)	Yield (dry wt/gdm^{-3})	$Y \cdot \frac{\Delta x}{\Delta s} / \%$
1	1.5	2.8	33
2	4.6	6.7	52
3	2.0	7.8	39
4	4.0	14.5	54
5	3.0	5.8	50

(TABLES 1 and 2). The aggregates seemed to be very healthy and green after 3 weeks' cultivation. The aeration rate had to be very high in order to maintain a homogenous culture (not reached in all experiments). Results from one cultivation are given in TABLES 1 and 2.

DISCUSSION

One difficulty when using plant cell cultures for production of metabolites is the analysis of products. We used both RIA and HPLC for analyzing cardenolides. RIA is easy and suitable for screening, but from TABLE 1 it can be seen that RIA gives quite a different picture than HPLC. Results obtained in RIA are affected by cross reactions. For example, digoxin antiserum also reacts with many other cardenolides. On the other hand, the HPLC method is not as exact as would be hoped because the plant cell extracts vary depending on cultivation conditions and many peaks may disappear under the background.

Cell yields in our cultivations were as high as in plant cell cultures normally: the maximum cell yield reached was 14.5 g/dm^3 and $\Delta x/\Delta s = 54\%$ in the Kluyver flask with gibberellic acid (5 mg/l). Maximum cardenolide concentrations (digitoxin equivalent 174 pmol/mg dry wt, RIA) were obtained in mechanically stirred fermentors although the cells appeared to be "sick."

Our aims are to optimize further the production conditions and to determine the regulation points in cardenolide synthesis. Many unresolved problems still remain in plant cell cultivation, particularly in the production of secondary metabolites with suspension cultures.

REFERENCES

1. WEILER, E. W. & M. H. ZENK. 1976. Radioimmunoassay for the determination of digoxin and related compounds in *Digitalis lanata*. Phytochemistry **15:** 1537–1545.
2. HAGIMORI, M., T. MATSUMOTO & Y. OBI. 1982. Studies on the production of *Digitalis* cardenolides by plant tissue culture. Plant Physiol. **69:** 653–656.
3. LUCKNER, M., C. KUBERSKI, H. SCHEIBNER, C. SCHWIEBODE & B. DIETTRICH. 1982. Principles regulating cardenolide formation in cell cultures of *Digitalis lanata*. Planta Med. **45**(3): 134.
4. ZENK, N. H. 1978. The impact of plant cell culture on industry. *In* Frontiers of Plant Tissue Culture. T. A. Thorpe, Ed. Int. Assoc. Plant Tissue Culture. University of Calgary. Calgary, Canada. pp. 1–13.
5. VOGEL, E. 1979. Die radioimmunologische bestimmung von Cardenoliden in pflanzlichen Zellmaterial. Thesis, Martin Luther Universität. Halle-Wittenberg, Federal Republic of Germany.
6. WICHTL, M., W. WICHTL-BLEIER & M. MANGKUDINJOJO. 1982. Hochleistungs-flüssigkeits-chromatographische analyse von Digitalis-Blattextrakten. II. Quantitative Analyse. J. Chromatogr. **247:** 359–365.

Noninvasive ^{31}P NMR Studies of the Metabolism of Suspended and Immobilized Plant Cells

PETER BRODELIUS

Institute of Biotechnology
Swiss Federal Institute of Technology
Hönggerberg
CH-8093 Zürich, Switzerland

HANS J. VOGEL

Physical Chemistry 2
Chemical Center, University of Lund
P.O. Box 740
S-220 07 Lund, Sweden

Whole cell immobilization has become an important area of research. Large parts of the metabolism of the immobilized cells may be utilized for biosynthetic purposes. Simple, inexpensive carbon sources can be converted to very complex, valuable compounds. This is true in particular for immobilized plant cells producing secondary metabolites such as alkaloids and steroids.

Preparations of immobilized whole cells have been characterized in various ways. However, noninvasive methods of studying the intact cell in the immobilized state have been lacking. *In vivo* ^{31}P NMR has recently emerged as a very useful method for the study of bioenergetics and metabolism of living systems.[1] Here we report on our comparative ^{31}P NMR studies of freely suspended and immobilized plant cells. The technique is also applicable to other types of immobilized cells.

Plant cells have been immobilized by entrapment in various matrices with viability and biosynthetic capacity preserved.[2,3] Such immobilized cell preparations have been employed for *de novo* synthesis, for synthesis from metabolically distant precursors, or for bioconversions. These studies indicate that the secondary metabolism in immobilized plant cells is not largely altered by entrapment in the polymer.

In this study, we have investigated several aspects of phosphate metabolism. *In vivo* studies of both free and immobilized cells of *Catharanthus roseus* and *Daucus carota* were performed. The cells were cultivated in submerged cultures in a modified growth medium (low concentration of manganese, which is paramagnetic and therefore can interfere in the NMR experiments), and they were entrapped in alginate or agarose according to standard procedures.[4,5]

Phosphorus-31 NMR has the advantage that the phosphorus nucleus has a spin of 1/2 and is 100% abundant, thus alleviating the need for expensive enrichment procedures. This, combined with the reasonable sensitivity of the ^{31}P nucleus for NMR experiments, has contributed to the recent popularity of ^{31}P NMR studies. Only a few low-molecular-weight, phosphorus-containing compounds are present normally in the cells, so chemical shift assignments can usually be made readily. Long accumulation times are, however, often necessary to obtain spectra with a reasonable signal to noise ratio.

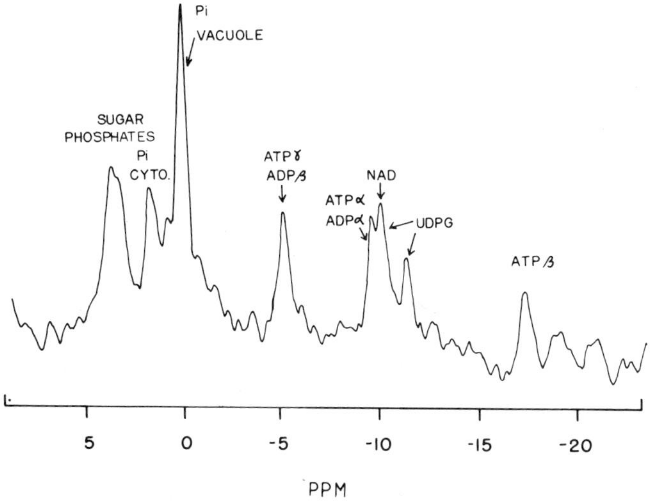

FIGURE 1. ^{31}P NMR spectrum (103.2 MHz) of *C. roseus* cells. For assignments of the resonances see TABLE 1.

The experiments reported here were carried out with a specially designed horizontal probe using a homebuilt 6T NMR instrument.[6] The immobilized cells are placed in this probe and can be perfused with oxygenated medium, which is a requirement for long-time experiments. In order to obtain satisfactory spectra, an accumulation time of 12–14 hours was required which corresponds to 40–50000 scans (1 sec per scan). For spectra where we studied only P_i levels, a considerably shorter accumulation time (1–2 hours) could be used.

FIGURE 1 illustrates the ^{31}P NMR spectrum of *C. roseus* cells. The corresponding spectrum of *D. carota* is very similar and the same phosphorylated metabolites can be found even though the peak corresponding to vacuolar P_i is much smaller. The chemical shifts of relevant compounds are given in TABLE 1. Since the chemical shifts of P_i and some of the phosphorylated metabolites are a function of pH at near physiological pH values, ^{31}P NMR may also be used to evaluate intracellular pH. When organelles have a different pH, one can study compartmentalization of metabolites in cells. To determine the intracellular pH, one can use titration curves of

TABLE 1. ^{31}P NMR Overview of Chemical Shifts of Some Biological Compounds[a,b]

Compound	Chemical Shift (ppm)	Compound[c]	Chemical Shift (ppm)
Glucose-6-P	5.1	NDPG	−10.6/−12.5
Glucose-1-P	3.0	NDP (α)	−9.3
AMP	4.4	NDP (β)	−5.3
P_i	3.3	NTP (α)	−9.9
NAD^+	−10.6	NTP (β)	−18.5
NADH	−10.5	NTP (γ)	−4.8

[a]All shifts are referenced to 85% phosphoric acid; upfield shifts are given a negative sign.
[b]All shifts are measured at pH 8. Those of NTP and NDP are measured in the presence of excess $MgCl_2$.
[c]N = nucleoside (mainly adenosine).

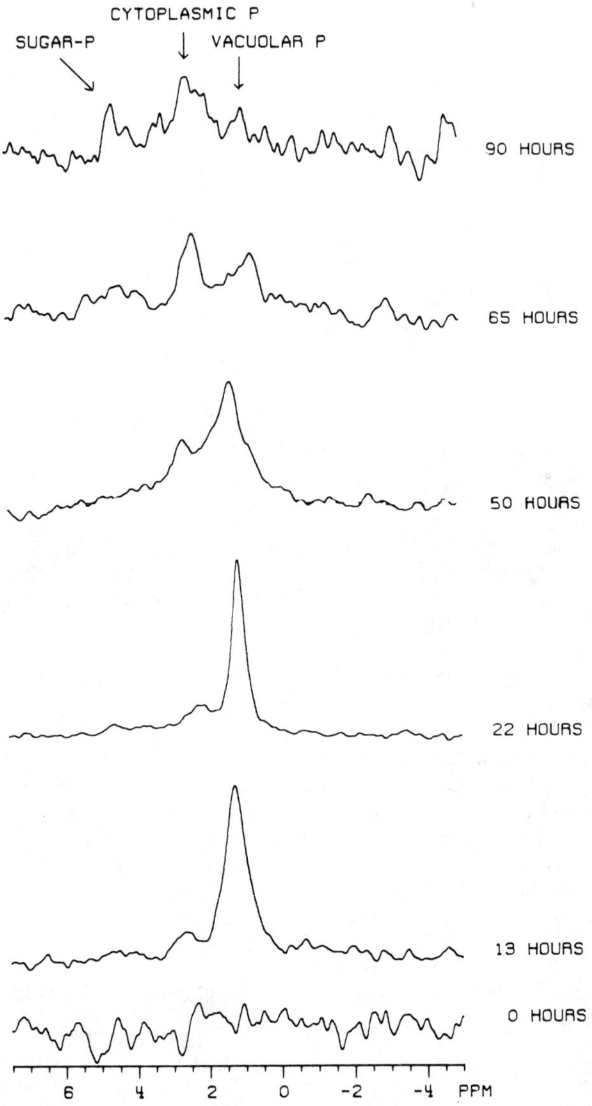

FIGURE 2. ^{31}P NMR spectra (145.7 MHz) of *C. roseus* cells at various times after inoculation. The spectra were obtained by accumulating the following number of scans (one scan per second): 0 hour, 5260 scans; 13 hours, 2900 scans; 22 hours, 7200 scans; 50 hours, 6700 scans; 65 hours, 6840 scans; 90 hours, 9100 scans.

chemical shifts versus pH as determined for standard solutions of P_i or glucose-6-phosphate. From the chemical shift position of the P_i-resonances, internal pH values of 7.0 and 6.0 (or below) were found for the cytoplasm and the vacuoles, respectively, for both freely suspended and immobilized cells of *C. roseus* or *D. carota*.

Phosphate uptake by the plant cells was followed by studying the increase and decrease of the vacuolar and cytoplasmic P_i as a function of time with ^{31}P NMR. It was confirmed by following the disappearance of ^{32}P from the medium. *C. roseus* quickly took up all phosphate from the medium into the vacuoles. There it was stored temporarily until it was used as indicated in FIGURE 2. *D. carota* took up phosphate from the medium only as it was needed. Also, when the phosphate concentration of the medium was increased fivefold, the *C. roseus* cells took it up completely within a few hours.

The uptake of P_i from the medium by agarose- or alginate-entrapped *C. roseus* cells was somewhat slower than for freely suspended cells. All the P_i was, however, taken up by the immobilized cells within a few hours, indicating that phosphate uptake was not affected to any great extent by the immobilization. Therefore it can be assumed that the phosphate uptake was not a limiting process.

The presence of phosphorylated metabolites in freely suspended and agarose-entrapped *C. roseus* cells has been studied by ^{31}P NMR. The vacuolar P_i reached its maximum within a few hours after inoculation of cells into fresh medium and subsequently it decreased continuously as the P_i was consumed and incorporated in various compounds, as is seen in FIGURE 2.

Comparison of the ^{31}P spectra of free and immobilized cells shows that the same metabolites were produced by all cell preparations during 4 days of incubation. In fact, the agarose-entrapped cells showed a spectrum very similar to that observed with freely suspended cells (see FIG. 1) indicating that the primary metabolism of these cells was not affected to any great extent by the immobilization. However, the ADP/ATP ratios appeared to be somewhat higher in the immobilized cells, suggesting slightly lower energization. Alginate-entrapped cells, on the other hand, appeared to be more affected. Plant cells immobilized in this polymer were somewhat restricted in growth,[3] but it is likely that this restricted growth was not due to a changed phosphate metabolism, since the same phosphorylated metabolites are found here as in free or agarose-entrapped cells.

The ^{31}P NMR studies carried out in this investigation showed that various plant species in culture have fundamentally different mechanisms for phosphate uptake. The P_i was either taken up from the medium and stored within the vacuoles before use (*C. roseus*) or it was taken up from the medium as it was needed (*D. carota*). The phosphate uptake mechanisms of these cells were not affected by immobilization. Our studies have also shown that the concentration of phosphorlyated metabolites in agarose-entrapped plant cells was not much different from that in freely suspended cells. Only a somewhat higher ADP/ATP ratio could be estimated for the immobilized cells.

The previously observed restricted growth of alginate-entrapped cells was most likely not due to a changed phosphate metabolism, but rather to limited diffusion of another nutrient (oxygen?) within this polymer.

The NMR technique for studying phosphate metabolism of whole immobilized cells should prove to be a valuable tool for characterizing biocatalysts.

REFERENCES

1. GADIAN, D. G. 1982. Nuclear magnetic resonance and its application to living systems. Oxford University Press. Oxford.

2. BRODELIUS, P. & K. MOSBACH. 1982. Immobilized plant cells. *In* Advances in Applied Microbiology. A. Laskin, Ed. Vol. **28:** 1–26. Academic Press. New York.
3. BRODELIUS, P. 1983. Immobilized plant cells. *In* Immobilized Cells and Organelles. B. Mattiasson, Ed. Vol. **1:** 27–55. CRC Press. Boca Raton, FL.
4. BRODELIUS, P. & K. NILSSON. 1980. Entrapment of plant cells in different matrices. A comparative study. FEBS Lett. **122:** 312–316.
5. NILSSON, K., S. BIRNBAUM, S. FLYGARE, L. LINSE, U. SCHRÖDER, U. JEPPSSON, P.-O. LARSSON, K. MOSBACH & P. BRODELIUS. 1983. A general method for the immobilization of cells with preserved viability. Eur. J. Appl. Microbiol. Biotechnol. **17:** 319–326.
6. DRAKENBERG, T., S. FORSEN & H. LILJA. 1983. ^{43}Ca NMR studies of calcium binding to proteins: Interpretation of experimental data by band shape analysis. J. Magn. Reson. **53:** 412–422.

Comparison of Immobilization Methods for Plant Cells and Protoplasts[a]

J. M. S. CABRAL,[b] P. FEVEREIRO,[c] J. M. NOVAIS,[b] AND
M. S. S. PAIS[c]

[b]*Laboratório de Engenharia Bioquímica*
Instituto Superior Técnico
1000 Lisboa, Portugal

[c]*Departamento de Biologia Vegetal*
Faculdade de Ciências de Lisboa
1200 Lisboa, Portugal

Plant cells have been immobilized by entrapment methods using polysaccharides from seaweeds and synthetic matrices for the biosynthesis and biotransformation of valuable plant cell products.[1] Protoplasts, which are obtained by the removal of the plant cell walls, have been mechanically stabilized by immobilization in calcium alginate gels.[2] In an attempt to develop inexpensive and simple immobilization techniques for these sensitive cells, several experiments with alternative methods were carried out with *Silybum marianum* cells and protoplasts. The flowers and leaves of *Silybum marianum* are used traditionally in Portugal for the small-scale production of Serpa cheeses. The coagulant properties of their petals or leaves are due to their high rennin content. Protoplasts were obtained both from the second pair of vegetative leaves of *Silybum marianum* as well as *callus* and cell suspension cultures.

TABLE 1. Clotting of Milk by Immobilized *Silybum marianum* Cells

Preparation	Clotting Time (h)	Clotting Activity (h^{-1} ml^{-1} cells)	Activity Retention (%)
Free cells	15	0.067	100
Agar (2%)	20	0.050	75
Calcium alginate (2%)	18	0.056	83
Gelatin (10%) + glutaraldehyde (5%)	—	0.0	0
Polyacrylamide (10%)	—	0.0	0
Hydrous titanium (IV) oxide	18	0.056	83

Slices (0.25 cm^2 area) of the second pair of leaves of *Silybum marianum* were used for the production of *callus* cell suspension cultures. After surface sterilization for 10 min in a 10% hypochlorite solution, the slices were inoculated in a solid medium using 2,4-D (0.5 mg/l), kinetin (0.02 mg/l), and benzyladenine (0.02 mg/l) as growth regulators. The *callus,* when two weeks old, was transferred to 50 ml of the liquid medium contained in 250-ml Erlenmeyer flasks. An inoculum of 25 ml of the cell

[a]We are grateful for the financial support received from the Instituto Nacional de Investigacao Cientifica, Lisboa, Portugal.

TABLE 2. Clotting of Milk by Immobilized *Silybum marianum* Protoplasts

Preparation	Clotting Time (h)	Clotting Activity (h^{-1} ml^{-1} protoplasts)	Activity Retention (%)
Free protoplasts from leaves of intact plant	1	1	100
Free protoplasts from suspension cell culture	2	0.50	50
Agar (2%)	4.5	0.22	22
Calcium alginate (1%)	1.3	0.77	77
Calcium alginate (2%)	1.5	0.67	67
Cellulose acetate (2%)	14	0.071	7.1
Gelatin (2%) + glutaraldehyde (5%)	—	0.0	0
Gelatin (10%) + glutaraldehyde (5%)	—	0.0	0
Polyacrylamide (10%)	—	0.0	0
Hydrous titanium (IV) oxide	3.3	0.30	30
Alkylamine derivative	4	0.25	25
Carbonyl derivative	—	0.0	0
Carboxyl derivative	4	0.25	25

suspension culture, obtained after two weeks, was transferred to 25 ml of new liquid medium, and the four-week-old cells were used for immobilization assays.

Protoplasts were obtained using 5 ml of the cell suspension culture or slices of leaves to which 5 ml of solution A were added. (Solution A: cellulase 2%, pectinase 2%, sorbitol 0.45 *M* and $CaH_4(PO_4)_2$ 0.004 *M*.) The pH of the solution was adjusted to 5. Incubation of this solution with the material was carried out at 25°C for four hours with constant shaking of 70 strokes/min. The protoplasts obtained were washed twice in a solution containing sorbitol 0.45 *M* and $CaH_4(PO_4)_2$ 0.004 *M*.

Plant cells and protoplasts were immobilized by the following methods: (1) gel entrapment in agar (2%), calcium alginate (1–2%), gelatin (2–10%) plus glutaraldehyde (5%) and cross-linked polyacrylamide (10%); (2) fiber entrapment in cellulose acetate (2%); and (3) chelation[3] on hydrous titanium(IV) oxide and its alkylamine, carbonyl and carboxyl derivatives. For the clotting assay, 1 ml of cells or protoplasts (10^6 protoplasts), free or immobilized, was added to 1 ml of milk at pH 6.0 and 30°C.

TABLES 1 and 2 show the clotting activity of the immobilized plant cells and protoplasts, respectively, as well as the comparison with their free counterparts. From these tables it can be seen that cell and protoplast entrapment in natural polysaccharides (agar and calcium alginate) led to active preparations, mainly with the mild calcium alginate procedure. The lower activity obtained with agar-entrapped protoplasts was due to the relatively high contact temperature (50°C) between the agar solution and the protoplasts. Cell and protoplast immobilization in synthetic (poly-

TABLE 3. Stability of Immobilized *Silybum marianum* Cells at 30°C

Time (days)	Clotting Activity (h^{-1} ml^{-1} cells)	
	Calcium alginate (2%)	Hydrous Ti(IV) Oxide
0	0.059	0.056
2	0.053	0.050
4	0.056	0.042
6	0.050	0.048

acrylamide) or natural (gelatin) gels using toxic chemicals (acrylamide or glutaraldehyde) and on glutaraldehyde-activated support (carbonyl derivative of titanium(IV) oxide) always led to inactive preparations. Protoplasts were partially inactivated by chelation on hydrous titanium(IV) oxide; however, chelated plant cells show similar activity and stability to that of calcium alginate-entrapped cells. Both calcium alginate and titanium(IV)-chelated plant cells show no decrease in clotting activity for six days, as can be seen in TABLE 3.

REFERENCES

1. BRODELIUS, P. & K. MOSBACH. 1982. Immobilized plant cells. *In* Advances in Applied Microbiology. A. Laskin, Ed. Vol. **28:** 1–26. Academic Press. New York.
2. SHEURICH, P., H. SCHNABL, U. ZIMMERMANN & J. KLEIN. 1980. Immobilisation and mechanical support of individual protoplasts. Biochim. Biophys. Acta **598**(4): 645–651.
3. CABRAL, J. M. S., M. M. CADETE, J. M. NOVAIS & J. P. CARDOSO. 1984. Immobilization of yeast cells on transition metal-activated pumice stone. Ann. N.Y. Acad. Sci. **434:**. This volume.

A Simple Inulin Assay for Renal Clearance Determination Using an Immobilized β-Fructofuranosidase

D. F. DAY AND W. E. WORKMAN

Audubon Sugar Institute
Louisiana State University
Baton Rouge, Louisiana 70803

Screening tests for renal function involve the determination of blood constituents such as urea nitrogen and creatinine. If the level of these substances increases, this implies impaired renal function. However, the sensitivity of these tests is limited, since kidney function may decrease as much as 50% before they yield abnormal results.

The clearance tests provide a more sensitive measure of renal function. Normal hospital practice involves use of the creatinine clearance test. The creatinine clearance test is a relatively accurate measure of glomerular filtration rate and has largely replaced the less accurate urea clearance test. However, when plasma creatinine levels increase considerably above normal, creatinine is also secreted by the kidneys, producing creatinine clearance values greater than the actual glomerular filtration rate.

The polysaccharide inulin has become the substance of choice for precise measurements of glomerular filtration rates because it is freely filtered by the glomeruli but is neither secreted nor reabsorbed by the tubules. Inulin clearance currently is not in routine clinical use, due in large part to the difficulty of accurately and routinely assaying for inulin in urine and plasma. We propose a simple assay system suitable for routine testing of inulin in urine and plasma. A two-step enzymatic process is used for the conversion of inulin to sorbitol with the oxidation of NADH to NAD^+. The first step is the conversion of inulin (polyfructose) in the sample (urine or plasma) to fructose (Eq. 1).

$$\text{Inulin} \xrightarrow{\beta\text{-fructofuranosidase}} \text{Fructose} \qquad \textbf{(1)}$$

Inulin is hydrolyzed, using the enzyme from *Kluyveromyces fragilis* prepared by the method of Workman and Day.[1] Hydrolysis is carried out at 50°C and is normally complete within 20 min (FIG. 1). This enzyme produces fructose as the main product with minor amounts of glucose.

The second step in the assay (Eq. 2) is the conversion of fructose and NADH to sorbitol and NAD^+ by the enzyme sorbitol dehydrogenase.

$$\text{Fructose} \xrightarrow[\text{NADH}^+\text{H}^+ \;\curvearrowright\; \text{NAD}^+]{\text{SDH}} \text{D-Sorbitol} \qquad \textbf{(2)}$$

This is simply a modification of the clinical sorbitol dehydrogenase (SDH) assay established for serum SDH.[2] In our case, fructose, not SDH, is the unknown compound. Because this enzyme does not have the same pH optimum (5.0) as the β-fructofuranosidase, the pH is adjusted with Tris buffer (pH 7.6) before the Equation 2 part of the assay. After incubation at room temperature, the decrease in absorbance

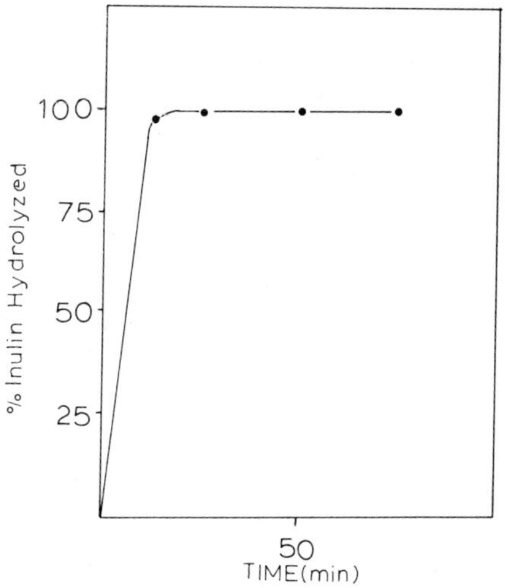

FIGURE 1. The hydrolysis of inulin with time by 300 IU of immobilized β-fructofuranosidase. The inulin concentration (7.5%) is at the upper limit of what would be found clinically.

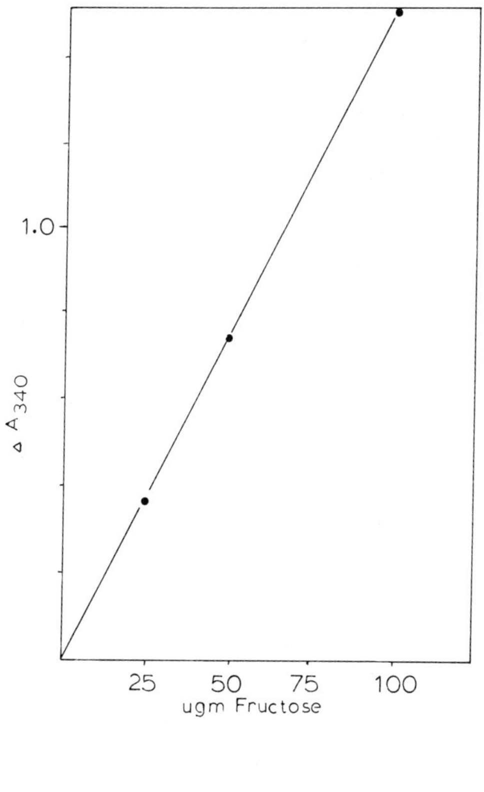

FIGURE 2. A typical standard curve for fructose using the sorbitol dehydrogenase assay.

TABLE 1. Assay System

1. Test sample, buffered with 0.01 *M* sodium phosphate pH 5.0 reacted with Con A-immobilized β-fructofuranosidase (300 IU enzyme) with stirring for one hour at 50°C.
2. Remove immobilized enzyme by decantation or filtration.
3. Withdraw 0.25 ml of hydrolyzed sample and add to 0.75 ml of fructose assay mix. Incubate for one hour at room temperature.
4. Clarify by centrifugation.
5. Read the absorbance at 340 nm. Take the difference in A340 from the control value (-sorbitol dehydrogenase) and determine the fructose concentration from a standard curve. The fructose assay mix contains 1.4 IU sorbitol dehydrogenase, 0.2 mg NADH, and 0.75 ml 0.1 *M* Tris buffer at pH 7.6.

at 340 nm (or increase in absorbance at 280) is measured, and the fructose (inulin) concentration is determined from a standard curve (FIG. 2). A summary of the assay is given in TABLE 1.

This assay will offer the clinical technician a simple wet chemistry method for measuring inulin in clinical samples. It should enhance the value of the inulin clearance test to the clinician and may lead to increased use of this procedure in clinical practice. This assay is not only a practical wet chemistry procedure but should be adaptable to automated clinical analyzers. Both because of limited availability of this enzyme and in order to test the suitability of the protein for use in automated analyzers, it was decided to use the enzyme in immobilized form. Agarose-bound Con A (Sigma Chemical Co., St. Louis, MO.) was used as the matrix for this work, although it is not necessarily the best available matrix. A summary of the properties of the immobilized β-fructofuranosidase are given in TABLE 2. The short half-life is probably due to lectin (Con A) instability as the native enzyme is extremely stable under these conditions.[1]

SUMMARY

Current hospital practice for testing renal function is to use the creatinine clearance test. Inulin clearance, an inherently more accurate procedure, currently is only carried out by specialized laboratories because there is not a simple biochemical assay for inulin, that is, an assay that could be carried out by any laboratory without special facilities.

We have developed a simple enzymatic assay system for measuring inulin in plasma and urine. The procedure uses a β-fructofuranosidase immobilized on Concanavalin A to convert inulin to fructose. Fructose is then measured by measuring the

TABLE 2. Properties of Con A Agarose-immobilized β-Fructofuranosidase

Loading Capacity	432 IU/mg lectin
Half-life	20 hours
K_m^a	14 m*M*
pH Optimum[a]	5.0

[a]The kinetic parameters of the enzyme after immobilization were identical to those of the free enzyme.

NADH→NAD conversion produced when fructose is converted to sorbitol by the enzyme sorbitol dehydrogenase. Kinetic parameters, binding capacities, and operating conditions for the immobilized β-fructofuranosidase were determined as well as general operating parameters for the complete assay system. This system offers the potential for replacing the creatinine clearance test as the assay of choice for renal function.

REFERENCES

1. Workman, W. E. & D. F. Day. 1983. Purification and properties of the β-fructofuranosidase from *Kluyveromyces fragilis*. FEBS Lett. **160**(1,2): 16–20.
2. Sigma Technical Bulletin 50-UV. 1974. Sigma Chemical Company. St. Louis, MO. pp. 1–10.

Hybrid Biosensor for Clinical and Fermentation Process Control

ISAO KARUBE, IZUMI KUBO, AND SHUICHI SUZUKI

Research Laboratory of Resources Utilization
Tokyo Institute of Technology
Nagatsuta, Yokohama, 227, Japan

The determination of urea and creatinine in biological fluids is a diagnostically important test. External dialysis (artificial kidney) has become so widely practiced that precise, rapid, and simple methods for urea and creatinine determinations are very desirable. The determination of urea is also important in various fields such as environmental and bioindustrial process analyses. Spectrophotometric methods have conventionally been used for measuring serum urea and creatinine. However, these methods are time-consuming, complicated and other substances in serum interfere seriously. By combining the advantages of an immobilized enzyme catalyst and the convenience of electrochemical ammonia determination, Guilbault *et al.*[1,2] reported the development of urea and creatinine electrodes. Ammonia gas or ammonium ion electrodes were used to measure ammonium ion produced by the enzymatic reactions. However, ions or volatile compounds such as amines sometimes interfere with the determination of ammonia and ammonium ion.

This paper describes new sensors for the amperometric determination of urea and creatinine. They are based on amalgamation of an enzyme reaction and bacterial metabolism.

UREA SENSOR

Urease (EC 3.5.1.5) hydrolyzes urea to ammonium ion and carbon dioxide. A microbial sensor consisting of immobilized nitrifying bacteria and an oxygen electrode has been developed for the amperometric determination of ammonia.[3]

The ammonia is successively oxidized to nitrite and nitrate by nitrifying bacteria. The bacteria have not been completely characterized, but are known to be a mixed culture of *Nitrosomonas* sp. and *Nitrobacter* sp. The sequence of reactions is:

$$NH_4^+ \xrightarrow[\textit{Nitrosomonas}\text{ sp.}]{O_2} NO_2^- \xrightarrow[\textit{Nitrobacter}\text{ sp.}]{O_2} NO_3^-$$

The nitrifying bacteria consume oxygen, so that the oxygen decrease can be detected by an oxygen electrode. Therefore, a microbial sensor for ammonia consisted of immobilized nitrifying bacteria and the oxygen electrode. A urea sensor consisted of immobilized urease, immobilized bacteria, and an oxygen electrode. Urease was immobilized on a porous polyvinylchloride (PVC) membrane by the method described previously.[4] Nitrifying bacteria were cultivated and immobilized on a porous acetylcellulose membrane as described previously.[3] The porous membrane retaining nitrifying bacteria was attached on the surface of the oxygen electrode. The immobilized urease membrane then was placed on the bacterial membrane. These membranes were covered with a dialysis membrane and fastened with a rubber O-ring. The sensor system consisted of the sensor inserted in a flow cell, a carrier (buffer) solution, a peristaltic pump, an injection port and a recorder. The temperature of the carrier solution and the flow cell was maintained at 35° ± 1°C.

The optimum pHs of the enzyme reaction and of bacterial metabolism are different. Two optimum responses were observed at pH 7 and 8. The determination was performed at pH 7.0 because the sensor was stable at this pH. The determination time of the urea sensor was three minutes. A linear relationship was observed between the current decrease and the urea concentration below 50 mg · dl^{-1} (FIG. 1). The minimum detection concentration was 10 mg · dl^{-1}. The selectivity of the sensor for urea was examined with solutions containing other organic compounds. The sensor did not respond to citrate, oxalate, aspartate, uric acid, creatinine, ascorbic acid, and glucose. The reproducibility of the sensor was examined with a 25 mg · dl^{-1} urea

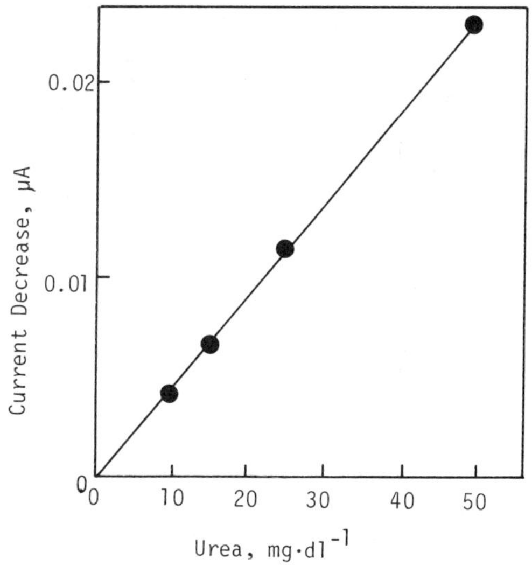

FIGURE 1. Calibration curve of urea sensor.

solution. The current output was reproducible within 4.8%. The sensor could be used for more than 2 weeks and 200 assays.

CREATININE SENSOR

Creatinine deiminase (EC 3.5.4.21) hydrolyzes creatinine to *N*-methylhydantoin and ammonium ion, and ammonia produced is determined with the microbial sensor described above. Therefore, a creatinine sensor consisted of immobilized creatinine deiminase, immobilized nitrifying bacteria, and an oxygen electrode. The creatinine deiminase was immobilized on a triamine membrane prepared by the method described previously.[5] The sensor consisted of a cellulose dialysis membrane, immobilized creatinine deiminase, immobilized nitrifying bacteria, and an oxygen electrode. The sensor system is composed of the sensor inserted in a flow cell, a carrier solution

TABLE 1. Influence of Immobilization Conditions on Characteristics of Creatinine Sensor

		A	B	C
Immobilization conditions:	Temp.	4°C	4°C	20°C
	pH	7.0	8.5	7.0
	t(h)	15	15	2
Current (μA) Creatinine concentration 50 mg · dl^{-1}		0.22	0.06	0.05
Minimum determination concentration		5 mg · dl^{-1}	50 mg · dl^{-1}	10 mg · dl^{-1}

(borate buffer, pH 8.5), a peristaltic pump, an injection port, and a recorder. The temperature of the carrier solution and the flow cell was maintained at 30 ± 1°C.

The immobilization of creatinine deiminase was performed under various conditions. TABLE 1 shows the influence of immobilization conditions on characteristics of the sensor. The optimum condition for creatinine deiminase immobilization was 4°C, pH 7.0, and 15 hours. The creatinine deiminase membrane prepared at these conditions was used for the sensor.

When a sample solution containing creatinine was applied to the sensor system, creatinine permeated through the dialysis membrane and was decomposed to ammonia and *N*-methylhydantoin. The ammonia was assimilated by the immobilized bacteria. At the same time, the bacteria consumed dissolved oxygen around the membrane, so that the current from the oxygen electrode decreased markedly and reached a minimum value within three minutes. There was a linear relationship between the current decrease and the concentration of creatinine below 100 mg · dl^{-1}, with a

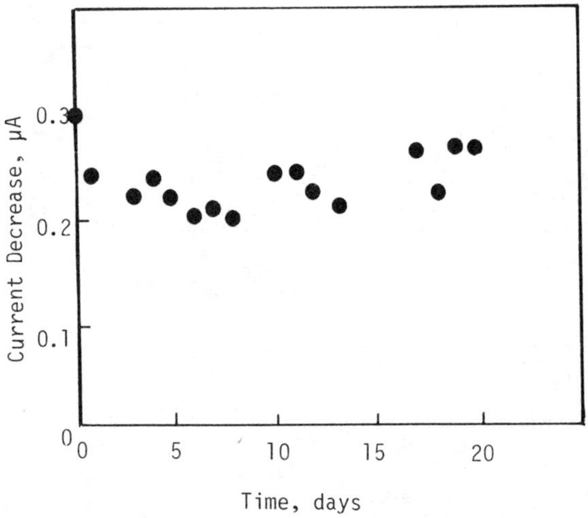

FIGURE 2. Stability of creatinine sensor (pH 8.5, 30°C, 1 ml · min^{-1}, 50 mg · dl^{-1} creatinine).

maximum current decrease of about 0.23 μA at 50 mg · dl^{-1}. The minimum detectable concentration of creatinine with this sensor was 5 mg · dl^{-1}.

The reproducibility of the sensor was examined with a 50 mg · dl^{-1} solution. The current decrease was reproducible within 6.7%. The selectivity of the creatinine sensor was satisfactory. The reusability of the creatinine sensor was examined with a sample solution containing 50 mg · dl^{-1} creatinine (FIG. 2). The current output of the sensor eventually decreased, but it could be used for more than 3 weeks and 300 assays.

In conclusion, the hybrid biosensor appears quite promising for the determination of creatinine and urea in biological fluids.

REFERENCES

1. GUILBAULT, G. G. & G. NAGY. 1975. Improved urea electrode. Anal. Chem. **45:** 417–419.
2. GUILBAULT, G. G., S. P. CHEN & S. S. KUAN. 1980. A creatinine-specific enzyme electrode. Anal. Lett. **13:** 1607–1624.
3. KARUBE, I., T. OKADA & S. SUZUKI. 1981. Amperometric determination of ammonia gas with immobilized nitrifying bacteria. Anal. Chem. **53:** 1852–1854.
4. HIROSE S., M. HAYASHI, N. TAMURA, S. SUZUKI & I. KARUBE. 1979. Polyvinylchloride membrane for a glucose sensor. J. Mol. Catal. **6:** 251–260.
5. WATANABE, E., K. ANDO, I. KARUBE, H. MATSUOKA & S. SUZUKI. 1983. Determination of hypoxanthine in fish meat with an enzyme sensor. J. Food Sci. **48:** 496–500.

Oxalate-sensing Enzyme Electrode

LELAND C. CLARK, JR.,[a] LINDA K. NOYES,[a]
THOMAS A. GROOMS,[b] AND PAMELA E. MOORE[a]

[a]*Children's Hospital Research Foundation*
Elland and Bethesda Avenues
Cincinnati, Ohio 45229

[b]*Yellow Springs Instrument Company*
Yellow Springs, Ohio 45387

The discovery of oxalate oxidases (E.C.1.2.3.4) in several plants has made possible the determination of oxalate in natural products by enzymatic analysis. Oxalate oxidase is found in usable amounts in moss,[1] barley sprouts,[2] and beet stems.[3] Previous methods for enzymatic analysis of oxalate depend upon the utilization of the released peroxide to oxidize a dye via a peroxidase reaction.[4] Oxalate has also been determined by the use of oxalate decarboxylase (E.C.4.1.1.2) and a conductometric electrode[5] to measure the CO_2 released as the formate is generated.

Our present method depends upon the polarographic measurement of hydrogen peroxide liberated according to the oxalate oxidase-catalyzed reaction between oxalate and oxygen to form carbon dioxide and hydrogen peroxide. The enzyme is immobilized between two membranes that are affixed to a platinum/silver-tipped electrode held in a thermostated cuvette. The use of amperometry to measure hydrogen peroxide has been reviewed earlier by Clark.[6] The measurement of oxalate in blood is difficult because of the low levels normally present, while the measurement of oxalate in urine is complicated by the fact that it is easily precipitated as a calcium or magnesium salt or in other insoluble forms. Urine may contain substances which affect the activity of the enzyme. Oxalate measurements in "synthetic urine" have been reported[7] but have very little relevance in the real world. The present report, a summary of work over the past six years, describes the way in which we have produced oxalate oxidase, immobilized it for use in an enzyme/peroxide electrode, optimized the conditions, and used the sensor to measure oxalate.

A YSI glucose sensor and Model 23A glucose analyzer were adapted to measure oxalate. Some measurements were made with two oxalate sensors in the same measuring cuvette. The instrument was adjusted to read in nanoamperes; and the amplifier output was fed to a strip chart recorder. Polarograms were run with a Princeton Applied Research polarograph. Succinate buffer, 11.81 g succinic acid/l, was adjusted to pH 3.5 using Na_2HPO_4 and stored at 4°C without preservatives. Oxalate standards, 1.522 g sodium oxalate/l, were diluted to 10 mg%, 20 mg%, 30 mg%, 40 mg%, and 50 mg%. The work reported here was conducted with oxalate oxidase prepared from barley sprouts, using a procedure similar to that described by Chiriboga.[8] The method of preparation is diagramed in FIGURE. 1.

The enzyme membranes were made by depositing a concentrated, buffered glutaraldehyde solution of the enzyme on cellulose ester membranes, covering these with Nuclepore, and sealing on O rings with cyano-acrylate cement. The membranes were stored at 4°C, without drying agents. The sensor tip (FIG. 2) was wetted with a drop of the succinate buffer, the O-ring membrane affixed, and the sensor placed in a cuvette with a thermostat. A threaded nut bearing on the electrode held the membrane on the tip of the electrode and sealed the tip into the cuvette. The electrode was

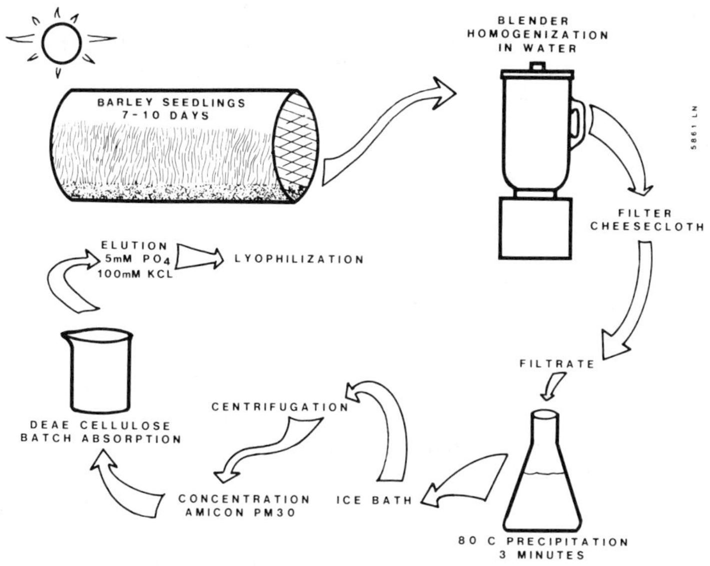

FIGURE 1. Preparation of oxalate oxidase from barley seedlings.[8]

continuously polarized at 0.7 volt. The current declined over a period of 20–60 minutes to reach a stable no-oxalate baseline current of about 2 nanoamperes. After a steady baseline was established, the instrument was calibrated by injecting 20 μl samples of standard in the expected range of analysis. Oxalate oxidase membranes that were stored at 4°C retained their activity over many weeks. The buffer that proved most satisfactory was based on succinate, with a pH optimum of 3.5. The current response of the oxalate sensor was linear to at least 50 mg/dl and passed through the origin.

Oxalate levels in blood are on the order of 100 to 300 μg/100 ml; in urine they are 1 to 114 μg/ml. In urine the oxalate is complexed in such a way as to require special handling and pretreatment before analysis. The many problems involved in the

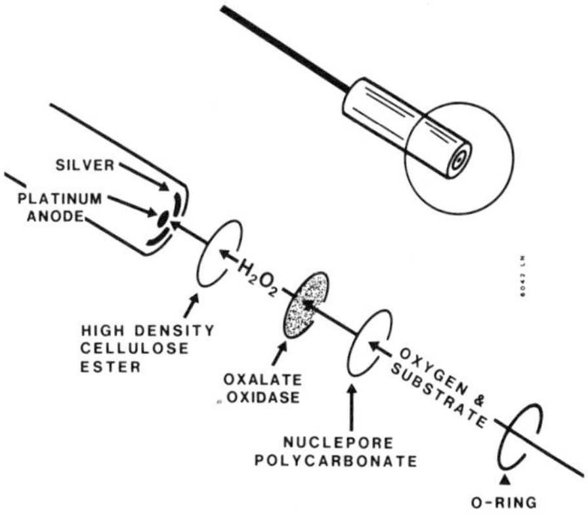

FIGURE 2. Spread out view of components that make up electrode tip. Arrow shows direction of movement of substrates and product to enzyme and platinum electrode, respectively.

measurement and interpretation of urinary oxalate have been reviewed.[17] Furthermore, there are substances in urine that apparently inhibit most oxalate oxidase.[4] The oxalate sensor we have described here is not sensitive enough for the direct analysis of oxalate in unprocessed blood. While it may be sensitive enough to be used in urine, at present it appears to have little, if any, advantage over other methods[1,2,9–16] in use today. Perhaps this oxalate sensor will be useful for calcium determination where oxalate titrations are used.

REFERENCES

1. LAKER, M. F., A. F. HOFMANN & B. J. D. MEEUSE. 1980. Spectrophotometric determination of urinary oxalate with oxalate oxidase prepared from moss. Clin. Chem. **26:** 827–830.
2. BAIS, R., N. POTEZNY, J. B. EDWARDS, A. M. ROFE & R. A. J. CONYERS. 1980. Oxalate determination by immobilized oxalate oxidase in a continuous flow system. Anal. Chem. **52:** 508–511.
3. OBZANSKY, D. M. & K. E. RICHARDSON. 1983. Quantification of urinary oxalate with oxalate oxidase from beet stems. Clin. Chem. **29:** 1815–1819.
4. POTEZNY, N., R. BAIS, P. D. O'LOUGHLIN, J. B. EDWARDS, A. M. ROFE & R. A. J. CONYERS. 1983. Urinary oxalate determination by use of immobilized oxalate oxidase in a continuous-flow system. Clin. Chem. **29:** 16–20.
5. BISHOP, M., H. FREUDIGER, U. LARGIADER, J. D. SALLIS, R. FELIX & H. FLEISCH. 1982. Conductivometric determination of urinary oxalate with oxalate decarboxylase. Urol. Res. **10:** 191–194.
6. CLARK, L. C., JR. 1979. The hydrogen peroxide-sensing platinum anode as an analytical enzyme electrode. Methods Enzymol. **LVI:** 448–479.
7. RUBINSTEIN, I., C. R. MARTIN & A. J. BARD. 1983. Electrogenerated chemiluminescent determination of oxalate. Anal. Chem. **55:** 1580–1582.
8. CHIRIBOGA, J. 1966. Purification and properties of oxalic acid oxidase. Arch. Biochem. Biophys. **116:** 516–523.
9. ZERWEKH, E., E. DRAKE, J. GREGORY, D. GRIFFITH, A. F. HOFMANN, M. MENON & C. Y. C. PAK. 1983. Assay of urinary oxalate: Six methodologies compared. Clin. Chem. **29:** 1977–1980.
10. HALLSON, P. C. & G. A. ROSE. 1974. A simplified and rapid enzymatic method for determination of urinary oxalate. Clin. Chim. Acta. **55:** 29–39.
11. COSTELLO, J., M. HATCH & E. BOURKE. 1976. An enzymic method for the spectrophotometric determination of oxalic acid. J. Lab. Clin. Med. **87:** 903–908.
12. SALLIS, J. D., M. F. LUMLEY & J. E. JORDAN. 1977. An assay for oxalate based on the conductometric measurement of enzyme-liberated carbon dioxide. Biochem. Med. **18:** 371–377.
13. KOHLBECKER, G., L. RICHTER & M. BUTZ. 1979. Determination of oxalate in urine using oxalate oxidase: Comparison with oxalate decarboxylase. J. Clin. Chem. Clin. Biochem. **17:** 309–313.
14. MAYER, W. J., J. P. MCCARTHY & M. S. GREENBERG. 1979. The determination of oxalic acid in urine by high-performance liquid chromatography with electrochemical detection. J. Chromatogr. Sci. **17:** 656–660.
15. KOBOS, R. K. & T. A. RAMSEY. 1980. Enzyme electrode system for oxalate determination utilizing oxalate decarboxylase immobilized on a carbon dioxide sensor. Anal. Chim. Acta **121:** 111–118.
16. YRIBERRI, J. & S. POSEN. 1980. A semi-automatic enzymic method for estimating urinary oxalate. Clin. Chem. **26:** 881–884.
17. ROBERTSON, W. G. & A. RUTHERFORD. 1982. Aspects of the analysis of oxalate in urine—a review. Scand. J. Urol. Nephrol. **53**(Suppl): 85–93.

Direct Rapid Electroenzymatic Sensor for Measuring Alcohol in Whole Blood and Fermentation Products

LELAND C. CLARK, JR.,[a] LINDA K. NOYES,[a]
THOMAS A. GROOMS,[b] AND PAMELA E. MOORE[a]

[a]*Children's Hospital Research Foundation*
Cincinnati, Ohio 45229

[b]*Yellow Springs Instrument Company*
Yellow Springs, Ohio 45387

At the present time, there are no rapid methods for the direct measurement of ethanol in *whole* unprocessed blood. One may wonder about the validity of tests based upon breath analysis, since they depend upon the accurate equilibration of the partial pressure of alcohol in the blood with that in expired air.[1] Yet the medical, ethical, moral, and legal consequences of excessive consumption of alcoholic beverages is enormous. A *stat.* whole blood alcohol method may prove equally valuable in the intensive care of trauma patients.[2]

The millions of blood gas and pH measurements made every month in laboratories and hospitals using whole blood samples of only 50 to 100 μl attests to the value of rapid on-the-site results. Blood gas, pH, and glucose measurements rely upon electrochemical principles. This report represents our progress towards a rapid easy-to-use blood alcohol sensor.

A number of years ago, we described[3] an alcohol sensor based on an alcohol oxidase from *Basidiomycetes*. Because of the lack of general availability of this enzyme, we have investigated alcohol oxygen oxidoreductases ("oxidases") from a number of other biological sources. We have looked at alcohol oxidase from *Basidiomycetes, Candida boidinii, Hansenula polymorpha,* and *Pichia pastoris.* Our efforts were directed toward finding an enzyme that could be immobilized in the sensor's membrane and that would yield peroxide in large enough amounts, and in reproducible enough ways to function in our system.[4]

The present sensor is modeled after the first enzyme electrode described.[5] This determination is conducted at a constant pO_2 in 350 μl of stirred, buffered solution maintained at a constant temperature in a cuvette. The sample, either 10 or 25 microliters, is injected into the cuvette and the ethanol diffuses through a polycarbonate membrane where it contacts the oxidase. The liberated peroxide diffuses through a cellulose ester membrane to a platinum anode (0.7 V) where it is oxidized with the generation cf a current. The reaction is:

$$RCH_2OH + O_2 \xrightarrow[\text{oxidase}]{\text{alcohol}} RCHO + H_2O_2$$

METHODS

Instrumentation

A YSI glucose sensor and Model 23A glucose analyzer were adapted to measure alcohol.

Alcohol Oxidase

Alcohol oxidase (E.C. 1.1.3.13) preparations were obtained largely from Sigma. The enzyme was also prepared in our laboratories by the procedure shown in FIGURE 1 based on a method described by Van Dijken.[6]

Buffer

Phosphate, chloride-containing buffers at pH 7.4 are compatible with the sensor. Azide and ethylenediamine (tetraacetate) are added in small amounts as preservatives.

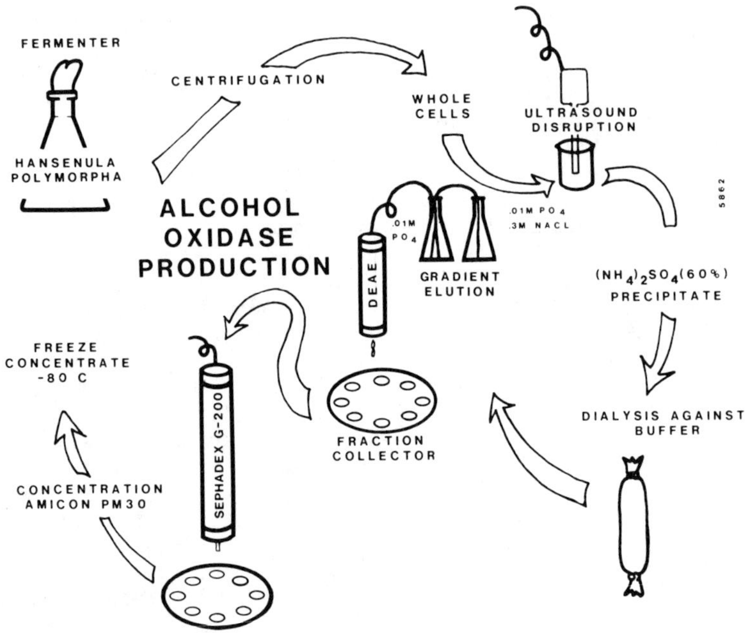

FIGURE 1. Procedure for alcohol oxidase production.

Alcohol Standards

An ethanol stock standard is prepared by weighing 10 grams of absolute ethanol and diluting to one liter with water. Working standards are prepared by diluting the stock with water or isotonic saline so as to have concentrations of 10, 30, 50, 100, 200, and 500 mg/dl as required. Standards of methanol and other alcohols are also prepared gravimetrically. Alcohol stocks and standards are stored at 4°C in stoppered Pyrex bottles. When small samples are prepared and aliquoted for daily bench working standards, great care must be taken to avoid evaporation.

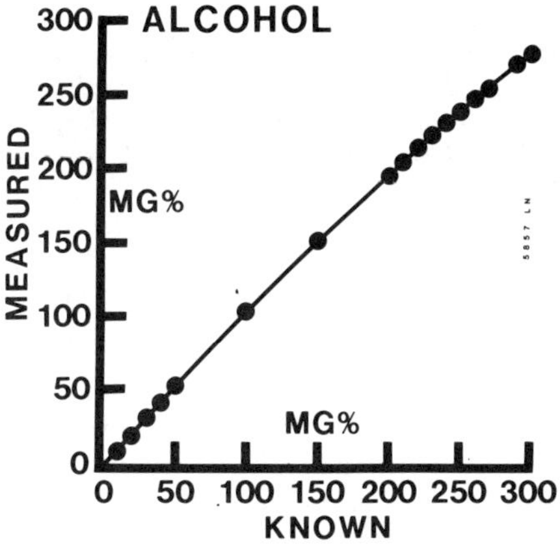

FIGURE 2. Calibration curve based on 150 mg% ethanol standard. Duplicate determinations were made in each of three instruments and the values averaged and plotted in the figure.

Alcohol Oxidase Membranes

Many of the enzyme membranes were obtained from The Yellow Springs Instrument Co. (Yellow Springs, OH 45387). We also prepared many enzyme membranes by placing a concentrated solution of alcohol oxidase on cellulose ester membrane, covering it with a Nuclepore membrane and sealing on O-rings with cyanoacrylate cement.

Preparation of Sensor

The sensor tip was wetted with a drop of the buffer, the membrane was affixed, and the sensor secured in the cuvette. A diagram of our sensor is published elsewhere in these proceedings.[7]

RESULTS

The current response of the alcohol sensor is linear to 200 mg% (FIG. 2). The relative response of the sensor to methanol, ethanol, and isopropanol is shown in TABLE 1.

Mice injected intraperitoneally with 0.5 cc of 10% ethanol attain maximum blood levels of nearly 200 mg%, while an identical dose of methanol gives blood levels as high as 800 mg% methanol. There is no competitive inhibition between methyl and ethyl alcohol; total current output is a function of the amount of methanol times its

TABLE 1. Current in Nanoamperes from 25 μl Samples of 10 m*M*/l Solutions

Methanol	25.7
Ethanol	7.6
n-Propanol	1.5
Isopropanol	0.1
Ethylene glycol	0.0
Propylene glycol	0.0

polarographic current plus ethanol times its current. For example, when 10 m*M* ethanol reads 10.3 nA and 10 m*M* methanol reads 31.6 nA, an equal mixture of the two solutions reads 21.9 nA. Addition of Triton X-100 to the buffer so as to make it hemolytic has very little, if any, effect on the response time or current of the alcohol sensor. Isotonic Gomori buffer with sodium chloride as well as succinate buffer (pH 6.0) gave only half the reading of the commercial phosphate buffer routinely used for glucose measurements. Mopso buffer (0.1 *M*, pH 7.4) gave the lowest reading of the buffers tested. The blood levels of a mouse that was breathing oxygen bubbled through absolute alcohol was between 50 and 116 mg%, depending on the bubbling rate and temperature of the alcohol.

DISCUSSION

All of the alcohol oxidases tested so far respond well when trapped or immobilized on our peroxide sensor. The catalase activity of whole blood does not interfere. No coenzyme needs to be added to the buffer. Selection of the best oxidase on a practical basis will depend on the stability of the enzyme in the instrument. Their relative response to various alcohols is similar.

Our work shows that it will be possible to perfect an instrument for the rapid measurement of whole blood samples using finger puncture samples and yielding immediate results.

We have previously measured alcohol diffusing through skin of animals using a $TCPO_2$ electrode and alcohol oxidase.[8,9] Guilbault *et al.* described an oxidase-based amperometric electrode and thermistor probe for measuring alcohol.[10]

Alcohol dehydrogenase (E.C. 1.1.1.1.) has been used in measuring alcohol where a Clark pO_2 electrode was used to measure NAD^+-dependent reactions.[11] The dehydrogenase procedure gives no response to methanol but gives 80% of the ethanol value for isopropanol and 60% for *n*-propanol.[12] Our present method gives nearly zero response for isopropanol but about three times the ethanol reading for methanol. If pretreatment of a blood sample with alcohol dehydrogenase/NAD does not lower the reading obtained with our sensor, the probability[13] is that the alcohol involved is methanol. In the comatose patient, therefore, neither our oxidase method nor the ADH/NAD method will reveal the exact situation with respect to alcohols, and gas chromatography or HPLC must be used. There are several possibilities for increasing the selectivity of enzyme-based alcohol sensors by the use of multiple enzymes.

CONCLUSION

We have made considerable progress in the understanding and development of a peroxide-based enzyme electrode for rapid analysis of whole blood for alcohol. As little as 10 μl of blood is required and the analysis can be completed in less than one minute. Optimum conditions for temperature, buffering, hydrogen ion activity, and enzyme entrapment have been found. This sensor responds to methanol, ethanol, and isopropanol in a ratio of 1, 0.30, and 0.004.

ACKNOWLEDGMENTS

Barbara Williams helped in the preparation and editing of this paper. Richard Hoffmann performed the alcohol injections and blood sampling in the mice. The paper was typed by Estelle Riley. Thanks are also due our murine corps. Without the rapid, reliable service of the Sigma Chemical Company, this work would not have been possible.

REFERENCES

1. DUBOWSKI, K. M. 1977. Collection and later analysis of breath alcohol after calcium sulfate sorption. Clin. Chem. **23:** 1371–1373.
2. BOURKE, D. L., M. B. ROSENBERG & K. F. SCHMIDT. 1978. Anesthesia for the trauma patient. Orthop. Clin. North Am. **9:** 661–678.
3. CLARK, L. C., JR. 1972. A family of polarographic electrodes and the measurement of alcohol. *In* Biotechnology and Bioengineering Symposium Series, Enzyme Engineering 3, L. B. Wingard, Jr., Ed. pp. 377–394.
4. CLARK, L. C., JR. 1979. The hydrogen peroxide-sensing platinum anode as an analytical enzyme electrode. Methods Enzymol. **LVI:** 448–479.
5. CLARK, L. C., JR. & C. LYONS. 1962. Electrode systems for continuous monitoring in cardiovascular surgery. Ann. N.Y. Acad. Sci. **102:** 29–45.
6. VAN DIJKEN, J. P. 1976. Oxidation of Methanol by Yeasts. Doctoral dissertation, University of Groningen, the Netherlands.
7. CLARK, L. C., JR. L. K. NOYES, T. A. GROOMS & P. E. MOORE. 1984. Oxalate-sensing enzyme electrode. Ann. N.Y. Acad. Sci. **434:** This volume.
8. CLARK, L. C., JR. 1979. Measurement of circulating alcohol with a $tcPO_2$ electrode. Birth Defects **15:** 37–38.
9. CLARK, L. C., JR. 1983. Cutaneous methods of measuring body substances. U.S. Patent No. 4,401,122.
10. GUILBAULT, G. G., B. DANIELSSON, C. F. MANDENIUS & K. MOSBACH. 1983. Enzyme electrode and thermistor probes for determination of alcohols with alcohol oxidase. Anal. Chem. **55:** 1582–1585.
11. CHENG, F. S. & G. D. CHRISTIAN. 1977. Amperometric measurement of enzyme reactions with an oxygen electrode using air oxidation of reduced nicotinamide adenine dinucleotide. Anal. Chem. **49:** 1785–1788.
12. VASILIADES, J., J. POLLOCK & C. A. ROBINSON. 1978. Pitfalls of the alcohol dehydrogenase procedure for the emergency assay of alcohol: A case study of isopropanol overdose. Clin. Chem. **24:** 383–385.
13. REDETZKI, H. M. & W. L. DEES. 1976. Comparison of four kits for enzymatic determination of ethanol in blood. Clin. Chem. **22:** 83–86.

FAD and Glucose Oxidase Immobilized on Carbon[a]

OSATO MIYAWAKI[b] AND LEMUEL B. WINGARD, JR.

Department of Pharmacology
School of Medicine
University of Pittsburgh
Pittsburgh, Pennsylvania 15261

Electroanalytical techniques can be used for direct characterization of electroactive prosthetic groups, such as flavin mononucleotide (FMN) and flavin adenine dinucleotide (FAD), in enzymes immobilized on electroconductive supports. Studies with these techniques will lead to a better understanding of the electrochemical state of the prosthetic groups as well as the development of new methodology in the field of enzyme electrodes. In the present work, glucose oxidase (GO) and FAD were immobilized by adsorption[1] or covalent attachment to the surface of two types of carbons, medium porosity spectroscopic graphite (SG) and low-porosity glassy carbon (GC).

FAD adsorbed strongly to SG (SG-FAD) and remained adsorbed for more than a week in aqueous and organic solutions at room temperature. Cyclic voltammetry showed an $E_{1/2}$ of -533 mV (ref. Ag/AgCl with 1 *M* KCl) for SG-FAD, as measured in 0.1 *M* Tris buffer, pH 8.0. This was in good agreement with literature results.[2] The surface concentration of FAD was 1.98×10^{-10} mol/cm^2, as determined by integration of the peak area for FAD on the cyclic voltammogram. The SG-FAD peak current from the same voltammogram was proportional to the scanning rate. This showed that the FAD was attached on the electrode surface, since the current would have been proportional to the square root of the scan rate if the FAD had been in solution.

GO, adsorbed on SG, showed only a slight peak for FAD by cyclic voltammetry but a large peak by the more sensitive differential pulse voltammetry. When the apoenzyme of GO (apoGO) was adsorbed on SG, no electrochemical activity was observed. However, the combination of SG-FAD and apoGO gave restoration of enzymatic activity (FIG. 1). The reconstituted glucose oxidase activity was estimated to account for only 0.1% of the total FAD adsorbed on the SG electrode. With either the reconstituted GO or the holoenzyme of GO adsorbed on the surface, the SG could function as an amperometric glucose sensor, provided that a potential of +0.9 V was applied to the enzyme electrode and oxygen was present in the system. The potential of +0.9 V was sufficient to oxidize the hydrogen peroxide. In this case, the reaction sequence at the electrode was as follows

$$\text{Glucose} + \text{FAD} \rightarrow \text{Gluconolactone} + \text{FADH}_2 \quad (1)$$

$$\text{FADH}_2 + O_2 \rightarrow \text{FAD} + H_2O_2 \quad (2)$$

$$H_2O_2 \rightarrow 2H^+ + 2e^- + O_2 \quad (3)$$

However, when the potential was changed to +0.2 V (sufficient to oxidize $FADH_2$) and oxygen was removed from the system, no current was observed. This showed that

[a]Supported by ARO contract DAAG2982K0064.

[b]On leave from the Department of Agricultural Chemistry of the University of Tokyo, Japan.

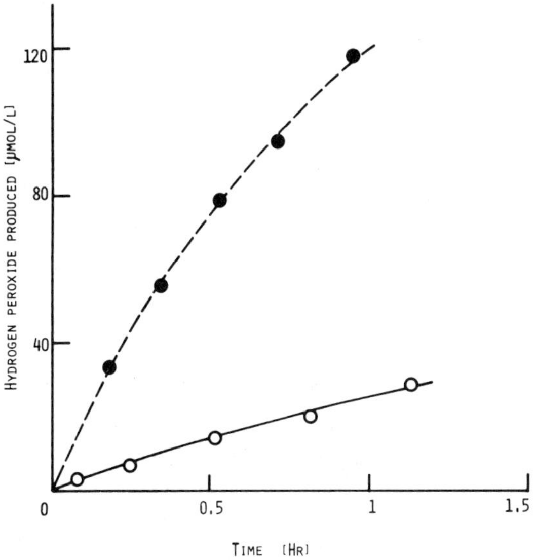

FIGURE 1. Production of hydrogen peroxide by adsorbed holoenzyme of GO (●) and by reconstituted enzyme (0) (adsorbed FAD and added apoGO) on SG. Enzyme activity, defined as initial rate of production of hydrogen peroxide, measured in oxygen-saturated 0.1 *M* glucose in 0.1 *M* potassium phosphate buffer at pH 6 and 25°C.

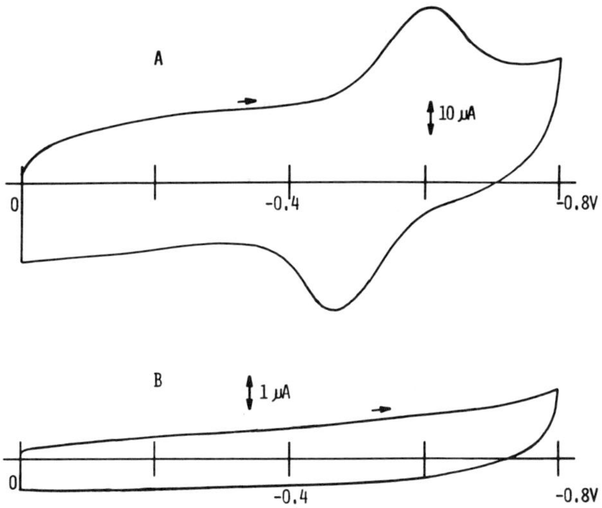

FIGURE 2. Cyclic voltammograms of (A) FAD covalently attached to hot acid-activated glassy carbon via carbodiimide coupling and (B) plain glassy carbon. Measurements made in 0.1 *M* Tris buffer, pH 8, 25°C, nitrogen blanketed, 50 mV/sec; potentials ref. Ag/AgCl (1 *M* KCl).

direct electron transfer from coenzymatically active FAD (Eq. 4) was not obtained by this method of immobilization-reconstitution.

$$FADH_2 \rightarrow FAD + 2H^+ + 2e^- \quad (4)$$

For covalent binding of FAD to GC, carboxylic groups were introduced onto the GC surface by one of three methods: oxygen plasma treatment,[3] electrochemical activation[4] (+2.2 V and 10% HNO_3), or hot acid treatment[5] (170°C, conc. H_2SO_4). The hot acid treatment gave the highest concentration of carboxylic groups on the surface. FAD was coupled, through either the adenine amino or the sugar hydroxyl groups, to the GC carboxylic acids, using a carbodiimide as the condensing agent. The covalently bound FAD showed very good electrochemical activity, as measured by cyclic voltammetry (FIG. 2). This activity was the result of covalently bound FAD since the adsorption of FAD on GC was not very strong or very stable. However, no coenzymatic activity was observed upon addition of apoGO to the covalently bound FAD. To reduce the effect of steric hindrance upon reconstitution, several types of spacers were introduced between FAD and the surface carboxylic groups. ϵ-Aminocaproic acid, ethylenediamine with glutaraldehyde, and hexamethylene diamine with formaldehyde were used for this purpose. After these spacers were introduced, the electrochemical activity of the immobilized FAD remained very good, but reconstituted enzyme activity on addition of apoGO still was not obtained. This work is continuing in order to elucidate which functional groups on the FAD molecule can be used for covalent binding of FAD to the support and still allow for reconstitution of enzyme activity. It did appear that the covalently immobilized FAD was capable of oxidizing NADH in solution; this interesting route for NAD regeneration will be reported elsewhere.

REFERENCES

1. MIYAWAKI, O. & L. B. WINGARD, JR. 1984. Electrochemical and enzymatic activity of flavine adenine dinucleotide and glucose oxidase immobilized by adsorption on carbon. Biotechnol. Bioeng. **26**:. In press.
2. GORTON, L. & G. JOHANNSON. 1980. Cyclic voltammetry of FAD adsorbed on graphite, glassy carbon, platinum, and gold electrodes. J. Electroanal. Chem. **113**: 151–158.
3. EVANS, J. F. & T. KUWANA. 1979. Introduction of functional groups onto carbon electrodes via treatment with radio-frequency plasmas. Anal. Chem. **51**: 358–365.
4. BOURDILLON, C., J. P. BOURGEOIS & D. THOMAS. 1980. Covalent linkage of glucose oxidase on modified glassy carbon electrodes: Kinetic phenomena. J. Am. Chem. Soc. **102**: 4231–4235.
5. WINGARD, L. B. JR. & J. L. GURECKA, JR. 1980. Modification of riboflavin for coupling to glassy carbon. J. Mol. Cat. **9**: 209–217.

Enzyme Electrode Based on a Fluoride Ion Selective Sensor Coupled with Immobilized Enzyme Membrane

MAT H. HO AND TAI-GUANG WU

Department of Chemistry
University of Alabama in Birmingham
Birmingham, Alabama 35294

Analytical measurements with oxidase enzyme electrodes in samples exhibiting high variability in oxygen content, such as plasma or whole blood present problems. In addition measurements with a peroxidase probe may suffer from interferences because species other than enzymatically generated hydrogen peroxide may diffuse through the membrane and be oxidized on a platinum electrode at 0.7 V. To circumvent these problems, we report a new immobilized enzyme probe that uses a fluoride ion-selective electrode as the sensor. The principle of the probe is based on the reaction of an organo-fluoro compound with H_2O_2 in the presence of peroxidase to produce fluoride ion.[1-3] The liberated fluoride ion, which can be used for the quantification of H_2O_2, is measured by the fluoride ion-selective electrode. This novel potentiometric method can also be applied to other peroxidase-coupled reactions.[4,5] In the work reported here glucose oxidase and peroxidase were coimmobilized onto a membrane, which in turn was attached directly to a fluoride electrode. The result was a self-contained probe for glucose.

Glucose oxidase (E.C.1.1.3.4, from *Aspergillus niger*), horseradish peroxidase (E.C.1.11.1.7), bovin serum albumin (BSA), glutaraldehyde (25% aqueous solution), and α-chymotrypsin (E.C.3.4.21.1) were obtained from Sigma. 2,3,4,5,6-Pentafluorophenol; 2,3,5,6-tetrafluorophenol; 4-fluorophenol; 4-fluoroaniline, and 3-fluoro-DL-tryosine were obtained from Aldrich. The organo-fluoro solutions were prepared in acetate buffer, pH 5. Glucose standard solutions were prepared from anhydrous β-D-glucose and acetate buffer containing 1 g/l benzoic acid. These solutions were allowed to mutarotate overnight before use.

FIGURE 1 shows the schematic diagram of the apparatus. A combination fluoride ion-selective electrode (Model 96-09, Orion Research Inc., Cambridge, MA) was used as a sensor. It was covered with pig small intestine membrane, obtained from Radelkis, Budapest. Glucose oxidase and peroxidase were coimmobilized onto the membrane using glutaraldehyde,[6] and enzyme probe was assembled, as shown in FIGURE 1. The electrode potential was measured with a 901 Orion Ionanalyzer and recorded.

The probe was fixed in a plastic thermostated cell, since fluoride ion reacts with glass. The probe was allowed to equilibrate in 5 ml of organo-fluoro solution (acetate buffer, pH 5) under constant stirring. After a stable reading was obtained, 100 μl of standard glucose was injected, and the slope was determined from the linear portion of the response curve.

The peroxidase-catalyzed reaction of 4-fluoroaniline with hydrogen peroxide to produce fluoride ion has been reported.[1-3] Recently, Siddigi[4] screened several organo-fluoro compounds for susceptibility to peroxidase-catalyzed C-F bond rupture. In our study, five organo-fluoro compounds (2,3,4,5,6-pentafluorophenol; 2,3,5,6-tetrafluorophenol; 4-fluorophenol; 4-fluoroaniline, and 3-fluoro-DL-tyrosine) appear to be suitable for analytical applications. These compounds have good linearity and a high reaction rate to achieve adequate sensitivity. The effects of variations in the amounts of enzymes, bovine serum albumin, and glutaraldehyde were investigated. A series of probes was prepared in which the amounts of peroxidase and glucose oxidase were varied, and the response of each probe was determined. The response rate of the probe increased with increasing enzyme concentration up to a limit imposed by the ratio of enzyme to BSA and glutaraldehyde and by the thickness of the immobilized enzyme layer. The optimum amounts of peroxidase and glucose oxidase were found to be 660 units (2 mg) and 400 units (3 mg), respectively. For BSA, the optimum volume was found to be 45 μl of 10% BSA. At BSA amounts lower than 45 μl, the ratio of enzymes to BSA was too high and a relatively smaller portion of the enzyme was immobilized. Larger volumes of BSA might have increased the degree of crosslinking, but it also may have led to blocking part of the active centers of the enzymes. Furthermore, larger volumes increased the thickness of the immobilized enzyme layer and therefore may have affected the diffusion rate. The amount of glutaraldehyde had a similar effect on the response of the probe as did BSA. The optimum amount of glutaraldehyde was 1 μl of 25% aqueous solution. Acetate buffer, pH 5 was used, and the optimum temperature was found to be 37°C. For glucose, 1m*M* *p*-fluorophenol was used as the reactant.

FIGURE 2 shows the calibration curve of the probe at optimum conditions. The linear regression line was ΔE (mv) = 25.96 log [glucose] − 8.18. A relative standard deviation of 3% was obtained for 20 repetitive measurements at 20 mg/100 ml glucose. With storage at room temperature, the activity of the enzyme decreased about 6% and

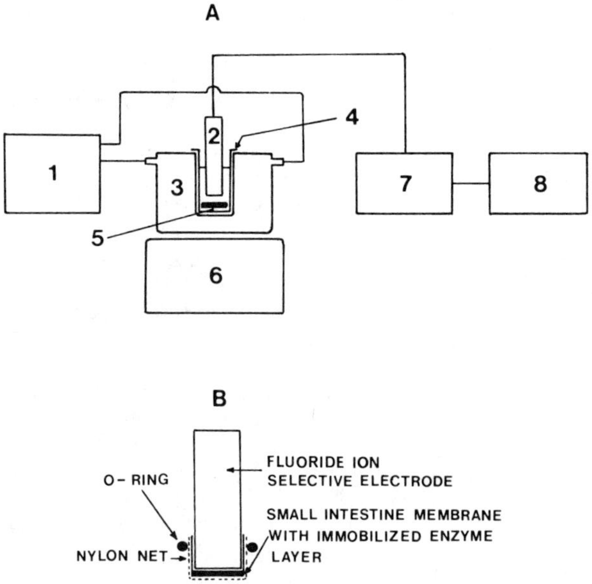

FIGURE 1. Schematic diagram of the apparatus (A) and immobilized enzyme probe (B) for glucose. (1) Temperature-controlled water bath circulator, (2) enzyme probe, (3) circulating jacket, (4) plastic cell, (5) magnetic stirring bar, (6) magnetic stirrer, (7) Orion Ionanalyzer 901, and (8) recorder.

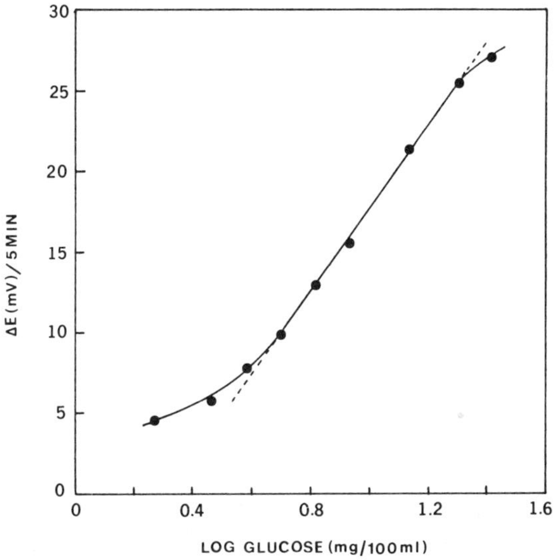

FIGURE 2. Calibration curve of the immobilized enzyme probe for glucose.

50% after 14 and 30 days, respectively. With storage in Tris buffer at 4°C, the activity remained almost the same after 30 days and 45 analyses. The selectivity of the probe depends on both the enzyme and the electrochemical detection method. Glucose oxidase is almost specific for β-D-glucose and the fluoride electrode is specific for fluoride ion. No interferences were observed with 10.4 mg/100ml uric acid, 40.8 mg/100 ml ascorbic acid, 35.2 mg/100 ml DL-cystine, 0.4 mg/100 ml bilirubin, 2.24 mg/100 ml sodium thiocyanate, and 87.0 mg/100 ml reduced glutathione. This immobilized enzyme probe, based on fluoride ion sensing, was highly specific for glucose and free of the interferences usually observed with the peroxidase probe.

REFERENCES

1. Hughes, G. M. K. & B. C. Saunders. 1954. Enzymatic rupture of a C-F bond. Chem. Ind. (London): 1265.
2. Hughes, G. M. K. & B. C. Saunders. 1954. Studies in peroxidase reaction. Part IX. Reaction involving the rupture of C-F, C-Br and C-I links in aromatic amines. J. Chem. Soc.: 4630–4634.
3. Yamazaki, I., H. S. Mason & L. Piette. 1960. Identification, by electron paramagnetic resonance spectroscopy of free radicals generated from substrates by peroxidase. J. Biol. Chem. **235:** 2444–2449.
4. Siddiqi, I. W. 1982. An electrochemical assays system for peroxidase and peroxidase-coupled reactions based on a fluoride ion-selective electrode. Clin. Chem. **28**(9): 1962–1967.
5. Ho, M. H. & T. G. Wu. 1984. Flow injection analysis of glucose using glucose oxidase coupled with potentiometric detection. In press.
6. Ho, M. H. & T. G. Wu. 1984. Immobilized enzyme electrode for glucose using fluoride ion sensor. In press.

Use of Immobilized Enzymes in Flow Injection Analysis

MAT H. HO AND MOORE U. ASOUZU

Department of Chemistry
University of Alabama in Birmingham
Birmingham, Alabama 35294

Flow injection analysis (FIA) is based on the injection of a precisely measured volume of sample into a continuously flowing, nonsegmented carrier stream.[1,2] The carrier stream transports the sample toward the flow-through detector. Necessary reagents needed for a particular analysis are either present in the carrier stream or can be added further downstream on the way to the detector. As it moves towards the detector, the sample disperses into the carrier stream both longitudinally and radially by a combination of controlled convection and diffusion processes. Controlled dispersion of the sample generates excellent, reproducible results. Since the reaction products are measured before "steady-state" conditions are established, the readout is available within seconds of introducing the sample, thus making high sample throughput possible. The advantages of FIA make it attractive for the enzymatic analysis of biologically important compounds. Unfortunately, the automated and continuous assays of these compounds in clinical laboratories require large and costly amounts of enzymes. This problem can be overcome by using immobilized enzyme reactors. In this paper, glucose was chosen as a model to illustrate the application of an immobilized enzyme reactor in FIA. Glucose oxidase catalyzes the oxidation of glucose to produce H_2O_2, which is detected electrochemically.

Glucose oxidase (E.C.1.1.3.4, from *Aspergillus niger*), β-D-glucose and glutaraldehyde (25% aqueous solution) were obtained from Sigma Chemical Co. Aminopropyl-controlled pore glass (AMP-CPG) of mean pore diameter 569 Å, pore volume 1.15 cc/g, surface area 43 m^2/g, and particle size 120/200, was from Electro-Neuclonics, Fairfield, NJ. The phosphate buffer (pH 7.0, 0.1 *M* phosphate) contained 3.0 g/l NaCl, 1.0 g/l sodium benzoate and 0.5 g/l dipotassium EDTA dihydrate.

Glucose oxidase was immobilized on AMP-CPG using the glutaraldehyde coupling method. The reactor was a Teflon tube (2 mm i.d.). Fifty mg of AMP-CPG was dried, treated with 2 ml of glutaraldehyde (1.0% in 0.1 *M* phosphate buffer, pH 7.0) for one hour at room temperature with shaking, and washed thoroughly with distilled water to remove excess glutaraldehyde. The derivatized AMP-CPG was incubated with 1,500 units of glucose oxidase in 500 μl phosphate buffer, pH 7.0, overnight at 4°C followed by extensive washing with buffer and water.

A schematic diagram of the flow injection system is shown in FIGURE 1. Buffer was used as a carrier and was delivered at a constant flow of 1 ml/min using a peristaltic pump. A Rheodyne valve (Model 7010, Rheodyne Inc., Cotati, CA) was modified with a bypass and was used as an injection valve. A Model 25 hydrogen peroxide electrode (Yellow Springs Instrument Co., Inc., Yellow Springs, OH) was covered with two layers of cellulose acetate membrane, prepared as described by Taylor *et al.*[3] A PAR-174A polarograph (Princeton Applied Research, Princeton, NJ) was used to apply a constant potential (0.7 V) to a two-electrode system. The output signal was fed into a strip chart recorder. Teflon tubing (0.8 mm i.d.) was used throughout the system.

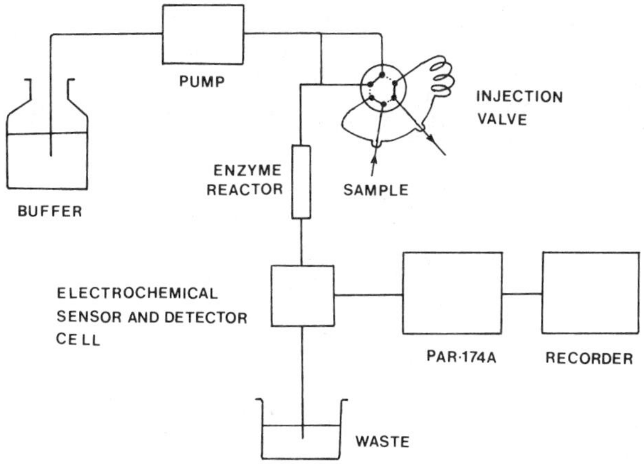

FIGURE 1. Schematic diagram of the flow injection system.

High conversion efficiency and low dispersion in the enzyme reactors are very important for analytical applications, particularly for flow injection analysis.[4] The length of the tubing and pump speed were chosen so that the time for the sample plug to travel through the reactor was long enough to allow a significant amount of reaction to occur but sufficiently short to minimize the dispersion. Better sensitivity was observed at low flow rates. This may be due to the fact that the reactor operates at less than 100% efficiency, and sensitivity is therefore inversely proportional to flow rate.[4] The packed-bed reactor was prepared using the glucose oxidase-derivatized AMP-CPG. The reactor was stored at 4°C in pH 7.0 phosphate buffer when not in use. Long-term stability of the reactor was studied by periodically testing its response to 48 mg/100 ml glucose. The activity of glucose oxidase increased slightly during the first

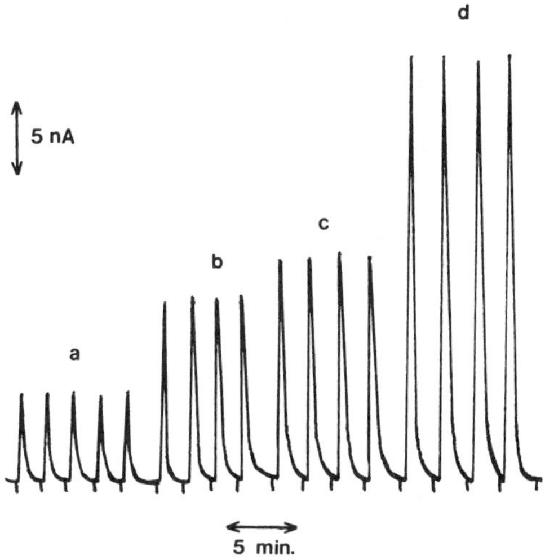

FIGURE 2. Typical response peaks in FIA system for glucose at (a) 15 mg/100 ml; (b) 33 mg/100 ml; (c) 40 mg/100 ml; and (d) 80 mg/100 ml.

few days and then decreased gradually to 92% of the initial value at 45 days. The selectivity of the system depends on both the enzyme and the electrochemical detection method. Glucose oxidase is almost specific for β-D-glucose, but the hydrogen peroxide electrode is capable of oxidizing other reducing agents, such as ascorbic acid, uric acid, and amino acids, which are common components in biological fluids. To improve the electrode selectivity, a cellulose acetate membrane with a molecular weight (MW) cutoff at about 100, was used. This membrane was permeable to H_2O_2 but presumably not to other medium MW reducing agents. In addition, the cellulose acetate membrane effectively prevented electrode poisoning arising from protein adsorption.

The FIA calibration plot was linear up to about 80 mg/100 ml. Above 80 mg/100 ml, deviation from linearity begins to occur. FIGURE 2 shows typical response peaks at several concentrations. Reproducibility of the assay showed a relative standard deviation for sets of 18 injections of 1.4 and 1.8% at concentrations of 16 mg/100 ml and 64 mg/100 ml, respectively.

REFERENCES

1. ROCKS, B. & C. RILEY. 1982. Flow injection analysis, a new approach to quantitative measurements in clinical chemistry. Clin. Chem. **28**(3): 409–421.
2. RUZICKA, J. & E. HANSEN. 1981. Flow Injection Analysis. Wiley Interscience. New York.
3. TAYLOR, P. J., E. KMETEC & J. M. JOHNSON. 1977. Design, construction, and applications of a galactose selective electrode. Anal. Chem. **49**(6): 789–794.
4. JOHANSSON, G., L. ORGEN & B. OLSSON. 1983. Enzyme reactors in unsegmented flow injection analysis. Anal. Chim. Acta **145:** 71–85.

Multifunctional Biosensor for the Determination of Fish Meat Freshness

ESTUO WATANABE AND KENZO TOYAMA

Department of Food Engineering and Technology
Tokyo University of Fisheries
4-5-7 Konan, Minato-ku
Tokyo, 108, Japan

ISAO KARUBE, HIDEAKI MATSUOKA, AND SHUICHI SUZUKI

Research Laboratory of Resources Utilization
Tokyo Institute of Technology
Nagatuta cho, Midori-ku
Yokohama, 227, Japan

The estimation of fish freshness is very important in the food industry for the manufacture of high-quality products. Indicators of fish freshness, such as nucleotides,[1] ammonia,[2] amines,[3] volatile acids,[4] catalase activity,[5] and pH,[6] have been proposed so far. However, the determination of these indicators requires complicated and time-consuming procedures. Recently, sensors consisting of immobilized enzymes and electrochemical devices have been developed for estimation of organic compounds[7-9] In the present study, a multifunctional enzyme sensor system was proposed for estimation of fish freshness. The freshness indicator K_I was represented by Equation 1.

$$K_I = \frac{\text{HxR} + \text{Hx}}{\text{IMP} + \text{HxR} + \text{Hx}} \times 100 \qquad \textbf{(1)}$$

K_I is based on changes in the concentrations of adenosine triphosphate (ATP), adenosine diphosphate (ADP), adenosine-monophosphate (AMP), inosine monophosphate (IMP), inosine (HxR), and hypoxanthine (Hx) in the fish meat. After the death of the fish, the ATP and ADP are decomposed rapidly, and IMP is formed. Changes in the AMP level are negligible, and the amount of AMP also is small. HxR and Hx gradually accumulate with decomposing IMP, because the degradations of HxR and Hx are rate-determining steps in this pathway. Therefore, there is a good correlation between K_I value and fish freshness. The enzyme reaction sequence is summarized as follows:

$$\text{ATP} \xrightarrow{\text{ATPase}} \text{ADP} \xrightarrow{\text{myokinase}} \text{AMP} \xrightarrow{\text{AMP-deaminase}} \text{IMP} \xrightarrow{\text{Nucleotidase}}$$

$$\text{HxR} \xrightarrow{\text{nucleoside phosphorylase, Pi}} \text{Hx} \xrightarrow{\text{xanthine oxidase, } O_2} \text{uric acid}$$

Various enzyme sensors for the determinations of Hx,[10] HxR,[11] IMP,[12] and AMP[13] have been prepared. Xanthine oxidase, xanthine oxidase-nucleoside phosphorylase, xanthine oxidase-nucleoside phosphorylase-nucleotidase, and xanthine oxidase-nucleoside phosphorylase-nucleotidase-AMPdeaminase were immobilized on a cellulose triacetate membrane containing 1,8-diamino-4-aminomethyloctane. These immobilized enzyme preparations were attached to an oxygen electrode, and the oxygen

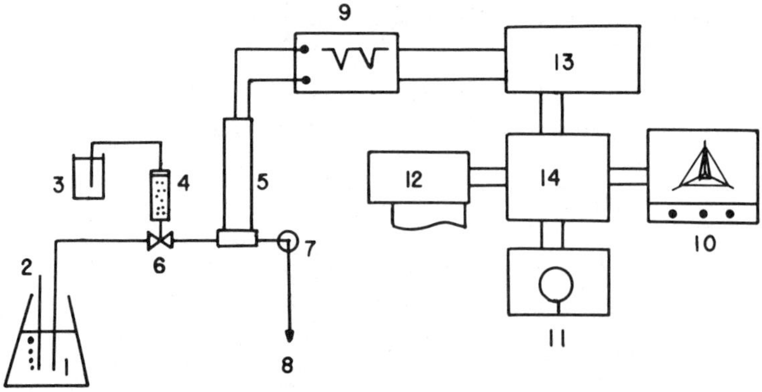

FIGURE 1. Schematic diagram of multifunctional enzyme sensor system. (1) buffer tank, (2) air, (3) eluate, (4) ion-exchange resin column, (5) enzyme electrode, (6) injection port, (7) peristaltic pump, (8) waste, (9) recorder, (10) disk, (11) monitor, (2) printer, (13) A/D converter, (14) Apple II.

consumption due to the oxidation of hypoxanthine was monitored. Each component was determined as the decrease in current. The multifunctional enzyme sensor system consisted of a combination of IMP sensor, ion-exchange resin column and computer (FIG. 1).

Optimum conditions for the enzyme sensor system were established as follows: pH 7.8; temperature, 30–32°C; flow rate of buffer, 1.5 ml; min^{-1}; flow rate of column, 1 ml · min^{-1}; and sample volume, 20 μl. One assay was completed in 25 minutes.

Fish freshness was determined simply and rapidly by this sensor system (TABLE 1). Moreover, fine differences among fish freshness, which could not be shown by any indicators so far proposed, were expressed in the freshness pattern (FIG. 2).

TABLE 1. K_I Values of Fish Meat

Fish	Storage[a] Time (days)	K_I Value Sensor	K_I Value Conventional Method
Sea bass	1	9.2	10.9
	2	19.4	19.0
	5	34.8	33.0
	10	55.3	56.0
Bream	1	4.7	6.6
	2	5.2	5.8
	5	12.4	10.9
	10	17.0	19.9
Flounder	1	10.0	12.9
	2	36.5	33.0
	5	58.2	55.0
	10	63.0	60.5
Yellowfin tuna	0	5.1	7.0
	1	8.6	8.0
	2	12.5	13.0
	3	24.6	25.0
	5	36.8	35.0

[a]Storage time after purchase from a retail store.

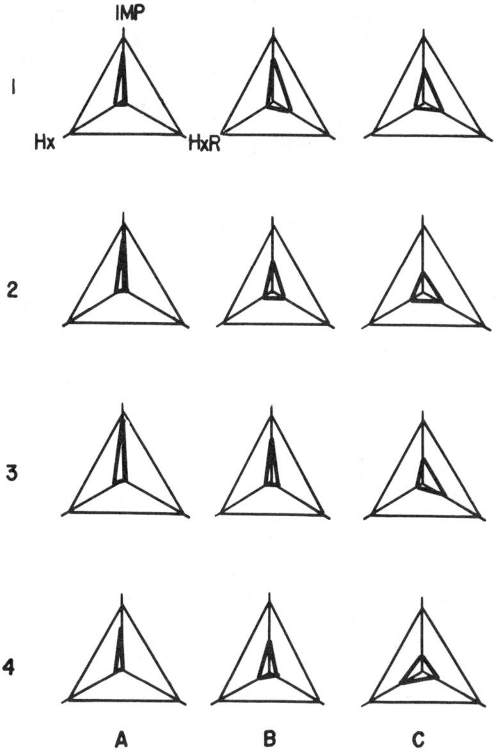

FIGURE 2. Freshness pattern. (1) yellowfin tuna, (2) sea bass, (3) bream, (4) flounder. (A) very fresh ($K_I < 10$), (B) fresh ($K_I < 40$), (C) not fresh ($K_I > 40$). Hx: hypoxanthine, HxR: inosine, IMP: inosine-5-monophosphate.

REFERENCES

1. SAITO, T., A. ARAI & M. MATSUYOSHI. 1959. A new method for estimating the freshness of fish. Bull. Jpn. Soc. Sci. Fish. **24:** 749–750.
2. OTA, F. & T. NAKAMURA. 1952. Variation of ammonia contents in fish meat by heating under pressure. Relation between the increase of ammonia and the freshness of fish. Bull. Jpn. Soc. Sci. Fish. **18:** 15–20.
3. KARUBE, I., I. SAITOH, Y. ARAKI & S. SUZUKI. 1980. Monoamine oxidase electrode in freshness testing of meat. Enzyme Microb. Technol. **2:** 117–200.
4. SUZUKI, T. 1953. Determination of volatile acids for judging the freshness of fish. Bull. Jpn Soc. Sci. Fish. **19:** 102–105.
5. MORI, T. & M. HATA. 1949. The new simplified method of determination for freshness of fishes. Bull. Jpn. Soc. Sci. Fish. **15:** 407–411.
6. KAWABATA, T., M. FUJIMAKI, K. AMANO & A. TOMIYA. 1952. Studies on pH of fish muscle. Variation in pH of fresh albacore muscle on the locality examined. Bull. Jpn. Soc. Sci. Fish. **18:** 32–40.
7. KARUBE, I., K. HARA, I. SATOH & S. SUZUKI. 1979. Amperometric determination of phosphatidyl choline in serum with use of immobilized phospholipase D and choline oxidase. Anal. Chim. Acta **106:** 243–250.

8. MELL, L. D. & J. T. MALOY. 1976. Amperometric response enhancement of the immobilized glucose oxidase enzyme electrode. Anal. Chem. **48:** 1597–1601.
9. GUILBAULT, G. G. & F. R. SHU. 1971. An electrode for the determination of glutamine. Anal. Chim. Acta **56:** 333–338.
10. WATANABE, E., K. ANDO, I. KARUBE, H. MATSUOKA & S. SUZUKI. 1983. Determination of hypoxanthine in fish meat with an enzyme sensor. J. Food Sci. **48:** 496–500.
11. WATANABE, E., K. TOYAMA, I. KARUBE, H. MATSUOKA & S. SUZUKI. 1984. Biofunctional enzyme sensor for hypoxanthine and inosine in fish meat. Eur. J. Appl. Microb. Biotechnol. **19:** 18–22.
12. WATANABE, E., K. TOYAMA, I. KARUBE, H. MATSUOKA & S. SUZUKI. 1984. Determination of inosine-5-monophosphate in fish meat with an enzyme sensor. J. Food Sci. **49:** 114–116.
13. WATANABE, E., T. OGURA, K. TOYAMA, I. KARUBE, H. MATSUOKA & S. SUZUKI. 1984. Determination of adenosine-5-monophosphate in fishes and shellfishes with an enzyme sensor. Enzyme Microb. Technol. **6:** 207–211.

Oxidase Enzyme: Enzyme and Immunoenzyme Sensor

J. L. ROMETTE AND J. L. BOITIEUX

Laboratoire de Technologie Enzymatique (U.T.C.)
B.P. 233
60206 Compiegne, France

ENZYME ELECTRODE

Classical analytical procedures require the precipitation or dialytic separation of proteins. In contrast, electrochemical monitoring with enzyme electrodes can be carried out on whole fermentation broth or biological media, thus eliminating preparation of the sample.

The literature reviewed shows numerous references on enzyme electrode production but only a few are commercially available. In fact, the problem is the application of theoretical procedures to practical measurements. For this important reason, experiments have been focused on the use of oxidase enzymes that are very specific and that do not need any cofactor added in the sample medium. But in this case, another difficulty exists: the oxygen content of the sample. Measurements in samples exhibiting high variability in oxygen content present an important problem. Different solutions exist, all based on the stabilization of the pO_2 sample before or during the measurement. However, how can one be sure that the enzyme incorporated inside the insoluble matrix of the electrode is not oxygen rate dependent? These enzymes usually are not the Michaelien type but of the ping-pong mechanism type. So there is not one Michaelis constant (K_m) but two: one for the main substrate, the other for the oxygen. We have solved the problem by using an enzyme membrane in which oxygen solubility is far higher than in aqueous solution.[1] In this way, the oxygen used during measurement is the oxygen dissolved in the membrane, not the oxygen present in the sample.

The use of gas-sensing membrane electrodes has led to the development of better enzyme electrodes that are free from ionic protein and electrochemically active compounds. The sensor in the electrode was a Clark electrode allowing an amperometric measurement of O_2 concentration. In order to get good mechanical stability, the active film containing the oxidase enzyme is coated directly on the gas-selective membrane of the pO_2 electrode.

We have previously described the technology that permits the development of these sensors.[2] Different enzyme electrodes are described for analysis applications in the fermentation field: glucose,[3] sucrose and lactose,[2] ethanol,[4] and L-lysine.[5]

An important characteristic of such sensors is their stability. For instance the stability of the lysine electrode is characterized by both its storage and operational conditions. When stored at 5°C in buffer, the enzyme membrane remains stable for at least six months. The operational stability was examined by repeatedly measuring the electrode response (a measurement every two minutes). The enzyme seems to be inactivated by the reaction. One molecule of enzyme can transform only a limited number of molecules of substrate. For 1 m*M* of lysine concentration, 200 measurements can be done; for 0.5 m*M* of lysine concentration, 400 measurements are performed. This type of inactivation is very similar to those observed for the glucose oxidase enzyme reaction.[6] It seems to be a constant for the FAD oxidase enzyme. When used in dynamic-state response, 3,000 measurements were assayed successfully for L-lysine determination and 5,000 for β-D-glucose determination.

A major problem for the in-line analysis and control of bioengineering systems is to significantly measure data such as substrate concentrations in real time. A computerized enzyme electrode[7] offers an efficient solution for this important problem. In this way, such electrodes can realize sixty analyses per hour in routine work, with an accuracy typically better than 1%.

IMMUNOENZYME ELECTRODE

We have previously described the fixation of immunoglobulin on gelatin membranes[8] and introduced an enzyme immunoassay (EIA) technique for the determination of hepatitis B surface antigen (HBsAg) in biological fluids.[9] We report the use of glucose oxidase as the enzyme label. The advantages of the glucose oxidase label have been recently demonstrated.[10] The purpose of our work was to study the possibilities of automating this type of antigen estimation in a semicontinuous flow mode, after processing the signals with the aid of a microprocessor. Thus, we demonstrated the applicability of this label to EIA by developing a computerized automatic system for the determination of HBsAg in biological fluids by quantitative evaluation of the enzyme reaction.

The active membrane is fixed on a pO_2 electrode, modified in our laboratory. The pO_2 consumption, due to the enzyme reaction, is measured in real time, when the electrode is in contact with a glucose standard solution. The signal of such an electrode is directly proportional to the O_2 consumption and to the antigen concentration. We also studied the effects of different variables such as incubation time, temperature, and pH on the formation of the antigen-antibody complex and on the amperometric kinetics before determining the sensitivity, reproducibility, and reliability of the procedure for the quantitative estimation of HBsAg by the computerized system. Although the immunological procedures take as long as other classical techniques, the enzymatic stage shortens the overall time used. This computerized enzyme immunosensor[11] allowed 30 sample measurements per hour. The measurement time included the washing step between samples, the sampling step and the pO_2 measurement step. The volume of the sample for each measurement is around 50 μl. The stability of the active membrane when stored at 4°C is very good (not more than 20% loss of initial activity after 6 months) so that the calibration curve can be reproduced easily with the same original membrane.

The calibration curve reveals the range of glucose activity, and consequently the antigen concentration estimated by this enzyme immunosensor. We have used a range of 0.1–10 μg/l for our assays and found the response to be perfectly linear. A good correlation ($\sigma = 0.970$) was observed between the values obtained by radioimmunoassay (RIA) and this enzyme immunosensor. The intra-assay coefficient of variation for samples (0.1–100 μg/l) was between 2 and 4% ($n = 10$). The interassay coefficient of variation was between 7 and 12%. The method gave a linear response for antigen concentrations up to 100 μg/l and the limit of sensitivity was approximately 0.1 μg/l.

CONCLUSION

It seems possible right now to anticipate a significant development of these techniques of analysis because they can be applied easily in the clinical analysis field as well as in biotechnology. The cost of the measurement is appreciably reduced, which is very important when one considers the financial problems that exist in the health-care economy. The other main advantage is the potentiality of such techniques. Today only a few enzymes are commercially available for analysis application. Microbiology and genetics will permit in the near future the production of more and more enzymes that have been improved for their use in analysis.

REFERENCES

1. Quenesson, J. C. & D. Thomas. 1977. French patent No. 7715,616.
2. Romette, J. L. 1980. Réalisation d'une électrode à enzyme pour le dosage du glucose. Ph.D. Thesis. University of Compiègne.
3. Romette, J. L., B. Froment & D. Thomas 1979. Glucose oxidase electrode. Clin. Chim. Acta **95:** 249–253.
4. Belghith, H. & J. L. Romette. 1984. Determination of ethanol by oxidase enzyme electrode. Biotechnol. Bioeng. To be submitted.
5. Romette, J. L., J. S. Yang, H. Kusakabe & D. Thomas. 1983. Enzyme electrode for specific determination of L-lysine. Biotechnol. Bioeng. **25:** 2557–2566.
6. Bourdillon, C., T. Vaughan & D. Thomas. 1982. Electrochemical study of D-glucose oxidase autoinactivation. Enzyme Microb. Technol. **4:** 175–180.
7. Kernevez, J. P., L. Konate & J. L. Romette. 1983. Determination of substrate concentrations by a computerized enzyme electrode. Biotechnol. Bioeng. **25:** 845–855.
8. Boitieux, J. L., G. Desmet & D. Thomas. 1978. Immobilization of anti HBsAg antibodies on artificial protein membranes. FEBS Lett. **93:** 133–136.
9. Boitieux, J. L., G. Desmet & D. Thomas. 1978. Determination potentiométrique de l'antigène de surface du virus de l'hépatite B dans les liquides biologiques. Clin. Chim. Acta **88:** 329–336.
10. Johnson, R. B., R. M. Libby & R. M. Nakamura 1980. Comparison of glucose oxidase and peroxidase as labels for antibody in enzyme-linked immunosorbent assays. J. Immunoassay 1. **1:** 27–37.
11. Boitieux, J. L., J. L. Romette, N. Aubry & D. Thomas. 1984. A computerized enzyme immunosensor: Applications for the determination of antigens. Clin. Chim. Acta **136:** 19–28.

Current Status of Activity Assays for Tissue Plasminogen Activator

JOHN C.-T. TANG, SHIRLEY LI, PAULA McGRAY, AND ANDREW VECCHIO

Biogen
14 Cambridge Center
Cambridge, Massachusetts 02142

Tissue plasminogen activator (TPA) is a serine protease involved in the fibrinolytic system that dissolves blood clots. The enzyme catalyzes the conversion of a zymogen. plasminogen, to the enzymatically active form, plasmin, by limited proteolysis. In the course of searching for specific activity assays that might be useful in monitoring the purification of TPA, we have developed several coupled photometric assays. In addition, radioactive, agarose-plate, and other activity assays have also been considered and investigated for this purpose (TABLE 1).

We have previously reported the one-step photometric procedure consisting of a thioester, thiobenzyl benzyloxycarbonyl-lysine (Z-Lys-S-Bzl) and fibrinogen-coated plates.[1] It is simpler and more sensitive than the old two-step method without using immobilized fibrinogen.[2] The new assay has been used successfully for protein purification and can be easily adapted to automated processes.

Recently, other chromogenic substrates, D-Val-L-Leu-L-Lys-*p*-nitroanilide (Val-Leu-Lys-*p*NA) and D-Ile-L-Pro-L-Arg-*p*-nitroanilide (Ile-Pro-Arg-*p*NA) were also used in the one-step assay. It is found that TPA activity is greatly enhanced by immobilized fibrinogen and free fibrinogen when either the thioester, Val-Leu-Lys-*p*NA, or Ile-Pro-Arg-*p*NA were used in the colorimetric assay (FIG. 1, A–D). Enzyme kinetics studies indicate that the K_m for plasminogen assayed on the thioester and fibrinogen-coated plates is 1.5 μg per ml, which is substantially lower than that observed in untreated plates (4.8 μg per ml). This is not due to the effect of fibrinogen on the second step of the coupled photometric assay because there is no change in the plasmin activity under these conditions (FIG. 1E). Similar results in TPA activation have also been observed, when fibrin-coated plates were used.

Free fibrinogen, which is an activator of TPA, has been included in the standard assay mixture. We are able to detect less than 1 ng of TPA activity within a one-hour incubation time at 20°C (FIG. 2A). In the thioester assay, however, high concentrations of reducing agents and nonspecific proteins cause significant background due to the interaction of DTNB with these reagents.

^{125}I-labeled fibrin-coated plates had been extensively used in the past for urokinase and TPA assays.[3] Although the sensitivity of the radioactive procedure is equivalent to that of the thioester photometric method, it appears that the kinetics of the enzyme are not easy to follow nor is the reproducibility great. In addition, the preparation and handling of radioactive substrates requires more work and training. We observed a narrow linear range of enzyme activity in the radioactive assay, usually between 10% to 50% of trypsin-releasable [^{125}I]fibrin in the reaction mixture (FIG. 2B).

The optimal pH of TPA activity assayed on the thioester, Z-Lys-S-Bzl (one-step or two-step assay) was in the range of pH 8.0 to pH 8.5, with a rapid decline of the activity below pH 6.0.[1] Therefore, the enzyme reaction with *p*-nitroanilide as a substrate can be stopped by adding acetic acid (0.2 *M*, final concentration). However, soybean trypsin

TABLE 1. Activity Assays for Tissue Plasminogen Activator

No.	Activity Assay	Substrate	Product	Reference
1	Radioactive assay	[^{125}I]Fibrin [^{125}I]Plasminogen	Soluble [^{125}I]fibrinopeptides Two-chain [^{125}I]plasmin (SDS-PAGE)	Unkeless, *et al.*[3] Roblin & Young[4]
2	Coupled photometric assay (One-step and two-step assays)	Thioester (Z-Lys-S-Bzl) *p*-nitroanilides (Val-Leu-Lys-*p*NA and Ile-Pro-Arg-*p*NA)	Free sulfhydryl group (Reacts with DTNB), *p*-Nitroaniline	Tang *et al.*,[1] Coleman & Green[2]
3	Agarose plate assay	Casein Fibrin	Soluble fibrinopeptides Soluble peptides (Lysis zone)	Failly-Crepin,[5] Marsh & Arocha-Pinazo,[6] Taylor *et al.*[7]
4	Fluorometric assay	CBZ-Gly-Pro-Arg-AEC CBZ-Gly-Gly-Arg-AMC tBOC-Val-Gly-Arg-β-naphthylamide	Fluorescent products	Nieuwenhuizen *et al.*,[8] Zimmerman *et al.*,[9] Obrenovitch *et al.*[10]
5	Direct photometric assay	Fibrin-suspension	Soluble fibrinopeptides (Decrease in tubidity)	Kanai et al.[11]

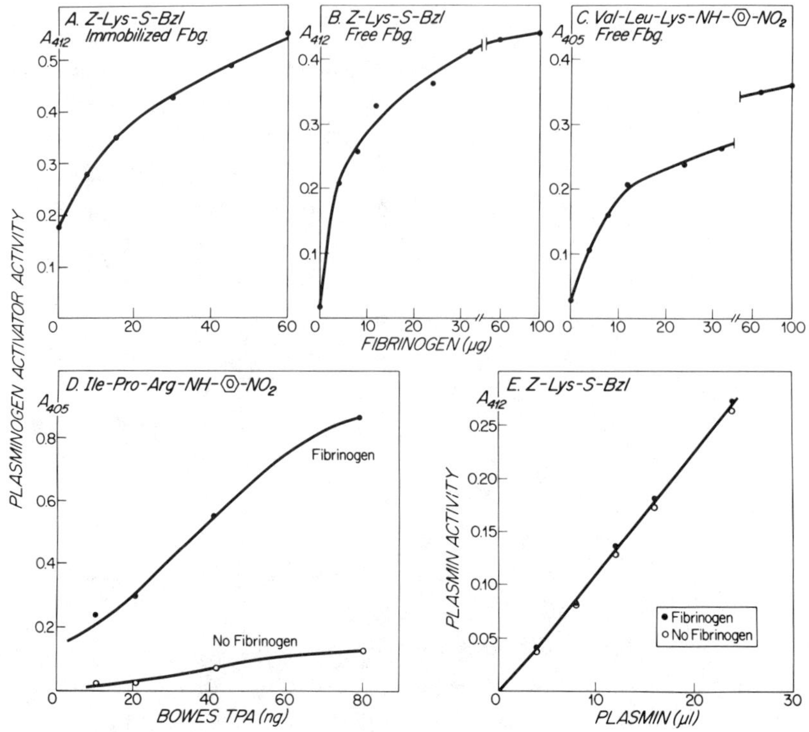

FIGURE 1. Effects of fibrinogen on activities of plasminogen activators and plasmin. The activity of plasminogen activators was assayed by a single-step photometric assay.[1] (A) The reaction mixture (total volume of 1.0 ml) contained 20 μmol Tris-HCl, pH 8.1; 0.001% Triton X-100; 5 μg plasminogen; 0.22 μmol DTNB, 0.20 μmol Z-Lys-S-Bzl, and enzymes. The assays were performed in the fibrinogen-coated plates. To prepare fibrinogen-coated plates, fibrinogen was diluted in 0.1× PBS to various concentrations. To each well of the tissue culture plates (Costar 24-well tissue culture clusters, well diameter, 16 mm), 0.5 ml of the fibrinogen solution was applied. The plates were dried at room temperature and then stored at 4°C until use. (B) The reaction mixture (total volume of 0.50 ml) contained 10 μmol Tris-HCl, pH 8.1; 0.001% Triton X-100; 2.5 μg plasminogen; the indicated concentrations of fibrinogen; 0.11 μmol DTNB; 0.10 μmol Z-Lys-S-Bzl; and enzymes. (C) Same assay mixture as that of B except that DTNB and Z-Lys-S-Bzl were replaced by 0.3 μmol Val-Leu-Lys-NH  NO_2 (Kabi substrate S-2251). (D) The assay mixture (total volume of 0.50 ml) contained 10 μmol Tris-HCl, pH 8.1; 0.001% Triton X-100; 50 μg fibrinogen; 0.3 μmol Ile-Pro-Arg-NH 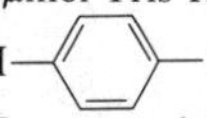 NO_2 (Kabi substrate S-2288), and enzymes. (E) Same assay mixture as that of B except that TPA was replaced by plasmin, and no plasminogen was added. The assays were performed in the presence or absence of 50 μg of free fibrinogen. Unless otherwise indicated, the enzymes used were purified from Bowes cells.

inhibitors were employed to stop TPA reactions using other substrates, since their enzymatic products were not stable under acidic conditions. The effects of natural and synthetic inhibitors on fibrinolytic activities of TPA are shown in TABLE 2.

The agarose-plate assay has been applied for a zymogram technique to detect TPA activity in polyacrylamide gels as well as an *in vivo* enzyme assay.[5–7] The diameter of

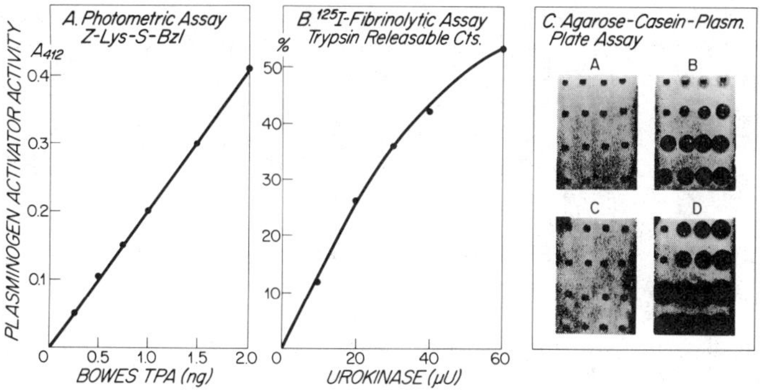

FIGURE 2. Activity assays for tissue plasminogen activators and urokinase. (A) The assay mixture contained 0.25–2.00 ng of purified Bowes TPA, 50 μg of free fibrinogen, and other standard components described in FIG. 1B. The enzyme reaction was assayed for 1 h at 20°C. (B) Fibrinolytic assays were performed in the tissue culture plates coated with [^{125}I]fibrin. Radioiodination of fibrinogen was carried out by the chloramine T method. To prepare fibrin-coated plates, [^{125}I]labeled fibrinogen was diluted in 0.1× PBS. To each well of the plates, 0.25 ml of the fibrinogen solution (30 μg and 40,000 cpm per well) was applied. They were dried at room temperature. Fibrinogen is converted into fibrin by the action of thrombin, which is present in serum. The fibrin-coated plates were prepared by further incubation of the fibrinogen-coated plates with 0.50 ml of 2.5% serum in PBS at 37°C for 3 hours. They were subsequently washed two times with distilled water and stored at 4°C. The assay mixture (total volume of 0.50 ml) contained 50 μmol Tris-HCl, pH 8.1; 125 μg BSA; 4 μg plasminogen and various urokinase concentrations (10 to 60 Sigma microunits). The enzyme mixtures were incubated at 20°C for one hour. Total releasable counts were determined by adding 400 μg trypsin to the reaction mixture. (C) The agarose gel consisted of 1% agar, 2% milk (Carnation nonfat dry milk), 0.02% sodium azide, and 0.05% plasminogen in PBS. For a 85 mm × 100 mm gel-bound film, 20 ml of the solution (warmed to 45°C before use) was poured onto the hydrophilic side of the film and spread evenly. Samples were placed into wells punched out of agarose gel. Gels A and C were control gels that did not contain the substrate plasminogen. Purified Bowes TPA (0.00, 0.005, 0.010, 0.020, 0.040, 0.064, 0.090, 0.120 plough units) was added to gels A and B in duplicate. Urokinase (0.00, 0.004, 0.008, 0.016, 0.032, 0.051, 0.077, 0.096 plough units) was added to gels C and D, also in duplicate. They were incubated at 37°C overnight in a humid environment.

TABLE 2. Effects of Inhibitors on Activity of Tissue Plasminogen Activator Using [^{125}I]Fibrinolytic Assay

Inhibitors	Concentration	Relative Activity (%)
None		100
Trypsin inhibitors		
a. Soybean	0.4 mg/ml	2
	1.6 mg/ml	0.5
b. Aprotinin	0.2 μg/ml	34
	20 μg/ml	2.0
	400 μg/ml	0.4
Lysine	20 m*M*	0.4
ε-Aminocaproic acid	20 m*M*	0.3
Arginine	20 m*M*	83
Serum	1%	64
	10%	15

the lysis zone is proportional to the activity of PA added and the lysis is fully dependent upon plasminogen. TPA activity monitored by the agarose-casein (or fibrin)-plasminogen plate assay shows a wider range of sensitivity than that of the radioactive method and provides semiquantitative results. We have used the casein-agar overlays as a routine procedure for screening TPA activity (FIG. 2C).

The sensitivity of a TPA assay using Z-Lys-S-Bzl is slightly higher than that using Val-Leu-Lys-*p*NA. However, the second *p*-nitroanilide, Ile-Pro-Arg-*p*NA, which is a substrate for a broad spectrum of serine proteases, shows TPA activity about an order of magnitude lower than that of the other two substrates tested. Both Z-Lys-S-Bzl and Val-Leu-Lys-*p*NA are suitable substrates for TPA and urokinase assays employing biological samples, since the enzyme reaction is fully dependent upon the natural substrate, plasminogen. Other TPA assay procedures indicated in TABLE 1 are not practical for enzyme kinetics studies or protein purification. This is due to either expensive substrates and equipment in fluorometric procedures,[8–10] or complicated methods with poor sensitivity in radioactive[4] and direct photometric[11] procedures.

ACKNOWLEDGMENTS

We thank Dr. Richard Flavell and Dr. John Smart for discussion. We also thank Miss Donna Roberto for editorial assistance.

REFERENCES

1. TANG, J. C.-T., P. MCGRAY & P. CHEN. 1983. Improved photometric assays for plasminogen activator (abstract). 83rd Annual Meeting of the American Society for Microbiology.
2. COLEMAN, P. L. & G. D. J. GREEN. 1981. A coupled photometric assay for plasminogen activator. Methods Enzymol. **80:** 408–414.
3. UNKELESS, J. C., A. TOBIA, L. OSSOWSKI, J. P. QUIGLEY, D. B. RIFKIN & E. REICH. 1973. An enzymatic function associated with transformation of fibroblasts by oncogenic viruses. J. Exp. Med. **137:** 85–126.
4. ROBLIN, R. & P. O. YOUNG 1980. Dexamethasone regulation of plasminogen activator in embryonic and tumor-derived human cells. Cancer Res. **40:** 2706–2713.
5. FAILLY-CREPIN, C. & J. URIEL. 1979. An electrophoretic assay for plasminogen activator. Biochimie **61:** 567–571.
6. MARSH, N. A. & C. L. AROCHA-PINANZO. 1972. Evaluation of the fibrin plate method for estimating plasminogen activators. Thromb. Diath. Haemorrh. **28:** 75–88.
7. TAYLOR, J. C., D. W. HILL & M. ROGOLSKY. 1972. Detection of caseinolytic and fibrinolytic activities of BHK-21 cell strains. Exp. Cell Res. **73:** 422–428.
8. NIEUWENHUIZEN, W. G., G. WIJNGAARDS & E. GROENEVELD. 1977. Fluorogenic peptide amide substrates for the estimation of plasminogen activators and plasmin. Anal. Biochem. **83:** 143–148.
9. ZIMMERMAN, M., J. P. QUIGLEY, B. ASHE, C. DORN, R. GOLDFARB & W. TROLL. 1978. Direct fluorescent assay of urokinase and plasminogen activators of normal and malignant cells: Kinetics and inhibitor profiles. Proc. Natl. Acad. Sci. USA **75:** 750–753.
10. OBRENOVITCH, A., C. MAINTIER, T. MAILLET, R. MAYER, C. KIEDA & M. MONSIGNY. 1983. Sensitive fluorometric determination of plasminogen activator in cell lysates and supernatants. FEBS Lett. **157:** 265–270.
11. KANAI, S., H. OKAMOTO, Y. TAMAURA, S. YAMAZAKI & Y. INADA. 1979. Fibrin suspension as a substrate for plasmin: Determination and kinetics. Thromb. Haemostas. **42:** 1153–1158.

PART VIII. ENZYMES AND HYDROCARBON SUBSTRATES AND ENVIRONMENTS

Oxidation of Gaseous Hydrocarbons by Methanotrophs: Heterogeneous Bioreactor

CHING T. HOU

Corporate Research Laboratories
Exxon Research and Engineering Company
Route 22 East, Clinton Township
Annandale, New Jersey 08801

Methane is the main ingredient of natural gas and oil field flare gas and is a by-product in several industrial processes. Microbial transformation of methane gas into liquid chemicals such as methanol, formaldehyde, and formic acid or into single cell protein has high industrial potential.

Methanotrophs are microorganisms that can grow on methane and other C_1 compounds and cannot utilize compounds having a C—C bond. Whittenbury and his coworkers[1] first systematically classified obligate methane utilizers into two groups based on their internal membrane structure type. Type I obligate methane utilizers have their membrane arranged in bundles of vesicular disks and form cysts in their resting stage. These organisms utilize the ribulose monophosphate carbon assimilation pathway and have an incomplete TCA cycle. The Type I membrane obligate methane utilizers consist of three genera: *Methylomonas, Methylobacter,* and *Methylococcus.*

Type II obligate methane utilizers have their membranes arranged in pairs around the cell periphery and form exospores or lipid cysts. The Type II organisms utilize the serine pathway for carbon assimilation and have a complete TCA cycle. Type II obligate methane utilizers have two genera: *Methylosinus* and *Methylocystis.*

Methane is oxidized by methylotrophic microorganisms through methanol, formaldehyde, formate and then into carbon dioxide and water (FIGURE 1). We have purified all of the enzymes in this methane catabolic pathway: methane monooxygenase,[2] methanol dehydrogenase[3] (or methanol oxidase in the case of yeast[4]) aldehyde dehydrogenase,[5,6,7] and formate dehydrogenase.[8]

Epoxides have become extremely valuable products because of their ability to undergo a variety of chemical reactions. The products of epoxidation are industrially important because of their ability to polymerize under thermal, ionic, and free radical catalysis to form epoxy homopolymers and copolymers. Ethylene oxide and propylene oxide are the two most important commercial epoxides. Over 90% of the current industrial output of propylene oxide is produced by either the chlorohydrin or the oxirane process. Both of these processes require multiple steps and depend heavily on the market price of their by-products, such as styrene. The only direct oxidation process known is still not far beyond the laboratory curiosity stage. It requires a metal oxide as the catalyst.

Several years ago, we isolated more than 20 methane-utilizing cultures and found that resting cell suspensions of these newly isolated cultures as well as known cultures all epoxidize gaseous alkenes to their corresponding 1,2-epoxides.[9] After the reaction, the cells were centrifuged, and the product, propylene oxide, was found totally in the supernatant fraction, that is, the product accumulated extracellularly. Control experiments with heat-killed cells indicated that the epoxide was produced enzymatically.

$$CH_4 \xrightarrow{E_1} CH_3OH \xrightarrow{E_2} HCHO \xrightarrow{E_3} HCOOH \xrightarrow{E_4} CO_2 + H_2O$$

- E_1: METHANE MONOXYGENASE
- E_2: METHANOL DEHYDROGENASE OR METHANOL OXIDASE
- E_3: ALDEHYDE DEHYDROGENASE
- E_4: FORMATE DEHYDROGENASE

FIGURE 1. Enzymes involved in the oxidation of methane.

More information on the epoxidation of propylene by methanotrophs was gathered in both *in vivo* and *in vitro* systems.

In batch experiments with whole cells, the rate of propylene oxide production was linear for the first 120 min (FIGURE 2). The optimum pH and temperature for the production of propylene oxide by resting cell suspensions of methanotrophs were pH 6–7 and temperatures around 35°C. Propylene oxide was found not to be further metabolized by methanotrophs. Although it is known that epoxidation of alkenes by cell-free methane monooxygenase (MMO) systems requires reducing power in the form of $NADH_2$, resting cell suspensions of methanotrophs do not require an exogeneous supply of reducing power for the epoxidation reaction. This is possibly due to the endogeneous storage of reducing power by the cells. The reason for the slower reaction rate after two hours of incubation was found to be mainly due to the depletion of the stored cofactor, $NADH_2$. In fact, we found that methane metabolites stimulated the epoxidation of propylene[10] (TABLE 1). The magnitude of stimulation depended on which compound was used. The addition of methanol or formate stimulated epoxidation of propylene.

The mechanism for the stimulation by methane metabolites is explained in FIGURE 3. Most steps in the methane metabolism, except the initial step, produce $NADH_2$ or a reducing power. Epoxidation of alkene by MMO requires $NADH_2$. The product epoxide is an end product and cannot be further metabolized to produce reducing

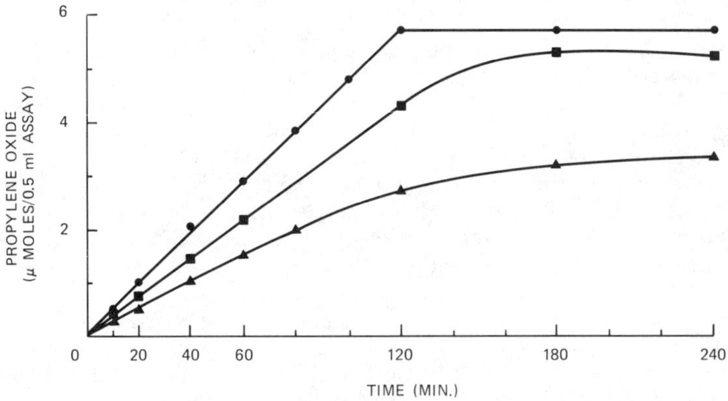

FIGURE 2. Time course of propylene oxide production by resting cell suspensions of methanotrophs.[9] *Methylosinus trichosporium OB3b* (●); *Methylococcus capsulatus* CRL M1 (■); and *Methylobacterium* sp. CRL 26 (▲).

TABLE 1. Stimulation of Epoxidation of Propylene in Whole Cell Suspensions of *Methlococcus capsulatus* CRL M1 by Methane Metabolites[10]

	Propylene Oxide Produced (μ Moles)		
	1 Hour	2 Hours	3 Hours
Control	2.0	3.5	3.6
+ CH_3OH (6 m*M*)	2.7	4.2	5.2
+ CH_3OH (48 m*M*)	1.8	4.2	5.2
+ HCHO (0.25 m*M*)	2.3	3.6	4.5
+ HCHO (4 m*M*)	1.9	3.5	3.6
+ HCHO (10 m*M*)	1.8	3.0	3.5
+ HCOOH (10 m*M*)	2.6	4.0	4.5
+ HCOOH (40 m*M*)	2.1	3.5	4.1

power. Therefore, the addition of methane metabolites to resting cell suspensions of methanotrophs provided additional reducing power in the form of reduced NAD^+.

The generation and/or regeneration of $NADH_2$ is a key problem in the design of a bioreactor to be used for the microbial epoxidation process. We have studied cofactor regeneration in a cell-free system and demonstrated the generation and regeneration of $NADH_2$ with various dehydrogenases.[11] $NADH_2$ regenerated from formate and formate dehydrogenase was found compatible in coupling with the MMO epoxidation system (FIGURE 4). For example, the production of propylene oxide was conducted with either excess or limited amounts of $NADH_2$. In the case of a limited amount of $NADH_2$ (—△—), the production of propylene oxide stopped within a short time and resumed with the addition of formate (↓), indicating the regeneration of cofactor. In addition, we also demonstrated that $NADH_2$ can be generated from NAD^+ and formate (the price of NAD^+ is six times lower than that of $NADH_2$) (FIGURE 5). NAD alone without formate showed no production of propylene oxide.[5] When formate was added (at the arrow point), the production of propylene oxide started. Both the complete system (with NAD^+ and formate) and the control system (with $NADH_2$) showed identical epoxidation activity. Other $NADH_2$-generation systems (TABLE 2), such as diol dehydrogenase, secondary alcohol dehydrogenase, primary alcohol dehydrogenase, and their substrates were found equally compatible in coupling with the MMO epoxidation system.[11]

However, because of the instability of the MMO epoxidation system in its cell-free form,[1,12] we concluded that any industrial applications of bio-epoxidation processes would probably use a whole cell system.[13] Therefore, we devised a laboratory-scale

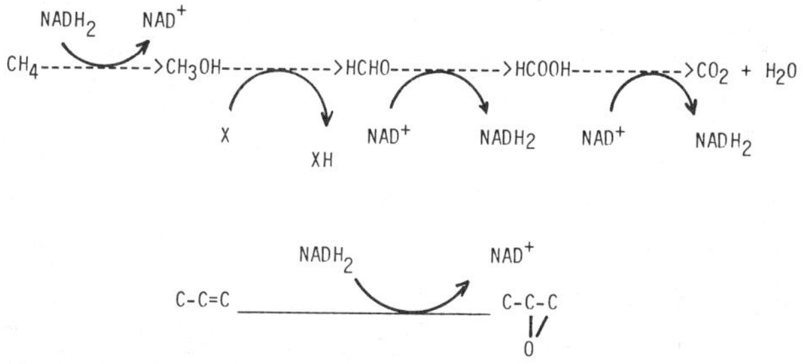

FIGURE 3. Mechanism for the stimulation of epoxidation by methane metabolites.

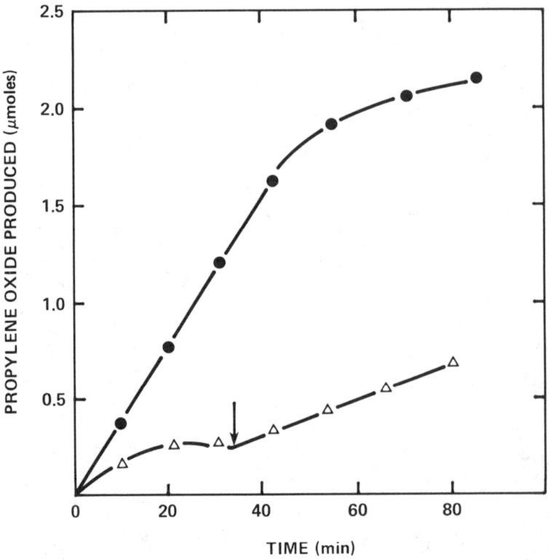

FIGURE 4. Epoxidation of propylene by soluble methane monooxygenase from *Methylosinus* sp. CRL 31 using $NADH_2$ regenerated from formate and formate dehydrogenase.[11] (■) Excess amount of $NADH_2$; (△) limited amount of $NADH_2$; ↓ addition of sodium formate.

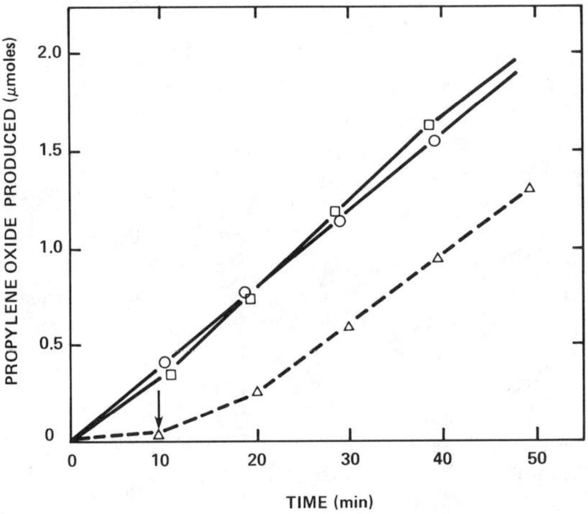

FIGURE 5. Epoxidation of propylene by soluble methane monooxygenase from *Methylosinus* sp. CRL 31 using cofactor generated from NAD^+ by formate and formate dehydrogenase.[11] (O) Control, with $NADH_2$; (△) with NAD^+; arrow point indicates the addition of formate; (□) NAD^+ and formate were added at the beginning of the reaction.

TABLE 2. Epoxidation of Propylene by Soluble Methane Monooxygenase from *Methylosinus* sp. CRL 31 Using Cofactor Generated by Various Dehydrogenases and Their Substrates[11]

Cofactor Generation System[a]	Rate of Epoxidation of Propylene (nmoles/min/mg protein)
Formate dehydrogenase, formate and NAD^+	70
Diol dehydrogenase, 1,2-propanediol and NAD^+	60
Secondary alcohol dehydrogenase, 2-propanol and NAD^+	60
Primary alcohol dehydrogenase, ethanol, and NAD^+	75
Control 1 $NADH_2$	75
Control 2 NAD^+	0

[a]Note: The specific activities for dehydrogenases added or in the soluble fraction (in nmoles/min/mg protein) were: formate dehydrogenase, 60; diol dehydrogenase, 40; secondary alcohol dehydrogenase, 40; and primary alcohol dehydrogenase, 80. Substrate for the dehydrogenase and NAD^+ added were 10 μmoles and 5 μmoles, respectively.

bioreactor using immobilized whole cells. The gaseous nature of the substrates makes this bioreactor system unique.

Immobilization of cells via covalent binding of the cells to long ligands of a macromolecule resulted in loss of epoxidation activity. This is possibly due to alteration of the membrane structure of the cells. Cells were also entrapped in polyacrylamide, and the cell-containing polyacrylamide was freeze-dried and was ground into powder. These powders were suspended in buffer solution and their activity was tested. Again, no epoxidation activity was detected. This is possibly due to loss of enzyme activity during the lengthy cell-manipulation procedure and the diffusion barrier created by polyacrylamide, which could prevent the diffusion of gaseous substrates.

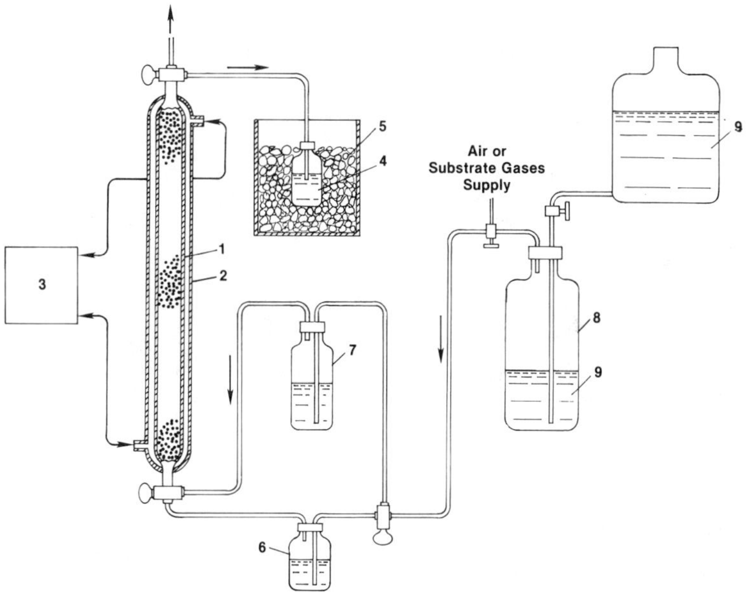

FIGURE 6. Schematic diagram of the gas-solid bioreactor system with biocatalyst regeneration capacity; (1) bioreactor; (2) jacket; (3) temperature control device; (4) liquid for product recovery; (5) ice; (6) water at 40°C; (7) methanol for biocatalyst regeneration; (8) gaseous substrate mixture; and (9) water.

Finally, we devised a simple gas-solid heterogeneous bioreactor system that provided good yields of propylene oxide (FIGURE 6). The capacity of this bioreactor was about 7 ml. Cell paste of methane-grown *Methylosinus* sp. CRL 31 in phosphate buffer pH 7.0 was coated on porous glass beads (2 mm diameter). About 20 mg of cells were physically adhered in a thin layer on these beads. The cell-coated glass beads were then packed into a glass column. The bioreactor was equipped with a jacket for circulating water to control the reaction temperature.

The gaseous substrate (a mixture of propylene and oxygen) was introduced through a water bottle (maintained at 40°C) to pick up moisture and then into the bottom of the bioreactor. The temperature of the bioreactor was kept at 40°C (higher than the boiling point of propylene oxide, 35°C). The product was recovered from the exhaust gas stream by cooling using ice as the coolant.

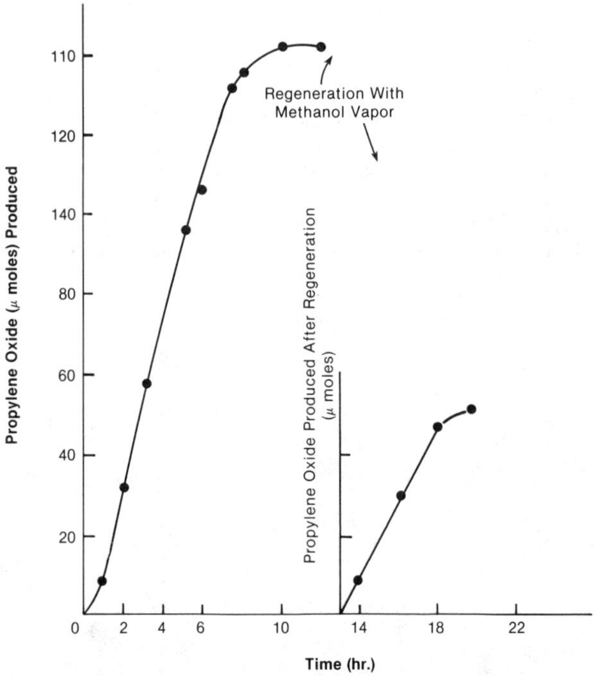

FIGURE 7. Production of propylene oxide from propylene by cells of methane-grown *Methylosinus* sp. CRL 31 packed in a gas-solid bioreactor. Regeneration of the biocatalyst was conducted after 12 hours of operation.

Initially, the gaseous phase inside the bioreactor was evacuated and the reactor was then filled with the gaseous substrates. The substrates were then introduced continuously into the bioreactor at a flow rate of 0.5 ml/min. The relative humidity inside the bioreactor was maintained at about 70%. The production of propylene oxide was at a constant rate of 18 μmoles/h for the first seven hours (FIGURE 7). The turnover rate for propylene was 2.7%.

After seven hours of continuous operation, the rate of propylene oxide production slowed down, possibly due to the depletion of endogeneous reducing power ($NADH_2$). After ten hours of operation, product formation essentially stopped. In batch experiments, the same cell suspension produced propylene oxide at 1.5 μmoles/h/mg protein for two hours.

After 12 hours, regeneration of the biocatalyst was attempted *in situ* in the

TABLE 3. Production of Propylene Oxide from Propylene in a Gas-Solid Bioreactor

	Propylene Oxide Produced (μmol/20 mg cells)		
Microbes	Initial	1st Regeneration	2nd Regeneration
Methylosinus sp. CRL 31	145	50	18
Methylococcus sp. CRL M1	120	18	6
Methylosinus trichosporium OB3b	135	40	10

bioreactor. The regeneration substrate used was methanol. The substrate gas mixture was replaced with air. The air was forced to pass through methanol (also maintained at 40°C) and then into the bioreactor, carrying methanol vapor with it. The regeneration of the biocatalyst was continued for about 30 min at a 3 ml/min flow rate. At the end of the regeneration, substrate gases were reintroduced. The production of propylene oxide immediately resumed at a constant rate of 12 μmoles/h. After an additional six hours of operation, the reaction rate slowed down again, indicating the need for renewed cofactor regeneration. Two repeat cofactor regeneration operations were performed.

Cells of other methanotrophs were also tested in this bioreactor for the production of propylene oxide. Results are shown in TABLE 3. It appears that this heterogeneous bioreactor can be applied to all types of methanotrophs.

A major problem in the biotechnological process is the toxicity of the reaction product for the biocatalyst. To avoid epoxide accumulation in the microenvironment of the biocatalyst and thus inactivation, rapid removal of this product is essential.

Although we have not optimized the reaction conditions for an individual strain of methanotroph for this heterogeneous gas-solid bioreactor, our data indicate the feasibility of using this new type of bioreactor for the production of chemicals using gaseous substrates. This type of bioreactor offers the following advantages: a simple step for loading and unloading of biocatalyst, a process for continuous production of product, a rapid and continuous removal of product from the reactor, simple cofactor regeneration, and little process water required.

REFERENCES

1. WHITTENBURY, R., K. C. PHILLIPS & J. F. WILKINSON. 1970. Enrichment, isolation, and some properties of methane-utilizing bacteria. J. Gen. Microbiol. **61:** 205–218.
2. PATEL, R. N., C. T. HOU, A. I. LASKIN & A. FELIX. 1982. Microbial oxidation of hydrocarbons: Properties of a soluble methane monooxygenase from a facultative methane-utilizing organism, *Methylobacterium* sp. strain CRL-26. Appl. Environ. Microbiol. **44:** 1130–1137.
3. PATEL, R. N., C. T. HOU & A. FELIX. 1978. Microbial oxidation of methane and methanol: Crystallization of methanol dehydrogenase and properties of holo- and apo-methanol dehydrogenase from Methylomonas methanica J. Bacteriol. **133:** 641–649.
4. PATEL, R. N., C. T. HOU, A. I. LASKIN & P. DERELANKO. 1981. Microbial oxidation of methanol: Properties of crystallized alcohol oxidase from yeast, *Pichia* sp. Arch. Biochem. Biophys. **210:** 481–488.
5. PATEL, R. N., C. T. HOU & A. FELIX. 1979. Microbial oxidation of methane and methanol: Purification and properties of aldehyde dehydrogenase from *Methylomonas methylovora*. Arch. Microbiol. **122:** 241–247.
6. PATEL, R. N., C. T. HOU, P. DERELANKO & A. FELIX. 1980. Purification and properties of a heme-containing aldehyde dehydrogenase from *Methylosinus trichosporium*. Arch. Biochem. Biophys. **203:** 654–662.
7. PATEL, R. N., C. T. HOU & P. DERELANKO. 1983. Microbial oxidation of methanol: Purification and properties of formaldehyde dehydrogenase from *Pichia* sp. NRRL Y-11328. Arch. Biochem. Biophys. **221:** 135–142.
8. HOU, C. T., R. N. PATEL, A. I. LASKIN & N. BARNABE. 1982. NAD-linked formate dehydrogenase from methanol-grown *Pichia pastori*, NRRL Y-7556. Arch. Biochem. Biophys. **216:** 296–305.
9. HOU, C. T., R. N. PATEL, A. I. LASKIN & N. BARNABE. 1979. Microbial oxidation of gaseous hydrocarbons: Epoxidation of *n*-alkenes by methylotrophic bacteria. Appl. Environ. Microbiol. **38:** 127–134.
10. HOU, C. T., R. N. PATEL, A. I. LASKIN, I. MARCZAK & N. BARNABE. 1980. Microbial

oxidation of gaseous hydrocarbons: Oxidation of lower *n*-alkanes and *n*-alkenes by resting cell suspensions of various methylotrophic bacteria and effect of methane metabolite. FEBS Lett. **9:** 267–270.

11. Hou, C. T., R. N. Patel, A. I. Laskin & N. Barnabe. 1982. Epoxidation of alkenes by methane monooxygenase: Generation and regeneration of cofactor, $NADH_2$, by dehydrogenases. J. Appl. Biochem. **4:** 379–383.
12. Colby, J., D. I. Stirling & H. Dalton 1977. The soluble methane monooxygenase of *Methylococcus capsulatus* (Bath). Its ability to oxygenate *n*-alkane, *n*-alkenes, ethers, and alicyclic, aromatic, and heterocyclic compounds.
13. Hou, C. T., R. N. Patel & A. I. Laskin. 1980. Epoxidation and ketone formation by C_1-utilizing microbes, Adv. Appl. Microbiol. **26:** 41–69.

Hydrocarbon Micellar Solutions in Enzymatic Reactions of Apolar Compounds

P. L. LUISI,[a] P. LÜTHI,[a] I. TOMKA,[a] J. PRENOSIL,[b] AND A. PANDE[a]

[a]*Institut für Polymere*
ETH-Zentrum
CH-8092 Zürich, Switzerland

[b]*Technisch-Chemisches Laboratorium*
ETH-Zentrum
CH-8092 Zürich, Switzerland

INTRODUCTION

There have been several attempts to use enzymes, which are generally stable in predominantly aqueous systems, in organic solvents. The aim of these studies was to set up methods to enzymatically convert water-insoluble, highly lipophilic substrates, such as steroids, aldehydes, or alcohols with long aliphatic chains, lipidic substances, waxes, and so forth. The best-known method is perhaps the one that uses two phases, an aqueous one containing the enzyme, and the nonaqueous one containing the lipophilic substrate.[1-3] The reaction takes place at the interface. In a more recent paper, Klibanov and coworkers describe a method whereby the enzyme is immobilized in the cavities of porous materials that can operate in organic solvents.[4]

A few years ago, a novel method was proposed, which is based on the use of reverse micelles. These are aggregates formed by certain surfactants when dissolved in apolar solvents; the polar heads of the surfactants tend to avoid the solvent and form a polar core, whereas the aliphatic chains of the surfactants are directed as schematized in FIGURE 1. The term "reverse" conveys the idea that the structure is reverse as compared to the one present in aqueous micellar solution, where in fact the polar heads are exposed to water. The polar core of reverse micelles can solubilize water, giving rise to what is defined as a "water pool"; consequently, hydrophilic molecules can be solubilized in this water pool. The literature on micelles and reverse micelles is abundant.[5-7]

It has been shown that enzymes can be solubilized in the water pool of reverse micelles without loss of activity. One of the most-used systems consists of isooctane as solvent, and *bis*(2-ethyl-hexyl) sodiumsulfosuccinate, abbreviated AOT (Aereosol-OT) as the surfactant. In this system, an enzymatically active hydrocarbon solution, containing as little as 1% water (vol/vol) and a few percent (by weight) of surfactant is readily obtained. The micellar hydrocarbon solutions are generally transparent and lend themselves to the same manipulations and studies (including spectroscopic investigations in the far UV-spectral range) as aqeuous solutions.

A number of papers have described the activity of hydrophilic enzymes, such as α-chymotrypsin, lysozyme, peroxidase, ribonuclease, and alcohol dehydrogenase, as a function of the micellar parameters, such as water content, concentration of and type of surfactant, and the reader is referred to such papers for detailed information.[8-12]

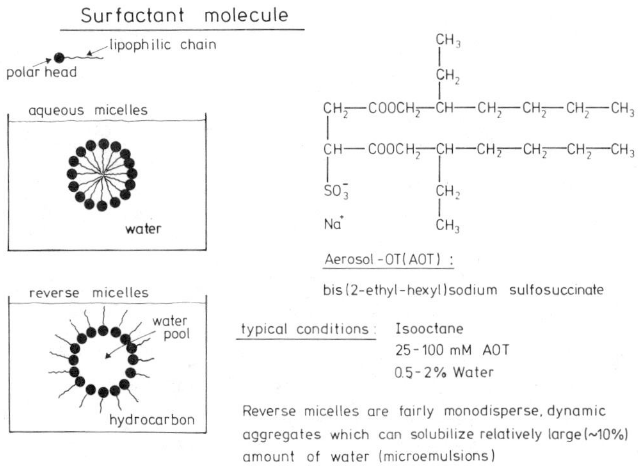

FIGURE 1. Schematic representation of reverse micelles (cross-section); and the structure of the surfactant AOT, with a typical set of conditions.

Here, we would like to review only the use of such enzymatic systems for apolar substrates. Although this is the main reason for interest in enzymes in reverse micelles as far as biotechnology is concerned, the major trust so far has been on water-soluble enzymes and substrates, and in dealing with the basic chemistry of the processes involved. This is understandable, since the field is quite new and this necessitates a basic screening with simpler components. However, it is time now to ascertain what is the possible biotechnological relevance of these systems. In this paper, we bring together the sparse data on apolar substrates, and address the question: What kind of generalizations are possible on the basis of available data?

APPLICATIONS

Oxidation of Polyunsaturated Fatty Acids—Linoleic Acid

Soybean lipoxygenase-l catalyzes the oxidation, with molecular oxygen, of polyunsaturated fatty acids containing a 1:4 diene system to give hydroperoxides.[13] Linoleic acid, the commonly used substrate in assaying this enzyme, presents a problem due to poor solubility in aqueous solutions, especially in the acidic pH range,[14] which makes the spectroscopic monitoring procedures difficult. This problem is eliminated in reverse micellar solutions (10 m*M* AOT/isooctane), as linoleic acid (and probably other similar substrates) is soluble in isooctane, permitting the reaction to be carried out in an acidic pH range.

The enzyme exhibits typical Michaelis-Menten type behavior (FIG. 2) and the product, from the absorption spectral characteristics, appears to be the same as in the aqueous medium. The kinetic parameters, in our preliminary studies, do not appear to compare favorably with those in the aqueous solution.[15] For example, $K_m{}^a$ is about

$^a K_m$ stands for the Michaelis constant, k_{cat} for the turnover number, and w_o for [water]/[surfactant].

fourfold higher and k_{cat} about an order of magnitude lower.[16] However, we have not yet optimized the reaction parameters and optimization (discussed later) may give better results. Indeed, an increase in w_o, for example, appears to give higher reaction rates until an optimum ($w_o = 30$) is reached.[17]

Lipoxygenase-1 in aqueous solutions shows a fluorescence emission maximum at 328 nm on excitation at 280 nm. Upon denaturation, the emission maximum shifts to 345 nm.[18] The emission maximum in reverse micelles remains at 328 nm.[16] Likewise, the far ultraviolet circular dichroic spectrum[19] is the same as in the aqueous solution (for example, at $w_0 = 16$). However, the negative ellipticity is lower at lower w_0 (for example at $w_0 = 6$), indicating a less-ordered secondary structure at lower w_0 values.[16] The important point is that the enzyme appears to maintain its structure and function in reverse micelles. It is by no means trivial, considering that it is a large (MW 100,000) metalloprotein. Heme proteins, for example, myoglobin and cytochrome P-450, do not appear to be so stable in this system.[20,21]

Reduction of Ketosteroids

It has been shown[17] that the reduction of ketosteroids can be achieved using NADH and horse liver alcohol dehydrogenase (LADH) in an AOT/isooctane reverse micellar system or, as shown by Laane and coworkers,[22] for progesterone and prednisone, by the use of NADH and 20β-hydroxysteroid dehydrogenase in a cetyltrimethylammonium bromide (CTAB)/hexanol or chloroform and octane reverse micellar system. The latter workers actually used a novel NADH-regenerating system

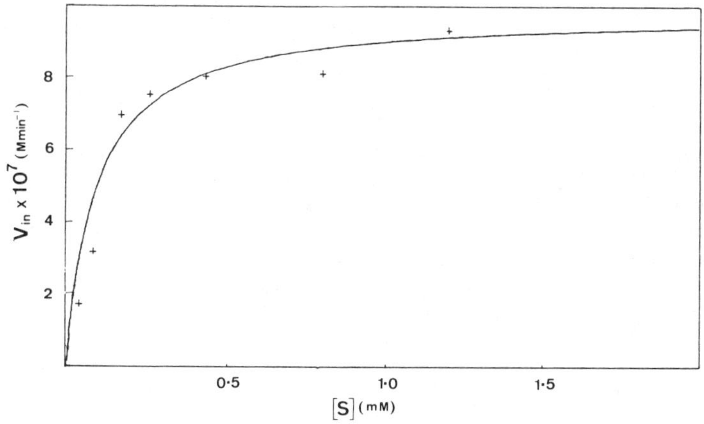

FIGURE 2. A plot of initial reaction velocity (for the conversion of linoleic acid by liponygenase-1), at various overall substrate concentrations (the term "overall concentration" refers to concentration in the total volume of the solution, hydrocarbon and aqueous). Experimental data (+) and the calculated curve (—) obtained by fitting the data to a linearized Michaelis-Menten plot using the weighted linear least square fitting procedure. The following kinetic parameters were obtained: $K_m = 100\ \mu M$, $k_{cat} = 2.0\ sec^{-1}$. Reaction medium: 10 m*M* AOT/isooctane ($w_o \simeq 15$; pH of the enzyme and substrate solution = 10.0, 50 m*M* borate buffer; overall concentration of lipoxygenase = 8.2×10^{-9} *M*. Note that while in aqueous solution, the pH optimum of this reaction is at pH 9, solutions at pH 10 were used because in AOT/isooctane reverse micelles, the pH optimum shifts to pH 10.[17]

consisting of a hydrogenase that uses hydrogen gas (H_2) to reduce methyl viologen (MV^{2+}) which in turn reduces NAD^+ in the presence of lipoamide dehydrogenase. Thus NAD^+ and MV^{2+} are recycled while H_2 is consumed. The complete system is depicted in FIGURE 3.

α-Chymotrypsin-catalyzed Synthesis of Peptide Bonds

The reverse micelle, as a microreactor, permits interesting chemical reactions when the product(s) and the reactants have opposite solubility characteristics in this microheterogenous system. Specifically, if the reactants are water-soluble and the product(s) more hydrocarbon-soluble, the latter, following synthesis, would be expelled from the microreactor. Even when the partition coefficient of the product is not highly in favor of the hydrocarbon phase, the large excess of the bulk hydrocarbon solvent with respect to water results in a significant transfer of the product to the hydrocarbon phase. One can achieve higher efficiency by using an enzyme reactor that permits spatial segregation of the product from the reaction mixture. This is shown in FIGURE 4, where a reactor with hollow fibers is employed for the purpose. The capillaries of the semipermeable tubular polyamide membrane (molecular weight cutoff of 10,000) permit the required segregation by preventing the enzyme-containing micelles to enter the bulk hydrocarbon phase. The reaction is represented in the figure as $A + B \rightleftharpoons C$. $C \rightleftharpoons C^*$ represents the partitioning of the product between the micellar water pool and the bulk hydrocarbon solvent.

We have tried to apply this concept to the synthesis of peptides catalyzed by proteases. We have been working in this area using enzymes in aqueous solution,[23] and the extension to hydrocarbon micellar solution appeared obvious, both because of the relatively small amount of water present in the reverse micellar system (excess water favors hydrolysis over synthesis) and because of the above-mentioned preferential solubility effects. As a first example, we studied the reaction

$$\text{Ac-Trp-OEt} + \text{H-Trp-NH}_2 \rightleftharpoons \text{Ac-Trp-Trp-NH}_2 + \text{EtOH}$$

catalyzed by α-chymotrypsin under the following conditions: reaction medium, 100 m*M* AOT in isooctane; $w_o = 9$; overall concentration of the enzyme and reactants, 1 μ*M* and 0.6 m*M*, respectively; 50 m*M* carbonate buffer, pH = 10.0. Most of these conditions are based on the work of Morihara and Oka, who studied the above reaction in aqueous solution.[24] Although the yield was lower than 40%, the approach appears promising and further studies towards optimizing the process are in progress.

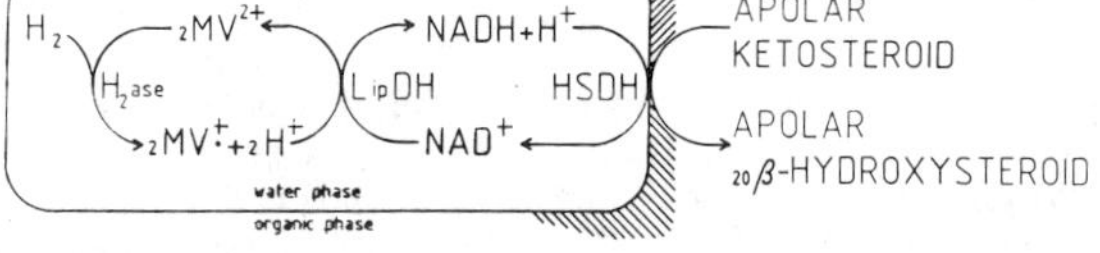

FIGURE 3. Scheme for the H_2-driven regeneration of NADH and the subsequent reduction of an apolar steroid in a reverse micellar medium. For clarity, all the water-soluble components of the system are drawn in one micelle. Abbreviations used are: H_2ase for hydrogenase, LipDH for pig heart lipoamide dehydrogenase, HSDM for 20β-hydroxysteroid dehydrogenase, NAD^+ and NADH for nicotinamide adenine dinucleotide and reduced nicotinamide adenine dinucleotide, respectively, and MV for methyl viologen. (Figure taken from Gibian and Galaway,[15] in which other details are also given.)

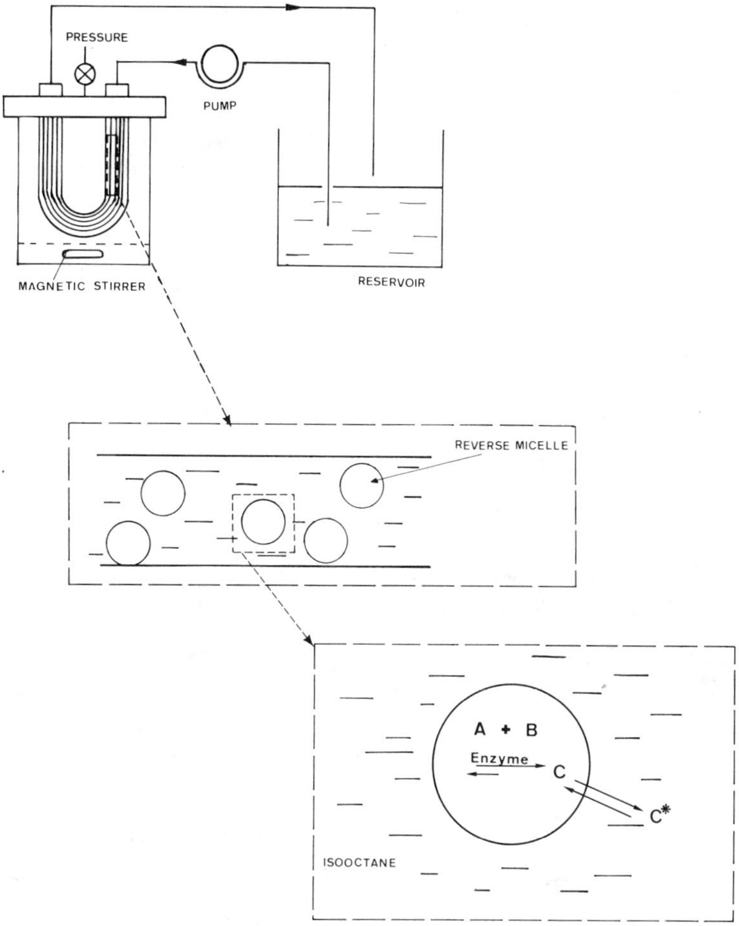

FIGURE 4. A schematic representation of a reactor (Berghof Miniconcentrator BMS) with the capillary system (internal volume 0.4 ml and surface area 20 cm^2); an enlarged view of a capillary of semipermeable tubular membrane and further magnification showing a reverse micelle in the organic solvent. The enzymatic coupling reaction takes place in the water pool of the reverse micelle, as shown.

Another reaction that appears to give better yield of the product is the synthesis of Ac-Phe-Leu-NH_2, starting from Ac-Phe-OEt and H-Leu-NH_2 and using α-chymotrypsin as the catalyst.

Miscellaneous Examples

Almost a decade ago, when reverse micellar systems were not yet as well characterized, Wells and coworkers published a series of papers[25–27] on the phosphatidylcholine/diethylether reverse micellar system. They showed that phospholipase A_2 was active in that system and the activity was a function of w_o and calcium ion concentration. This study, incidentally, provides an interesting case in which the surfactant lipid also serves as a substrate.

Substrate may also serve as a cosurfactant and probably does so, for example, in the oxidation of aliphatic alcohols of $H(CH_2)_nOH$ (n = 2–10) type using LADH.[10] Martinek *et al.* observed in this case that while the apparent second-order rate constant (k_{cat}/K_m) was maximal for octanol in aqueous solutions, it was maximal for butanol in AOT/octane reverse micellar solution. Indeed, if aliphatic alcohols function as cosurfactants, the structural properties of micelles may alter,[28] and the effective substrate concentration must also be lowered. For example, if octanol functioned as a cosurfactant and butanol could not, the effective concentration of the former as a substrate may be lowered, raising the apparent K_m and lowering its k_{cat}/K_m value compared to butanol. This would appear to result in altered specificity in the micellar medium.

Not all enzymes are stable in a reverse micellar system. Liver microsomal cytochromes P-450 and b_5 fall in this group.[21] It appears that in such cases, the stability can be improved by using a mixture of surfactants, instead of one surfactant. However, these systems are not well characterized in terms of structure, and as such, it is not strictly correct as yet to refer to them as reverse micelles.

OPTIMIZATION OF REACTION PARAMETERS

The most common parameter varied in the enzyme reactions in reverse micelles is w_o. It has been observed in several cases that k_{cat} or the reaction rate for the enzymatic reactions first increases with increasing w_o, reaches an optimum value (at w_o of 10 to 15) and then decreases when w_o further increases.[10,29] As has been indicated earlier in this paper in the case of soybean lipoxygenase-1, the effect of w_o on k_{cat} is rather atypical. The rate increases up to a w_o of about 30 and then appears to level off.[17] In any case, optimization with respect to w_o is perhaps the simplest and most common.

The nature of reverse micelles can be altered by certain additives, for example, cosurfactants. Laane and coworkers[22] observed that depending on the polarity of their ketosteroid substrate, they had to use either chloroform or hexanol as cosurfactant. The nature of the substrate may indeed dictate its location in the microheterogenous medium of the reverse micelle,[30] resulting in an alteration in the course of a reaction.

The nature of the surfactant, especially the charge of the head group, is another factor that can play an important role. Laane and coworkers found that the chain of enzymes they used[31] was apparently inactive in AOT reverse micelles while it was active in those of CTAB. Similarly, Erjomin *et al.*[21] found a mixture of surfactants better for stabilizing cytochrome P-450 than a single surfactant.

Temperature can also play a significant role in the stability of an enzyme in reverse micelles. The freezing point of water is lowered in reverse micelles[20] and enzymes that are not stable at normal room temperatures in such systems can be stabilized at lower temperatures. Hilhorst *et al.*[22] conclude from their studies that the rate of enzyme reaction in reverse micelles is proportional to the substrate flux and if indeed this is true, then the substrate concentration may need to be optimized for an apolar substrate. However, their conclusion is based on their observation that the initial reaction rate was directly proportional to the substrate concentration, even when the latter was several orders of magnitude higher than K_m (observed in aqueous solutions). But since K_m may and indeed does increase in some cases in the micellar solutions, their results could simply be explained by a large effective increase in K_m.

CONCLUDING REMARKS

We indicated in the first section that three different systems have been used for the conversion of apolar substances by water-soluble enzymes. Of these, the reverse micellar system has an inherent advantage over the simple biphasic system, as it provides an extremely large interfacial area. This, in turn, alleviates the problems of diffusion-limited reactions. Thus, as pointed out by Laane and coworkers,[22] some enzymatic reactions that apparently do not take place in the biphasic system do so in reverse micelles.

It is, however, difficult to say at present how the reverse micellar system would compare with the enzyme immobilization procedure. The latter technology has been successful in normal aqueous enzymatic reactions and as a result, now enjoys far greater investment. From the practical viewpoint, micellization of enzymes is trivial compared to immobilization and yet it affords similar flexibility in terms of reaction parameters. For example, in the case of immobilized enzymes, one has the choice of selecting the hydrophobicity of the support and the polarity of the solvent system.[32] For the same purpose in reverse micellar systems, the polarity of water pools and the nature of the interface can be varied, depending on the solvent(s) and cosurfactant(s). This flexibility, in both systems, permits the use of a whole range of apolar substrates with varying degrees of polarity.[22,32] Although efforts are under way (as indicated in FIG. 4), the reverse micellar system, unlike the immobilized enzyme system, has only been used in a batch-type process. It therefore requires procedures, such as those proposed by us[33,34] and used successfully by Laane and coworkers,[22] to isolate the product and recover the enzyme. It should be noted that the use of water-soluble enzymes in reverse micelles inherently requires a large amount of the organic solvent. In conclusion, both of these systems in their application to apolar substrates are in a rapid stage of development and one would have to wait to make a reliable comparison.

From the standpoint of basic enzymology however, the homogeneous reverse micellar system appears to have a definite edge as it can be monitored by most of the common spectroscopic techniques permitting the study of mechanisms as well as the structure-function relationships. In addition, it has also become a medium of choice for cryoenzymology.[20]

From the mechanistic point of view, we do not yet know where in the microheterogenous system the reaction between a water-soluble enzyme and an apolar substrate takes place. The experiments of Hilhorst *et al.*[22] indicate that the polarity of the cosurfactant (when a cosurfactant is used) and the hydrophobic chain length of the surfactant may play a significant role in determining whether the substrate is allowed access to the enzyme. Penetration of the apolar substrate through the surfactant interface, they suggest, is the rate-limiting factor in such reactions. Since a reverse micellar system provides a microheterogenous solution with regions of dramatically different polarity,[30] one would first need to know, for each set of reactants, where these reactants are likely to reside and this may vary depending, among other factors, on their polarity. Thus, there may not be a general mechanism for such reactions.

REFERENCES

1. BUTLER, L. G. 1979. Enzymes in nonaqueous solvents. Enzym. Microb. Technol. Vol. **1**(4): 253–259.
2. MARTINEK, K. & A. N. SEMENOV. 1981. Enzymes in organic synthesis: Physiochemical means of increasing the yield of end product in biocatalysis. J. Appl. Biochem. **3**(2): 93–126.

3. CREMONESI, P., G. CARRERA, L. FERRARA & E. ANTONINI. 1979. Immobilized hydroxysteroid dehydrogenases for the transformation of steroids in water-organic solvent systems. Biotech. Bioeng. **22**(1): 39–45.
4. KLIBANOV, A. M., N. N. IVANOV, A. A. BOGDANOV & V. P. TORCHILIN. 1984. Protein immobilization on liposomes. Ann. N.Y. Acad. Sci. **434:** This volume.
5. FENDLER J. H. 1982. Membrane Mimetic Chemistry. John Wiley & Sons. New York.
6. MITTAL, K. L. & E. J. FENDLER. 1982. Solution Behaviour of Surfactants Vol. 1 and 2. Plenum Press. New York.
7. LUISI, P. L. & B. STRAUB. 1984. Reverse Micelles. Plenum Press. New York.
8. BARBARIC, S. & P. L. LUISI. 1981. Micellar solubilization of biopolymers in organic solvents. 5. Activity and conformation of α-chymotrypsin in isooctane-AOT reverse micelles. J. Am. Chem. Soc. **103:**(14): 4239–4244.
9. GRANDI, C., R. E. SMITH & P. L. LUISI. 1981. Micellar solubilization of biopolymers in organic solvents. J. Biol. Chem. **256**(2): 837–843.
10. MARTINEK, K., A. V. LEVASHOV, Y. L. KHMELNITSKY, N. L. KLYACHKO & I. V. BEREZIN. 1982. Colloidal solution of water in organic solvents: A microheterogenous medium for enzymatic reactions. Science **218:** 889–891.
11. MEIER, P. & P. L. LUISI. 1980. Micellar solubilization of biopolymers in hydrocarbon solvents. II The case of horse liver alcohol dehydrogenase. J. Solid-Phase Biochem. **5**(4): 268–82.
12. WOLF, R. & P. L. LUISI. 1979. Micellar solubilization of enzymes in hydrocarbon solvents. Enzymatic activity and spectroscopic properties of ribonuclease in *n*-octane. Biochem. Biophys. Res. Commun. **89**(1): 209–17.
13. VLIEGENHART, J. F. G., G. A. VELDINK & J. BOLDINGH. 1979. Recent progress in the study on the mechanism of action of soybean lipoxygenase. Agric. Food Chem. **27**(3): 623–626.
14. GROSSMAN, S. & R. ZAKUT. 1979. Determination of the activity of lipoxygenase (lipoxidase). Methods Biochem. Anal. **25:** 303–329.
15. GIBIAN, M. J. & R. A. GALAWAY. 1976. Steady-state kinetics of lipoxygenase oxygenation of unsaturated fatty acids. Biochemistry **15**(19): 4209–4214.
16. PANDE, A. & P. L. LUISI. Unpublished results.
17. MEIER, P. 1983. Dissertation (ETH No. 7222).
18. FINAZZI-AGRÒ, A., L. AVIGLIANO, G. A. VELDINK, J. F. G. VLIEGENHART & J. BOLDINGH. 1973. The influence of oxygen on the fluorescence of lipoxygenase. Biochem. Biophys. Acta **326:** 462–470.
19. SPAAPEN, L. J. M., G. A. VELDINK, I. J. LIEFKENS, J. F. G. VLIEGENHART & C. M. KAY. 1979. Circular dichroism of lipoxygenase-1 from soybeans. Biochem. Biophys. Acta **574:** 301–311.
20. BALNY, C. & P. DOUZOU. 1979. New trends in cryoenzymology: II. Aqueous solutions of enzymes in apolar solvents. Biochimie **61:** 445–452.
21. ERJOMIN, A. N. & D. I. METELITZA. 1983. Catalysis by hemoproteins and their structural organization in reversed micelles of surfactants in octane. Biochem. Biophys. Acta **732:** 377–386.
22. HILHORST, R., C. LAANE & C. VEEGER. 1983. Enzymatic conversion of apolar compounds in organic media using an NADH-regenerating system and dihydrogen as reductant. FEBS Lett. **159**(1,2): 225–278.
23. PELLEGRINI, A. & P. L. LUISI. 1978. Pepsin-catalyzed peptide synthesis. Biopolymers **17**(11): 2573–2580.
24. MORIHARA, K. & T. OKA. 1977. α-Chymotrypsin as the catalyst for the peptide synthesis. Biochem. J. **163**(2): 531–542.
25. MISIOROWSKI, R. L. & M. A. WELLS. 1974. The activity of phospholipase A_2 in reversed micelles of phosphatidylcholine in diethyl ether: Effect of water and cations. Biochemistry **13**(24): 4921–4927.
26. POON, P. K. & M. A. WELLS. 1974. Physical studies of egg phosphatidylcholine in diethyl ether-water solutions. Biochemistry **13**(24): 4928–4936.
27. WELLS, M. A. 1974. The nature of water inside phosphatidylcholine micelles in diethyl ether. Biochemistry **13**(24): 4937–4942.

28. EICKE, H.-F. 1980. Surfactants in nonpolar solvents. Aggregation and micellization. Topics Curr. Chem. **87:** 85–145.
29. LUISI, P. L., P. MEIER & R. WOLF. 1980. Properties of enzymes solubilized in hydrocarbons via reversed micelles. *In* Enzyme Engineering. H. H. Weetall & G. P. Royer, Eds. Vol. **5:** 369–371. Plenum Press. New York.
30. SCHANZE, K. S. & D. G. WHITTEN. 1983. Solubilization in surfactant media: Use of an isomerizable solute probe to determine microheterogeneity in microemulsions. J. Am. Chem. Soc. **105**(22): 6734–6735.
31. LAANE, C. Personal communication.
32. SODA, K. 1983. Bioconversion of lipophilic compounds by immobilized biocatalysts in organic solvents. TIBS **8**(12): 428.
33. LUISI, P. L., F. J. BONNER, A. PELLEGRINI, P. WIGET & R. WOLF. 1979. Micellar solubilization of proteins in aprotic solvents and their spectroscopic properties. Helv. Chim. Acta **62**(3): 740–753.
34. LUISI, P. L., V. E. IMRE, H. JAECKLE & A. PANDE. 1983. Microemulsions: Proteins and nucleic acids as quest molecules. *In* Topics in Pharmaceutical Sciences. D. D. Breimer & P. Speiser, Eds. Elsevier Science Publishers. Amsterdam.

Continuous Synthesis of Glycerides by Lipase in a Microporous Membrane Bioreactor

TSUNEO YAMANÉ, MOHAMMAD MOZAMMEL HOQ, AND SHOICHI SHIMIZU

Laboratory of Bioreaction Engineering
Department of Food Science and Technology
Faculty of Agriculture
Nagoya University
Nagoya, Japan

SHIRO ISHIDA AND TADASHI FUNADA

Amagasaki Works
Nippon Oil and Fats Co., Ltd.
Amagasaki
Hyogo, Japan

Glycerol and polyglycerol esters of fatty acids are prepared industrially and used commercially as food, cosmetics, pharmaceuticals, and emulsifiers. The main industrial process is based on direct esterification of fatty acids with glycerol and polyglycerol of various chain lengths in the presence of an inorganic catalyst at 200–250°C. The chemical reaction is tedious, nonselective, and energy consuming, and the crude products usually need further purification and bleaching. On the other hand, the enzymatic process, if it is feasible, has certain advantages over the chemical process in that it involves mild reaction conditions, specificity of the reaction, and conservation of energy.

In this context, several researchers have worked on the synthesis and hydrolysis of fat with free or immobilized lipase in emulsion systems.[1–4] Such conventional emulsion systems of lipase-catalyzed reactions have certain drawbacks:

(1) Emulsification of oil needs surfactant and/or a large power input such as stirring at high speed.
(2) The separation of emulsified oily products is complicated and needs powerful centrifugation.
(3) The reaction system of free lipase plus emulsified substrate is hard to operate continuously from an economic viewpoint. Fresh lipase must be continuously supplemented into the bioreactor because the lipase adsorbed on the surface of the emulsified oil droplets is lost from the bioreactor together with the droplets of product.
(4) Immobilization of lipase involves restricted diffusion of substrate and product. One can easily imagine that particulate droplets have limited access to the enzyme, resulting in a very low reaction rate. It is common that the activity of immobilized lipase is only several percent of the original activity of the free lipase.
(5) Control of water content of the reaction system is difficult. This is very

important because the catalytic action of lipase is reversible. Lower water content favors the synthetic reaction; at higher water content, hydrolysis prevails.

These disadvantages of the emulsion system led us to develop a nonemulsion system. We have attempted continuous operation of the reaction catalyzed by lipase in a membrane bioreactor, initially for the synthesis of glycerides of fatty acids from glycerol and liquid-free fatty acids.

EXPERIMENTAL SETUP

FIGURE 1 shows the setup of the membrane bioreactor. One unit of a flat plate-type dialyzer made of plastic material (Hospal Hemodialyzer RP-6) was used as the membrane bioreactor. First, a frame was covered on both sides with membranes, making a narrow compartment. The effective size of one sheet of membrane was 11.6 cm × 31.3 cm. Outside the membranes were placed two plates, the upper and lower

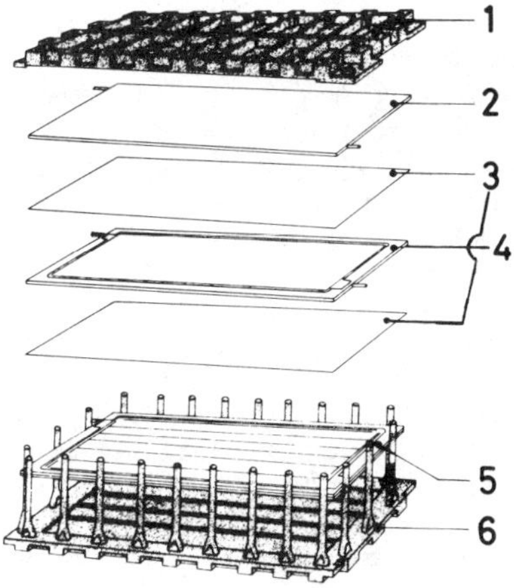

FIGURE 1. The setup of the microporous membrane bioreactor. (1) upper support; (2) upper plate; (3) microporous membrane; (4) frame; (5) lower plate; and (6) lower support.

plates. The surfaces of the plates adjacent to the membranes were ditches (0.5-mm deep and 0.8-mm wide bifurcated grooves; total number of ditches was 115). The plates and frame together with membranes were placed between upper and lower supporters and bolted tightly together. Thus, the bioreactor system comprised a closed chamber with no room inside for air when filled with reactants. The bioreactor is unique in the membrane used. It was made of polypropylene film with numerous micropores (porosity = 45%, maximum pore is 0.04 × 0.4 μm^2, 25 μm thick; JURAGARD 2500, Polyplastic Co., Ltd., Tokyo).

The flow of liquids in the bioreactor can be understood from the frontal and lateral views in FIGURE 2. In the lateral view of the bioreactor, the ditches are illustrated. The glycerol-water-lipase solution passes through the narrow compartment between two sheets of membrane, while fatty acid passes on the other side of the membrane in the ditches of the plates.

FIGURE 3 shows a conceptual cross-sectional view of the bioreactor and site of the reaction. Owing to the hydrophobic nature of the membrane, the fatty acid can readily

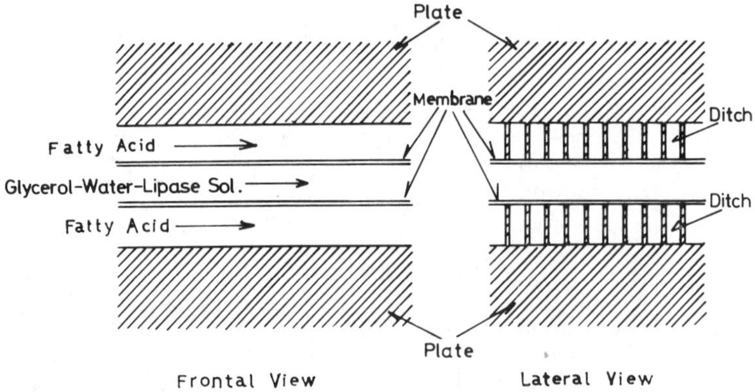

FIGURE 2. Schematic picture to show flows of fatty acid and glycerol-water-lipase solution in the bioreactor.

penetrate the micropores while the glycerol-water-lipase cannot. However because of the high viscosity of the glycerol-water-lipase solution, the pressure of the solution, P_G, is a little greater than that of the fatty acid, p_F. Hence, the fatty acid cannot leave the membrane and cannot mix with the glycerol-water-enzyme solution. The reaction therefore takes place at the interface of the membrane and glycerol-water-lipase solution. The glycerides produced diffuse back into the bulk flow of the fatty acid phase while the other product, water, diffuses into the bulk of the glycerol-water-lipase solution.

A process flow diagram of the reaction system is shown in FIGURE 4. The fatty acid was fed into the reactor with a plunger pump (Micro pump, Model SP-100, Sibata Chemical Apparatus Manufacturing Co., Ltd., Tokyo) and the glycerides formed were obtained at the outlet in a pure state. That is, the glycerides obtained were never mixed with the glycerol solution. The other reactant, glycerol, and a small quantity of water with solubilized lipase was fed into the bioreactor with a peristaltic pump (Mini-pulse 2, Gilson France SA Villiers Le Bel, France) through another path formed by the hydrophobic membrane. The glycerol solution from the outlet passed into the controlled dehydration system and was then recycled into the bioreactor. In the short-term experiments, no glycerol was supplemented, but in the long-term experiments, glycerol was supplied at an appropriate feed rate. The membrane bioreactor was placed in a thermostated water bath at 40°C.

For the controlled dehydration system, several methods were tried, of which two proved to be suitable. As a continuous dehydrator, we used a molecular sieve (bead type, 2–3 mm diameter, 5A-50, Union Showa Co., Ltd., Yokkaichi, Mie) column through which glycerol solution was passed. Alternatively, a vacuum dehydrator (a

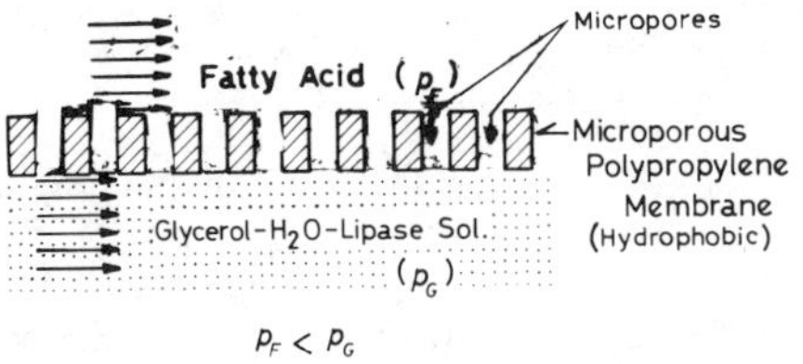

FIGURE 3. Illustrative figure of the neighborhood of the membrane. Micropores are drawn very schematically.

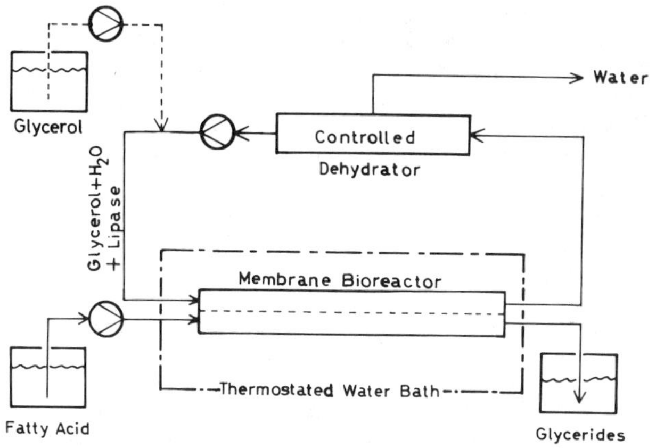

FIGURE 4. Process flow diagram of the reaction system. Glycerol was supplemented only during the long-term continuous experiments.

vacuum pump reduced the pressure inside the desiccator in which the glycerol-water-lipase solution reservoir was placed) was used for intermittent dehydration in some short-term experiments.

EXPERIMENTAL PROCEDURE

Materials

Microbial lipases were gifts from their manufacturers (these are listed in TABLE 1). Their hydrolytic activities were determined by the olive oil emulsion method.[5] One unit of activity is defined as the amount of the enzyme that liberates 1 μmol equivalent of fatty acid from olive oil in one minute under the analytical conditions (37°C). Most experiments were carried out with the lipase from *Chromobacterium viscosum* var. *paralipolyticum*[6–10] manufactured by Toyo Jozo Co., Ltd., Tokyo. This lipase has a weak positional specificity for the α position. As liquid fatty acids, oleic and linoleic acids of high purity were used without any dilution. Two kinds of oleic acids were used:

TABLE 1. Glyceride Synthesis Activities of Various Lipases

Source of Lipase	Manufacturer	Enzyme Activity (unit/mg)[a]	Conversion (%)[b]
Candida cylindracea	Meito Sangyo	290	3.8
Chromobacterium viscosum	Toyo Jozo	56	78
Mucor miehei	Novo Industri Japan	35	74
Pycomyces nitens	Takeda Yakuhin	2300	2.5
Pseudomonas fluorecens	Amano Seiyaku	2000	55
Rhizopus delemer	Tanabe Seiyaku	350	32
Rhizopus japonicus	Osaka Saikin Kenkyujyo	60	5.3

[a]Hydrolysis activity measured according to Yamané *et al.*

[b]Initial water content in glycerol-lipase solution was 3%. Initial enzyme activity (hydrolysis) was 6000 units/ml glycerol solution. Flow rate of oleic acid was 2.1 ml · h^{-1}.

Extra Olein 90 (the content of oleic acid was about 92%) and Extra Olein 99 (the content of oleic acid was over 99.9%). The purity of linoleic acid used was 99.3%. Extra Olein 90, Extra Olein 99, and linoleic acid were produced by Nippon Oil and Fats Co., Ltd., Tokyo.

Glycerol used was of reagent grade.

Start-up of Continuous Experiments

Glycerol solution having the required water content was prepared in a beaker and a defined amount of lipase was dissolved in it. The beaker was placed in a desiccator which contained silica gel so as to prevent moisture absorption. The glycerol-water-lipase solution was fed into the bioreactor with the help of a peristaltic pump. After the bioreactor was filled with the glycerol-water-lipase solution, the fatty-acid feeding was started. When the fatty acid came out of the outlets, the bioreactor was dipped into a thermostated water bath. Initially both fatty-acid and glycerol-water-lipase solution was fed at higher flow rates to fill them up rapidly in the bioreactor to save time. After the bioreactor was dipped in the thermostated water bath, the flow rates were adjusted. The recycle flow rate of the glycerol-water-lipase solution was kept at 10 $ml \cdot h^{-1}$ throughout all the experiments. The flow rate of fatty acid was adjusted and kept constant at a defined value during each run.

Analytical Procedure

The extent of conversion of fatty acid to glycerides was estimated by measurement of acid values,[11] and was calculated by the following equation:

$$\text{Conversion } (\%) = \frac{(AV)_{in} - (AV)_{out}}{(AV)_{in}} \times 100$$

$(AV)_{in}$ is the acid value of the fatty acid supplied into the bioreactor and $(AV)_{out}$ is the acid value of the product obtained from the outlet of the bioreactor.

Water content of the glycerol-water-lipase solution (or glycerol concentration) was monitored by measurement of refractive index with an Abbé refractometer (Type 3, ATAGO Co., Ltd., Tokyo, at 38.5°C). A calibration curve was constructed for a glycerol-water-lipase solution.

Composition of glycerides produced was analyzed quantitatively by gas-liquid chromatograph after silylation. The details of the quantitative analysis of glycerides by gas-liquid chromatography will be published elsewhere.

RESULTS AND DISCUSSION

Comparison of Glyceride Synthesis Activities of Various Lipases

The conversions of glyceride synthesis were compared in experiments using the amounts of lipases that gave the same activity of olive oil hydrolysis (6000 units/ml glycerol sol.), which were analyzed by the standard method,[11] because an analytical method to measure glyceride synthesis activity of lipase has not yet been established. Water content was not controlled in these experiments. The result is shown in TABLE 1. It is clear that, of the microbial lipases tested, the one produced by *Chromobacterium viscosum* gave the highest conversion. Hence, this lipase was used in all the subsequent experiments.

Effectiveness of Water Content Control

It was anticipated that water that accumulated in the recycled glycerol-water-lipase solution would affect the reaction rate and hence the outlet conversion. Therefore, two experiments were carried out, one with and one without control of water content. FIGURE 5 indicates that water formed resulted in a decrease in the conversion, and that a higher conversion level could be maintained when water content was controlled. These results demonstrate that the control of water is essential to maintain the steady-state conversion at a constant level for a prolonged time. This figure also shows typical time courses at the startup period of the continuous operation. In this particular case, it took about five hours to reach a steady state, and the steady state was

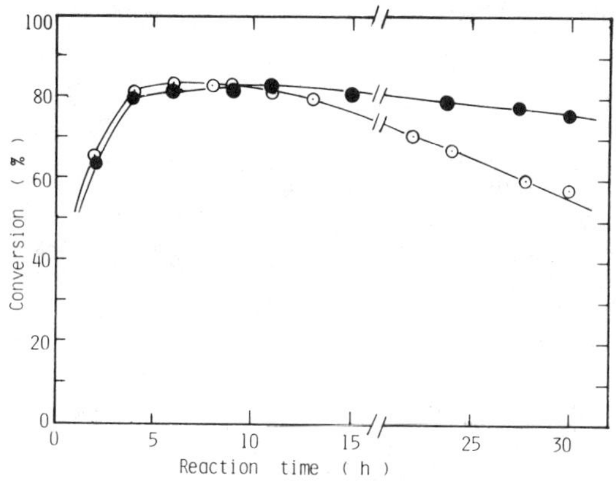

FIGURE 5. Effectiveness of water content control. The fatty acid fed was oleic acid (92% purity) whose flow rate was 3.2 ml · h^{-1}. —○—, no elimination of water formed; —●—, water content was controlled in the range of 7–9%.

maintained for more than 20 hours without supplementation of fresh glycerol when the water content was controlled.

Effect of Water Content on Conversion

The effect of water content on conversion was examined by keeping the water content of glycerol-water-lipase solution in the ranges shown in FIGURE 6. It was found that the conversion was very low when water content was low and that the enzyme dissolved in pure glycerol exhibited no synthesis activity. This loss of synthesis activity is probably because some amount of water is essential for the enzyme's catalytic activity. On the other hand, at higher water contents, the conversion decreased gradually. This is because the hydrolysis reaction, the backward reaction of ester synthesis, becomes significant. In between, at three to four percent of water, the conversion was the highest. We therefore employed an initial water content of three percent in the glycerol solution.

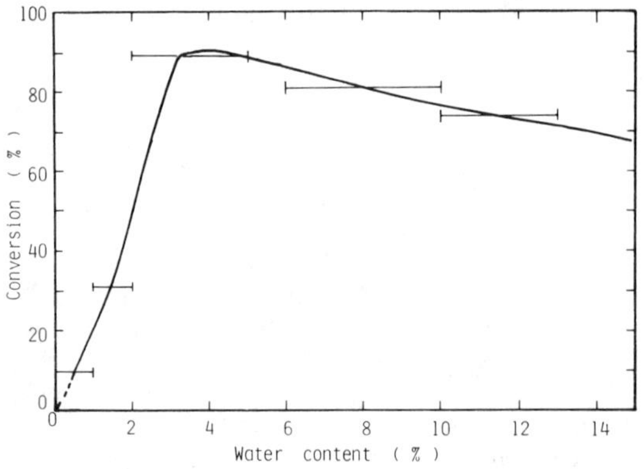

FIGURE 6. Effect of water concentration on conversion of oleic acid to glycerides. The line (——) indicates the range of water-content control. The flow rate of oleic acid (92% purity) was 3.2 ml · h^{-1}.

Effect of Enzyme Concentration on Conversion

FIGURE 7 shows the effect of initial concentration of lipase in the glycerol-water solution on conversion. At lower enzyme concentration, the conversion was almost proportional to the enzyme concentration; at higher concentrations, the dependence reached a plateau. This could be attributable to enzyme adsorption at the interface. The plateau at the higher enzyme concentration would be due to saturation of enzyme adsorbed on the interface.

Effect of Flow Rate of Fatty Acid on Conversion

The effect of flow rate of oleic acid on conversion was investigated at the two levels of water content (FIG. 8). It can be easily understood that the conversion increased

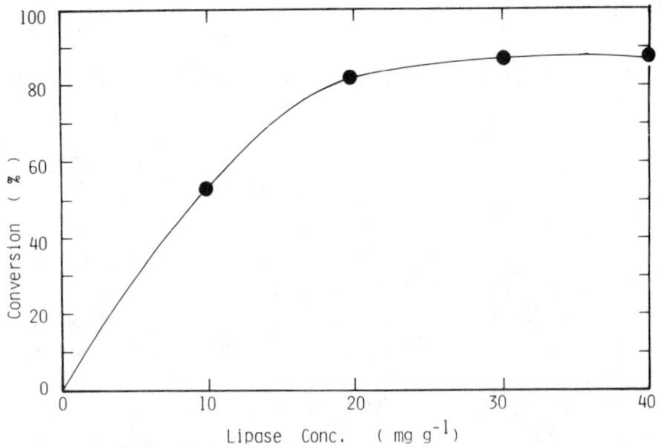

FIGURE 7. Effect of initial concentration of lipase in the glycerol-water solution on the conversion of oleic acid to glycerides. The flow rate of oleic acid (92% purity) was 3.2 ml · h^{-1}. Initial water concentration was 3–4%.

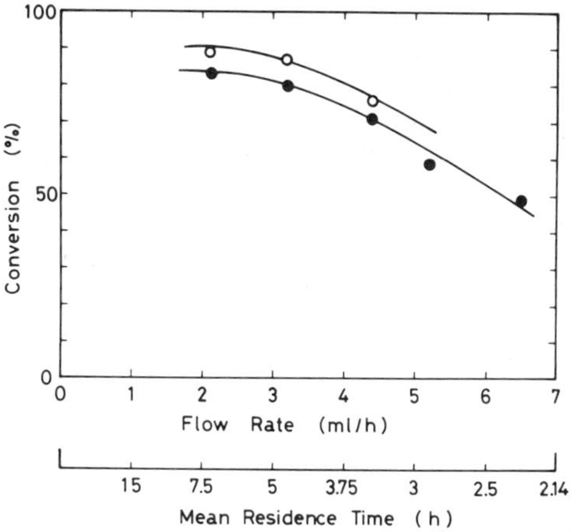

FIGURE 8. Effect of flow rate of oleic acid (92% purity) on conversion. —●—, 5–10% of water content; —O—, 3–5% of water content.

with decrease in the flow rate, because of longer residence times of fatty acid in the bioreactor. At a lower flow rate, the conversion approached its maximum value, which would be close to the true chemical equilibrium with the water content. If one wanted to increase the conversion further, one would have to carry out the reaction at lower water content and lower flow rates.

Operational Stability of Lipase

Long-term experiments were carried out to examine the operational stability of the lipase with and without 1% calcium chloride in the glycerol-water-lipase solution (FIG. 9). A molecular sieve column was used for controlled dehydration. Owing to the loss of

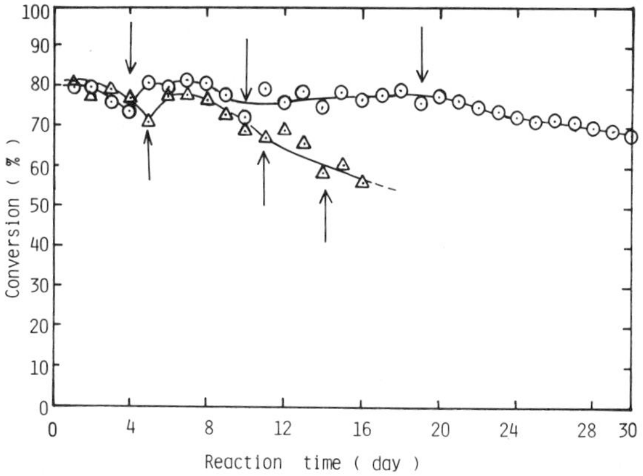

FIGURE 9. Operational stability of the lipase. The fatty acid fed was oleic acid whose flow rate was 2.1 ml · h^{-1}. The arrows indicate the times of replacement with new molecular sieve columns. —Δ—, without $CaCl_2$; —O—, with 1% $CaCl_2$ in the glycerol-water-lipase solution. Water content was kept 3–5% in the solution.

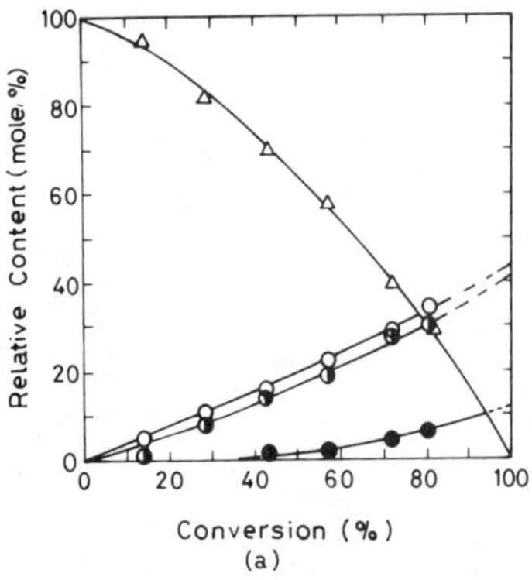

FIGURE 10. Compositions of glycerides produced. Oleic acid (99.9% purity) was supplied at the flow rate of 2.1 ml · h^{-1}. Initial water content was 3–4%. Symbols: —Δ—, free fatty acid; —O—, monoglyceride; —◐—, diglyceride; —●—, triglyceride.

dehydration ability of the molecular sieves, the column was replaced at the times indicated by vertical arrows in FIGURE 9. Glycerol was supplemented in an equivalent amount to that removed by the reaction under the experimental conditions. It can be seen from FIGURE 9 that the calcium ion is effective in preventing inactivation of the lipase. The estimated half-life of the enzyme in the presence of calcium ion is about 52 days by extrapolation of the curve. The prolonged high activity of the enzyme may be due to a combination of the protective effect of high glycerol concentration and the stabilizing effect of the calcium ion. It is known that polyhydric alcohols such as glycerol and sorbitol protect the enzymes from denaturation[12] and that some lipases are stabilized or even activated by calcium ion.[13]

Composition of Glycerides

In order to determine the composition of glycerides having various acid values, the output samples were analyzed by gas-liquid chromatography. The results of analysis of oleic acid glyceride and linoleic acid glyceride are shown in FIGURES 10 and 11, respectively. The results indicate that almost equimolar amounts of mono- and diglycerides were produced even when the conversion was more than 80%, and that the quantities of triglycerides were negligible or very small. This phenomenon may be in

FIGURE 11. Linoleic acid (99.3% purity) was supplied at the flow rate of 2.1 ml · h^{-1}. Initial water content was 3–4%. Symbols: —Δ—, free fatty acid; —O—, monoglyceride; —◐—, diglyceride; —●—, triglyceride.

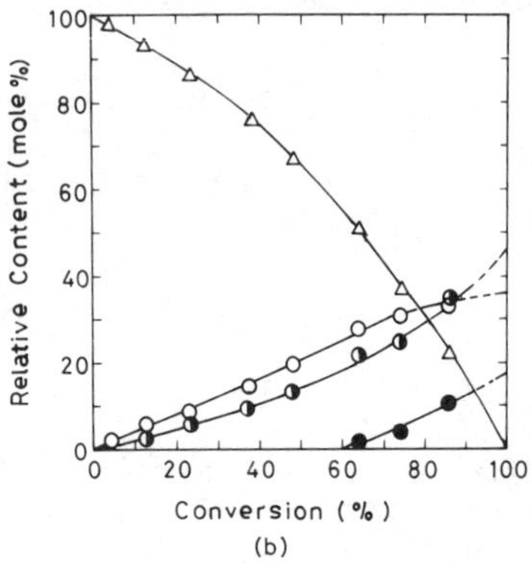

accord with the fact that the lipase used has a weak specificity of hydrolysis toward the α position.

CONCLUSION

A new bioreactor with a microporous membrane was developed for continuous synthesis of glycerides by lipase in a nonemulsion system. The advantages of the new microporous membrane bioreactor can be summarized as follows:

(1) The bioreactor does not require the making of an emulsion. This obviates the need for surfactant and stirring.
(2) The oily product can be obtained in a pure state with no other phase.
(3) The control of the water content is easier in this nonemulsion system than in the conventional emulsion system.
(4) Auto-oxidation of fatty acid is avoidable in the bioreactor simply by freeing the fed substrate from oxygen, because the bioreactor has no free space to hold air inside.
(5) The flow of the fatty acid in this microporous membrane bioreactor is close to plug flow while the emulsion system is always operated in a perfect mixing mode.

These advantages might be exploitable to push forward to industrial application. Unit sets of the bioreactor could be piled up to make a bigger set, resulting in a large surface area in a compact system. Alternatively, one might be able to use hollow-fiber modules as a potentially more effective bioreactor configuration.

ACKNOWLEDGMENTS

The authors wish to thank the companies listed in TABLE 1 for kindly providing gifts of lipase samples.

REFERENCES

1. TSUJISAKA, Y., S. OKUMURA & M. IWAI. 1977. Glyceride synthesis by four kinds of microbial lipase. Biochim. Biophys. Acta **489:** 415–422.
2. BELL, G., J. A. BLAIN, J. D. E. PATTERSON, C. E. L. SHAW & R. TODD. 1978. Ester and glyceride synthesis by *Rhizopus arrhizus* mycelia. FEMS Microbiol. Lett. **3:** 223–225.
3. LIEBERMAN, R. B. & D. F. OLLIS. 1975. Hydrolysis of particulate tributyrin in a fluidized lipase reactor. Biotechnol. Bioeng. **17:** 1401–1419.
4. KIMURA, Y., A. TANAKA, K. SONOMOTO, T. NIHIRA & S. FUKUI. 1983. Application of immobilized lipase to hydrolysis of triacylglyceride. Eur. J. Appl. Microbiol. Biotechnol. **17:** 107–112.
5. YAMANÉ, T., T. FUNADA & S. ISHIDA. 1982. Repeated use of lipase immobilized on amphiphilic gel for hydrolysis of a small amount of glycerides included in liquid crude fatty acid. J. Ferment. Technol. **60:** 517–523.
6. YAMAGUCHI, T., N. MUROYA, M. ISOBE & M. SUGIRA. 1973. Production and properties of lipase from newly isolated *Chromobacterium*. Agric. Biol. Chem. **37:** 999–1005.
7. SUGIRA, M., M. ISOBE, N. MUROYA & T. YAMAGUCHI. 1974. Purification and properties of a *Chromobacterium* lipase with a high molecular weight. Agric. Biol. Chem. **38:** 947–952.

8. Sugira, M. & M. Isobe. 1974. Studies on the lipase of *Chromobacterium viscosum.* III. Purification of a low-molecular-weight lipase and its enzymatic properties. Biochim. Biophys. Acta **341:** 195–200.
9. Sugira, M. & M. Isobe. 1975. Studies on lipase of *Chromobacterium viscosum.* IV. Substrate specificity of a low-molecular-weight lipase. Chem. Pharm. Bull. **23:** 1226–1230.
10. Horiuchi, Y., H. Koga & S. Gocho. 1976. Effective method for activity assay of lipase from *Chromobacterium viscosum.* J. Biochem. **80:** 367–370.
11. Paquot, C. 1979. Standard Method for the Analysis of Oils, Fats and Derivatives, 6th Ed. Pergamon Press. Oxford, United Kingdom. pp. 52–54.
12. Butler, L. G. 1979. Enzymes in nonaqueous solvents. Enzyme Microb. Technol. **1:** 253–259.
13. Brockerfoff, H. & R. J. Jensen. 1974. Chapter 4: Lipases. *In* Lipolytic Enzymes. Academic Press. New York. pp. 25–175.

The Enzymatic Synthesis of Esters in Nonaqueous Systems

I. L. GATFIELD

Haarmann & Reimer GmbH
P.O. Box 12 53
3450 Holzminden, Federal Republic of Germany

The replacement of water by an organic solvent as an enzymatic medium is almost invariably accompanied by a dramatic decrease in the catalytic activity of the enzyme and a decline in its substrate specificity. In the most successful attempts, it has been found that enzymes become inactive if the concentration of the organic component is greater than 90%.[1,2] Organic cosolvents nearly always exert deleterious effects on catalysis by both free and immobilized enzymes and significant activities in non-aqueous media are the exception rather than the rule.

As part of our research program on the application of enzyme systems for preparation purposes, we have investigated certain nonaqueous systems and have been able to show that the lipase-esterase preparation obtained from the mold *Mucor miehei* is capable of synthesizing esters in such systems.[3] Contrary to literature reports that describe the enzymatic synthesis for which the presence of a considerable quantity of water is essential,[4–6] the *Mucor miehei* enzyme system performs best under nonaqueous conditions. In fact, the degree of esterification achieved is inversely proportional to the water content of the system.[7]

RESULTS AND DISCUSSION

The enzymatic synthesis of ethyl oleate under standard conditions is shown in FIGURE 1. Equimolar quantities of oleic acid and ethanol were stirred at room temperature in closed vessels with 3% by weight, relative to the acidic component, of the *Mucor miehei* enzyme.

The enzymatic nature of the esterification reaction was demonstrated by boiling an aqueous solution of the enzyme for 30 minutes and freeze drying. The lyophilized product gave a 4% degree of esterification when checked under standard conditions. It had been shown previously that the freeze-drying process itself had no adverse effects upon the synthetic capabilities of the enzyme preparation.

Increasing the ethanol concentration of the heterogeneous system caused a slight inhibition of the enzyme such that a yield of some 75% was obtained after three days when a sevenfold molar excess of ethanol was employed. Other organic solvents such as toluene, diethyl ether, tetrahydrofuran, methylene chloride, and chloroform gave rise to differing degrees of inhibition, the magnitudes of which depended upon the amount of solvent added.

The enzyme system is remarkably stable and can be reused a number of times. Thus the same enzyme was used for some 12 cycles over a 36-day period in the oleic acid-ethanol model system without any apparent loss of activity. The esterification yields varied in a random fashion between 83% and 87% during the 12 cycles.

When certain hydroxy acids were treated with this enzyme preparation, lactones were formed. In the case of 4-hydroxybutyric acid, the corresponding γ-butyrolactone was isolated and its identity was proved with the acid of IR and NMR spectroscopy. Similarly, gas chromatographic analysis showed that ω-hydroxypentadecanoic acid underwent conversion to the macrocyclic lactone pentadecanolide when treated in toluene with the enzyme system.

The enzyme exhibits a pronounced substrate specificity and will only esterify

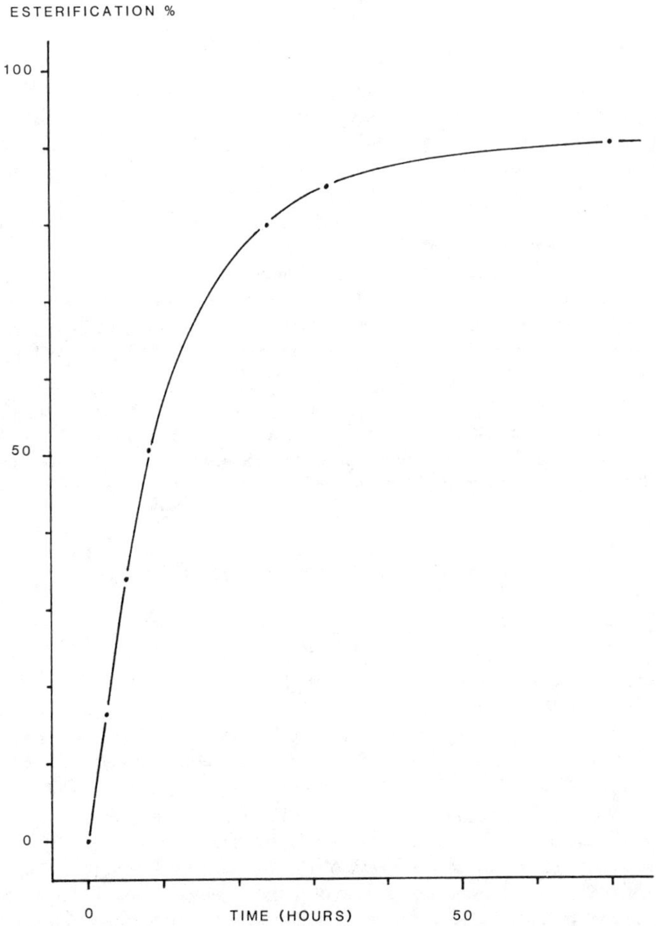

FIGURE 1. Esterification profile for the model system oleic acid–ethanol.

straight-chain carboxylic acids. The degree of esterification achieved depends upon the chain length of the acid as can be seen in FIGURE 2. Furthermore, propionic and acetic acids do not undergo esterification at all.

Some substituents can cause an increase in the degree of esterification as is shown in TABLE 1. Thus both the keto- and the hydroxy-derivatives of propionic and butyric

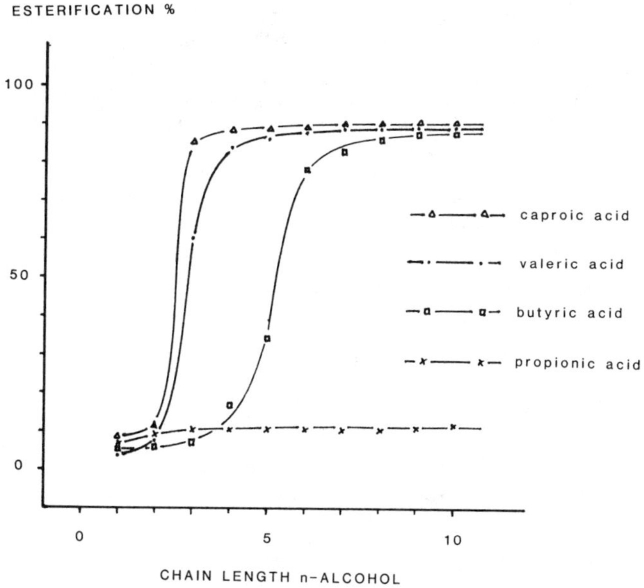

FIGURE 2. Effect of chain length of carboxylic acid upon the degree of esterification obtained with *n*-alcohols.

TABLE 1. Effect of Certain Substituents upon the Degree of Esterification of Short-chained Carboxylic Acids with Ethanol

Acidic Component	Degree of Esterification (%)
Butyric acid	9
α-Keto butyric acid	31
3-Hydroxy butyric acid	21
Propionic acid	6
α-Keto propionic acid[a]	27
2-Hydroxy propionic acid[b]	38

[a]Pyruvic acid.
[b]Lactic acid.

TABLE 2. Effect of Structure of Alcohol Component upon Degree of Esterification with Oleic Acid

Alcohol Component	Degree of Esterification (%)
Aliphatic alcohols:	
n-Butanol	92
sec-Butanol	65
tert-Butanol	14
Terpene alcohols:	
Geraniol	90
Menthol	20
Linalool	5

acids are more readily converted into their ethyl esters than the parent acids. On the other hand, aromatic and α, β-unsaturated acids do not undergo esterification.

The structural limitations concerning the alcohol component are not as stringent. Thus both primary and secondary aliphatic alcohols can be esterified efficiently, but not tertiary alcohols (TABLE 2). This does not apply, however, to terpene alcohols, since those containing secondary alcohol groups such as menthol undergo only slight esterification.

One of the driving forces of this reaction is, without doubt, the removal of the reaction product water from the system, probably via adsorption by the enzyme. The considerable change in the physical appearance of the enzyme during the course of the reaction would support this hypothesis.

ACKNOWLEDGMENT

The technical assistance of Ms. Anni Kloidt is gratefully acknowledged.

REFERENCES

1. BUTLER, L. G. 1979. Enzymes in nonaqueous solvents. Enzyme Microb. Technol. **I:** 253–259.
2. KLIBANOV, A. M., G. P. SAMOKHIN, K. MARTINEK & I. V. BEREZIN. 1977. A new approach to preparative enzyme synthesis. Biotechnol. Bioeng. **XIX:** 1351–1361.
3. GATFIELD, I. L. & T. SAND. 1981. German Patent Application 3,108,927.
4. HOSONO, A. & J. A. ELLIOTT. 1974. Properties of crude ethyl ester-forming enzyme preparations from some lactic acid and psychrotrophic bacteria. J. Dairy Sci. **57:** 1432–1437.
5. HOSONO, A., J. A. ELLIOTT & W. A. MCGUGAN. 1974. Production of ethyl esters by some lactic acid and psychrotrophic bacteria. J. Dairy Sci. **57:** 535–539.
6. IWAI, M., S. OKUMURA & Y. TSUJISAKA. 1980. Synthesis of terpene alcohol esters by lipase. Agric. Biol. Chem. **44:** 2731–2732.
7. GATFIELD, I. L. 1980. Unpublished results.

Cholesterol Degradation by Polymer-entrapped *Nocardia* in Organic Solvents[a]

JOSÉ M. C. DUARTE[b] AND M. D. LILLY

Department of Chemical and Biochemical Engineering
University College London
London, United Kingdom

Some of the work presented in this volume indicates an increasing interest in the use of two-liquid-phase biocatalytic reactors.[1] In some cases, such as steroid transformations, the second organic phase consists of a water-immiscible solvent in which the substrate is dissolved.[2] Both Buckland *et al.* and Cremonesi *et al.*[3] used a range of organic solvents for this purpose. A more systematic approach was made by Duarte and Lilly[4] and Duarte.[5]

One potential and real disadvantage of the technique using water-immiscible organic solvents is the deleterious effect that the solvent may have on the structure and function of the microorganisms. Cell entrapment was an obvious approach to this problem and Duarte and Lilly[4] showed that the half-life for the cholesterol oxidase activity of *Nocardia rhodochrous* could be extended by entrapment in calcium alginate or polyacrylamide. Entrapment of cells for steroid transformation in the presence of organic solvents was also exploited by Fukui *et al.*[6] Immobilization of a yeast has also been found to increase its operational stability for stereoselective hydrolysis of methyl esters dissolved in heptane.[7]

In none of these cases was there a need for maintenance of a whole degradative pathway. We have therefore examined the complete oxidation of cholesterol in such two-phase systems. Some microorganisms can metabolize cholesterol as a carbon and energy source yielding carbon dioxide and water. One of these is *Nocardia rhodochrous,* but its capacity to degrade cholesterol as sole carbon and energy source in aqueous systems seems to be limited (unpublished observation). When this organism was used in the presence of high proportions of organic solvents, cholestenone accumulated,[8] and no further degradation was observed. Here we show that by entrapping the cells in polymers, *Nocardia rhodochrous* was capable, if an additional energy source was added, of completely degrading cholesterol dissolved in a water-immiscible organic solvent.

MATERIALS AND METHODS

Nocardia rhodochrous (NCIB 10554) grown on cholesterol and a mineral medium was entrapped in two different types of supports, alginate and polyacrylamide.[4] Two

[a]We thank the Engineering Foundation (New York) for subsidizing our participation in this conference. We also thank the Calouste Gulbenkian Foundation for the financial support it gave to one of us (J.M.C. Duarte) during this work.

[b]Address reprint requests to Dr. Duarte at DTIQ/LNETI, National Laboratories of Industrial Engineering and Technology, Est. das Palmeiras, Queluz de Baixo, 2745 Queluz, Portugal.

TABLE 1. Shake Flask Experiments on the Conversion of Cholesterol by Immobilized *Nocardia rhodochrous*

Experimental Conditions	Flask Number 1	2	3
Support	4% Alginate	4% Alginate	8% PAA
Energy source	2% Glycerol	1% Acetate	YE/Gly[a]
Substrate (mg/ml of TCE)	50	50	50
Reaction rate (mg/ml/h)	0.05	0.018	0.3

[a]YE/Gly: 1.2% yeast extract + 0.6% glycerol in Tris buffer.

hundred milliliters of the fermentation medium were centrifuged at 29°C and the harvested cells were immediately immobilized.

The substrate, cholesterol, was dissolved in 1,1,1-trichloro-ethane (TCE), which was previously shown to be a good solvent for use with this system.[4]

The transformation reaction was conducted in shake flasks on an orbital shaker at a temperature of 29°C. The total volume of reaction was 100 ml:50 ml of TCE (containing the substrate) + 20 ml of an aqueous solution (containing a carbon source) + 30 ml of gel with the entrapped cells. The flasks were agitated at 200 rpm.

The reaction was followed by measuring the decrease in the cholesterol concentration and the appearance of cholestenone and of other possible intermediates by HPLC.

RESULTS AND DISCUSSION

It was previously observed[9] that cells grown on cholesterol initially had a high cholesterol degradation activity when used in the absence of organic solvent. This

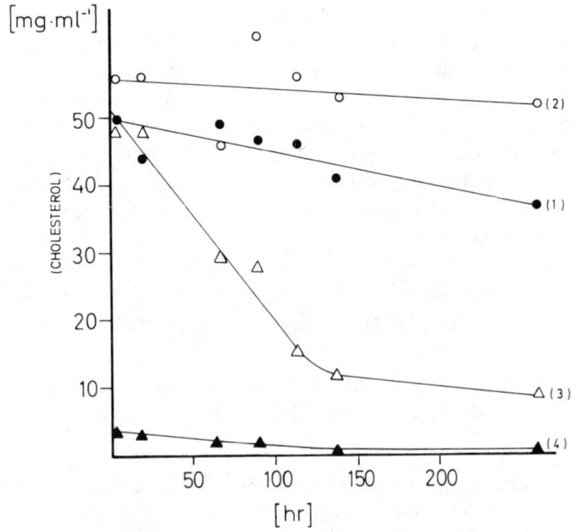

FIGURE 1. Cholesterol consumption by gel-entrapped *Nocardia* cells grown on cholesterol. Numbers in parentheses refer to the flask number (cf. TABLE 1) Symbols: flask 1, ●; flask 2, O; flask 3, △; flask 4, ▲.

activity was increased when a carbon energy source was added to the reaction medium. Glycerol supplemented with yeast extract was used with the polyacrylamide-entrapped cells, but this medium was not used with the alginate-entrapped cells to avoid any adverse effects on the gel structure. Glycerol and acetate were tried as potential carbon energy sources with the alginate gel.

When the free cells were used in the presence of the organic solvent, cholestenone accumulated in practically stoichiometric amounts with minor quantities of androstenedione (AD) and androstdienedione (ADD) clearly visible after separation by TLC. We believe the AD and ADD were formed during the production of the cells and remained adsorbed to the cells. On exposure to trichloroethane, the two compounds were slowly extracted.

When the cells, entrapped either in alginate or polyacrylamide, were used in the presence of various carbon energy sources (TABLE 1) a different behavior was observed (FIGS. 1 and 2). In each case, the cholesterol concentration decreased almost linearly with time (FIG. 1). Flask 4 represented a control experiment in which no cholesterol was added; apparently harvested cells had some undegraded cholesterol still associated

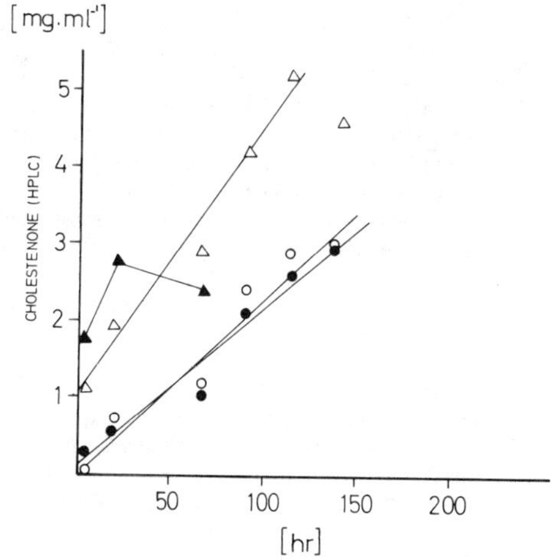

FIGURE 2. Cholestenone formation by gel-entrapped *Nocardia* cells grown on cholesterol. Numbers in parentheses and symbols as in TABLE 1 and FIGURE 1, respectively.

with them. There were large differences in the rates at which the entrapped cells oxidized cholesterol (TABLE 1).

These rate differences could be a mass transfer effect resulting from the smaller size of the polyacrylamide particles (0.5-mm diameter) compared to the alginate gels (2.0 mm). A fourfold difference in rate had been observed for the initial step of cholesterol to cholestenone conversion,[4] with similar preparations. When intraparticle mass transfer limitation is present, the radius that is proportional to the Thiele modulus has a significant effect on the effectiveness factor. Klein and Wagner,[10] for example, showed that for phenol degradation by immobilized microorganisms, doubling the particle radius caused a 57% decrease in observed activity. It is also likely that the reaction rates for cholesterol oxidation were affected by the different hydrophobicities of the two gels in a manner similar to that described by Omata *et al.*[11] Glycerol seemed to have a beneficial effect on the transformation using alginate-entrapped cells. This could be due either to it being more readily utilized by the *Nocardia* or because it increased the solubility of cholesterol in the aqueous/solid matrix.

For each of the reactions shown in FIGURE 1, only a small part of the cholesterol that disappeared could be accounted for as accumulated cholestenone. In the case of the polyacrylamide-entrapped cells, only 12% of the cholesterol was recovered as cholestenone. No other steroids could be detected in substantial amounts. With the polyacrylamide-entrapped cells the level of AD had reached about 0.4 mg/ml after 114 hours. The rate of cholesterol consumption slowed down at this time and an unknown steroid appeared, reaching levels of about 0.7 mg/ml after 138 hours. This compound is probably an intermediate between cholestenone and AD which accumulated as complete side-chain degradation began to slow down. Much lower levels of AD and the unknown intermediate were found in flasks 1 and 2, but a higher concentration of ADD (0.4 mg/ml) was reached in flask 1 than in the other flasks.

Our results indicate that immobilization of *Nacardia rhodochrous* protects the microorganism from the deleterious effects of the organic solvent so that the cells are capable of maintaining for some time an active degradation pathway. These results are encouraging, especially if other microbial systems can be protected in the same way.

REFERENCES

1. LILLY. M. D. 1982. Two-liquid-phase biocatalytic reactions. J. Chem. Technol. Biotechnol. **32:** 162–169.
2. BUCKLAND, B. C., P. DUNNIL & M. D. LILLY. 1975. The enzymatic transformation of water-insoluble reactant in nonaqueous solvent. Conversion of cholesterol to cholest-4-ene-3-one by a *Nocardia* sp. Biotechnol. Bioeng. **XVIII:** 815–826.
3. CREMONESI, P., G. CARREA, L. FERRARA & E. ANTONINI. 1974. Enzyme dehydrogenation of testosterone coupled to pyruvate reduction in a two-phase system. Eur. J. Biochem. **44:** 401–405.
4. DUARTE, J. M. C. & M. D. LILLY. 1980. The use of free and immobilised cells in the presence of organic solvents: The oxidation of cholesterol by *Nocardia rhodochrous. In* Enzyme Engineering. H. H. Weetall & G. P. Royer, Eds. Plenum Press. New York. Vol. **5:** 363–367.
5. DUARTE, J. M. C. 1982. Enzyme kinetics and mass-transfer on two-liquid-phase heterogeneous systems. *In* Enzyme Engineering. I. Chibata, S. Fukui & L. B. Wingard, Jr., Eds. Plenum Press. New York. Vol. **6:** 157–158.
6. FUKUI, S., S. A. AHMED, T. OMATA & A. TANAKA. 1980. Bioconversion of lipophilic compounds in nonaqueous solvent. Effect of gell hydrophobicity on diverse conversions of testosterone by gel-entrapped *Nocardia rhodochrous* cells. Eur. J. Appl. Microbiol. Biotechnol. **10:** 289–301.
7. OMATA, T., N. IWAMOTO, T. KIMURA, A. TANAKA & S. FUKUI. 1981. Stereoselective hydrolysis of *dl*-menthyl succinate by gel-entrapped *Rhodotorula minuta* var. *texensis* Cells in Organic Solvent. Eur. J. Appl. Microbiol. Biotechnol. **11:** 199–204.
8. LEWIS, D. J. 1981. Ph.D. Thesis. University of London.
9. DUARTE, J. M. C. Unpublished results.
10. KLEIN, J. & F. WAGNER. 1978. Immobilised Whole Cells. Dechema-Monograph. Vol. **82:** 142–164. Verlag Chemie. Weinheim.
11. OMATA, T., T. IIDA, A. TANAKA & S. FUKUI. 1979. Transformation of steroids by gel-entrapped *Nocardia rhodochrous* cells in organic solvent. Eur. J. Appl. Microbiol. Biotechnol. **8:** 143–155.

Catalysis by Enzymes Entrapped in Reversed Micelles of Surfactants in Organic Solvents

I. V. BEREZIN

A. N. Bach Institute of Biochemistry
The USSR Academy of Sciences
117071 Moscow, USSR

KAREL MARTINEK

Department of Chemistry
Moscow State University
117234 Moscow, USSR

Enzyme engineering is reaching its maturity.[1] Nevertheless, stabilization of enzymes remains one of the most challenging areas of biotechnology, and its importance will increase as more and more enzymes are used commercially.[2] Great advances have been made toward solving the problem of enzyme stabilization against the denaturing action of nonnative environments.[3–10]

Denaturation of enzymes in organic solvents can be avoided if the reaction is carried out in a water-*water-immiscible* organic solvent biphasic system;[11,12] see also reviews.[13–15] For systems with an extremely low content of water, an enzyme should be entrapped in reversed micelles of surfactants;[16] for review, see Martinek *et al.*[17] and Levashov *et al.*[18]

Colloidal solution of water in an organic solvent as a microheterogeneous medium for enzymatic reaction has gained credence in chemical and clinical analyses. This technique is used to enhance the sensitivity of bioluminescent assay,[19] in biophotolysis of water,[20] and especially in fine organic compound syntheses such as the following: (1) to convert high concentrations of substrates poorly soluble in water, such as steroids;[21] and (2) proteins may be hydrophobized by their chemical modification with water-insoluble compounds, for example by their acylation with $C_{16}-C_{18}$ acyl chlorides.[22]

Recently we worked out another application. As a model reaction, we selected the oxidation of isobutanol to isobutyraldehyde

$$RCH_2OH + NAD^+ \rightleftharpoons RCHO + NADH \quad (1)$$

which is catalyzed by alcohol dehydrogenase in a water-octane system with the addition of 0.5 *M* Aerosol OT as a surfactant. The results are presented in FIGURE 1 as the dependence of equilibrium constant (K) on the ratio between the volumes of the organic and aqueous phase (V_{org}/V_w), where the equilibrium constant in water is $K_w = 9\ 10^{-4}$. As the water content in the system decreases, the equilibrium is shifted to the right (toward the aldehyde). The main result is the specific effect appearing when the surfactant is added to the water-organic solvent reaction medium. For details see Martinek *et al.*

There is no doubt that this approach to the regulation of chemical equilibrium will find application in preparative organic synthesis.

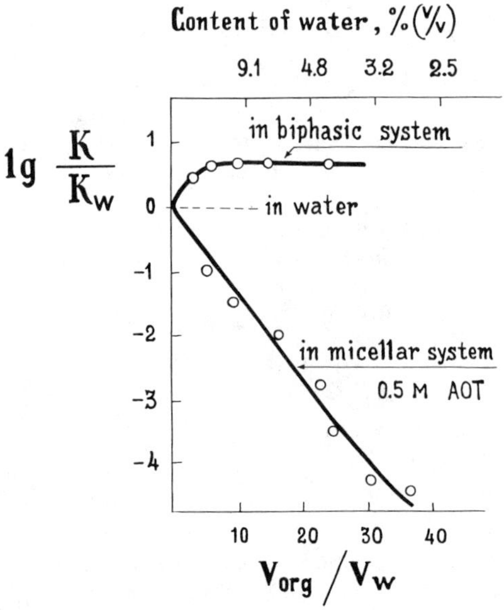

FIGURE 1. Plots of the dependence of the equilibrium constant of reaction[1] on the ratio between the volumes of the organic (octane) and aqueous phases in biphasic (without surfactant) and micellar (0.5 *M* Aerosol OT) systems. Conditions: pH 8.8 (0.02 *M* phosphate buffer), 20°C. For details see Martinek *et al.*[23]

REFERENCES

1. Katchalski, E. & A. Freeman. 1972. Trends Biochem. Sci. **7:** 427.
2. Klibanov, A. M. 1983. Science **219:** 722.
3. Martinek, K., A. M. Klibanov & I. V. Berezin. 1977. J. Solid-Phase Biochem. **2:** 393.
4. Klibanov, A. M., N. O. Kaplan & M. D. Kamen. 1978. Proc. Natl. Acad. Sci. USA **75:** 3640.
5. Marshall, J. J. 1978. Trends Biochem. Sci. **3:** 79.
6. Klibanov, A. M. 1979. Anal. Biochem. **92:** 1.
7. Schmid, R. D. 1979. Adv. Biochem. Bioeng. **12:** 41.
8. Martinek, K., V. V. Mozhaev & I. V. Berezin. 1980. *In* Enzyme Engineering: Future Directions. L. B. Wingard, Jr., A. A. Klyosov & I. V. Berezin, Eds. Plenum Press. New York. p. 3.
9. Mozhaev, V. V., V. A. Šikšnis, V. P. Torchilin & K. Martinek. 1983. Biotechnol. Bioeng. **25:** 1937.
10. Mozhaev, V. V. & K. Martinek. 1984. Enzyme Microb. Technol. **6:** 50.
11. Klibanov, A. M., K. Martinek & I. V. Berezin. 1977. Biotechnol. Bioeng. **19:** 1351.
12. Martinek, K., A. M. Klibanov & I. V. Berezin. 1977. Biorg. Khim. (Russ.) **3:** 696.
13. Antonini, E., G. Carrea & P. Cremonessi. 1981. Enzyme Microb. Technol. **3:** 291.
14. Martinek, K. & A. N. Semenov. 1981. J. Appl. Biochem. **3:** 93.
15. Lilly, M. D. 1982. J. Chem. Technol. Biotechnol. **32:** 162.
16. Martinek, K., A. V. Levashov, N. L. Klyachko & I. V. Berezin. 1977. Dokl. Akad. Nauk SSSR (Russ.) **236:** 920. Engl. Ed. 1978. **236:** 951.
17. Martinek, K., A. V. Levashov, Yu. L. Khmelnitsky, N. L. Klyachko & I. V. Berezin. 1982. Science **218:** 889.
18. Levashov, A. V., Yu. L. Khmelnitsky & K. Martinek. 1984. *In* Surfactants in Solution K. Mittal, Ed. Plenum Press. New York. **2:** 1069.
19. Brovko, L. Yu., N. L. Klyachko, A. V. Levashov, N. N. Ugarova, K. Martinek & I. V. Berezin. 1983. Dokl. Acad. Nauk SSSR (Russ.) **273:** 494.
20. Hilhorst, R., C. Laane & C. Veeger. 1982. Proc. Natl. Acad. Sci. USA **79:** 3927.

21. Hilhorst, R., C. Laane & C. Veeger. 1983. FEBS Lett. **159:** 225.
22. Levashov, A. V., A. V. Kabanov, K. Martinek & I. V. Berezin. 1984. Dokl. Acad. Nauk SSSR (Russ.) **278:** 246.
23. Martinek, K., Yu L. Khmelnitsky, A. V. Levashov & I. V. Berezin. 1981. Dokl. Acad. Nauk SSSR (Russ.) **256:** 1423. Engl. Ed. 1981. **256:** 143.

Protein Immobilization on Liposomes

A. L. KLIBANOV, N. N. IVANOV, A. A. BOGDANOV, AND V. P. TORCHILIN

USSR Cardiology Research Center
Institute of Experimental Cardiology
Moscow, USSR

During the past few years several methods of drug targeting have been actively elaborated. One promising method is a liposome-containing therapeutic preparation with water in the inner space or in the structure of the membrane and with a molecule immobilized on the outer surface so as to give recognition and binding to the target zone.[1,2] Thus, the problem arises to develop methods of attaching vector molecules, mainly immunoglobulins, enzymes, or other proteins, on liposome surfaces. Two principal methods of immobilization have been developed in our and other laboratories. These are (1) covalent binding of the protein with the liposome surface via a spacer group, and (2) incorporation of the protein into the membrane using a protein previously modified with a hydrophobic agent (for a review see Torchilin and Klibanov).[3]

We have studied the interaction of enzymes and immunoglobulins, modified with fatty acids and natural phospholipids, with liposome surfaces.[4,5] Cheap and commercially available hydrophobic derivatives of different sugars also can be easily incorporated into liposome membranes. The sugar moiety can be activated by oxidation with periodate to make the sugar reactive towards protein amino groups. This method combines the advantages of covalent binding via a spacer group (the sugar moiety serves as a spacer) and hydrophobic modification of the protein. Here, the fatty acid residues work as anchors for fixing the protein on the surface of the liposome membrane.

For incorporation into liposomes, we have used sucrose stearate-palmitate 15 (SSP 15) (Serva). Liposomes were prepared from egg lecithin, SSP 15, and dicetylphosphate in the molar ratio 6:2:2 by the reverse-phase evaporation method.[6] To follow the integrity of the liposomes during further manipulations 40 m*M* carboxyfluorescein was included in the inner water space of the liposomes. At this concentration, self-quenching of carboxyfluorescein takes place. If liposome integrity is disturbed, carboxyfluorescein is released into the surrounding medium, and the resulting dilution causes the fluorescence to increase. The increase in fluorescence intensity corresponds to the degree of liposome destruction or leakage. The liposomes obtained were activated by oxidation of sucrose residues with sodium periodate at pH 5.8. During this procedure, the liposomes remained intact. The activated liposomes were incubated with $1.5 \times 10^{-5} M$ α-chymotrypsin in 0.01 *M* borate buffer, pH 8.8. The Schiff bases formed between the oxidized sucrose aldehyde groups and the protein ϵ-amino groups of lysine residues were reduced with sodium cyanoborohydride (FIG. 1). Unbound protein was separated from the liposome by gel filtration on Sepharose CL4B. The quantity of bound enzyme was determined according to its catalytic activity towards *N*-acetyl-L-tyrosin ethyl ester.

The data permitted the following conclusions to be made: (1) SSP 15 was easily incorporated into liposomes which remained intact; (2) sucrose residues on the liposome surface could be oxidized with sodium periodate without apparent increase in liposome permeability; (3) enzymes could be bound to the activated liposomes and the

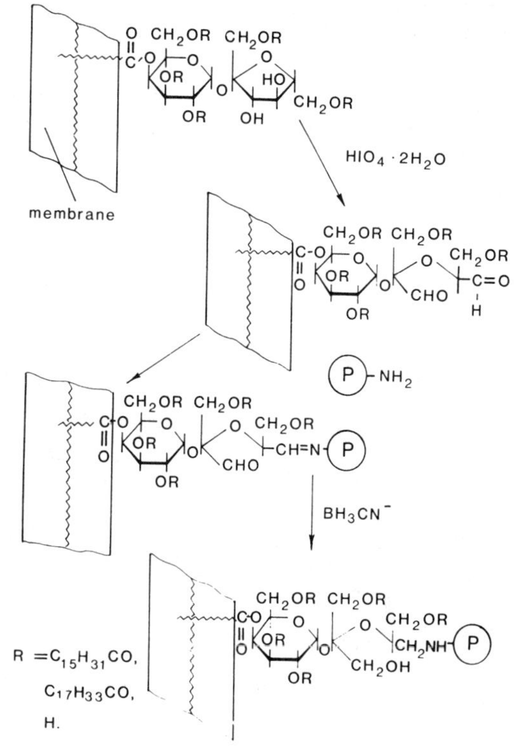

FIGURE 1. The scheme of protein immobilization on liposome via SSP 15.

preparation reduced while maintaining the liposomes intact; and (4) from kinetic studies, the quantity of liposome-bound α-chymotrypsin was as high as 1.2×10^{-4} moles of protein per mole of lipid.

REFERENCES

1. GREGORIADIS, G. & A. S. ALLISON, Eds. 1980. Liposomes in Biological Systems. John Wiley and Sons. Chichester-New York-Brisbane-Toronto.
2. TORCHILIN, V. P., V. N. SMIRNOV & E. I. CHAZOV. 1982. Problems and perspectives of liposome use for drug targeting. Vopro. Med. Khim. (Problems in Medical Chemistry, Russ.) **28:** 3–14.
3. TORCHILIN, V. P. & A. L. KLIBANOV. 1981. Immobilization of proteins on liposome surface. Enzyme Microb. Technol. **3:** 297–304.
4. TORCHILIN, V. P., V. G. OMEL'YANENKO, A. L. KLIBANOV, A. V. MIKHAILOV, V. I. GOLDANSKII & V. N. SMIRNOV. 1980. Incorporation of hydrophilic protein modified with hydrophobic agent into liposome membrane. Biochim. Biophys. Acta **602:** 511–521.
5. TORCHILIN, V. P., A. L. KLIBANOV & V. N. SMIRNOV. 1982. Phosphatidylinositol may serve as the hydrophobic anchor for immobilization of proteins on liposome surface. FEBS Lett. **138:** 117–120.
6. SZOKA, F. Jr. & D. PAPAHADJOPOULOS. 1978. Procedure for preparation of liposomes with large internal aqueous space and high capture by reverse-phase evaporation. Proc. Natl. Acad. Sci. USA **75:** 4194–4198.

Index of Contributors

Acuña, M. E., 110–114
Agathos, S. N., 44–47
Alfani, F., 39–43
Andersson, E., 115–118
Aoki, K., 278–281
Asouzu, M. U., 526–528
Auriol, D., 267–270
Azuma, M., 394–405

Bader, J., 171–185
Bailey, J. E., 31–38
Ballesteros, A., 296–298
Berezin, I. V., 577–579
Berke, W., 257–258
Blanch, H. W., 373–381
Boccù, E., 127–130
Bogdanov, A. A., 580–582
Boitieux, J. L., 533–535
Bourdillon, C., 343–346
Bourne, J. R., 136–139
Branner, S., 340–342
Brodelius, P., 382–393, 487–490, 496–500
Bückmann, A. F., 87–90
Bungay, H. R., 155–157

Cabral, J. M. S., 483–486
Cadete, M. M., 483–486
Cambou, B., 214–218, 219–223
Cantarella, M., 39–43
Cardoso, J. P., 483–486
Castner, J. F., 224–231
Catapano, G., 123–126
Chen, C-S., 186–193
Chibata, I., 450–453
Chotani, G. K., 347–362
Chao, W.-S., 334–339
Christensen, M., 340–342
Clark, D., 31–38
Clark, L. C., Jr., 512–514, 515–519
Clemente, A., 131–135
Combes, D., 48–60, 61–63
Constantinides, A., 347–362

Dal, A., 461–464
Dalbadie-McFarland, G., 232–238
Day, D. F., 504–507
Demain, A. L., 44–47
Drioli, E., 123–126
Duarte, J. M. C., 131–135, 573–576

Eigtved, P., 340–342
Ermolin, G. A., 289–291
Esclade, L., 214–218

Fernández, V. M., 296–298
Fevereiro, P., 501–503
Fiolitakis, E., 91–94
Flaschel, E., 70–77
Freeman, A., 418–427
Fukui, S., 479–482
Fukushima, S., 148–151
Funada, T., 558–568
Furui, M., 450–453
Furusaki, S., 168–170

Gainer, J. L., 465–467
Gallifuoco, A., 39–43
Gatfield, I. L., 569–572
Ge, S., 282–284
Gianfreda, L., 7–19, 127–130
Girdaukas, G., 186–193
Gopalan, A. S., 186–193
Greco, G., Jr., 7–19, 127–130
Grizzuti, N., 7–19
Grooms, T. A., 512–514, 515–519
Günther, H., 171–185
Guillochon, D., 214–218
Guo, J., 275–277
Gustafsson, J.-G., 285–288

Hagi, N., 95–98
Hahn-Hägerdal, B., 115–118, 152–154, 161–163, 475–478
Haq, M. M., 558–568
Harada, T., 95–98
Harju, M., 406–417
Hashimoto, Y., 206–209
Hatton, K. S., 334–339
Haufler, U., 99–105
He, B., 275–277
Hediger, T., 136–139
Hedman, P., 285–288
Heikonen, M., 406–417
Henley, J., 64–69
Hirata, T., 206–209
Ho, M. H., 523–525, 526–528
Høјer-Pedersen, B., 271–274
Holst, O., 472–474
Hou, C. T., 541–548
Huie, T., 363–372
Huitron, C., 110–114
Hummel, W., 87–90

Idman, T., 491–495
Ignashenkova, G. V., 289–291
Inuzuka, K., 394–405

Iorio, G., 123–126
Irino, S., 95–98
Ishida, S., 558–568
Ivanov, N. N., 580–582

Ji, X., 264–266
Jiang, Z., 275–277
Johansson, A.-C., 115–118
Jones, C. K. S., 119–122
Jonsson, B., 152–154

Kaiser, E. T., 321–326
Kajiwara, K., 427–436
Kamikubo, T., 158–160
Karrenbauer, M., 78–86
Karube, I., 427–436, 508–511, 529–532
Kasche, V., 99–105
Kauppinen, V., 491–495
Kautola, H., 454–458
Kimura, K., 206–209
Kirwan, D. J., 465–467
Klein, J., 437–449
Klibanov, A. M., 20–26, 219–223, 363–372, 580–582
Klotz, I. M., 302–320
Kohn, J., 254–256
Koistinen, T., 406–417
Kominek, L. A., 106–109
Kubo, I., 508–511
Kula, M.-R., 87–90, 91–94, 194–205, 257–258

Landcrt, J.-P., 70–77
Larsson, M., 144–147, 475–478
Laskin, A. I., xiii–xv
Lee, Y. Y., 164–167
Legoy, M.-D., 259–263
Leuchtenberger, W., 78–86
Li, S., 536–540
Li, H., 264–266
Lilly, M. D., 573–576
Linko, P., 406–417, 454–458
Linko, Y.-Y., 406–417, 454–458
Linse, L., 487–490
Liu, S., 264–266
Lohmeier-Vogel, E., 152–154
Lorenz, J. W., 363–372
Lüthi, P., 549–557
Luisi, P. L., 549–557
Lundbäck, H., 472–474

Maeda, M., 427–436
Maksimenko, A. V., 289–291
Markkanen, P., 491–495
Marrese, C., 224–231
Marrucci, G., 7–19
Martin, W. G., 292–295
Martinek, K., 577–579
Matsuno, R., 158–160
Matsuoka, H., 427–436, 529–532
Mattiasson, B., 144–147, 161–163, 472–474, 475–478
Mauz, O., 251–253
McGray, P., 536–540
Minard, P., 259–263
Miron, T., 254–256
Miyawaki, O., 224–231, 520–522
Monsan, P., 48–60, 61–63, 267–270, 468–471
Morr, M., 257–258
Moore, P. E., 512–514, 515–519
Mosbach, K., 239–248
Murata, H., 427–436

Nagashima, M., 394–405
Nam, D. H., 210–213
Nanmori, T., 278–281
Neitzel, J., 232–238
Neumann, S., 171–185
Nishira, H., 278–281
Noetzel, S., 251–253
Noguchi, S., 394–405
Novais, J. M., 483–486
Noyes, L. K., 512–514, 515–519
Núñez, M., 296–298
Nuti, M., 461–464

Ogasa, T., 206–209
Oriol, E., 267–270
Oyama, K., 95–98

Pais, M. S. S., 501–503
Pande, A., 549–557
Paul, F., 267–270
Perlot, P., 468–471
Peruffo, B., 461–464
Plöcker, U., 78–86
Prenosil, J. E., 136–139, 549–557

Rella, R., 123–126
Remy, M. H., 343–346
Renken, A., 70–77
Richards, J. H., 232–238
Riechmann, L., 99–105
Riggs, A. D., 232–238
Roda-Santos, L., 131–135
Rodriguez, R., 363–372
Rollings, J. E., 140–143
Romette, J. L., 533–535
Rossi, M., 123–126
Royer, G. P., 334–339
Ryu, D. D. Y., 210–213

Sadana, A., 64–69
Saito, H., 206–209

Samejima, H., 394–405
Sauber, K., 251–253
Saval, S., 110–114
Scardi, V., 39–43
Schiavon, O., 127–130
Schülein, M., 271–274
Schütte, H., 87–90, 194–205
Scouten, W. H., 249–250
Seissler, P., 363–372
Shieh, W-R., 186–193
Shimizu, S., 558–568
Shinke, R., 278–281
Sih, C. J., 186–193
Simon, H., 171–185
Skoog, K., 161–163
Song, S. K., 164–167
Sonomoto, K., 479–482
Spettoli, P., 461–464
Spiesser, D., 70–77
Stellwagen, E., 1–6
Stenroos, S. L., 406–417
Stuker, E., 136–139
Sundaram, P. V., 299–301
Suzuki, S., 427–436, 508–511, 529–532

Takamatsu, S., 450–453
Takasawa, S., 206–209
Tanaka, A., 479–482
Tanaka, M., 158–160
Tang, J. C.-T., 536–540
Taniguchi, M., 158–160
Thanos, J., 171–185
Thøgersen, H., 340–342
Thomas, D., 214–218, 259–263, 343–346
Thompson, R. W., 140–143
Threefoot, S. A., 363–372
Tischenko, E. G. 289–291
Tomka, I., 549–557
Torchilin, V. P., 27–30, 289–291, 580–582
Torres, V., 296–298
Tosa, T., 450–453
Toyama, K., 529–532
Trubetskoy, V. S., 27–30
Tsao, G. T., xiii–xv

Ueyama, K., 168–170

Van der Tweel, W., 249–250
VanMiddlesworth, F., 186–193
Vecchio, A., 536–540
Veronese, F., 127–130
VickRoy, T. B., 373–381
Vogel, H. J., 496–500
Vorlop, K.-D., 437–449

Wandrey, C., 87–90, 91–94, 257–258
Wagner, F., 437–449
Watanabe, E., 529–532
Weaver, J. C., 363–372
White, E. T., 119–122
Wichmann, R., 87–90
Wichmann, U., 91–94
Wilchek, M., 254–256
Wilke, C. R., 373–381
Williams, R. E., 292–295
Wingard, L. B., Jr., xiii–xv, 224–231, 520–522
Wolf, H. J., 106–109
Workman, W. E., 504–507
Wu, T.-G., 523–525
Wulff, G., 327–333

Yamade, K., 148–151
Yamanaka, H., 278–281
Yamané, T., 558–568
Yang, L.-W., 459–460
Yang, R. Y. K., 119–122
Yang S., 282–284
Yuan, Z., 264–266
Yuki, S., 278–281

Zale, S. E., 20–26
Zamorani, A., 461–464
Zhang, S., 275–277, 282–284
Zhong, L-C., 459–460
Zhou, B.-N., 186–193
Ziomek, E., 292–295

Subject Index[a]

Acetamide, 8 M, ribonuclease stabilization and, 21–22
Acetobacter oxydans, 396
Acetone-dried cells, dehydrogenase activity in, 107
Acetylcholine, affinity isolation of, 299–301
N-acetyl-DL-amino acids, acylase-catalyzed stereospecific hydrolise of, 78
N-acetyl-L-tyrosine ethyl ester, 31
Acid hydrolysis, ribonuclease thermoinactivation and, covalent changes responsible for, 24–25
Acid phosphatase
  as deactivation material, 9–10
  temperature dependence for, 11, *12*
Acremonium chrysogenum, 465
Active-site design, chemical mutation approach to, 322
Active-site tritration, as enzyme immobilization technique, 31
Acylamino-β-lactam acylhydrolase, phenoxymethylpenicillin synthesis using, 210–13
Acylase
  amino acid production using, 80–83
  systems, model conditions for, *82*
ADH, yeast extraction of, 286–88
Aerobacter aerogenes, 271
Affinity chromography, polysaccharide activation for, 254–56
Agar, plant-cell entrapment use of, *384*
Agarose, plant-cell entrapment use of, *384*
Agroindustrial by-products, microbial enzymes produced from, 110–14
Ajmalicine, price of, *383*
Alanine
  industrial production methods for, 78, *79*
  lactate transformation to, 91–92
  production of, 450–53
  residue parameters of, *4*
Alcohol
  fermentation of starch using, with immobilized cell/enzyme bioreactor, 148–51
  glucose conversion to, by *Saccharomyces cervisiae,* 421–22
  measurement of, electroenzymatic sensor for, 515–19
Algae, biochemical produces from, *240*
Alginate
  in butanediol production, 455
  plant-cell entrapment use of, *384*
Amino acids
  composition of, *272*
  N-acetyl-DL-derivative preparation of, 80–83
  dehydrogenases of, *196*
  groups, crosslinking of, 5
  hydrophobic, catalyzation reactions of, 95
  from hydroxy acid racemic mixture, 91–94
  industrial production methods for, *79*
  from keto acids, 84–85
  Mucor miehei analysis with, 340–42
  NADH-dependent, production of, 195
  polypeptide chains and sequence differences of, 3
  racemization of, 25
  replacement for, identification difficulties and, 3
  residue parameters of, *4*
  synthetase stability from, *46*
6-aminopenicillanic acid
  production of, 437–40
  semisynthetic penicillin produced from, 127
Aminopyridines, hydrolysis of, 310–15
Ammonia, 25
Ampicillin, semisynthesis of, 99, 102–3
Amylase
  aqueous two-phase systems for production of, using *Bacillus subtilis,* 115–18
  producing strains for, fermentation conditions of, 275–77
  production improvement for, polypepton fraction effective and, 278–81
Aniline, hydroxilation of, 215–16
Anionic-acrylamide, 9
Antibiotics
  lactam, kinetically controlled semisynthesis of, 99–105
  Streptomyces clavuligerus production of, 420
Apolar compounds, enzymatic reactions of, 549–58
Aqueous two-phase systems
  for amylase production, 115–18
  for starch to ethanol, 144–47
Arginine
  industrial production methods for, *79*
  residue parameters of, *4*

[a]An italicized page number refers to a table or figure.

Arrhenius law, 10
Arthrobacter simplex, 106–9
steroid transformation using, 459–60
Asparaginase, stabilization of, 27–30
Asparagine
deamidation of, 25
replacement of, 3
residue parameters for, *4*
Aspartame, enzymatic production of, 95–98
Aspartate
industrial scale synthesis reaction with, 194
residue parameters of, *4*
Aspartic acid
industrial production methods for, *79*
peptide bonds involving, 25
Aspergillus sp., 110
Aspergillus niger, 39, 282
ATP, regeneration of, for enzymatic synthesis, 257–58
ATPase F_1, stability of, *2*
Aureobasidium sp., 110
Azide concentrations, xylose utilization and, 153
Aztobacter vinelandii, 465

Bacillus acidopullulyticus, thermophilic pullulanases from, 271–74
Bacillus brevis, 44–47
Bacillus cereus, 278
Bacillus macerans, 70
Bacillus megaterium, 271
Bacillus mycoides, 271
Bacillus subtilis
amylase production using, aqueous two-phase system for, 115–18
biochemical products from, *240*
production, purification, and immobilization of, 486–71
Bacteria
biochemical products from, *240*
enantioselective synthesis of PG-Nor by, *208*
heat-dried, steroid-1-dehydrogenation using, 106–9
lactic acid, applications of, 406–17
photosynthetic, fuel cell using, 427–36
storage stability of, 433
thermophilic
enzyme stability from, 3
stress adaptation of, 1
See also specific bacteria
Bakers' yeast
ADH and hexokinase extraction from, 286–88
keto ester reductions by, 186–93
Basidiomycetes, 131
Benzisoxazolcarboxylic acids, decarboxylation reactions of, 304–7
Bifunctional reagents, intramolecular cross-linking with, subunit enzyme stabilization by, 27–30
Binding, covalent, of biomolecules, 241–43
Binding groups, function of, 331
Biocatalysts, advantages and limitations of, 334–35, *335*
Biochemicals
immobilized plant cells in production of, *386*
production of, *240*
Bioconversion
fermentation methods, *108*
non-fermentation methods, *109*
Biofuels, production of, 481
Biohydrogenation reduction methods, in chiral compound preparation, 171–85
Bioligands, immobilization of, chromophoric sulfonyl chloride agarose for, 249–50
Biomolecules, covalent binding of, 241–43
Bioreactors
cross-flow, 352–53
fluid-agitated bed-type fermenter, reproductive immobilized cell study using, 349–50, *351,* 352
heterogeneous, 541–48
immobilized cell/enzyme
alcohol fermentation with, 148–51
comparison of, *360*
microporous membrane, 558–68
See also Reactors
Biosensors
for fish freshness determination, 529–32
hybrid, in fermentation, 508–11
Blood, alcohol in, electroenzymatic sensor for measurement of, 515–19
Boehringer Glucose Kit, 39
Brevibacterium, 197
Butanediol, *Enterobacter* sp., production of, 454–58

Calcitonin, as semisynthetic enzyme, 321
Calcium alginate gel, *Saccharomyces cerevisiae* immobilization in, 394–405
Calf alkaline phosphatase, residual activity in, *68*
Candida boidinii, 87
Candida cylindracea, 223
Candida tropicalis, xylitol to ethanol formation using, in pentose fermentation, 152–54
Carbecephem compounds, optically pure, synthesis of, 206–9
Carboniimide, 27

Carbohydrates, in chemical and fuel production, 161–63
Carbon
  FAD and glucose oxidase immobilization on, 520–22
  metabolism shift in, microbial enzyme activity and, 45
Carbonyl group compounds
  chiral alcohol hydrogenation to, *175*
  asymmetric reduction of, 186–93
Carboxyl esterases, catalyzation by, 219
Carboxyl groups, crosslinking of, 5
Cardenolide production, by *Digitalis lanata,* 491–95
Carnitine
  cholinesterase-catalyzed production of, 219
  fatty acid trasport role of, 186
Carrageenan, plant-cell entrapment use of, *384*
Carriers, synthetic, for enzymes, 251–53
Catalysts, synthetic hydrogenation, 334–39
Catharantus roseus, biochemical products from, *240*
Cell entrapment, 239–40
Cell homogenates, enzyme recovery from, 285–88
Cellobiose hydrolysis, mathematical model of, 41–42, *42*
Cells
  biocatalytic processes with, reaction kinetics and, 437–49
  gel entrapment of, 418–26
  growth of, product inhibition effect on, 353–55
  immobilization of, microbiological measurements by, 363–72
  living, applications of, 479–82
  metabolic behavior of, 475–78
  microbial
    ethanol production using, 394–405
    immobilization of, 420–22
    steroid transformation by, 459–60
  oxygen supply to, 472–74
  plant
    entrapment of, 481
    immobilized viable, 382–93
    immobilization methods for, 501–3
    metabolism of, 496–500
    polymers used for entrapment of, *384*
  procaryotic, growth of, 373–81
  reproductive immobilized, 347–62
Cellulase
  economics of recycling of, 155–57
  kinetics of, 158–60
Cellulose
  acid–catalyzed, kinetics of, 164–67
  hydrolysis of
    acid and enzymatic, 161–63
    data and model prediction in, *165*
    enzyme fate after, *155*
    kinetic parameters in, 166
  microcrystalline, hydrolysis of, 158–60
  in rice husks, 131
  saccharification of, glucose inhibition in, 39–43
Cellulose diacetate, in butanediol production, 455
Cephalosporinase, coupling of, 251
Ceramide glucosyl transferase, residual activity in, *68*
Cereals, hydrogenase activity in, 296–98
Chaotropic agents, enzyme stability to, 1
Chiral compounds
  enzymes for synthesis of, 194–205
  reduction methods for, 171–85
Chloramphenicol, synthetase and, 46, *46*
Chlorella vulgaris, biochemical products from, *240*
Cholesterol degradation, by *Nocardia,* 573–76
Cholinesterase, catalyzation with, 219
Chromography, affinity, polysaccharide activation for, 254–56
Chromophoric sulfonyl chloride agarose, for immobilizing bioligands, 249–50
Chymotrypsin
  coupling of, *251*
  deactivation of, kinetic and spectroscopy studies of, 31–38, *32*
  immobilized, properties of, *35*
  modification of, hexanal and, 343–46
  specific activity change in, 32–33, *32*
  spin-labeled, spectroscopy of, 33–35, *33, 34*
Chymotrypsinogen, cystine destruction in, 24
Citronellol, carboxyl esterase-catalyzed resolution of, *220*
Clostridia, hydrogenase catalyzation in, *172*
Clostridium sp., 174
Codeine, price of, *383*
Cofactor regeneration, enzyme-coenzyme system use of, 259–63
Compaction, gel particle effectiveness factor effects from, 168–70
Conformational processes, irreversibility of, noncovalent forces and, 21
Continuous-flow reactors, fermentation by *Leuconostoc oenos* in, 461–64
Cornynebacterium simplex, 106–9
Corticoid-responsive diseases, steroids used for, 106
Covalent binding, of biomolecules, 241–43
Covalent mechanisms, enzyme inactivation and, 20
Creatinine, sensors for, 509–11
Cross-flow bioreactor, 352–53

Crosslinking, intramolecular, subunit enzyme stabilization and, 27–30
Cyclic backflushing, membrane life prolonged by, 79
Cyclodextrin
pH activity profile of, *72*
production of, by enzymatic starch degradation, 70–77
Cysteine
industrial production methods for, *79*
residue parameters of, *4*
synthetase activity in, 45
Cystine
elimination reactions in, 23–24
industrial production methods for, *79*
Cytidine-5′-monophospho-N-acetyl-neuraminic acid galactosyl glucosylceramide sialyltransferase, residual activity in, 68
Cytochromes
renaturation rate of, 1
stability of, 4–5, *4*

Dairy industry, whey disposal in, 136
Daucus carota, 487
Deactivation kinetics, immobilization, chemical modifiers, enzyme aging and, 64–69
Decanol
cyclodextrin thermal denaturation with, 71–73, *72, 73*
mass transfer retarding effect of, at high enzyme concentrations, 75
Denaturation reactions, 1
Desulfovibrio desulfuricans, 292
Deuterated compounds, preparation of, 174
Dextrans, invertase stability and, 56
Dextransucrase, production and purification of, from *Leuconostoc mesenteroides,* 267–70
Dicarboxylic acids, intramolecular crosslinking tested with, 27–30
Digitalis lanata, cardenolide production by, 491–95
Digitoxin, price of, *383*
Dihydrolipoamide, thiol interchange with, 46
Dihydronicotinamide dinucleotide, oxidation of, 307–9
Diimidoesters, 27
Dimethyl pimelimidate, as crosslinking reagent, 28
Disease, cortocoid-responsive, steroids used for, 106
Disulfide bonds, reduction of, 21
DMSO, 389–90
DNA, mutagenic strategies of protein function and, 232–38

E. coli
growth speed of, 363
phenyl sepharose extraction of, 288
Electrical conductance, in microbiological measurement, 365–68
Electrodes
oxalate-sensing, 512–14
rotating ring disc, mass transport resistance studies using, 224–31
Electroenzymatic reduction methods, in chiral compound preparation, 171–85
Electroenzymatic sensors, for alcohol measurement, 515–19
Electromicrobial reduction methods, in chiral compound preparation, 171–85
Enantiomeric compounds, manufacture of, enzyme membrane reactor process for, 78–86
Enantioselective reductions, of keto esters, by bakers' yeast, 186–93
Enoate reductase, 176–79
Enoates, hydrogenation of, to chiral carboxylates, *173*
Enterobacter sp., butanediol production by, 454–58
Environmental paramaters, alteration of, in xylitol to ethanol formation, 152–54
Enzymes
aging of, deactivation kinetics influence of immobilization, chemical modifiers and, 64–69
catalytic properties of, 219–23
in chiral compound synthesis, 194–205
coenzyme complexes, 243–47
coenzyme systems, cofactor regeneration in, 259–63
conformational inactivation of, *28*
coupling of, *251*
electrodes for, sensors for, 523–25
entrapped, catalysis by, 577–79
gel entrapment of, 418–26
immobilized, in flow injection analysis, 526–28
immobilized cell reaction with, 450–53
irreversible thermoinactivation of, mechanisms of, 20–26
isolated, in NAD(P)H regeneration, 171–85
membrane reactor scale-up process of, for enantiomeric compound manufacture, 78–86
microbial, agroindustrial by-product production of, 110–14
oligomeric, properties of, *28, 29*
oxidase, 533–35
recovery from microbial cell homogenates by, 285–88
semisynthetic, 321–26

soluble and stabilized, thermal and chemical deactivation of, 7–19
stability of
and glucose inhibition in cellulose saccharification, 39–43
reaction rate reduction of, 17
strategies for increasing, 1–6
water activity effect on, 48–60
starch degradation of, cyclodextrin production by, 70–77
stereospecificity of, 199–200
subunit, intramolecular crosslinkings of, 27–30
support matrices for, mass transport resistance studies through, 224–31
synthesis of, ATP regeneration for, 257–58
synthetic carriers for, 251–53
thermal deactivation kinetics for, 7–9
thermal stability of, *40*
thrombolytic, substrate affinity for, 289–91
time-decaying activities of, effectiveness factors of, 13
Erwinia aroideae, acylamino-β-lactam acylhydrolase from, phenoxylmethylpenicillin synthesis using, 210–13
Erythritol, stabilizing effect of, 53–54
Esters, enzymatic synthesis of, 569–72
Ethanol
production of, immobilized cell use in, 394–405, 442–47
starch conversion to, 144–47
xylitol formation to, with *Candida tropicalis,* 152–54
Ethylene glycol, stabilizing effect of, 53–54
Eupenicillium javanicum, 158
Extraction, definition of, 78

Fatty acids, transport role of, carnitine role in, 186
Fermentation
air-life, 494
draft tube, 494
amylase producing strains and, 275–77
bioconversions for, *108*
of lactic acid, *376*
malolactic, by *Leuconostoc oenos,* 461–64
mechanically stirred, 494
process control in, hybrid biosensor for, 508–11
Fibril, cellulose, 159
Fish, freshness of, biosensor for, 529–32
Flavin, semisynthetic enzyme studies with, 322–24
Flavin adenine dinucleotide, immobilization on carbon of, 520–22
Fluoride ion selective senors, enzyme electrode based on, 523–25
Flor injection analysis, immobilized enzymes in, 526–28
Fluid-agitated bioreactor, 349–50, *351,* 352
Fructofuranosidase, inulin assay using, 504–7
Fuel cell systems, photosynthetic bacterial, 433–35
Fumeric acid, malic acid from, 83–84
Fungus, rice husk growth of, 133

Galactose oxidase, catalyzation of, 219
Galactosidase, immobilization of, 136
Gastrin, aspartame as methyl ester of, 95
Gelatin, in butanediol production, 455
Gel entrapment, of cells and enzymes, 418–26
Gel microdroplets, in microbiological measurement, 368–71, *369, 370*
Gel particles, effectiveness factor of, compaction effects on, 168–70
Globular domains, enzyme structural feature and, 3
Gluconobacter oxydans, 472
biochemical products from, *240*
Glucose
alcohol conversion from, by *Saccharomyces cervisiae,* 421–22
concentration effects of, on reactor productivity, *377*
electroactivity of, 228
inhibition of, in cellulose saccharification, 39–43
Glucose oxidase
catalyzation of, 219
immobilization on carbon of, 520–22
ring-disc electrode attachment using, 227
Glucosidase, thermal deactivation of, 39–43
Glutamate, residue parameters of, *4*
Glutamic acid, industrial production methods for, *79*
Glutamine
deamidation of, 25
residue parameters of, *4*
Glutaraldehyde
enzyme crosslinkings by, 160
tyrosimase immobilization and, 66
Glutaredoxin, thiol interchange with, 46
Glutathione, thiol interchange with, 46
Glyceraldehyde-3 phosphate dehydrogenase, stabilization of, 27–30
Glycerides, synthesis of, by lipase, 558–68
Glycerol, stabilizing effect of, 53–54
Glycine, residue parameters of, *4*
Glycosidases, microbial, 282–84
Glyoxal, polyacrylamide-hydrazide crosslinking by, 419–20, *419*
Gramicidin S synthetase, stabilization of, 44–47
Guanidine hydrochloride, enzyme stability to, 1

6 M GuHCl, ribonuclease thermoinactivation effects of, 21–22

Hemicellulose, in rice husks, 131
Hemoglobin-containing systems, hydroxylation by, 214–18
Herpes simplex, 239
Hexanal, chymotrypsin modification and, 343–46
Hexokinase, yeast extraction of, 286–88
Histidine
  industrial production methods for, *79*
  residue parameters for, *4*
Hollow-fiber reactors, 373–81
  lactose hydrolysis in, 119–22
Homogenates, microbial cell, enzyme recovery from, 285–88
HPLC, cyclodextrin analysis by, 71
Hybrid biosensors, in fermentation, 508–11
Hydrocarbon micellar solutions, 549–58
Hydrocarbons, oxidation of, 541–48
Hydrogen
  bond distribution in, sucrose concentration and, 50
  organic acid production of, 428–31
  photosynthetic bacteria production of, 432–32
  production and uptake reactions of, enzyme kinetic parameters for, *294*
Hydrogenase, activity of, cereal expression of, 296–98
Hydrogenase catalyzed reactions, viologen mediator for, 292–95
Hydrogen peroxide, oxygen supply to cells using, 472–74
Hydrolysis
  cellobiose, mathematical model of, 41–42, *42*
  enzymatic, of lactose, 119–22
  of microcrystalline cellulose, by cellulase, 158–60
  of sucrose, 48-52, *49, 50, 51*
Hydrophobic molecules, chymotrypsin modification with, 343–46
Hydroxamate, 310
Hydroxy acids
  optically active, enzymatic production of, 87–90
  racemic mixture of, amino acids from, 91–94
2-hydroxyisocaproate dehydrogenases, 200–4
Hydroxylation
  by hemoglobin containing systems, 214–18
  protoplast immobilization and, 489–90
Hydroxylic groups, invertase stability and, 52–53

Imidazole, 310
Immobilization, new techniques of, 239–48
Immunoenzyme sensors, 533–35
Imprinting, molecular, 327–33
Indole solutions, chymotrypsin in, 33, *34, 36*
Industrial catalysts, enzymes as, limitations of, 20–26
Inoculum, extracellular enzyme production effects of, *113*
Intramolecular crosslinking, with bifunctional reagents, subunit enzyme stabilization and, 27–30
Inulin assay, for renal clearance, 504–7
Invertase stability
  dextrans and, 56
  hydroxylic groups and, 52–53
  polyhydric alcohols and, 61–63
  poly(1,2-ethanediol) and, 54–56
Ionic detergents, enzyme stability to, 1
Isoleucine
  industrial production methods for, *79*
  residue parameters of *4*

Jasmine, price of, *383*

Keto acids, amino acids from, 84–85
Keto esters, enantioselective reductions of, by bakers' yeast, 186–93
Ketosteroids, reduction of, 551–52
Kidney acylase, amino acid production using, 80–83
Kinetics, chymotrypsin deactivation studied with, 31–38
Klebsiella pneumoniae, 70, 153, 454
Kluyvera citrophilia, 207–8

Lactam antibiotics, kinetically controlled semisynthesis of, 99–105
Lactamase, protein function study use of, 232–38
Lactase, kinetic parameters of, *120*
Lactate, alanine transformation from, 91–92
Lactate dehydrogenase, 198–99
Lactic acid
  fermentation systems for, *376*
  immobilized bacteria of, applications of, 406–17
  Lactobacillus delbrueckii production of, 374–79
Lactobacillus casei, 87
Lactobacillus confusus, 87, 200
Lactobacillus delbrueckii, lactic acid production by, 374–79
"Lactohyd," whey production using, 136–39

Lactose, enzymatic hydrolysis of, in hollow-fiber reactor, 119–22
LDH, residual activity in, *68*
Leucine
  industrial production methods for, *79*
  residue parameters of, *4*
  synthetic stabilization and, 46
Leuconostoc sp., 201
Leuconostoc mesenteroides, dextransucrase production from, 267–70
Leuconostoc oenos, malolactic fermentation by, 461–64
Levan sucrase, 468–71
Lignin, in rice husks, 131
Linoleic acid, oxidation of, 550–51
Lipase, glyceride synthesis by, 558–68
Liposomes, protein immobilization on, 580–81
Lithospermum erythrorhizon, 382
Liver microsomes, immobilization of, 423
Low water condition, acid-catalyzed cellulose hydrolysis under, 164–67
LSP 21, biochemical products from, *240*
Lysine
  industrial production methods for, *79*
  residue parameters of, *4*
Lysis, 489

Magnesium chloride, sucrose concentration and, 50
Malic acid
  decomposition of, 123
  from fumaric acid, 83–84
  industrial scale synthesis reaction with, 194
Mass transport resistances, rotating ring-disc electrode studies of, 224–31
Measurement, microbiological, by cell immobilization, 363–72
Meicelase, 158
Memory effect, origin of, 329–31
Mentha, monoterpene reduction by, 422–23
Mercaptan, 310
Methane, microbial transformation of, 541–48
Methantrophs, hydrocarbon oxidation by, 541–48
Methemoglobin, hydroxylation by, 214–18
Methionine
  acylase use production of, 80–83
  industrial production methods for, 78, *79*
  residue parameters of, *4*
Micelles, catalysis of enzymes entrapped in, 577–79
Michaelis constant, 41
Microbial acylase, amino acid production using, 80–83
Microbial cell homogenates, enzyme recovery from, 285–88
Microbial glycosidases, 282–84
Microorganisms, immobilization of, 465–67
Microsomes, liver, immobilization of, 423
Milk, prefermentation of, 412–15
MLA 144, biochemical products from, *240*
Molecular imprinting, 327–33
Molecules, hydrophobic, chymotrypsin modification with, 343–46
Monascus sp., 282
Monoterpene, *Mentha* reduction of, 422–23
Mucor miehei, amino acid and NMR analysis of, 340–42
Mushroom tyrosinase
  deactivation kinetics of, 66–69
  residual activity in, *68*
Mutagenic strategies, protein function studies by, 232–38
MUTMAC, 31
Mycobacterium sp.
  biochemical products of, *240*
  steroid reduction by, 420–21

NAD, reduction of, 179
NAD kinase, immobilization of, 264–66
Neurospora crassa, 3
Nitro-phenyl phosphate hydrolysis, 9–17
4-nitro-3-carboxytriflouroacetanilide, hydrolysis of, 314–15
NMR analysis, of *Mucor miehei,* 340–42
Nocardia sp., cholesterol degradation by, 573–76
Nylon affinity tubes, affinity isolation of acetylcholine with, 299–301
Nylon net, in butanediol production, 455

Oligonucleotide synthesis, 3
Oligosaccharides
  solid superacids for hydrolyzing of, 161–63
  starch conversion to, 140
Optical measurements, in microbiological measurement, 368
Ornithine, synthetase stabilization and, 46
Oxalate-sensing electrodes, 512–14
Oxidase, enzymes, 533–35
2-oxo-acid reductase, 176–79
Oxygen, supply to cells of, 472–74
Oxyhemoglobin, hydroxylation by, 214–18
Oxytetracycline, production of, 480–81

Palladium, adduct of, biocatalyst use of, 337
Papain, semisynthetic enzyme studies using, 322–24
Pectinase activity, 111

Pellicularia filamentosa, 158
Penicillin acylase
coupling of, *251*
immobilization of, in polarized ultrafiltration membrane reactor, 127–30
penicillin G hydrolytic deacylation by, 206
Penicillin amidase, semisynthesis of, 99–102
Penicillin digitatum, 465
Pentose sugars, fermentation of
in rice husks, 131
xylitol to ethanol formation in, 152–54
Peptide bonds
chymotrypsin-catalyzed synthesis of, 552–53
cleavage of, 25
Peptides, kinetically controlled semisynthesis of, 99–105
Perfluorinated resin-sulfonic acid catalysts, chemical production utilization of, 161–63
Peroxidase, catalyzation with, 219
Phenols, isomeric, sulfatase-catalyzed separation of, 219
Phenoxymethylpenicillin, synthesis of, using α-amino-acylamino-β-lactam acylhydrolase, 210–13
Phenylacetic acid, in penicillin production, 128
Phenylalanine
acylase use production of, 80–83
industrial production methods for, 78, *79*
residue parameters for, *4*
synthetase stabilization and, 46
Phenyl lactic acid, production of, 87
Phenyl sepharose, *E. coli* protein extracted with, 288
pH extremes, enzyme stability to, 1
Plant protoplasts, immobilization of, 487–90
Plants
cells of, immobilized viable, 382–93
secondary products of, prices for, *383*
Plasminogen activators, activity assays for, 536–40
Polarized ultrafiltration cells, enzyme stabilization technique using, 7
Polyacrylamide hydrazide, gel entrapment of cells and enzymes in, 418–26
Polyacrylamides, 49
in penicillin production, 129
plant-cell entrapment use of, *384*
Polyethylenimines
catalytic effectiveness of, *306*
modified, conformation of, 315–18
Polyhydric alcohols, stabilizing effect of, 53–54, 61–63
Polymeric carriers, enzyme stabilization and, 27–30
Polymerization, enzyme structural feature and, 3
Polymers
preparation of cavities in, by molecular imprinting, 327–29
synthetic, 302–20
Polymer stability, 11–13, *12*
Poly(1,2-ethanediol), invertase stability and, 54–56
Polypeptide chains
amino acid sequence differences and, 3
length of, enzyme structural feature and, 3
thermoinactivation mechanisms in, 20
Polypeptide entropy, 5
Polypepton fraction effective, amylase production and, 278–81
Polysaccharides
activation of, and protein immobilization, 254–56
renewable, utilization of, 140–43
solid superacids for hydrolyzing of, 161–63
Polyunsaturated fatty acids, oxidation of, 550–51
Potassium ferricyanide, 33
Potassium ferrocyanide, mass transfer for, parameter values for, *230*
Potato starch, decanol substrate concentration with, 73–74
Prepolymerized polyacrylamide, plant-cell entrapment use of, *384*
Prepolymer methods, applications of cells immobilized by, 479–82
Procaryotic cells, growth of, 373–81
Proline
industrial production methods for, *79*
residue parameters of, *4*
synthetase stabilization and, 46
Propanol, chymotrypsin deactivation and, 31
Protein A, extraction of, 285–86
Protein conformational transition, analysis of, *2*
Protein denaturants, enzyme stability to, 1
Proteins
coupling of, to tresyl chloride activated supports, *242*
covalent change in, enzyme inactivation as result of, 20
crystallographic structures of, 3
surface analysis of, 5
function of, by mutagenic strategies, 232–38
immobilization of
on liposomes, 580–81
polysaccharide activation and, 254–56
Proteus mirabilis, 177–78
Proteus vulgaris, 177–78
Protoplasts
immobilization of, 487–90

immobilization methods for, comparison of, 501–3
Pseudomas putida, biochemical products from, *240*
Pseudomonas dacunhae, alanine production using, 450–53
Pseudomonas melanogenum, 206–7
Pullulanases
properties of, *283*
thermophilic, *Bacillus acidopullulyticus* and, 271–74
Pumice stone, transition metal-activated, yeast cell immobilization on, 483–86
Pyrethrins, price of, *383*
Pyridine, 310
Pyruvate, as intermediary in lactate to alanine transformation, 91–92

Quinine, price of, *383*

Racemization, of amino acid residues, 25
Reaction kinetics, biocatalytic processes with immobilized cells and, 437–49
Reactors
capillary membrane, characterization of, 123–26
continuous-flow, fermentation by *Leuconostoc oenos* in, 461–64
in ethanol production, 395–96
hollow-fiber, 373–81
lactose hydrolysis in, 119–22
polarized ultrafiltration membrane, penicillin immobilization in, 127–30
See also Bioreactors
Reductases, in chiral compound preparation, 171–85
Reductions, enantioselective, of keto esters, 186–93
Renal clearance, inulin assay for, 504–7
Reproductive immobilized cells, 347–62
Residue parameters, *4*
Resistance, mass transport, ring-disc electrode studies of, 224–31
Rhodospirillum rubrum, 427
Ribonuclease A
bovine pancreatic, enzyme thermoinactivation study use of, 20–26
irreversible thermoinactivation of, 21–25, *22, 24*
Rice husks, pretreatment influence on, by *Sporotrichum pulverulentum,* 131–35
Ring-discelectrodes, mass transport resistance studied with, 224–31

Saccharification, of cellulose, glucose inhibition in, 39–43
Saccharomyces cerevisiae, 148, 163, 186
ethanol production using, 394–405
glucose to alcohol conversions by, 421–22
reproductive immobilized cellular study with, 347–62
Secondary structural elements, enzyme structural feature and, 3
Sensors
electroenzymatic, for alcohol measurement, 515–19
fluoride ion selective, enzyme electrode based on, 523–25
immunoenzyme, 533–35
SEPARAN AP273, 9
Serine
industrial production methods for, *79*
residue parameters of, *4*
Shikonin, price of, *383*
Silica, in rice husks, 131
Spearmint, price of, *383*
Sodium chloride, superacid use with, 161–62
Sodium hydroxide, ligocellulosic pretreatment with, 133
Sorbitol, stabilizing effect of, 53–54
Soybean trypsin inhibitors, cystine destruction in, 24
Spectroscopy
chymotrypsin deactivation studied with, 31–38
nuclear magnetic resonance, 341
Sporotrichum pulverulentum, rice husk utilization by, pretreatment influence on, 131–35
Starch
alcohol fermentation of, with immobilized cell/enzyme bioreactor, 148–51
enzymatic degradation of, cyclodextrin production by, 70–77
ethanol conversion from, 144–47
hydrolysate of, *276*
liquefaction of, 140–43
Steroid-1-dehydrogenation, heat-dried bacteria for, 106–9
Steroids
hydroxylation of, 479–80
microbial cell transformation of, 459–60
Mycobacterium sp. reduction of, 420–21
Streptomyces clavuligerus, antibiotic production by, 420
Succinic acid, as crosslinking reagent, 28
Succinic anhydride, ribonuclease isoelectric point altered by, 22–23, *22*
Sucrose, hydrolysis of, 48–52, *49, 50, 51*
Sugars, starch conversion to, 144–47
Sulfatase, catalyzation with, 219
Sulfhydryls, intracellular, time course of, *45*
Sulfite, waste liquor from, lactic acid produced from, 411

Superacids, solid, in oligo- and polysaccharide hydrolyzation, 161–63
Synthetic carriers, for enzymes, 251–53
Synthetic hydrogenation catalysts, synzymes as, 334–39
Synthetic polymers, 302–20
Synzymes, as synthetic hydrogenation catalysts, 334–39

Temperature, cyclodextrin production influence from, *72*
Thermal deactivation kinetics, for enzymes, 7–9
Thermoanaerobacter cervisiae, 152
Thermoinactivation, irreversible, of enzymes, 20–26
Thermophilic pullulanases, *Bacillus acidopullulyticus* and, 271–74
Thiol, interchange of, 46
Thioredoxin, thiol interchange with, 46
Threonine
  industrial production methods for, *79*
  residue parameters of, *4*
Tissue plasminogen activators, activity assay for, 536–40
Tresyl chloride, support activation with, protein coupling and, *242*
Trichoderma koningii, 282
Trichoderma viride, 39, 41, 158
Triticum durum, 296
Trypsin, coupling of, *251*
Tryptophan
  industrial production methods for, 78, *79*
  production of, 440–42
  residue parameters of, *4*
Tyrosinases, deactivation kinetics of, 66–69
Tyrosinase, residual activity in, *68*
Tyrosine
  industrial production methods for, *79*
  residue parameters of, *4*

UDP glucose, residual activity in, *68*
Urea
  chemical denaturation by, 9
  enzyme reduction in, 22
  enzyme stability to, 1
  sensors for, 508–9
Urease, coupling of, *251*

Valine
  industrial production methods for, 78, *79*
  residue parameters of, *4*
  synthetase stabilization and, 46
Vincristine, price of, *383*
Vinylphenylboronic acid, ester linkages of, 328, *329*
Viologen mediators, for hydrogenase-catalyzed reactions, 292–95
Vitamin B_{12}, production of, 480

Water
  activity of, enzyme stability effect of, 48–60
  amylase production with, 115–18
  pectinolytic activity effect from, 111
Whey
  enzymatic hydrolysis of, 136–39
  as substrate in lactic acid production, 411–12
Wine, malic acid in, 123

Xanthan gum, plant-cell entrapment use of, *384*
Xanthine oxidase, catalyzation with, 219
Xylanase, 111, 112, *112*
Xylitol
  ethanol formation from, with *Candida tropicalis,* 152–54
  stabilizing effect of, 53–54
Xylose, butanediol effects of, 457, *457*

Yeast
  bakers', ADH and hexokinase extraction from, 286–88
  biochemical products from, *240*
  galactosidases of, 138
  immobilization of cells of, 483–86
  improvement of viability of, 397
  keto ester reductions by, 186–93

ZAPM, formation rate of, substrate concentration effects on, *96*

PUBLISHED BY THE NEW YORK ACADEMY OF SCIENCES

$135.00